FIELD GUIDE TO THE WILD FLOWERS
of the
EASTERN MEDITERRANEAN

FIELD GUIDE TO THE WILD FLOWERS
of the
EASTERN MEDITERRANEAN

Chris Thorogood

Kew Publishing
Royal Botanic Gardens, Kew

© The Board of Trustees of the Royal Botanic Gardens, Kew 2019
Text and illustrations © Chris Thorogood
Photographs © Chris Thorogood unless otherwise stated

The author has asserted his right to be identified as the author of this work in accordance with the Copyright, Designs and Patents Act 1988.

All rights reserved. No part of this publication may be reproduced, stored in a retrieval system, or transmitted, in any form, or by any means, electronic, mechanical, photocopying, recording or otherwise, without the written permission of the publisher unless in accordance with the provisions of the Copyright Designs and Patents Act 1988.

Great care has been taken to maintain the accuracy of the information contained in this work. However, neither the publisher, the editors nor the authors can be held responsible for any consequences arising from use of the information contained herein. The views expressed in this work are those of the individual authors and do not necessarily reflect those of the publisher or of the Board of Trustees of the Royal Botanic Gardens, Kew.

First published in 2019 by
Royal Botanic Gardens, Kew
Richmond, Surrey, TW9 3AB, UK
www.kew.org

ISBN 978 1 84246 691 9
eISBN 978 1 84246 692 6

Distributed on behalf of the Royal Botanic Gardens, Kew in North America by the University of Chicago Press, 1427 East 60th Street, Chicago, IL 60637, USA.

British Library Cataloguing in Publication Data
A catalogue record for this book is available from the British Library

Copy-editing: Matthew Seal
Design and page layout: Christine Beard
Proofreading: Sharon Whitehead
Production management: Georgina Hills

COVER ILLUSTRATIONS
Front: *Tulipa saxatilis*
Back, clockwise from top left: *Iris bismarckiana*; *Viola scorpiuroides*; *Anacamptis papilionacea* subsp. *aegaea*; *Anemone coronaria*; *Onobrychis venosa*; *Gladiolus atroviolaceus*

FRONTISPIECE: *Dracunculus vulgaris*, Crete

Printed in Spain by GraphyCems S.L.

For information or to purchase all Kew titles please visit
shop.kew.org/kewbooksonline or email publishing@kew.org

Kew's mission is to inspire and deliver science-based plant conservation worldwide, enhancing the quality of life.

Kew receives approximately one third of its running costs from Government through the Department for Environment, Food and Rural Affairs (Defra). All other funding needed to support Kew's vital work comes from members, foundations, donors and commercial activities including book sales.

Acknowledgements

The eastern Mediterranean is one of the richest floristic regions in the world. Distilling such a rich and complex flora into a single usable guide has been a challenge assisted greatly by many botanists over the years. First and foremost, my thanks to Simon Hiscock, a great source of inspiration, with whom I have spent many happy years teaching Mediterranean field botany to undergraduate students. Special thanks to Ori Fragman-Sapir and Yuval Sapir for their hospitality and for the wonderful excursions to survey the flora of Israel and Palestine, which has an exceptional beauty of all of its own; thanks also to Dar Ben-Natan for showing me some remarkable Israeli desert sites.

I am very grateful to Nick Turland, Yuval Sapir and Arne Strid for critically reviewing the manuscript and offering many useful suggestions. Many photographs were kindly provided by Yuval Sapir, Banan Al Sheikh, Oron Peri, Zuheir Shater, Nick Turland, Eleftherios Dariotis and Arne Strid, which together make this book a much more complete representation of the flora. Experts on particular regions were generous with their advice; for example, Yuval Sapir and Zuheir Shater provided information on the floras of Israel and Palestine and of Syria, respectively. Stephen Harris was an invaluable source of knowledge on the history of plant hunting in the Levant. Finally, I am grateful to botanists and enthusiasts for their recommendations of locations; for example, Zissis Antonopoulos recommended site locations in northern Greece, and Eleftherios Dariotis, who happened to be in Crete at the same time as me, recruited an army of local shepherds and hunters to keep an eye out for rare plants in flower!

Like all books of this nature, this guide is a synthesis that builds upon existing floras. Seminal works including the *Flora Europaea*, *Flora of Turkey & the East Aegean Islands*, *Flora Palaestina* and *Flora of Cyprus* created an early foundation for interpreting the eastern Mediterranean flora. However, profound changes in classification led by the APG (Angiosperm Phylogeny Group), along with numerous new species records, have driven a reassessment of the region's flora. Exciting recent developments that have been instrumental in this process have been the *Vascular Plants of Greece: An annotated checklist* — a collective effort to compile the entire flora of Greece; the comprehensive *Atlas of the Aegean Flora*; *Flowers of Crete*; and online resources created by numerous authors, such as *The Flora of Israel*, *Flora Ionica*, *Flora of Cyprus* and *greekflora.gr*. The *Flora of Greece Web* project will advance our knowledge of the flora of Greece and adjacent territories even further. Together, these resources will create a robust platform that will shape how we interpret and conserve the eastern Mediterranean flora. The ambitions of this guide are to form a modest addition to this canon and to inspire a new generation of field botanists to appreciate the beauty and importance of the wild flowers of the eastern Mediterranean.

Contents

ACKNOWLEDGEMENTS   v

INTRODUCTION   1
This book's coverage   2
How to use this book   3
The plants   4
  Sequence of plant families   4
  Plant descriptions   4

A HISTORY OF PLANT HUNTING IN THE LEVANT   7

PLANT IDENTIFICATION   9
Common flower types deconstructed   10
Common flower forms   11
Common leaf forms   12
Common fruit forms   13
How to identify plant families   14

PLACES TO SEE WILD FLOWERS IN THE EASTERN MEDITERRANEAN   17
Ionian Archipelago   17
Mainland Greece   19
  Rhodope   19
  Mount Olympus   20
  Mount Ossa and Mount Pelion   22
Peloponnese   23
Aegean Islands   24
Crete   26
Cyprus   28
Mediterranean Turkey   30
The Southeast Mediterranean   31
  Israel and Palestine   31
  Lebanon   33
  Syria   33

WILD FLOWER HABITATS 35
Agricultural land 35
Maquis, garrigue and phrygana 37
Evergreen and deciduous forests 41
Gorges 43
Coasts 44
Offshore islets 45
Arid habitats 46
Seasonal interest 47

SPECIES DESCRIPTIONS 49
Gymnosperms 50
Basal angiosperms 57
Monocots 64
Eudicots 182

GLOSSARY 611
SELECTED REFERENCES 618
INDEX OF ENGLISH NAMES 619
INDEX OF SCIENTIFIC NAMES 631

Cistanche violacea, Israel

Introduction

The Mediterranean Basin boasts one of the world's richest floras. Indeed, up to 10% of all vascular plants occur in the Mediterranean Basin, which represents just 1.6% of the Earth's surface. The area covered in this guide encompasses mainland Greece, the Peloponnese, the Ionian and Aegean archipelagos, Crete, Karpathos, Cyprus, Turkey, and the southeast, including the coasts and adjacent deserts of Syria, Lebanon, and Israel and Palestine. This large area has a complex and varied geology and topography but is united by its typically Mediterranean climate of hot, dry summers and mild, wet winters. The eastern Mediterranean phytogeographic region is especially rich, has an exceptionally high number of endemic species, and has a flora quite distinct from that of the western Mediterranean Basin, which is covered in a separate guide.

Hippocrepis emerus subsp. *emeroides*, a common shrub in the pea family (Leguminosae) in the eastern Mediterranean.

Map of the geographical coverage of this book, courtesy of NASA Worldview. Coverage across this vast area is weighted by geography, habitat, and classification.

West of region:

AA: Aegean Archipelago
CR: Crete
CY: Cyprus
GR: Mainland Greece
IA: Ionian Archipelago
KP: Karpathos (mainly grouped with CR in the descriptions — the largest neighbouring island)
PN: Peloponnese
TR: Turkey

Southeast Mediterranean:

IP: Israel & Palestine
LB: Lebanon
SY: Syria

THIS BOOK'S COVERAGE

This book's mission is to be the most comprehensive single-volume field guide of the eastern Mediterranean, which gives the reader the best possible chance of finding and identifying a given plant that they have encountered. However, it would be impossible to condense all of the many thousands of species that occur in the region into a single easy-to-use volume. Coverage is therefore balanced by geography, habitat, and classification.

Mastic (*Pistacia terebinthus*) is common across the region.

Geography. All species commonly encountered across the countries covered are included, for example, the widespread mastic (*Pistacia terebinthus*). Local endemics are also covered because these make up a significant proportion of the island flora in the region; for example, Crete and Cyprus each have a particularly rich endemic flora. Fewer species from the conventionally less touristic regions are included as most readers are unlikely to encounter them, although some are featured that are of particular local interest. Although Egypt and Libya fall within the broader area covered, they are mainly desert with very narrow Mediterranean zones, and are not included here specifically.

Calicotome villosa is typical of the thorny vegetation on maquis, garrigue, and phrygana.

Habitat. Coverage spans all the major habitats, including maquis and garrigue plant communities, which are dominant across most of the region, and which are dominated by thorny shrubs such as *Calicotome villosa*; as well as aquatic and woodland habitats, each of which have their own specific floras; and arid and semi-desert zones, which are a feature of much of the extreme southeast of the region. A lesser focus is placed on montane habitats, which have a unique flora of their own that is not typically Mediterranean. Most species covered are found in maquis, garrigue, and phrygana habitats as these are the most common forms of vegetation.

Brugmansia cultivars are commonly planted across the region.

Classification. All major flowering plant families are included, although some families are represented in more detail than others. This is because some families are of wider interest or more conspicuous, and better suited to a field guide. For example, a greater coverage of bulbs and orchids is given than of grasses and sedges. Commonly planted ornamentals and naturalised exotics are covered because while they are not native to the eastern Mediterranean, they nevertheless form a significant contribution to the flora; in any case, it is often not immediately obvious whether a plant is native or not.

HOW TO USE THIS BOOK

This book is designed to be a highly visual guide, with as many photographs and illustrations as possible in a single volume that are useful to both the amateur and the professional botanist alike. Plants are sequenced in order of relatedness, which is the most objective form of classification. Readers with a more advanced knowledge will typically head straight to the relevant plant family and search for the species in question. Beginners' approach may take a little longer and will typically follow the sequence below:

1. Identify the relevant section of the book by checking whether the plant belongs to one of the major plant families (*see* How to identify plant families, p. 14).
2. Scan through the photographs for this family for a species that is similar to the plant in question.
3. For larger or more complex groups, find the relevant subgroup headed (a), (b), (c), etc., which summarises the salient features.
4. Use the plant descriptions and illustrations associated with the photographs to confirm the identification.
5. Finally, verify the identification by cross-checking descriptions for closely related species that are described in the same section and have key distinguishing characteristics highlighted.

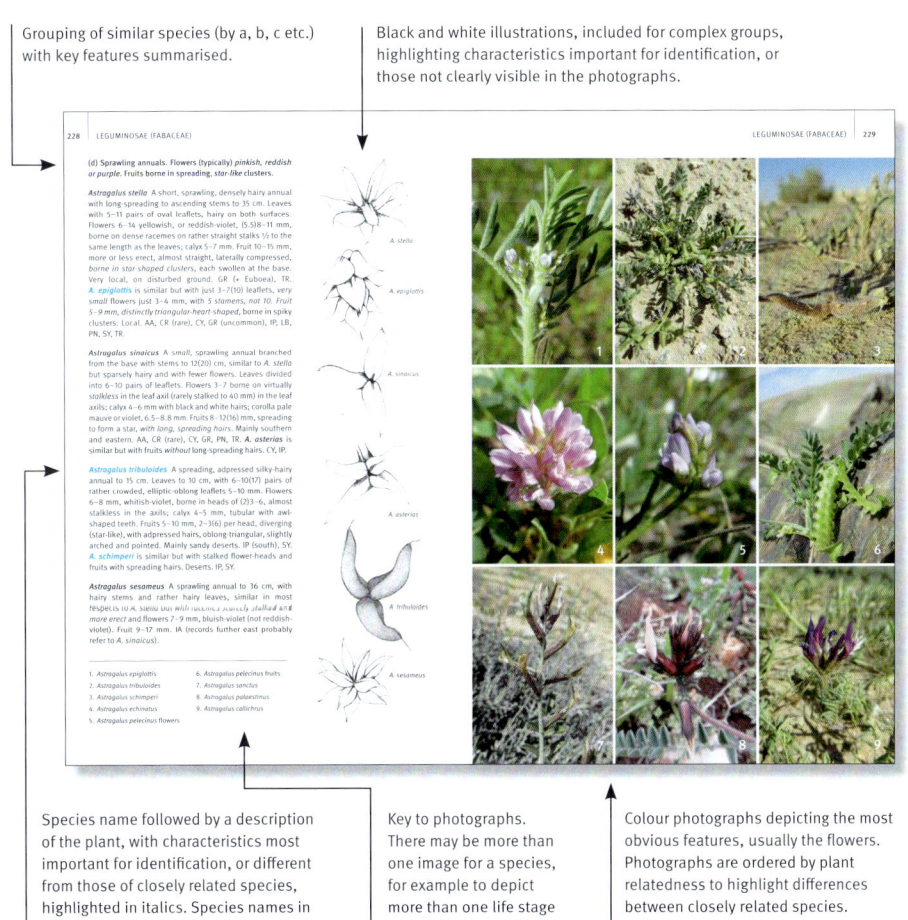

Grouping of similar species (by a, b, c etc.) with key features summarised.

Black and white illustrations, included for complex groups, highlighting characteristics important for identification, or those not clearly visible in the photographs.

Species name followed by a description of the plant, with characteristics most important for identification, or different from those of closely related species, highlighted in italics. Species names in blue are accompanied by a photograph.

Key to photographs. There may be more than one image for a species, for example to depict more than one life stage such as flower and fruit.

Colour photographs depicting the most obvious features, usually the flowers. Photographs are ordered by plant relatedness to highlight differences between closely related species.

THE PLANTS

SEQUENCE OF PLANT FAMILIES

The sequence of plants is based on plant relatedness, which is the most objective form of classification. All plants in this book are assigned to a hierarchy that begins with four groups: the gymnosperms (non-flowering seed plants), the basal angiosperms (an ancient lineage), the monocots (a group including bulbs, orchids and grasses) and the eudicots (representing the majority of plants). Within each of these four groups are families, which in turn, are made up of genera and species (see A plant family tree, p. 5). The relatedness of families on which this sequence is based follows the APG 4 (Angiosperm Phylogeny Group) system, which is the globally accepted classification of plants. The assignment of species to a family in this book will sometimes differ from that of traditional floras, in which plants were classified according to their morphology (appearance), rather than their genetic relatedness, an obvious example being the snapdragon (*Antirrhinum*), which is now assigned to the plantain family (Plantaginaceae). The sequence of families may also deviate subtly from texts that followed earlier versions of APG, which has been updated four times, but is now relatively stable. To navigate the sequence of plant families, the reader should consult How to identify plant families, p. 14.

PLANT DESCRIPTIONS

Similar plants within a large or complex family have been grouped by shared important features under headings (a), (b), (c), etc. These replace traditional keys, for which there is no space in this book and which can lead to an incorrect identification if all the species in a given family are not included. The descriptions include details of dimensions, habitat and distribution, with the most important features highlighted in italics. Not all descriptions are written in exactly the same way. For example, it is useful to highlight a particular set of characteristics (such as leaf stalk length) that is more important in distinguishing species in one genus than those in another. To describe species precisely, the use of technical terms is often necessary, so a list of definitions is included in the Glossary.

Names. Each plant is assigned an English name (if it has one) and a scientific name, which is made up of two parts: the first indicates the genus to which the species belongs, and the second is the specific epithet (see A plant family tree, p. 5). Some are further split into subspecies, which differ subtly but are not distinct enough to be treated at the species rank. Plant species are frequently revised as more information becomes available about their relatedness. This can lead to confusion, especially where botanists do not agree on the status of a species, subspecies or variety. For these reasons, plant names may vary from one guide to another as well as over time. Therefore to avoid confusion, commonly used synonyms are included for species that are widely known by another name, indicated by '(Syn.)'. Subspecies are included where they are geographically important or particularly distinct. Although it is conventional to include all subspecies that occur in the region covered, it would be beyond the scope of this guide to do so, particularly for very variable species for which numerous subspecies exist.

INTRODUCTION | 5

A plant family tree

FAMILY. A group (name usually ending '-aceae') of related genera. Every family has a number of defining characteristics useful for identification (*see* How to identify plant families, p. 14).

FAMILY APIACEAE

GENUS (plural: genera). A group of related species. All species in a given genus share the same common generic name (the first word in a name made up of two parts). Species in the same genus share many characteristics and are sometimes difficult to tell apart.

Daucus   *Ferulago*

SPECIES. A unit within a genus. Each species has a unique specific epithet (the second word in a name made up of two parts). Every species has uniquely defining traits necessary for identification.

Daucus carota   *Ferulago nodosa*

SUBSPECIES. Some species are further divided into subspecies. These normally differ subtly from the typical form of the species, but are not distinct enough to be described at the species rank.

Daucus carota subsp. *drepanensis*   *Daucus carota* subsp. *major*

Dimensions. Dimensions are based on the author's observations and on scientific papers and floras. Dimensions vary under different environmental conditions and with plant age, so should be viewed as an approximate guide unless emphasised specifically.

Habitat. The habitat is described because this is often useful for identification. For example, a plant typically found on sea cliffs is unlikely to be encountered in an oak woodland. Where several similar species that share a habitat are described together, the habitat is only described for the first, to avoid repetition.

Flowering time. The precise flowering time is not given in the descriptions because this varies from year to year, with location and elevation, and will be most probably affected by climate change. Many species flower in March and April in the eastern Mediterranean, and for those that are not spring-flowering, the approximate flowering time is highlighted in the description.

Distribution. It would be impossible to include maps for every species, and there is no other precise yet simple means of indicating distribution. The distributions given are based on regional floras and databases, and on the author's own observations. It is possible that a species may be encountered outside the distribution given in this guide. The countries or areas in which a species is likely to occur are listed in abbreviated form. Abbreviations do not follow international convention, to allow discrimination between islands (*see* map on p. 1 for main area of coverage).

Photographs and illustrations. Because photographs cannot be included for all species, a balance must be struck between detailed descriptions and photographs. Photographs have been selected to reflect coverage based on geography, habitat and classification (*see* This book's coverage, p. 2). Names of plants for which photographs are included are indicated in blue bold. The majority of photographs depict the flower, which is usually the most useful for identification. Black and white illustrations are provided to complement the descriptions and to illustrate characteristics that are useful for identification.

Nerium oleander

A history of plant hunting in the Levant

The Levant (which includes all countries along the Eastern Mediterranean shores in its widest, historical sense) has a very rich history of plant collecting and a significant botanical legacy. Collections in the region were made by William Sherard (1659–1728), a key figure in early eighteenth-century European botany, who was appointed Consul at Smyrna (Izmir), Turkey in 1703. Sherard was a travelling companion and tutor to various wealthy individuals and made European grand tours between 1694 and 1698. In his will, Sherard bequeathed his herbarium, library, and notes to the University of Oxford. Today 20,000 specimens that he collected make up the Sherard Herbarium.

Reproductions of some of the illustrations produced by Ferdinand Bauer published in the *Flora Graeca*: (A) *Cyclamen persicum*, now a popular garden plant; (B) *Asphodelus ramosus*; (C) *Arum dioscoridis*; (D) *Ferula communis*.

Among the most important treatments of the region was the *Flora Graeca*, published between 1806 and 1840. This work was based on surveys across Greece, Turkey and Cyprus by botanist John Sibthorp (1758–1796) and illustrator Ferdinand Bauer (1760–1826). Sibthorp and Bauer followed the footsteps of the French botanist Joseph Pitton de Tournefort (1656–1708), who between 1700 and 1702 collected plants in the islands of Greece, Constantinople, the borders of the Black Sea, Armenia and Georgia. Sibthorp and Bauer were the first botanical explorers of Cyprus. The spectacular quality of the botanical illustrations, the size of the publication (10 volumes) and its significant cost upon publication together make *Flora Graeca* one of the most extraordinary botanical publications of all time. The voyages also bestow a rich horticultural legacy, including the cultivation of popular garden plants such as *Crocus* hybrids derived from species first collected in Turkey, and *Cyclamen persicum* collected in Cyprus. From March to December 1786, Sibthorp collected and described while Bauer prepared dried specimens and produced colour-coded pencil sketches. Bauer's watercolours prepared in Oxford from these annotated sketches are now regarded as some of the finest examples of botanical illustration, and many of them represent new species. Today 2,462 specimens from *Flora Graeca* are housed in the Sibthorpian Herbarium at the University of Oxford.

One of the 20,000 specimens that make up the Sherard Herbarium, *Dracunculus vulgaris*.
COPYRIGHT (©) THE OXFORD UNIVERSITY HERBARIA, REPRODUCED WITH PERMISSION.

Some of the 2,462 herbarium specimens collected during Sibthorp's voyages that make up the Sibthorpian Herbarium. Left: *Crocus flavus*; right: *Anemone coronaria*.
COPYRIGHT (©) OXFORD UNIVERSITY HERBARIA, REPRODUCED WITH PERMISSION.

Plant identification

A grasp of the basic plant anatomy including flower, leaf and fruit form is essential for wild flower identification. The photographs in this book act as a visual guide, but for accurate identification, checking the shape and dimensions of key features included in the descriptions is necessary. Beyond the obvious diagnosis of grass, bulb, shrub or tree, flower arrangement is often the most important indicator of the family to which a plant belongs (*see* How to identify plant families, p. 14). For example, an umbel (an umbrella-shaped inflorescence) is characteristic of the carrot family (Apiaceae), whereas a capitulum (a daisy-like inflorescence made up of many tiny flowers called florets) is typical of the daisy family (Asteraceae). Within a family, other characteristics such as leaf shape or level of hairiness may then be important to identify to genus or species level. A hand lens is necessary to observe minute features of some plants, such as level of hairiness of flower parts not visible with the naked eye.

The next few pages offer a brief introductory guide to plant identification in the field, including the most common flower, leaf and fruit forms encountered in the region, along with a quick reference guide to the most common families.

Malva punctata

COMMON FLOWER TYPES DECONSTRUCTED

Here are some flower types common in the east Mediterranean flora with the main parts annotated.

A regular (actinomorphic) flower in the mallow family (*Malva*, Malvaceae)

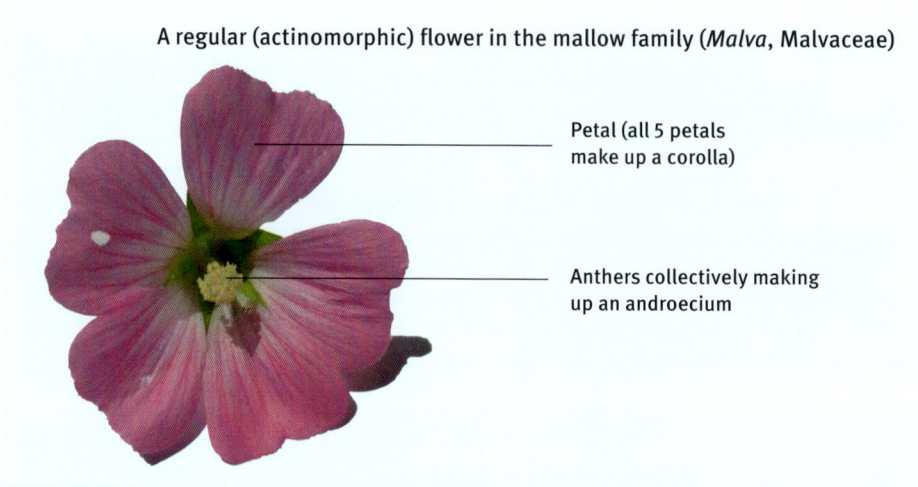

Petal (all 5 petals make up a corolla)

Anthers collectively making up an androecium

A zygomorphic flower in the mint family (*Phlomis*, Lamiaceae)

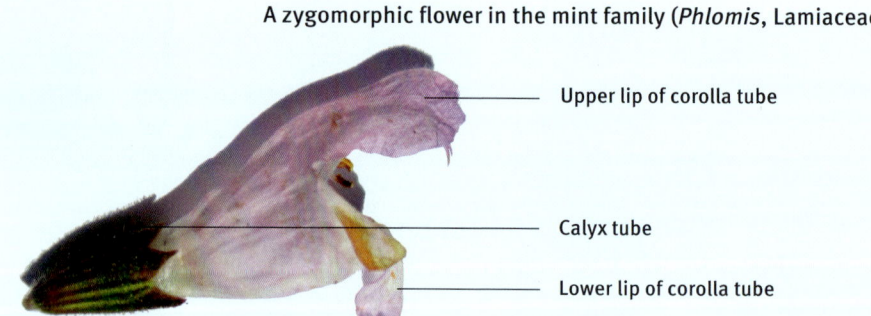

Upper lip of corolla tube

Calyx tube

Lower lip of corolla tube

A zygomorphic flower in the pea family (*Spartium*, Leguminosae)

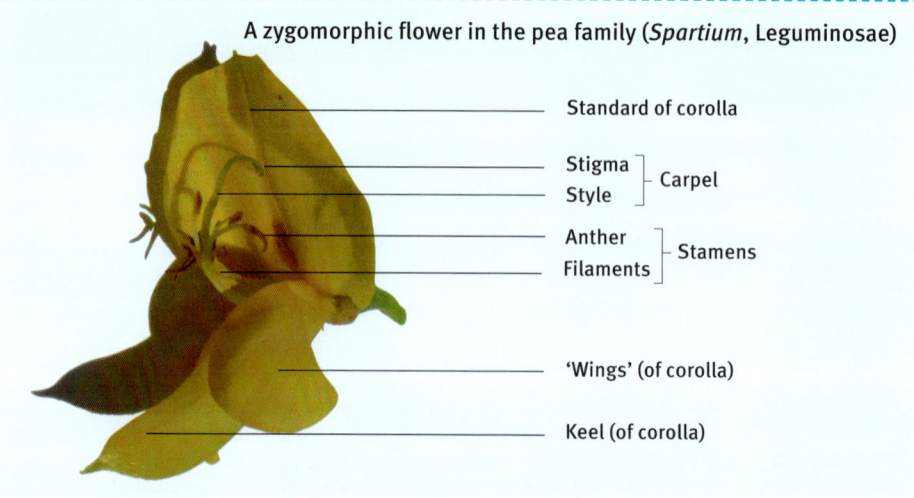

Standard of corolla

Stigma ⎤
Style ⎦ Carpel

Anther   ⎤
Filaments ⎦ Stamens

'Wings' (of corolla)

Keel (of corolla)

Some common flower forms in the eastern Mediterranean flora: (A) inflorescence composed of **spikelets**, typically seen in Poaceae (*Hordeum murinum*); (B) inflorescence a **lax raceme** (*Ornithogalum narbonense*); (C) inflorescence a dense **terminal raceme** (*Bituminaria bituminosa*); (D) inflorescence composed of **whorls** (*Salvia fruticosa*); (E) inflorescence a **capitulum**, typically seen in Asteraceae (*Pallenis spinosa*); (F) inflorescence a **spathe and spadix** typically seen in Araceae (*Dracunculus vulgaris*); (G) inflorescence a lax terminal raceme (*Spartium junceum*); (H) **male** (left) and **female/cosexual flowers** (right) borne on separate plants (**dioecious**) (*Ceratonia siliqua*).

Some common leaf forms in the eastern Mediterranean flora: (A) whorled (*Rubia peregrina*); (B) linear (*Gladiolus italicus*); (C) alternate, elliptic (*Pittosporum tobira*); (D) opposite, elliptic (*Olea europaea*); (E) pinnately divided into elliptic leaflets (*Pistacia lentiscus*); (F) arrow-shaped (*Arum italicum*); (G) circular-triangular and cordate (heart-shaped) at the base (*Cyclamen hederifolium*); (H) 3-pinnately divided into linear leaflets (*Foeniculum vulgare*); (I) round and palmately lobed (*Ficus carica*); (J) simple, oval-elliptic, shallow-lobed (*Salvia fruticosa*); (K) deeply pinnately lobed and toothed (*Sonchus oleraceus*); (L) succulent (*Opuntia maxima*).

PLANT IDENTIFICATION | 13

Some common fruit forms in the eastern Mediterranean flora: (A) non-splitting legume (pod) typical of Leguminosae (*Ceratonia siliqua*); (B) 3-parted, splitting capsules (*Leopoldia comosa*); (C) Schizocarp (dry fruit splitting into 2 1-seeded mericarps) *Malabaila aurea*; (D) globose, valved capsule (*Cyclamen hederifolium*) (E) 3-parted, ovoid capsule (*Gladiolus italicus*); (F) drupe (succulent fruit with 1 stony seed) (**Prunus dulcis**); (G) an aggregate of drupes (*Morus alba*); (H) discrete spikelets, typical of grasses (*Briza maxima*); (I) capitulum of achenes each with 'fluffy' pappus, typical of **Asteraceae** (*Urospermum picroides*); (J) stalked berries (*Smilax aspera*); (K) stalkless berries (*Arum italicum*).

HOW TO IDENTIFY PLANT FAMILIES

An understanding of the most important features of the main plant families will enable the reader to pick out the relevant part of this book even in what may at first appear to be a bewildering number of flowers. Related families often share characteristics, so if a plant does not belong to any of the families given below, this list should at least guide the reader to the relevant section. Relative representation of the different families varies by region, but the selection below is a guide to the most important based on abundance, diversity, and conspicuousness. In the Species Descriptions section, plant families are sequenced by their genetic relatedness.

Asteraceae, daisy family

Individual flowers minute, tubular and 5-lobed or flat and strap-like, aggregated into daisy-like, and dandelion-like or thistle-like flower-heads (capitula) surrounded by bracts. Fruit single-seeded, often with a 'parachute' (pappus) attached.

Poaceae, grass family

Leaves alternate, usually linear, often in 2 ranks and sheathing the stem. Flowers complicated: usually cosexual with 1, 3 or 6 stamens and a pistil enclosed in 2 bracts, the lemma, and palea, the whole making a floret; florets arranged in spikelets with empty bracts (glumes) at the base; styles normally 2. Fruit grain-like (a grain).

Lamiaceae, mint family

Plants often glandular and aromatic with square stems, leaves opposite. Flowers borne in (often) congested whorls towards the stem apex. Calyx with 5 teeth, often 2-lipped; corolla normally 2-lipped; stamens 2 or 4. Fruit made up of nutlets hidden at the calyx base.

Apiaceae, carrot family

Leaves alternate, simple or 2–4 times divided (feathery). Flowers tiny, petals 5, often uneven, borne in umbrella-like heads (umbels); stamens normally 5. Fruit 2-parted, often flattened.

Leguminosae (Fabaceae), pea family

Leaves alternate, often pinnate, with stipules. Calyx 5-toothed, sometimes 2-lipped; flowers zygo-morphic with 5 petals, the upper a standard, the laterals as 2 wings and the lower fused into a keel; stamens 10. Fruit often pea-pod-like, splitting when ripe, sometimes coiled.

Caryophyllaceae, pink family

Leaves opposite. Sepals 5, separate or fused into a tube. Petals 5, separate but sometimes spilt. Fruit a capsule with 6, 8 or 10 teeth.

Brassicaceae, cabbage family

Leaves usually opposite. Flowers often white or yellow, with 4 separate sepals; petals 4, separate and forming a cross. Fruit typically a 2-parted pod.

Liliaceae, lily family

Traditionally a single family now split into several closely related families characterised by leaves with parallel veins, often strap-like. Tepals (perianth segments) 6, petal-like, separate or fused; stamens 6; ovary superior. Fruit a 3-parted capsule or berry.

Orchidaceae, orchid family

Leaves in rosettes or alternate, narrow and untoothed. Flowers in upright spikes or racemes; sepals 3, petals 3, the laterals spreading or united with one or all sepals, the lower forming a variably shaped lip; pollen in a paired sticky mass (pollinia); ovary inferior. Fruit a capsule.

Amaranthaceae, amaranth family

Now including many species formerly in the Chenopodiaceae. Leaves alternate or opposite. Flowers regular, borne in cymes or panicles, usually with 5 (1–8) tepals, often fused; stamens 1–5; ovary superior. Flowers often enlarging in fruit; fruit 1-seeded.

Cistaceae, rock rose family

Leaves normally opposite, often with stipules. Flowers cosexual and regular, often showy; sepals 3 or 5, petals 5; stamens numerous. Fruit a 1–10-valved capsule.

Euphorbiaceae, euphorbia family

Plant produces a white latex when cut, leaves opposite or alternate. Flowers variously arranged (genera very distinct) but all with 2–3-parted, often warty ovary, and 2–3 or branched stigmas.

Papaveraceae, poppy family

Plant produces a watery sap or white or yellow latex when cut, leaves usually lobed, without stipules. Flowers cosexual, regular or zygomorphic; sepals usually 2, soon-falling; petals 4(–6), showy. Fruit an achene or capsule.

Plantaginaceae, plantain family

Leaves alternate or opposite. Flowers borne in cymes or racemes; sepals 4–5, petals 4–5, often fused into a corolla tube, sometimes 2-lipped; stamens 2, 4 or 5. Fruit a capsule. The Orobanchaceae is a related family of parasites many of which lack green pigment (chlorophyll).

Boraginaceae, borage family

Leaves normally alternate and bristly. Flowers borne in coiled cymes; calyx 5-toothed; petals 5, fused into a variably long tube, often pink, white or blue; stamens 5. Fruit made up of 5 nutlets.

Malvaceae, mallow family

Leaves alternate, often palmately lobed. Flowers regular, often with an epicalyx; sepals and petals 5; stamens numerous. Fruit a capsule, or nutlets.

Rosaceae, rose family

Leaves stalked and usually with stipules. Flowers regular with typically 5 sepals and petals and 2–4 x as many stamens; carpel 1–many, each with a separate style. Fruit an achene, drupe, capsule or fleshy.

Campanulaceae, bellflower family

Plants often with a white latex, leaves alternate, without stipules. Flowers often showy, cosexual, zygomorphic or regular; sepals and petals 5; petals often blue and fused; stamens borne on style-base (not on petals). Fruit a capsule or berry.

Lathraea rhodopea

Places to see wild flowers in the eastern Mediterranean

Local floras are very much shaped by their geological history and by climate. Local climates are influenced by nearby land masses. For example, cold air from continental Europe and Turkey to the north, and hot air from the Sahara to the south have each led to very different plant assemblages in nearby regions. A geographically widespread selection of locations from across the eastern Mediterranean is described here, broadly from northwest to southeast, along with a handful of the species most likely to be found there. The best time to visit is indicated, though this varies with the weather from year to year.

IONIAN ARCHIPELAGO

The Ionian Archipelago is a series of over a hundred islands located parallel to the western coast of mainland Greece. From north to south, there are six major islands: Corfu, Paxos, Lefkada, Ithaca, Cephalonia and Zakynthos, as well as a string of smaller, largely uninhabited islets. Geologically, the islands are mainly characterised by limestone and schist sediments covered with terra rossa soil. The Archipelago is typically Mediterranean in climate and sheltered from cool winds by the mainland Greek massifs, but with a higher precipitation than the more easterly Aegean Islands. Owing to the islands' westerly location, the vegetation shows elements typical of both the western and eastern Mediterranean floristic regions.

The only way to see many of the rarer plants on the Ionian Archipelago is by boat.

Map of the Ionian Archipelago. COURTESY OF NASA WORLDVIEW

Furthermore because the islands span some 300 km, there is a cline from green, *Cupressus*-dominated communities in the north, to sparser and scrubbier, more xerophytic vegetation in the south. These geographic aspects, combined with island endemism, mean that the Ionian Archipelago has a rich and diverse flora, comprising almost 2,000 species. Some of the Ionian endemics are restricted to remote coastal cliffs, so that many of the rarer plants on the islands can be accessed only by boat. April and May are the best times to visit.

The rare island endemic *Centaurea paxorum* (top right) and sea lavender *Limonium antipaxorum* (centre right) occur only on a few remote sea cliffs on the islands of Paxos and Antipaxos, accessible only by boat. Ionian Island endemic *Stachys ionica* (bottom right, photograph by Eleftherios Dariotis) overlooking the Porto Katsiki beach in Lefkada.

MAINLAND GREECE

Mainland Greece forms part of the Balkan Peninsula. This land mass was once a peninsula of Turkey, and this geological history has given rise to a rich legacy in Irano-Turanian species. The area boasts world-class mountain floras including numerous ancient relict species and endemics. Indeed the northeast region of Greece alone has over 3,000 species. It is therefore well beyond the scope of this guide to cover the mountain floras comprehensively, and most strictly alpine species are excluded here.

Map of Greece. COURTESY OF NASA WORLDVIEW

Rhodope

The Rhodope mountains cross the Bulgarian–Greek border, and mark the northern limit of the region covered by this flora. The mountains are mainly non-calcareous, and form a bridge between north Aegean and Pontic species from Turkey to the south and east and Bulgarian and Macedonian species to the north and west. This, combined with their isolation and geological history, means that these mountains have a rich and unique sub-Mediterranean forest flora, as well as woods with a more Nordic feel, including the southernmost forests of *Picea abies*. Warmer slopes have an open, thermophylous vegetation of *Quercus coccifera*, *Q. pubescens* and *Juniperus oxycedrus*, scattered with *Pinus nigra*, which is more abundant in cooler areas at higher altitudes. Rocky beech forests are home to the endemic *Haberlea rhodopensis*, which has survived for millions of years, long since its nearest living relatives became extinct. Another rarity and arguably one of the most spectacular of all the species described in this book is the poorly known parasitic toothwort, *Lathraea rhodopea*, which lives off the roots of deciduous trees and grows only very locally in the forests of northern Greece and Bulgaria.

Sub-Mediterranean forests of the Rhodope mountains (bottom left) are home to spring-flowering species such as *Corydalis solida* (bottom right) and the spectacular *Lathraea rhodopea* (page 16).

Mount Olympus

Mount Olympus is the highest mountain in Greece at 2,917 m and boasts a uniquely diverse flora in a relatively small area. The mountain's geographical isolation, proximity to the sea, and juxtaposition of central European and Mediterranean vegetation have resulted in a flora of exceptional diversity with numerous endemics. The eastern and northern slopes are more densely forested than the drier western and southern slopes. Predominantly alpine species descend far below their normal range in the summits to the sheltered, humid limestone valleys and canyons. Here, Mediterranean, Central European and Balkan species all exist in unusual proximity. The forested foothills (300–700 m) are dominated by *Quercus coccifera*, *Juniperus oxycedrus* (occasionally parasitised by the mistletoe relative *Arceuthobium oxycedri*), *Phillyrea latifolia*, and, in damper areas, *Quercus pubescens* and *Platanus orientalis*. Lower slopes are home to irises such as *Iris reichenbachii* and about 50 species of orchid, including numerous forms of the variable *Ophrys sphegodes*. Rocky crags between 400 m and 1,400 m in the mountain's gorges and ravines host many endemic chasmophytes, including *Genista sakellariadis*, *Aubrieta thessala* and, on damp boulders and shady cliffs, rosettes of the extraordinary endemic 'living fossil' *Jankaea heldreichii*. Above 1,400 m lies a Balkan mountain vegetation zone with many more interesting alpine species not covered in this flora. The mountain and its foothills are worth visiting from mid-April through to August (depending on the altitude).

The foothills of Mount Olympus are a refuge for Balkan-Mediterranean *Iris reichenbachii* (bottom left) and numerous forms of the highly variable *Ophrys sphegodes* (bottom right). Canyons and gorges display a rare mix of Mediterranean, Central European and Balkan species including many rarities and endemics such as (facing page, top to bottom): *Genista sakellariadis*, *Astragalus monspessulanus*, *Saxifraga scardica* and *Aubrieta thessala*.

Mount Ossa and Mount Pelion

Like Mount Olympus, these mountains lie close to the sea and receive moist coastal breezes. Similarly the mountains' flanks are densely forested and a mixing pot of Mediterranean, Central European and Balkan plant species. Coastal plains around the mountains are parched white in summer but in spring host swathes of annuals and bright red anemones (*Anemone pavonina*). The higher slopes of Ossa, above 1,600m, host a sub-alpine Mediterranean vegetation dominated by *Quercus coccifera* and *Juniperus oxycedrus* and, here and there, interesting spring-flowering rarities including attractive clumps of *Viola rausii* and small cushions of *Saxifraga scardica* on bare rock and limestone scree. Interestingly, horse chestnut occurs in forested ravines on the eastern slopes. Pelion supports extensive humid forests of sweet chestnut and beech, as well as more open hillsides which are home to *Iris tuberosa* and locally, the Balkan endemic *Campanula incurva*. The upper slopes of Pelion are covered with *Fagus* forests and are the southernmost locality for several Central European woodland species, such as enchanter's nightshade (*Circaea lutetiana*). Both mountains are worth visiting from April through to August (depending on the altitude).

The foothills of Mount Ossa (foreground) support rare pure stands of *Quercus coccifera*, as well as beech, Macedonian fir and sweet chestnut. Anemones (*Anemone pavonina*) (right inset) are common on lower plains; the summit (background) supports a Mediterranean sub-alpine zone including rarities such as *Viola rausii* (left inset).

PELOPONNESE

The Peloponnese region borders west Greece to the north and Attica to the northeast. The region comprises a series of rugged limestone highlands fringed by sandstone foothills and narrow coastal plains, ending in a wild and windswept peninsula at the mainland's southernmost point. Isolation from the mainland and its southerly position have contributed to the evolution of a unique flora of nearly 3,000 species and a high degree of endemism. The highest and most isolated mountain ranges in the north, such as Aroania and Killini, support dense forests of *Abies cephalonica* and *Pinus nigra* subsp. *pallasiana*, cut with deep gorges rich in endemic spring-flowering species including many orchids. The mountains' sunny slopes, cliffs and screes host a very rich and diverse flora all of their own, which is not covered exhaustively in this guide, and host remnant populations of species otherwise found mainly in Asia, such as *Biebersteinia orphanidis*. Mount Parnon is the highest of a lower, more southerly range of limestone mountains, which is known for its rare alpine species. Terraces, olive groves, orchards and plains surrounding the rugged mountain ranges of the Peloponnese support a plethora of wild flowers in the spring. The central plains of Arcadia are ablaze with scarlet anemones in April, while bulbs including *Bellevalia ciliata*, *Tulipa undulatifolia* and *Muscari neglectum* flower in synchrony in pastures and orchards around Argolida in the east. The Peloponnese is best visited in April, or from May onwards for the higher mountains.

The Peloponnese. COURTESY OF NASA WORLDVIEW

Valeriana italica overlooking the Styx Valley of the Aroania mountain range in the northern Peloponnese.
PHOTOGRAPH BY ELEFTHERIOS DARIOTIS.

AEGEAN ISLANDS

For simplicity, in this guide the Aegean Islands as a collective include: the North and East Aegean Islands; the Sporades; Euboea; the Saronic Islands; Cyclades and the Dodecanese Islands (including Rhodes and Karpathos). Crete and Cyprus each have their own diverse floras, so are described separately. The Aegean Islands are much drier than Greece's other island archipelago, the Ionian Islands (which occupy a similar latitudinal range); they have little or no summer rainfall, and consequently host a lower and sparser vegetation.

The Aegean Islands. COURTESY OF NASA WORLDVIEW

The northernmost islands are relatively cool owing to their proximity to the neighbouring continental land masses. By contrast, the easternmost and southernmost islands are influenced by continental Anatolian and North African coastal climates, respectively. The islands collectively have about 3,420 species (excluding Crete), and a high number of endemics, owing to their long isolation from the mainland. The East Aegean Islands are especially rich, home to many Anatolian species and botanically quite distinct from the westernmost islands. Typically, the islands are best visited in April and May.

Carlina tragacanthifolia, native to the Dodecanese and the Marmaris peninsula of southwest Turkey; seen here on Karpathos. PHOTOGRAPH BY ELEFTHERIOS DARIOTIS.

Asphodeline lutea

CRETE

Crete is the largest of the Aegean Islands and has a famously rich flora of about 2,000 species, of which approximately 10% are endemic. Most species have a circum-Mediterranean or Balkan distribution, many with a centre of distribution further east, while some on the drier southern coast have North African affinities. Crete is a central remnant of an arc of mountains that once formed part of an Aegean land mass that stretched from the Peloponnese to southwest Turkey. This historic geography, combined with the island's complex and varied geology, have contributed to its exceptionally diverse flora. The gorges, cliffs and limestone mountains, particularly in the west, have provided a refuge for ancient endemic chasmophytes, such as *Eryngium ternatum*. The central plains, such as those around Spili, are home to swathes of endemic tulips and a diversity of orchid species quite without compare, including the endemic *Ophrys kotschyi* subsp. *cretica*. Striking stands of dragon arum (*Dracunculus vulgaris*) and yellow asphodel (*Asphodeline lutea*) are common on rocky coasts along Crete's southern shores. The east of the island is less mountainous but is dissected by gorges and is an important habitat for the peculiar *Aristolochia cretica*. Also in the east, valley bottoms and banks of rivers and coastal streams host stands of the rare palm *Phoenix theophrasti*. Crete is best visited in spring from late March to mid-April.

Map of Crete. COURTESY OF NASA WORLDVIEW

Gorges in Crete such as Imbros (below left) support species such as *Valeriana asarifolia* (below right). Opposite, clockwise from top left to centre, some of Crete's extraordinarily diverse flora: *Ferula glauca*, *Dracunculus vulgaris*, *Asphodeline lutea*, *Ophrys kotschyi* subsp. *cretica*, *Tulipa saxatilis*, *Tulipa cretica*, *Iris unguicularis* subsp. *cretensis*, *Leopoldia spreitzenhoferi*, and *Arum creticum*.

CYPRUS

Cyprus, which is the third largest Mediterranean Island, has a floral diversity similar to that of Crete, comprising about 1,600 species of which 7–9% are endemic. Cyprus lies 150 km from Syria and 75 km from Turkey, with which the island shares floristic links, to the north.

Map of Cyprus. COURTESY OF NASA WORLDVIEW

The island's coastline ranges from steep cliffs covered in coastal maquis dominated by *Juniperus phoenicea*, to extensive dune systems and associated wetlands, such as the salt lakes of Akrotiri and Larnaka. Rare geophytes in coastal stations include numerous forms of bee orchid (*Ophrys* spp.) and an endangered tulip (*Tulipa cypria*). Forests on the island extend from outposts in the hills of Akamas to the east, across the island's two main mountain ranges: the east-central Troodos and the northern Pentadaktylos. The foothills of these mountains are flanked by maquis and thermophilous pine forests of *Pinus brutia*, which occur from sea level to 1,400 m. The Troodos range is the most cool and humid area of the island. Hill forests here are covered with a dense evergreen mantle, particularly in the Paphos forest, which is home to rare species such as the endemic Cyprus cedar (*Cedrus libani* var. *brevifolia*), perennials including *Gagea troodi* and *Arum rupicola*, as well as little patches here and there of endemic thyme (*Thymus integer*). The endemic evergreen golden oak (*Quercus alnifolia*) occurs in the Troodos, as does an isolated forest of *Pinus nigra* subsp. *pallasiana* on the highest peaks. The northern slopes of the massif are populated with semi-deciduous oaks (*Quercus infectoria* subsp. *veneris*), which have a distribution predominantly east of Cyprus. The more northerly Pentadaktylos range is dominated by pines mixed locally with cypress (*Cupressus sempervirens*) on limestone outcrops. Spring arrives early in Cyprus, and the island is best visited from early March to early April.

The Akamas Peninsula (below) hosts a rich coastal maquis flora. Facing page, clockwise from top left to centre, notable examples of Cyprus's flora: the endangered *Tulipa cypria*, *Arum dioscoridis*, *Ophrys fuciflora* subsp. *bornmuelleri*, the endemic *Orchis troodi*, and from the Paphos Forest in the Troodos Mountains, *Vicia lunata* and *Thymus integer*.

MEDITERRANEAN TURKEY

Turkey has the richest flora of all of the Mediterranean countries and the highest number of endemics, an estimated 25–30%. Up to 10,000 plant species are recorded in Turkey, about half of which are in the country's Mediterranean zone. The country's incredible species richness is in part due to its vast range of habitats and geology. Habitats span Mediterranean and Black Sea coasts, coastal and interior mountains, gorges, steppes, alluvial plains and arid areas. Here, only the Mediterranean seaboard is covered, which makes up a relatively small area of Turkey's total landmass. The mountains of this area, for example the Taurus Mountains, which divide the Mediterranean coastal region from the central Anatolian Plateau, are particularity rich in endemics. These mountains are home to large stands of black pine (*Pinus nigra*), Taurus fir (*Abies cilicica*) and Lebanon cedar (*Cedrus libani*). Nearer the coast where summers are hot and dry and winters wet, vegetation is dominated by *Quercus coccifera* and *Pinus brutia*. The ultramafic rock floras of Antalya and Içel, and the hills and gorges of Muğla host particularly diverse floras. Some of the species of note in the hill forests around Turkey's Mediterranean coastline include *Aristolochia guichardii*, an unusual species found on rocky screes and cliffs, *Cephalanthera epipactoides* subsp. *kurdica*, an orchid of lime-rich *Quercus* forests, and *Arum rupicola*, which grows on rocky slopes and screes. *Eremurus spectabilis* is a spectacular species of rocky hillsides with a distribution predominantly further east. Turkey is best visited from early April to late May, depending on the altitude.

Map of Turkey. Only the Mediterranean seaboard is covered in this guide. COURTESY OF NASA WORLDVIEW

Eremurus spectabilis (left), *Arum rupicola* (centre) and *Cephalanthera epipactoides* subsp. *kurdica* (right). PHOTOGRAPHS BY NICK TURLAND.

THE SOUTHEAST MEDITERRANEAN

The southeast Mediterranean encompasses Israel and Palestine, Lebanon and western Syria. The region has strong desert and Irano-Turanian climatic influences, and a very different flora from the northern and western territories covered in this guide. Much of this territory falls within the Fertile Crescent, an area defined by its diverse climates and the area of origin of Neolithic founder crops.

Israel and Palestine

The flora of Israel and Palestine is diverse for a land area of just over 22,000 km², comprising about 2,800 species. The rate of endemism is about 7%, but many species are rare or absent further west in the region. Habitats are rich and varied, ranging from coastal plains along the eastern seaboard, through rolling hill and mountain vegetation on Mount Hermon, the Upper Galilee, and the Lebanon transition zone, to deserts (such as the Negev) and mountain deserts in the hot, dry south. The region's Mediterranean vegetation is the focus of this guide. However, strong Irano-Turanian, Middle-Eastern and Saharo-Arabian influences intermix throughout to form mosaic plant communities, so some desert species are also included. Interestingly, a small number of Sudanian species also occur at the northernmost limits of their range in the Dead Sea Rift Valley – a series of cliffs, flood plains and saline depressions that extends from Syria to East Africa. A rare assortment of plants occurs here that are absent elsewhere in the region, including many rare Apocynaceae, such as the cactus-like *Caralluma sinaica*. Meanwhile deserts in the south harbour an interesting and unique flora of their own. The vegetation here is sparse and dominated by drought-tolerant shrubs such as *Salsola vermiculata* and *Anabasis articulata*, and, here and there in the western Negev, the imposing yellow parasite, *Cistanche tubulosa*.

Map of the Southeast Mediterranean, encompassing Israel and Palestine, Lebanon and western Syria. COURTESY OF NASA WORLDVIEW

Israel and Palestine are particularly rich in rare and endangered iris species. *Iris atropurpurea*, for example, is an endemic of the coastal Sharon Plain, while *I. bismarckiana* and *I. lortetii* are restricted to northern districts around the Upper Galilee, Lower Galilee, Mount Hermon and adjacent areas. Spring comes early in the region; the best times to visit are March and April in the south and north, respectively.

The flora of Israel and Palestine, clockwise from top left: *Iris atrofusca*, *I. bismarckiana*, *Tulipa agenensis*, *Aristolochia bottae*, the northern Negev, the Upper Galilee, *Fritillaria persica* and *Allium rothii*.

The Golan Heights, at the crossroads between Israel and Syria, are an open basaltic landscape with sparse trees and shrubs. The area has a different vegetation from the surrounding sedimentary landscape.

Lebanon

Lebanon has a complex topography and geology. Over 70% of the country's 10,500 km² falls upon vast, rugged mountain systems, which host important and species-rich remnant forests, contrasting with the surrounding arid environment. Lebanon is home to over 2,600 species, about 3.5% of which are strictly endemic, and the country is recognised globally as an 'important plant area'. Many rare and endemic species are restricted to isolated mountain summits such as those of Mt Makmel, Mt Sannine, Qamouaa, Ehden and Mt Hermon. These mountain ranges harbour a rich mosaic of relic forests, including stands of cedar (*Cedrus libani*), fir (*Abies cilicica*) and juniper (*Juniperus foetida*), as well as steep limestone outcrops dominated by pine (*Pinus brutia*) and oak (*Quercus calliprinos*, *Q. infectoria* and *Q. cerris*). Understorey vegetation in these mountain forests hosts geophytes such as cyclamens (*Cyclamen coum*) and the endemic *Ornithogalum libanoticum*.

Syria

Syria covers a large land area of 185,000 km² and is home to over 3,000 plant species. Syrian vegetation at low altitudes is typically Mediterranean, dominated by shrubs such as *Sarcopoterium spinosum* and *Lavandula stoechas*, and home locally to the unusual southwest Asian endemic *Michauxia campanuloides*. This Mediterranean garrigue vegetation extends across the low-lying Syrian Saddle, which separates the Turkish Anti-Taurus and Syro-Palestine ranges to the north and south, respectively. The western Syrian mountain vegetation, on terra rossa-limestone substrates, is particularly rich and has a strong Irano-Turanian influence. Here, a sub-humid Mediterranean mosaic of evergreen and non-deciduous trees and shrubs is dominated by various shrubby Rosaceae (*Amygdalus* spp., *Crataegus* spp., and *Prunus* spp.) and oaks (*Quercus calliprinos*). The vegetation of the northwest Syrian mountains, on magnesium-rich serpentine substrates, has a plant diversity of rare distinction. Numerous endemic species grow here, such as *Iris nusairiensis*. To the east, the arid Syrian Desert is sparsely vegetated with xerophytic scrub. Floristic surveys of Syria are scarce and its native flora deserves much further attention.

The coastal magnesium-rich northwest Syrian mountains (top left) harbour a plant diversity of unusual distinction, including rare and endemic species such as *Iris nusairiensis* (top right). Garrigue on the western seaboard of Syria, such as near Latakia (bottom left), is home to eastern Mediterranean perennials such as the spectacular *Michauxia campanuloides* (bottom right). PHOTOGRAPHS BY ZUHEIR SHATER.

Wild flower habitats

AGRICULTURAL LAND

Traditional agriculture in the eastern Mediterranean is based on four major crops that have been cultivated since antiquity: olive, carob, fig and almond, as well as a plethora of other fruits and nuts including citrus, mulberries, grapes, dates, prickly pear, loquat and pistachios. The latter crops are often grown in smallholdings and gardens as multipurpose plants for farming, soil conservation and ornament or as field boundaries (e.g. prickly pears, *Opuntia* spp.). The four ancient crops do not require intensive farming and dot the landscape across the region, forming an important constituent of the flora. The olive (*Olea europaea*) has the greatest agricultural importance, as a source of olive oil. Traditionally farmed olive groves are very important habitats for wild flowers, including bulbs and perennials such as orchids and cyclamens, as well as a multitude of late-spring flowering annuals. The carob (*Ceratonia siliqua*) is a small evergreen tree in the pea family (Leguminosae). Its seeds have been used for millennia as a source of food and fodder. These trees, along with ancient, gnarled figs (*Ficus carica*) and almonds (*Prunus dulcis*), are a common feature of pastures and terraces across the region; they all harbour a rich and abundant flora similar to that of olive groves. Finally, although citrus crops originate in Asia, orange and lemon groves have also become symbolic of the Mediterranean landscape and culture, and they too represent an important refuge for wild flowers. Traditionally managed fields, in addition to the smallholdings of the major crops described above, are very significant habitats for wild flowers, including perennials with bulbs or tubers below the traditional depth of tilling, such as *Leontice leontopetalum*. These important habitats are under threat from agricultural intensification where modern farming regimes realise faster profits; thankfully, however, in many parts of the region they are still managed sympathetically, and are an excellent place to observe the local flora.

Olive groves are an important habitat for a range of wild flowers, including the cyclamen (*Cyclamen hederifolium*, inset)

The five most important cultivated trees of the eastern Mediterranean maquis vegetation: (A) **citrus**, for example the bitter orange (*Citrus × aurantium*), and (B) the more unusual 'fingered citron' (*Citrus medica* var. *sarcodactylis*); (C) **carob**, *Ceratonia siliqua*; (D) **fig**, *Ficus carica*; (E) **almond**, *Prunus dulcis*; and (F) **olive**, *Olea europaea*.

MAQUIS, GARRIGUE AND PHRYGANA

The dominant form of vegetation in the eastern Mediterranean is a dense, shrub-dominated plant community 1.5–3.5 m high. It typically forms a dark green mantle on the *terra rossa* soil with exposed rocky ribs. The shrubs are tough, woody, and drought-tolerant, a habit known as sclerophyllous ('hard-leaved'), which has evolved across all Mediterranean climate regions in response to water stress. This vegetation, known locally by the French name *maquis*, includes a multitude of culinary herbs and is typically resinous and strongly aromatic. Often dominant are mastics (*Pistacia lentiscus* and *P. terebinthus*), which form dense, green thickets, as well as shrubs and small trees, such as the strawberry tree (*Arbutus unedo*) and tree heath (*Erica arborea*). The understorey is dotted with myriad geophytes, which survive the hot dry summers in the form of dormant bulbs, for example, the sea squill (*Drimia maritima*).

Mastic (*Pistacia lentiscus*) is a major component of maquis vegetation.

The low-lying, stunted form of this vegetation less than 1 m high, which grows on shallow soils such as coastal slopes or limestone screes inland, is known as *garrigue*. Garrigue has a similar composition to maquis, but extreme water stress, weather fronts, and soil erosion lead to a stunted and degraded shrub community that has fewer of the species associated with deep acidic substrates, such as heathers (*Erica* spp.) and strawberry trees (*Arbutus unedo*). These are replaced by stunted *Quercus coccifera*, and cushion-like shrubs of *Genista acanthoclada* and *Euphorbia acanthothamnos*. The sparser thicket and increased availability of light supports a rich variety of smaller annuals, perennials, and geophytes. The lowest, open form of this vegetation is often referred to as *phrygana* (Greece) and *batha* (Israel and Palestine). This is often bare in places, and dominated by shrublets of *Sarcopoterium spinosum* and *Thymbra capitata*. Because it intergrades with garrigue, it is described as such in this book.

WILD FLOWER HABITATS

Right: maquis vegetation 1.5–3.5 m high is dominated by small trees and shrubs such as *Pistacia terebinthus*. Below: the low-lying, stunted form of this vegetation, less than 1 m high, that grows on shallower soils such as coastal slopes is known as garrigue. The open, partially bare form shown is often referred to as phrygana or batha, and dominated by shrublets of *Sarcopoterium spinosum* or the occasional small tree (*Ceratonia siliqua*).

Common plants of the eastern **Mediterranean maquis**: **shrubs**, such as (A) *Asparagus aphyllus*, (B) *Paliurus spina-christi*, (C) *Myrtus communis* and (D) *Morus alba*; **conifers**, such as (E) *Juniperus phoenicea*; **legumes**, such as (F) *Coronilla valentina*, (G) *Anthyllis hermanniae*, (H) *Calicotome villosa* and (I) *Spartium junceum*; [J–Q overleaf] **perennials**, such as (J) *Stachys cretica*, (K) *Pallenis spinosa*, (L) *Salvia fruticosa*, (M) *Clinopodium vulgare* and (N) *Teucrium capitatum*; **creepers**, such as (O) *Rubia peregrina*; **bulbs**, such as (P) *Drimia maritima*, and **grasses**, such as (Q) *Aegilops neglecta*.

EVERGREEN AND DECIDUOUS FORESTS

Mediterranean evergreen forests are a common feature across coastal lowlands in the region. Many are dominated by Aleppo pines (*Pinus halepensis*), which form extensive, open stands on a variety of substrates from sea cliffs up to about 1,000 m. Large stands of Aleppo pine are scarcer in the drier south and east of the region, where the similar Turkish pine (*P. brutia*) is more common, along with olive and carob (see below). Stone pines (*P. pinea*) are widespread and commonly planted, although they rarely occur in large stands except locally in parts of the western and central Peloponnese. Patchy, open pine forests of all of the above species have an understorey of scrub and maquis vegetation (*see* Maquis, garrigue and phrygana, p. 37) in which juniper (*Juniperus oxycedrus*, *J. phoenicea*), mastic (*Pistacia lentiscus*), heathers (*Erica arborea*, *E. manipuliflora*) and butcher's broom (*Ruscus aculeatus*) are among the most common shrubs. At higher altitudes, these pines are frequently mixed with oaks. The Kermes oak (*Quercus coccifera*) is widespread and rarely forms pure stands except around some of the mountains of mainland Greece. It is a common feature of mixed evergreen forest intergraded with maquis vegetation. Holm oaks (*Quercus ilex*) favour higher altitudes and damper conditions, and are often associated with an understorey of large shrubs or small trees, such as *Laurus nobilis* and *Myrtus communis*.

The Aleppo pine (*Pinus halepensis*) (left) is common across the region; the umbrella pine (*P. pinea*) (right) only forms natural stands in the western and central Peloponnese, but is also frequently planted throughout the area.

WILD FLOWER HABITATS

In hotter, drier parts of the south and east, including the Peloponnese, the southern Aegean region and Crete, olive and carob are often the dominant tree species. These form scattered stands rather than extensive natural woods, and are cultivated on hillsides and terraces; they are an important feature of the landscape and have their own associated flora (*see* Agricultural land, p. 35). More locally in the south and east, evergreen forests are dominated by a handful of other species, including *Cupressus sempervirens* in parts of Crete, Syrian juniper (*Juniperus drupacea*) on Mount Parnon in the Peloponnese, and more frequently in southern Turkey and Syria, and the Eastern strawberry tree (*Arbutus andrachne*) on Aegean coasts.

Forested habitats can be dominated by coniferous evergreen species, such as (A) Aleppo pine (*Pinus halepensis*), and (B) *Cupressus sempervirens*; deciduous trees, such as (C) chestnut (*Castanea sativa*) and (D) oak (*Quercus pubescens*); their common understorey shrubs, (E) *Arbutus unedo* and (F) *Ruscus aculeatus*; and alien plantation species, such as (G) *Eucalyptus camaldulensis*.

GORGES

Rocky cliffs, gorges and canyons are important habitats across the eastern Mediterranean. Their geology, exposure, aspect and altitude together affect the diverse assemblages of plants they host. Species that are highly specialised for growing in the crags and crevices of sheer rock surfaces are called *chasmophytes*. Gorges and cliff faces are a refuge from grazing and man's activities, and some species grow exclusively in these habitats, the so-called obligate chasmophytes, for example the rare Cretan endemics *Campanula saxatilis* subsp. *saxatilis* and *Linum arboreum*. Gorges that span from mountains to coast, for example the famous Samaria Gorge in Crete, often host particularly diverse plant communities in which mountain species descend below their normal range and grow in rare proximity to lowland species. Meanwhile, deep ravines and gorges in ho t, dry areas provide islands of damp, shaded microhabitats with their own distinct floral assemblages. For example, gorges in Cyprus, such as the Avakas Gorge, are home to stands of trees that are absent elsewhere in the arid lowland plains, such as *Platanus orientalis*, which grows alongside rare perennials such as *Arum hygrophilum* among boulders by running water.

Cyprus's Avakas Gorge (top left) is home to rarities such as *Arum hygrophilum* (top middle) and the critically endangered *Centaurea akamantis* (top right). Cliffs and gorges, including Crete's famous Samaria gorge (bottom left), are home to rare chasmophytes such as the endemic *Campanula saxatilis* subsp. *saxatilis* (bottom centre) and *Linum arboreum* (bottom right).

COASTS

The eastern Mediterranean coastline, with its hundreds of islands and islets, is vast. It is quite distinct from the western Mediterranean, which has a stronger Atlantic influence, and has fewer extensive alluvial dune systems. This, coupled with weak tides and an absence of large rivers, means the coastline is on the whole much more uniform. Much of the rocky coastline is flanked by stands of Aleppo (*Pinus halepensis*) or umbrella pine (*P. pinea*), or maritime garrigue, characterised by small shrubs such as *Plocama calabrica* and *Anthyllis hermanniae*, perched on rocky sea cliffs and stunted by coastal fronts. Owing to this habitat uniformity, littoral and halophytic plant communities are often rather scarce, and there are fewer rarities and endemics on sea-cliffs than on the cliffs and gorges inland. Nevertheless, coastal inlets, dune systems and saltmarshes, where they occur, are still a rich hunting ground for interesting plants, including some local endemics, particularly sea lavenders (*Limonium* spp.).

Sea daffodils (*Pancratium maritimum*) are common on dunes (left) while small shrubs like *Plocama calabrica* (right) favour coastal garrigue.

Many of the most common coastal species are also frequent across most of the Mediterranean Basin. Species that are widespread across coastal dune systems include the dominant marram grass (*Ammophila arenaria*), the striking, summer-flowering sea daffodil (*Pancratium maritimum*), and patches here and there of the succulent sea rocket (*Cakile maritima*). On shingle beaches, *Salsola kali*, *Matthiola tricuspidata* and *Crithmum maritimum* are common. Salt marshes are characterised by halophytes tolerant of saline conditions and even salt water submersion, such as *Atriplex portulacoides*, succulent shrubs such as *Arthrocnemum fruticosum* and *Suaeda vera*, and annuals such as *Spergularia marina*.

OFFSHORE ISLETS

The flooding of the Mediterranean basin some five million years ago dissected the land into nearly 5,000 islands and islets, 4,000 of which have a surface area of less than 10 km². Offshore islets are especially numerous in the eastern Mediterranean, many of which are too small and remote for habitation, rarely grazed, and therefore an important refuge for wild flowers. The floral diversity of offshore islets depends on their size, their proximity to other landmasses, and their topography. Many are home to islet specialist species that are scarce or extinct on their nearest associated island or on the mainland. For example, the broomrape *Orobanche sanguinea*, bulbous perennial *Androcymbium rechingeri* and knapweed *Centaurea pumilio* all grow in abundance on Crete's southwesterly offshore islet of Elafonisi and its associated dunes, yet all three species are rare or absent on adjacent mainland Crete. Limited population size and distribution, together with habitat sensitivity, mean that many islet specialist species are highly vulnerable to extinction.

The offshore islet of Elafonisi is home to species that are rare or absent on adjacent mainland Crete, such as the broomrape *Orobanche sanguinea* (inset).

ARID HABITATS

Mediterranean plant communities in the east (Israel, Palestine and Syria) lie in close proximity to arid habitats. Where these phytogeographic regions meet, a mixture of Mediterranean and desert species transitions into true desert vegetation. *Loess* – a powdery sediment on which rainwater stands – is a common substrate in these transition zones, which enables annuals to prosper seasonally in otherwise barren areas. A unique species assemblage in the east comprises Irano-Turanian elements in the Syrian Desert and Anatolia; Saharo-Arabian elements, from the Sahara, Sinai and Arabian deserts; and Sudano-Zambesian elements, typical of the subtropical savannas of Africa. Silt, clay and sand support sparse shrub vegetation of *Anabasis articulata*, *Salsola tetrandra* and *S. vermiculata*. Wadis dissect less vegetated sand sheets; here annuals occur such as *Aaronsohnia factorovskyi*, tufts of *Stipa capensis* and patches of *Retama raetam*. Wet saline areas occur locally in the Dead Sea Valley, as does savannoid vegetation, such as flat-topped stands of *Vachellia tortilis* subsp. *raddiana*. Adjacent arid plains and gullies host interesting species that are rare elsewhere, such as parasitic *Cistanche tubulosa*, the cucumber relative *Cucumis prophetarum* and the peculiar grass-like *Gymnarrhena micrantha* of the daisy family (Asteraceae).

Arid habitats adjacent to the Dead Sea Valley (bottom) are home to the cucumber relative *Cucumis prophetarum* (left), parasitic *Cistanche tubulosa* (centre) and grass-like *Gymnarrhena micrantha* (right).

SEASONAL INTEREST

Spring interest

Spring is the best time of year to see a wide variety of wild flowers. Flowering time is strongly influenced by seasonal weather and by altitude. Depending on the region, this can be from early February to the end of April (often later on higher mountains), and typically one or two weeks earlier than in the western Mediterranean. Planning a visit between the last two weeks of March and the first two of April is a good rule of thumb.

Typically, the eastern Mediterranean spring is warm, often cloudy or showery, and the nights cool. Following the winter rains, the landscape is often lush. Many shrubs are in flower, as are a plethora of interesting geophytes such as orchids, alliums and arums, numerous annuals, herbaceous perennials and curiosities including parasitic broomrapes (*Orobanche* and *Phelipanche* spp.).

Clockwise from top left, spring-flowering plants include shrubs such as *Cistus creticus*, orchids (*Ophrys scolopax* subsp. *heldreichii*), arums (*Arum concinnatum*), perennials (*Anemone coronaria*), geophytes (*Asphodelus ramosus*) and parasitic broomrapes (*Orobanche sanguinea*).

Summer interest

Summer typically lasts from mid-June to mid-September. Apart from the occasional thunderstorm, precipitation is sparse and the landscape is typically brown, barren and parched. Annuals and herbaceous perennials die back, surviving the hot summer months in the form of seeds and root storage organs, respectively.

The most conspicuous plants are the summer-flowering geophytes such as *Pancratium maritimum* and *Drimia maritima*, which start to appear in July and August, respectively. Also of note are late-flowering Asteraceae such as *Carlina* and *Cynara* spp., and the diagnostic fruits of many Apiaceae and Brassicaceae.

Bottom of previous page, left: *Pancratium maritimum*; right: *Drimia maritima*; opposite left: *Cynara cornigera*; right: fruits of *Thapsia garganica*.

Autumn and winter interest

The onset of cooler nights and the first significant rainfall starts in late September and early October, when the first signs of vegetation appear in the landscape parched brown by the summer sun. The most conspicuous plants are the autumn-flowering bulbs, for example crocuses (*Crocus* spp., *Colchicum* spp.), daffodils (*Narcissus* spp.) and bright yellow sternbergias (*Sternbergia* spp.). Curiosities include the unusual, foetid, trap-pollinating arums (*Biarum* spp.), which are often smelled before they are seen, in October. Cyclamens such as *Cyclamen persicum* flower throughout the winter months, and in late winter and early spring, turban buttercups (*Ranunculus asiaticus*) begin to appear. By December or January, all but the most barren areas are lush with winter vegetation, quite unlike the brown summer landscapes seen by most tourists. The Mani Peninsula in the southern Peloponnese is a good area for many rare and endemic autumn-flowering bulbs, particularly in early November.

Clockwise from top left, autumn-flowering plants include *Colchicum bivonae*, *Narcissus obsoletus*, planted and naturalised belladonna lilies (*Amaryllis belladonna*), and *Sternbergia lutea*; late winter- and early spring-flowering plants include cyclamens (*Cyclamen persicum*) and turban buttercups (*Ranunculus asiaticus*).

Species descriptions

Cyclamen persicum

GYMNOSPERMS

Seed plants without true flowers, in which ovules are 'naked' rather than enclosed within an ovary.

PINACEAE | PINE FAMILY

Evergreen or deciduous trees with needle-like leaves borne spirally and vegetative buds with brown scales. Male cones with copious pollen; female cones borne on the same tree and more substantial, with woody scales, taking years to mature.

Cedrus CEDAR

Evergreen trees with needle-like leaves scattered on long shoots, whorled on lateral spurs. Cones erect; seeds winged.

Cedrus libani CEDAR OF LEBANON An evergreen conifer 20–30(40) m with a large, forked trunk and spreading canopy (narrowly pyramidal in dense forests). Leaves short and needle-like, to 17 mm long, in clusters of 15–35. Female cones 80 mm–12 cm; seeds winged. Mountain forests above 500 m. CY, LB, SY, TR. **Var. *brevifolia*** is the form on Cyprus, variably treated as a variety, subspecies or true species. It has shorter, thicker leaves, typically <15 mm long and incurved. Rare, in mountains above 900 m. CY.

Abies FIR

Evergreen trees with stalkless, needle-like leaves borne singly (not clustered) on long shoots. Species freely hybridising in the region.

Abies cephalonica GREEK FIR A pyramidal fir tree 25–30(40) m with spreading branches, *resinous buds* and *hairless,* cylindrical young twigs. Leaves 2–3 mm wide, needle-like, spine-tipped, with 2 white lines on the undersurface, surrounding the twigs. Female cones 10–20 cm long, cylindrical and blunt; seeds winged. Mountain forests above 800 m, primarily in the Peloponnese. GR (Pindus), IA (Cephalonia), PN. **A. alba** is a similar tree to 50 m but with *densely hairy* young twigs and obtuse needle-like leaves. Female cones 90 mm–17 cm. Mountain forests. GR (north). **A. × borisii-regis** MACEDONIAN FIR is of hybrid origin between *A. cephalonica* and *A. alba* and is intermediate in characteristics. Twigs with short, brown hairs, appearing flat with leaves arranged in lateral rows. Mountain forests. GR, PN (rare).

1. *Cedrus libani*
2. *Cedrus libani* var. *brevifolia*
3. *Abies cephalonica* male cones
PHOTO: NICK TURLAND

Pinus PINE

Evergreen trees with stiff, often long, needle-like leaves borne in clusters of 2, 3 or 5 on very small shoots.

(a) Cones large, >(50)70 mm long.

Pinus pinea UMBRELLA PINE A large, distinctive tree to 30(40) m, *parasol-shaped* when mature. Bark grey-brown with peeling red patches. Leaves green, 70 mm–15(20) cm long, 1.2–1.7 mm wide, with minute, forward-pointing teeth along the margins. Female cones 70 mm–14 cm long, broadly ovoid to spherical, shiny, red-brown with large, *scarcely winged* or unwinged seeds; scales more or less all the same size. Common on hills and in coastal habitats, often in large stands, frequently planted for the edible seeds. Throughout, except the north.

Pinus halepensis ALEPPO PINE A medium-sized tree to 20 m with an *irregular, rounded crown*, and distinctly twisted branches; branches and twigs *greyish*, bark becoming reddish and fissured; buds not sticky. Leaves needle-like, borne in pairs, 60 mm–12(15) cm long and *narrow*, just 0.7–1 mm wide, straight or twisted. Female cones egg-shaped, 60 mm–12 cm long, borne on *recurved* stalks; seeds *winged*. *P. brutia* TURKISH PINE is similar but *without grey twigs* and with *broader paired leaves* >1 mm wide, 10–16 cm long. Female cones 50 mm–11 cm. Seeds winged. Maquis below 600 m. Throughout.

(b) Cones small, <70(90) mm long.

Pinus sylvestris SCOTS PINE A tall tree to 36 m. Bark dark grey-brown below, *pale orange-red and flaking* above; *crown flat-topped and lopsided when mature*. Leaves *small and twisted*, 20–80 mm long, 3 mm wide, distinctly stalked, and in pairs with a persistent grey basal sheath; *blue-grey* and twisted. Female cone more or less spherical, acute, deflexed and yellowish-brown, 25–75 mm long. Mountain slopes and planted for timber. GR (north), TR (north).

Pinus nigra CORSICAN PINE A variable (with subspecies described), tall, robust tree to 42 m with a *straight trunk and open, pyramidal crown*; bark dark grey above and deeply fissured; twigs dark brown with slightly sticky buds. Leaves borne in pairs, dark green (not bluish), 80 mm–18 cm long, and 2 mm wide, scarcely twisted. Female cones solitary or clustered, long, egg-shaped, to 30–90 mm, yellowish-brown, shiny and more or less *unstalked*. Mountain forests above 500 m. **Subsp.** *nigra* has straight or incurved, spine-pointed leaves 70 mm–10 cm, and female cones 50–80 mm. GR (north). **Subsp.** *pallasiana* (also treated as a species) has rigid, twisted or curved leaves 12–18 cm long (to 12 cm, thick and rigid in populations on CY), and female cones 50 mm–12 cm. CY (Troodos), GR, TR.

1. *Pinus pinea*
2. *Pinus pinea* male cones
3. *Pinus halepensis* habit
4. *Pinus halepensis* female cones
5. *Pinus brutia* female cones
6. *Pinus brutia* habit
7. *Pinus nigra* subsp. *pallasiana*; (inset) detail

CUPRESSACEAE | JUNIPER FAMILY

Evergreen trees or shrubs with resin, leaves opposite or in whorls, scale-like or needle-like, usually in groups of 3; vegetative buds without bud-scales. Female cones with woody or succulent scales.

Juniperus JUNIPER

Evergreen shrubs with twigs spreading in 3 dimensions; leaves needle-like or opposite, scale-like when mature, borne on spreading (not flat) branches. Typically dioecious; female cones berry-like, with *fused, succulent* scales. Seeds not winged.

Juniperus communis COMMON JUNIPER A dense, greyish shrub to 26 m. Leaves needle-like, 4–20 mm long, borne in *groups of 3* with a *single* white band above. Female cones egg-shaped to spherical, 5–10 mm, green and later bluish-black when ripe (cones take years to ripen); male cones solitary. Mainly mountain forests above 1,000 m. Almost restricted to the mainland (regional subspecies exist). GR, TR.

Juniperus oxycedrus PRICKLY JUNIPER Similar to *J. communis*: a greyish dioecious shrub to 15 m with gradually narrowing, sharply spine-tipped needle-like leaves 8–25 mm long and 1 mm wide, with *2 pale bands* above, borne in whorls of 3. Female cones 8–11(15) mm across, ripening red-brown. Dunes and maquis. Throughout. *J. macrocarpa* is similar (also considered to be a subspecies of *J. oxycedrus*) but with leaves tapered at the base and *larger* cones 12–16 mm across. AA, CR (+ KP), LB, SY, TR.

Juniperus phoenicea PHOENICEAN JUNIPER A shrub to 8 m, with grey-brown bark peeling in narrow strips. Leaves 5–14 mm long (just 0.7–1 mm on older branches) and *scale-like, with membranous margins,* adpressed to the stem. Female cones 8–14 mm, spherical, red-brown when mature. Maquis and evergreen scrub. Throughout. **Var. *turbinata*** is the typical coastal form with elongated (turbinate), rather than spherical, female cones. Throughout. *J. foetidissima* has *strongly 4-angled stems,* leaves that lack membranous margins and female cones almost black when mature. Mountain slopes. AA (Euboea + Lesbos), GR, TR.

Cupressus CYPRESS

Evergreen trees with twigs spreading in 3 dimensions and opposite, scale-like leaves when mature. Female cones woody, separating into scales; seeds winged.

Cupressus sempervirens CYPRESS A tree to 30 m with either a conical or slender, columnar canopy. *Leaves dark green*, scale-like, just 0.5–1 mm long and closely overlapping. *Female cones spherical, 25–40 mm across* and yellowish. The slender form is widely planted in towns and gardens. Throughout, except IP. *C. macrocarpa* MONTEREY CYPRESS has dark green leaves 1–2 mm long and female cones 20–35 mm long. Native to California, USA, widely planted. *C. arizonica* has *pale blue-grey*, scale-like leaves. Native to Arizona, USA, widely planted.

---

1. *Juniperus communis*
2. *Juniperus oxycedrus*
3. *Juniperus phoenicea*
4. *Juniperus phoenicea* var. *turbinata*
5. *Cupressus sempervirens*
6. *Cupressus sempervirens* female cones

ARAUCARIACEAE | MONKEY-PUZZLE FAMILY

Very large, evergreen trees with columnar trunks and needle-like or flattened leaves borne spirally; vegetative buds without bud-scales. Male cones large, catkin-like and drooping. Seeds not winged.

Araucaria

Evergreen trees native to Australasia and South America; leaves broad and with many veins or awl-shaped and incurved.

Araucaria heterophylla NORFOLK ISLAND PINE A tall, erect tree to 60 m, with horizontal spreading branches arranged somewhat symmetrically in whorls around the trunk; leaves awl-shaped and incurved. Female cones squat and spherical, with woody, spirally arranged scales and wingless seeds. Native to Norfolk Island in the Pacific; widely planted.

EPHEDRACEAE | JOINT PINE FAMILY

Small shrubs with rush- or broom-like stems, small opposite or whorled leaves, often reduced to scales. Dioecious; male cones in small axillary clusters; female cones with several pairs of bracts. 'Fruit' (in fact a female cone) berry-like with 1–2 seeds.

Ephedra

Broom-like shrubs with branched stems and inconspicuous, scale-like leaves. Male cones with several 2–3-celled stamens.

Ephedra distachya JOINT PINE *A low, scrambling shrub to 1 m* (depending on exposure) with erect stems from creeping rhizomes, forming a thicket. Stems rigid and broom-like, leafless except for ash-coloured scale-like leaves to 2 mm at the joints, *thin, 0.7–1 mm across, greenish-yellow, and not easily broken at the nodes*. Male and female cones separate; greenish yellow and small, the male in clusters of 4–8 pairs, the female in solitary pairs. Mature female cone 5–7 mm, berry-like and reddish, the seed protruding. Local on dry rocky slopes, sea cliffs and river banks. AA, GR, TR.

Ephedra foeminea A sprawling shrub, similar to *E. distachya* but with dense, pendent stems with a whitish pith. Leaves to 2.5 mm, with *hairless* margins. Male and female cones separate; female cones borne on *conspicuously curved, narrowly cylindrical stalks*. Seeds 1–2 per cone. Woods, ravines and cliffs. Throughout. *E. major* (syn. *E. nebrodensis*) is similar but almost erect, with mature female cones with 1 protruding seed. AA (north), GR, PN.

Ephedra aphylla A sub-erect shrub, similar to the above species, but with leaves with minutely *fringed* (ciliate) margins and female cones borne on *straight stalks*. Semi-arid to arid areas in the southeast only. IP, LB.

BASAL ANGIOSPERMS

An ancient flowering plant lineage.

ARISTOLOCHIACEAE | BIRTHWORT FAMILY

Perennial herbs and woody climbers with creeping rhizomes. Leaves simple, alternate, untoothed, with heart-shaped bases. Flowers very distinctive, usually zygomorphic; stamens 6–12 in 1–2 whorls; ovary inferior; styles 6. Fruit a capsule.

Aristolochia BIRTHWORT

Herbs with stems >10 cm. Flowers *zygomorphic and tubular*, swollen at the base (pitcher-like); stamens 6. Root and micro-morphology are important but difficult to observe in the field. A diverse and complex group in the region.

(a) Flowers solitary; tube *scarcely curved*; lower lip without distinct lobes.

Aristolochia rotunda ROUND-LEAVED BIRTHWORT A scrambling or ascending perennial to 70 cm, with oval to heart-shaped, virtually *stalkless leaves*. Flowers 25–50 mm long, more or less straight-tubed, yellowish striped with brown, with a *large limb that droops* when mature. Fruit 20–25 mm. Dry, rocky and scrubby places, mainly on the mainland. AA (west), GR, TR. ***A. pallida*** is a similar, ascending perennial to 50 cm with broad, oval to kidney-shaped leaves 23–72 mm long, with *short stalks* to 15(25) mm. Flowers 24–57 mm, greenish-yellow with brownish stripes, the limb darker, short and slightly curved. Capsule ovoid, to 20 mm. Scattered on the mainland, absent from islands. GR, TR. ***A. elongata*** is similar, with brownish flowers 25–50 mm, with a straight tube, broadest above the middle and shorter limb. GR, IA, PN.

Aristolochia parvifolia A sprawling or ascending perennial to 25(50) cm with hairless, *small,* short-stalked (<10 mm) leaves, just 10–20(35) mm long. Flowers 30–50(60) mm, rather large relative to leaf size, ascending, *borne on very short stalks*; tube greenish-brown to dusky-purple with prominent dark brown veins in the throat; limb *extended, linear-lance-shaped*. Fruit *spherical*, 7–14(15) mm. Rocky slopes and woods. CY, GR, IP, LB, SY, TR.

(b) Flowers solitary; tube *strongly curved*; lower lip shallowly lobed.

Aristolochia sempervirens A *climbing, woody-based* perennial to 5 m with stalked, triangular-heart-shaped, *leathery, evergreen* and hairless leaves; lobes shallow. Flowers 20–50(60) mm long, borne on hairy stalks, dull purplish-brown with a *distinct yellow throat;* ovary hairy. Fruit 10–35(40) mm. Woodland, thickets and damp, shady habitats. AA (rare), CR, CY, IA (Cephalonia), IP, PN, TR.

Aristolochia cretica An erect or spreading *distinctly hairy* perennial to 60 cm with blunt, kidney-triangular-shaped, hairy leaves 20–55(60) mm long; leaf stalks <30 mm. Flowers *large*, 50 mm–12 cm long, dull purple, with *long white hairs* within; lower lip shallowly lobed. Gorges near the sea. Seeds 3–5 mm. An endemic of Crete, mainly in the east. CR. ***A. hirta*** is similar but with *acute* oval-triangular leaves and grey-brown flowers with shallow, *rounded* lower lobes. Seeds *larger*, 7–10 mm. AA (east), GR (northeast), TR.

Aristolochia guichardii A simple or branched perennial to 30 cm, similar to *A. cretica*. Leaves heart-shaped, 35–55 x 30–50 mm, with stalks 10–30 mm long. Flowers borne on hairy stalks, *small*, 20–35 mm long, with an *abruptly curved tube*; interior dark purple and *densely white-hairy*; lower lip shallowly lobed. Capsule 20–30 mm. Deciduous and pine woods and maquis. AA (Rhodes), TR (southwest).

(c) Flowers solitary; tube *scarcely curved*; lower lip *prominently lobed and heart-shaped* (southern Turkey only).

Aristolochia auricularia A short perennial with leafy shoots. Leaves borne on stalks 1–6 mm or stalkless with an oval-triangular blade; leaves small, 10–22 x 8–20 mm, and much wider than long. Flowers 25–40 mm long, borne on stalks exceeding the leaf stalks; *tube straight*, widened at the top; limb markedly expanded, slightly exceeded by the tube; lip heart-shaped at the base. Fruit spherical, to 30 mm. TR (southwest Anatolia). The following species are all similar: **A. isaurica** has *larger leaves,* always >26 mm, wider than long. Flowers with small limb to just 13 mm, greatly *exceeded* by the tube. TR (south-central Anatolia). **A. geniculata** has flowers with slightly bent tube, not widened at the top, and densely hairy within. Mountains. TR (Anatolia). **A. rechingeriana** has larger flowers >44 mm with a straight tube, distinctly widened at the top, sparely hairy within. TR (southwest Anatolia).

1. *Aristolochia rotunda*
2. *Aristolochia rotunda* (cross section)
3. *Aristolochia sempervirens* in fruit
4. *Aristolochia cretica* PHOTO: ARNE STRID

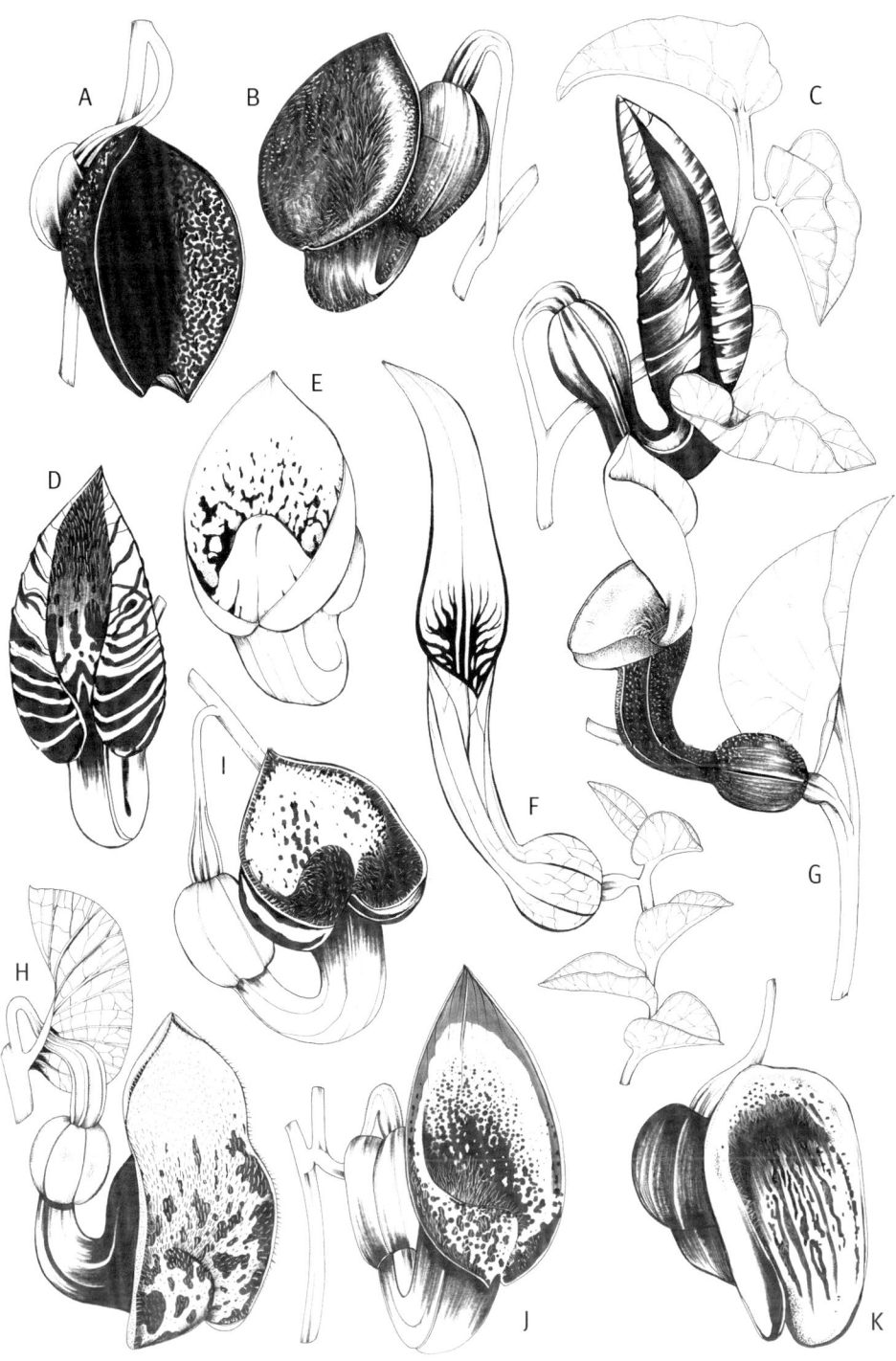

Aristolochia species: **A.** *A. poluninii*; **B.** *A. guichardii*; **C.** *A. incisa* (habit); **D.** *A. incisa*; **E.** *A. paecilantha*; **F.** *A. parviflora*; **G.** *A. krausei*; **H.** *A. bottae*; **I.** *A. maurorum*; **J.** *A. pontica*; **K.** *A. bodamae*. Illustrations not to scale.

(d) Flowers solitary; tube *strongly curved*; lower lip *prominently lobed and heart-shaped* (many similar species).

Aristolochia paecilantha A simple, ascending, hairy perennial to 40(50) cm. Leaves triangular to oval, 40 mm–10(12) cm x 30–80 mm, with *incurving lower lobes*; stalks 3–20 mm. Flowers 35–65(70) mm, borne on nodding stalks (shorter than flowers); *virtually hairless within*; tube *greenish-yellow*; throat with blackish-purple marbling; opening incurved heart-shaped at the base; lower portion of tube ovoid. Fruit oblong, 30–55 mm. Shady, rocky habitats; common in Lebanon. IP, LB, SY, TR. **A. scabridula** is similar but with shorter, rougher leaves, shorter flower-stalks and small, *tawny-reddish flowers*. IP (north), LB, SY.

Aristolochia maurorum A simple, ascending, *short-hairy* perennial to 40 cm. Leaves small, 25–80 x 15–50 mm, *short-stalked* (1–5 mm) and distinctly *linear-lance-shaped* with oblong-spatula-shaped lower lobes. Flowers 20–60(85) mm, borne on (equally) long, nodding stalks; tube yellowish-green to *rusty red*, rather hairless, with blackish-purple markings; strongly *expanded with heart-shaped lower lip, lobes often divergent*; throat white-hairy. Fruit oblong, 25–35(60) mm. Hedges and fields. IP, LB, SY, TR. **A. bottae** is similar but *branched*, with *broader*, triangular-arrow-shaped leaves 25–70 x 15–60 mm. Flowers yellowish with discrete brown-purple markings with sparse white hairs. Fields and gullies. IP, LB, SY, TR. **A. poluninii** is branched, with *longer leaves* to 11 cm. Flowers often *entirely brown* and *densely hairy within*, with a markedly expanded opening and broad round-heart-shaped lobes to lower lip. Pine forests. TR. **A. cilicica** has *unbranched stems* to 30 cm and yellowish flowers, variably marked brown on the limb and white-hairy in the throat. Mountains. TR (Taurus). **A. incisa** has slightly blue-grey, wavy-edged leaves 51–35 mm and flowers shiny blackish-purple outside, 25–45 mm, the opening with *strongly incurved margins*. AA (east), GR (northeast), TR.

Aristolochia pontica An ascending, shortly branched perennial to 50 cm. *Leaves large, about as long as wide*, up to 12.5 x 16 cm, round-oval to kidney-shaped and virtually hairless. Flowers greenish-purple to reddish, pale yellow internally with darker spots and purplish hairs in the throat; lower lip with deflexed lobes 10–20 mm long. Fruit to 25 mm. Riverbanks and damp, shady habitats. IP, LB, SY, TR.

Aristolochia bodamae Similar to *A. hirta* in form (but note the lobed lower lip). Leaves 12–14 cm long and 80 mm–12 cm across (*distinctly longer than wide*), triangular-oval with stalks, 30–50 mm. Flowers dull *blackish or brownish-maroon*, the interior hairy and yellowish with darker spots towards the margin; the *lower lip lobed* at the base (not rounded). Shady habitats. TR.

Aristolochia krausei Similar to *A. guichardii* in form (but with lower lip lobed to at least 2–2.5 mm deep). Leaves broadly oval-triangular, slightly longer than broad. Flowers purplish with a *distinctly expanded*, yellow opening; the lobes curved inwards and the limb often deflexed; throat with sparse white hairs. TR.

1. *Aristolochia paecilantha* PHOTO: YUVAL SAPIR
2. *Aristolochia scabridula* PHOTO: YUVAL SAPIR
3. *Aristolochia bottae*
4. *Aristolochia poluninii* PHOTO: NICK TURLAND

(e) Flowers solitary; *opening reduced to an apical pore* (without distinct limb or lip).

Aristolochia microstoma An ascending, simple or branched perennial to 40 cm with oval-heart-shaped leaves 10–30 mm long. Flowers very distinctive: short, yellowish, flushed purple towards the top, 15–30 mm long, with the *opening reduced to a small pore* (limb absent). Fruit 10–20 mm. Rocky screes. AA, PN.

(f) Flowers all yellow, borne in *clusters* of 2 or more.

Aristolochia clematitis BIRTHWORT A vigorous, more or less hairless *erect or ascending* (not climbing or scrambling) perennial to 1 m with numerous unbranched stems. Leaves to 15 cm, bluntly heart-shaped, with stalks 15–50 mm. Flowers to 20–35 mm long and entirely *yellow*; tube straight, borne on very short stalks. Capsule 20–50 mm. Fields, waste places and damp thickets in the north. GR (north, northeast – common), TR (north).

MAGNOLIACEAE | MAGNOLIA FAMILY

Trees with flowers arranged in rings, with stamens and pistils in spirals on a conical receptacle (an arrangement seen in fossil flowers).

Magnolia

Large trees with broad, leathery leaves. Flowers large and regular with 9–13 tepals; stamens and carpels numerous. Fruit an aggregate of follicles.

Magnolia grandiflora MAGNOLIA An exotic tree to 20 m (often shorter) with simple, oval, leathery dark green leaves to 20 cm long and large, white, scented flowers to 30 cm across. Native to the southeast United States; planted in towns and gardens. Throughout.

LAURACEAE | LAUREL FAMILY

Aromatic, evergreen shrubs and trees with simple, alternate leaves. Flowers small and cosexual, regular, with 8–12 stamens; style 1. Fruit a 1-seeded berry.

Laurus BAY LAUREL

Aromatic trees and shrubs, mainly tropical. Leaves with shining oil glands.

Laurus nobilis BAY LAUREL A monoecious shrub or small tree to 10(18) m with blackish bark and hairless shoots. Leaves 50 mm–10 cm long, alternate, oblong to lance-shaped, untoothed with slightly wavy edges and dotted with glands. Flowers to 10 mm across, rather inconspicuous: greenish yellow, borne in clusters in the leaf axils, with 4 petals. Fruit a black, oval berry. Rocky habitats and ravines, also frequently planted. Throughout, except smaller islands.

Persea AVOCADO

Trees and shrubs with large, fleshy, 1-seeded fruits.

Persea americana AVOCADO A tree to 15 m with large, oval-elliptic leaves, leathery and pointed at the apex. Flowers small, white, with 5 petals. Fruit an avocado pear. Native to Mexico and Central America; frequently planted in gardens in the region. Throughout.

1. *Aristolochia clematitis*
2. *Magnolia grandiflora*
3. *Laurus nobilis*
4. *Persea americana*

MONOCOTS

A major group of plants, typically with a single seed leaf, leaves with parallel (not net) veins, and flower parts in multiples of 3. Mostly herbaceous and with underground storage organs such as bulbs and corms.

ARACEAE | ARUM FAMILY

Perennial, usually hairless, tuberous herbs with leaves all basal and stalked. Individual flowers tiny, borne in a compact spike (spadix), enfolded in a large, often leafy bract (spathe). Fruit a berry.

spathe

spadix (limb)

Arum

Tuberous, usually spring-flowering perennials with arrow-shaped, untoothed leaves. Flowers unisexual, borne at the base of a spadix, which is enveloped by the spathe. Fruit a red berry, produced generally after the leaves have withered.

(a) Spadix cylindrical and *yellow*, spathe greenish with some, or no purple pigmentation.

Arum italicum ITALIAN ARUM A perennial with a large tuber and long-stalked leaves appearing in autumn; leaves large, 15–35 cm long, arrow-shaped with pointed lobes, semi-erect on long stalks; mid to dark green with variable paler marbling. Visible portion of the spadix *narrowly* club-shaped and pale yellow, ¼–½ the length of the spathe limb; spathe to 40 cm long, pale yellow-green; male flowers yellow. Berries red, borne in dense spikes. Damp hedgerows and ditches. Throughout. *A. concinnatum* is similar but with a distinct, *robustly* club-shaped spadix, *½ to equal* the length of the spathe limb. Woods and scrub at various altitudes. AA, CR (+ KP), PN, TR (southwest). *A. byzantinum* is a local species, often confused with *A. concinnatum*, but with narrow leaves without pale markings, a much *narrower* spadix appendage, and *purplish male flowers*. TR (northwest).

Arum creticum A distinctive perennial with shiny arrow-shaped leaves appearing in autumn; lobes divergent. Inflorescence borne above the leaves; spathe 70 mm–12 cm, *cream, whitish or pale green, reflexing at maturity*; spadix narrow, cylindrical and long-protruding, typically bright *yellow-orange*; sterile flowers few or absent; inflorescence sweetly scented. Stony hillsides. AA (Rhodes – rare), CR (+ KP), TR (southwest).

(b) Spadix cylindrical and *brown*, spathe greenish-white.

Arum idaeum A perennial with arrow-shaped leaves, similar in form to *A. creticum*. Spathe erect, greenish outside, *white within*, with contrasting *dark brown spadix*; spadix long-cylindrical, almost as long as the spathe; sterile flowers absent; inflorescence scentless. Mountain woods only; a rare island endemic. CR.

(c) Spadix cylindrical and *brownish* (rarely pale), spathe marked with at least some purple.

Arum hygrophilum A perennial with a discoid tuber, leaves long-stalked and erect, with basal lobes ⅓ of the length of the blade. Inflorescence borne beneath the leaves, long-stalked; spathe elongated ovoid, white-green with a distinct *purple border*; spadix purple and *very narrow, <3 mm across*; inflorescence not foetid. Gorges and streamsides, strongly eastern. CY, IP, LB, SY, TR.

1. *Arum italicum*
2. *Arum italicum* fruits
3. *Arum concinnatum* (cross section)
4. *Arum concinnatum*
5. *Arum creticum*
6. *Arum idaeum*   PHOTO: NICK TURLAND
7. *Arum hygrophilum*
8. *Arum hygrophilum* (cross section)

Arum maculatum WILD ARUM A perennial with a large tuber and long-stalked leaves appearing in *spring*, 70 mm–20 cm long, wrinkled, sometimes spotted, the lower lobes not markedly divergent. Inflorescence foetid, the visible portion of the *spadix chocolate-brown* (sometimes grey or yellow), ‹½ the length of the spathe limb; spathe, if flushed purple, darker towards the edges (not centre). Cool, damp woods; absent from most islands and hot, dry areas. GR (north), TR.

Arum orientale A perennial with a discoid tuber from which leaves appear in the autumn; leaves arrow-shaped and deep green, often with purple flushed stalks (but without dark spots). Inflorescence with a white–green spathe *evenly suffused with purple*, especially along the margin, but white within the tube; visible portion of spadix *dark* purple–black *long and narrow* (not club-shaped), 3–6 mm across and ›½ the length of the spathe limb; male flowers in 2 whorls; smelling of horse dung. GR (north).

(d) Spadix cylindrical and *brownish* (rarely pale), spathe strongly marked with, or entirely purple.

Arum nigrum A perennial with leaves to 21 cm, appearing in autumn. Spathe *elliptic* to boat-shaped, entirely *dark purplish-black* internally (unique in the region covered), 15–20 cm long; spadix robust, purplish-grey, about ½ the length of the spathe, *club-shaped with a long stipe*; sterile flowers spine-like; inflorescence smelling strongly of dung. Stony ground, rare and local. AA (northwest), GR (north). ***A. palaestinum*** is similar to *A. nigrum* but with a lance-shaped spathe 10–23 cm long and blackish, *long, cylindrical spadix with a short stipe,* ½ to almost equalling the length of the spathe; inflorescence weakly scented. Open hillsides in the east of region only. IP, LB, SY.

Arum gratum A perennial with arrow-shaped leaves to 18 cm. Spathe small, 12–18 cm (the base often partially buried), purple above, whitish below; spadix short, 55–65 mm and club-shaped; sterile flowers thread-like; inflorescence *sweetly scented. Cedrus* and deciduous woods. LB, SY, TR. ***A. elongatum*** is similar but with a much longer, stouter spadix, *almost equalling the spathe*; strongly smelling of dung. Rare and local. GR (north).

Arum cyrenaicum A perennial with arrow-shaped leaves to 20 cm. Spathe 12–16 cm, *strongly flushed pale red–purple on the inner surface*, typically *paler towards the centre*; spadix narrowly cylindrical, ½ to ¾ the length of the spathe and pale brown; sterile flowers thread-like; inflorescence smelling faintly of dung. Among rocks in shady scrub. Very rare, on Crete (also Libya). CR (southwest). ***A. purpureospathum*** is similar but with an intensely and *uniformly blackish-purple spathe interior* and mauve (not yellow) sterile flowers. CR (southwest + KP – rare).

Arum cylindraceum (syn. *A. alpinum*) A perennial, similar to the species above. Inflorescence borne on long stalks (equalling or exceeding the leaf stalks), rather small, unscented; spathe *small*, 90 mm–14 cm, greenish to whitish, flushed pale purple; spadix ½ the length of the spathe, gradually tapering into the stipe, purplish. CR, GR, TR.

1. *Arum maculatum*
2. *Arum nigrum* PHOTO: ARNE STRID
3. *Arum palaestinum*
4. *Arum cyrenaicum*
5. *Arum purpureospathum*
6. *Arum dioscoridis*
7. *Arum dioscoridis* (cross section)
8. *Arum rupicola* habit, in rocky scree
9. *Arum rupicola* PHOTO: NICK TURLAND

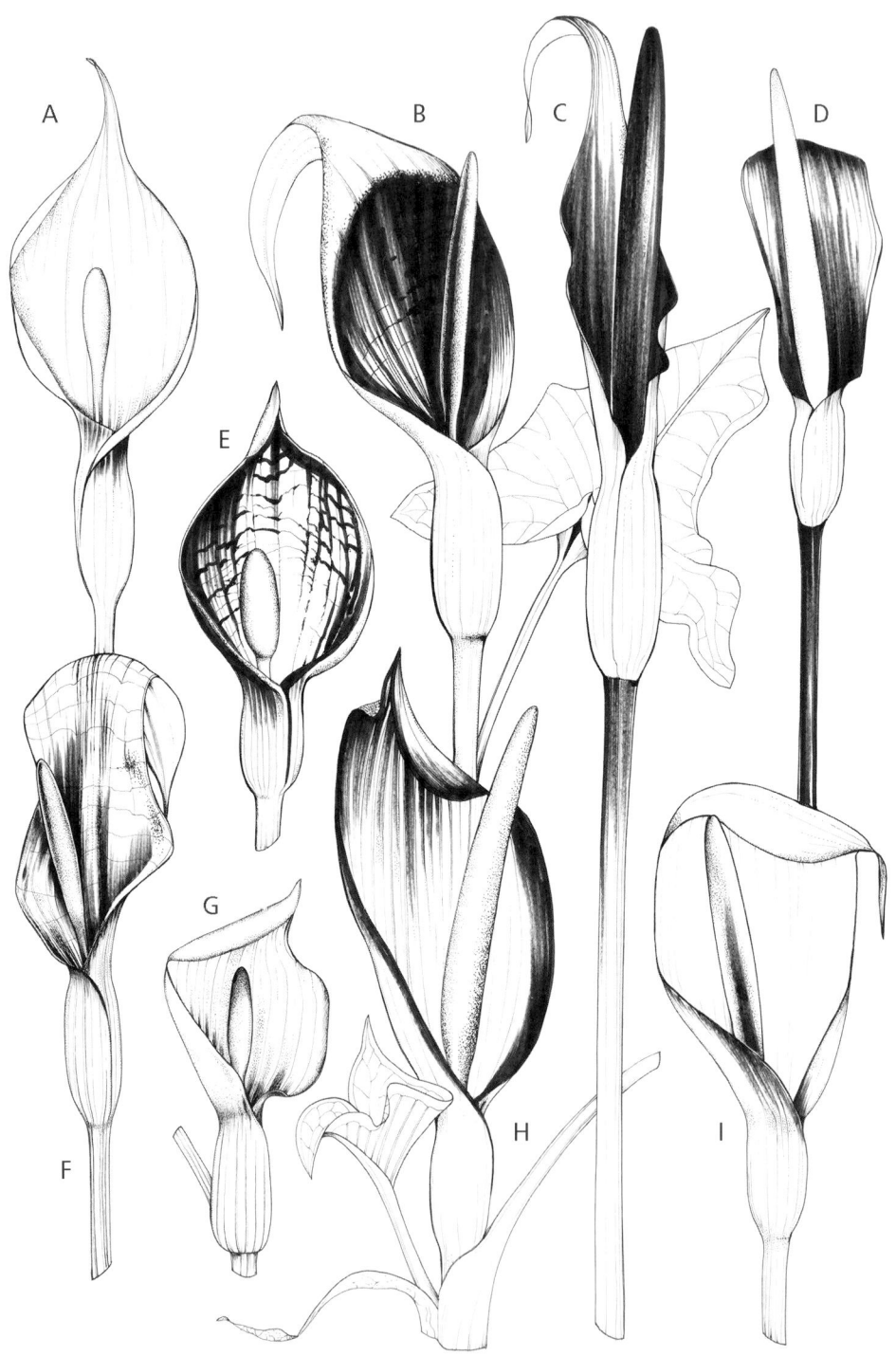

Arum species: **A.** *A. byzantinum*; **B.** *A. palaestinum*; **C.** *A. rupicola*; **D.** *A. elongatum*; **E.** *A. nigrum*; **F.** *A. cyrenaicum*; **G.** *A. gratum*; **H.** *A. orientale*; **I.** *A. concinnatum*. Illustrations not to scale.

(e) Spadix cylindrical, *brownish, purple or black*; spathe conspicuously *spotted* with purple.

Arum dioscoridis A perennial with leaves to 30 cm long with strongly divergent basal lobes. Spathe to 30 cm long, *strongly suffused with purple spots*; spadix cylindrical, brown, 70 mm–15(20) cm; inflorescence *strongly* smelling of dung. Streamsides, field borders and rocky hillsides; strongly eastern. AA (east), CY, IP, LB, SY, TR (south).

(f) Spadix brownish and *fusiform*.

Arum rupicola A large perennial with leaves to 30 cm with slightly divergent basal lobes. Inflorescences borne on *long stalks to 50 cm – exceeding the foliage*; spathe large, to 50 cm, dull purple; spadix slightly shorter than the spathe, with a scarcely defined stipe, *robustly conical–cylindrical in shape* (fusiform). Rare on scree and mountainsides. AA (Lesbos), CY (mainly Paphos Forest – Troodos), IP, SY, TR.

Arisarum

Small, tuberous perennials with heart-shaped leaves. Inflorescences like those of *Arum* but smaller and the spathe fused over most of its length, thus tube-like. Fruit green, several-seeded.

Arisarum vulgare FRIAR'S COWL An autumn-flowering, short perennial with basal leaves 20 mm– 17 cm long, broadly heart-shaped, with rounded basal lobes. Flower-stalk more or less equal to the leaf stalks (55 mm–35 cm), sometimes spotted. Spadix greenish or brown, slightly protruding from the white, green to to chocolate brown-striped, *hooded spathe*; spathe tubular–cylindrical below and drooping forward at the apex. Berries greenish to blackish. Widespread and common on garrigue and maquis. Throughout.

Biarum

Small tuberous perennials with narrow *unlobed* (not arrow-shaped) leaves. Inflorescences virtually stalkless above ground. Fruits greenish, whitish or purplish, 1-seeded. Closely related and difficult to distinguish in the field: careful observation of sterile flowers is important. Mostly autumn- to winter-flowering.

(a) Spathe long-tapering; spadix *brown* (many similar species in the east of range).

Biarum tenuifolium An autumn-flowering perennial to 20 cm with leaves appearing after flowering. Leaves 11–29 cm (from the base), oblong to lance-shaped (linear-lance-shaped late in season). Inflorescence borne at ground level; spadix long and slender, brown, and often *much-exceeding* the purplish spathe, with sterile, thread-like flowers *above* and below the fertile male flowers, 1.5–3 mm long. Fruit a *whitish* berry. Garrigue, rocky slopes and stony pastures. AA, CR, GR (south), IA, PN. The following are very similar but *without* sterile male flowers above the fertile male flowers: **B. rhopalospadix** is *spring-* (not autumn)-*flowering* with oblong to spatula-shaped leaves and sterile male flowers *hooked and thickened*. GR (Attica); PN. **B. angustatum** is autumn-flowering with narrowly elliptic leaves and *markedly downward-pointing* sterile male flowers 3–4 mm long. IP, SY. **B. carduchorum** is late autumn-flowering with broadly elliptic leaves and *upward-pointing* sterile male flowers 3–7 mm long. Strongly eastern. SY (south); TR (south). **B. syriacum** has linear to linear-ellipctic leaves and spreading to erect, thread-like sterile male flowers 3–6 mm long. Rare on red clay in Syria only. SY (northwest). **B. eximium** has linear to linear-ellipctic leaves and purple sterile male flowers spread evenly between the fertile male and female flowers. TR. **B. olivieri** is distinguished by its markedly swollen spathe tube, exceeding the width of the inrolled limb. IP (south).

(b) Spathe long-tapering; spadix *yellow*.

Biarum auraniticum A small, winter-flowering bulb with long-stalked, channelled leaves 60 mm–12 cm long. Inflorescence similar in form to that of *B. tenuifolium* but the spathe *yellow to greenish-white* (sometimes flushed maroon); *spadix yellow*. Open hillsides and volcanic habitats. IP (Golan), SY.

(c) Spathe small, white and *flask-shaped*; spadix brown.

Biarum davisii A very distinctive autumn-flowering perennial with 5–9 elliptic to oval-shaped leaves 30–40 mm long. Spathe 50–60 mm long, flask-shaped, whitish, spotted with pink–brown; spadix equalling or just exceeded by the spathe, brown and forward-pointing, 2 mm thick. Fruit a dirty-white berry. A rare island endemic. CR. *B. marmarisense* is similar but with a spathe 70–80 mm long and *thread-like spadix* just 0.5 mm thick. Red limestone garrigue. AA (Simi), TR (west).

(d) Spathe *much-reduced*; spadix yellow.

Biarum ditschianum An unmistakable spring-flowering perennial with linear to lance-shaped leaves to 20 cm. Spathe *much-reduced*, typically <20 mm above ground, purplish-red within; spadix robust, *much exceeding* the spathe, 70–80 mm long, yellow, the very base purplish with numerous *white hairs*. Limestone garrigue and rocky crevices. AA (Kastellorizo – rare), TR (southwest).

Eminium

Similar to *Biarum* but with *lobed* leaves and 2-seeded fruits.

Eminium spiculatum A spring-flowering perennial with long-stalked, lobed leaves. Inflorescence borne at ground level simultaneously with the leaves; spathe deflexed, blackish-purple and *warty and wrinkled within*; the tube greenish and faintly purple-spotted outside; spadix blackish or yellowish, exceeded by the spathe. Fruit a 2-seeded berry. East of region only. IP, LB, SY, TR.

Zantedeschia ARUM LILY

Rhizomatous herbaceous perennials with long-stalked, broadly oval leaves with heart-shaped bases. Spadix without an appendix. Fruit a yellow berry, several-seeded. Native to southern Africa.

Zantedeschia aethiopica ARUM LILY A hairless perennial with rather fleshy, arrowhead-shaped leaves 10–45 x 10–25 cm, borne on long stalks to 75 cm. Flower-stalks somewhat exceeding the leaves; *spadix yellow*, to 15 cm long, and about ½ the length of the spathe; individual flowers indistinct; *spathe pure white*. Berries yellow (often not developing). Planted and naturalised in ditches. Throughout.

---

1. *Arisarum vulgare*
2. *Biarum tenuifolium*
3. *Biarum tenuifolium* (cross section)
4. *Biarum auraniticum*   PHOTO: YUVAL SAPIR
5. *Biarum davisii*   PHOTO: NICK TURLAND
6. *Biarum marmarisense*   PHOTO: ARNE STRID
7. *Biarum ditschianum*   PHOTO: NICK TURLAND
8. *Eminium spiculatum*

Dracunculus DRAGON ARUM

Large, robust tuberous perennials with deeply palmately lobed leaves. Inflorescences with a long, prominent spadix (to 40 cm). Fruit a several-seeded berry.

Dracunculus vulgaris DRAGON ARUM A robust, erect perennial to 1 m with a solitary, purple-spotted stem. Leaves green with white streaks, *deeply divided into lance-shaped, more or less equal segments to 20 cm long*. Spathe maroon–purple and hairless within, large, 20–40 cm long, tapering to the tip, soon drooping; spadix blackish and shiny; inflorescence foul-smelling. Fruit a red berry, borne numerously in a large, compact head. Locally common in gorges and garrigue at a range of altitudes (absent from CY, the southeast and most of the west). AA, CR (+ KP, common), GR (rare in the west), TR (west).

ALISMATACEAE

Annual or perennial, hairless, aquatic herbs, rooted in mud, with erect or floating leaves; leaves stalked, with linear to rounded blades. Flowers cosexual and regular, arranged in panicles or racemes; sepals and petals 3; stamens 6–numerous; style absent, ovary superior. Fruit a head of achenes or few-seeded follicles.

Baldellia LESSER WATER-PLANTAIN

Aquatic herbs with all basal leaves in a rosette. Flowers pale mauve or white, long-stalked, borne in umbels; stamens 6; carpels spiralled.

Baldellia ranunculoides A perennial herb to 20(50) cm. Leaves all basal, with stalks 70 mm–33 cm; leaves lance-shaped. Flowers cosexual, borne on long, spreading stalks, with 3 pale pink petals. Fruit an aggregate of 35–55 achenes. Pond margins and seasonally wet fields. Absent from much of the south and east; rare. AA (rare), GR (west), PN, TR (west).

Damasonium THRUMWORT

Aquatic herbs with all basal, usually submerged or floating leaves. Flowers with white petals each with a basal yellow blotch; stamens 6; carpels 6–10 in 1 whorl. Fruits with *carpels long-beaked and spreading in a star*.

Damasonium bourgaei A submerged aquatic, annual herb to 35 cm with floating oval to lance-shaped leaves 20 mm–30 cm long. Flowers cosexual with 3 pinkish-white petals 25–35 mm long. Fruits star-like, follicles with 2 seeds; seeds <1.5 mm. Still waters. GR, PN (northwest – rare), TR. *D. polyspermum* is similar to the previous species but with fruits with *follicles containing 5–16 seeds*. GR, IP (north), SY.

Sagittaria ARROWHEAD

Aquatic monoecious perennials with leaves all basal. Flowers *unisexual*, borne in whorls on leafless stems; male above, female below; stamens 7–numerous; carpels spiralled in a rounded head. Leaf shape variable (sometimes misleading).

Sagittaria sagittifolia ARROWHEAD A monoecious, rather large-flowered perennial with leaves all basal, of 3 forms: submerged and linear; floating and elliptic; aerial and arrow-shaped. Flowers 20–30 mm across, white with a *dark purple centre and purple anthers*; *bracts free at the base*. Stagnant or slow-moving water. Absent from hot, dry areas. GR, TR.

---

1. *Dracunculus vulgaris*; (inset) cross section
2. *Dracunculus vulgaris* fruits
3. *Baldellia ranunculoides*
4. *Sagittaria sagittifolia*

Alisma

Aquatic perennials with leaves all basal, linear or narrowly elliptic to oval. Flowers many, each with 3 petals, white or pale mauve with a yellow basal blotch; stamens 6; carpels numerous in 1 whorl, 1-seeded and curved inwards.

Alisma plantago-aquatica WATER-PLANTAIN An erect, hairless, aquatic perennial with leaves and stems above the water, 20 cm–1.2 m. Leaves all basal, long-stalked and *oval–elliptic* with *rounded lobes at the base*; blade 36 mm–29 cm long. Numerous whorls of pale flowers form a branched, pyramidal inflorescence; flowers 7–12 mm across, long-stalked; anthers 2 x as long as broad. Seeds without protuberences. Muddy and damp habitats or shallow water. Throughout, in suitable habitats. *A. lanceolatum* is similar but with leaves with blades graduating into their stalks; blade 50 mm–30 cm long. Fruits widest in the middle, with straight, erect styles. Throughout. *A. gramineum* has *linear to linear-lance-shaped leaves*; blade just 35–55 mm long. Fruits widest in the upper ½ and with recurved styles. Rare. GR, PN.

BUTOMACEAE | FLOWERING RUSH FAMILY

Perennial, aquatic, hairless herbs rooted in mud, with linear, all basal leaves. Flowers borne in *Allium*-like umbels with flowers in parts of 3; stamens 9; style 1, carpels 6. Fruit a group of 6 splitting follicles.

Butomus

Hairless aquatic perennials with basal leaves. Sepals and petals 6, all similar. Follicles virtually free.

Butomus umbellatus FLOWERING RUSH A distinctive, erect, aquatic perennial to 1.5 m. Leaves all basal, linear, pointed and 3-angled, almost as tall as the flowering stems. Flowers pale pink–white, with similar petals and sepals 10–15 mm long, borne in umbels with brownish bracts on tall, rounded stems. Local, in still or sluggish water. Throughout.

JUNCAGINACEAE | ARROWGRASS FAMILY

A small family of hairless, aquatic herbs with linear, sheathing basal leaves. Flowers small and green, borne in erect spikes or racemes; tepals and stamens 6; carpels either 3 or 6, joined to a superior ovary. Fruit 3 or 6 1-seeded units splitting when ripe.

Triglochin

Aquatic, hairless herbs, superficially *Juncus*- or *Plantago*-like but with 3 or 6 1-seeded fruit segments.

Triglochin barrelieri (syn. *T. bulbosa* subsp. *barrelieri*) A short perennial to 35 cm with a bulbous rootstock with fibres 0.8 mm wide. Developed leaves 4–7(18), linear, tapered to a point and 70 mm–35 cm long. *Flowers borne in spring*, along with the leaves, virtually stalkless, tiny and greenish, to 2.5 mm, borne in *slender spikes* 25 mm–25 cm long. Fruit elliptic and spreading, 6–10 mm long. Local in saltmarshes and on river banks. Throughout. *T. laxiflora* (syn. *T. bulbosa* subsp. *laxiflora*) is similar but with leaves developing *after flowering* and fibres 0.3 mm wide. *Flowers borne in autumn*. Similar habitats. GR.

ZOSTERACEAE | EELGRASS FAMILY

A small family of marine perennial grass-like herbs. Flowers unisexual or cosexual, borne in 2 rows in a congested inflorescence enclosed in sheath-like bracts; perianth absent; stamen 1; style 1 with 2 stigmas. Fruit a stalkless drupe.

Zostera

Grass-like marine herbs distinguished by their complex, *congested* inflorescence of *unisexual* flowers enclosed in a leafy sheath.

Zostera marina COMMON EELGRASS A submerged marine perennial with rhizomes to 2 mm across with 1–2 bundles of roots at each node, and leaves 50 cm–1.2 m long and *broad*, (2)4–20 mm wide, with 5–11 veins, and broad, rounded, *bristle-tipped points*; *basal sheaths entire, not split*. Flowers greenish, borne in terminal, much-branched inflorescences, enclosed in the base of the sheath; stigma 2 x the length of the style; female flowers with *1 style*; male flowers with *1 stamen*. Locally abundant in suitable habitats; fine sandy or silty substrates submerged to 10 m. Widely recorded in the area but possibly in error; rare or absent in most areas. *Z. noltii* is similar but with a *smaller rhizome*, <1.5 mm across, *with up to 4 bundles of roots* at each node and *narrower leaves,* just 0.5–1 mm wide with 3 veins, to 22 cm long; *leaf sheaths open down 1 side*. AA (north).

POSIDONIACEAE

Submerged marine perennials with flowers in branched, few-flowered spikes borne in the axils of long bracts; flowers regular; male flowers on the upper spikes and cosexual flowers below; stamens 3; stigma disk-like. Fruit fleshy and about the size of an olive.

Posidonia

Grass-like marine herbs distinguished by their inflorescence of male and *cosexual* flowers borne in *cymes*.

Posidonia oceanica SEAGRASS A grass-like perennial with stout rhizomes *densely covered in numerous brown fibres formed from old leaf bases*. Leaves strap-shaped, 40–70 cm long, dark green with a blunt, rounded tip and 13–17 longitudinal veins. Flowers greenish (rarely produced), 2–3 in cymes, cosexual below, male above. Fruit fleshy and ovoid, 20–32 mm. Common in the northern Aegean, rare or absent further south.

1. *Alisma plantago-aquatica*
2. *Posidonia oceanica* (commonly seen balls formed from fibres)

CYMODOCEACEAE

A small, mainly tropical family of marine perennials with narrow, flat leaves with 3–numerous veins and serrated edges. Flowers inconspicuous, unisexual, with petals absent, borne solitary or in pairs, or in a cyme; male flowers with 2 anthers; styles 1–3. Fruit stony and compressed.

Cymodocea

Marine, grass-like perennials distinguished by their *solitary, unisexual* flowers.

Cymodocea nodosa A grass-like perennial with leaves in groups of 2–7 on short shoots and annual leaf scars at the nodes. Leaves dark green, to 40 cm long, *narrow*, to just 4 mm wide, 7–9 veined and spiny-toothed towards the tips. Flowers solitary, without a perianth; male flower with a filament-like stalk, *stamens 2*; female flower stalkless with *2 styles* and *2 thread-like stigma lobes*. Fruit laterally compressed. Local on submerged marine sandy substrates. Throughout – the most common eastern Mediterranean sea grass.

DIOSCOREACEAE

A primarily tropical family including the yam. Leaves simple and alternate. Plants dioecious; flowers small and greenish borne in spikes or racemes in the leaf-axils; stamens 6 (vestigial in female flowers); style 1, stigmas 3. Fruit a several-seeded berry.

Dioscorea

A twining herb unusual in having dioecious, inconspicuous flowers and red berries (*Bryonia* shares these characteristics but with hairy, palmately lobed leaves and tendrils).

Dioscorea communis (syn. ***Tamus communis***) BLACK BRYONY A twining perennial climber to 5 m, dying back annually to a tuber; superficially similar to *Smilax* but spineless. Leaves glossy dark green, heart-shaped and long-stalked, 50 mm–15 cm x 40 mm–11 cm. Flowers greenish-yellow, 3–6 mm across, borne in loose racemes; male flowers with 6 stamens, female flowers with 6 minute lobes and a conspicuous ovary. Fruit a rounded red berry 10–13 mm across. Locally common in damp woods and thickets. Throughout.

MELANTHIACEAE

Lily-like herbaceous perennials with bulbs or rhizomes. Leaves typically with sheathing bases. Flowers often borne in branched inflorescences. Fruit a capsule or berry.

Paris

Hairless perennial herbs with horizontal rhizomes. Leaves 3–6 in a terminal whorl. Flowers solitary. Fruit a berry.

Paris quadrifolia HERB PARIS A distinctive herbaceous perennial 10–50 cm, with *leaves borne in a whorl of 4*, 50 mm–16 cm long. Flower solitary and terminal, borne on a stalk 20–60 mm; tepals 4(6); anthers 8(10). Fruit a berry to 18 mm long. Cool mountain woods; absent from arid areas. GR.

COLCHICACEAE

Herbaceous perennials with rhizomes or corms, previously included in the Liliaceae. Leaves few, alternate and sheathing below. Flowers 1–few, conspicuous; tepals and stamens 6; styles 3. Fruit a capsule.

Colchicum AUTUMN CROCUS

Cormous perennials with basal leaves appearing with, or after the flowers. Flowers crocus-like with 6 spreading tepals; stamens 6; styles 3. Fruit a capsule, borne centrally in the leaves. A variable and difficult genus in the region with numerous species (many not described here).

(a) Leaves partially or well-developed during flowering, in autumn.

Colchicum cupanii MEDITERRANEAN MEADOW SAFFRON A low, autumn-flowering perennial, normally with *2 leaves*, linear to linear-lance-shaped, sometimes with hairy margins at the base; *leaves present at flowering-time* but initially short, to 15 cm long when mature. Flowers pale pink to lilac with markedly *narrowly* elliptic tepals 10–25 mm long; anthers purplish-black and styles yellow. Rocky habitats and maquis. AA, CR (rare), GR. The following species are similar: *C. pusillum* has *3–8 narrow leaves* just 1–4 mm wide, typically developed during flowering (except at high altitudes). Flowers 1–4, white to dark pink with blunt tepals 10–20 mm long. Anthers 1.5–3 mm, grey–brown to dark purple. Rocky habitats and pastures. AA, CR (common), CY GR, PN. *C. cretense* has large anthers 2.5–4.5 mm long. CR (mountains). *C. stevenii* has *bright purplish-pink* tepals, 20–30 mm long; anthers 2.5–4.5 mm, *yellow*. CY, SY, TR.

Colchicum baytopiorum A low, autumn-flowering perennial typically with *3* partially developed leaves; leaves *recurved*, narrowly lance-shaped, 22–32 cm long when mature. Flowers 1–2, funnel to bell-shaped, mid-pink with tepals 25–35 mm; anthers 5–7 mm, lemon yellow. Maquis and open limestone woods. AA (Rhodes), TR.

1. *Dioscorea communis*
2. *Paris quadrifolia*
3. *Colchicum cupanii*
4. *Colchicum baytopiorum*

(b) Leaves partially or well-developed during flowering in winter to spring.

Colchicum triphyllum A low, spring-flowering perennial with 3 partially developed linear to lance-shaped, erect to spreading leaves during flowering, sometimes rough along the margins, 11–15 cm long when mature. Flowers white to dark pink with blunt tepals 6–12 mm long; anthers 2.5–3.5 mm, purplish or greyish. Open, stony habitats. GR (mountains), TR.

1. *Colchicum autumnale*
2. *Colchicum graecum*
3. *Colchicum graecum* (cross section)
4. *Colchicum bivonae*
5. *Androcymbium rechingeri*
6. *Androcymbium rechingeri* in fruit
7. *Androcymbium palaestinum* PHOTO: ORON PERI

(c) Leaves undeveloped during flowering; flowers not chequered.

Colchicum autumnale MEADOW SAFFRON An autumn-flowering perennial with a dark brown long-necked corm with a *thin, papery* tunic. Leaves 3–4(5), *broadly lance-shaped*, 30–48 mm wide, often twisted. Flowers 1–3 with pink tepals, not chequered, 30–50 mm long. Damp meadows and other grassy habitats. Woods and pastures. GR (north). *C. parnassicum* has arching leaves and a *membranous* corm tunic. PN.

Colchicum graecum A perennial similar to the previous species, with a long-necked corm with a thicker, *leathery*, reddish to blackish tunic. Leaves more or less erect, to 25 cm long and 60 mm wide, usually *scarcely or not twisted*. Flowers 2–5 with pale pink tepals, not chequered, to 60 mm long. Mountain slopes. Rare. GR, PN.

Colchicum troodi A low, autumn-flowering perennial with 3–4(6) leaves borne in spring, 12–20 cm long. Flowers 2–6, white to pink with narrowly oblong tepals 28–45 mm long. Anthers 6–7 mm, yellow. Rocky maquis and pine forests. CY, IP (north), TR.

Colchicum parlatoris A late summer- to autumn-flowering perennial with 4–9(10) leaves developing after flowering. Flowers 1(2), light purplish-pink (to white) with narrowly elliptic lobes to 50 mm long; anthers 4–6 mm, yellow. IA, PN.

(d) Leaves undeveloped during flowering; flowers normally *chequered*.

Colchicum bivonae A low, autumn-flowering perennial. Leaves borne in spring in clusters of 5–9, linear-lance-shaped or lance-shaped, hairless and sub-erect, 15–25 cm long. Flowers cup-shaped with tepals 50–60 mm long, borne in clusters, pink to lilac and *strongly chequered*; tepals not twisted; anthers brownish-purple. Rocky slopes and maquis. AA (rare), GR (+ adjacent islands), IA, PN, TR. The following local species are similar: *C. euboeum* has dark brown outer corm tunics with long necks to 13 cm, and flowers often appearing in summer, with yellowish anthers. AA (Euboea), GR (north). *C. boissieri* has smaller, rhizome-like corms and uniformly pink tepals. *C. chalcedonicum* has shorter, broader, grey–green leaves, short-necked corms and 1–2(3) flowers with tepals 30–55 mm long. GR (north and east + adjacent islands), TR.

Colchicum macrophyllum An autumn-flowering perennial. Leaves borne in spring, *large, broad and channelled*, 22–44 cm. Flowers 1–5, chequered pink (rarely all white) with tepals 45–70 mm long; anthers yellow. Rocky habitats. AA (east + Euboea), CR, TR (southwest).

Androcymbium

Herbaceous, cormous perennials. Inflorescences congested with enlarged floral bracts; stamens 6; carpels 3. Species all similar; capsule dehiscence is a diagnostic character (to be observed in mature fruits only). Also classified in the genus *Colchicum*.

Androcymbium rechingeri A small, bulbous, late winter-flowering perennial to just 50 mm. Leaves 6–10, in a *rosette at ground level*, 60 mm–10 cm, linear-lance-shaped, glossy green and long-tapering to a point. Flowers 1–few, *white*, 20–25 mm across with 6 *short-pointed* tepals with thin purplish stripes; stamens 6; styles 3, all free. Fruit a 3-valved, *non-splitting*, pear-shaped capsule. Coastal sands, a very rare island endemic of just 4 locations on Crete and associated islets. CR (west + islets of Imeri Gramvousa and Elafonisi). *A. palaestinum* is a similar species to 10 cm with a lowermost pair of lance-shaped, tapering leaves, the others *oblong-lance-shaped*; leaves overtopping the scapes. Flowers whitish-*lilac* with prominent darker stripes; tepals 16–18 mm. Deserts and arid cliff tops in the southeast of the region. IP (eastern).

SMILACACEAE

Climbing lianas with heart-shaped leaves, hooked spines and paired tendrils at the leaf bases. Flowers unisexual; tepals and stamens 6; style absent, stigmas 3. Fruit a berry with 1–several seeds.

Smilax

Woody vines with spines and tendrils. Flowers greenish with 6 tepals. Berries black when ripe, with 1–3 seeds.

Smilax aspera COMMON SMILAX A variable, creeping, scrambling or climbing shrub to 6 m with angled, smooth or prickly stems. Leaves dark shiny green and leathery, triangular to heart-shaped, often with prickles on the margins, and with a pair of tendrils at the base of the leaf stalk; blade 42–68 mm (10.3 cm). Flowers unisexual, green–white or yellowish or pinkish, scented, to 5 mm, borne in branched clusters. Berry red, ageing black. Common on garrigue and coastal scrub. Throughout.

LILIACEAE | LILY FAMILY

Bulbous, tuberous or rhizomatous, mostly hairless perennials with leaves with parallel veins. Flowers usually with 6-parted perianth, often with star-like tepals not fused; stamens 6, style 1 or 3. Fruit a capsule or berry. Traditionally a much larger family, now divided into smaller familes following DNA-analysis-based revision.

Lilium LILY

Large perennials with bulbs consisting of overlapping, fleshy scales and leafy flowering stems. Flowers conspicuous with 6 spreading tepals and long-protruding stamens, or trumpet-shaped with stamens included. Fruit a capsule.

Lilium chalcedonicum A tall perennial to 1.2 m. Leaves alternate, lance- to oval-shaped, *hairy* beneath and along the margins, 3–5-veined; lower leaves *spreading*, *abruptly* changing to *adpressed* stem-leaves. Flowers borne in loose terminal clusters of 1–7, *orange-red to bright red, unspotted*; anthers red. Open, rocky mountain woods. AA (Euboea), GR, PN. *L. carniolicum* is smaller, to 80 cm, with *3–9-veined* leaves *gradually* merging from spreading basal leaves to adpressed stem-leaves; flowers yellow, orange or red, sometimes spotted; tepals 30–65 mm. Mountain scree. GR (north). *L. rhodopeum* is taller with longer, narrower leaves and large flowers borne in clusters of 3–5, *bright lemon-yellow* and unspotted; tepals 80 mm–12 cm. Mountains of the Rhodope. GR (north).

Lilium candidum MADONNA LILY A tall, stout, hairless perennial to 1.2 m. Leaves alternate, lance-shaped, shiny mid-green with 3–5 veins. Flowers sweetly scented, *white, trumpet-shaped* with deflexed tepals, borne in loose, terminal clusters of 5–6; anthers yellow. Rocky mountain habitats and woods. Widespread (often doubtfully native, except in the east), in remote habitats across the region. AA, CR, GR, IP, LB, PN, SY, TR.

Gagea STAR-OF-BETHLEHEM

Small perennials with 1–2 bulbs and few basal leaves, and 0–few stem-leaves. Flowers 1–5, white or yellow and bell-shaped or star-like with tepals not fused. Fruit a capsule. Many similar species occur in the region that are difficult to distinguish in the field.

(a) Plants with a *single basal leaf*; stems often angular. Flowers yellow.

Gagea pratensis A short, slender perennial 40 mm–20 cm with 2–3 bulbs and a solitary, broadly linear, flat basal leaf 80 mm–30 cm long and 2–4(5) mm wide, and a single opposite pair of lance-shaped stem-leaves with hairy margins. Flowers 2–6, with tepals 9–16(18) mm long, borne on hairless, *angled* stems; tepals yellow, tinged greenish towards the base, narrowly lance-shaped; stigmas 3, yellow. Grassy habitats. AA (Lesbos, Thasos – rare), GR, IA. The following species are similar: **G. pusilla** is smaller, to just 30–80 mm with a single basal leaf 1–2 mm wide and starry flowers with narrow tepals. GR. **G. lutea** has a *wider* basal leaf, 4–11(13) mm, with a hooded tip, and flowers with a *single, green stigma*; bracts 0–1(2); tepals 9–14(18) mm long. GR. **G. minima** has long-pointed, *very narrow*, often deflexed sepals. GR.

1. *Smilax aspera*
2. *Smilax aspera* fruits
3. *Lilium candidum*
4. *Lilium chalcedonicum*
   PHOTO: ARNE STRID
5. *Gagea pratensis*

Gagea reticulata A small *tufted, desert* perennial with a fibrous, netted tunic; sheath elongated (above ground). Basal leaf 1, narrowly linear, 1(2) mm wide, exceeding the inflorescence; stem-leaves 3–5, *whorled beneath the umbel*, linear, exceeding the flowers. Flowers 2–5, borne in umbellate clusters, with thickish stalks, on stems to 12 cm; tepals green outside, yellow within, linear-lance-shaped, 15–18(20) mm. Capsule 13–15 mm, spherical, exceeded by the tepals. Dry fields and deserts in the east. IP, SY, TR.

(b) Plants with *(1)2(3)* basal leaves; stems typically cylindrical to semi-cylindrical. Flowers yellow.

Gagea bohemica EARLY STAR-OF-BETHLEHEM A *very small* bulbous perennial with stems <40 mm. Basal leaves 2, very narrow, 1.5–4.5 mm and thread-like, often curled and prostrate, stem-leaves shorter, wider (2.5–12 mm) and alternate. Flowers 1(–4), yellow, with a greenish exterior, with *blunt tepals* 10–14(15) mm long, borne on hairy or hairless stalks. Dry grassy and rocky habitats or in woods. Throughout. **G. peduncularis** is similar but with erect, stout flower-stalks that elongate as they mature. Common on garrigue and in woods. Throughout.

Gagea juliae A short, sparsely hairy perennial 20 mm–10 cm with 2 basal leaves, much exceeding the stems, 90 mm–25 cm long and 1–2 mm wide. Flowers *numerous,* 3–15(25), borne in loose umbels, yellow, 18–20 mm across, borne on slender, hairless to thinly woolly stalks to 40 mm long, *with a dense tuft of hairs* where the flowers join. Rare and local, in shaded maquis and open pine and oak hill forests. CY, TR.

Gagea villosa A short, greyish, slightly hairy bulbous perennial with 2 basal, narrowly linear, channelled leaves with a rounded keel, 12–25 cm x 1–2.5(3) mm, and a single opposite pair of stem-leaves (rarely 3) just below the flowers. Flowers usually *numerous (3–21)*, yellow with a greenish exterior, star-shaped with narrow, pointed tepals 10–14(17) mm long (>4 x as long as broad), the outer ones often deflexed. GR, PN.

Gagea amblyopetala A short, hairless, bulbous perennial to 16 cm; bulbs with rather shiny, brown tunic. Leaves 2, almost equal, *flat* and solid, 1–1.25 mm wide, shiny and green; stem-leaves channelled. Flowers rather numerous 3–10(15), nodding in bud; tepals 6–10 mm. AA, CR, GR, IA, PN, TR. **G. omalensis** is similar but with fewer (1–5) flowers and *angular* leaves. CR (west – rare). **G. heldreichii** has *erect* (not nodding) buds, leaves 2–3 mm wide and more pointed tepals 10–16 mm. Rare and local CR, GR (south), PN.

Gagea rigida A bulbous perennial with 1–2 narrowly linear basal leaves exceeding the flowers. Stems branching at the very base, the first stem-leaves similar to the basal ones. Flowers 1–3, rather large and opening widely, borne on stalks to 25 mm (elongating in fruit); tepals 15–20 mm. Confused records for the species across the region. AA (mainly southern), CR, CY, PN, TR. **G. fibrosa** is similar with 1–2 linear leaves, 1–4 flowers borne on woolly stalks to 80 mm (the longest exceeding the scape); tepals 15–25 mm with transparent margins. Dunes, garrigue and pine forests. CY, IP, TR.

(c) Plants with 2–4 leaves. Flowers *white*.

Gagea graeca A short perennial 50 mm–10 cm, with 2–4 narrowly linear leaves 30 mm–10 cm long and 1–2 mm wide. Flowers 1–5, borne in loosely racemose clusters, *white, with purple-striped tepals,* widely *bell to funnel-shaped* (unlike the previous species), 7–10 mm across. Garrigue and rocky habitats. AA, CR (common), CY, PN, TR.

---

1. *Gagea reticulata*
2. *Gagea bohemica*
3. *Gagea juliae*
4. *Gagea graeca*

Tulipa TULIP

Bulbous perennials with solitary stems, and few, rush-like leaves. Flowers with tepals all alike; stamens 6. Fruit a 3-parted capsule containing fairly large, flat seeds.

(a) Filaments hairy at the base. Flowers constricted (waisted) near the base; flowers yellowish.

Tulipa sylvestris (syn. *T. australis*) WILD TULIP A hairless, bulbous perennial to 50 cm. Leaves 2–3(4), strap-shaped and channelled, 8–37 cm long and 10–18 mm wide. Flowers 1–2, *nodding in bud*; *yellow* tinged with orange and or green, tepals 33–70 mm, becoming recurved with age; stamens hairy at the base. Fruit a capsule (often not produced). Garrigue, rocky slopes, scrub and roadsides. AA (rare), GR, PN, TR.

(b) Filaments hairy at the base. Flowers constricted (waisted) near the base; flowers pink, white or orange.

Tulipa orphanidea A hairy to hairless, bulbous perennial. Leaves 2–5(6), narrow, strap-shaped, channelled, dark green with purplish margins. Flowers solitary (rarely 2), bowl-shaped when fully open; tepals rather dull coppery *orange–red or brownish*, often flushed green outside, 25–57 mm long; anthers greenish to brownish-orange; filaments hairy beneath. Damp meadows and mountain slopes; mainly in the Peloponnisos. GR, PN (recorded in Turkey). Eastern forms often referred to as *T. whittallii*. *T. goulimyi* is similar but with bulbs with thick, woolly brown hairs within the bulb tunic and more (5–7), *wavy-edged l*eaves. Flowers bright orange to *brownish-red*, without a basal black spot. Damp, sandy meadows, rare and local. CR (west), PN. *T. doerfleri* is very similar to *T. orphanidea* (and often considered a form of it) but with *darker red* flowers; tepals elliptic and rather blunt. Stony fields in west-central Crete only. CR. *T. bithynica* resembles the species listed above but with *narrowly elliptic* and rather *sharply pointed* tepals.

Tulipa saxatilis A short, hairless, bulbous, patch-forming perennial. Leaves 2–3, broadly lance-shaped, green and *shiny*. Flowers 1–4, *large* and opening widely in the sunshine, *pale pink to lilac* with a large yellow centre; tepals 38–54 mm; anthers >5 mm long; filaments hairy beneath. Seldom setting seed. Rocky crevices and fields. CR. *T. bakeri* is similar but with *darker pink–purple* flowers, and frequently setting seed. Local in rocky fields and slopes inland. An island endemic. CR.

Tulipa cretica A small, hairless, bulbous perennial, similar to *T. saxatilis* but *smaller,* to just 11 cm. Leaves narrowly lance-shaped, *prostrate*, broad, green and shiny, with reddish margins. Flowers 2–3, *small and white*, sometimes flushed lilac, with a yellow centre; tepals 15–32 mm; anthers yellowish, 1.5–3 mm long; filaments hairy beneath. Rocky habitats; widespread from sea level to mountains. CR.

(c) Filaments *hairless* at the base. Flowers *not* constricted (waisted) near the base; flowers whitish-pink.

Tulipa clusiana A bulbous perennial to 30 cm. Leaves 3–5, linear, channelled, 20–25 cm long and 10–17 mm wide. Flowers solitary, whitish-pink with darker pink stripes externally, and dark purple anthers. Native to central Asia from Iraq eastwards, very locally naturalised. AA (Chios), GR.

(d) Filaments *hairless* at the base. Flowers *not* constricted (waisted) near the base; flowers yellow, orange or red.

Tulipa undulatifolia A bulbous perennial to 40 cm with hairy stems. Leaves 3–4, grey–green, broadly lance-shaped with *markedly undulating margins*. Flowers solitary, *scarlet with long-pointed tepals* 30–70 mm long, each with a basal black spot edged with yellow. Local in stony fields and pastures. AA (local), GR, PN, TR.

1. *Tulipa goulimyi* PHOTO: NICK TURLAND
2. *Tulipa sylvestris*
3. *Tulipa orphanidea*
4. *Tulipa saxatilis*
5. *Tulipa cretica*
6. *Tulipa clusiana*
7. *Tulipa undulatifolia* PHOTO: ARNE STRID

LILIACEAE

Tulipa praecox A bulbous perennial to 65 cm with hairless, or slightly hairy stems. Leaves 3–4, *greyish blue–green*, the lowermost large (to 70 mm wide), lance-shaped; upper leaves linear-lance-shaped. Flowers large, *bright orange*, flushed green outside; tepals 36–82 mm, *2 x as long as wide*, each with a brownish-green basal spot edged with yellow, the innermost with a prominent central yellow–green stripe. AA (Chios), GR, TR.

Tulipa agenensis A bulbous perennial similar to *T. praecox* with *green* (not greyish) leaves 20–60 mm wide and *bright orange–red* flowers with pointed tepals 48–85 mm long, *3 x as long as wide*, all without a prominent central yellow–green stripe, and with a *conspicuous dark, yellow-bordered basal blotch*. Field margins, garrigue and stony pastures. AA (Chios), IP, LB, SY, TR. ***T. cypria*** A bulbous perennial very similar to *T. agenensis,* to 35 cm with a bulb tunic woolly-hairy within. Flowers *dark purplish crimson* (not bright red); tepals 30–60 mm, each with a *small or obscure, rounded* blue–black basal spot edged yellow. A rare and endangered island endemic. Garrigue and *Juniperus* forest. CY.

Tulipa systola A predominantly *desert* bulb 15–20(35) cm, similar to *T. agenensis* but with (typically) 4 greyish leaves with conspicuously undulate margins, 20–35 mm wide. Flowers red; tepals with a subtle grey–purple bloom on the outer surface, and the black spot on the base of the inner surface often *without* a yellow margin; outer tepals (32)45–70(80) mm, oval-diamond-shaped; inner tepals nearly as long to slightly shorter than the outer. Rocky hills and deserts in the far southeast and east only. IP (mainly southern), SY.

Fritillaria FRITILLARY

Bulbous perennials with solitary, unbranched stems. Leaves alternate, and mostly on the stems. Flowers tubular or bell-shaped, nodding; tepals 6, all petal-like; stamens 6; style 3-lobed. Fruit an erect, 3-parted capsule containing many flattened seeds. A diverse genus, well-represented in the region.

(a) Flowers solitary or 2–3, *broadly* or squarely bell-shaped, about as long as wide, (20)25–50 mm across.

Fritillaria graeca A short, bulbous perennial to 25 cm. Leaves grey–green, lance-shaped, the *lowermost pair opposite*, the rest alternate, >10 mm wide. Flowers 20–30 mm long, *deep purple–brown, chequered*, often with a green stripe along each tepal; style 3-lobed; nectaries 5 mm long. Light woods and alpine scree. GR, PN. ***F. davisii*** is similar to *F. graeca* but with *shiny-* (not grey-) green leaves and flowers *without* a green stripe along the tepals. Rocky fields. PN (Mani Peninsula). The following are similar: ***F. epirotica*** is small, to *just 10 cm*, with the upper 3 leaves sometimes whorled, and *deep purple–brown* flowers *chequered inside but not outside*, 20–25 mm long; nectaries 10 mm long. GR (northwest). ***F. pontica*** is larger, 15–45 cm, with a *whorl of grey leaves near the top of the stem* and 1–3 *green* flowers 25–30 mm long, stained pale brown (*not chequered*). AA (Lesbos), GR (north), TR. ***F. gussichiae*** has broad, alternate, distinctly grey leaves, which *clasp* the stem at the base. Flowers 1–3, to 30 mm long, *green* flushed brown, *not* chequered. Deciduous woods. GR (north).

1. *Tulipa agenensis*
2. *Tulipa cypria*
3. *Tulipa systola*
4. *Fritillaria messanensis* PHOTO: ELEFTHERIOS DARIOTIS
5. *Fritillaria spetsiotica* PHOTO: ELEFTHERIOS DARIOTIS

LILIACEAE

Fritillaria messanensis A short, bulbous perennial 20–35 cm. Leaves grey-green, linear, the *upper 3 whorled*, the rest alternate, <10 mm wide. Flowers solitary, with a flared mouth, 25–30 mm long, *green or brownish, scarcely chequered*, often with a green stripe along each tepal; style 3-lobed; nectaries lance-shaped, 6–10 mm long. Light woods and alpine scree. CR, GR, PN. **F. orientalis** is a similar, bulbous perennial 20–50 cm. Leaves grey-green, opposite or whorled. Flowers 1–3(5), with a flared mouth, 25–30 mm long, yellowish-green, *markedly chequered* with deep maroon-brown; nectaries 10–15 mm long. Light woods and scrub. GR.

Fritillaria acmopetala A short, bulbous perennial 20–40 cm. Leaves linear, all alternate. Flowers solitary (rarely 2–3), broadly bell-shaped, 25–40 mm long, green marked reddish-brown, not chequered; tepals recurved; style 3-lobed; nectaries lance-shaped, 6–10 mm long. Limestone woods and pastures. CY, LB, TR (southwest).

Fritillaria rhodocanakis A very short, bulbous perennial 50 mm–15 cm. Leaves green and lance-shaped, mostly alternate, the lowermost opposite. Flowers solitary, broadly bell-shaped, (15)20–22(25) mm long, bioloured deep purplish-brown; tepals with *prominently yellow*, outward-pointing tips; style 3-lobed. Rocky hillsides. PN. **F. spetsiotica** is very similar (and hybridises with *F. rhodocanakis*); tepals with yellow along the margins and tips only (not in a prominent band). A rare endemic. PN.

1. *Fritillaria latakiensis*
2. *Fritillaria latakiensis* (cross section)
3. *Fritillaria acmopetala*
4. *Fritillaria bithynica*
5. *Fritillaria persica*

(b) Flowers solitary or 2–3, conical or *narrowly* tubular bell-shaped, about 15–25 mm across, with at least some purplish or blackish markings.

Fritillaria latakiensis A bulbous perennial 12–25 cm. Leaves 5–7, all alternate, linear, 30–70 mm long and 4–7 mm wide. Flowers 1–2, *narrowly* bell-shaped with linear tepals, purplish outside, greenish-yellow within, *not* chequered; style 8–9 mm, *divided*. Capsule not winged. Deciduous forest and scrub. Rare and local. SY, TR.

Fritillaria stribrnyi A bulbous perennial 10–50 cm. Leaves grey–green narrowly linear, alternate, the upper 3 in a whorl. Flowers 1–3, narrowly bell-shaped, 15–20 mm long, green with a metallic purple tint and greyish bloom, not chequered; style *undivided*. Rocky hillsides. TR.

Fritillaria ehrhartii A bulbous perennial 20–35 cm. Leaves alternate, the lowermost pair opposite. Flowers 1–3, almost *conical*, 15–20 mm long, *deep purple–brown*, tipped with yellow, with a grey bloom outside; style undivided. Rocky hillsides. AA (Euboea + adjacent islets).

Fritillaria drenovskii A short, bulbous perennial 15–30 cm. Leaves grey–green, alternate and narrowly lance-shaped. Flowers 1–4, narrowly bell-shaped, 15–20 mm long, not flared at the mouth, deep *reddish-maroon* to purple, not chequered; style 3-lobed. Light mountain woods. GR (northeast).

Fritillaria obliqua A short, bulbous perennial 10–20 cm. Leaves grey–green, lance-shaped and alternate. Flowers 1–2, *blackish with a grey bloom*, conical, 20–30 mm long; style 3-lobed. AA, GR (Attica). Large forms to 35 cm on Kythnos with numerous leaves clothing the stems and 1–5 flowers are referred to by some authors as *F. tuntasia*.

(c) Flowers solitary or 2–3, conical or narrowly tubular bell-shaped, about 15–25 mm across, *entirely yellow* or greenish (without purple or blackish markings).

Fritillaria bithynica A short, bulbous perennial 70 mm–20 cm. Leaves green and oblong to lance-shaped, the lowermost paired, the uppermost in a whorl of 3. Flowers solitary, *narrowly* bell-shaped, 17–27 mm long, *yellowish-green throughout* (rarely with purplish stripes); nectaries small, brown or green; style slender and undivided. Capsule winged. Pine and oak forests. AA (Chios, Samos), TR (west).

Fritillaria conica A short bulbous perennial 10–20 cm. Leaves *shiny green*, alternate, the lowermost more or less opposite, lance-shaped. Flowers 1–2, *yellowish-green throughout,* conical, 10–20 mm long; style 3-lobed. Limestone hills and scrub. PN (southwest). *F. euboeica* is similar but just 50 mm–10 cm with *grey–green* leaves and *yellow*, narrowly bell-shaped (not conical) flowers. Limestone hill slopes. AA (Euboea).

Fritillaria sibthorpiana A slender, bulbous perennial 15–25 cm. Leaves 2(3), alternate, *elliptic*, *to 50 mm wide*, green; the upper much smaller and narrower. Flowers solitary, bell-shaped, lemon to buttery yellow, 18–22 mm. Capsule unwinged. AA (Simi), TR (Marmaris). *F. rhodia* is similar but with middle leaves linear, and *very narrow*, just 1–2 mm wide. Flowers 1–2, 15 mm, bell-shaped and *flared* at the mouth, greenish, later yellow. Capsule unwinged. *F. forbesii* is very similar with thread-like leaves just 0.5 mm wide, and flowers 15–25 mm. Pine forests and maquis. TR.

(d) Flowers borne *numerously* in long racemes.

Fritillaria persica A *tall, robust* bulbous perennial 20 cm–1.5 m. Leaves numerous, lance- to oval-shaped, all alternate. Flowers typically 7–20 borne in elongated racemes; flowers greyish or greenish (sometimes blackish-purple), narrowly to broadly bell-shaped, 15–20 mm long; nectary small, triangular, to 2 mm; style slender and undivided. Capsule winged or angled. Various grassy, open habitats, in the southeast only. IP, LB, SY, TR (south).

ORCHIDACEAE | ORCHID FAMILY

One of the largest families of flowering plants, with about 26,000 species worldwide. Perennials with flowers borne in spikes or racemes; flowers each with a bract, usually conspicuous, zygomorphic; perianth with 6 tepals in 2 whorls of 3: the outer 3 tepals ('sepals') all similar, the inner 3 ('petals') with the central the largest and distinct, known as the lip or labellum; ovary inferior; anthers and stigma together form a central column. Fruit a capsule. Some groups were taxonomically exaggerated before the availability of DNA data, so nomenclature is inconsistent in floras.

Epipactis HELLEBORINE

Perennials with numerous stem-leaves. Flowers borne in racemes, often twisted so that flowers face one way; inner perianth whorl of 2 similar upper segments with a 2-parted lip consisting of an inner hypochile and an outer, triangular epichile; ovary not twisted. Most species occur in woods at higher altitudes; absent from many islands.

(a) Hypochile concave but not distinctly cupped. Typically in damp or wet habitats.

Epipactis palustris MARSH HELLEBORINE An erect, perennial to 45(60) cm with leaves oblong to lance-shaped, pointed. Flowers borne in loose, few-flowered racemes (to 14); lower bract *as long as flower*; the outer tepals oval to lance-shaped and brownish or purplish green, the inner upper 2 segments shorter and whitish with purple markings; lip with a *heart-shaped epichile with frilly margins* and a yellow spot at the base; flowers open widely. Damp habitats on the northern mainland. GR, TR.

Epipactis veratrifolia A large, clump-forming, rhizomatous perennial to 1.5 m with lance-shaped leaves to 25 cm, which are largest around the centre of the stem. Flowers large, *exceeded by their lower bracts*; horizontal to nodding, bell-shaped and green, flushed with maroon; lip with a narrowly triangular epichile 9–11 mm long, yellowish to reddish in the centre and with 2 paler triangular lobes at the base. Wet, rocky cliffs. East and southeast only. CY, IP, LB, SY, TR.

(b) Hypochile concave and *markedly cupped*. Typically in pine or oak forest habitats or dry scrub.

Epipactis microphylla SMALL-LEAVED HELLEBORINE A slender perennial 20–40(60) cm with *few (<12), short leaves those on the stems 30–60 mm (12 cm), the largest leaf shorter than its internode*. Inflorescence lax; outer surface of perianth covered in short hairs; flowers nodding and scented; sepals and petals oval, incurved and green–white within, reddish outside; lip 5–7.5 mm long with a shallow cup-shaped hypochile pinched sharply at the front; epichile heart-shaped with crinkled bosses at the base. Shady, deciduous forests and scrub. AA, CR, CY, GR, IA, PN, TR.

Epipactis condensata A rhizomatous perennial 20–75 cm with 4–10 leaves all on the stem, oval-lance-shaped, 40–60 mm long, bract-like above. Flowers large, horizontal, borne in dense, elongated inflorescences; petals whitish to pale pink; lip with a heart-shaped epichile, *white*, flushed dark pink, with scalloped margins. Open woods, rare, in the east only. CY, LB, SY, TR.

Epipactis atrorubens DARK-RED HELLEBORINE A slender perennial to 30(60) cm with up to 11 oval leaves usually >50 mm and forming 2 rows. Inflorescence long, 1-sided and thickly covered in short hairs at the top; flowers *purplish to brownish-red with contrasting bright yellow anthers*, nodding and vanilla-scented with pointed tepals; epichile frilly and with wrinkled bosses at the base. Rare, in high-altitude scrub and woodland. AA, GR, PN.

Epipactis troodi A rhizomatous perennial 20–40 cm, green flushed violet, with 1–4 oval leaves to 70 mm. Flowers yellowish-green, borne in weakly 1-sided inflorescences; sepals and petals 10–12 mm long; lip with a purplish, heart-shaped epichile with 2 pinkish, wrinkled bosses at the base. Mainly pine woods; rare. CR, CY. Cretan populations have smaller sepals (<9 mm) and are often referred to as ***E. cretica***.

Epipactis helleborine BROAD-LEAVED HELLEBORINE A variable (with many forms described), erect perennial forming clumps of 1–3 stems. Stem-leaves larger than the basal leaves, *oval–elliptic*, strongly veined and spirally arranged; *longer than their internodes*. Raceme many-flowered (up to ~100) and rather 1-sided, greenish – rather dull. Locally widespread in a range of habitats, mostly in woodland scrub and hill forests. AA (local), CY, GR, IA, IP, PN.

Epipactis turcica A slender perennial 25–60 cm with a thick, pale green stem and *leaves arranged spirally*, clustered at the base, to 10 cm long. Flowers large and bright, borne in weakly 1-sided, lax inflorescences; flowers only half opening with olive–green sepals, petals pinkish; lip with an epichile broadly heart-shaped, whitish or yellowish flushed dark pink with wrinkled bosses at the base. Very rare in pine and oak forests. AA (northeast; local), TR.

Cephalanthera HELLEBORINE

Differs from *Epipactis* in that flowers have a less clear demarcation of the hypochile and epichile, and perianth folded forward with tepals forming a bell-like structure.

(a) Flowers white, cream or pink.

Cephalanthera epipactoides A robust, clumped perennial 20–70 cm with 2–4 short, elliptic leaves. Flowers 10–30(50), borne in dense spikes with leafy bracts; white, yellowish or cream with rather *open*, narrowly lance shaped sepals and petals; lip 15–22 mm long; epichile narrowly heart-shaped with cream to brownish ridges; spur 3–4 mm. Local in forests. AA (northeast), GR (north), TR. **Subsp. *kurdica*** has bright *pink flowers*; epichile whitish with cream or brownish ridges, heart-shaped, to 10 mm. TR, LB, SY. ***C. cucullata*** is a similar plant, *smaller* in all parts, with flowers remaining *semi-closed*, with tepals ≤ 20 mm and spur just 1–2 mm. A rare local endemic, mainly on the Psiloritis massif. CR.

1. *Cephalanthera epipactoides* subsp. *kurdica*   PHOTO: NICK TURLAND
2. *Cephalanthera longifolia*
3. *Cephalanthera damasonium*

Cephalanthera rubra RED HELLEBORINE An erect perennial to 60 cm with short, straight, lance-shaped leaves; stems flexuous and with glandular hairs above. Inflorescence lax with 3–8(15) *bright pink flowers*, which open widely; epichile dark pink with numerous orange–yellow ridges. Shaded woods and scrub, usually on limestone. AA (northeast), CY, GR, IA (Cephalonia), PN, TR.

(b) Flowers white with *conspicuous yellow–orange* markings.

Cephalanthera longifolia SWORD-LEAVED HELLEBORINE An erect, hairless perennial to 60 cm with stems with whitish scales below, and long, *narrow, linear leaves* around the stem, with the tips somewhat drooping. Flowers *pure white* and *open*, with orange markings inside the lip, borne in dense spikes of 2–15(20); tepals pointed; bracts shorter than their ovaries. Local in damp, rocky woods. Throughout.

Cephalanthera damasonium WHITE HELLEBORINE An erect, hairless perennial with angled stems to 60 cm. Leaves oblong-lance-shaped at the tip, to 10 cm. Flowers 3–11(16), each with a leafy bract below; perianth *creamy white*, to 20 mm long, *partially open* and erect, with blunt segments; hypochile with an orange blotch; epichile with orange ridges; bracts longer than their ovaries. Mountain woods, mainly in the west; local. CR, GR, IA, PN, TR.

Limodorum LIMODORE

Distinctive perennials that lack green pigment and true leaves, in pine forests.

Limodorum abortivum VIOLET LIMODORE A distinctive, *purplish* mycoheterotroph (plant lacking chlorophyll and parasitising a fungus) 10–47 cm (1.1 m) tall, *without any green leaves*. Flower spike lax with 10–45(65) flowers; lateral segments 20–37 mm long, violet or whitish; lip yellowish or white with violet veins; bracts exceeding the ovaries; *spur 10–20 mm long*. Locally common in pine forests or mixed evergreen and deciduous woods. Throughout.

1. *Limodorum abortivum*
2. *Limodorum abortivum* fruits
3. *Neottia ovata*
4. *Neottia nidus-avis*

Neottia

Perennials either with green, functional leaves or brown, scale-like leaves devoid of chlorophyll. Flowers with lip greatly exceeding the other 5 tepals and divided into 2 terminal lobes; pollinia stalkless.

(a) Brownish perennials devoid of chlorophyll (mycoheterotrophic) and with highly reduced, scale-like leaves.

Neottia nidus-avis BIRD'S NEST ORCHID A brown plant *devoid of chlorophyll* (superficially like an *Orobanche*). Stems 10–52 cm with sheath-like scales; inflorescence with most or all flowers in the upper ½; dense and many-flowered (15–70). *Flowers brown–yellow* and short-stalked; sepals and petals similar, 5–7 mm; lip 9–11 mm long, divided into 2 basal, divergent lobes. Fairly common in deciduous or mixed evergreen and deciduous mountain woods. Throughout, except the southeast.

(b) Leaves normally 2 in an opposite pair on the stem; flowers yellow–green or dull reddish; lip deeply divided into 2 apical lobes.

Neottia ovata (syn. *Listera ovata*) TWAYBLADE A greenish, hairy-stemmed perennial 20–60(75) cm, normally with 2 oval leaves low on the stem. Inflorescence lax but many-flowered (up to ~100); *flowers yellowish-green* with equal tepals that are curved forwards, the sepals 2 x the width of the petals; lip to 10 mm long with 2 blunt, divergent lobes. Rather rare and absent from hot, dry areas. AA (local), CR, GR, PN, TR.

Orchis

Tuberous perennials with 2 ovoid tubers and several leaves in basal rosettes, those on the stem often sheath-like. Flowers often borne in short, dense spikes; upper 5 tepals incurved, the 2 lateral sepals incurved or erect to spreading; lip with 2 lateral lobes and 1 terminal lobe, the latter often larger and 2–3-lobed. Spur absent, short or long.

(a) Flowers greenish with upper tepals incurved to form a 'helmet'; lip with 3 lobes, the terminal lobe with 2 sub-lobes; spur absent. Still widely described within the genus *Aceras*.

Orchis anthropophora (syn. *Aceras anthropophorum*) MAN ORCHID A short, slender orchid 12–38(50) cm with 4–9 oval-shaped, blunt, shiny green leaves forming a basal rosette, and sheathing the stem. Inflorescence many-flowered, cylindrical and slender; flowers green–yellow streaked with dull red; sepals and petals forming a loose hood above the lip; lip 11–12(15) mm long and pendent, with 2 slender 'arms' and 2 shorter, spreading 'legs' and with 2 swellings near the base; spur absent. Locally frequent on maquis and in light woodland scrub or grassland. Throughout.

(b) Flowers white, pink or purple with tepals forming a hood; *lip characteristically human-shaped* (anthropomorphic) with narrow lateral lobes and a 2-lobed terminal lobe, often toothed.

Orchis simia MONKEY ORCHID A perennial to 30(45) cm with 2–5 unspotted basal leaves and 1 or 2 stem-leaves. Inflorescence dense and many-flowered, the flowers opening from the top downwards; bracts to 4 mm long; tepals curved to form a hood with upturned points; white to lilac outside; lip deeply 3-lobed with a narrowly rectangular central area spotted with red, all lobes ending in spindly, bright magenta points (to just 1 mm wide), the secondary lobes of the *middle lobe as narrow as lateral lobes*. Throughout, except IP and much of AA.

Orchis militaris **MILITARY ORCHID** A perennial to 45(60) cm with 3–5 unspotted basal leaves and 1–2 sheathing the stem. Inflorescence dense above, laxer below, with up to 40 flowers; tepals curved to form a hood that is lilac–red (sometimes pale) outside and strongly veined within. Lip pink with a pale centre with groups of dark hairs; deeply 3-lobed the laterals curved inwards and the middle lobe divided into 'legs'; *secondary lobes of central lobe much broader than the laterals*. GR (north and centre).

Orchis italica **ITALIAN MAN ORCHID** A robust perennial 18–43(50) cm with 5–10 leaves in a rosette, and 2–4 sheathing the stem (not reaching the flowers). Leaves oblong-lance-shaped, wavy-edged, often flecked with brown. Flowers whitish-pink with darker veins, borne in dense, many-flowered inflorescences; tepals forming a loose hood; lip 12–21(25) mm long, tipped and spotted with purple, with *slender, pointed* 'arms and legs' with a short 'tail' in the middle; spur down-curved and ½ the length of the ovary; bracts tiny, 1-veined and much exceeded by the ovaries. Locally frequent to abundant on base-rich garrigue. Throughout (rare in the north).

Orchis galilaea A slender perennial 15–50 cm with 3–8 basal, unspotted leaves to 12 cm and 1–3 smaller stem-leaves. Flowers musk-scented, borne numerously (12–90) in dense, ovoid to cylindrical clusters; flowers greenish, yellowish, whitish or pink; tepals forming a hood, typically with dark veins within; lip deeply 3-lobed, to 12 mm long, pale with *large, dark purple markings*; spur 3–4 mm long, curved downwards. Oak woods and olive groves, southeastern only. IP, LB, SY.

Orchis purpurea **LADY ORCHID** An erect, robust perennial to 50 cm (1 m) with up to 6 unspotted, shiny basal leaves and 1–2 sheathing the stem. Inflorescence rather tall with many flowers densely crowded in the upper part only. Bracts to 3 mm long. Tepals curved to form a *strongly brownish-purple spotted green 'helmet'*; lip paler with numerous tufts of brownish hairs; lip 3-lobed and almost entire with an apical notch forming a 'skirt'. Mainly mountains. GR, PN (rare), TR.

(c) Flowers white, pink or purple with lateral sepals spreading to erect; petals folded over the column; lip 3-lobed, the laterals broad and the middle lobe longer and divided into 2 further lobes.

Orchis quadripunctata A perennial 10–30 cm with purplish stems and 2–6 spotted or unspotted leaves to 12 cm long. Flowers small, pink to lilac with whitish centres (rarely all white), borne in lax to dense, cylindrical spikes; tepals bluntly oval, the outer 3 spreading, the inner 2 converging to form a tight hood; lip 3-lobed, the *centre pale with 4–8 dark spots;* spur 8–14 mm, long and slender, equalling the ovary. Gorges, hill slopes and maquis. AA (west), CR (common), CY, GR, PN, TR (local).

Orchis troodi A slender to robust perennial 15–50 cm with broadly oval, sometimes spotted leaves to 20 cm; basal leaves almost erect. Inflorescence lax to dense with 3–18 flowers, initially *pale pink*, ageing crimson; tepals broadly oval, the lateral *greenish centrally*, the dorsal nearly erect; *lip large and wedge-shaped*, to 22 mm long, 3-lobed, with sparse dark spots; spur 20–25 mm long, horizontal to nearly vertical. Rare in mountain woods of Cyprus. CY. **O. anatolica** is very similar but shorter, to 40 cm, with darker pink flowers *without greenish markings* on the lateral tepals; spur straight. Garrigue and pine woods, strongly southern and eastern. AA, CR, CY, IP, LB, SY, TR. **O. sitiaca** is very similar but larger, with silvery (not fresh green) leaves and an *upward-pointing* spur. Garrigue. An island endemic. CR (mostly eastern).

1. Orchis anthropophora
2. Orchis simia
3. Orchis militaris
4. Orchis italica
5. Orchis galilaea
6. Orchis purpurea
7. Orchis quadripunctata
8. Orchis troodi
9. Orchis pauciflora

Orchis mascula EARLY PURPLE ORCHID A variable, slender perennial with broadly lance-shaped, shiny-green leaves with darker spots. Bracts lance-shaped and shorter than the ovary. Inflorescence dense, with up to 20 sweet-smelling flowers; tepals mauve or reddish, the middle sepal curving with the petals to form a hood, the laterals deflexed; *lip 8–16 mm long, 3-lobed and convex from the centre, the centre pale and spotted*, the middle lobe notched, sometimes with white spots towards the base; lateral lobes scarcely deflexed; spur *at least as long as the ovary* (as long or 2 x length of lip), pointing upwards and thickened at the apex. Open woods. Probably only the mainland (distribution poorly known). AA (east; local), GR, TR.

Orchis spitzelii A slender perennial 21–41 cm, with 2–5(7) basal leaves, spreading to erect and *unmarked*, those on the stem sheathing. Flowers with dorsal sepals and petals slightly curved to form a hood, lateral sepals erect or curving inwardly; sepals dark crimson or pink outside and tinted *olive green within*; *lip very convex, appearing folded*: pink, spotted red *without* a white basal area; 3-lobed, the middle lobe indented and with a slightly wavy margin; spur conical and *pointing downwards*, 6–10 mm long (*just* shorter than ovary, and more or less equalling lip). High altitudes; rare. CR (**subsp. *nitidifolia***), LB, TR.

(d) Flowers cream, yellowish or white.

Orchis provincialis PROVENCE ORCHID A perennial 14–39 cm with 4–9 densely spotted leaves below, and up to 3 sheathing the stem. Bracts narrow with 1–3 veins, equalling or exceeding the ovaries. Inflorescence lax and cylindrical with *pale yellow flowers*; dorsal sepals erect, laterals spreading, petals curved inwards to form a partial hood; lip 3-lobed, the middle lobe bend downwards in the centre and red-spotted; spur curved upwards and blunt at the tip. Maquis and deciduous or evergreen woods. AA, CR, GR, PN. ***O. pallens*** is similar but with unspotted leaves and an unmarked lip. GR, TR. ***O. pauciflora*** has unspotted, channelled leaves and lax inflorescence with only up to 7 (not 20) flowers; lip much darker in the centre than the edges (to sulphur yellow) with small, dark specks, and scalloped to toothed along the margin. Meadows and garrigue. CR, GR.

Neotinea

Perennials with 2 ovoid tubers and leaves 2–4 at the base; those on the stem reduced. Flowers with upper 5 tepals incurved; lip with 2 large lateral lobes and a larger terminal lobe; spur short and rounded at the tip.

Neotinea tridentata (syn. ***Orchis tridentata***) A short perennial 55 mm–26 cm with a dense, cylindrical inflorescence. Flowers with dark pink to pinkish-white tepals, greenish at the base and with darker spots, curving to form a hood, the sepals green below and terminating in long, fine points, and a 3-lobed, *flat* or slightly concave lip. AA, CR, GR, IA (rare), IP, TR (common).

Neotinea lactea (syn. ***Orchis lactea***) MILKY ORCHID A small perennial to 20 cm with up to 8 foliage leaves and a further 1–3 sheathing the stem. Inflorescence very dense and ovoid-elongated with a whiskered appearance. Bracts as long as ovaries. Tepals dull whitish-pink with dark veining and green centre, curved to form a hood; *lip convex*, 3-lobed with a further-divided middle lobe, pale with darker spots; spur cylindrical and curving downwards. AA, CR, GR, TR.

Neotinea ustulata (syn. ***Orchis ustulata***) BURNT ORCHID A small perennial to 15(30) cm with unspotted leaves, increasing in size up the stem. Inflorescence many-flowered, opening from the base upwards and dense and *dark-coloured at the top*; tepals curve to form a blackish-red hood (unopened buds form the 'burnt' tip to the flower spike); lip to 8 mm long and white with red spots, deeply 3-lobed with spreading, linear laterals and a rectangular central lobe. Mountain woods and scree; rare in the area. GR (north).

ORCHIDACEAE | 97

Neotinea maculata (syn. *Orchis intacta*) DENSE-FLOWERED ORCHID A small, pale perennial 10–25(40) cm, normally with densely spotted leaves (sometimes unspotted) forming a rosette. Inflorescence *small and dense; flowers very small* (appearing dwarfed by bracts) and scented; dull pinkish-white with purplish markings; *sepals forming a pale hood*; lip 3-lobed, 3–5 mm long only, the middle lobe rectangular and the laterals pointed. Fairly common on maquis, open pine and deciduous woods. Scattered throughout.

Anacamptis

Perennials with 2 rounded tubers and several leaves, decreasing in size up the stem. Upper tepals incurved; lip 3-lobed, the middle lobe largest; spur long, slender to thick.

(a) Tepals forming a tight hood; lip 3-lobed, the middle lobe entire and longer than the laterals; spur thick, sequestering nectar.

Anacamptis coriophora (syn. *Orchis coriophora*) BUG ORCHID A small perennial 14–37(60) cm. Leaves 4–11, lance-shaped and folded. Inflorescence ovoid and dense with many flowers; flowers pinkish-brown, scented; tepals converging to form a beaked hood; lip spotted; spur downward-pointing and ½ the length of the ovary; spur usually shorter than the lip. Mountains and damp habitats. Scattered throughout.

1. *Neotinea tridentata*
2. *Neotinea lactea* (typical form)
3. *Neotinea lactea* (white form)
4. *Neotinea maculata*
5. *Anacamptis coriophora*

Anacamptis sancta **(syn.** ***Orchis sancta*)** A perennial 15–45 cm with 5–15 unmarked, linear-lance-shaped basal leaves to 12 cm (often withered during flowering); stem-leaves 3–6, smaller and clasping. Flowers purplish, slightly scented, borne in lax, cylindrical spikes; hood elongated, *long-pointed*; lip *unmarked*, 3-lobed, to 15 mm, grooved at the base; spur conical, 6–10 mm. Rare, local and strongly eastern. AA, CR (rare), CY, IP, LB, SY, TR.

(b) Leaves narrow and unmarked, not in a basal rosette but arranged on the stem; bracts leaf-like. Damp habitats.

Anacamptis laxiflora (syn. *Orchis laxiflora*) LOOSE-FLOWERED ORCHID A tall perennial to 50(80) cm with up to 8 channelled, unspotted leaves. Inflorescence *lax*; stems red and flowers rather uniformly purplish-red with outwardly spreading sepals; petals incurved to form a loose hood; lip 3-lobed and *strongly convex*, the central lobe shorter than the down-folded laterals (or absent), usually forming a tooth between them; centre of the lip white and *unspotted*. Spur to 2/3 the length of the ovary, and *as long or longer than the lip* and thickened at the end. Widespread but local, in wet grassland, marshes and near water. AA, CR, CY GR, IA, PN, TR.

(c) Sepals oval, often with green veins, the upper tepals forming a hood; lip broad and weakly 3-lobed, the laterals rounded.

Anacamptis morio (syn. *Orchis morio*) GREEN-WINGED ORCHID A perennial 12–37(50) cm. Leaves mostly in a basal rosette, with several sheathing the stem and crowded, unspotted. Flowers 4–16, usually pink (sometimes white), with tepals forming a hood, the lateral sepals strongly *veined and suffused in green*; lip bluntly 3-lobed and almost folded in 2, with a central paler patch *with red spots*; spur *thick*, equalling the ovary and horizontal or pointing upwards. Throughout (except CR + smaller islands). **Subsp.** *syriaca* is the eastern form of the species, generally less robust, 10–30 cm, typically with a pale lip (with lateral lobes as pale as the centre), *unmarked*. CY, IP, LB, SY, TR.

Anacamptis israelitica A perennial 10–25 cm with 3–6 basal leaves to 12 cm and 1–2 long-clasping stem-leaves. Flowers 5–15, borne in lax inflorescences, opening from the top first; sepals and petals whitish to pinkish or lilac; lip 3-lobed and convex, whitish, with *a parallel row of large purple spots;* margins scalloped; spur 11–13 mm. Endemic to northern Israel. IP. *A. boryi* is similar but with a dark pink (not pale) hood and *longer spur, 12–18 mm*. Open woods. AA (west), CR, GR, PN.

(d) Lip entire, fan-shaped to diamond-shaped.

Anacamptis papilionacea (syn. *Orchis papilionacea*) PINK BUTTERFLY ORCHID A variable perennial 18–38(55) cm with 4–9 leaves in a basal rosette, and up to 5 sheathing the stem; stems reddish above. Inflorescence robust with few (6–22) large flowers; bracts reddish and exceeding the ovary; tepals pink–red, pointing forward and not forming a hood; lip not divided, *with a crinkly margin* with streaks of darker coloration, often upturned at the sides; spur 8.7–13.5 mm long and *slender* (1.4–2.5 mm). Poor grassland and garrigue. Throughout. Many forms exist that are not regarded as distinct by many authors. **Subsp.** *aegaea* is the broad-lipped form, common in the Aegean. AA. Narrower-lipped eastern forms (IP) are often referred to as **subsp.** *palaestina*. *A. collina* (syn. *Orchis collina*) is similar (sometimes co-occurring), but with a large, *expanded, sack-like spur* (4.6–7.8 x 3–4.7 mm). AA, CR (common), CY, IP, LB, SY, TR.

(e) Lip with 2 ridges at the base.

Anacamptis pyramidalis PYRAMIDAL ORCHID An erect perennial to 60 cm. Leaves lance-shaped, grey–green, unspotted. Flowers pink–purple (rarely white), borne in distinctly *cone-shaped or dome-shaped* dense spikes; tepals all pink, the lip broad and deeply 3-lobed, the spur long and slender (12–14 x 1 mm). Readily distinguished by the shape of the inflorescence. Widespread and common in sunny, grassy habitats. Throughout.

---

1. *Anacamptis sancta* PHOTO: ARNE STRID
2. *Anacamptis morio* subsp. *syriaca*
3. *Anacamptis papilionacea* subsp. *aegaea*
4. *Anacamptis papilionacea* subsp. *palaestina*
5. *Anacamptis collina*
6. *Anacamptis pyramidalis*

Himantoglossum

Tall, robust perennials with 2 ovoid tubers. Flowers with a very long, narrow lip with 2 lateral lobes; spur short.

(a) Lateral lobes of the lip long and thread-like with straight margins; middle lobe divided into 2 long, thread-like divisions.

Himantoglossum comperianum A distinct perennial 25–60 cm with numerous leaves decreasing in size along the stem. Flowers 5–20, large, whitish-violet, borne in a lax, cylindrical spike; hood *elongated and bell-shaped;* sepals oval; petals linear, 12–14 mm; lip hairless, wedge-shaped and 3-lobed, often spotted pink, the lateral and middle lobes all terminating in *long, thread-like, slightly twisted extensions* to 60 mm; spur cylindrical and curved, 12–18 mm. Rare and local. AA (Lesbos, Samos, Kos, Rhodes), LB, SY, TR (south).

(b) Lateral lobes of the lip short, with undulate margins; middle lobe divided.

Himantoglossum robertianum (syn. *Barlia robertiana*) GIANT ORCHID A robust perennial with stout stems to 80 cm. Leaves 5–10, matt green, to 30 cm long, sometimes with faint markings. Inflorescence dense and many-flowered, the lower bracts prominent and exceeding the flowers; lateral sepals forming a hood with the petals; lip distinctly 3-lobed, the laterals with wavy edges, the middle with 2 parallel ridges; flowers greenish to purplish brown; spur cone-shaped, to 6 mm. Frequent on scrub and open woodland on base-rich substrates. Throughout, except for the far southeast.

(c) Lateral lobes of the lip with strongly crimped margins; *middle lobe long* and ribbon-like, spirally wound in bud.

Himantoglossum hircinum LIZARD ORCHID A very robust and *tall* perennial to 70(90) cm with numerous foliage leaves decreasing in size up the stem. Inflorescence dense (smelling of goat), with up to ~100 flowers with very long lips; tepals form a hood; lip 3-lobed with a whitish, spotted central area, the lateral lobes linear-pointed, the middle lobe *markedly long and twisted, 30–65 mm long*, with a shallow notch 2–4 mm deep; spur cone-shaped. Northern mainland Greece, doubtfully elsewhere. GR. The following species are all similar, but geographically fairly disjunct: *H. caprinum* has a lip with a *very long* middle lobe 45–90 mm long with a *deep notch* at the tip, 10–50 mm; spur thick, 5–15 mm. IP, TR. *H. affine* has a lip with a middle lobe 30–50 mm long with a notch 3–15 mm deep; *spur short* and sack-like, 1–3 mm. Rare. PN, TR. **Subsp.** *samariense* is a rare island endemic form, found in three mountain massifs. CR.

Spiranthes LADY'S TRESSES

Perennials with tuberous roots and several leaves (all basal or with some stem-leaves). Flowers arranged spirally in dense spikes, usually white; lip virtually unlobed and close to the other tepals forming an almost tubular perianth; pollinia unstalked.

Spiranthes spiralis AUTUMN LADY'S TRESSES A small, tuberous, autumn-flowering perennial 10–30 cm with a basal leaf rosette at one side of the stem; stem with 3–7 overlapping scale-like leaves. Flowers 6–30, tiny, white, stalkless, borne in a *spirally twisted* spike; lip 6–7 mm, green with a white, frilly margin, just exceeding the petals and sepals. Dry, grassy habitats. Throughout.

Platanthera

Tuberous perennials with paired or few leaves at the stem base. Flowers white or greenish with spreading lateral lobes; lip entire; spur long and slender.

Platanthera chlorantha GREATER BUTTERFLY ORCHID A tuberous perennial with stems to 60(80) cm with 2 (rarely 3) basal leaves and up to 4 smaller stem-leaves. Inflorescence lax and many-flowered; flowers white (often greenish or yellowish), scented; central sepal broad and heart-shaped, forming a helmet with the 2 petals; lip tongue-shaped, to 18 mm; *anther lobes divergent downwards*; spur short and thick, 19–28 mm long. Open woods and meadows on limestone. AA, CY, GR, IA, TR. *P. bifolia* is similar but more slender with smaller, sweetly scented flowers with *anthers running parallel and close together*; spur 15–20 mm. Open woods and meadows on acid substrates. Absent from hot, dry areas. GR, TR. *P. holmboei* has *entirely green* (not white) flowers. East of region only. CY, IP, LB, SY.

Gymnadenia

Perennials with several tapering tubers and several leaves decreasing in size up the stem. Flowers fragrant, with 3-lobed lip and horizontal marginal petals; spur long and slender.

Gymnadenia conopsea FRAGRANT ORCHID An erect perennial 16–81 cm with 4–8 glossy, linear, unspotted basal leaves to 19 cm, and few stem-leaves. Flowers numerous (17–82), borne in *dense, cylindrical spikes*; flowers pink to lilac (rarely white) with lateral segments spread out and the remaining 3 forming a hood; lip with 3 blunt, equal lobes, *spur long, slender and downward-pointing*, 11.4–15(18) mm, almost 2 x the length of the ovary. Mountain woods and meadows. AA, GR, TR.

1. *Himantoglossum robertianum*
2. *Spiranthes spiralis*
3. *Platanthera chlorantha*

Dactylorhiza

Perennials with tubers lobed or clustered, leaves several, often spotted. Flowers with all tepals separate, equal or with inner lobes smaller; lip shallowly 3-lobed with a spur; leaf-like bracts *as long, or longer, than the flowers*. Highly variable and hybridising; a difficult group to distinguish in the field (only a small number of this complex genus are described here).

Dactylorhiza romana ROMAN ORCHID A tuberous perennial 15–35 cm with 3–7 foliage leaves in a basal rosette, and 1–3 stem-leaves. Inflorescence dense, ovoid-cylindrical, the bracts rather longer than the flowers; flowers variable in colour – *cream, yellow or magenta*; lateral sepals erect and turned outwards, to 10 mm, the central sepal slightly shorter and curving over the lateral petals, which are broader; lip rather flat and 3-lobed, the middle lobe raised; *spur cylindrical and bent to point upwards, exceeding the ovary* (to 25 mm). Higher-altitude maquis and open woodland to 2,000 m on base-rich substrates. Throughout.

Dactylorhiza sambucina ELDER-FLOWERED ORCHID A tuberous perennial with hollow stems 17–26(39) cm. Leaves 4–7, along the stem or in a loose rosette. Inflorescence dense, ovoid and many-flowered; flowers variably yellow to magenta, scented of elderflower; lateral sepals upright, the central sepal inclined over the petals; lip to 8–9 mm long and 8–12(14) mm wide: elliptic, folded and scarcely 3-lobed. Spur cylindrical-tapering, 12–15(17) mm long and *curved downwards, parallel with ovary*. Mountains, woods and meadows; local, on the mainland. GR.

Dactylorhiza saccifera A robust perennial 25–90 cm with solid stems, grooved above; leaves 10–20 cm. Flowers 15–80, borne in a dense, conical–cylindrical spike, whitish to dark pink; lateral sepals 7–14 mm, sometimes spotted; lip 9–16 mm long, deeply 3-lobed, kidney- to wedge-shaped at the base, with bold dark markings; spur robust, straight and cylindrical to conical, 9–15 mm long. Damp habitats in mountains. GR, IA, TR.

Ophrys BEE ORCHID

Perennials with 2 ovoid tubers. Flowers dupe male insects into attempting to mate with them to bring about cross-pollination (pseudocopulation). Sepals large, often greenish or pink; petals 2, smaller and hairy; lip large, hairy and variously patterned. Taxonomy complex and discordant across floras; DNA shows the genus has been excessively split and many of the forms previously described are not genetically distinct species; some of these are shown here to enable to reader to cross-reference other floras.

(a) Column rounded; stigmatic cavity about as wide as the anther.

Ophrys speculum MIRROR ORCHID A short perennial 7 mm–50(65) cm. Basal leaves oblong, blunt-tipped, stem-leaves pointed. Flowers borne in spikes of 1–15(18); sepals green or yellowish, usually striped brown; petals dark purple, hairy, and 1/3 the length of the sepals; lip to 10–16 mm long, *broad*, 7–14 mm, 3-lobed, the middle lobe with a *blue, shiny mirror* framed with yellow, and fringed with brown or blackish hairs. Throughout, except for CY and the southeast; rare on CR. **Subsp.** *speculum* has a lip with scarcely recurved margins. Throughout the species range. **Subsp.** *regis-ferdinandii* has a lip with a long middle lobe with markedly recurved margin, appearing *cylindrical*. AA (east), TR.

(b) Column rounded; stigmatic cavity about as 2 x as wide as the anther.

Ophrys insectifera FLY ORCHID A slender perennial 10–60 cm, with few (2–15), sparse flowers. Sepals pale green, petals dark brown or yellowish-green; lip brown (sometimes with yellow margin) with recurved margins, 3-lobed, the *middle lobe much longer than the side lobes* and with a deep cleft; mirror distinct and dull blue–grey; column rounded. Grassland and open woodland. Predominantly central European, extending to northern Greece. GR (north).

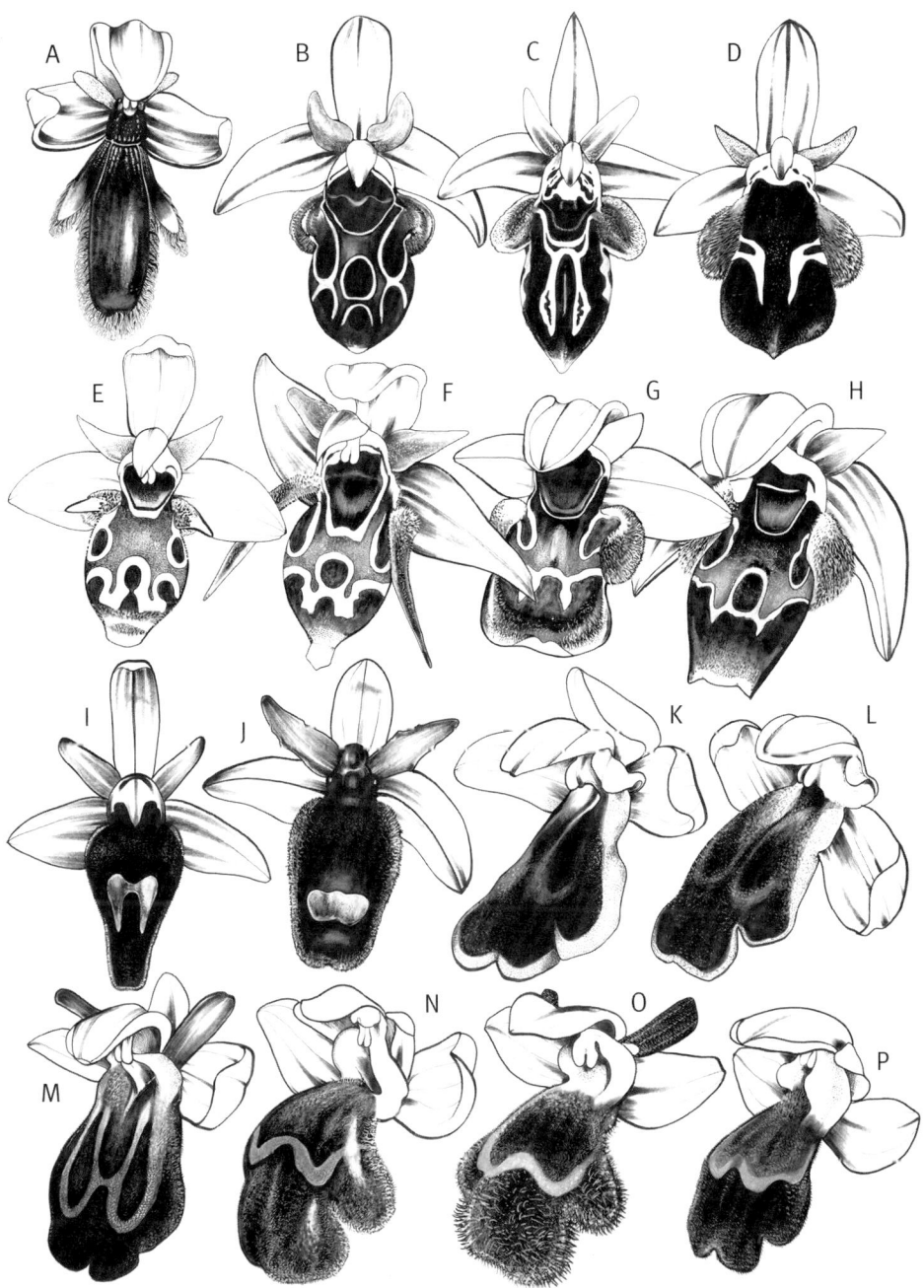

Ophrys species: **A.** *Ophrys speculum* subsp. *regis-ferdinandii*; **B.** *O. kotschyi* subsp. *kotschyi*; **C.** *O. kotschyi* subsp. *ariadnae*; **D.** *O. reinholdii*; **E.** *O. scolopax* subsp. *rhodia*; **F.** *O. scolopax* subsp. *cornuta*; **G.** *O. umbilicata* subsp. *flavomarginata*; **H.** *O. umbilicata* subsp. *bucephala*; **I.** *O. ferrum-equinum* subsp. *gottfriediana*; **J.** *O. bertolonii*; **K.** *O. fusca* subsp. *blitopertha*; **L.** *O. fusca* subsp. *cinereophila*; **M.** *O. fusca* subsp. *iricolor*; **N.** *O. omegaifera*; **O.** *O. omegaifera* subsp. *fleischmannii*; **P.** *O. omegaifera* subsp. *israelitica*. Illustrations not to scale.

Ophrys bombyliflora BUMBLEBEE ORCHID A short, loosely clump-forming perennial 50 mm–35 cm. Basal leaves oval to lance-shaped, forming a flat rosette; stem-leaves erect and clasping the stem. Flowers few, borne in short spikes of 1–6; sepals green; petals green with a purplish base, triangular, and 1/3 the length of the sepals; *lip small*, 6–10 mm long, 3-lobed with the lateral lobes deflexed, brown with a central bluish, shield-shaped mirror. Throughout, except the southeast (rare on Lesbos, Samos and Rhodes).

Ophrys lutea YELLOW OPHRYS A short perennial perennial 70 mm–50 cm. Basal leaves oblong and pointed, stem-leaves smaller and narrower. Flowers borne in spikes of 3–12; sepals green, petals greenish or yellowish and ½ the length of the sepals; lip 9–18 x 10–15 mm, 3-lobed, with a broad, flat, *yellow margin* surrounding a red–brown area with a blue–grey mirror; common. Throughout, with many regional forms (some doubtfully distinct). **Subsp.** *lutea* has a bent lip 14–19 mm wide with a yellow margin 3–6 mm wide. Throughout; rare in the far east and southeast. Forms with a *lip held upright* without basal prominences often described as ***phryganae*.** **Subsp.** *galilaea* has a *smaller, straighter lip* 5–10(12) mm wide with a *narrow* yellow margin 2–3 mm across; side lobes mostly yellow. Throughout. Forms with *raised basal prominences* on the lip, and the lip held upright are often described as ***sicula*.** **Subsp.** *melena* has a lip with side lobes mostly brown, yellow margins often absent. CR (east), IA (Corfu), GR.

Ophrys tenthredinifera SAWFLY ORCHID A short perennial 10–60 cm. Basal leaves oval to lance-shaped, blunt or pointed; stem-leaves similar. Flowers borne in short spikes of 3–8(11); sepals *purplish-pink* (rarely green or white), the central erect; petals similar in colour, 1/3 the length of the sepals; *lip broad and square, scarcely lobed or unlobed*, 9–18 mm long, brownish-purple with a broad yellowish margin, and a small, brown-spotted, 2-parted mirror. Throughout, except for the southeast. Forms with a prominent beard of *white hair* around the lip base are sometimes described as the form ***villosa.*** AA, CR.

Ophrys fusca SOMBRE BEE ORCHID A variable perennial 80 mm–44 cm. Basal leaves oblong to lance-shaped, blunt-pointed; stem-leaves smaller and narrower. Flowers borne in spikes of 3–6(9); sepals green or pinkish; petals green and ½ the length of the sepals; lip 8–20 mm long, horizontal to down-curved, 3-lobed and purplish or yellowish-brown, often yellow-edged, and with a bluish or greyish W-shaped mirror. Throughout. **Subsp.** *fusca* has a *down-turned* lip, *pale green underneath*. Throughout, with numerous regional forms, many continuously variable and doubtfully distinct, for example: forms with an arched lip deeply grooved at the base and prominent yellow border are sometimes described as ***cressa***. CR. Forms with distinctly down-turned lateral lip lobes and a prominent omega pattern are described as ***creberrima***. CR. **Subsp.** *blitopertha* has a *straight, almost flat* lip, often with a broad yellow margin. AA. **Subsp.** *cinereophila* is very similar but with a lip with a prominent knee-like bend close to the base. AA, CR, CY, SY, TR. **Subsp.** *iricolor* has a lip *wine red* underneath and sharply delineated mirror. Probably throughout.

Ophrys omegaifera A variable, slender to robust perennial to 50 cm. Flowers with pale green–yellow and oval–elliptic sepals; petals yellowish, greenish or brownish along the margins, almost flat at the margins; lip grey–brown or blackish, straight to abruptly curved, 3-lobed and velvety to hairy; *mirror fish-tail-shaped* and violet to grey (sometimes marbled), delineated by a distinct white to *blue omega-shaped band*; column rounded. AA, CR (+ KP), CY, IP, TR. **Subsp.** *omegaifera*

---

1. *Ophrys speculum*
2. *Ophrys insectifera*
3. *Ophrys bombiliflora*
4. *Ophrys lutea* subsp. *lutea* (*phryganae* form)
5. *Ophrys lutea* subsp. *galilaea*
6. *Ophrys lutea* subsp. *galilaea* (*sicula* form)
7. *Ophrys tenthredinifera* (typical form)
8. *Ophrys tenthredinifera* (*villosa* form)
9. *Ophrys fusca* subsp. *fusca* (*cressa* form)
10. *Ophrys fusca* subsp. *fusca* (*creberrima* form)

has a lip abruptly *downcurved* at the base and with mid- and side-lobes of the lip *short-hairy*. AA (Rhodes, scattered elsewhere), CR (east, centre), TR. **Subsp.** *fleischmannii* is similar but with the mid- and side-lobes of the lip *long-hairy*. AA, CR. **Subsp.** *israelitica* has a *straight* (or scarcely downcurved) lip base. AA (Naxos, Paros, Siros – rare), CY, IP, TR.

(c) Column extended into an S-shaped tip; pollinia with drooping stalks.

Ophrys apifera BEE ORCHID A very variable perennial 15–50 cm. Basal leaves oval to lance-shaped, blunt or pointed; stem-leaves similar but smaller. Flowers borne in short spikes of 5–10(15); sepals bright pink with a green mid-vein (rarely green or white); petals green or purplish and <1/3 the length of the sepals; lip 3-lobed, the *central lobe curved backwards*, with a shield-shaped brown or violet mirror with a yellowish margin; stamen elongated into a snout-like appendage. Common and widespread. Throughout.

(d) Column acute; pollinia with firm (not drooping) stalks.

Ophrys scolopax WOODCOCK ORCHID A perennial 10–50 cm. Basal leaves lance-shaped, pointed; stem-leaves narrower and more pointed. Flowers borne in rather long spikes of 3–15; *sepals pink* (rarely green or white), the *central erect*; petals similar in colour and ½ the length of the sepals; lip 8–14 mm, oval, 3-lobed, and brownish-purple and velvety with a large brownish-blue, H-shaped

or spotted mirror with a narrow yellow margin. Absent from CY and the southeast. AA, CR, GR. **Subsp. *cornuta*** has long, *horn-like* side lobes on the lip, 6–12(20) mm long. AA, GR, PN, TR. **Subsp. *heldreichii*** has a *long* lip, 6–13(15) mm, with *very distinctive appendage*, 1.5–2.5 mm. AA (south), CR (common), GR. **Subsp. *rhodia*** has petals *as long as wide* and a mirror covering the entire mid-lobe of the lip; lip 6–13 mm. AA (KP, Rhodes). Easily confused with forms of *O. umbilicata*. **Subsp. *scolopax*** has triangular-lance-shaped petals *longer than wide* and mid-lobe of lip *not* markedly narrowed at the base. AA, GR, PN.

Ophrys umbilicata A variable perennial 10–45(60) cm with 2–12 flowers; similar to, and easily confused with, forms of *O. scolopax*. Sepals pink, white or green, the *dorsal sepal boat-shaped, forming a roof over the column*; petals greenish or pinkish, velvety, recurved; mirror often complicated, blue to violet with a white border. AA (south), CY (common), GR (south), IP (common), PN. Numerous forms described, the following often being regarded as distinct. **Subsp. *umbilicata*** has a slender habit and a lip 8.5–15 mm wide. Throughout the species range. Forms with white to pink sepals are sometimes described as ***umbilicata***. AA, CY. Forms with uniformly green sepals are sometimes described as ***attica***. **Subsp. *bucephala*** has a compact habit and a lip 15–19 mm wide. AA (Lesbos, Chios, Samos). **Subsp. *flavomarginata*** has a strongly recurved lip with a yellow margin and woolly-hairy lateral lobes. CY, IP (rare).

Ophrys fuciflora A variable, tall, many-flowered perennial to 60 cm. Sepals purplish, white or green, the dorsal petal boat-shaped or flat; petals bright pink (less often purplish or green), triangular with recurved margins; *lip broad and square, 18–23 mm across*, yellowish-brown and entire to weakly 3-lobed with a distinct H-shaped, often complicated mirror that is dull blue or reddish-brown with a *conspicuous cream border*; column acute. Absent from most of the mainland. AA, CR, CY, GR, IP, PN, TR. **Subsp. *candica*** has a *marbled* mirror with a clear, unbranched, broad cream border. AA (Rhodes), CR, PN, TR. **Subsp. *andria*** has a *reduced* and fragmented mirror of *isolated spots*. AA (Andros, Naxos, Tinos, Kimolos). **Subsp. *fuciflora*** is the widespread and variable form with sepals 10–16 mm long, and leaves remaining fresh during flowering. AA (scattered), CR, GR, IP, PN. **Subsp. *bornmuelleri*** has smaller flowers with a *horizontally oriented* lip and petals just 1.5–2.2 mm long. CY, IP, TR. ***O. heterochila* (syn. *O. holoserica*)** is similar to *O. fuciflora* but with smaller flowers; petals 2.5–3 mm long; lip 8–11 mm wide, slightly vaulted; sepals strongly deflexed. AA (east), IP, TR.

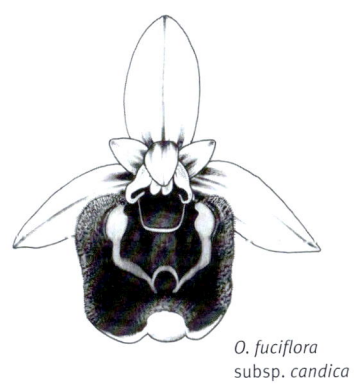

O. fuciflora subsp. *candica*

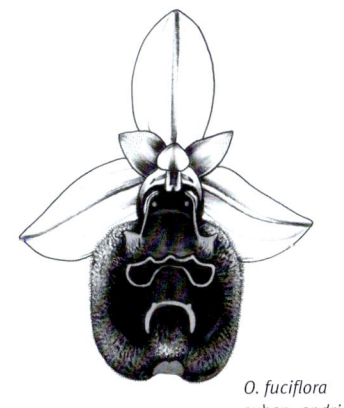

O. fuciflora subsp. *andria*

1. *Ophrys apifera*
2. *Ophrys scolopax* subsp. *heldreichii*
3. *Ophrys umbilicata* (*attica* form)
4. *Ophrys umbilicata* (*umbilicata* form)
5. *Ophrys umbilicata* (*umbilicata* form)
6. *Ophrys fuciflora* subsp. *bornmuelleri*

Ophrys argolica A variable, compact to slender perennial to 50 cm with large flowers. Sepals green, pink, white or violet and rather flat; petals green or pink and elliptic with flat margins; lip reddish or yellowish-brown, *straight and flat*, entire to weakly 3-lobed with a mirror consisting of *horseshoe-shaped figure of 2 central spots*, which are grey–blue, not distinctly shiny, with or without a paler border; column acute to obtuse. **Subsp. *argolica*** has *dense white hairs* along the basal lip margin. PN. **Subsp. *aegaea*** has a lip reddish-brown in the centre and whitish stigmatic cavity *intersected by a dark brown line*. Rare. AA (KP, Kasos, Amorgos, Iraklia). **Subsp. *lesbis*** is similar but with the lip centre and edges *yellowish* and stigmatic cavity brown with a pale centre. AA (Lesbos), TR (southwest Anatolia). **Subsp. *lucis*** has a *3-lobed* lip (similar to in shape to *O. sphegodes* subsp. *spruneri*). AA (Rhodes, Tilos, Nisiros), TR (southwest Anatolia). ***O. bertolonii*** is similar to *O. argolica* but shorter, to 35 cm, with normally bright pink sepals, pink petals, and a broad blackish-brown lip that is *bent forwards* at the base; mirror nearly square and shining blue–grey. GR (west only).

Ophrys sphegodes A widespread and very variable perennial (many forms described, considered by some authors to be a distinct species). Slender, 10–70 cm, with 3–10 flowers in a lax spike. Sepals green or yellowish, sometimes violet-flushed but *not pink*; petals also greenish or yellowish; lip brownish or blackish, often with a yellowish or reddish paler margin, rather *round, straight and flat* and entire or very weakly 3-lobed; mirror usually H-shaped and variably complicated and bordered. Column normally acute. Very widespread and common. Throughout, except arid areas. **Subsp. *sphegodes*** is the common form, but highly variable: lip *10–20 mm long* (large), with or without a narrow yellow margin, and lateral sepals all of 1 colour (which varies). Throughout, except the southeast. **Subsp. *mammosa*** has bicoloured sepals (green and purple–brown) and *prominent protruberances on the lip,* at least half as large as the stigmatic cavity. More or less throughout. Numerous regional forms exist that are variably distinct: forms with a lip with smaller protruberances and broad paler margin, and dorsal sepal curved are sometimes referred to as ***alasiatica***. CY. Forms

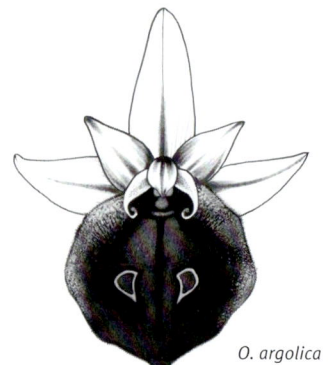

O. argolica

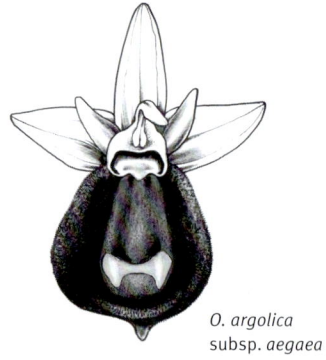

O. argolica subsp. *aegaea*

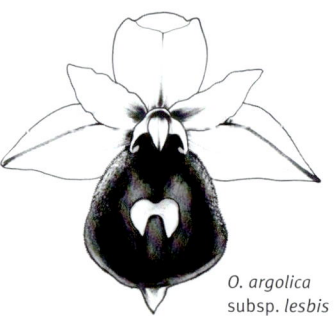

O. argolica subsp. *lesbis*

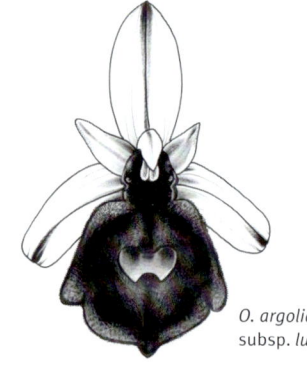

O. argolica subsp. *lucis*

---

1. *Ophrys sphegodes* subsp. *sphegodes*
2. *Ophrys sphegodes* subsp. *mammosa* (mainland Greek form)
3. *Ophrys sphegodes* subsp. *mammosa* (Cypriot form)
4. *Ophrys sphegodes* subsp. *mammosa* (*morio* form)
5. *Ophrys sphegodes* subsp. *mammosa* (*alasiatica* form)
6. *Ophrys sphegodes* subsp. *mammosa* (*herae* form)
7. *Ophrys sphegodes* subsp. *spruneri*
8. *Ophrys ferrum-equinum*

with a prominently beaked column, elongated sepals and petals, and dorsal sepal with recurved margins are sometimes referred to as *morio*. CY. Forms with greenish petals and sepals and lip with orange stigmatic cavity and basal field are sometimes referred to as *herae*. CR. **Subsp.** *spruneri* has bicoloured lateral sepals and a *prominently 3-lobed lip*. AA, CR, GR. **Subsp.** *helenae* has a lip with *no mirror*. GR. **Subsp.** *epirotica* has a lip with a yellowish margin, brown stigmatic cavity and greyish eyes on the column. GR (northwest). **Subsp.** *aesculapii* has a stigmatic cavity speckled green or brown and pale yellow–green eyes on the column. AA, GR, PN. **Subsp.** *cretensis* has a lip just 5–9 mm long, typically without a yellowish margin. AA (Paros), CR (+ KP). **Subsp.** *gortynia* has a lip distinctly wedge-shaped at the base. AA (Naxos, Paros, Antiparos), CR.

O. sphegodes subsp. *helenae*

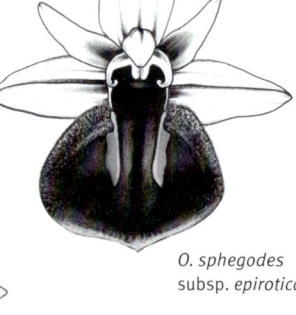

O. sphegodes subsp. *epirotica*

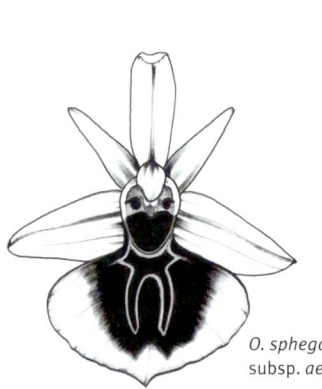

O. sphegodes subsp. *aesculapii*

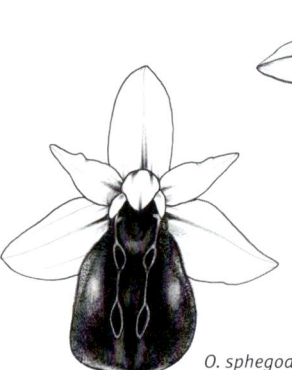

O. sphegodes subsp. *gortynia*

1. *Ophrys kotschyi* subsp. *cretica*
2. *Ophrys kotschyi* subsp. *kotschyi*

Ophrys ferrum-equinum A variably compact to slender perennial 10–35 cm with 2–8 flowers in a lax spike. Sepals violet to greenish-white; petals pink, violet or white, recurved or spreading, nearly *hairless;* lip *dark blackish-brown and velvety*, entire to slightly 3-lobed, rarely with lateral protuberances; mirror distinct, horseshoe-shaped or 2 drop-shaped bands, *not* connected to the lip base; column acute. AA, GR, PN, TR (west). **Subsp.** *ferrum-equinum* has pink sepals and spreading to recurved lip margins. Throughout the range of the species. **Subsp.** *gottfriediana* has green, white or muddy-pink sepals and lip with strongly deflexed margins. GR, IA.

Ophrys reinholdii A slender perennial 15–50 cm with 2–10 flowers in a lax spike. Sepals pink, violet or greenish; petals pink or green; lip 12–16 mm, *blackish and deeply 3-lobed*, the side lobes *shaggily-hairy*; mirror distinct, white or grey with a clear white border, horseshoe-shaped or 2 drop-like markings, not connected to the lip base; column acute. AA, CY, GR, PN (possibly further east). *Ophrys kotschyi* is similar, also with a black and white lip, variably 3-lobed, with an H-shaped or *complicated mirror, connected to the lip* base by 2 broad bands. AA, CR, CY, PN. **Subsp.** *kotschyi* is the endemic form found locally on Cyprus. CY. **Subsp.** *cretica* has a *circular* outline to the stigmatic cavity and prominent, obliquely *conical lateral lip lobes*. AA, CR, PN. **Subsp.** *ariadnae* has a *square* outline to the stigmatic cavity and often less prominent, slightly vaulted lateral lip lobes. AA (KP, Naxos, Paros, Chios), CR.

Serapias TONGUE ORCHID

Tuberous perennials with 2(–5) ovoid tubers, similar to *Ophrys* but with an elongated lip that is downward-pointing and tongue-like with 2 short, upturned lateral lobes. Potentially hybridising with other genera including *Ophrys*, the extent of which is not fully known.

(a) Base of flower lip with a *single* (sometimes deeply channelled) blackish protuberance.

Serapias lingua TONGUE ORCHID A short, tuberous, clump-forming perennial 10–30 cm; stems sometimes spotted. Leaves narrowly lance-shaped and grey–green. Bracts *shorter* than the hoods. Flowers borne in lax spikes of 2–6(9); sepals and petals purple; lip to 32 mm long, maroon-coloured with a *single*, coffee bean-like blackish protuberance at the base; middle lobe oval-lance-shaped, to 12 mm wide, *pale pink* to reddish. Common in a range of habitats including garrigue, woodland and wet meadows. Scattered across the region (apparently absent east of Rhodes).

(b) Base of flower lip with 2 distant protuberances; petals pointed and drop-shaped.

Serapias parviflora A slender perennial 15–30 cm. Leaves linear-lance-shaped, erect. Bracts *exceeding* the hoods. Flowers *small,* sepals 13–16 mm, petals drop-shaped and slightly shorter; lip *only just as long as the hood,* narrow with 2 brownish-red (not blackish) ridges at the base; middle lobe short, narrow, pointed, 6–10 mm, pale brown–pink to yellowish-green. Limestone areas, damp grassland and marshes. Probably throughout, except for the southeast.

(c) Base of flower lip with 2 distant protuberances; petal bases distinctly rounded.

Serapias cordigera HEART-FLOWERED ORCHID A perennial 12–40 cm. Leaves narrowly lance-shaped, channelled, sharply pointed; the lowermost leaves sheath-like, purplish and spotted at the base. Bracts *broad*, just shorter than the hoods. Flowers borne in compact clusters of 2–12; sepals and petals reddish; lip to 30–45 mm long, reddish, the prominent central lobe *broad and heart-shaped, 2 x as large as the sepals*, hairy at the mouth, with 2 dark ridges at the base. Damp grassland and marshes. Scattered throughout, except the southeast and smaller islands.

Serapias vomeracea LONG-LIPPED SERAPIAS A slender to robust perennial 15–60 cm, the base of the stem rarely spotted. Leaves to 19 cm, linear-lance-shaped, channelled, grey–green, sharply pointed; lowermost leaves sheath-like, green (not purple). *Bracts markedly longer than the hoods.*

1. *Serapias lingua*
2. *Serapias parviflora*
3. *Serapias cordigera*
4. *Serapias orientalis*
5. *Serapias orientalis* subsp. *levantina*
6. *Serapias aphroditae*
7. *Serapias bergonii*

Flowers borne in short spikes of 3–10; sepals and petals purplish; lip 21–45 mm – much exceeding the sepals; pale yellowish-red to maroon, with 2 parallel, similar-coloured ridges at the base; the prominent middle lobe triangular, to 30 mm. Damp grassland and marshes. CY, GR (+ adjacent islands), IP, PN.

Serapias orientalis A robust perennial 10–30 cm, with stem base often streaked purple. Leaves lance-shaped, channelled, to 14 cm. Bracts equal to the hoods. Flowers 2–3, borne in compact clusters; sepals 20–32 mm, petals slightly shorter, the bases rounded and purple; lip 28–40 mm, pink to brick-red, with 2 parallel orange ridges at the base and a *prominent tuft of long, whitish hairs*; middle lobe heart-shaped and deflexed, 9–14(20) mm across. AA, CR, CY, IP, PN, TR. **Subsp. *levantina*** has a narrow middle lobe of the lip, 8–12 mm across. CY, IP, TR.

Serapias aphroditae A short perennial 11–24 cm. Leaves 50–90 mm long. Bracts more or less equal to the hoods. Flowers borne on *dark purplish stems*, 3–6(10); tepals 10–15 mm, petals slightly shorter and rounded, crimson at the base; lip 16–21 mm, with 2 shiny dark purple ridges at the base and short, white hairs; middle lobe variably pointed forward or deflexed and red to deep purple, *to just 11 mm*. A rare island endemic restricted to the Akamas and Akrotiri peninsulas. CY.

Serapias bergonii A slender perennial 15–42 cm, the base of the stem and leaves often streaked red. Leaves lance-shaped, to 14 cm, the uppermost bract-like. Flowers 3–12, bracts 18–55 mm, exceeding the hoods; sepals 15–21 mm, petals 19–28 mm, the bases rounded and crimson; lip 18–29 mm, pale red to deep purple, with 2 small, parallel white to purple ridges at the base; *middle lobe to 20 mm*, orange, pinkish or greenish with short hairs. Widespread and common. AA, CR, CY, IP, GR, PN.

IRIDACEAE | IRIS FAMILY

Bulbous, tuberous or rhizomatous perennials, usually with linear leaves, all basal or alternate. Flowers with 6 tepals, the outer 3 often different from the inner 3, enclosed in 1–2, often papery, spathes when in bud; stamens 3; style often with 3 branches, ovary inferior. Fruit a 3-parted capsule.

Iris IRIS

Perennials with rhizomes or bulbs with flat, vertical leaves (or 4-angled). Flowers with the outer 3 tepals (falls) horizontal to down-turned, and the inner 3 (standards) erect; styles rather petal-like and arching over the falls, each with a stamen beneath. Some classifications recognise several genera, which here are all described under *Iris*, consistent with most floras.

(a) Beardless irises with rhizomes.

Iris pseudacorus (syn. *Limniris pseudacorus*) YELLOW FLAG A tall, erect spring-flowering rhizomatous perennial to 1.5 m. Leaves sword-shaped, 10–30 mm wide, green or greyish. Flowers large, 70 mm–10 cm across, *bright yellow* with faint red-purple veins, borne 1–3, each with a papery spathe below; unbearded, with standards markedly smaller than the falls. Capsule large and splitting into 3 segments when ripe, exposing large brown seeds. Damp habitats, river banks and near permanent water. Throughout (rare or absent on many islands; very rare in the far southeast).

Iris species: **A.** *I. hermona*; **B.** *I. haynei*; **C.** *I. mesopotamica*; **D.** *I. bismarckiana*; **E.** *I. atrofusca*; **F.** *I. petrana*; **G.** *I. lortetii*; **H.** *I. mariae*; **I.** *I. histrio*; **J.** *I. edomensis* (an endemic of Jordan, an area not covered in the text). Illustrations not to scale.

Iris unguicularis A late winter–spring-flowering, virtually stemless, clumped perennial with tough, branching rhizomes and persistent dead leaves. Leaves grass-like, ‹8 mm wide, 10–30(50) cm long, often deflexed. Flowers solitary, fragrant; variable in colour from pale blue (e.g. SY) to dark blue or purple; falls whitish at the base with violet veins, beardless. Garrigue and maquis. **Subsp.** *carica* has falls and standards ›55 mm. AA (mainly Paros, Cos and Rhodes), GR, LB, PN, SY, TR. **Subsp.** *cretensis* has falls and standards ‹55 mm. CR (+ KP – common).

(b) Beardless irises with corms, bulbs or spreading tubers (including *Iris* and *Moraea*).

Iris nusairiensis A compact, bulbous perennial to 10 cm with a *prounounced stem*, sheathed by the lower leaves; lower leaves 15 mm wide at the base; upper leaves erect, pale glossy green, inflated at the base and bract-like, concealing the true bracts. Flowers 1–3, pale blue to whitish; falls with a *winged* claw, the blade round to elliptic and with undulating margins; standards spreading to slightly deflexed; style branches with obtuse lobes. Rocky mountain habitats. SY.

Iris tuberosa (syn. ***Hermodactylus tuberosus***) SNAKE'S-HEAD; WIDOW IRIS   A spring-flowering perennial 20–40 cm with spreading, finger-like tubers (unlike other species described here). Leaves 1.3–4 mm wide, rush-like, 4-angled (square in cross section) and *long*, exceeding the flowering stems (which reach 40 cm). Flowers solitary with a broad, leafy sheath to 20 cm; sweetly scented, yellow–green, usually with velvety brown–purple falls 40–50 mm long, the standards about ½ as long, and narrow. Grassy and rocky habitats, garrigue and roadsides. AA, CR, GR, IA, PN, TR.

Moraea sisyrinchium (syn. ***Gynandriris sisyrinchium***) BARBARY NUT   A spring-flowering, cormous perennial 10–30(40) cm with slender stems. Leaves typically 2, sheathing at the base, the free portion longer than the flowering stems, linear and channelled, to 40 cm. Flowers (opening in the afternoon sun) borne in groups of 1–4, *violet-blue*; outer tepals 20–40 mm. Common on pathways, roadsides and other open habitats near the coast; often patch-forming. Throughout. ***M. mediterranea*** (syn. ***Gynandriris monophylla***) is shorter, to just 60 mm, has paler, *grey–blue* flowers with outer tepals 11–20 mm and typically just *1 leaf* per corm. AA (local), CR (west), GR (far southeast), IP.

1. *Iris pseudacorus*
2. *Iris unguicularis* subsp. *cretensis*; (inset) *I. unguicularis* subsp. *carica*

Iris planifolia A late winter-flowering bulb with broad, all basal, recurved leaves to 35 cm. Flowers 1–4, *pale blue–violet*, borne on short stalks, fragrant; falls to 75 mm with whitish markings and a central yellow band; standards to 35 mm. CR (southwest). *I. palaestina* is a similar, small winter–spring-flowering, bulbous perennial 50–75 mm (20 cm) high; bulbs with membranous sheaths. Leaves borne 8–10 cm, in 2 rows, folded and recurved, long-tapering, lance-shaped to linear. Flowers sweet-smelling, *whitish-yellow* with a pale lilac tinge; lobes acute, with darker blue and yellow markings; falls 45–55 mm; standards 15–20 mm. Dry hillsides and coastal or desert scrub. IP, LB, SY (rare), TR (south).

Iris histrio A small bulbous perennial to 80 mm (15 cm). Leaves to 60 mm, 4-angled and squarish in cross section. Flowers 50–70 mm across, pale blue–violet with paler marbling and *conspicuous* darker blue–purple spots; falls (45)50–60 mm; claw of falls 9–14 mm wide, not green-veined outside, the blade broadly oval. Stony limestone hillsides and maquis scrub. IP (north), LB, SY (west), TR (south – Taurus). *I. vartanii* is similar, the flowers typically exceeded by the leaves; falls (35)40–45(50) mm; limb veined but *not conspicuously spotted*; claw green-veined outside. Open, dry, rocky habitats. Southern and eastern. IP (central), SY.

Iris pamphylica A small bulbous perennial 15–25 cm. Leaves 1–2, 4-angled, 17–24 cm during flowering, elongating to 55 cm in fruit. Flowers 35–40 mm across with a *short tube* (20 mm), *bicoloured*; falls *brownish-purple,* yellow at the base with small dark spots; standards to 40 mm, *pale to deep blue with darker veins*; style branches to 35 mm with narrowly lance-shaped, acute lobes. Fruit capsule at first erect, later *pendulous*, with a *long-tapering* beak. *Pinus* and *Quercus* forest and fields. A rare endemic of southern Anatolia. TR.

(c) Bearded irises with rhizomes; flowers white, yellow or violet, not heavily suffused with darker markings.

Iris × *germanica* BEARDED IRIS A robust, spring-flowering perennial herb 40–90 cm (1.2 m) with branched stems and a thick rhizome, visibly spreading above ground. Leaves broad and sword-shaped, grey–green, (20)30–60 mm wide; shorter than the flowering stems. Spathes beneath the flowers green at the base but papery at the tips, often tinged purple. *Flowers large*, to 10 cm across, borne in *groups of 4*; falls 40–60 mm wide, typically *uniformly* violet–blue with a *conspicuous yellow beard* along the centre; standards paler. Various rocky, dry and grassy habitats. Cultivated and widely naturalised. White-flowered, cultivated forms are referred to as *I. florentina*. *I. hellenica* is very similar to *I.* × *germanica* but with *dark violet–purple falls, whitish at the base*. Open forests. PN (Mount Saitas).

Iris albicans A short, erect, spring-flowering, rhizomatous perennial herb, 40–75(80) cm; rhizomes thick and spreading partially above ground, similar in most respects to *I. germanica*. Leaves broad (13–50 mm) and sword-shaped, 30–45 cm. Flowers borne in clusters along the stems, the *lowermost stalkless*, each with a spathe, green during flowering, or papery at the tip; *flowers large and pure white*; falls 60–80 mm, with yellow beards; standards equalling the falls. Native to the Middle East; locally commonly naturalised and patch-forming on roadsides. CR, CY, IP, LB, SY (probably elsewhere).

---

1. *Iris nusairiensis* PHOTO: ZUHEIR SHATER
2. *Iris tuberosa*
3. *Moraea sisyrinchium*
4. *Iris planifolia*
5. *Iris vartanii* PHOTO: BANAN AL SHEIKH
6. *Iris florentina*
7. *Iris albicans*
8. *Iris reichenbachii*
9. *Iris attica* PHOTO: ARNE STRID

Iris mesopotamica An erect, spring-flowering perennial to 1.2 m with a very robust, creeping rhizome. Leaves developing early but *not* evergreen, slightly bluish, 45–60 cm. Flowers very large, borne in terminal *heads of 3–4 with 2–3 lateral branches*, the lowermost the largest; falls to 90 mm, mostly violet, *whitish with thick orange–purple veins* at the base with a yellow beard; standards to 85 mm, blue–purple (lighter than the falls). Old cemeteries, stony limestone slopes in high mountains, and widely cultivated. IP, LB, SY.

Iris reichenbachii A short, erect, spring-flowering perennial to 30 cm with leaves persistent through the winter; leaves sickle-shaped, 5–15 mm wide. Flowers 1–2, borne on slender stems to 30 cm, *greenish-yellow or brownish-purple throughout*, sometimes with violet veins; standards and falls 40–60 mm long, much exceeding the tube. Maquis and pine forests, mostly in mountains. GR (mainly northern).

Iris attica A short, erect, spring-flowering perennial to 25 cm. Leaves grey–green, strongly curved and *sickle-shaped*, 4–7 mm wide. Bract green, often keeled; bracteole thin and transparent. Flowers solitary, often yellow, sometimes blue, purple or bicoloured; falls 35–45(60) mm, with a yellowish or bluish beard. Rocky and grassy habitats and *Quercus* scrub on hill slopes. GR, TR. ***I. suaveolens*** is very similar but <25 cm, with 1–2 flowers and *boat-shaped*, sharply keeled, green bracts and *similar* (slightly smaller) bracteoles (unless when 2-flowered in which innermost bracteole membranous). TR (northwest, west).

(d) Bearded irises with rhizomes; flowers *dark* purple to blackish, not heavily suffused with darker markings.

Iris atropurpurea An erect, spring-flowering perennial herb, 25–35 cm, similar in form to *I. atrofusca* but less compact, with longer rhizome nodes. Leaves 7–11, linear-sickled-shaped; typically 5 basal (½ as long as scapes) + 2 stem-leaves. Falls 35–60 mm, *uniformly* blackish-maroon with a beard of scattered yellowish hairs tipped with purple–black; standards 55–80 mm, circular, dark red–purple, undotted with inconspicuous veins. Sandy coastal habitats. A very rare endemic. IP (mainly coastal plains).

Iris atrofusca A perennial 20–30 cm, with stout rhizomes. Leaves about 6, weak, sword-shaped, tapering, pale green. Flowers borne singly, the stalks concealed by *2 sheathing leaves*; falls 60–75 mm, (usually) nearly *black* with a velvety black–brown beard, mixed with some yellow hairs at the very base; standards 70–90 mm, *dark claret-brown* to blackish with radiating black veins; style branches convex and keeled. Hillsides and deserts; rare. IP. ***I. haynei*** is very similar (continuously variable in some areas), with short-creeping rhizomes; taller, to 50 cm. Leaves 5–6(8), typically all basal. Flowers borne singly with a *pair of bract-like leaves* near the base; falls 70–80 mm, dark lilac-purple with a yellowish beard; standards 90 mm–10 cm, similar in colour or paler, strongly deflexed from the middle. Hillsides; a rare endemic. IP (mainly Mount Gilboa).

Iris petrana An erect, spring-flowering perennial herb, 20–25 cm, with a *creeping* rhizome. Leaves 6–10, slightly sickle-shaped, narrow (3–6 mm). Flowers large; falls 35–55 mm, dark lilac-maroon with a yellowish beard of purple-tipped hairs; standards of similar size and variably maroon to purple, strongly veined dark purple (flower rarely all yellow); style branches arched and keeled. Semi-desert plains. IP (Negev Highlands; Edom). ***I. mariae*** is very similar but with more strongly curved sickle-shaped leaves (4–6 mm wide), and lilac to violet flowers with dark purple to *blackish* (not yellowish) hairs in the flower throat; falls 50 mm; standards 60–65 mm. Loose sands in the Negev and Sinai. IP (south).

1. *Iris atropurpurea*
2. *Iris atrofusca*
3. *Iris atrofusca* (pale form)
4. *Iris haynei*
5. *Iris mariae*
6. *Iris bismarckiana*
7. *Iris hermona*
8. *Iris lortetii* PHOTO: YUVAL SAPIR

(e) Bearded irises with rhizomes; flowers with fall petals heavily suffused with contrasting darker markings.

Iris bismarckiana An erect, spring-flowering, rhizomatous perennial herb to 40 cm. Leaves 6–7, sickle-shaped, pale bluish-green, to 40 cm. Flowers borne on short stems, not exceeding the leaves; falls cream, to 60 mm, *dotted and marbled strongly with brown–purple*, blackish at the base; standards, 70–90 mm, almost circular, whitish to pale sky-blue with lilac veins, becoming brown below; style branches arched. Hillsides and maquis; rare and local. IP (north), LB (south). The following two species are similar: *I. hermona* has stout rhizomes without stolons. Leaves about 8, *partially evergreen* (appearing early), nearly erect. Flowers similar to *I. bismarckiana* but larger, borne on scapes much *exceeding* the leaves; falls 65–85 mm cream, intricately veined with dark lilac-brown and covered in small, embossed pink spots, purple–black at the base; standards 65–85 mm, whitish, circular, lined with faint purple but virtually without dots; style branches the same colour as the falls. Rocky ground. Very rare. IP (Mount Hermon; Golan), SY (south). *I. lortetii* has a compact rootstock. Leaves linear, more or less *equalling* the stems (to 50 cm). Flowers borne singly (rarely 2 per stem); falls 50–80 mm, not lined, deflexed, pale peach to lead-coloured with numerous *very fine maroon spots*, brownish-red near the base, sparsely red-hairy; standards 90 mm–11 cm, circular, erect, pinkish-white with dark pink veins; stigmas maroon, horizontal with crests deflexed. Garrigue. Rare. IP (Upper Galilee + Samaria Hills), LB (south).

Crocus

Small perennials with corms. Leaves basal, with a central whitish channel, enclosed beneath by sheaths. Flowers cup-shaped, with 1–2 spathes; tepals 6, typically all similar; stamens 3; style solitary with 3 (or more) branches, each with variably divided stigmas. Corm tunic, leaf number and style divisions are all important. Numerous species in the region (especially Turkey), of which just a few are described.

(a) Winter- to spring-flowering. Flowers mainly white, lilac, pink or blue.

Crocus sieberi A spring-flowering bulb with netted, fibrous tunics and 2–7(10) leaves present during flowering. Flowers white with a yellow throat, striped externally; tepals 15–45 mm; anthers yellow; stigma yellow to red. An island endemic of mountain scrub and woods. CR. The following are similar. *C. veluchensis* has flowers *without yellow* in the throat; filaments white. Woods and grassland near snow in mountains. GR. *C. nivalis* has 2–7(10) leaves present during flowering. Flowers 1–3, *lilac–blue* with a yellow throat; tube 30 mm–10 cm with egg-shaped segments; anthers yellow; style red, scarcely divided into 3 branches. AA, GR, IA, PN. The following mainland forms are variably recognised as true species, or as forms of *C. sieberi*, but may be better considered to be forms of *C. nivalis*. *C. sublimis* **(syn.** *C. sieberi* **subsp.** *sublimis***)** has bulbs with finely netted tunics and lilac–blue flowers with conspicuous hairs in the throat of the flower. Mountain turf or open woods in mountains. GR, PN. *C. atticus* **(syn.** *C. sieberi* **subsp.** *atticus***)** has bulbs with coarsely netted tunics and lilac–blue flowers with a yellow throat. GR, PN.

Crocus biflorus A spring-flowering bulb with 3–8 leaves appearing with the flowers, with narrow white stripes. Flowers 1–4, *white within*, with purplish-brown stripes on the outer surface of the outer 3 tepals; style short with 3 orange branches; anthers yellow to maroon. AA (east), GR, PN, TR.

---

1. *Crocus sieberi*
2. *Crocus veluchensis*
3. *Crocus sublimis* PHOTO: ELEFTHERIOS DARIOTIS
4. *Crocus atticus*
5. *Crocus baytopiorum*
6. *Crocus kotschyanus* (cross section)
7. *Crocus gargaricus*

Crocus baytopiorum An early spring-flowering bulb with a coarsely netted tunic, often bristly at the neck, and 4–5 leaves present during flowering. Flowers pink with a *pale blue tinge* and darker veins; tepals 20–30 mm; anthers yellow; styles orange, divided. Rare, in rocky limestone woods and screes. TR (southwest).

Crocus kotschyanus A spring-flowering bulb with a very *thin*, papery tunic and 4–6 leaves developing *after flowering*. Flowers lilac with a yellow throat; tepals 16–43 mm; anthers *white* to pale yellow; style yellow with slender branches. Open scrub and mountain pastures. A variable species with regional subspecies described. LB, SY (northwest), TR (central, south).

Crocus hadriaticus A late winter-flowering bulb with 5–9 greyish, very slender leaves <1 mm wide, present during flowering. Flowers 1–2, *white*, usually with a yellow throat; style reddish-orange with 3 branches above the throat. IA, PN.

(b) Winter to spring-flowering. Flowers yellow.

Crocus gargaricus A spring-flowering bulb with a *finely* netted tunic and 3–4 leaves just developing during flowering. Flowers funnel-shaped, bright yellow with tepals 25–45 mm; anthers yellow; styles orange and divided. Rare, in damp pine woods and grassland. TR (northwest).

Crocus vitellinus A winter- to early spring-flowering bulb with a thick papery tunic splitting into parallel fibres, and 5–8 dark shiny green leaves present during flowering. Flowers *deep yellow–orange* (occasionally marked purple–brown externally); tepals *rounded*; anthers yellow; styles orange and divided. Stony habitats and open woods. SY (west), TR (south).

Crocus flavus A spring-flowering bulb with papery tunics with parallel fibres and a sheath of dead leaf bases, and 4–8 erect leaves present during flowering. *Flowers pale to deep yellow throughout*; tepals 15–35 mm; stamens with yellow filaments and anthers; stigma slightly branched into 3 yellow to orange branches. Woods and scrub; common. GR.

Crocus chrysanthus is similar to *C. biflorus*, except for the bright yellow flowers. TR.

(c) Late summer- to autumn-flowering.

Crocus boryi An autumn-flowering bulb with papery to leathery tunics with parallel fibres, and 3–7 leaves present during flowering. Flowers *white* with a yellow throat, sometimes with external stripes; tepals 15–50 mm; anthers *white*, on slightly hairy filaments. CR, GR (west, and Mount Menikio), IA, PN. ***Crocus niveus*** is similar but with yellow anthers. PN.

Crocus goulimyi An autumn-flowering bulb with smooth, leathery tunics, splitting at the base, and 6–12 leaves present during flowering. Flowers fragrant, white to violet, often the *3 outer segments darker and larger*; tepals 16–38 mm; style white to orange and *branched*; anthers yellow. Olive groves and beneath old walls. PN (south).

Crocus laevigatus An autumn-flowering bulb with smooth, leathery tunics splitting at the base, and 3–4 leaves present during flowering. Flowers fragrant, lilac or white, with a yellow throat and usually with *1–3 prominent external stripes*; tepals 13–30 mm; anthers *white*. AA (Euboea, Ikaria, southern islands), CR, GR (Attiki), PN (southeast).

Crocus cartwrightianus An autumn-flowering bulb with a finely netted tunic and 5–11 leaves present during flowering. Flowers white to lilac, veined dark purple; tepals 15–32 mm long; *style long, exceeding the flowers and divided into 3 orange–red branches* from the throat. Stony habitats. AA (south), CR (west), PN (local). The following are similar: ***C. sativus*** is the sterile cultivated saffron, which has larger flowers and similar, long stigmas *collapsing against the inner flower*. Native to Turkey, widely cultivated. ***C. hadriaticus*** has flowers with a *yellow throat* and style branched from above the throat. GR (west), IA, PN. ***C. oreocreticus*** has short or absent leaves during flowering and narrower tepals. CR (east and central).

Crocus asumaniae An autumn-flowering bulb with a weakly netted, fibrous tunic, and 6 leaves *just* present during flowering. Flowers white to pale lilac with darker basal veins; anthers yellow; styles divided into *3 long orange–red branches*, exceeding the tepals. Rare, in open *Quercus* woods and hill forests. TR (south).

---

1. *Crocus vitellinus*
2. *Crocus goulimyi* (cross section)
3. *Crocus laevigatus*
4. *Crocus sativus*
5. *Crocus sativus* (cross-section)
6. *Crocus asumaniae*

Romulea

Low, cormous perennials with basal, linear leaves, often 4-grooved. Flowers normally enclosed in bud by 2 spathes; flowers *Crocus*-like, *borne on slender green stems*; tepals 6, contracting into a very short perianth tube; stamens 3, style solitary and 3-branched.

Romulea bulbocodium A low, cormous spring-flowering perennial. Leaves 3–7, deep green and curved or erect-spreading. Flowers crocus-like, borne on scapes 30 mm long that often have ›1 flower; white to lilac, *always with a yellow throat*, and striped purple; tepals elliptic and pointed, to 30 mm; stamens about ½ the length of the tepals, stigmas variable, usually *overtopping the anthers*. *Spathe papery almost throughout and weakly green-veined*. The most common species in the region, in a range of habitats from mountain rocks to coastal sands. Throughout. The following are similar. *R. linaresii* has 2 basal leaves, typically recurved-spreading. Flowers borne 1–2(3) on stems with 1–4 leaves; flowers violet–purple, all of one colour, 17–33 mm long; filaments purple; stigmas exceeded by the anthers. Throughout, except the southeast. *R. tempskyana* has 3–6 recurved leaves 80 mm–16 cm long. Flowers 1–6, *deep violet* (rarely white), always with a yellow throat. AA, CY, IP, TR (south).

Romulea columnae SAND CROCUS A low, cormous perennial, similar to the species above, with 3–8 erect to spreading, recurved leaves 0.6–1 mm wide. Flowers solitary or in groups of 2–3, *small*, to 12 mm across; corolla tube to 5 mm; tepals 10–19 mm, narrowly lance-shaped (9–15 mm wide), long-pointed; flowers mauve to white, usually with a whitish-yellow to yellow throat, sometimes greenish externally; anthers to just 2.5 mm. Damp coastal turf and sand; widespread across the region. Throughout.

Romulea ramiflora A low, cormous spring-flowering perennial with 4–6 recurved to erect leaves. Flowers borne on short stalks, pale to deep blue–violet, often with darker veins and greenish on the outside; throat normally yellowish; tepals oblong and blunt, 14–21 mm long. *Spathes almost entirely green with a narrow papery margin*. Coastal sandy habitats. Virtually throughout, except the southeast mainland.

Gladiolus

Cormous perennials with fans of flat, sword-shaped, ribbed leaves. Flowers borne on long, rigidly erect stems, each flower with a green bract; tepals 6, unequal, fused into a short tube at the base; stamens 3; style slender with 3 short branches. A confused genus with conflicting keys in regional floras, possibly owing to high levels of hybridisation; reliable identification in the field may not be possible.

(a) Anthers longer than their filaments (or aborted). Flowers pinkish-red.

Gladiolus italicus A tall, cormous perennial to 80 cm, with *broad, sword-shaped leaves* 40–80 cm long, 17 mm wide, arising from a spherical corm with netted fibres. Flowers 6–15, borne in a 2-sided inflorescence, each flower with an equally long leafy bract; flowers 40–50 mm long, pinkish-red, the segments very unequal, the upper longer and broader than the laterals; anthers *longer* than the filaments. Seeds winged. Common on garrigue, maquis, roadsides and cultivated land. Throughout, except some small islands.

(b) Anthers equalling, or shorter than, their filaments. Flowers pink or pinkish-red.

Gladiolus communis var. *byzantinus* A bulbous perennial 50 cm–1 m, similar to *G. italicus* in form, with leaves 30–70 cm long and 5–22 mm wide. Flowers bright pink, borne 10–20 in a lax, weakly 2-sided inflorescence. Anthers spear-shaped, the *same length or shorter than their filaments*. Seeds winged. Stony fields, woods, maquis and cultivated places. Recorded throughout, but

1. *Romulea columnae*
2. *Romulea bulbocodium*
3. *Gladiolus italicus*; (inset, top) fruits; (inset, bottom) cross section
4. *Gladiolus communis* var. *byzantinus*; (inset) cross section

possibly confused with for related taxa. *G. illyricus* has traditionally been differentiated from *G. communis* in being shorter, 25–50 cm, with leaves 10–40 cm long and 4–10 mm wide and with fewer (3–10) flowers. Seeds brown and winged. AA, GR, IA, PN, TR. *G. anatolicus* is very similar but with seeds *not winged*. AA (east), TR.

Gladiolus triphyllus A *short*, bulbous perennial, 15–50 cm (often less) with 3–4 linear leaves, the lowermost 10–30 cm long, <5 mm wide and markedly 3–4-veined. Flowers few (1–7), stalkless, sweet-smelling in the evening, *pale to dark pink* (not magenta) with segments 25–30 mm long; anthers *shorter* than the filaments. Seeds not winged. An island endemic. CY.

(c) Flowers *purple*.

Gladiolus atroviolaceus A perennial 30–60 cm similar to the previous species but with *deep purple–lilac* flowers. Leaves 3, 3.5–5 mm wide, bluish, with 6 parallel veins. Flowers *purple*, borne in spikes of 4–7(8); spathes unequal; anthers usually just *longer* than their filaments. Capsule ellipsoid, 18 mm; seeds ovoid, dark brown, 2.5 mm, not winged. Fallow fields and plains. IP (mainly Judean Mountains; rare elsewhere), SY (west), TR.

ASPHODELACEAE (including Xanthorrhoeaceae)

Many genera formerly included in the Liliaceae (in its broader, traditional description), characterised by dense tufts of long narrow leaves and stout, woody spikes of flowers; perianth of 6 tepals, often conspicuous; stamens 6; style 1 with minute or 3-lobed stigma. Fruit a capsule.

Eremurus

Robust perennials with long spike-like racemes of white, yellow, pink or brownish flowers. Predominantly Asian (poorly represented in the Mediterranean).

Eremurus spectabilis A distinctive, robust hairless perennial 1–2 m. Leaves grey–green, linear, all in a basal tuft, channelled with a sharp keel. Flowers whitish to yellowish, cup-shaped, 15–20 mm across, borne on *long, very dense racemes* of about 100; stamens conspicuous, orange. Fruit a rounded capsule with angular seeds. Stony hillsides. TR.

Aloe

Succulent shrubs with robust, spiny-margined leaves forming a rosette. Flowers in racemes; corolla tubular, often orange or red. Fruit a capsule. Native to Africa and Madagascar, widely cultivated in the region.

Aloe vera A stoloniferous perennial succulent with numerous basal leaf rosettes with grey-bluish or reddish-tinged leaves to 40–50 cm. Inflorescences to 1 m, with *yellow flowers* 25–30 mm with protruding stamens. Planted and locally naturalised. Throughout. *A. maculata* has dark grey–green leaves, often flushed orange–red, with distinct *pale markings*. Planted and occasionally naturalised in hot, dry areas.

Asphodelus ASPHODEL

Robust, hairless, herbaceous and tuberous perennials (rarely annuals) with leafless stems and linear leaves in basal tufts. Tepals free, all similar; flowers borne on simple or branched stems; filaments hairless. Many species very similar; fruit shape an important diagnostic.

(a) Leaves flat. Flower-stalks *thickened in fruit* (1–1.3 mm wide).

Asphodelus ramosus A tall, robust herbaceous perennial 50 cm–1.6(2) m with numerous fleshy roots. Leaves flat, strap-shaped and grey–green. Flowers white and star-like, with tepals 11–20 mm, with brownish mid-veins, borne in tall, erect, much-branched inflorescences in which the *lateral branches are almost as long as the terminal*. Capsules ovoid, large, 6–12 x 4–9 mm. Stony pastures and goat-grazed maquis. Probably throughout, except the north.

---

1. *Gladiolus illyricus*
2. *Gladiolus triphyllus*
3. *Gladiolus atroviolaceus*
4. *Eremurus spectabilis* PHOTO: NICK TURLAND
5. *Aloe maculata*
6. *Aloe vera*

(b) Leaves flat and keeled. Flower-stalks *not* thickened in fruit (<1 mm wide).

Asphodelus albus WHITE ASPHODEL A perennial 60 cm–1.5 m. Leaves linear and grey–green, with a central keel, 60 cm–1.1 m x 25 mm. Inflorescence a *narrow, spike-like raceme*, rarely branched at the base, and usually *unbranched* above or with few, short lateral branches; flowers white, to 30 mm across; tepals with greenish mid-veins. Capsules 7–13 x 6–11 mm, borne densely along the spike. Rocky ground and grazed maquis. GR (north and central).

(c) Leaves hollow.

Asphodelus fistulosus HOLLOW-LEAVED ASPHODEL A small, tufted *annual* or short-lived perennial, 30–60 cm (1.5 m) with numerous fleshy roots. *Leaves linear and slender, 1–35 mm wide, hollow, and cylindrical*. Flowers white, borne in a lax raceme of 10–15 on a long, smooth stalk with membranous, whitish bracts; tepals 8.5–12.5 mm, with brownish mid-veins. Capsule 5–6 x 4–6 mm. Common in grazed pastures, stony ground and waste places across the region. Throughout. *A. tenuifolius* is similar but with fibrous roots, narrower leaves to 2.5 mm and smaller flowers with tepals 5.5–7.5(8) mm. Capsule 3–4 mm long. Arid habitats. CY, IP, PN (rare).

Asphodeline

Similar to *Asphodelus* but rhizomatous (not tuberous) and flowers borne on *leafy stems*.

Asphodeline lutea YELLOW ASPHODEL A herbaceous perennial to 1 m with fleshy rhizotomous roots. Leaves linear, blue-green, 2-3 mm wide, untoothed, both basal and *along the flowering stems*. Flowers *yellow*, to 40 mm across, borne in a dense raceme; tepals with green mid-veins; stamens 6, of 2 different lengths, the longest with curved filaments. Capsule to 12 mm. Rocky slopes, garrigue and mountain slopes. Throughout. *A. liburnica* is similar but much more slender, with leaves *borne only on the lower third of the stem* (not its whole length). Flowers yellow, borne in loose racemes. Rocky hills and mountains. AA (rare), CR, GR, PN (rare), TR.

AMARYLLIDACEAE

Bulbous or rhizomatous (or cormous) perennials with leafless stems, similar to Liliaceae. Flowers solitary or in umbels, enclosed in papery bracts in bud; tepals 6, usually all petal-like; stamens 6; style 1, ovary 3-parted. Fruit a 3-parted capsule, often succulent.

Galanthus SNOWDROP

Hairless, tufted, bulbous perennials with 2 basal leaves per bulb. Flowers solitary, pendulous, borne on slender stems; tepals 6, the outer 3 petal-like, the inner 3 smaller and notched; stamens 6. Absent from hot, dry areas.

(a) Winter- to spring-flowering.

Galanthus elwesii A small, winter-flowering bulb with strap-shaped grey-blue, mat-surfaced leaves 60 mm-32 cm long, 6-32 mm wide, appearing *after the flowers*. Flowers white with 3 outward-spreading tepals 20-27 mm long; inner tepals 8-15 mm, flared and marked green at the tips. Dry rocky garrigue and pine forests. AA (north), GR (north), TR. *G. samothracicus* is similar but with inner tepals with green marks only near the apex (not also in the lower half). Damp woods. An island endemic. AA (north).

Galanthus woronowii A small, winter-flowering bulb with rather broad, mid-green leaves 12.5 mm-6 cm long, 10-20 mm wide. Flowers white with 3 outward-spreading tepals 18-27 mm long; inner tepals 7-12 mm, with *inverted green v-shaped markings* on the outer surface. TR. *G. ikariae* is very similar, with bright, shiny-green leaves 15-25 mm wide and flowers with outer tepals 15-25 mm long. AA (Andros, Ikaria), TR.

(b) Autumn-flowering.

Galanthus reginae-olgae A small, *autumn-flowering* bulb with narrow, dark green leaves with recurved margins, and a central silver mid-vein, 11-15 cm long, 3-8 mm wide. Flowers white with 3 outward-spreading tepals 15-35 mm long; inner tepals 9-12 mm, not flared, variably marked with green. Damp woods. GR (west), IA (Corfu), PN. *G. peshmenii* is a similar, very local, autumn-flowering species without leaves during flowering. AA (Kastellorizo), TR (south).

---

1. *Asphodelus ramosus*
2. *Asphodelus albus*
3. *Asphodelus fistulosus*
4. *Asphodelus tenuifolius*
5. *Asphodeline lutea*
6. *Asphodeline luburnica*

Amaryllis

Large, bulbous perennials with several strap-shaped leaves arranged in 2 rows. Flowers borne in terminal clusters on leafless stems; tepals 6, all similar.

Amaryllis belladonna A large, clumped, *autumn-flowering* bulbous perennial with several strap-shaped leaves 30–50 cm long, 20–30 mm wide, arranged in 2 rows, *appearing after the flowers*. Flowers about 10 cm across, bright pale pink with a yellowish centre, borne in terminal clusters on leafless stems 30–60 cm; tepals 6, all similar. Native to South Africa, planted and occasionally naturalised.

Sternbergia WINTER DAFFODIL

Bulbous perennials, not smelling of garlic. Flowers with funnel-shaped, crocus-like perianth with 6 yellow, all similar tepals.

Sternbergia lutea WINTER DAFFODIL An autumn-flowering bulbous perennial with strap-like leaves 70 mm–10 cm x 4–15 mm, which appear before or during flowering. *Flowers crocus-like, bright yellow* and held erect on stems 25 mm–10(20) cm; flowers short-tubed with 6 oval and pointed tepals; stamens 6, the filaments greatly exceeding their anthers. Rocky maquis. Throughout, except the southeast. **S. colchiciflora** is similar but with much *narrower leaves* (2–5 mm) appearing *after flowering*. Flowering stems just 10–20 mm long, largely concealed by their bulb tunics; flowers bright yellow, funnel-shaped with linear tepals, broadest above the middle, to just 4 mm wide. Stony and rocky pastures at high altitude; almost exclusively on the mainland. AA (Chios), GR, IP (north – rare), LB, SY, TR. **S. clusiana** is similar to the previous species but always flowering *before* the leaves emerge, and with *broader* tepals at least 15 mm wide. AA (east – rare), IP, SY, TR.

Acis

Bulbous perennials in dry habitats with *narrow leaves*. Flowers borne in umbels of 1–5 flowers; corolla bell-shaped with 6 tepals, pink or white; anthers greatly exceeding their filaments. Recently separated from *Leucojum*.

Acis ionica A small, autumn-flowering bulb with 2–3(5) thread-like leaves appearing after the flowers, 12–22 cm long, 2–3 mm wide. Flowers white, borne singly on slender, solid stems to 20 cm long; tepals 9–13(15) mm long, the outer three oblong, the inner three broadly oval, narrower at the base; anthers bright yellow. Fruit a 6-lobed capsule; seeds black. Coastal rocky slopes and garrigue. GR (west Sterea Ellas), IA.

Leucojum

Bulbous perennials in wet habitats with *hollow stalks and broad leaves*. Flowers borne in umbels of 1–5 flowers; corolla bell-shaped with 6 tepals, pink or white; anthers greatly exceed their filaments.

Leucojum aestivum SUMMER SNOWFLAKE A *robust, clump-forming*, bulbous, spring- to summer-flowering perennial 31–61 cm with strap-shaped leaves 50 cm x 5–15 mm. *Flowering stems flattened, twisted and 2-winged*, with up to 8 flowers 11–13(15) mm long; corolla bell-shaped and white, each tepal with a green spot at the tip. River banks, ditches and other damp or wet habitats. GR, TR.

Narcissus DAFFODIL

Bulbous, usually early-blooming perennials with basal leaves and hollow, leafless scapes. Flowers with 6 tepals; stamens 6, surrounded by a cup- or trumpet-like corona. Capsule 3-parted. A difficult genus with high variability and hybridisation.

(a) Flowers with white tepals, flowering in autumn.

Narcissus papyraceus PAPERWHITE NARCISSUS A hairless, bulbous perennial with broad, strap-like leaves, blue–green and slightly channelled, 6.9–9(13) mm wide. *Flowers pure white*, borne in umbels of 4–11(20); flowers to 40 mm across and strongly fragrant. Dry rocky places and fallow ground. GR (and naturalised elsewhere).

Narcissus obsoletus A hairless, slender, bulbous perennial with 1–2, bluish-green, rush-like leaves to just 1 mm wide. Flowers 20–30 mm across, solitary or paired; tepals white, greenish-yellow at the very base, oblong and pointed; corona *very short*, orange, 6-lobed. Dry slopes, sandstone hills (IP) and pathsides. Throughout (populations in the region were previously described under *N. serotinus*, which is similar but genetically distinct, and occurs further west).

(b) Flowers with cream, white or green tepals, flowering mainly in late winter to early spring.

Narcissus tazetta BUNCH-FLOWERED DAFFODIL A hairless, bulbous perennial 23–60 cm, with blue–green flat, broad, strap-like leaves 6.2–12 mm wide, equalling the stems, present during flowering; flowers appear in autumn in rocky mountains, and in winter to early spring in seasonal lakes. Flowers borne in umbels of 3–many (15); fragrant, white (sometimes cream to yellow), with a prominent yellow corona 3–6 mm long that is 2 x as wide as high, on a 2-edged, flattened scape. Spathe at the base of the inflorescence membranous. Meadows and pastures. Throughout, except the north.

Narcissus poeticus PHEASANT'S-EYE DAFFODIL A hairless, bulbous perennial to 60 cm with blue–green flat, linear leaves 5–9(13) mm wide, equalling the stems, present during flowering. Flowers rather large and showy, *normally solitary*, nodding and scented; corolla white with a greenish tube; tube to 30 mm long, tepals wide-spreading, to 22–38 mm long; corona to 2.3–5.2 mm long, yellow with a *bright red, frilly rim* and 3 stamens protruding. A high-altitude species, locally abundant in deciduous woods and mountain pastures. GR (northwest, north).

Pancratium SEA DAFFODIL

Bulbous perennials with flowers borne on long-stemmed, terminal umbels; corolla funnel-shaped with 6 narrow tepals and a corona with 12 teeth.

Pancratium maritimum SEA DAFFODIL A distinctive, summer-flowering, clump-forming, bulbous perennial. Leaves fleshy, grey–green, hairless and long 16–37(75) cm. Flowers large and white, to 15 cm long, the perianth tube ‹2 x *as long as the segments*; fragrant, borne in umbels of 4–9(19); tepals all similar; stamens 6, borne on the rim of the cone-like corona. Fruit a 3-parted capsule. Characteristic of the Mediterranean dune flora. Throughout. *P. sickenbergeri* is similar but smaller with *markedly curled leaves*. Deserts. IP.

1. Narcissus obsoletus
2. Narcissus tazetta
3. Narcissus papyraceus
4. Pancratium maritimum
5. Pancratium sickenbergeri

Vagaria

A small genus of bulbous perennials, closely related to *Pancratium*, with strap-shaped leaves. Flowers with 6 tepals, borne in loose umbels; stamens 6.

Vagaria parviflora A clumped, bulbous autumn-flowering perennial; bulbs with black tunics. Leaves 47, 30–60 cm long. Flowers 6–9, borne in loose terminal umbels; flowers 55–75 mm across; tepals hooded at the tip, white with a greenish vein. Rocky crevices. IP, SY.

Allium

Distinctive bulbous perennials smelling of onion or garlic when crushed. Flowers borne in terminal, often spherical umbels; tepals 6 all similar; stamens 6. Fruit a 3-parted capsule. Some species produce *bulbils* in the flower-heads (and/or *bulbets* — offsets of the bulb). Numerous species occur in the eastern Mediterranean; a small subset is described here.

(a) Innermost stamens with conspicuously *3(5)-parted filaments* (the anther borne on the central cusp).

Allium ampeloprasum WILD LEEK A *robust, leek-like* perennial with a membranous bulb and stout flowering stems to 1(2) m with *numerous yellow–brown bulblets*. Leaves 4–10, pale blue–green, V-shaped in cross section (at least below), rough along the margins and with a central keel beneath; leaves often withered during flowering. Flower-heads large and spherical with numerous lilac flowers with slightly protruding stamens, borne on long stalks; innermost stamens with 3-parted filaments, the *lateral teeth greatly exceeding the central (fertile) part*; spathe usually soon-falling. Common, particularly near the coast. Throughout. The following are similar: **A. atroviolaceum** is less robust and with *dark purple flowers* borne in smaller umbels 30–60 mm across. Dry fields. GR, LB, TR. **A. scorodoprasum** has crowded *blackish-crimson* bulblets and stamens not protruding. Distribution unclear owing to confusion (possibly throughout).

Allium commutatum A leek-like bulbous perennial 40 cm–1.2 m, similar to the previous species group. *Leaves many* (6–12), *not rough* along the margins. Flowers whitish-pink or dull purple, with greenish or purplish keels along the tepals, to 3.3–4.2 mm long; *margin of inner tepals finely toothed*. Coastal sands and cliffs. Throughout except CY and the far southeast.

Allium sphaerocephalon ROUND-HEADED LEEK A bulbous perennial 20–70(90) cm with scapes circular in cross section. Leaves 2–4(5) semi-cylindrical with a groove along the surface, sheathing the stem below, 1–2(3) mm wide; spathes 2-valved, shorter than the umbels; leaf sheaths concealing bulblets. Flowers dark purple–red (rarely pale), each tepal with a greenish keel, borne in *dense, spherical heads* to 40 mm across; lateral teeth of 3-parted filaments only just *equalling* the central (fertile) part; stamens clearly *protruding*; flower-stalks 5–20(30) mm long. Throughout. **A. junceum** is similar but with stamens *not* protruding and flower-stalks <4 mm long; bulblets absent. Inner filaments 5-parted. Garrigue and semi-deserts. AA (east), CY, SY, TR. **A. guttatum** is similar to both the previous species but with often, sparser heads of paler *greenish or whitish* flowers, in which the lateral teeth of the 3-parted filaments *much exceed* the central (fertile) parts. Throughout except the far southeast.

Allium amethystinum A rather robust, bulbous perennial to 1.2 m. Leaves 3–7, narrowly cylindrical, about 8 mm wide, sheathing the lower part of the stem, often *withered* during flowering. Flower-stalks reddish above, cylindrical in cross section, with purple, tubular-bell-shaped flowers, to 4.5 mm long, borne on *unequal* stalks, in *large, regular* spherical to drop-shaped umbels to 50 mm across; stamens protruding. Rocky and disturbed habitats. Throughout except the far southeast.

---

1. *Allium ampeloprasum*
2. *Allium commutatum*
3. *Allium sphaerocephalon*
4. *Allium junceum*
5. *Allium subhirsutum*
6. *Allium schubertii* PHOTO: BANAN AL SHEIKH
7. *Allium neapolitanum*
8. *Allium palaestinum*

Allium vineale CROW GARLIC A bulbous perennial 22 cm–1 m with rounded (not angular) scapes. Leaves 2–4(6), cylindrical, channelled and sheathing the base of the scape. Spathe 1-valved and beaked, papery, *not longer than the flowers* and soon-falling; flowers reddish or greenish and bell-shaped, 3–4 mm long, often few: *most or all of the flowers replaced by bulbils, often sprouting green shoots before falling*; stamens of flowers *protruding*. Dry grassy habitats. GR, IA, PN, TR.

Allium schoenoprasum CHIVES A *tufted* bulbous perennial 12–46 cm with many leaves and flowering stems. *Leaves cylindrical and hollow,* grey–green, 10–29 cm long and 1.5–2.6(4) mm wide. Flower-stalks cylindrical, with dense, small umbels to 40 mm across, with 7–50 *pink–purple flowers*, and no bulbils; anthers *yellow and not protruding*; spathe *2-valved*, shorter than the umbel. Generally mountain-dwelling in rocky habitats, or cultivated as a culinary herb. GR, TR.

(b) Innermost stamens typically entire (without teeth or with 2 small teeth either side of the base) and leaves *flat or folded*.

Allium chamaemoly A *small*, bulbous perennial just 10–85 mm. Leaves 4–8, *spreading flat* on the ground, sheathing only the very base of the flower scape, linear with hairy margins; spathes 1-valved and 2–4 lobed. Flowers white and star-like, 10–16 mm across, borne in umbels of 7–18; tepals with greenish mid-veins. Flower-stalks *curving in fruit*. Strongly western. GR, IA, PN.

Allium subhirsutum A short, bulbous perennial 10–30 cm. Leaves linear, 2–7 mm wide, sheathing the stem at the very base, distinctly *hairy along the margins*. Flowers 7–78, white, flat and starry with tepals 5–7 mm long, borne in loose umbels about 30–60 mm across; stamens included. Dry, stony places; common. AA, CR, GR, IA.

Allium schubertii A large, bulbous perennial. Leaves flat, spreading, lance- to strap-shaped, slightly wavy, with rough margins. Spathe short, 2–3-valved. Flowers 20–200, borne on *long* stalks many times the length of the flowers and *very unequal* – the fertile shorter, the sterile 3–4 x as long, all club-shaped at the tips; corolla pink–mauve. IP, LB, SY, TR.

Allium neapolitanum NAPLES GARLIC A bulbous perennial to 50 cm. Leaves few (2–3), linear, 11–24(36) mm wide, and sheathing the stem at the base and *keeled* along the back. Flower stems triangular in cross section (at least at the apex) and smooth, with white flowers in loose clusters to 70 mm across; *spathe 1-valved* and shorter than the flower-stalks; flowers broadly cup-shaped. Locally common in grassy habitats and pastures. Widespread. Throughout (rare in the west). The following species are similar: ***A. palaestinum*** has *cylindrical* (not angled) stems. Rocky deserts. IP (south). ***A. trifoliatum*** has *hairy* leaves and sheaths (at least along the margins), cylindrical stems and white flowers with pink mid-veins along the tepals, or flushed pink with age; flowers broadly cup-shaped with spreading lobes. Rocky habitats. Throughout. Related species in the Israel–Palestine area include: ***A. erdelii***, which has leaves with *long, soft hairs* and *narrowly bell-shaped* white to cream flowers borne on long, erect stalks of similar length. Arid and semi-desert habitats. ***A. negevense*** has yellowish-white flowers with stalks 10–15 mm long. Semi-desert highlands. ***A. longisepalum*** has leaves with *short* hairs and white to pink flowers 10–15 mm long borne in rounded clusters on stalks 15–25 mm. ***A. akirense*** has pinkish-white flowers just 3–5 mm long. Coastal hills. ***A. papillare*** has leaves covered by backward-pointed tiny hairs; flowers 5–7 mm long, cream, with purplish veins and blunt lobes. Desert sands. ***A. qasyunense*** is very similar, with pointed corolla lobes without purplish veins. Maquis and semi-desert.

1. *Allium trifoliatum*
2. *Allium erdelii*
3. *Allium negevense*
4. *Allium israeliticum*
5. *Allium rothii*
6. *Allium tel-avivensis*

Allium nigrum A bulbous perennial, 60–90 cm. Leaves 2–5, all basal, *broadly linear*, to 25–33(90) mm wide, tapered to a point, much exceeded by the scapes, which are rounded in cross section. *Spathes 2–3(4)-valved* with pointed tips, free to the base. Flowers 30–90, white or very pale lilac with green mid-veins, and *blackish ovaries*, borne in dense, spherical umbels; anthers yellow. Various dry, grassy habitats; local. AA, CR (common), CY, GR, PN, SY. ***A. cyrilli*** is similar but with tepals narrow (1–1.5 mm, not 1.5–3 mm wide) with incurved tips. AA, CR, GR, TR. ***A. orientale*** is similar to *A. nigrum* but much shorter, 10–40 cm, with narrower leaves ‹25 mm wide, tapering at the tips. CY, IP, LB, SY, TR.

Allium israeliticum A bulbous perennial to 25(40) cm. Leaves 2–6, thick, flattish to recurved, 15–30 cm, blue–grey. Flower-heads borne on cylindrical scapes; spathe papery, whitish-brown; flowers whitish; tepals 6–8 mm; filaments just exceeded by the tepals, narrowly triangular with yellow anthers; ovaries persistently green. Capsule 3-angled, 6–8 mm. Sparse garrigue, sandy plains and fallow fields. IP (mainly the north).

Allium rothii A *desert-dwelling*, short, robust bulbous perennial to 15 cm. Leaves numerous, broad, flat and prostrate, equalling or exceeding the scapes. Umbels hemispherical, borne on thick stems; spathe short, lobed; tepals 4–4.5 mm, white with *dark blackish-purple* mid-veins; stamens and ovaries deep purple. Capsule 5 mm, spherical. Rare, in dry, disturbed areas and rocky deserts. IP (Negev Highlands; rare elsewhere), SY (Syrian Desert).

Allium tel-avivense A bulbous perennial similar to the *A. nigrum* group, to 25(40) cm. Leaves spreading, folded with slightly undulating margins, mostly exceeding the stems, rather narrow (10–20 mm). Flowers pink–lilac, borne in umbels 50–80 mm across, with prominent green ovaries; tepals 7–9 mm. A rare endemic of sandstone hills. IP (mainly Sharon and Philistine plains).
A. aschersonianum is a similar *desert-dwelling* bulbous perennial, 30–70(80) cm, with broader (20–50 mm) leaves *shorter* than the scapes. Flowers pink to purple; filaments and anthers deep purple; tepals 6–7 mm. Rare and local, in sandy and rocky deserts. IP (northern Negev, Negev Highlands, Jordan Valley), SY, TR.

1. *Allium aschersonianum*
2. *Allium roseum*
3. *Allium siculum*
4. *Allium pallens*

Allium roseum ROSY GARLIC A bulbous, hairless perennial 40–65(85) cm, with 2–4, linear, keeled leaves, 4–12 mm wide, similar to *A. neapolitanum* in most respects but with *cylindrical* (not angled) stems and *pink* (sometimes whitish), bell-shaped flowers to 12 mm long, borne in loose umbels that are sometimes mixed with bulbils; stamens not protruding; *spathe with 2–4 valves, fused at the base*. A range of dry habitats; common. Throughout.

Allium siculum A tall, hairless, bulbous perennial 50 cm–1.5 m. Leaves mostly basal, strongly keeled beneath, to 50 mm wide, often withered during flowering, the innermost sheathing the flowering stems. Flowers greenish-white to greenish-maroon, *bell-shaped*, 14–16 mm long, borne on long, arching stalks in *loose umbels;* stalks becoming erect in fruit. Damp shady woods and maquis. GR, TR.

(c) Innermost stamens entire (without teeth) and leaves *cylindrical, subcylindrical or filiform.*

Allium paniculatum A variable, bulbous perennial 26–73 cm. Leaves 2–4, 40 mm–20 cm long, semi-cylindrical, *very narrow, 0.9–1.8 mm wide*, sheathing the stem at the base. Flowers borne in lax to compact umbels to 35 mm across; perianth *bell-shaped to cylindrical and open at anthesis*; tepals 4–6 mm, white or pale pink, often on drooping stalks; stamens *included* within the perianth (or the yellow anthers just protruding); spathe persistent, *2-valved*, and greatly exceeding the flower-stalks. Dry, open habitats; common. Throughout (not CR or KP). **A. stamineum** is similar but shorter, 50 mm–30(35) cm, with *brownish to purplish* flowers with *stamens distinctly protruding* (by at least 0.5 mm). Scattered throughout.

Allium pallens A bulbous perennial 20–80 cm, similar to *A. paniculatum*. Leaves 3–4, sheathing the stem to almost halfway, hairless. Flowers borne in rounded, usually compact heads 20–40 mm across; whitish or yellowish-green with unequal spathes, the longest much exceeding the flower-heads; tepals 3.5–5 mm; anthers exserted, yellow. Disturbed habitats. Scattered throughout; rare on some islands (e.g. CR).

ASPARAGACEAE

A large and variable family comprising many genera of shrubs or large perennials, formerly included in an even larger broadly defined family, the Liliaceae. Flowers cosexual (rarely dioecious, e.g. *Asparagus* and *Ruscus*); perianth with 6 tepals; stamens 6; styles 1–3. Fruit a capsule or berry with 1–numerous seeds.

Dracaena

Robust, rather succulent, tree-like perennials with stout trunks. Flowers with 6 tepals and stamens. Fruit a fleshy berry or capsule. Native mainly to tropical and sub-tropical Africa and Macaronesia.

Dracaena draco A slow-growing tree-like perennial 8–12 m with a very *robust, swollen trunk*, eventually densely branched above. Leaves sword-shaped, thick, grey–green and flat, 40–90 cm long, 30–45 mm wide. Flowers borne in numerous clusters of 4–7 in large, hairless inflorescences to 1.2 m long; flowers 7–9 mm across, whitish-yellow with 6 tepals and 6 thickened, flattened stamens. Fruit a yellowish to reddish-brown berry. Native to western Africa and Macaronesia, widely planted throughout.

Asparagus ASPARAGUS

Shrubby, hairless, rhizomatous perennials with tough stems (becoming woody) and with *cladodes* (rather than true leaves). Flowers small and inconspicuous, bell-shaped with 6 tepals. Fruit a small berry.

Asparagus officinalis WILD ASPARAGUS   A tall, hairless, *herbaceous* perennial with erect to spreading, smooth (not spiny) stems, to 1.5(2) m, with or without clusters of small green cladodes in the axils, with a *feathery appearance*. Flowers very small, to 6 mm long, borne in the leaf axils; unisexual, greenish-yellow and bell-shaped. Fruit a spherical, red berry 10 mm across. Various habitats; common. Throughout.

Asparagus acutifolius   A *climbing* subshrub with much-branched, woody stems to 1.5(3) m. Cladodes dark green, *more or less equal*, borne *numerous clusters, 10–35(70)*; *small, just 2–8 mm long*, and *spine-tipped*, to 0.5 mm thick. Flowers borne in clusters of *2–4*, 4 mm long, *pale yellowish-green*, sweetly scented and mixed among the cladodes; flower-stalks surrounded by bracts at the base. Berry black when ripe. Throughout, except some islands (not CR). **A. aphyllus** is similar to A. acutifolius but *shorter, and non-climbing*, to 60 cm (1 m) tall with hairless green stems. *Cladodes markedly unequal*, in clusters of 3–10(15), to 20 mm long and 0.6 mm thick. Flowers in groups of 3–8(10). Berries black when ripe; flower-stalks surrounded by bracts at the base. Rocky scrub and garrigue; very common in the Aegean. Throughout (not CY). **A. horridus (syn. A. stipularis)** is similar to the previous species but with grey–green *solitary* cladodes, in rigid clusters of 2–3, 5–40 mm long. Hot, dry areas. Absent from the Greek mainland. AA, CR, CY, IP, SY, TR.

1. *Dracaena draco*
2. *Asparagus acutifolius*
3. *Asparagus horridus*

ASPARAGACEAE

Cordyline

Trees and shrubs with dense, strap- or sword-shaped leaves and numerous white or cream flowers.

Cordyline australis A small tree up to 10 m (usually less) with a stout grey trunk, which is simple or branched. Leaves sword-shaped, erect to 1 m long. Flowers white, scented and borne in dense panicles to 1 m long. Native to New Zealand, widely planted in towns and gardens. Throughout.

Ornithogalum STAR-OF-BETHLEHEM

Bulbous, spring-flowering perennials with basal leaves in rosettes. Flowers normally white and star-shaped, borne in racemes; each flower with 1 bract; tepals free and petal-like; stamens 6. Fruit a 3-valved capsule with many seeds. A complex group in the area and identification often challenging (some species cannot be identified reliably in the field).

(a) Inflorescence an elongated raceme; leaves without central white mid-veins.

Ornithogalum narbonense A spring-flowering perennial, 25–57 cm. Leaves 4–6, linear with sheathing bases, to 16 m wide, persistent until after flowering. Flowers 20–30 mm across, white, scentless, borne erect on stalks the same length in a *tall, many-flowered, slender raceme* of up to 50, with tepals wide and spreading, with prominent external green stripes; anthers yellow; ovary to 5 mm with a flattened top, *distinctly exceeded by the narrow style*. Common in rocky pastures; one of the most widespread species in the region. Throughout. *O. brachystylum* is smaller with tepals white above, green beneath, inwardly rolled soon after flowering. AA (Rhodes + adjacent islets). *O. prasinantherum* has a cone to ovoid-shaped ovary to 4 mm, tapering into the style. AA (rare), GR (west), IA, PN. *O. spetae* has reddish anthers. AA (Andros, Siros).

Ornithogalum pyrenaicum BATH ASPARAGUS A spring-flowering perennial to 80 cm (1 m) with slender green stems, somewhat asparagus-like in bud. Leaves 5–8, linear and slightly channelled, without a white mid-vein, often withered during flowering. Flowers many (>20), *yellow–green*, 18–23 mm across, star-shaped, borne in *long, slender, spike-like racemes*, faintly scented; anthers yellow; ovary *egg-shaped to cylindrical*. Various habitats including woods and scrub. AA (local), GR. *O. creticum* is similar but with a *spherical* ovary. AA, CR.

(b) Inflorescence an elongated raceme; leaves with white mid-veins.

Ornithogalum nutans DROOPING STAR-OF-BETHLEHEM A perennial to 60 cm with 4–6, strap-like leaves to 15 mm wide. Flowers white, broadly bell-shaped, to 45 mm, nodding, borne in *1-sided racemes*; tepals somewhat recurved, with an external grey–green stripe; bracts longer than their flower-stalks. Open woodland, scrub and pastures. AA, CR, GR, IA, TR.

(c) Inflorescence *corymb-shaped*; leaves typically *without* white mid-veins.

Ornithogalum pedicellare A spring-flowering perennial with 5–12 or more leaves. Leaves linear to thread-like, 40 mm–20 cm long, 1–3 mm wide, spreading, typically (though not always) without a paler mid-vein. Flowers 2–8(15), white, borne in *lax clusters* on stalks initially erect, later spreading, the lowermost 20–70 mm long; anthers yellow. A rare island endemic; garrigue and maquis. CY. *O. trichophyllum* is similar but with fewer (4–8) leaves and *crowded* flower clusters, the lowermost stalks 10–15 mm. CY, IP.

Ornithogalum neurostegium A small bulbous perennial, 20 mm–25 cm, with erect scapes exceeding the leaves. Leaves several, linear, 7–9 mm wide without a white line on the upper surface, with rough, reddish, undulate margins, rough-hairy or hairless beneath. Flowers white, 10–20, borne in corymbs; tepals 12–15 mm, green outside; flower-stalks stiff and spreading at an angle in fruit, 25–45 mm. Capsule unwinged. Maquis, fields and hillsides. IP, LB, SY, TR. Two subspecies are widely recognised but their distinction and distributions are unclear: **Subsp. *neurostegium*** has leaves rough-hairy beneath, and bulbs with or without bulblets. **Subsp. *eigii*** has *hairless* leaves (except the margins) and numerous bulblets.

Ornithogalum montanum A short perennial to 20 cm. Leaves 3–6 in a basal rosette, strap-shaped, 10–15(20) mm wide, *long-tapering to a pointed tip*. Flowers 5–15, white with a broad green stripe on the outer surface of each tepal, 20–30 mm across, borne in wide, flat-topped racemes; flower-stalks 30–80 mm long, horizontal to erect-spreading in fruit. Damp mountain habitats. GR, IP, SY, TR. *O. atticum* is similar but with grey–green, twisted leaves with undulating edges. Rocky habitats. GR, PN. *O. oligophyllum* has typically just *2(3) grey–green* leaves 3–15 mm wide, becoming broader towards the tips, which in turn are abruptly narrowed. Alpine grassy habitats. GR, TR.

(d) Inflorescence *corymb-shaped*; leaves *with white mid-veins* (many similar species).

Ornithogalum umbellatum COMMON STAR OF BETHLEHEM A short, bulbous perennial 15–30 cm, often with numerous offsets forming leafy clumps. Leaves 2–5 mm wide, dark green with a central white stripe. Flowers 5–20, white with a bright green band on the outer surface of each tepal, 15–25(40) mm across, borne in wide, flat-topped racemes; *flower-stalks horizontal in fruit*. Fruit a 6-lobed capsule. Common in woods and grassy habitats. Throughout. The following are similar: *O. gussonei* lacks numerous bulblets and has horizontal to semi-erect flowering stalks. AA (Rhodes), GR, TR. *O. wiedemannii* has flower-stalks horizontal to *down-turned in fruit*. Fruit capsules with 6 distinct wings on the lobes. GR (north), TR. *O. armeniacum* has *very narrow, hairy leaves* to just 1–2 mm wide and 7–11 erect flowers 20–26 mm across; flower-stalks *down-turned in fruit*. Mountain slopes. AA, GR (north), TR. *O. exaratum* is smaller, lacks offsets and has tepals 17–23 (not 18–30) mm long. GR (southeast), AA.

Ornithogalum collinum A short perennial to 20 cm. Leaves 4–15, linear, *channelled*, to 8 mm wide, with a central white stripe, sometimes with minutely hairy margins. Flowers white, 20–32 mm, borne in broad corymbs *at ground level* on very short scapes (<40 mm); tepals white with a bright green stripe on the external surface; anthers yellow or reddish. Garrigue and rocky slopes. AA, CR, GR, IA, PN. The following are similar: *O. exscapum* has hairless leaves with central white stripes. Flowers white, borne in flat-topped clusters, small, <30 mm across (tepals to 15 mm); *flower-stalks clearly deflexed and rigid in fruit*. Considered by some authors to be a form of *O. collinum*. AA, CR, GR, IA, PN. *O. sibthorpii* has rather soft flower-stalks, *grooved* immediately below the first flowers. AA, CR, GR, PN. *O. sphaerolobum* has short style and shorter, broader capsules. AA (Kastellorizo), TR.

(e) Inflorescence corymb-shaped, with star-like flowers; leaves without white mid-veins; *fruiting stalks ascending*.

Ornithogalum arabicum A stout, bulbous perennial 40–60 cm with numerous offsets from the main bulb. Leaves 5–8, flat or channelled, plain green and sheathing at the base, 20–45 mm wide. Flowers white or cream, 40–50 mm across with a *conspicuous violet–black ovary* in the centre, borne in a lax raceme of up to 25; tepals broad, *without a central green stripe*. Capsule cylindrical. Dry, rocky places. Virtually throughout except the north (absent from CY, and patchy in the west).

1. *Ornithogalum narbonense*
2. *Ornithogalum nutans*
3. *Ornithogalum pedicellare* (poor specimen)
4. *Ornithogalum neurostegium* subsp. *eigii*
5. *Ornithogalum umbellatum*
6. *Ornithogalum exscapum*
7. *Ornithogalum sibthorpii*

Drimia SEA SQUILL

Similar to *Ornithogalum* but with very large bulbs and numerous flowers borne in a long, terminal spike, in late summer to autumn.

Drimia maritima (syn. *Urginea maritima*) SEA SQUILL A stout, late summer-flowering perennial 60 cm–1.2(1.5) m with a very large bulb to 6–15(18) cm across, sitting close to the soil surface; bulb greenish with a red tunic. Leaves borne in winter–spring, all basal, large, broadly lance-shaped, at least 20 mm wide, wavy-margined and shiny green. Flowers white and star-like, to 16 mm across, borne in a long, stout, spike-like inflorescence arising erect from the ground; filaments 4.5–5.5 mm long. Common in a range of dry habitats including coastal scrub and garrigue. Recorded throughout, but some populations correspond with the following. *D. aphylla* is an eastern (diploid) form with smaller green to pink bulbs. CY, IP, LB, SY, TR. *D. numidica* has ovoid, pink bulbs and filaments 6–7 mm long. CR, GR, IA, PN. Both of these taxa are probably indistinguishable from *D. maritima* in the field.

Scilla SCILLA

Bulbous spring-flowering perennials with basal leaves and leafless stalks. Flowers star-shaped and usually blue, borne in racemes or solitary, without, or with *1 bract at the base of each flower-stalk*; tepals 6, separate and spreading. Fruit a 3-parted capsule.

(a) Small bulbous perennials, typically <20 cm.

Scilla bifolia (syn. *S. subnivalis*) A short, early spring-flowering bulbous perennial to 15 cm. Leaves typically *2*, linear or broader towards the tip, 5–10 mm wide. Flowers *bright purplish-blue*, borne 1–7 in loose, rather 1-sided racemes on stalks 10–30 mm (shorter above); bracts absent; tepals 6–9 mm; anthers dark blue. Common in mountain woods and open grassy areas in the north. AA, GR, TR. *S. cilicica* has *more* (3–4), *larger*, linear leaves 15–30 cm long and 7–13 mm wide. Flowers bright purple–blue. Shaded limestone rocks. CY, LB, SY, TR (south). *S. nana* is similar to *S. bifolia* but with larger, paler blue tepals (to 16 mm), fused (rather than free) below; anthers yellow. CR (mountains – common).

Scilla morrisii A short, early spring-flowering bulbous perennial. Leaves lax, 15–30 cm long, strap-shaped. Flowers pale, *milky pinkish-lilac*, bell-shaped, borne 1–3 in loose, rather 1-sided racemes on stalks 10–30 mm (shorter above); bracts very small or absent. A rare island endemic remaining in just a few places in ancient oak forests. CY (west).

Scilla cilicica A small, bulbous perennial with 3–4(6) broadly linear leaves 15–25(38) cm long, 8–15 mm wide, *appearing in autumn, withering by spring*. Flowers borne in short racemes of 5–15, 18–32 mm across, pale to mid-blue or violet, bell-shaped, the tepals later deflexed, 9–16 mm. Seeds black, ovoid, 3 mm. Rare, in limestone woods. IP (mainly around the Upper Galilee and Mount Hermon – rare), LB, SY.

Scilla messeniaca A small, winter- to spring-flowering bulb to 15 cm with 5–7 *broadly* linear-lance-shaped leaves 6–15 mm wide, appearing with the flowers and sheathing at the base. Flowers (7)10–20, with stalks 4–8 mm, rather drooping, *pale* blue to lilac, borne in lax racemes. An endemic of the Peloponnese, in grassy habitats and garrigue. PN. *S. bithynica* is a similar, larger perennial to 30 cm with 3(4) leaves and 6–12(15), erect mauve–blue flowers on stalks 3–12 mm, in rather dense spikes. Damp, grassy habitats at low altitude. TR (northwest).

ASPARAGACEAE

(b) Large bulbous perennials, to >1 m.

Scilla hyacinthoides A large, *robust* bulbous perennial to 1.5 m. Leaves 8–12, linear-oblong, narrowed at both ends, with rough margins, 13–46 cm long, 15–30(40) mm wide. Flowers borne numerously (60–110) in long scapes opening from the bottom upwards, much exceeding the leaves; corolla 8–9 mm across, violet to light blue, star-shaped; tepals 5–7 mm; flower-stalks horizontal, prominently long and straight in fruit. Seeds 3 mm, black. Damp habitats, grassy slopes and hillsides. IP (Mediterranean region), SY (west), TR.

1. *Drimia aphylla*
2. *Drimia maritima*
3. *Scilla bifolia*
4. *Scilla morrisii* (atypical pale specimen)
5. *Scilla cilicica*
6. *Scilla hyacinthoides*
7. *Scilla messeniaca*

Prospero

Bulbous perennials previously grouped with *Scilla*, differing in having strap-shaped to filiform leaves, and scapes of flowers appearing without, or adjacent to, the leaves in autumn.

Prospero autumnale (syn. *Scilla autumnalis*) AUTUMN SQUILL A short *autumn-flowering* perennial with stems to 14–30(45) cm, and 5–10(14) hairless, narrowly linear, *channelled* leaves only 1–2(4) mm wide, absent during flowering. Flowers 10–25(66), usually *pink–purple* (sometimes lilac or blue) and star-like, borne in a spike-like raceme; bracts absent. Locally common in a range of dry, rocky and grassy habitats; widespread in the region. Throughout.

Bellevalia

Bulbous perennials with lax spikes of dull-coloured flowers; corolla with 6 deeply divided teeth; bell-shaped and tubular but not constricted into a throat. Fruit a 3-parted capsule. Hybridisation is apparent among some species.

(a) Flowers *bright blue or violet* in bud, <10 mm long and mostly blue–purple when mature.

Bellevalia dubia A small, bulbous perennial 13–35(40) cm with 4–5(6) leaves. Leaves erect or recurved, equalling or slightly exceeding the flower-stalks, linear and tapered towards the tips. Inflorescences cylindrical with 15–30(40) rather long-stalked (4–8 mm) flowers; perianth to 5–8 mm, bell-shaped and *dark blue* when open, the lobes with a whitish margin. Various grassy and scrubby habitats. CR (west – rare), GR, IA, PN, TR (local).

(b) Flowers *dull purplish* in bud, ≥10 mm long and mostly blue–purple when mature.

Bellevalia trifoliata A bulbous perennial to 60 cm. Leaves 2–4, about as long as the flowering spike. Flowers 12–14 mm long, borne in *cylindrical,* rather lax flower-heads of 10–40(60); corolla *long-tubular*, violet–purple in bud, and pale yellow–brown at the apex when fully open, borne on *short stalks*, 4–8(16) mm long. AA (mainly east), CR (west – rare), GR, IA, PN (rare), TR.

(c) Flowers tinged light mauve, green or yellow in bud, but *mostly cream or white* when mature.

Bellevalia warburgii A bulbous perennial to 60 cm. Leaves 3–6, exceeded by the inflorescence. Flowers 10–13 mm, whitish-green, flushed brownish-purple upon opening, borne in lax, cylindrical racemes up to 11 cm wide, longer than their scape; flower-stalks ascending in bud, slightly recurved in flower, straight and *conspicuously elongated* in fruit (25–50 mm). Capsule 10 mm. Fields and plains; rare and endangered. IP (Galilee to the southern West Bank; especially the Esdraelon Plain).

Bellevalia nivalis A small bulbous perennial to 12(25) cm, the stems exceeded by the leaves. Leaves (2)3–5, linear and spreading, 50 mm–25 cm long, 3–12 mm wide, not long-sheathing at the base, with membranous margins. Flowers 7–14, borne on 1–3 *condensed, cylindrical* racemes per bulb; flowers borne on *erect* stalks, which are *short to virtually absent*; corolla dirty white or cream, the uppermost tinged violet, (2)4–(5)7.5 mm long; anthers purple–violet. Coastal garrigue and rocky hillsides (CY) or alpine habitats (SY). *B. flexuosa* is a similar bulbous perennial with 5–6 lance-shaped, prostrate leaves tapered at both ends with membranous, wavy margins. Flowers borne in *lax* ovoid clusters on slender, *longer, ascending* stalks *just* shorter than the corolla; anthers blue. Dry, stony habitats. IP, LB. *B. densiflora* is similar to the previous species but with flowers borne in dense clusters with slender stalks typically half as long as the corolla (or longer in some forms); anthers light blue but *yellow* when mature. Swamps and waterlogged habitats. IP, LB, SY.

ASPARAGACEAE 147

Bellevalia desertorum DESERT HYACINTH A small, bulbous perennial with (3)4–5(7) fleshy, broad, strap-shaped and rather twisted leaves, *prostrate* on the ground, greatly exceeding the flower-stalks; sometimes pale-spotted. Flowers pale blue–lilac, slightly constricted, tubular-bell-shaped, virtually stalkless, borne in ovoid clusters; teeth shorter than the tube; filaments triangular, just longer than the anthers. Arid and semi-arid bare ground, particularly the north and central Negev and Judean Desert. IP, LB, SY.

Bellevalia eigii A bulbous perennial in similar habitats to the previous species but with a *more erect* rosette. Flowers borne on *long, slender, spreading* stalks in elongated, *lax racemes*; corolla brown, cream in the distal half. Semi-arid steppe habitats in the north and central Negev. IP.

1. *Prospero autumnale*
2. *Bellevalia trifoliata*
3. *Bellevalia warburgii*
4. *Bellevalia nivalis*
5. *Bellevalia flexuosa*
6. *Bellevalia desertorum*
7. *Bellevalia eigii*

Bellevalia ciliata A small, bulbous perennial to 50 cm with linear-lance-shaped leaves with *long-hairy margins*. Flowers blue–violet in bud, with greenish teeth, maturing dull brown, borne in broad, *conical-pyramidal* (not cylindrical) flower-heads of 30–50 flowers; flowers 9–11 mm long, borne on slender, very *long, spreading stalks*, 30–40 mm long; stalks horizontal and rigid in fruit. Cultivated land and field margins. GR, PN (east), TR.

Bellevalia romana A small bulbous perennial to 35 cm, the stems exceeded by the leaves; leaves 3–6, linear-lance-shaped, 5–15 mm wide, with smooth margins. Flowers 20–30, borne in lax, cylindrical racemes on spreading to erect stalks; perianth *whitish*, becoming brown upon opening, 7–10 mm long. Damp meadows and cultivated land. GR (mainly west), IA.

Leopoldia TASSEL HYACINTH

Bulbous perennials with dense to lax racemes. Fertile flowers constricted at the throat, often brownish or greenish; apical flowers blue, violet or pink, sterile, often highly reduced. Capsule 3-lobed, splitting and often compressed.

(a) Corolla teeth whitish, pale brown or yellow.

Leopoldia comosa (syn. ***Muscari comosum***) TASSEL HYACINTH A bulbous perennial, 20–60(90) cm. Leaves 2–3(5), linear, recurved, grooved, tapered gradually to the tip, to 40 cm long and 6–25 mm wide. Flowers borne in a lax raceme with small, blue, sterile flowers held *erect, on long, slender blue stalks* to 15 mm long, *forming a tuft at the apex*; fertile flowers brownish-green, *shortly tubular-bell-shaped* with 6 outwardly curved short, cream-coloured or whitish teeth. Common in open rocky and grassy habitats; the most widespread species. Throughout. The following are similar:

1. *Leopoldia comosa*
2. *Leopoldia spreitzenhoferi*; (inset) detail
3. *Leopoldia longipes*

L. cycladica has fertile flower-stalks <1 mm long, and sterile flower-stalks spreading (not erect). Limestone hills and cliffs. AA. ***L. weissii*** has sterile flowers numerous, *not* erect, and fertile flowers with yellow teeth. Dunes and garrigue. CR, GR. ***L. dionysicum*** has broader leaves, *large* feritile flowers 7–12 mm, long-stalked, and teeth of fertile flowers yellow or cream. AA (small offshore islets), CR (islets off the north coast).

Leopoldia spreitzenhoferi A bulbous perennial to 15(20) cm. Leaves 10–15 cm long, 2–3 mm wide. Flowers borne in dense racemes *becoming lax*; lower fertile flowers narrowly tubular, 5 mm long, deep brownish or yellowish with yellow, spreading to recurved teeth; flower-stalks ascending; sterile flowers *absent* or small (reduced to a few tiny threads). A local island endemic of dunes, among rocks and coastal garrigue. CR (+ adjacent islets).

(b) Corolla teeth blackish.

Leopoldia longipes A variable, bulbous perennial to 50 cm. Leaves 15–40 cm, folded, rolled and rather twisted. Flowers numerous, oblong to cone-shaped, 7–12 mm with stalks 20 mm, rather recurved when mature (later elongating and straightening in fruit); corolla purple in bud, cream to brownish upon opening, with blackish teeth; sterile flowers few, small and violet. Plains, fields and desert gullies; rather rare and local. IP (south-central), SY, TR.

Leopoldia tenuiflora A bulbous perennial to 30(60) cm, similar in form to *L. comosa* and relatives. Leaves 3–7, linear and channelled. Flowers light grey–brown with recurved, *blackish teeth*; sterile flowers bright violet, borne on fleshy, violet, erect or ascending stalks 3–16 mm long. Dry habitats. GR, TR.

Muscari GRAPE HYACINTH

Bulbous perennials similar to *Leopoldia* with dense racemes (laxer in fruit). Fertile flowers usually blue, purple or blackish, constricted at the throat, or not; sterile flowers smaller and fewer (0–10). Capsule 3-lobed and splitting.

(a) Spring-flowering. Corolla tube blue to blackish with *paler to whitish* teeth.

Muscari neglectum COMMON GRAPE HYACINTH A short, bulbous perennial 12–25 cm. Leaves 3–9, linear, channelled and all basal, 1.5–3.5(6) mm wide. Inflorescence a short, dense raceme of touching ovoid, *dark to blackish-blue* flowers with *whitish teeth,* 4–8 mm long, borne on spreading or recurved stalks; the uppermost flowers sterile and paler. Common in open rocky and grassy habitats; a widespread and variable species. Throughout (populations in AA may be referable to *M. puchellum*). ***M. armeniacum*** is similar but with *bright azure–blue* flowers to 5.5 mm long with paler to whitish teeth, borne in dense ovoid to cylindrical clusters; sterile flowers few, the *same colour* as (or slightly paler than) the fertile flowers. Grassy mountain habitats. GR, TR. ***M. botryoides*** has ribbed, greyish leaves, hooded at the tips and *very small* bright blue flowers just 3–4 mm long with white recurved teeth; sterile flowers few, the same colour or paler. Various habitats. GR, TR.

Muscari pulchellum A short, bulbous perennial, scarcely distinct from *M. neglectum* but always short (<15 cm) with 3–8 very narrow leaves *just 0.75–4 mm wide* and 0 (to very few) offsets. Raceme *cylindrical, lax* and few-flowered (8–18 fertile flowers *not* in dense and conical clusters); flowers small, 4–6.6 mm long, and mostly *dark magenta* (not dark grey–violet). AA, GR (Attica), PN (east). ***M. kerkis*** is a very similar, small perennial with 6–7 leaves 1.5–2 mm wide, with a *whitish line*, and dark magenta flowers 4.9–5.2 mm long with white recurved teeth. An endemic of cliffs and pine forests on Mount Kerkis. AA (Samos).

1. *Muscari neglectum*
2. *Muscari armeniacum*
3. *Muscari botryoides*
4. *Muscari kerkis*
5. *Muscari commutatum*
6. *Agave americana* inflorescence
7. *Agave americana* (vegetative form)
8. *Agave sisalana*
9. *Agave attenuata*

(b) Spring-flowering. Corolla tube and teeth *both* blackish-blue.

Muscari commutatum A short, bulbous perennial 60 mm–30 cm with 2–5 channelled leaves about as long as the flowering stalks. Flowers *violet–black* with *tiny black* (not whitish), *recurved* teeth, borne in dense, ovoid clusters; corolla 4–7 mm. Grassy and rocky habitats; garrigue. AA (common), CR, GR.

(c) Spring-flowering. Corolla tube *yellow*; sterile flowers normally absent.

Muscari macrocarpum A short, bulbous perennial with leaves 10–20 cm long, exceeding the flower-stalks. Flowers borne in dense clusters becoming lax, at first purple then *bright yellow*, sweetly scented; corolla egg-shaped with minute spreading, brownish teeth. Limestone cliffs. AA (mainly Samos, Fournoi, Simi + adjacent islets).

(d) Autumn-flowering.

Muscari parviflorum A short, bulbous perennial 10–25(35) cm with 2–4 linear, thread-like leaves just 1.5–3 mm wide, exceeded by the flowering stem. Flowers sky-blue and bell-shaped, to 5 mm long, borne in *very lax* racemes; sterile flowers few or absent, tiny if present. Coastal garrigue or dry, rocky and grassy habitats. AA (local), CR, GR, IA (Zykanthos), IP.

Pseudomuscari

Very similar to *Muscari* (and formerly included in the genus) but with bell-shaped corolla *not* constricted at the mouth.

Pseudomuscari inconstrictum A short bulbous perennial 50 mm–17 cm with 2–4 *narrow to thread-like leaves* 50 mm–20 cm long and just 1–3 mm wide, reddish at the base. Flowers 3–15, borne in long, rather lax racemes; corolla oblong-*bell-shaped*, *not constricted* at the tip, 3–4 mm long, dark indigo-blue with *similar coloured teeth,* borne on slender stalks to 2 mm; sterile flowers smaller and paler. Dry rocky garrigue in the east. CY, IP, SY, TR (south).

Agave

Large, imposing succulents with fibrous, very robust leaves in rosettes. Flowers cosexual and regular; stamens 6; ovary superior or inferior and 3-parted. Fruit a splitting capsule; seeds numerous. Native to the Americas.

Agave americana CENTURY PLANT A very large and imposing succulent 6–8 m during flowering; plants take at least 10 years to flower, after which they die, but perennate by offsets. Leaves very large and succulent, to 2 m long and 15–22(30) cm wide, with a stout *black spine at the tip 30–50 mm long*; blue–green or variegated with yellow stripes. Flowers borne in summer on a *tree-like inflorescence* that persists for years, greenish-yellow, to 90 mm long, borne on at least ½ the length of the inflorescence scape. Fruit an oblong capsule. Common, particularly near developed areas; cliffs and roadsides. Throughout. *A. sisalana* has straight, rigid leaves that are much narrower, 5–11 cm wide, which when mature are spineless along the margins; terminal spine 15–25 mm long. Flowers blue–green, and *mixed with leafy bulbils*, which detach and form new plants. Planted, sometimes naturalised, in semi-arid areas.

Agave attenuata is similar to the above species but smaller, with less succulent leaves with a *pale blue–white waxy bloom,* and *flowers borne in an extremely dense, fox-tail like inflorescence, drooping to the ground when mature*. Planted in towns and resorts in hot, dry parts; rarely far from urban areas.

Yucca YUCCA

Perennial succulents with sword-shaped leaves in rosettes. Inflorescence an erect panicle of creamy-white flowers; perianth bell-shaped with lobes longer than their tube. Fruit a non-splitting capsule. Garden escapes, native to North America. Naturalised plants in the region are generally infertile.

Yucca gloriosa YUCCA A tough, erect, succulent shrub 1–2(3) m with a thick, woody trunk, branching with age and a terminal leafy rosette. Leaves *blue–grey* green with reddish margins, rigid and tough, tapering gradually into a spine-tip; *margins entire* or inconspicuously toothed. Flowers cream–white, bell-shaped and pendent, 45–60 mm long, borne in stout, terminal panicles; tepals 6, all similar. Fruits dry, not fleshy. Widely planted in towns, parks and gardens. Naturalised on dunes. *Y. filamentosa* is similar to the previous species but with leaves with *whitish margins from which fibrous filaments are shed*. Widely planted.

Anthericum

A genus of about 300 rhizomatous perennials with long, narrow leaves and simple or branched stems carrying starry white flowers with 6 tepals. Mainly in tropical and southern Africa and Madagascar.

Anthericum liliago ST BERNARD'S LILY A semi-erect perennial to 60 cm with a short root stock of fleshy roots. Leaves more or less equalling the scapes, linear and narrow, 5–7(8) mm wide. Flowers white and rather with tepals 16–23 mm long (larger than other species), which markedly *exceed the stamens*, borne in *loose, erect racemes* of up to 10(30); tepals 3-veined and the style curved. Capsules ovoid, 7–9 mm wide. Stony and scrubby ground, woodland and mountain slopes.

Ruscus BUTCHER'S BROOM

Woody evergreen shrubs with stems flattened into leaf-like blades (cladodes) on which the flowers and fruits are borne directly; dioecious; tepals free. Fruit a berry.

Ruscus aculeatus BUTCHER'S BROOM A sclerophyllous, shrubby perennial 20–80 cm (1 m) with ribbed, *intricately branched stems* (especially above), with cladodes 40–65 mm long arranged in 2 rows, *ending in a spiny tip*. Dioecious; flowers inconspicuous, greenish-white, borne on the middle of the upper suface of the cladodes. Fruit a large red berry 10–15 mm across. Common and widespread in woods at higher altitude. Throughout. *R. hypophyllum* is similar but rather shorter, 10–80 cm, and less sclerophyllous, with simple stems (or with few branches) with large, flexible cladodes 50 mm–10 cm long *without spine-tips*; flowers borne on the upper or lower surface of the cladodes; bracts *papery*. Native to the western Mediterranean but planted and naturalised. GR. *R. hypoglossum* A perennial to 40 cm, similar to the previous species with generally unbranched stems and cladodes without spine-tips bearing flowers on the upper surface; bracts *green and leathery* (not papery). Often cultivated, rarely naturalised.

ARECACEAE | PALM FAMILY

Trees with large, pinnately or palmately divided leaves. Flowers unisexual or cosexual, small and greenish, usually borne in spikes or panicles with papery sheaths; petals 3; stamens 6. Fruit a berry or drupe. Most species in the region are planted.

Chamaerops

Trunk short; leaves fan-shaped and palmately cut into stiff segments. Native in the region.

Chamaerops humilis DWARF FAN PALM A bushy, clumped, normally stemless dwarf palm to 4(7) m; trunk covered in grey fibres. Leaves green or grey–green and fan-like, tough and divided ⟩2/3 their length into 20 pointed, narrowly lance-shaped segments; stalk long and with a spine-toothed margin. Flowers yellow, borne in dense panicles 12–25 cm long, with sheathing bracts. Fruit oblong, 7–16 mm, yellow or brown when ripe. Common and widespread on garrigue and in rocky ravines or pastures. Throughout.

---

1. *Yucca filamentosa*
2. *Yucca gloriosa*
3. *Ruscus aculeatus*
4. *Anthericum liliago*
5. *Chamaerops humilis*

Washingtonia

Trunk long and slender; leaves with long bare stalks terminating in a rounded fan of segments. Native to North America.

Washingtonia robusta A robust palm with a very stout, smooth trunk with diagonal furrows to 25 m tall and 40–70 cm wide, the upper part *clothed with the long-persistent withered leaves*. Leaves large and fan-like. Native to Mexico. Planted throughout. **W. *filifera*** is similar but shorter, to 15 m, with a trunk to 1 m across and leaves with white threads pendent from the leaf intersections. Native to California. Planted throughout.

Phoenix

Trunk long and stout; leaves pinnately cut into numerous leaflets. DNA-analysis shows species are freely hybridising and accurate identification of true species in the field may not be possible.

Phoenix canariensis CANARY PALM A tall palm to 20 m with a solitary trunk 50 cm–1.2 m across and crown with fronds 1.8–6 m long, with pinnately divided leaflets 21–89 cm. Fruits 15–23 mm, orange ripening to purple–brown. Native to the Canary Islands, planted as an ornamental in towns and gardens in the region. Throughout. *P. dactylifera* is often similar (especially when immature) but taller, to 30 m with a *narrower trunk 20–50 cm across* and smaller crown. Fruit (the edible date) larger, 25–75 mm long, orange–brown and sticky when ripe. Widely cultivated and native in desert oases in the far southeast. Hybrids of the above species (*P. canariensis* × *P. dactylifera*) are intermediate in characteristics, and also frequently planted. *P. theophrasti* grows to 15 m, is very like the previous species, but is typically a *dune-dwelling* tree with small, dry, inedible fruits 15 mm, and *short-stalked* flower clusters. Rare and local on coasts and in gorges. Considered to be native to the region. AA (south), CR, PN (Epidaurus), TR (southwestern coasts – rare).

Jubaea

Trunk very long and smooth (not covered in fibrous leaf bases).

Jubaea chilensis CHILEAN WINE PALM A palm similar to *Phoenix canariensis* but with shorter leaves and a smooth, grey trunk that is rather swollen in appearance; diamond-shaped markings left when leaves fall. Native to Chile, planted along roadsides. Throughout.

STRELITZIACEAE

Herbaceous (though often tree-like) perennials similar in form to plants in the banana family. Leaves sheathing, sometimes forming a 'trunk'. Leaves simple. Flowers borne in infloresences, zygomorphic; stamens 5 or 6; style 1, stigmas 1 or 3. Fruit a capsule.

Strelitzia BIRD OF PARADISE

Evergreen perennials with long-stalked leaves arranged in 2 ranks (fan-like) and flowers borne in horizontal infloresences emerging from robust, folded, boat-like spathes.

Strelitzia reginae BIRD OF PARADISE An erect, evergreen, rhizomatous perennial to 2 m. Leaves borne in 2 ranks, appearing fan-like, broadly oval, 25–70 cm long, 10–30 cm wide, borne on *long stalks* to 1 m. Flowers distinctive in shape, borne above the leaves; spathe held horizontally, greyish, ending in a long point; sepals *bright orange*; petals *purple–blue to white*. Commonly planted in gardens throughout (not naturalised).

1. *Washingtonia robusta*
2. *Phoenix canariensis*; (inset) fruits
3. *Phoenix dactylifera*
4. *Phoenix theophrasti* PHOTO: NICK TURLAND
5. *Strelitzia reginae*

MUSACEAE | BANANA FAMILY

Herbaceous (though tree-like) perennials with leaves with persistent overlapping basal sheaths forming a 'trunk'. Monoecious with flowers borne in clusters (effectively a raceme of cymes); stamens 5–6; carpels 3. Fruit a berry.

Musa BANANA

Herbaceous, robust perennials with trunk-like stems and 6–20(–numerous) leaves in a canopy. Flowers borne 12–20 per cluster, unisexual. Fruit the familiar banana.

Musa cultivar BANANA Herbaceous (tree-like), tropical-looking perennials 2–9 m. Leaves very large and broad, to 2 m long and 50 cm across with a prominent midrib, sometimes splitting at the margins; pinnately veined. Flowers unisexual, borne in large pendulous clusters. Fruit an elongated berry (a banana), 30 mm–40 cm long. Cultivars widely cultivated.

Musa cultivar

TYPHACEAE | BULLRUSH FAMILY

Herbaceous, aquatic (typically marginal) perennials with stout rhizomes and rush-like, linear leaves sheathing at the base. Flowers unisexual, borne in distinctively dense, cylindrical spikes, the male flowers above and the female below; perianth of bristles or scales (difficult to distinguish from stamens); stamens 1–8; styles and stigmas 1. Fruit a 1-seeded spongy drupe or capsule.

Typha BULLRUSH

Unmistakable aquatic perennials with flowers borne in cylindrical spikes.

Typha latifolia BULLRUSH, GREAT REEDMACE A robust, herbaceous, marginal perennial to 3 m with stout, creeping rhizomes and rush-like, erect, thick and flat, linear leaves 8–20 mm broad, sheathing at the base. Flowers unisexual, borne in dense, cylindrical spikes: male flowers borne above in a narrow, yellow spike, later falling; female flowers below, forming a broad, cylindrical spike that turns brown when mature, 18–30 mm wide; *female flowers lack bracteoles*. Fruits 1-seeded. Widespread and locally common in suitable habitats (pond and river margins, swamps and ditches). Absent from most islands. AA (rare), GR, IA (Corfu), IP, PN (probably elsewhere).

Typha angustifolia A perennial similar to *T. latifolia* but with narrower leaves 3–6(10) mm wide, sheaths with few or no mucilaginous glands and female flowers *with abruptly pointed bracteoles, and separated from the male flowers by a naked section of the stem 30–80 mm (12 cm) long*. Similar habitats. Throughout, except most islands. *T. domingensis* is similar, with leaf sheaths internally covered with *numerous brown, mucilaginous glands*, and rather slender, honey-brown cylindrical spikes of female flowers with *tapering bracteoles*. Similar habitats. Virtually throughout (probably the most common species of *Typha*).

Sparganium BUR-REED

Aquatic, herbaceous perennials rooted in mud with simple or branched stems and erect or floating leaves. Flowers unisexual, crowded into spherical heads, the female below and subtended by leafy bracts; perianth with 1–6 membranous scales. Fruit dry, not splitting.

Sparganium erectum BRANCHED BUR-REED An erect, hairless perennial to 1.5 m, usually in shallow water. Leaves all erect (sometimes floating), triangular in section, *keeled* and linear. Inflorescence branched, with the spherical, yellow male heads borne above the female on each branch; flower clusters unstalked; tepals thick with dark tips. Waterlogged and aquatic grassy habitats. Throughout, except many of the smaller islands.

JUNCACEAE

Erect, reed-like annuals or perennials with white, pith-filled stems. Differing from grasses and sedges in having regular flowers with 6 similar tepals and (3)6 stamens; style 0 or 1; stigmas 3. Fruit a capsule with 3–numerous seeds. Relative anther to filament length important, and should be measured in late flower or in fruit. Only the more widespread and common species in the region are described.

Juncus

Annuals or perennials with 1–2-faced leaves; hairless. Flowers with 6 tepals (unlike in other grass-like families); stamens 6. Capsule with numerous seeds.

(a) Leaves (normally) all basal; flower clusters *lateral*.

Juncus effusus SOFT RUSH A perennial with almost smooth stems (30–50 obscure ribs) and acute but *soft-tipped* leaves, and inflorescences of pale brown flowers with the lowermost bract *long, with a narrow sheath*, and anthers 0.4–0.7 mm, *equalling or scarcely longer* than the 3 filaments; tepals more or less equal. Seasonally wet habitats and pool margins. Throughout, except much of the southeast. *J. inflexus* is a similar, densely tufted perennial with *blue–grey, leafless, strongly ribbed stems* (12–18 clear ribs) with an interrupted central pith, and 1-sided, brownish inflorescences, and with purple–black basal sheaths; inflorescence with many ascending, unequal, lax branches; anthers 0.8–0.9 mm, equal to 1.5 x the filament length. Pastures and damp habitats. Throughout. *J. conglomeratus* differs in that the stems are dull green (not blue–grey) with 10–30(35) ridges and with a continuous internal pith; bract adjacent to inflorescence flat and opened out. Throughout, except much of the southeast.

(b) Leaves (normally) all basal; flower clusters *terminal*.

Juncus capitatus DWARF RUSH A dwarf, tufted, firmly rooted rhizomatous annual just 17–50 mm (18 cm) with all basal bristle-like leaves 0.2–0.6 mm wide and several stiff, unbranched, leafless (except for leafy bracts at the top) stems with compact heads of flowers, overtopped by 1 or 2 bracts. Outer perianth segments greenish, later reddish with curved segments, exceeding the inner segments. Damp, sandy habitats. Throughout.

(c) Leaves all basal or stems with 1 or more leaves; *maritime habitats*.

Juncus maritimus SEA RUSH A stiffly erect perennial with stems 30 cm–1.5 m, and 2 mm wide, and a short, creeping rhizome. Leaves sharp, almost all basal, generally shorter than the stems, and *sharply pointed*. Inflorescence an interrupted panicle with 2–3 flower-heads. Flowers straw-yellow and 6-parted with unequal tepals; anthers 2 x as long as the filaments. Bracts sharply pointed, the largest longer than the inflorescence. *Capsules triangular-ovoid and pointed, 2.5–3 mm long, not as long as or slightly longer than the tepals*. Saltmarshes, dunes and coastal grassland. Throughout.

Juncus acutus SHARP RUSH Similar to *J. maritimus* but taller, 70 cm–1.8 m, with more *densely tufted stems*; leaves sharply pointed and longer than the stems, flowers reddish-brown, borne in dense, rounded inflorescences exceeded by their bracts; anthers up to 5 x longer than the filaments; tepals more or less equal. *Capsule 3.2–6 mm long, much longer than the tepals; inner tepals with membranous margins extended into lobes*. Coastal sands and saltmarshes. Throughout.

(d) Stems with 1 or more leaves; (often) *inland habitats*.

Juncus bufonius A small perennial with usually *solitary flowers* (1–3) each with 3 stamens, with anthers *shorter* than or as long as the filaments and unequal tepals; inner tepals longer than the capsule; capsule 3–5 mm. Damp places. Throughout. *J. hybridus* is very similar but with fascicles of 3–6(10) flowers, each with 6 stamens (not 3), with anthers often to only ½ the length of the filaments, and inner tepals equalling or scarcely exceeding the capsules. Damp habitats, including saline areas. Throughout, except much of the southeast.

CYPERACEAE

A large family of hairless, herbaceous sedges with rhizomatous root systems; leaves grass-like but stems solid, often triangular. Flowers wind-pollinated, borne on spikelets, highly reduced, the perianth often in the form of bristles; monoecious or sometimes dioecious; stamens 1–3; style 0 or short; stigmas 2–3. Fruit a small, 2–3-angled nut. Only a small subset is described here.

Scirpoides

Strongly rhizomatous perennials with cylindrical stems. Leaves semicircular in cross section. Infloresence made up of (1)5–many stalked or stalkless, *spherical* heads; lowest bract stem-like, making inflorescence appear lateral; stamens and stigmas 3.

Scirpoides holoschoenus An erect, clump-forming perennial to 1.5 m. Leaves semicircular in cross section, appearing stem-like, 0.5–2 mm wide. Flower-heads *spherical*, 2–12 mm wide, brown, with spikelets 2.5–4 mm. Local among damp rocks and seeps near the sea. Throughout, except for some islands.

Carex SEDGE

Rhizomatous, spreading or tufted perennials with stems triangular in section. Inflorescence of 1-flowered spikelets grouped in spikes; lowest bract leaf-like or glume-like; flowers *unisexual*, either mixed in 1 spike or dioecious (often the upper spikes male and the lower female); perianth bristles 0; stamens and stigmas 2–3.

Carex divisa A *creeping*, hairless perennial with *slender, wiry* stems 15–50(70) cm and flattened leaves; stems often clustered. Flowers male above, female below, in scarcely interupted spikes; lowest bract bristle- or leaf-like and just exceeding the inflorescence. Damp grassy and sandy places. Throughout.

C. divisa

Cyperus

Tufted annuals or perennials with stems triangular in section and grass-like leaves. Inflorescence an umbel or umbel-like raceme with many-flowered grass-like spikelets clustered in dense heads; lowest 2–10 bracts leaf-like and exceeding the inflorescence; flowers cosexual without a perianth; stamens 1–3; stigmas (2)3.

(a) Inflorescence a *very dense capitulum*.

Cyperus capitatus A small, tough, hairless, blue–grey rhizomatous perennial with few, wiry leaves; stems solitary, 30(40) cm. Leaves blue–green, becoming yellow with age, with inrolled margins. Bracts leaf-like, erect and exceeding the inflorescence; later brown and withered. *Flowers borne in dense, brown, terminal spikelets;* stamens 3. Common, sometimes abundant, on coastal dunes. Throughout.

1. *Juncus acutus*
2. *Scirpoides holoschoenus*
3. *Cyperus capitatus*
4. *Cyperus involucratus*
5. *Cyperus laevigatus*
6. *Cyperus papyrus*

(b) Inflorescence rather compact to lax; stamens (2)3.

Cyperus laevigatus A short, tufted perennial to 50 cm with mat-forming rhizomes; stems erect and bunched or solitary, rounded or triangular in section. Leaves few, reduced, 30–60 mm. Inflorescence a fascicle of 2–4(9) stemless spikelets with greenish-brown flowers; stamens 3; *bracts 2, erect, and forming an apparent elongation of the stem*, and exceeding the inflorescence. Local in marshes, stream margins and other wet habitats. Throughout.

Cyperus involucratus A robust, clumped perennial to 1.3 m with tough, woody rhizomes to 13 mm wide and numerous erect stems. *Leaves reduced to brownish sheaths* (sometimes extended into a spine-tip). Inflorescence terminal with numerous radiating leafy rays to 20 cm long and greenish to brownish spikelets; stamens and stigmas 3. Cultivated and occasionally naturalised in disturbed damp, coastal habitats, possibly throughout.

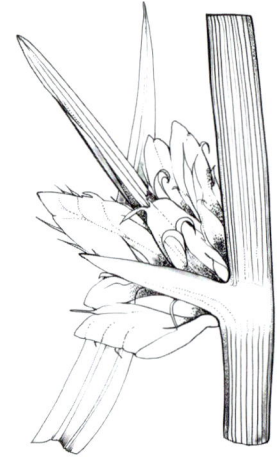

C. laevigatus

Cyperus papyrus PAPYRUS A very *robust and tall*, clumped perennial to 5 m with woody rhizomes and erect stems triangular in cross section. Inflorescence terminal, *large, 30–60 mm across, sub-spherical* with numerous (50–150) radiating and arching green *thread-like* rays on which spikelets are borne in 1–4 umbel-like clusters 20–30 mm long; stamens and stigmas 3. Commonly cultivated in aquatic habitats as an ornamental throughout; sometimes naturalised (e.g. north IP).

Cyperus longus A rather robust perennial 37–78 cm (1 m) with thick, far-spreading rhizomes 3–10 mm wide. Leaves to 2–5 mm wide, shorter than or equalling the stems. Inflorescence diffuse: a simple or compound umbel with 6–10(12) rays *of brownish or reddish spikelets*, 12–30 cm; stamens 3; bracts 2–6, the outer exceeding the inflorescence. Damp places and pool margins. Native to Africa, naturalised throughout.

C. longus

Cyperus esculentus EDIBLE CYPERUS A perennial to 60 cm (1 m) with *underground tubers* and leaves to 2–10 mm wide. Inflorescence an umbel of 3–8 *straw-coloured* spikelets 10–12 cm long, forming lax clusters on the branches; glumes densely overlapping; stamens and stigmas 3; bracts 2–3 × longer than the inflorescence. Cultivated in Mediterranean Europe.

(c) Inflorescence rather compact to lax; stamens 1.

Cyperus eragrostis A shortly rhizomatous perennial with erect, often solitary stems to 60(80) cm. Flowers borne in greenish-yellow to brownish spikelets 8–13 mm long, rather compact; *stamen 1*. Native to tropical America, naturalised in the region, sometimes abundantly.

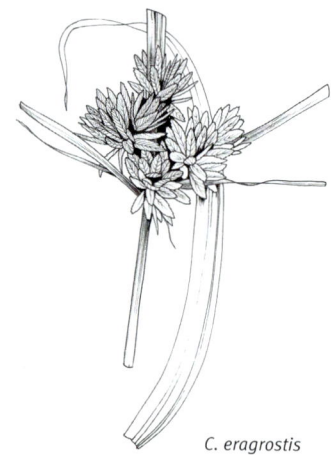

C. eragrostis

Eleocharis

Perennials with rounded to ridged stems and 0 leaf blades. Inflorescence a terminal spikelet, the lowest bract glume-like; flowers cosexual; perianth of 0–6 bristles; stamens 3; stigmas 2 or 3.

Eleocharis palustris COMMON SPIKE-RUSH A hairless, aquatic perennial with creeping rhizomes with solitary (1st year) and later numerous, *tufted*, leafless stems to 60(75) cm, bearing cylindrical spikelets 5–30 mm long; *stigmas 2*. Stems reddish at the base with *leafless sheaths*; sheaths pale brown. Stems with approximately equal air canals in cross section. Marshes and seasonally flooded habitats. Throughout. *E. multicaulis* is very similar, differing mainly in having *3 stigmas*. Nuts 3-angled. Similar habitats, mainly western. CR, GR.

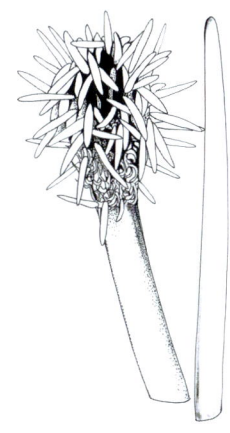

E. palustris

Schoenus

Tufted perennials with leaves crescent-shaped in cross section. Inflorescence with several spikelets in dense terminal heads of 1–4-flowered flattened spikelets; lowest bract leaf-like or glume-like; flowers cosexual with 0–6 perianth bristles; stamens and stigmas 3.

Schoenus nigricans BLACK BOG-RUSH A densely tufted perennial to 75 cm with all leaves basal, and terminal inflorescences. Leaves shorter than, or roughly equalling the stems, dark grey–green, hard and wiry. Inflorescence with 5–10 rather flattened spikelets, the lowest bract usually clearly exceeding it; stamens and stigmas 3. Nuts 3-sided and creamy-white. Locally common on maritime sands and peat. Throughout.

Isolepis CLUB-RUSH

Typically slender annuals. Inflorescence a head of 1–several spikelets with numerous florets; bracts if present exceeding the inflorescence, leafy and falling early; glumes spirally arranged; stamens 1–3, style 2–3-parted.

Isolepis setacea BRISTLE CLUB-RUSH A small, slender, tufted, sedge-like herb wth narrow stems <0.5 mm wide and 10–15(30) cm long. Leaves few, and shorter than the stems. Inforesence *much exceeded by the bract*, which is 5–20(35) mm long; spikelets 1–4, 2–4 mm long; empty bracts (glumes) purple–brown with green midribs and translucent margins. Damp, marshy habitats. Throughout, except parts of the southeast.

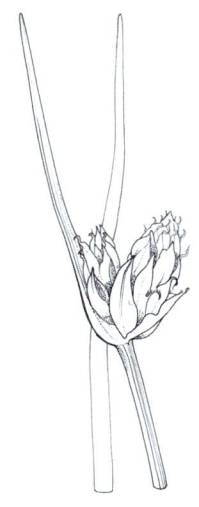

I. setacea

POACEAE | GRASS FAMILY

Annual or perennial, often rhizomatous or creeping plants. Leaves alternate, linear and sheathing the stem, generally with a membranous ligule at the base of limb. Inflorescences very variable, often a spike or panicle; flowers not brightly coloured, wind-pollinated and with (1)3(6) stamens; pistil with normally 2 styles, enclosed within 2 bracts; the whole called a floret; florets arranged into spikelets with 2 empty bracts (glumes) at the base. A large family divided into several subfamilies; it is well beyond the scope of this book to describe all species in the region; only a very small subsection is included.

SUBFAMILY POOIDEAE

The largest subfamily of grasses with numerous genera. Stems usually hollow. Spikelets normally cosexual with 1–many female florets; lemmas with or without awns; stamens 1–3; styles and stigmas 2.

(a) Spikelets with florets surrounded by long, white, silky hairs.

Ampelodesmos

Hairless perennials with much-branched, spike-like inflorescences; spikelets lance-shaped with 2–5 florets; glumes equal, long-pointed; lemma hairy below, leathery, 5-veined.

Ampelodesmos mauritanicus A very robust, densely clump-forming, hairless perennial 1–3 m with rush-like, rigid leaves with inrolled margins. Inflorescences borne on long, slender, arching stems exceeding the leaves, rather 1-sided and interrupted, light purplish-green, later straw-coloured; spikelets numerous, 10–15 mm, shortly awned; glumes purplish, 9–12 mm, long-pointed; lemmas with long silky hairs at the base and on the keel, with papery margins. Rocky habitats, often in ditches. Absent from most islands and much of the east. GR, IA, TR.

(b) Spikelets rounded (not flattened), arranged in 2 alternating rows.

Brachypodium

Annuals or rhizomatous perennials with membranous ligules. Inflorescence a raceme with 1 stalked spikelet at each node; spikelets with many cosexual florets; glumes unequal, 3–9-veined, much shorter than the spikelets; lemmas mostly 7-veined; stamens 3; stigmas 2.

Brachypodium retusum An ascending rhizomatous perennial to 60 cm. Leaves virtually hairless, 20 mm–10 cm long, 1–3 mm wide, stiff, bluish. Inflorescence a raceme with 1–5 alternately diverging spikelets at each each node; spikelets 20–30 mm long, laterally compressed, with 8–15 fertile florets; lower glumes 4–8 mm, 3-veined, upper glumes 8–9 mm, membranous, 5–7-veined. Maquis and garrigue. Virtually throughout, except CY, IP, LB.

(c) Spikelets with many overlapping florets, flattened from 1 side, often long-stalked. Numerous similar species.

Bromus BROME

Annuals. Inflorescence with flattened, long-stalked spikelets with many overlapping florets; glumes unequal, often awned, 3–7(9)-veined; lemma 7–9(11)-veined, minutely split at the apex. Many species in the region, all variable; dwarf specimens in dry areas are not reliable for identification.

1. *Ampelodesmos mauritanicus*
2. *Brachypodium retusum*
3. *Bromus fasciculatus*
4. *Bromus hordeaceus*
5. *Melica ciliata*
6. *Melica transsilvanica*

Bromus madritensis COMPACT BROME A short, tufted annual with erect stems. Leaves tapered, to 5 mm wide, softly white-hairy. Inflorescence a rather lax, erect, wedge-shaped panicle with short, bunched branches; spikelets to 60 mm; lemma with a long bristle-tip to 16 mm, hairy or hairless; glumes 1–3-veined; lemma 5–7-veined, 10–19 mm. Common on fallow and cultivated ground. Throughout. ***B. fasciculatus*** is a much more slender species with narrow, linear lemma 3–5-veined, 10–15 mm long and just 1–1.5(2) mm wide. Virtually throughout coastal areas. ***B. diandrus*** has a long, very lax, nodding inflorescence (not wedge-shaped) with spreading spikelets; lemma 22–45 mm, 7-veined. Throughout. ***B. hordeaceus*** has an erect, rather short and dense inflorescence with spikelets on short stalks (exceeded by their spikelets), with a slightly inflated appearance; lemma 6.5–11 mm long, 2.5–5 mm wide, 7–9-veined. AA, CR, GR, IA, PN, TR. ***B. squarrosus*** has a *deflexed* awn, at right angles to the lemma; glumes 3–9-veined; lemma 8–11 mm long, 5–9 mm wide, 7–11-veined. Dry ground and wasteland. CR, GR, IA, PN.

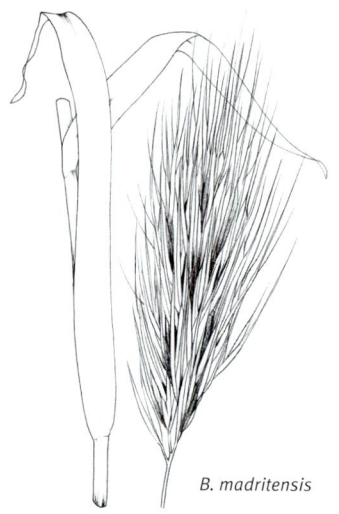

B. madritensis

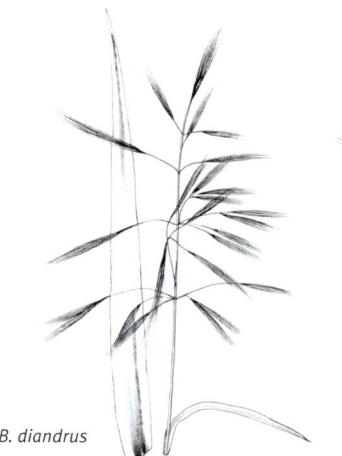

B. diandrus

B. hordeaceus

(d) Spikelets awnless, rounded (not flattened), with 2–4 florets, the upper sterile and club-shaped.

Melica

Clumped, perennial grasses with short rhizomes with spike-like or sparsely branched panicles; spikelets with 1–3 cosexual florets; glumes 3–5-veined; lemmas 7–9-veined.

Melica ciliata A clumped, rhizomatous perennial. Leaf-sheaths tubular; leaves flat, stiff, 50 mm –15 cm long, 1–3 mm wide, hairless, tapering at the tips. Flowers borne in long, cylindrical, straw-coloured, terminal panicles, 40 mm–20 cm, eventually nodding, continuous (or slightly interrupted); glumes membranous, 4–5 mm, 5-veined; lemma 7–9-veined. Probably throughout, except the southeast or casual only there. ***M. transsilvanica*** is similar, but with unequal glumes, denser panicles and lower leaf-sheaths with long, soft, downward-directed hairs. Rarer than the previous species and absent from much of the south and east.

(e) Spikelets numerous, compressed, awned with 2–4 florets.

Avena OAT

Annuals. Inflorescence a compound, diffuse panicle, branched with large, long-stalked and drooping spikelets with 2–3 florets; glumes papery and exceeding the florets, 7–11-veined; lemma leathery and 7–9-veined with a stout, bent, long awn. A variable and difficult genus.

Avena barbata BEARDED OAT An erect grass with solitary or clumped stems to 1 m. Leaves linear, to 15 mm wide and hairless or slightly hairy on the margin; ligule membranous, to 5 mm. Inflorescence a 1-sided, very lax panicle with spikelets drooping on slender stalks; lowest lemma 12–18 mm with 2 short bristles at the tip, 3–5 mm long. Fallow and cultivated ground and roadsides. Throughout. *A. sterilis* is similar but the lowest lemma 16–25 mm with 2 short points <1.5 mm, *strongly bent* below, forming a distinct dog-leg. Cultivated and fallow land. Throughout.

A. barbata

1. *Avena barbata*
2. *Gaudinia fragilis*
3. *Rostraria cristata*

(f) Spikelets stalkless, arranged alternately and overlapping on the axis in a slender, spike-like inflorescence.

Gaudinia FRENCH OAT-GRASS

Annuals to short-lived perennials with long, spike-like inflorescences; spikelets with 4–11 florets, all fertile; lower glume half as long as the upper, 3–5- and 5–11-veined, respectively; lemmas 5–9-veined.

Gaudinia fragilis FRENCH OAT-GRASS A softly-hairy annual to 40 cm (1 m) with flat leaves with hairy margins; lower leaves with spreading hairs on the sheaths. Inflorescences long, slender and green, about 10 cm; spikelets 9–20 mm, borne in 2 alternate rows, arranged edgeways to the axis; lower glume 3 mm, 3-veined, the upper 7 mm, 7–9-veined; lemma 7–11 mm, toothed, with a twisted, bent awn 5–13 mm long. Grassy habitats. Virtually throughout (not the far southeast).

(g) Annuals with dense, rounded, very soft-hairy inflorescences.

Lagurus HARE'S TAIL

Annuals. Inflorescence very compact, compound and spike-like, densely silkily hairy; spikelets with single florets, falling as a unit when ripe; glumes bilobed, 1-veined and awned, longer than the obscurely 5-veined lemma.

Lagurus ovatus HARE'S TAIL A softly hairy, grey–green annual to 60 cm. Leaves linear-lance-shaped and flat; ligule hairy and membranous, to 3 mm. Inflorescence distinctive: egg-shaped, dense, 'fluffy' soft and white, 5–20 mm long; lemma semi-transparent and with awns 8–20 mm. Common in sandy coastal environments and rocky slopes inland. Throughout.

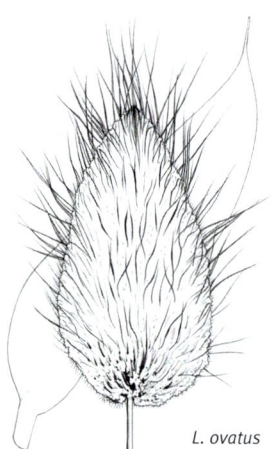

L. ovatus

(h) Panicles spike-like with very short branches.

Rostraria

Small annuals with spike-like panicles with very short branches; florets cosexual; glumes unequal (the lowermost 1-veined, the uppermost 3-veined); lemmas 5-veined, toothed.

Rostraria cristata MEDITERRANEAN HAIR GRASS A short, erect grass to 20(60) cm with hairless to hairy stems. Flowers borne in spikelets 3–8 mm long with awns 1–3, in green, cylindrical, spike-like panicles 10 mm–10 cm long and 4–10 mm wide. Disturbed habitats. Throughout.

(i) Spikelets ovoid, heart-shaped, awnless with overlapping florets.

Briza

Annuals or perennials. Inflorescence distinctive: spikelets flattened and inflated, ovoid- to heart-shaped and awnless, pendulous with 4–30 overlapping all cosexual florets.

Briza maxima LARGE QUAKING GRASS A hairless, low annual grass with often solitary, erect stems. Leaves flat and linear, to 4 mm wide. Inflorescence a lax panicle, with up to 15 large, drooping papery spikelets appearing inflated, on slender stalks, green then purplish ripening to pale brown, 8–25 mm long. Common on fallow ground, roadsides and maquis; widespread. *B. minor* is similar but with numerous (>20), smaller spikelets 2.5–5 mm long. Open habitats. Throughout.

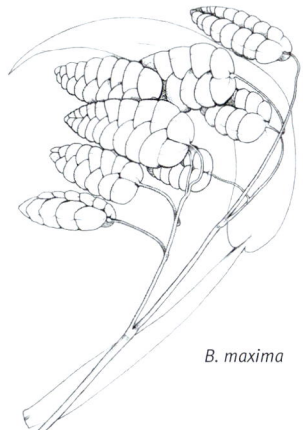

B. maxima

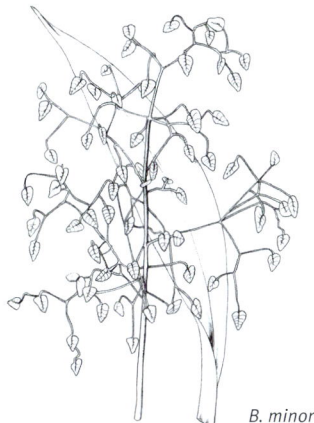

B. minor

(j) Spikelets large (10–16 mm), lance-shaped, with 1 cosexual floret.

Ammophila MARRAM

Rhizomatous perennials. Inflorescence spike-like, dense, cylindrical with large spikelets with single florets; glumes papery and keeled, 1–3-veined; lemma also papery, 5–7-veined, lance-shaped and hooded at the tip.

Ammophila arenaria MARRAM GRASS A tough, clump-forming rhizomatous perennial with smooth stems to 1.2 m. Leaves to 5 mm wide but appearing narrower owing to the inrolled margins; ligules narrow and pointed, 10–30 mm long. Inflorescence a slender, spike-like panicle with dense, erect spikelets to 16 mm; lemma with a very short, stiff bristle-tip with a ring of hairs at the base. Very common to dominant on coastal dunes. Throughout.

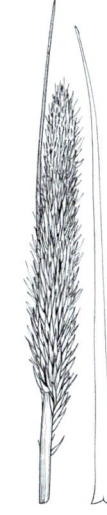

A. arenaria

(k) Spikelets small (3–8 mm) with 1 cosexual floret.

Gastridium NIT-GRASS

Annuals with compact, spike-like inflorescences with ascending to adpressed branches; spikelets laterally compressed, solitary with a single cosexual floret; glumes unequal and exceeding the florets, 1-veined and papery at the tips; lemma membranous, 5-veined and awned or not (awns shorter than the lemmas).

Gastridium ventricosum NIT-GRASS A small annual to 50(90) cm with flat, hairless leaves, often withered during flowering; ligules to 3 mm long, and pointed. Flowers borne in green, strongly erect (at least at first), more or less bilaterally symmetrical and laterally compressed panicles 5 mm–10(16) cm long; spikelets short, 3–5 mm long with a single floret. Probably throughout. **G. phleoides** is native to Asia but naturalised as a casual weed in the region, and distinguished by its very dense panicles with longer spikelets 5–8 mm long; lemma densely pubescent, with an awn 4–7(8) mm, often exceeding the glumes. Probably throughout.

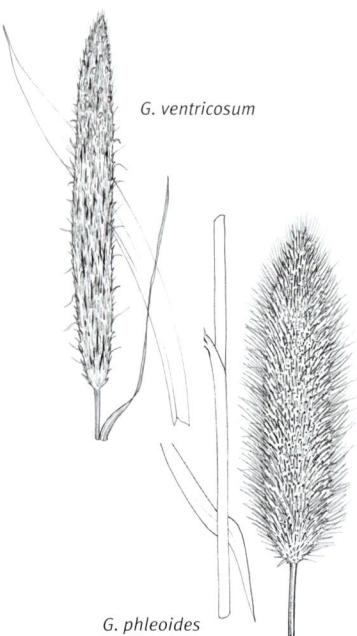

G. ventricosum

G. phleoides

(l) Spikelets small and stalkless, strongly compressed, with 1 floret.

Alopecurus FOXTAIL

Annuals or perennials without rhizomes. Inflorescence a dense, narrow, spike-like panicle; spikelets strongly compressed, with 1 cosexual floret; glumes 2, equal, 3-veined; lemmas transparent, 4-veined; stamens 3.

Alopecurus arundinaceus An erect, slender, hairless perennial 30 cm–1.2 m with flat, rough leaves 3–10 mm wide. Inflorescences long, dense and cylindrical, 20 mm–13 cm long and 10 mm wide; spikelets 4–6 mm, stalkless, arranged all around the axis; glumes parallel with fine hairs on the keel; lemma as long, oval and long-awned. Damp habitats. Throughout (except for many islands).

Alopecurus arundinaceus

(m) Spikelets stalked with 2 cosexual florets; inflorescence a compact panicle.

Corynephorus

Densely tufted perennial grasses with compact panicles; spikelets with 2 florets; glumes almost equal, 1-veined; lemmas 5-veined with a bent awn club-shaped at the apex.

Corynephorus divaricatus (syn. *C. articulatus*) A small, tufted annual to 42 cm; ligule a membrane 5–8 mm long; leaves 20–45 mm long, just 0.5 mm wide. Inflorescence an open panicle 20 mm–10 cm long with spikelets clustered towards the branch tip; branches spreading, 30–60 mm long; spikelets stalked with 2 fertile florets, laterally compressed, 4–4.5 mm long; glumes similar, shiny and 1-veined; lemmas 1.5–2 mm long, 5-veined. Dunes. Probably throughout.

(n) Spikelets flattened, stalkless, alternately arranged edgeways onto the axis.

Lolium RYE GRASS

Annuals or perennials. Inflorescence simple, unbranched and spike-like with stalkless spikelets alternately arranged edgeways onto a jointed axis, flattened; glumes solitary in lateral spikelets (2 in terminal); lemma 5–9-veined and awned or not.

Lolium perenne RYE GRASS A hairless, tufted, wiry perennial to 50(90) cm. Stems smooth and slender, bent below. Leaves narrow, up to 3 mm wide, and folded until mature. Ligule to 1 mm, abruptly pointed. Inflorescence simple, to 15 cm long with compressed, oval spikelets to 15 mm; lower lemmas 3.5–9 mm and almost always awnless. Cultivated as a fodder crop. Throughout. *L. multiflorum* is similar but an annual with awned lemmas (awns to 15 mm); glume shorter than the spikelet. Throughout.

L. perenne

(o) Spikelets crowded into dense bunches at the ends of side branches.

Dactylis COCK'S-FOOT

Perennials. Inflorescence a more or less 1-sided panicle or compound with spikelets crowded into dense clusters at the ends of the side branches; spikelets flattened, short-stalked with 2–5 florets; glumes keeled and 3-veined; lemma keeled and 5-veined, very shortly awned or awnless.

Dactylis glomerata COCK'S-FOOT A perennial, bluish clump-forming grass with erect or spreading stems to 1.4 m. Leaves rough, with ligules 2–10 mm. Inflorescence an erect, rather unequal and 1-sided tufted panicle of laterally compressed spikelets borne in dense clusters on lateral branches, often with prominent yellow stamens. Common in grassy habitats. Throughout.

D. glomerata

(p) Spikelets of 2 distinct kinds: fertile with 1 cosexual and 1 vestigial floret or sterile and made up of lemmas.

Lamarckia GOLDEN DOG'S TAIL

Annuals. Inflorescence with spikelets of 2 kinds: the upper with 1 fertile floret and 1 rudimentary floret, the lower with several pairs of overlapping, blunt, sterile lemmas in 2 ranks.

Lamarckia aurea GOLDEN DOG'S TAIL A more or less hairless, low annual grass with tufted, erect stems to 20(30) cm. Leaves linear, 2–6 mm wide with hairy margins; ligule membranous, 5–10 mm long, pointed or blunt. Inflorescence 30–90 mm long, dense, rather 'fluffy' and 1-sided with the outer spikelets sterile, greenish and later golden. Common on fallow and cultivated ground and roadsides. Throughout.

L. aurea

(q) Spikelets paired and of 2 kinds: outer sterile and comb-like; inner with fertile florets.

Cynosurus

Annuals or perennials. Inflorescence a compact spike-like panicle of fertile spikelets with (1)2–5 cosexual florets and sterile spikelets with sharp-pointed lemmas in a herring-bone arrangement.

Cynosurus echinatus ROUGH DOG'S-TAIL A short to tall hairless annual with erect or spreading stems to 75 cm (1 m) and flat leaves 2–10 mm wide. Inflorescence a dense, plume-like, 1-sided, oblong panicle 10–40(80) mm long of shiny green or purplish spikelets; the outer spikelet of each pair comb-like with several pairs of spreading, long-awned, sterile lemmae; inner spikelet fertile and wedge-shaped. Dry rocky and grassy scrub. Throughout.

(r) Spikelets compressed, of many florets arranged in 2 ranks.

Catapodium

Annual grasses with a simple or little-branched, stiff, 1-sided, spike-like inflorescence; spikelets rather compressed with many (3–14) florets in 2 ranks; glumes 2, nearly equal and papery; lemma blunt and leathery, 5-veined and awnless.

Catapodium rigidum (syn. *Desmazeria rigida*) FERN-GRASS A small, stiff, hairless, tufted, bluish annual with erect stems to 15(60) cm with several to numerous erect or spreading stems. Leaves often purplish, fine-pointed, to 2 mm wide and flat or with inwardly rolled margins. Inflorescence a more or less 1-sided panicle to 80 mm long, often branched below, with sparse, tiny spikelets to 7 mm, each with 5–10 minute florets; glumes 1.3–2 mm and pointed; lemmas longer, 2–2.6(3) mm and blunt. Coastal sands, dry, bare habitats and walls and rocks. Throughout.

1. *Lamarckia aurea*
2. *Catapodium rigidum*
3. *Stipa capensis*
4. *Oryzopsis miliacea*
5. *Aegilops neglecta*

(s) Spikelets alternate, arranged broadside onto the axis.

Parapholis HARD-GRASS

Annuals with very slender, whip-like inflorescences with alternate spikelets arranged broadside and set into hollows of the axis, each with a single floret; glumes equal and 3–5-veined; lemma finely 3-veined.

Parapholis incurva CURVED SEA HARD-GRASS A distinctive, short, tufted annual with spreading, curved stems 10–20 cm long. Leaves flat or inrolled, linear and pointed, to 2 mm wide, rough above and along the margins with reddish sheaths. Inflorescence 10–80 mm (15 cm), often not exserted from its sheath, slender, rigid, cylindrical, strongly curved and jointed with spikelets adpressed to the stem; spikelets to 7 mm long, a little longer than the joints of the axis; glumes equal, closing the cavities of the axis. Saline coastal habitats, particularly saltmarshes and cliffs. Throughout.

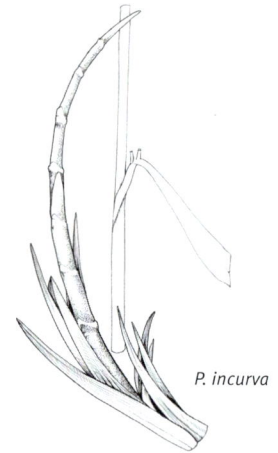

P. incurva

(t) Spikelets with single florets; lemmas ending in a sharp point or awn.

Piptatherum (including *Oryzopsis*)

Shortly rhizomatous perennial grasses with transparent ligules. Spikelets short and dorsally compressed; lemma with a long, straight, often falling terminal awn. A much-confused genus; *Oryzopsis miliacea* now established to be genetically distinct from *Piptatherum*.

Oryzopsis miliacea (syn. *Piptatherum miliaceum*) A tall, erect, perennial to 1.5 m with hairless stems, and leaves rough above. Flowers borne in light green to straw-coloured panicles to 40 cm long; 1-sided, drooping and lax, with several (to 20) branches at each node along the stalk; glumes 3–4 mm, lemma 2–2.5 mm, hairless, stiff, membranous (not leathery) with awns 3–5 mm. Dry open habitats among shrubs and other vegetation. Throughout.

O. miliacea

Piptatherum coerulescens (syn. *Oryzopsis coerulescens*) A tall, erect, perennial to 70 cm (1 m) with hairless stems. Leaves 15–31 cm long and 1–12 mm wide, rough on both surfaces, with long ligules to 11 mm long, transparent. Flowers borne in panicles 30 mm–15 cm long with spikelets 5–14 mm; glumes sub-equal, exceeding the florets, 3–9 veined; lemma 2.6–6.5 mm long, leathery, with awns 1–15 mm long, falling. Virtually throughout, except the far southeast.

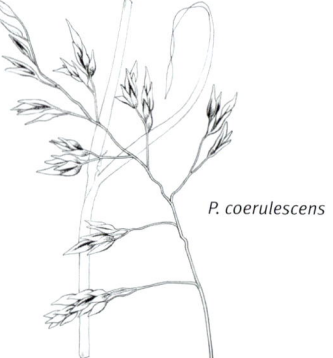

P. coerulescens

Stipa

Annuals or perennials with ligule a membrane fringed with hairs; lemma with forward-pointing bristles and a long terminal, persistent awn.

Stipa capensis An annual or biennial to 20 cm, with leaves to 15 cm long with revolute margins; blue–grey and hairy or hairless. Inflorescence a dense, slender panicle; enclosed at the base by a subtending leaf, to 15 cm long and 10 mm wide; spikelets solitary, the fertile spikelets stalked; glumes persistent and all more or less similar, 15–20 mm long and exceeding the florets and 3-veined; awns 70 mm–10 cm. Most common in deserts and semi-deserts. Throughout, except cool northern areas. ***S. parviflora*** is similar, but a perennial to 60 cm, with dissimilar glumes: those below to 15 mm and those above to 7.5 mm. Steppe habitats, mainly in the southeast. CR (southeast), IP, LB.

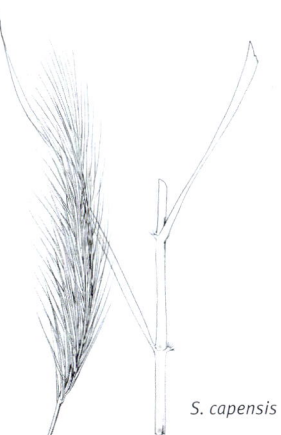

S. capensis

Stipa pennata A densely tufted, blue–grey perennial grass to 60 cm, with distinctive narrow, feathery inflorescences with very long, trailing awns with many conspicuous silvery-white hairs; spikelets yellow–green; glumes to 20 mm with a hairless slender awn 2–3 x as long; lemma to 25 mm with a very long, feathery and twisted awn to 35 mm long. Grassy habitats in the north only (if considered to be distinct from the following species). GR, TR. ***S. pulcherrima*** has lemma with a line of hairs *reaching the base* of the awn on the undersurface (rather than ending 3–8 mm below the tip). Probably widespread but distribution unclear owing to confusion with the previous species (It may not be possible to distinguish *S. pennata* and *S. pulcherrima* in the field).

(u) Spikelets stalkless, arranged broadside on the axis in a compact, ovoid inflorescence.

Aegilops

Inflorescence with stalkless spikelets arranged broadside along the axis in a distinctive compact ovoid or cylindrical cluster; glumes large and tough and strongly veined with 2–4 awns at the apex; lemmas also awned. Many similar species co-occur.

Aegilops geniculata A low, tufted annual with erect stems to 40 cm. Leaves with a flat blade 2–3 mm wide, finely hairy on the upper surface; ligule very short. Inflorescence congested and an inverted cone-shape, *not more than 2 x as long as broad*, with 1–2 vestigial spikelets at the base of the fertile ones; fertile spikelets often just 2–4, *broadest at, or just below the middle*; lemma with a long bristle 15–25(30) mm, *as long as those of the glumes*. Common on bare and fallow ground and in olive groves. Probably throughout. ***A. neglecta*** is similar but with glumes of lateral spikelets with 2–3 (not 3–5) awns, the spike (excluding awns) *at least 5 x as long as broad*, and the awns of the lemmas about *half as long* as those of the glumes. Probably throughout (though not widely recorded in the far east and southeast).

(v) Spikelets 1 per node with several to many florets with all but the apical 1–2 cosexual, flattened broadside onto the axis.

Elytrigia

Perennials. Inflorescence with 1 spikelet per node, with several to many florets, flattened broadside onto the rachis; glumes 3–11-veined, rarely awned; lemmas 5-veined, unawned or short (rarely long)-awned.

Elytrigia juncea (syn. *Elymus farctus*) A tough, clump-forming, rhizomatous perennial grass to 60(80) cm, often co-occurring with *Ammophila arenaria*. Leaves inrolled at the margins, 2–6 mm wide, and minutely hairy above, hairless beneath; ligules short. Inflorescence a slender spike with dense, erect, virtually stalkless, laterally compressed and hairless spikelets; lemma unawned. Common on coastal dunes. Probably throughout, except for many islands.

E. juncea

(w) Spikelets in clusters of (2)3, arising from each joint in the axis.

Hordeum

Annuals, sometimes perennials. Inflorescence spike-like, dense and long-awned with spikelets in clusters of 3 arising from each joint in the axis; glumes narrow, long-awned and 1–3-veined; lemmas 5-veined with long awns. Other, very similar species occur.

Hordeum murinum WALL BARLEY An annual 10–60 cm with tufted, erect, smooth stems. Leaves linear, to 4 mm wide and hairy on both surfaces with shiny sheaths; ligule membranous and small, to 1 mm long. Inflorescence a more or less bilaterally symmetrical, bristly, dense, spike-like panicle, lemmas with awns 10–45 mm long; glumes with awns 10–30 mm long. A common grass in a range of habitats. Throughout. *H. vulgare* subsp. *spontaneum* is a similar annual with *rough, robust* lemma with *longer* awns, 70 mm–15 cm. Fields, roadsides, olive groves; the wild progenitor of cultivated barley. AA and CR eastwards.

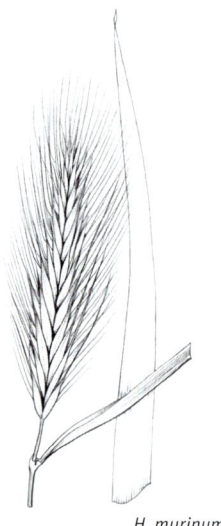

H. murinum

(x) Spikelets numerous, each with 1 floret.

Oryza

Annuals. Inflorescence with many spikelets, each with a single floret; glumes much shorter than the lemma, 1-veined; lemma compressed and keeled, strongly 5-veined; stamens 6.

Oryza sativa RICE A hairless, aquatic annual with leafy stems to 1.3 m. Leaves flat and smooth, to 10 mm wide. Inflorescence large, lax and erect or curved with numerous long lateral branches bearing numerous spikelets; glumes very small and equal, to 2 mm; lemma 7–9 mm, hairy above and short-pointed. Native to the tropics, cultivated in suitable areas (deltas).

SUBFAMILY ARUNDINOIDEAE

Stems hollow, plants often reed-like. Spikelets cosexual with 1–many female florets; lemmas with 1–3 awns; stamens 1–3; styles and stigmas 2.

Arundo

Robust perennials. Inflorescence large, compound, feathery with numerous spikelets, each 1–7-flowered; glumes lance-shaped, papery and keeled, 3-veined; lemma papery with dense silvery hairs.

Arundo donax GIANT REED The largest grass in the region: an extremely robust, rhizomatous perennial with bamboo-like stems to >6 m. Leaves grey–green, to 60 mm wide. Inflorescence a large panicle to 60 cm long; silky and silvery, with spikelets 12–20 mm long, each with usually 3(4) florets; lemma notched at the apex with a short bristle-tip; glumes papery, keeled and 3-veined, usually (not always) longer than florets. Common in a range of habitats. Throughout. *A. plinii* is rhizomatous, shorter and less robust, typically <2 m, with slender stems, and with smaller spikelets <8 mm long with fewer (1–2) florets; lemma entire with a short awn. Along verges and creeks or in saltmarshes. Scattered across the region (absent from many islands and most of the north).

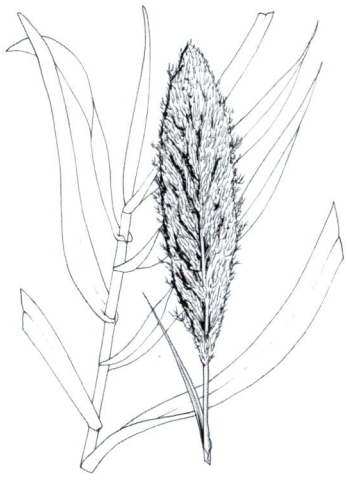

A. donax

Phragmites

Spreading perennials. Inflorescence a large, feathery and compound panicle with numerous slender spikelets with many 2–6(10) florets with hairy stalks; glumes unequal, 3–5-veined; lemma hairless and 1–3-veined, awnless.

Phragmites australis (syn. *P. communis*) A bed-forming large, reed-like grass, rather similar to *Arundo donax* with tall, rather slender stems to 3.5 m, which do not overwinter. Leaves to 50 cm long and 50 mm broad, grey–green and tapered to the tip; sheaths smooth and hairless, surrounding the leaf nodes. Flowers borne in drooping, 1-sided, more or less cylindrical, bunched greenish or purplish panicles with spikelets 8–16 mm long, each with up to 10 florets. Aquatic habitats such as lake margins. Throughout.

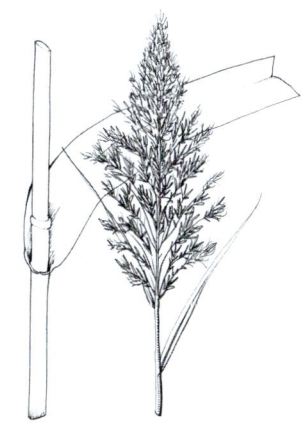

P. australis

SUBFAMILY CHLORIDOIDEAE

Stems hollow or solid; spikelets cosexual with 1–many female florets; lemmas with 1 or more awns; stamens 1–3; styles and stigmas 2.

(a) Inflorescence compound, comprising a whorl of branches of stalkless spikelets.

Cynodon BERMUDA GRASS

Perennials. Inflorescence a compound umbel of 3–6 slender branches with stalkless spikelets arranged in 2 rows; glumes 1-veined; lemma 3-veined.

Cynodon dactylon BERMUDA GRASS A spreading, short perennial with creeping stems to 30 cm. Leaves linear and flat, hairless or hairy along the margins. Inflorescence distinctive and star-like: 3–6 spikes 20–50 mm long, outwardly spreading from a single central axis; spikelets 2–3 mm long, stalkless. Native to tropical Africa but a commonly naturalised weed in a range of disturbed habitats. Throughout.

C. dactylon

(b) Inflorescence with side branches.

Tragus

Annuals. Inflorescence a spike, or spike-like, with 2–5 spikelets on very short branches at each node, each with 1 cosexual floret; glumes unequal, the upper longer and 5–7-veined, each vein with hooked bristles; lemma 3-veined.

Tragus racemosus STALKED BUR GRASS A creeping, branched, spreading annual, rooting at the nodes with erect stems to 40 cm and short, flat leaves to 3 mm wide with spines along the margins. Inflorescences spike-like, long, cylindrical, purple; spikelets 3–5 per node, the upper glumes with with 7 rows of fine-crooked bristles on the backs. Dry sandy areas, dunes, waste places and olive groves; local. GR.

(c) Inflorescence a lax, diffuse panicle.

Eragrostis LOVE GRASS

Annuals or tufted perennials with ligues a ring of hairs. Inflorescence a lax, diffuse panicle; spikelets with 3–many florets, narrow; glumes usually 1 veined; lemmas 3-veined, keeled and blunt.

Eragrostis barrelieri A tufted annual 10–60 cm with flat or inrolled leaves 20 mm–10 cm long, 20–55 mm wide, bluish; ligule a fringe of hairs. Inflorescence an open, diffuse panicle 30 mm–20 cm, the stiff branches with spikelets almost to the base; spikelets 5–20 mm long with 5–30 fertile florets; all glumes 1-keeled, 1-veined; lemma 3-veined, keeled, membranous with distinct lateral veins. Bare, dry habitats. Probably scattered throughout, except many islands.

1. *Oryza sativa*
2. *Arundo donax*
3. *Arundo plinii*
4. *Cynodon dactylon*
5. *Eragrostis barrelieri*

SUBFAMILY PANICOIDEAE

The second largest subfamily of grasses with numerous divergent species. Stems often solid. Spikelets unisexual or cosexual; lemmas sometimes awned; stamens 3; styles and stigmas 2.

(a) Spikelets in 1-sexed clusters, dissimilar.

Zea MAIZE

Leaves *broad*, 50 mm–12 cm across. Male flowers in slender terminal panicles, female spikelets in simple racemes (cobs) among the lower leaves.

Zea mays MAIZE A robust annual to 5 m with stout stems and many broad leaves 50 mm–12 cm wide. Male flowers borne in a terminal inflorescence of spikelets to 20 cm; female spikelets borne below on swollen, lateral branches and enclosed in overlapping leafy bracts, with long, projecting styles (to 25 cm). Fruit very distinctive: hard, shiny, yellow, white or purple, arranged around a very swollen axis ('corn on the cob'). Native to South America, widely cultivated.

(b) Spikelets comprising 1 fertile floret and 1 or more sterile florets.

Sorghum

Annuals or perennials. Inflorescence large and much-branched with paired, shiny spikelets with 1 stalkless fertile floret and 1 or more stalked, sterile or male florets; glumes all compressed and 3-pointed at the tip.

Sorghum halepense JOHNSON GRASS A large, erect, deeply rhizomatous perennial to 1.5 m with stems silkily-hairy at the nodes. Leaves hairless with rough margins, <20 mm wide. Inflorescence a large terminal, rather lax, pyramidal panicle to 30 cm long; spikelets shiny, to 5.5 mm long; lemma notched at the apex and with an awn to 12 mm long; sterile floret stalked, violet, hairy and lance-shaped. Locally common in damp, disturbed habitats. Throughout. *S. bicolor* is larger (to 2 m), with broader leaves (>20 mm) and lacks creeping stems; inflorescence dense, ovoid, to 50 cm long. Cultivated for livestock, sometimes a casual weed (native to Asia and Africa). Throughout.

S. halepense

Saccharum

Large, rhizomatous perennials, often several metres tall with ligule a fringed membrane or ring of hairs. Inflorescence large and plumose, with a robust stalk; spikelets enclosed by long, silky hairs; glumes equal; lemma with a straight awn or awnless.

Saccharum ravennae RAVENNA GRASS A very robust, pampas grass-like perennial 1–4.5 m. Leaves 30 cm–1 m long, with hairy sheaths; ligule a ciliate membrane, the leaf blade tapering into the midrib at the base. Inflorescence a plumose, rather open, straw-coloured panicle 20–70 cm long borne on very tall, erect stalks; spikelets 3–6 mm, in pairs, each with 1 fertile floret; glumes exceeding the florets, the lower 2-keeled, the upper membranous, 1-keeled and 3-veined; lemma 3 mm long, 1–3-veined and awned. Ditches and other damp habitats. Throughout.

(c) Spikelets paired, arising from each node of the branches, overlapping, the lower stalkless and fertile, the upper stalked and sterile.

Hyparrhenia

Inflorescence branched with paired, slender, spike-like clusters arising from leaf-like bracts.

Hyparrhenia hirta A tufted perennial to 1.2 m with smooth stems. Leaves linear, to 3 mm wide, more or less hairless. Inflorescence a panicle of paired racemes, each enclosed in leaf-like bracts; lemma to 4.5 mm with a stout bristle-tip to 20 mm, twisted and hairy below. Fairly frequent in dry, rocky habitats. Throughout.

H. hirta

(d) Spikelets paired and dissimilar: 1 typically stalkless and the other stalked.

Andropogon

Annuals or perennials. Inflorescence usually with paired racemes; spikelets compressed, paired, with 1 fertile floret; lower glumes with 1–11 veins, upper glumes with 1–3 veins; lemmas less firm, 1–3-veined.

Andropogon distachyos A clumped, rhizomatous perennial to 20 cm. Leaves 70 mm–20 cm long, 1–5 mm wide. Flowers borne in erect to divergent, slender paired racemes 40 mm–14 cm; spikelets in pairs, adpressed; glumes dissimilar, exceeding the florets, the lower with 7–11 veins; the upper 1-keeled and awned; lemma 2-veined and awned. Throughout.

(e) Spikelets each subtended or enclosed by stiff bristles.

Cenchrus SANDBUR

Annuals or tufted perennials. Ligules a fringe of hairs. Inflorescence spike-like with groups of 1–few spikelets on short stalks enclosed by a spiny bur of fused spines and bristles; spikelets with 2 florets, the lower sterile; glumes unequal; lemma 5-veined, awnless.

Cenchrus ciliaris A tufted perennial with short rhizomes and keeled leaves 50–85 mm. Inflorescence a dense, cylindrical, bristly panicle; spikelets with 1 basal sterile floret and 1 fertile floret, 2–5.5 mm long; lower glume 1-veined, the upper glume 1–3-veined; lemma 5-veined; anthers 3, stigmas 2. Disturbed and semi-arid habitats. Scattered probably throughout.

Setaria BRISTLE GRASS

Annuals or perennials. Inflorescence a cylindrical and spike-like panicle with numerous, densely clustered, stalkless spikelets with 2 florets; glumes unequal; lemma 5-veined, awnless.

Setaria viridis GREEN BRISTLE GRASS A loosely tufted annual to 50 cm (1 m) with wide, flat, hairless leaves; ligule a ring of hairs. Inflorescence dense, very densely hairy, cylindrical and erect; bright green, up to 12(17) cm long; spikelets 2–2.5(2.7) mm long; upper glume as long or almost as long as the spikelet. Common on sandy, fallow and waste ground. Throughout. *S. adhaerens* is similar but with the main axis of the inflorescence sparsely bristly with hairs <0.2 mm (not densely hairy with hairs >0.5 mm). Common in dry, disturbed habitats. Throughout.

1. *Sorghum halepense*
2. *Saccharum ravennae*
3. *Hyparrhenia hirta*
4. *Andropogon distachyos*
5. *Cenchrus ciliaris*
6. *Setaria viridis*
7. *Setaria adhaerens*

(f) Spikelets stalked, not subtended by bristles, borne in clustered into spike-like inflorescences.

Panicum

Annuals. Inflorescence a compound, diffuse and much-branched panicle with slender branches; spikelets flattened, with 2 florets, the upper fertile and the lower male or sterile; glumes unequal; lower lemma 5–11-veined, awnless.

Panicum repens TORPEDO GRASS A perennial with creeping underground stems and stiff erect stems to 80 cm. Leaves in 2 ranks, blue–grey and stiff, to 6 mm wide, the uppermost more or less equalling the inflorescence. Inflorescence erect and narrow, with slender ascending branches with numerous whitish branches of spikelets, each to 2 mm, without awns; glumes unequal, the upper exceeding the lower. Damp, sandy habitats, particularly on the coast. Throughout.

Panicum miliaceum A robust, annual grass, similar to *Zea* or *Sorghum* before flowering: leaves to 20 mm wide and sheaths with long hairs. Inflorescence rather dense and flopping, to 20 cm long, with numerous slender branches of plump (bead-like), often purplish spikelets each 4.5–5.5 mm long; glumes unequal, the upper exceeding the lower. Widely cultivated as a cereal and forage crop.

SUBFAMILY DANTHONIOIDEAE

Very robust, often woody-based perennials; stems often solid. Spikelets unisexual or cosexual with 1–many female florets; lemmas with 1 awn; stamens 3; styles and stigmas 2.

Cortaderia PAMPAS GRASS

Densely tufted perennials. Inflorescence a large, plume-like, spreading panicle; spikelets laterally compressed, with 2–7 florets; glumes slightly unequal and 1-veined; lemma silky-hairy, 3–5-veined, acuminate and long-awned.

Cortaderia selloana PAMPAS GRASS A very tall, dense perennial to 3 m tall and >1 m across. Leaves long and slender, to 2 m long and 1 cm wide, with sharp serrated edges. Inflorescence plume-like, overtopping the leaves; panicle open; ovoid and dense, 25 cm–1 m; spikelets solitary, lance-shaped and stalked, laterally compressed, 12–16 mm, each comprising 3–7 fertile florets with diminished florets at the apex; glumes all similar, exceeded by their spikelets. Native to South America, widely planted in towns and occasionally naturalised.

CERATOPHYLLACEAE | HORN-WORT FAMILY

A family with a single genus with few species, characterised by a submerged aquatic habit and minute unisexual flowers with a superior ovary; stamens 10–25; style 1. Fruit an achene.

Ceratophyllum

Submerged, aquatic, rootles perennial herbs with *whorled, forked leaves*.

Ceratophyllum demersum HORN-WORT A submerged aquatic perennial with slender, flexible stems to 1 m with whorls of brittle, dark green leaves *forked x 1 or x 2*. Flowers minute, arising from the axils though rarely formed; unisexual, green and stalkless. Fruit a 1-seeded nut 4–5 mm with a pair of spreading basal spines and a solitary terminal spine. Common in slow-moving or still water, often in man-made aquatic habitats. Throughout. *C. submersum* SPINELESS HORN-WORT is very similar but with leaves *forked x 3–4* and fruits that lack a basal spine and have a short or non-existent apical spine. In suitable habitats throughout.

EUDICOTS

A major group of plants, typically with 2 cotyledons (seed leaves), netted veins radiating from a central main vein (rarely parallel veins), flower parts in multiples of 4 or 5 (rarely 7), and unlike most Monocots, often with secondary growth, forming trees and shrubs.

PAPAVERACEAE | POPPY FAMILY

Annuals or perennials with milky or watery sap. Leaves shallowly 1–2-pinnately lobed. Flowers solitary or in racemes. Sepals 2(3), petals 4(6), often crumpled when newly opened; stamens numerous; style 1. Fruit a splitting capsule.

Papaver POPPY

Annuals or perennials often with a white latex. Flowers solitary with red, mauve or white petals; stigma a stalkless 4–20-rayed, flat disk.

(a) Fruits lacking prominent bristles.

Papaver somniferum OPIUM POPPY A vigorous, erect, *whitish* to blue–grey annual to 50 cm (1 m) with pinnately divided, oval leaves. Lower leaves with a short stalk, the upper leaves clasping the stem. Flowers large, petals 25–50 mm long, pale purple with a dark centre, anthers yellow. Capsule hairless. A relic of cultivation and widely naturalised in waste places. **Subsp.** *setigerum* has stems *with sparse long, fine bristles*. Leaves often more deeply lobed, and *ending in a bristle*. Capsule rather narrow. Virtually throughout.

Papaver rhoeas COMMON POPPY An erect, bristly annual to 60(80) cm. Leaves pinnately lobed, to 15 cm with pointed segments, often 2-pinnately divided. Flowers solitary on long stalks with long, bristly, *spreading* hairs; petals bright red–crimson with or without a dark centre, 30–45 mm long; anthers bluish. Capsule more or less *ovoid (narrower at the base), and hairless,* <20 mm (not >2 x as long as wide). Common on cultivated and disturbed ground. Throughout. The following are very similar: ***P. humile*** has upper leaves with short but *distinct, broadly winged, clasping stalks* (not stalkless). Mainly in dry lowland or coastal plains and deserts. IP, SY. ***P. umbonatum*** has flower-stalks with *adpressed* (not spreading) bristles and small *dark red* petals, always with a black basal spot. Capsule *narrowly tubular-oblong*. IP.

Papaver dubium LONG-HEADED POPPY An erect annual similar to *P. rhoeas* but with *adpressed* hairs on the upper parts of the stem, the leaf segments blunt, not pointed, flowers a paler or more orange–red, usually without a dark centre; petals 15–35 mm long; capsule *oblong*, somewhat widened towards the apex (2–4.5 x as long as wide); anthers *violet*, exceeded by the stigmatic disk. Common on cultivated and disturbed ground. Throughout. ***P. purpureomarginatum*** is similar to *P. dubium* but with upper stem-leaves typically entire, the outer *sepals with dark violet markings*, yellow anthers and stigmatic disk with *dark marks between the shallow lobes*. Capsule ellipsoidal with a slightly concave disk. AA, CR, CY, TR. ***P. lecoqii*** is similar to *P. dubium* but less hairy, with yellow sap and *yellow* anthers, equalling or exceeding the stigmatic disk. Possibly throughout.

1. *Papaver rhoeas*
2. *Papaver humile*
3. *Papaver dubium*
4. *Papaver purpureomarginatum*
5. *Papaver nigrotinctum*
6. *Papaver hybridum*

(b) Fruits with prominent bristles.

Papaver argemone PRICKLY POPPY A bristly annual to 45 cm with adpressed hairs on the stem, and pinnately divided leaves. Flowers scarlet, petals 15–25 mm long, often with a dark centre, and the *petals not overlapping*; anthers bluish. Capsule <25 mm, shortly *cylindrical*, ribbed, and *sparsely bristly*. Sandy waste ground, often coastal. GR (north), TR. ***P. nigrotinctum*** is similar to *P. argemone* (also considered to be a subspecies thereof) but *smaller* (often to just 10 cm) with clearly *hairy* flower buds, petals with a dark spot at the base and a conical stigmatic disk with teeth between the rays; anthers blue, greenish or yellow. Capsule 10–15 mm, elongated. Common on mainland Greece and PN. Widespread (except CY, IA, IP or much of the east). ***P. virchowii*** is similar but with cylindrical capsules ≥15 mm long, borne on *swollen* stalks. AA (east), TR.

Papaver hybridum An annual similar to *P. argemone* and relatives, but with darker crimson-red flowers with petals 10–25 mm long, and an ovoid to spherical capsule *densely covered in pale, stiff bristles*. Similar habitats. Throughout.

Papaver apulum An ascending to erect annual 10–40 cm. Leaves pinnately divided, slightly bristly, the lowermost with oblong lobes, those above with linear lobes. Buds nodding, *broadly ovoid* and *hairless*. Flowers with brick-red petals, darker towards the base but *without* a dark spot; stigmatic disk with 4–6 rays, *vaulted between the rays when mature*. Capsule 6–12 mm, not ribbed, 4–5 mm wide with *adpressed* bristles below and thin, spreading bristles above. AA, CR, GR, PN.

Roemeria

Annuals. Flowers with 4 petals; stamens numerous; stigmas 2–4. Similar to *Papaver* but with *linear fruits* that split at the base into 2–4 parts.

Roemeria hybrida (syn. *Chelidonium hybridum*) ROEMERIA A short, slightly hairy annual (10)20–40(50) cm with yellow sap. Leaves alternate, 3-pinnately divided into linear segments. Flowers poppy-like, *violet to purple* with a darker centre; petals 15–30 mm long; sepals 10–13 mm; anthers pale blue or cream. *Capsule long and linear,* bristly and 4-parted, 50 mm–10 cm long. Virtually throughout except the far west (not IA).

Adonis

Annuals with 1–3-pinnately divided leaves, often with linear segments. Flowers with 5(–8) petal-like sepals; petals 3–20, glossy; stamens numerous. Fruit a head of numerous achenes, elongating when mature.

(a) Flowers not more than 35 mm across, deep purplish-red and cup-shaped. Achenes <6 mm long with *no dorsal hump or crest*.

Adonis annua An erect annual 10–40 cm with simple stems or few erect to spreading branches. Leaves 3-pinnately lobed, the terminal lobes linear. Flowers solitary, *cup-shaped* with petals 6–12 mm long, narrowly oval, *deep purplish-red*, often with a *blackish spot* at the base; sepals spreading to deflexed. Fruiting head 8–15 mm across, ovoid, dense; achenes 3–4(5.5) mm, with a stout, outward-curving beak. Populations in the eastern Mediterranean are referable to **subsp. *cupaniana***, which has woolly sepals and achenes 4–5.5 mm (other forms have also been described but are doubtfully distinct). Throughout.

1. *Roemeria hybrida*
2. *Adonis annua*
3. *Adonis cretica*
4. *Adonis aestivalis*
5. *Adonis microcarpa*
6. *Adonis dentata*
7. *Adonis palaestina*

PHOTO: BANAN AL SHEIKH

(b) Flowers not more than 35 mm across, bright red, orange or yellow and flat. Achenes <6 mm long, typically with a dorsal hump and *crested*.

Adonis cretica An erect annual similar to *A. annua* but with larger, rather flat *yellow* flowers with *spreading* petals 12–17 mm long, *without* a black spot at the base. Achenes 3.5–4.5 mm with outwardly curving beaks and *prominently humped*. CR.

Adonis aestivalis A divergently branched annual to 50 cm, virtually hairless or sparsely hairy below. Leaves linear. Flowers 10–30 mm across; sepals pale, hairless or sparsely hairy, spreading; petals red, orange, yellow or whitish, with darker bases. Achenes 3–5 mm, rough, netted, with a dorsal keel and broadly triangular projection; transverse crest prominent and toothed; beak ascending, short, greenish or bluish. Fallow fields and plains on the mainland. GR, IP, PN, SY, TR.

Adonis microcarpa An erect annual 10–30 cm, similar to *A. annua* in form, but branched from the base. Leaves 3-pinnately lobed, the terminal lobes linear. Flowers solitary, flat and *small* with petals 4–10 mm long, *yellow or bright red*, often with a *blackish spot* at the base. Fruiting head 8–25 mm across, cylindrical to slightly conical; achenes *small*, 2.3–3 mm long with a prominent convex hump. **A. flammea** is similar but with small, scarlet flowers with petals not more than 1/3 as long as the sepals borne in a lax inflorescence. Achenes *lax*, rather larger and weakly crested, with a shallow dorsal hump addressed to the *dark-tipped* beak. Throughout (rarer in the southeast). **A. dentata** has densely crowded achenes with a prominent, often *toothed*, crest. Damp habitats in or near deserts. IP.

(c) Flowers *large*, 40–50(60) mm across, crimson. Achenes 5–7 mm long without a dorsal hump.

Adonis palaestina An erect annual similar to *A. microcarpa* but with *large* red flowers *40–50(60) mm across*. Fruits with large achenes (4)5–7 mm long, strongly ridged with a *long*, prominent beak equalling or *exceeding* the body, and no dorsal hump; crest scarcely developed. Rare; often co-occurring with *A. microcarpa*. IP, LB, SY (south). **A. aleppica** is very similar, also with long-beaked achenes but weakly ridged and the beak only *just* exceeding the body and with a *well-developed crest*. Similar distribution.

Glaucium

Annuals or perennials with a watery latex. Flowers solitary with 4 petals; stamens numerous. Fruits *very long and narrow,* splitting into 2 parts.

(a) Flowers typically orange to red; ovary more or less hairy or bristly.

Glaucium corniculatum RED-HORNED POPPY An erect, hairy, blue–grey annual 30–40 cm. Leaves 10–25 cm long with oblong, toothed lobes, those below stalked, the uppermost stalkless. Flowers borne on rather *short* stalks 25–40 mm long, *orange to red flowers* (sometimes yellow); petals 30–40(50 mm long; sepals hairy. Capsule cylindrical, long and narrow, 10–22 cm, *with appressed hairs*. Dry, rocky habitats. More or less throughout. **G. grandiflorum** is similar but with clasping upper leaves and *large* flowers with orange to red petals with a violet (to blackish) spot at the base, borne on *stalks exceeding the leaves*; sepals hairy. Capsule 10–18 cm long, densely hairy with *long, adpressed bristles*. IP. **G. aleppicum** is similar but green (not with a grey–blue hue) and only *sparsely* hairy. Flowers *crimson*; sepals often *hairless*. Capsule with adpressed hairs. Steppe, deserts and scrub. IP, SY. **G. arabicum** has smaller, *yellow–orange* flowers, the petals darker at the base; sepals hairy. Capsule 80 mm–15 cm long, swollen at the base, with scarce, scattered bristles. Deserts. IP.

(b) Flowers bright *yellow*; ovary with white tubercles (not bristly or hairy).

Glaucium flavum YELLOW HORNED-POPPY A blue–grey, branched biennial to perennial, 30–90 cm. Leaves 15–30 cm with oblong, wavy and pinnately lobed leaves, the upper leaves clasping the stem; stems with a yellowish latex when cut. Flowers *bright yellow*; petals 30–40 mm long. Fruit narrowly cylindrical, long and narrow, 15–30 cm; curved, and *hairless* but with small whitish tubercles. Coastal sands and shingle, or disturbed habitats inland. Throughout.

Hypecoum

Annual herbs. Flowers small with 4 rather unequal petals; stamens 4; stigmas 2. Fruit a capsule, usually curved.

Hypecoum dimidiatum A small, grey–green annual 50 mm–40 cm with leaves 2–3-pinnately divided. Flowers borne 1–24 in erect inflorescences, yellow; sepals oval to lance-shaped and toothed; outer petals 7–12 mm, rather 3-lobed, the inner 5.5–8.5(10) mm, deeply 3-lobed, the lateral lobes oblong-linear and the central *with ciliate margins*; filaments 4, membranous. Fruits cylindrical, slightly swollen at the joints. Dry, rocky habitats and deserts. IP (mainly east-central).

1. *Glaucium corniculatum*
2. *Glaucium grandiflorum*
3. *Glaucium flavum*
4. *Hypecoum dimidiatum*
5. *Corydalis solida*

PAPAVERACEAE

Hypecoum procumbens A delicate, short, hairless, blue–grey annual with wide-spreading stems to 15 cm. Leaves 2-pinnately lobed, segments lance-shaped or linear. *Bracts leaf-like.* Flowers borne in small, *sparsely branched* clusters, flowers to 15 mm across, with 4 *pale yellow* petals: 2 large 3-lobed petals, and 2 small, lateral petals. Capsule erect and jointed. Local in maritime habitats. Virtually throughout (not IA). *H. imberbe* is similar but more erect and with *linear bracts*. Flowers *orange–yellow* with more or less *evenly lobed large outer petals,* borne in *repeatedly branched* clusters. Fruits scarcely jointed. Cultivated land and waste places. AA (local), CY, GR, PN. *H. pendulum* is similar to *H. procumbens* but with the *outer 2 petals unlobed or scarcely lobed*. Capsule straight and scarcely jointed, borne on *sharply deflexed stalks*. Various disturbed habitats, mainly on the mainland. CY, GR, IP (probably elsewhere).

Corydalis

Hairless perennial herbs, often with thick, tuberous roots. Leaves divided. Flowers zygomorphic, borne in racemes; upper petal spurred, the upper group of stamens with the base prolonged into a spur. Fruit a capsule.

(a) Stem with a conspicuous *scale below the lowermost leaf.*

Corydalis solida A small, erect, tuberous perennial 60 mm–15(20) cm, often with one branch from the axil of a *scale-like leaf* at the base. Leaves divided into lance-shaped segments, blue–grey beneath. Flowers 8–20, borne in rather dense racemes, 15–25(30) mm long, uniformly pinkish-white to mauve, or darker towards the tips; flowers subtended by *deeply* divided bracts. Capsule 3–5 x as long as broad (10–25 mm), as long as its stalk, with 5–11 seeds. Woods and bushy habitats, widespread on the mainland, absent from many islands. Variable, with several regional forms described. GR, IA, PN, LB, TR. *C. thasia* is a similar, slender perennial 10–25 cm with a *lax raceme* of fewer (3–12) flowers, the outer petals *darker* in the distal part, and *narrower capsules*, 5–7 x as long as broad. AA (Andros, Thasos).

Corydalis integra A sub-erect, tuberous perennial 10–20 cm, typically with a single stem with 1–3 lateral branches arising from the axil of a scale-like leaf at the base. Leaves divided with oblong lobes, blue–grey. Flowers 5–15, borne in rather lax racemes, pale pink, the inner petals externally (abruptly) *tipped with dark purple;* flowers subtended by *entire* (or sometimes slightly divided) bracts. Capsule 15–22 mm, linear-lance-shaped with 5–8 seeds. AA (east), TR.

(b) Stem *without* a conspicuous scale below the lowermost leaf.

Corydalis uniflora An ascending, tuberous perennial *without* a scale-like leaf at the base, with long-stalked, *opposite* basal leaves and several flexuous flowering stems just 20–50 mm. Flowers *few*, 1–3(4), whitish, pale pink or mauve, 15–25 mm long, with a strongly curved spur 8–12 mm long; flowers subtended by entire diamond-shaped bracts. Capsule short and broad with 3–6 seeds, spreading when mature. CR.

Corydalis cava A tuberous perennial 10–30 cm without a scale-like leaf at the base, with *alternate* greenish to greyish divided leaves with blunt lobes. Flowers *8–20*, reddish-purple, yellow or cream, subtended by entire oval bracts. Capsule oblong, 18–24 mm long with 5–10 seeds. Mainly mainland Greece. AA (west), GR, IA, PN.

1. *Fumaria kraliki*
2. *Fumaria macrocarpa*
3. *Fumaria officinalis*
4. *Fumaria parviflora*
5. *Fumaria densiflora*
6. *Fumaria bracteosa*

PAPAVERACEAE

Fumaria

Trailing or scrambling annual herbs with 2–4-pinnately divided leaves and distinctive leaf-opposed racemes of tubular 2-lipped flowers with 2 small sepals, 2 outer petals and 2 narrower inner petals; stamens 2. Fruit a more or less spherical 1-seeded achene. Many similar species, which are difficult to distinguish grow in the region; just a few common species are described here. A hand lens is essential for identification in the field.

(a) Fruiting stalks *recurved*. Flowers large, ≥9 mm long.

Fumaria capreolata WHITE RAMPING-FUMITORY A hairless, blue–green, scrambling annual with stems 30 cm–1(2) m. Leaves wedge-shaped with blunt, narrowly oblong terminal lobes. Flowers 10–25(30), *held sub-erect*, on racemes equalling or shorter than their stalks; corolla 2-lipped, 10–13(14) mm, the *upper petal compressed with upturned margins not concealing the keel, creamy white*, often tinged with pink, and *tipped with reddish black*. Capsule 2–2.3 mm, its stalk *strongly curved*. Disturbed ground. Throughout.

Fumaria petteri A slender annual with oblong leaf segments. Flowers *reddish-pink* with darker tips, 10–15 mm, borne in dense racemes of 15–20; lower petal with short, erect margins; sepals large, oval, weakly toothed to entire; bracts equalling the slightly recurved flower-stalks. Fruit 2.2 mm. Throughout.

(b) Fruiting stalks *recurved*. Flowers small, <9 mm long.

Fumaria kraliki A slender annual to 80 cm, very similar to *F. petteri* but with narrower (almost linear) terminal leaf segments and *small*, pale to reddish-pink, darker tipped flowers 5–7 mm long, typically with more conspicuously *toothed* sepals (not slightly toothed to entire) and with lower petals with *spreading* (not narrow and erect) margins. Fruiting stalks strongly recurved with equal bracts. Fruit 1.75 mm. Throughout.

(c) Fruiting stalks more or less straight. *Flowers large*, ≥(8)9 mm long.

Fumaria macrocarpa A hairless, sprawling annual with stems 20 cm–1 m long. Leaves divided, the terminal lobes narrowly oblong and pointed. Flowers borne in racemes of 4–10(15), the raceme shorter than its stalk. Flowers 8–11 mm, white or pale pink, *tipped greenish* (not purple), the inner petals the same colour (or tipped dark purple on the upper side only). *Fruits large*, 3.5(4) mm, with tubercles, scarcely keeled but with a large apical pit. AA, CR, IP.

(d) Fruiting stalks more or less straight. *Flowers small* to minute, <9 mm long (many similar species).

Fumaria officinalis COMMON FUMITORY A delicate, hairless, blue–green, scrambling annual with broad, flat, oval-lance-shaped leaf segments. Flowers 7–8(9) mm, *numerous*, 10–45, *mauve*, tipped with blackish purple on the wings of the upper petal and apex of the inner petals, borne in a raceme longer than the *short stalk*. Fruits 1.8–2.5 mm, much *wider than long*, warted (*not* shiny). Very common on disturbed ground. Throughout. **F. parviflora** has *channelled leaf segments*. Flowers small, 5–6 mm long and *pallid* (white, flushed very pale pink) in almost stalkless racemes; sepals minute, 0.5–1.5 mm long; *bracts at least equalling fruiting stalks*. Fruit scarcely keeled. Most common in dry to arid habitats. Throughout. **F. vaillantii** is similar to *F. parviflora* but with *pink flowers* and bracts *shorter* than fruiting stalks. Scattered throughout; absent from most islands and the far southeast.

Fumaria densiflora A delicate, erect, later spreading annual, with channelled leaflets, similar to the species above but with smaller flowers; flowers borne on *very short-stalked* racemes of 15–30; corolla 5.5–6.7 mm long, dark reddish-pink with *large, round, toothed sepals* 2.5–3(3.5) mm long; bracts at least as long, often *longer* than the fruiting stalks. Fruit 2–2.25 mm, spherical, *shiny*, with rounded apical pits, borne on thick, erect to spreading stalks. Throughout, in various habitats; common in the south and east. **F. bracteosa** is very similar but with much *smaller flowers* just 3.5–5 mm, deep pink; sepals much smaller and less conspicuous, 1.5–2 mm long. Fruit shiny and strongly *keeled*. Dry and arid habitats. AA (east), CY, IP.

BERBERIDACEAE

Perennials or shrubs with alternate, simple to pinnate leaves without stipules. Flowers regular, cosexual, solitary or in racemes; petals in 4–6 whorls of 2–3, usually yellow; stamens 4 or 6 in 2 whorls; style short or absent. Fruit a capsule or berry.

Berberis

Shrubs with spiny stems. Leaves simple. Flowers yellow; stamens 6. Fruit a few-seeded red to black, bloomed berry.

Berberis cretica A *spiny shrub* to 1 m. Leaves deciduous, shorter than the spines. Flowers bright yellow, with 4–6 sepals and petals, broadly cup-shaped, borne in short racemes. Fruit an ellipsoidal blackish-blue berry. Forms spiny thickets on stony slopes inland. AA (local), CR, CY, GR (local), TR. **B. libanotica** is very similar, differing only in subtle leaf characteristics, but geographically distinct. LB, SY.

Leontice

Tuberous perennials. Leaves divided into 3 or pinnately divided. Flowers yellow; stamens 6. Fruit a *strongly* inflated capsule.

Leontice leontopetalum A *hairless perennial* 25–60 cm arising from a coarsely warted tuber. Basal leaves 2–3, large, blue-grey, with 3 lobes. Inflorescence *pyramidal*, 15–30 cm across, made up of 3–7 racemes in the axils of reduced stem-leaves; flowers bright yellow, about 16 mm across, the petals exceeded by the sepals. Fruit 30–40 mm long, *strongly inflated*, containing 1–2 large seeds 5–8 mm long. Fields and agricultural land; mainly southern and eastern. AA, CR, CY, GR, IP, PN, TR.

Bongardia

Tuberous perennials. Leaves *pinnately* divided. Flowers yellow; stamens 6. Fruit a slightly inflated, papery capsule.

Bongardia chrysogonum A hairless, tuberous perennial 20–30 cm, with leaves all basal, arising directly from the tuber. Leaves *pinnately divided* with 9–16 crowded pairs of leaflets. Flowers with petals longer than the sepals. Fruits ovoid and pleated, 12–16 mm long. Fields and stony slopes. Strongly eastern; rare or extinct in the far west of the region. AA (Chios), CY, IP, LB, PN (rare), SY, TR.

1. *Berberis cretica* PHOTO: ARNE STRID
2. *Berberis cretica* PHOTO: NICK TURLAND
3. *Leontice leontopetalum*
4. *Leontice leontopetalum* in fruit
5. *Bongardia chrysogonum*

RANUNCULACEAE | BUTTERCUP FAMILY

Herbaceous annuals and perennials or woody climbers with alternate, simple or compound leaves. Flowers typically regular (sometimes zygomorphic); sepals and petals 5; stamens numerous. Fruit an aggregate of achenes or follicles (or a berry or capsule).

Helleborus

Herbaceous perennials with leaves arranged spirally or all basal, with long, toothed leaflets. Flowers borne in *winter–spring* in branched clusters; sepals 5, petals 5–12 in the form of small nectaries; stamens numerous; carpels 2–5. Fruit an aggregate of follicles.

Helleborus odorus A robust perennial to 40(60) cm with a short, stout rhizome and large, overwintering (often solitary) basal leaves appearing with or shortly after the flowers; leaves round in outline, divided palmately into (5)9–11(14) leaflets. Flowers 2–5(7), 50–70 mm across, clear yellowish-green, large, nodding, bell-shaped and long-persistent; follicles 3–6, scarcely united, with a long beak; seeds 4 mm. GR, IA (Corfu).

Helleborus vesicarius A robust perennial 30–50 cm with many large, basal leaves dissected into wedge-shaped, toothed lobes; stem-leaves stalkless with 3–5 toothed lobes, overtopping the flowers. Flowers greenish, nodding. Follicles *inflated*, forming a light green, *spherical* capsule. Shady woods. SY, TR (south).

Anemone ANEMONE

Perennial herbs with palmate or palmately lobed basal leaves and a whorl of leafy bracts beneath the solitary (or few) flowers. Perianth with 1 whorl of 5–20 petal-like sepals; stamens numerous. Fruit an aggregate of many-seeded carpels, spirally arranged.

(a) Stem-leaves *short-stalked* and similar to basal leaves. Carpels *shortly hairy*.

Anemone apennina A tuberous perennial 80 mm–25 cm with an irregularly lobed rhizome. Basal leaves with 1–2(5) long-stalked leaves; leaf blade round–triangular in outline with stalked, lobed segments; stem-leaves in a whorl of 3, *similar to the basal leaves*, though short-stalked. Flowers 25–35 mm across with *many* (9–18) petals 12–22 mm long, *hairless*, white to pale or deep blue; anthers pale yellow. Fruiting head *erect*. Woods, hill forests. AA (local), CY, GR, IA, PN, SY. **Subsp. *blanda*** is the form in the region (also considered to be a distinct species), and has larger blue (rarely white) flowers 30–45 mm across with 10–15(18) petals *hairless underneath*, and *nodding fruiting heads*. Rocky woods and mountains.

Anemone nemorosa WOOD ANEMONE A delicate perennial 50 mm–30 cm with creeping rhizomes and palmately lobed basal leaves with oval, toothed or lobed segments, and similar stem-leaves borne on an erect stem bearing solitary, white, drooping flowers; palmately lobed leaf-like bracts groups of 3 below the flowers. Sepals 6–7(9) flushed pink or purple; anthers yellow. Fruits drooping. Deciduous woods in cooler areas. GR (north), TR.

1. *Anemone apennina* subsp. *blanda*
2. *Anemone coronaria* (scarlet form)
3. *Anemone coronaria* (violet form)
4. *Anemone pavonina* (scarlet form)
5. *Anemone pavonina* (violet form)
6. *Anemone hortensis* subsp. *heldreichii*

(b) Stem-leaves *stalkless* and disimilar to basal leaves. Carpels *densely woolly*.

Anemone coronaria CROWN ANEMONE A hairy, tuberous perennial 80 mm–20 cm with woody rhizomes and basal leaves 2–3-times dissected into narrow segments; stem-leaves in a whorl of 3, *deeply dissected* into numerous narrow, toothed segments and *stalkless*. Flowers large, 35–65(75) mm across, *variably pink, red, blue, purple or white*, often with a paler centre, bowl-shaped, with usually *few (6)* elliptic, overlapping petals; anthers bluish to *dark purple*. Fruiting head erect. Common on cultivated land. Throughout. The following are similar: **A. *pavonina*** has basal leaves 3-lobed nearly to the base, the segments wedge-shaped and shallowly divided; *stem-leaves undivided or occasionally lobed*. Flowers 30–60 mm across, with 7–11 narrow petals variable in colour but often pinkish-mauve or scarlet. AA, GR, IA, PN, TR. **A. *hortensis*** has basal leaf segments divided less than halfway into almost entire, to toothed lobes; stem-leaves entire or occasionally with 1–3 teeth. Flowers 25–45 mm across with 7–12 petals 11–19 mm long, narrowly elliptic, pink to mauve; anthers deep purple. Olive groves and abandoned pastures. **Subsp. *heldreichii*** is the form in the region (also considered to be a distinct species), and has smaller, paler flowers, which are *pale above* (whitish to pale mauve), and *darker beneath* (brownish-purple or pink); anthers deep blue. Rocky pastures; a rare island-group endemic. CR (+ KP + Kasos).

Pulsatilla group (now widely classified under *Anemone*)

Herbaceous perennials with leaves all basal, 1–2-pinnately divided, without stipules. Flowers solitary, with a whorl of leafy bracts below, regular with petal-like sepals; stamens and carpels numerous. Fruit an achene.

Anemone rhodopaea (syn. *Pulsatilla rhodopaea*) PASQUEFLOWER A *short, densely white woolly-hairy* perennial to just 50 mm without branches and with 1-pinnately divided leaves with numerous (50–100) lobes. Flowers with 6 petal-like, *deep violet* sepals 20–50 mm, silky-hairy on the undersurface; stamens numerous, yellow; style *elongated in fruit*. Mountains in the north only. GR (northeast).

Thalictrum MEADOW-RUE

Herbaceous perennials with divided leaves spirally arranged, with stipules. Flowers borne in racemes of compound inflorescences with a whorl of bracts below, regular; sepals petal-like, 4; stamens numerous; carpels 2–15. Fruit an achene.

Thalictrum orientale A patch-forming perennial 10–30 cm with fibrous roots. Leaves with 2–3 toothed leaflets 20 mm, *evenly spaced* along the stems. Flowers few, erect, with white to lilac petal-like sepals to 10(12) mm; *filaments slender*, much narrower than the anthers. Achenes 2–6, narrowly oblong, virtually stemless. Rocky crevices in gorges and hillsides; widespread on the mainland, absent from the islands. PN, SY, TR.

Thalictrum aquilegiifolium An erect perennial to 1.5 m, usually unbranched. Leaves with 2–4 toothed leaflets. Flowers whitish to lilac–pink, borne in *dense clusters* in compound panicles; *filaments thickened*, wider than the anthers. Achenes several, to 7 mm, lance-shaped, *with 3 winged angles*, pendent upon slender stalks. AA, GR. *T. lucidum* is similar but with more slender filaments and narrowly ellipsoidal, *erect, stalkless* and *ribbed* achenes, 2 mm long. AA, GR.

Ranunculus BUTTERCUPS

Terrestrial or aquatic herbs with entire or lobed leaves. Flowers borne solitary or in cymes; sepals and petals normally 5, the petals often shiny and white or yellow; stamens numerous (or 5–10); carpels numerous. Fruit a head of 1-seeded achenes. Many species occur; not all are described.

(a) Leaves entire to palmately divided, flowers yellow and *carpels spiny or warty*.

Ranunculus muricatus ROUGH-FRUITED BUTTERCUP A short, usually hairless annual with a stout, much-branched stem 50 mm–30(40) cm and *kidney-shaped* lower leaves with 3–7-shallow-toothed lobes; upper leaves wedge-shaped with up to 5 lobes or occasionally entire. Flowers pale yellow, 6–16 mm across; sepals deflexed. Achenes *strongly keeled, with spines to 1 mm on the faces* (not the grooves) and tapered into an abruptly curved beak 2–2.5 mm. One of the most common species across the region in ditches and in other damp, grassy places. Throughout.

1. *Thalictrum orientale* PHOTO: ELEFTHERIOS DARIOTIS
2. *Ranunculus creticus*
3. *Ranunculus asiaticus* (red form)
4. *Ranunculus asiaticus* (cream form)
5. *Ranunculus asiaticus* (white form)
6. *Ranunculus asiaticus* (yellow form)
7. *Ranunculus paludosus*
8. *Ranunculus gracilis*
9. *Ranunculus flammula*
10. *Ranunculus peltatus*

(b) Leaves *entire to shallowly palmately divided*, flowers yellow and *carpels smooth*.

Ranunculus creticus An erect perennial 20–60 cm with spindle-shaped tubers and stout stems, hairy at the base, sparingly branched above. Basal leaves long-stalked, 60 mm–14 cm across, kidney-shaped and shallowly (3)–5-lobed and toothed (*not* deeply divided). Flowers yellow, 24–36 mm across; sepals adpressed. Achenes 3.5–4 mm, compressed with a short, hooked beak. Garrigue. AA (south), CR.

Ranunculus bullatus A low, hairy perennial with leaves 20–60 mm across, all basal, stalked, oval to rounded and toothed but *not lobed*. Flowers borne solitary on leafless stalks, yellow, to 26 mm across with 5–10(12) petals, and *sweetly scented*; sepals spreading. Achenes 1.2 mm with a short, curved beak. Local on rocky slopes. AA, CR (+ KP), CY, PN (Elafonisos, Kithira, Andikithira).

(c) At least some leaves (usually) *deeply palmately lobed*, flowers yellow; carpels smooth, wrinkled or spiny.

Ranunculus asiaticus An erect perennial 10–30 cm with a cluster of spindle-shaped tubers and solitary stems. Leaves 20–50 mm across, divided into 3 lobes, variously dissected or toothed; stem-leaves few and reduced. Flowers *large and showy*, with 5–9 petals, 30–60 mm across, *variably white, pink, yellow or cream or crimson* – a particular colour often regionally dominant; anthers *dark purple*. Achenes 4.5–5.8 mm with a curved beak. Garrigue and grassy habitats, a characteristic feature of the eastern Mediterranean. AA (south), CR (fairly common + KP), CY (common), IA, IP (common), PN, SY, TR.

Ranunculus gracilis An erect perennial 10–30 cm with ovoid root tubers mixed with fibrous roots and a scarcely swollen stock. Basal leaves long-stalked with a triangular blade, deeply divided, the lobes dissected into blunt segments; stem-leaves few and reduced. Flowers yellow, 20–24 mm across with *deflexed sepals*. Achenes 1.8 mm, compressed, with a straight or slightly hooked beak. AA (local), CR (+ KP), CY, GR (southeast), IA, PN, TR.

Ranunculus paludosus JERSEY BUTTERCUP An erect perennial 10–25(60) cm with some roots developed into spindle-shaped tubers and a swollen stock. Basal leaves with hairy stalks, the innermost *narrowly and deeply divided* with saddle-shaped lobes, the outermost shallowly toothed; stem-leaves 1–2, small. Flowers pale lemon yellow, 20–30(36) mm across; sepals not deflexed. Achenes 2.5–3 mm, more or less smooth and slightly hairy, with a straight or slightly hooked beak 1–1.5 mm, borne on an elongated receptacle. Cool, damp, grassy habitats. Throughout.

Ranunculus arvensis CORN CROWFOOT A (typically) hairless annual 10–45(60) cm with spatula-shaped, simple or more commonly toothed to dissected leaves with narrowly lance-shaped to linear lobes. Flowers *pale* greenish to lemon-yellow, 4–12 mm across, borne in branched clusters on slender stalks. Achenes 5–6(8) mm with *prominent, long, rigid spines >1 mm long*; beak 2–3 mm. Common on cultivated and disturbed land. Virtually throughout.

Ranunculus sceleratus CELERY-LEAVED CROWFOOT A rather stout and fleshy, more or less hairless annual with *shiny green leaves*, deeply divided, those above into narrow segments, on much-branched, hollow, grooved stems. Flowers numerous, borne in branched clusters, yellow, 5–10 mm across, with deflexed sepals. Achenes hairless, to 1 mm, borne in cylindrical heads. Wet, marshy habitats. GR.

Ranunculus sardous A hairy *annual* with a scarcely swollen underground stem base. Leaves 3-lobed and shiny. Flowers 12–25 mm across and *pale yellow*; sepals with dark markings along the margins (best seen in bud). Achenes smooth except for a row of small tubercles surrounded by a green border. Coastal grasslands, damp places, meadows and other grassy habitats. Throughout.

Ranunculus velutinus A rather tall, silkily downy, hairy perennial 40–70 cm, with fibrous roots. Basal leaves 30–70 mm across, rather geranium-like, *broadly oval* with 3 wedge-shaped toothed to lobed divisions; stem-leaves fewer and smaller. Flowers yellow, rather small to 25 mm across, borne on *slender stalks*; sepals *strongly* deflexed. Achenes 3 mm, rounded, smooth with short beaks (0.5 mm). Throughout, except the southeast (not CY or IP).

1. *Ficaria verna* subsp. *verna*
2. *Ficaria verna* subsp. *chrysocephala*

(d) Leaves entire and markedly *longer than broad*, flowers yellow.

Ranunculus flammula LESSER SPEARWORT A rather fleshy, hairless (or nearly so), hollow-stemmed perennial; stems to 50 cm and rooting at the base. *Leaves narrow and entire*; broader and more heart-shaped at the base. Flowers rather few, borne on slightly furrowed and hairy stalks, small, shiny and yellow, 7–20(25) mm across; sepals hairless and spreading. Carpels numerous, hairless and pitted, 1–2(2.3) mm long. Locally common in damp and waterlogged habitats. GR (north). *R. lingua* is similar but more robust, to 1.2 m, erect and with much larger flowers, 20–50 mm across borne on unfurrowed stalks. Similar habitats. Rare. PN.

(e) *Flowers pink or white* and plant *aquatic* (species similar and difficult to distinguish).

Ranunculus peltatus POND WATER-CROWFOOT An annual or perennial aquatic herb with floating and submerged leaves with divergent leaf segments with *rounded tips*. Flowers to 30 mm across. Fruit stalk to 15 cm long (*longer* than stalk of opposite leaf). Locally common in aquatic habitats. Throughout. *R. trichophyllus* is similar but with petals *not touching when mature* (widely spaced). Still or slow-moving water. Throughout (except smaller islands; not on CY).

Ficaria

Tuberous perennial herbs with heart-shaped leaves. Flowers yellow, with 7–12(13) petals; stamens and carpels numerous. Fruit a head of achenes.

Ficaria verna (syn. *Ranunculus ficaria*) LESSER CELANDINE A variable (with several subspecies described), hairless perennial 50 mm–25 cm with long-stalked, triangular *heart-shaped*, wavy-margined to shallowly lobed, fleshy, dark green leaves. Flowers 10–30 mm across, shiny and yellow, turning white on ageing with 7–12(13) narrow petals and *3 sepals*. Common in damp, grassy places. The following subspecies are all similar, and their distributions poorly known. **Subsp.** *verna* has flowers 13–30 mm across; petals 6–15 x 2–5 mm. North of range only. GR (north). **Subsp.** *chrysocephala* is more robust and erect, with flowers 40–60 mm across; petals *broad*, to 9–18 mm wide. Strongly southern. CR, CY (probably also mainland Greece and Turkey). **Subsp.** *ficariiformis* has flowers 35–55 mm in diameter; petals 15–26 x 4–12 mm. Widespread; probably throughout. **Subsp.** *calthifolia* is a short plant with *small* flowers just 23–50 mm across. Mainly northern. AA, GR (north).

Nigella LOVE-IN-A-MIST

Annual herbs with solitary flowers and rather feathery, pinnately divided leaves with narrow segments. Flowers solitary (or few); sepals petal-like, petals in the form of clawed nectaries; stamens numerous; carpels usually 5, variably fused along their inner margins and many-seeded. Fruit a capsule.

(a) Carpels joined over most of their length (only the styles free). Flowers with bluish sepals.

Nigella damascena LOVE-IN-A-MIST An erect hairless annual to 15–40(50) cm with alternate, finely 2–3-pinnately divided leaves with narrow segments, the *uppermost in a feathery whorl just below the flowers*. Flowers solitary, and sky-blue, 15–30(35) mm across with 5 petals and a central cluster of stamens and carpels; carpels joined over most of their length. Fruit capsule spherical, *strongly inflated*, 10-celled when dissected. Common on cultivated, sandy, disturbed and waste ground. Throughout. **N. elata** is similar but often taller (40–80 cm) with a 5- (not 10-) celled when dissected, scarcely inflated capsule 14 x 12 mm. AA (Lesbos), TR.

(b) Carpels joined for <80% of their length. Flowers with whitish to bluish sepals (many similar species).

Nigella arvensis A slender, *ascending to erect* annual 15–55 cm with pinnately divided leaves with linear to oblong segments. Flowers solitary, pale blueish to greyish-white (often veined green), long-stalked, usually *not* surrounded by a feathery whorl; sepals usually 5, the limb *abrupt to heart-shaped* at the base; carpels 3-veined, *fused to about ½* (30–65%) *their length*, smooth to densely tuberculate; *anthers pale blue to yellowish*. Throughout (rare on CR). A variable species with numerous regional forms. The following are all similar: **N. degenii** has flowers 15–20 mm across with sepal limb abrupt to slightly heart-shaped at the base, about *as long as wide*; anthers pale blue to *vivid red*; carpels united to 45–60%. An Aegean endemic. AA. **N. icarica** is very similar to N. degenii with flowers 14–20 mm across with white sepals with greenish veins; anthers *pale to mid-blue*; carpels to 55–65%. AA (Ikaria). **N. carpatha** has a *spreading* habit with several prostrate branches; flowers 12–17 mm across; sepal limb *wedge-shaped* at the base and 1.5 x as long as wide; *anthers cherry red* (pollen yellow). Carpels united to 50–60%. AA (KP, Kasos). **N. doerfleri** has small flowers just 9–15 mm across with sepal limb 1.5 x as long as wide; anthers *yellowish*; carpels untited to 65–80%. AA, CR. **N. stricta** has a *spreading* habit with short, rigid branches; sepal limb wedge-shaped at the base; anthers vivid red; carpels *outward-curving* and *widely divergent*, united to 50–80%. CR (southwest), PN (Kithira). **N. deserti** has long-beaked carpels united to just 30% at the base. Deserts. IP.

(c) Carpels joined for <80% of their length. Flowers with *yellowish* sepals.

Nigella orientalis An erect annual with divided leaves with long, slender, *rigid* lobes. Flowers with *yellow*, hairless sepals; petals *oval*-shaped and divided into *2 terminal lobes*; carpels large, strongly compressed, divided to 50%, with long, outward-curving beaks, becoming straighter in fruit; seeds disk-shaped. Eastern only. IP, LB, SY, TR. ***N. oxypetala*** has upper leaves with oblong-lance-shaped lobes (not linear) and yellow sepals; petals minute, broader than long with *4 terminal lobes*, the lateral lobes short, the middle lobes bristle-like; carpels united to >50%. SY, TR. ***N. ciliaris*** has sepals with *long, white, spreading hairs*; petals with *2, long lobes*; carpels united to just 30%. CY, IP.

Delphinium

Annuals or perennials with broadly palmately divided leaves and (typically) blue flowers borne in erect racemes, each with 5 petal-like outer segments, the uppermost spurred at the back, and 4 inner petal-like segments, the 2 uppermost with spurred nectaries. Fruit made up of 3–5 follicles.

(a) Plant a stout biennial to 1 m. Seeds *large*, 6 mm long.

Delphinium staphisagria A *hairy biennial* 30 cm–1 m with alternate leaves along the stem, 1-pinnately divided into 5–7 lobes (sometimes 2-pinnate); leaves 10–15 mm across. Flowers borne in lax racemes, deep blue, large to 25 mm across; upper sepal with *short, blunt, down-turned spur* 30–50 mm long (shorter than the petals). Fruit with inflated follicles to 22 mm, bearing few (3–6) large seeds to 6 mm. Local on rocky slopes and on garrigue and maquis inland. Throughout, except the far southeast.

(b) Plant a tuberous perennial to 1 m. Seeds small, <6 mm long.

Delphinium fissum A *tuberous perennial* with erect, solitary (or sparingly branched) stems 30 cm–1 m. Leaves round, 30 mm–10 cm across, several times divided into broadly *linear*, pointed segments. Flowers numerous, borne in a long raceme, on stalks 4–8 mm long, swollen at the apex; corolla purplish-bue; upper sepal with a straight or scarcely curved spur 14–17 mm. Fruit with 3 variably hairy follicles 14–20 mm long; seeds 2 mm. AA (rare), GR, IA, PN, TR. ***D. ithaburense*** has palmately lobed leaves with *wedge-shaped* segments divided into *oblong*-linear, pointed lobes. Flowers *pinkish* (or bluish) *white*; sepals with *long, white, spreading hairs*. Follicles very hairy, tapering into a long beak. Rocky habitats. IP (north and centre).

(c) Plant an annual, typically <1 m. Seeds small, <6 mm long (many similar species).

Delphinium peregrinum VIOLET LARKSPUR A slender, hairy *annual* typically <1 m; stems with a white bloom of spreading hairs. Lower leaves greyish and palmate with narrow segments, the upper leaves unlobed. Flowers borne in slender, rather dense racemes, dull bluish-*violet* (or dirty white; sometimes bicoloured), *small* to 18 mm across, the spur *upturned and longer* (x 1.5) *than the petals*; the nectariferous petals hairless, *elliptic, equal to and gradually tapering into the claw*. Follicles 3, 5–8 mm, variably hairy. Dry scrub. Throughout (rare on CR). ***D. hellenicum*** is similar but with lax racemes of tawny–blue to violet flowers with nectariferous petals rounded to square, shorter than the claw, and abruptly contracted into it. IA, PN. ***D. balcanicum*** is similar to *D. hellenicum* but with *dense* racemes of *deep blue–violet* flowers. Follicles 5.5–8 mm long with *long, spreading hairs*; seeds just 1.2 mm. AA (Skiros), GR. ***D. bovei*** has blue flowers with nectariferous petals just 0.3–0.5 x the length of their claw and ascending spurs greatly exceeding (x 3) the petals. Desert habitats. IP.

---

1. *Nigella damascena*
2. *Nigella arvensis*
3. *Delphinium staphisagria*
4. *Delphinium peregrinum*

RANUNCULACEAE

Consolida LARKSPUR

Annuals similar to *Delphinium* with palmately divided leaves with *numerous thread-like segments*. Flowers in terminal racemes; sepals 5, the upper long-spurred, petals 4. Fruit a *solitary* follicle.

(a) Flowers typically purplish or bluish. Follicles variably hairy but *not* with adpressed hairs.

Consolida ajacis LARKSPUR A downy annual to 1 m with a simple or branched stem and deeply dissected lower leaves; leaves persistent (not withered) in flower. Flowers rather large and few, borne in lax inflorescences, typically *bright blue* (pink, white and pale blue forms are cultivated), the upper petal with a *long* backwardly projecting spur 13–18 mm long; flower-stalks *long*, 15–40 mm. Follicles tapered at the apex. Local on disturbed ground, sandy places and field margins. AA (local), CR (local), GR, IA, PN. *C. hispanica* is similar but with *short* flower-stalks (8–12 mm) equalling or exceeding the spur, and follicles *abruptly* contracted at the apex. GR, IA, IP. *C. phygria* is similar but with *very lax* inflorescences; lower flower-stalks *shorter than the spur* (and remaining short in fruit). AA (north), GR. *C. orientalis* has *long, dense* inflorescences of *vivid violet* flowers that retain their colour when dry; fruiting stalks curving upwards. Follicles abruptly pointed at the apex and sparsely glandular. TR. *C. glandulosa* is glandular-hairy on the leaf stalks and has inflorescences with *golden* hairs. Racemes lax; spur 2 x as long as the petals. Follicles oblong, long-beaked, rather flattened and incurved, shorter than their stalks. Dry fields. TR.

Consolida regalis FORKING LARKSPUR A slender, rather downy, widely branching annual to 50 cm. Leaves divided into linear lobes, *withered* during flowering. Flowers violet–blue to dark blue, to 28 mm, borne in lax *panicles*; spur 12–18(25) mm. Follicles 7–11 mm, hairless; seeds black. Arable and disturbed habitats. AA (north), GR, IA, TR.

(b) Flowers typically purplish or bluish. Follicles with *adpressed hairs*.

Consolida arenaria A *short*, ascending annual 50 mm–12 cm with adpressed hairs, and leaves divided into linear-lance-shaped lobes. Flowers deep purple, few (2–4) on stalks to 12 mm; spur 12–16 mm. Follicles to 10 mm, with adpressed hairs. Coastal sands. AA (Rhodes, Limnos). *C. samia* is very similar but with 1–2 *pale* blue flowers and with more spreading hairs. Limestone screes; very rare. AA (Samos), TR. *C. tomentosa* has short, lax racemes of pink or violet flowers; spur scarcely longer than the petals. Follicles *pendulous* with adpressed hairs, cylindrical, tapering to a beak. Dry fields. SY.

(c) Flowers s*ulphur-coloured*.

Consolida sulphurea A small annual with adpressed hairs. Flowers 3–5, borne in short racemes, *Sulphur to yellow-coloured*; spur 2 x as long as the petals. Follicle borne on recurved stalks, pendulous, with dense *white hairs*. Grassy habitats in the east. IP, SY.

Clematis

Woody climbers with opposite, pinnately divided leaves. Flowers borne in branched cymes. Fruit a cluster of achenes, each usually with an elongated, often feathery style.

Clematis vitalba TRAVELLER'S JOY A woody, vigorous climber 3–10(30) m with *1-pinnately divided leaves* with pointed leaflets 50–70 mm long. *Flowers dull greenish white or cream* with segments *hairy on both sides*, fragrant, borne in terminal or lateral loose compound panicles; styles feathery in fruit. Maquis. GR, IA, PN, TR. *C. flammula* FRAGRANT CLEMATIS is a similar woody climber with slender stems 2–5(6) m, and with *2-pinnately divided leaves* with stalked, oval to circular leaflets. *Flowers white*, to 20 mm across, fragrant, with 4 pointed segments *hairy only on the margins and under-surface*. Maquis. AA (local), GR, IA, IP, PN, TR. *C. elisabethae-carolae* is very similar to *C. flammula* but herbaceous and *scrambling* (*not* climbing) with *undivided* lower leaves; upper leaves with 3 leaflets. Flowers scented of orange blossom. A very rare and endangered endemic of Crete. CR (west).

Clematis cirrhosa VIRGIN'S BOWER An evergreen climber to 4 m with shiny green leaves, variably divided, 3-lobed or entire, toothed or not. Flowers cream, often red-spotted within, nodding and *bell-shaped*, to 20 mm long, and silky-hairy outside. Woods and scrub; common in the Aegean. AA, CR (+ KP), CY, GR (south), IA, IP, PN.

Clematis viticella A deciduous climber to 4 m with pinnately divided leaves with oval, untoothed segments. *Flowers blue or purple, 30–60 mm across*, opening widely; styles not feathery. Scrub and thickets; cultivated forms are also widely grown. CY, GR, SY, TR.

Aquilegia COLUMBINE

Herbaceous perennials with leaves spirally arranged. Flowers regular with 5 sepals and petals, each with a backward-pointing spur; stamens numerous; carpels 5(10). Fruit a follicle.

Aquilegia vulgaris COLUMBINE A variable, erect, often branched, hairy perennial to 60 cm (1 m). Leaves stalked and toothed or lobed. Flowers large, to 50 mm long, *blue–purple* (sometimes white or pink) with 5 similar tepals, the *petal-like segments elongated into erect, curving spurs* 15–22 mm long. Follicles 15–20 mm. Grassy, shady habitats and wetlands in the north only. GR.

1. *Clematis vitalba*
2. *Clematis vitalba* fruits
3. *Clematis flammula*
4. *Clematis cirrhosa*

Ceratocephala

Small annuals with trifoliate leaves with *narrow, forked segments*. Flowers yellow; stamens 5–10. Fruit an aggregate of achenes.

Ceratocephala falcata A short, hairy, rather tufted annual 20 mm–12 cm with all basal, long-stalked trifoliate leaves divided into narrow, forked segments. Flowers regular, solitary, yellow, (8)10–16 mm across; petals 1.5 x as long as sepals. Fruits borne in a teasel-like heads of long, upward-curving, spine-like achenes. Rather local in aquatic habitats. AA (rare), CR (west – rare), CY, GR, IA (Corfu), IP, LB, PN, SY, TR.

PLATANACEAE | PLANE TREE FAMILY

A small family of trees (with 1 genus), with peeling bark and simple, palmately lobed leaves. Flowers unisexual, regular; stamens 3–4; carpels 5–8. Fruit a hairy achene.

Platanus × hispanica (syn. *P. × hybrida, P. × acerifolia*) PLANE TREE A large, deciduous tree to 35(44) m with bark peeling in flakes, and 5–7 sharply lobed palmate leaves. Flowers borne in dense heads, often in pairs; fruit spherical and pendent, 20–35 mm across. Not native (of uncertain horticultural origin), widely planted on roadsides in the area. ***P. orientalis*** is similar, with the base of the leaves with numerous smaller lobes, the lobes longer than wide and toothed, and usually 3–6 fruits per cluster. Gorges and ravines; also widely planted. Throughout.

1. *Platanus orientalis* habit; (inset) leaves
2. *Buxus sempervirens*

BUXACEAE | BOX FAMILY

Evergreen trees and shrubs with simple, opposite leaves. Flowers borne in early spring, unisexual, small, borne in spikes or lateral clusters; sepals and stamens 4; styles 2–3. Fruit a capsule.

Buxus BOX

Shrubs with small, leathery, opposite leaves and unisexual flowers with 4 stamens. *Fruit a 2–3-horned capsule.*

Buxus sempervirens BOX An evergreen shrub to 5 m with downy, 4-angled shoots. Leaves small, oval, 13–25(30) mm long and 7–12(15) mm wide, often upward-swept, pale, sometimes hairy below. Monoecious; flowers borne in axillary and terminal inflorescences, greenish and inconspicuous. Fruit a shiny, green–yellow, ovoid capsule 8–11 mm long with 3 horns. Maquis and mountains on the northern mainland; dense thickets occur in northern Greece. GR, TR.

PAEONIACEAE | PEONY FAMILY

Perennial herbs with alternate, divided leaves. Flowers solitary, large and showy, with (normally) 5 sepals and 5–8(13) petals; stamens numerous. Fruit consisting of 2–8(9) fleshy follicles, with large black or brown seeds.

Paeonia PEONY

Robust, leafy perennials. Flowers large and conspicuous, regular, solitary; stamens ~140. Follicles variably hairy.

(a) Leaves with 9–21 broad leaflets. Flowers *reddish or purplish* (sometimes whitish).

Paeonia mascula An erect, bushy, usually hairless perennial to 60(80) cm with large leaves with *leaflets usually not further divided*; lower leaves normally with 6–13(16) broadly elliptic to oval leaflets 25 mm–12 cm cm wide with untoothed margins, rather thin; hairless beneath. Flowers soliatary, *red* and opening widely, 10–13(15) cm across with 5–7 petals; anthers yellow. Follicles in groups of 3–4(5), usually white-hairy. Rather rare, in light woodland and scrub. AA (local), CY, GR (southeast), IP, PN, SY, TR. **Subsp.** *mascula* has green stems, elliptic leaflets and *white* (rarely pink) flowers. Throughout the range of the species. **Subsp.** *hellenica* has pinkish stems *broadly* elliptic leaflets and dull *purplish-red* flowers. AA, GR, PN. *P. parnassica* is a similar, erect perennial with hairy stems and leaves typically divided into 9–13 elliptic to oval, *broad* segments (30–60 mm wide). Flowers *dark purple*. A very rare endemic of Mount Parnassos and Mount Ellkonas. GR.

(b) Leaves with 15–80 lance-shaped leaflets. Flowers *whitish*.

Paeonia clusii An erect perennial 20–35 cm, with typically hairless, purple-tinged stems. Leaves deeply dissected into 15–80 pointed, light green segments. Flowers 60 mm–10 cm across, rather open with 6–8 almost round petals, *white* or pinkish. Follicles 2–5, hairy. AA (Rhodes), CR (+ KP). **Subsp.** *clusii* has numerous (40–80), narrow, firm leaf divisions. CR (+ KP). **Subsp.** *rhodia* has fewer (15–40), broader and thinner leaf divisions. AA (Rhodes).

(c) Leaves with 15–20 leaflets. Flowers *bright ruby red*.

Paeonia peregrina An erect perennial 30–50 cm, typically with *hairless*, greenish stems. Leaves dissected into 15–20 glossy green segments, some *deeply toothed* (teeth to 10 mm long). Flowers 70 mm–10 cm across, cup-shaped, *bright ruby red; stigmas yellow*. Follicles 2(3–4), hairy. Deciduous oak forests. AA (Athos – rare, Thasos), GR (northeast), IA (Lefkada). *P. saueri* is similar but with *entire* leaflets (rarely lobed) and *red stigmas*. Oak woods on Mount Pangeon only; very rare. GR (north).

SAXIFRAGACEAE | SAXIFRAGE FAMILY

Annual or perennial herbs with alternate or all basal leaves (rarely opposite). Flowers regular, cosexual with 4–5 petals and 5–10 stamens; carpels 2(3). Fruit 2 follicles fused to form a capsule. Many species occur at higher altitude in the region that are not described here.

Saxifraga

Hairy, often rosette-forming annuals or perennials with simple or almost compound leaves. Flowers with 5 petals and 10 stamens; carpels 2. Fruit a capsule.

(a) Ovary superior (or virtually so).

Saxifraga rotundifolia A fragile, slightly glandular biennial to perennial with stalked basal leaves in a lax rosette; leaves 20–70 mm wide, round-kidney-shaped, heart-shaped at the base, with 11–17 lobes. Flowers white, often with purplish or yellowish dots, borne in lax panicles of 1–2 flowers per branch. AA (local), CR, GR, IA, PN, TR. **Subsp. *rotundifolia*** has a long rhizome, leaves with transparent margins and *distinctly purple-spotted flowers*. Woodland habitats. GR, IA, PN. **Subsp. *chrysospleniifolia*** is a biennial without a short (or absent) rhizome and leaves with stalks gradually expanded at the top and flowers with faint or absent dots. Rocky gorges. Throughout the range of the species.

Saxifraga hederacea A delicate, hairless or sparsely hairy winter annual 40 mm–10 cm. Leaves mainly basal with slender stalks, small, 6–13(20) mm across, slightly fleshy, oval to kidney-shaped, with 3–7 shallow lobes. Flowers *solitary* (or 2–4) borne in lax, leafy cymes on *delicate stalks* 5–12 mm; *petals 2–3 mm*, **w**hite or cream. Throughout except the north.

(b) Ovary inferior (or virtually so).

Saxifraga scardica A small, evergreen perennial, which forms *dense, hard cushions on bare rock*. Basal leaves borne in dense rosettes, rigid with rough margins, sharply keeled. Flowers 3–8(15) borne in *cymes* on reddish stems 50 mm–15 cm; petals 10–15 mm, oblong to spatula-shaped, white or pale pink, greatly exceeding the sepals. Predominantly an alpine, occasionally extending locally into rocky gorges at lower altitudes. GR, PN.

Saxifraga tridactylites An erect, *slender annual* 40 mm–12 cm, reddish and sparsely hairy. Leaves few, alternate or in lax rosettes, *withered during flowering*, divided into 3(5) lobes. Flowers 3–6 borne in diffuse cymes with bracts; sepals 1.5 mm; petals 2.5–3 mm, white. Throughout.

1. *Paeonia mascula*
2. *Paeonia parnassica* PHOTO: ARNE STRID
3. *Paeonia clusii* subsp. *clusii* PHOTO: NICK TURLAND
4. *Paeonia clusii* subsp. *rhodia* PHOTO: ELEFTHERIOS DARIOTIS
5. *Saxifraga rotundifolia*
6. *Saxifraga rotundifolia* subsp. *chrysospleniifolia*
7. *Saxifraga scardica*

CRASSULACEAE

Typically succulent annuals and perennials with alternate, opposite or whorled leaves. Flowers regular, star- or bell-shaped with 3–18(20) sepals and petals and an equal number of stamens, or 2 x as many. Fruit a cluster of follicles.

Crassula

Aquatic or terrestrial annuals and perennials, usually hairless with succulent leaves, often in fused pairs. Flowers with 3–5–numerous sepals, petals and stamens.

Crassula tillaea **MOSSY STONECROP** A tiny, often *dark red* moss-like annual just 10–50 mm. Leaves 1–2 mm, oval and crowded. Flowers borne in small groups in the leaf axis, *virtually stalkless*, 1–2 mm across with petals shorter than the sepals; white or pale pink; *sepals and petals 3* (rarely 4). Capsules with 1–2 seeds. Rocky or stony habitats. Throughout, except IP. *C. vaillantii* is similar but with rather sparser leaves and *flowers with 4 sepals and petals*; petals longer than the sepals. Capsules with 8–12 seeds. Seasonally wet habitats; rare and local. Throughout. *C. alata* is similar to *C. tillaea* but taller (30 mm–10 cm) with lance-shaped, spine-pointed leaves and distinctly *stalked* flowers. Rocky habitats in the south and east. AA, CR, CY, IP.

Sedum STONECROP

Succulent annuals or perennials with flat or cylindrical leaves. Flowers with 4–9–numerous petals and 2 x as many stamens. Numerous similar species; many rare or local species not included here.

(a) Leaves not, or scarcely, flattened; flowers greenish or yellow with (5)*6–9 petals*.

Sedum sediforme A very fleshy perennial to 60 cm with flowering and non-flowering shoots, woody at the base, reddish throughout. Leaves grey–green, later red, borne in spiralled rows. Flowers yellowish-white to bright yellow with 5–8 petals, borne in dense corymbs on tall stalks, the branches recurved when mature, concave in fruit. Common in dry rocky or sandy habitats. Throughout. *S. amplexicaule* (syn. *S. tenuifolium*) is a similar, slender perennial to 25 cm with small grey–green leaves closely overlapping and clasping the stem at the base, and *few flowers borne in 1-sided clusters*; petals yellow with reddish mid-veins; flowers with 6–8 parts. Throughout. *S. ochroleucum* is similar to *S. sediforme* but shorter, to 30 cm, and less robust. Leaves rounded and spurred at the base. Flowers *larger, to 20 mm across*, pale yellow; sepals 5–7 mm long, narrowly triangular and glandular-hairy; petals to 10 mm long; inflorescence flat-topped in bud. Ravines and cliffs, mainly in the north. AA (north), GR, IA, PN (north), TR.

(b) Leaves not, or scarcely, flattened; flowers greenish or yellow, normally with *5 petals*.

Sedum acre **WALLPEPPER** A succulent, tufted, bright yellow–green, hairless perennial with ascending to erect flowering shoots to 29 cm. Leaves small, 3–5 mm, oval and broadest at the base, blunt and overlapping. Flowers *bright yellow*, borne in small clusters; petals 5, 6–8 mm. Rocky and sandy habitats. GR, PN.

Sedum litoreum A short annual similar in general appearance to *S. rubens* (p. 209) with short, reddish stems, but with *slightly flattened, broadly oblong to spatula-shaped leaves* and pale *yellow flowers*; petals 2.5–3.5 mm; stamens 5. Rocky outcrops. Damp, sandy habitats, often coastal. Throughout.

1. *Crassula alata*
2. *Sedum sediforme*
3. *Sedum ochroleucum*
4. *Sedum acre*
5. *Sedum litoreum*
6. *Sedum album*

(c) Leaves not, or scarcely, flattened; *flowers white* (or flushed with pink, rarely blue) normally with 5(6) petals.

Sedum album WHITE STONECROP A short, hairless, tufted and loosely *mat-forming* perennial with creeping stems, with flowering stems to 20(25) cm. *Leaves alternate*, linear-cylindrical to egg-shaped, 4–12 mm long, *rather shiny and often reddish*. Flowers white (sometimes pink-tinged), borne in flat-topped clusters (cymes) of >20; with 5 petals 2–4(6) mm long; stamens 10. Rocky habitats and old walls in the north and centre. AA (rare), CR, GR, PN.

Sedum dasyphyllum THICK-LEAVED STONECROP A fleshy, grey–pink-tinged perennial with *glandular-hairy* stems to 10 cm. Leaves ovoid to almost spherical, 3–6(10) mm long, *mostly opposite*, and slightly flattened above. Flowers pink–white, with 5–6 petals 2.5–4 mm long. Rocks and walls. AA (north), GR, PN (north), TR.

Sedum caespitosum A tiny, *virtually hairless annual* 20–60(90) mm. Leaves *alternate*, stalkless, oblong to egg-shaped or cylindrical, blunt-tipped. Flowers few, borne in cymes, 4–5-parted; petals 2.5–4 mm, elliptic to lance-shaped, whitish or reddish with a green or red keel; sepals *fused*; stamens 4 or 5; anthers red or yellowish. Fruiting heads spreading and *star-like* when mature. Maquis. Throughout.

Sedum eriocarpum A small, often clumped, more or less *glandular* annual 40 mm–15 cm with leaves in *whorls* at the base, alternate above, oblong to narrowly elliptic-cylindrical, blunt and greyish or reddish (similar in general appearance to *S. rubens* but see stamens). Flowers borne in *glandular-hairy* terminal cymes, often with 2–3 branches; flowers 5-parted; petals white to pale pink; *stamens 10*, anthers dark red. Ripe fruiting heads spreading. Throughout. Populations on Crete usually referred to as the endemic **subsp. *spathulifolium***, which has slightly flattened, *broadly spatula-shaped leaves ending in a slight point*. Coastal sands and islets. CR + islets.

Sedum rubens RED STONECROP A *very small*, *erect annual*, usually to just 30 mm–9(15) cm, often sticky and glandular above, similar to *S. caespitosum* and *S. eriocarpum* but *leaves linear*, alternate, 7–16(20) mm long, greyish-green, tinged with red. Flowers whitish or pink with darker mid-veins, borne in small, compact clusters with erect-spreading to recurved branches; *petals 5, 4–5(6) mm long, 2 x the length of the sepals; stamens usually 5*. Fruiting heads spreading. Sometimes hybridising with *S. eriocarpum*. Rocky habitats and walls. Throughout.

(d) Leaves *flat* and fleshy, more or less in the form of *basal rosettes*; flowers borne numerously in *long*, cylindrical to narrowly pyramidal racemes.

Sedum cepaea A small, *hairless*, fleshy annual 15–28(40) cm, all leaves opposite or whorled, with white or pink flowers borne in long, lax, leafy inflorescences; petals 3–4(5) mm. Shady habitats on higher ground. AA (north), CR (rare), GR, IA, PN.

Sedum creticum A small, *glandular* annual to short-lived perennial, 50 mm–15 cm. Basal leaves in a lax rosette, spatula-shaped, green, greyish or reddish, 4 mm long. Flowers borne on most of the stem, forming *lax, narrowly conical panicles*; flowers 6–5-parted with petals 5–7 mm, tinged red; stamens 10, anthers red. Fruiting heads erect. CR (+ Antikythera and KP).

Sedum cyprium An erect, fleshy herb 90 mm–15(35) cm with basal leaves with entire (not papillose) margins, 30–60 mm long, 10–20 mm wide, forming rather loose, purplish rosettes. Flowers borne numerously in dull pink, *long, narrowly conical panicles*; petals 5, 2.5 mm long. Rocky crevices; an island endemic. CY. ***S. lampusae*** is similar but with basal leaves with minutely papillose margins, *large*, 40 mm–10 cm long, 20–40 mm wide. An island endemic. CY (especially the north). ***S. microstachyum*** is similar to the previous two species but with basal leaves thinly glandular-hairy, linear-spatular-shaped. Habitat and distribution as for *S. cyprium*. CY.

Telmissa

Differing from *Sedum* in having 3–5-parted flowers and *1-seeded capsules* (treated by many authors as indistinct).

Telmissa microcarpa (syn. ***Sedum microcarpum***) A hairless, *minute, reddish*, patch-forming annual 50 mm–12 cm. Leaves alternate, linear, semi-cylindrical, 4–10 mm, stalkless. Flowers 3–5-parted, borne in 2–4-branched cymes with 8–20 flowers per branch; petals pinkish-white, 1 mm; sepals 0.5 mm, blunt. Fruit capsule 1–1.5 mm, angular, with a single ovule. Rocky limestone hills and maquis. CY, IP (mostly the north), SY (especially the Syrian Desert), TR.

---

1. *Sedum dasyphyllum*
2. *Sedum caespitosum*
3. *Sedum eriocarpum* subsp. *spathulifolium*
4. *Sedum rubens*
5. *Telmissa microcarpa*

Umbilicus NAVELWORT

Perennial herbs with *round leaves joining their stalks at the centre of the blade* (peltate). Flowers borne in spike-like racemes; petals 5, stamens 2 x as many, fused to the corolla.

(a) Corolla whitish, pale yellow, greenish or reddish, with lobes much *shorter* than their tube.

Umbilicus rupestris NAVELWORT A fleshy, hairless perennial 15–30(60) cm with distinctive basal circular leaves with a central hollow, borne on long stalks; upper leaves smaller, more kidney-shaped and with rounded teeth. Flowers cylindrical with 5 petals and 2 x as many stamens, whitish-green or yellowish, sometimes pink, *cylindrical* and *drooping*, borne in long, tapered racemes in which the flowers occupy 2/3 the stem; flower-stalks 3–5 mm; corolla lobes equalling or longer than than their tube. Damp woods and rocks. AA, CR, CY, GR, IA, PN.

Umbilicus intermedius An erect, hairless perennial 10–50 cm. Basal leaves circular, 20–50 mm across, with scalloped margins; upper leaves linear, toothed. Flowers borne in dense racemes, simple or branched from the base; flowers spreading horizontally or drooping, with stalks 0.5–2 mm; sepals 1–1.5 mm; corolla greenish-white, often reddish; lobes 5–7 mm, broadly lance-shaped to oval, 1/4 the length of their tube. Limestone rocks, crevices, rocky deserts and woods in the east. IP, LB, SY, TR.

Umbilicus horizontalis A fleshy perennial, similar to *U. rupestris*, but with flowering stems with numerous crowded, linear leaves, and flowers held *horizontally* (not drooping); inflorescence stalk 13–70 cm, the flowers occupying 1/3 (to 1/2) the stem; flower-stalks 1–2 mm; corolla lobes considerably *shorter* than their tube. Various habitats including coastal. The commonest species on the islands; rare on the northern mainland. Throughout except the north and southeast.

(b) Corolla *bright yellow* with lobes more or less *equalling* their tube.

Umbilicus luteus A perennial 20–80 cm with round leaves to 80 mm across, withered during flowering. Inflorescence stout, unbranched (or with 1–3 basal branches); flowers borne on stalks 1–5 mm, almost *erect to horizontal*; corolla *large*, 9–14 mm, bright yellow, the lobes more or less equalling the tube. Mainly mountain woods. AA, CR, GR, IA, PN. ***U. albido-opacus*** is similar, with leaves to 25 mm across, with *spreading* to slightly drooping flowers 5–6.5 mm long on stalks 3–4 mm, the corolla lobes more or less equalling the tube. A rare island endemic. AA (Rhodes + Chalki).

(c) Corolla whitish, pale yellow, greenish or reddish, with lobes much *longer* than their tube.

Umbilicus chloranthus A slender perennial 10–30(70) cm. Leaves round. Inflorescence a narrow, *lax raceme with flowers occupying most of the stem* with few short lateral branches to 10 cm; flowers just 3–5 mm, borne on stalks 1–2.5 mm, horizontally spreading or slightly nodding, pale yellow; corolla lobes *much longer* than their tube. Rather rare. AA, GR, IA, PN (west), TR. ***U. parviflorus*** is similar but stouter, shorter and with flowers in dense panicles with numerous dense lateral branches. AA, CR.

CYNOMORIACEAE

A family (with 1 genus) of root parasites of members of the Amaranthaceae. Flowers unisexual or cosexual, borne in a dense, brush-like inflorescence; tepals (1)3–6(8); stamen 1. Fruit a small, 1-seeded nut.

Cynomorium coccineum MALTESE FUNGUS A highly distinctive blackish-red plant sprouting as a club-shaped structure to 25(30) cm, from an extensive underground rhizome system with lance-shaped scale leaves. Inflorescence cylindrical with very dense, tiny flowers; tepals 3–6(8); stamens solitary, exserted. Parasitic on shrubs in deserts and semi-deserts. Rare and local. IP, SY.

VITACEAE | GRAPE FAMILY

A large, primarily tropical family of climbers with leaf-opposed tendrils. Flowers small, borne in clusters; petals and stamens 5; style 1. Fruit a 1–4-seeded berry.

Vitis

Leaves simple, palmately lobed. Flowers with fused petals, falling as the flowers open.

Vitis vinifera GRAPE A climbing shrub to >10 m with alternate leaves and tendrils opposite; tendrils branched. Leaves long-stalked and palmately 5–7 lobed, coarsely toothed. Flowers small, greenish, cosexual, borne in clusters, pendent when mature. Fruit a berry (grape). Cultivated throughout. **Subsp. *sylvestris*** is the wild form, which differs in being dioecious (also hybridising with cultivated vines). Scattered across the region, mainly northern; absent from hot, dry areas.

---

1. *Umbilicus intermedius*
2. *Umbilicus horizontalis*
3. *Cynomorium coccineum*

ZYGOPHYLLACEAE

A mostly tropical family of plants with pinnately divided leaves with stipules. Flowers regular, with a 5-parted perianth and 8 or 10 stamens. Fruit fleshy or a capsule (sometimes splitting into 5 mericarps).

Tetraena

Shrubs and herbs (succulents in the region covered). Leaves opposite with 1–2 pairs of leaflets. Flowers solitary or in pairs, regular; stamens 10; ovary 3–5-parted. Fruit dry, splitting when ripe (a *schizocarp*).

Tetraena dumosa A fleshy *desert shrub* 40–90 cm, the young branches with adpressed, white hairs. Leaves opposite, with 2 fleshy, oblong-cylindrical, soon-falling leaflets 5–10 mm, nearly as long as their stalks. Flowers white, borne solitary in the axils; sepals 5, hairy, oblong; petals 5, oblong-spatula-shaped, 2.5 x as long as the calyx with circular-oval scales. Fruit with 4–5 angles or wings, often somewhat broader than the capsule. Dry hillsides and deserts. IP.

Tetraena alba A much-branched, *fleshy* shrub to 30 cm (1 m) tall. Leaves opposite, succulent, greyish, with 2 cylindrical leaflets 5–10 mm long and a similar stalk (resembling a leaflet). Flowers solitary, virtually stalkless, 5-parted; petals *white*, to 5 mm, borne on stalks shorter than the calyx. Fruit a fleshy spherical capsule 8–10 mm. Coastal habitats. Rare and strongly southern in the region. AA (Rhodes), CR (+ KP), CY. *T. coccinea* has flowers borne on stalks *longer* than the calyx and white to pink flowers. Capsule cylindrical. Deserts. IP.

Zygophyllum

Shrubs and herbs. Leaves opposite with 1–10 pairs of leaflets. Flowers solitary or in pairs, regular; stamens 5–10; ovary 3–5-parted. Similar to *Tetraena* but with fruit a *capsule* with chambers (locules) and undivided stamen appendages.

Zygophyllum fabago SYRIAN BEAN CAPER An erect, hairless, branched perennial 20–60 cm, woody at the base. Leaves stalked, alternate, greyish, often with a rudimentary stalk (rachis) projecting beyond the leaflets; leaflets *2*, rather fleshy, egg-shaped to elliptic, 20–35(40) mm. Flowers solitary in the upper leaf axils, seemingly in pairs; erect, stalked; sepals greenish and blunt, the inner with papery margins, 7–8 mm; petals equalling or slightly exceeding the petals, white or white and orange; stamens exserted. Fruit a pendent, broadly cylindrical, slightly 5-angled capsule 22–30(25) mm. Dry waste places, rocky and sandy habitats in the east. IP, LB, SY, TR.

Tribulus

Prostrate, annual herbs with leaves with 5–8 pairs of leaflets. Flowers yellow. Fruits generally spiny.

Tribulus terrestris SMALL CALTROPS A prostrate, hairy annual with long, trailing stems to 50(80) cm with pinnately divided leaves with 5–8 pairs of elliptic leaflets 6–10 mm, somewhat silver-hairy. Flowers yellow, with 5 petals 8–14 mm long, borne in the leaf axils borne on short stalks (4–5 mm). Fruits 8–10 mm, *with prominent, robust spines*. Dry, bare areas and disturbed sand dunes; frequent. Throughout.

Fagonia

Perennial herbs, often woody-based with opposite, often trifoliate leaves. Flowers borne solitary in the leaf axils with 5 free sepals and petals and 10 stamens. Fruit a 5-parted capsule.

Fagonia cretica A short, hairless to hairy spreading, prostrate perennial 60–70 cm, woody at the base. Leaves 6–25 mm, trifoliate, leathery, with spine tips. Flowers bright reddish-purple with 5 free petals 8–9.5 mm, borne solitary between pairs of spine-tipped stipules. Fruit a 5-angled, egg-shaped capsule 7–9 mm. Dry habitats, rare in the region. AA (south), CR, CY.

Fagonia mollis A variably hairy, woody-based subshrub 20–40 cm with numerous branched, angled and grooved stems with *spines*. Leaves with 3 bristle-tipped, oval-elliptic leaflets 8–12 mm, the terminal larger and broader than the laterals. Flowers pink; petals 20 mm (2–3 x as long as the sepals). Fruit an ovoid capsule, tapered at the base, 6–7 mm. Seeds oval, brown. Deserts. IP (south and east).

1. *Tetraena dumosa*
2. *Tribulus terrestris*
3. *Fagonia cretica*
4. *Fagonia mollis*

LEGUMINOSAE (FABACEAE) | PEA FAMILY

The third largest family of flowering plants with 6 subfamilies. Herbs or trees, usually with trifoliate or pinnately compound leaves. Flowers zygomorphic, with an upper petal (standard), 2 lateral wings, which lie on the side of the 2 lower, typically fused petals (keel), concealing the 10 stamens and single style; stamens 9 (sometimes all 10), fused into a basal tube. Fruit (legume), highly variable, often splitting and pod-like, or a nut.

SUBFAMILY CAESALPINIOIDEAE

Trees and shrubs, often with tendrils and 2-pinnately divided leaves. Flowers borne in racemes; calyx 2–5-lobed or sepals free; petals 5 (rarely 0, 2 or 6); stamens 10. Fruit a splitting pod.

Albizia

Trees with pinnately divided leaves. Flowers *regular* with *numerous*, free, conspicuously *protruding* stamens. Fruit a pod-like legume.

Albizia julibrissin ALBIZIA A deciduous tree 5–12(15) m with a broad crown. Leaves large, 20–45 cm, and 2-pinnately divided with numerous (20–30) oblong leaflets, hairy beneath. Flowers borne in large spherical heads of up to 50, long-stalked; corolla tubular with 5 even teeth; greenish-white with a *conspicuous fringe of long, pink stamens* 20–30 mm. Fruit oblong, 40 mm–20 cm, with prominent seeds. Native to Asia; commonly planted as an ornamental in towns.

Acacia ACACIA

Trees or shrubs with either 2-pinnately divided true leaves, or leaves reduced to a single blade (phyllode) and flowers borne in dense clusters; stamens numerous, free. Fruit a splitting or non-splitting pod-like legume. Native to South Africa and Australia; numerous forms and cultivars, often difficult to distinguish.

(a) Leaves 2-pinnately divided.

Acacia farnesiana A *spiny, deciduous* tree 1.5–4 m with *short spines* 10–30 mm. Leaves 2-pinnately divided into 10–21 pairs of leaflets 2–7 mm long and just 0.75–1.75 mm wide. Flowers bright yellow, 2.5–3 mm long, borne in *dense, spherical clusters* 10–15 mm across, solitary or in groups of 2–5 on stalks 35 mm long. Fruit *broadly* sub-cylindrical, rather straight and rigid, 40–70 mm. LB, SY.

Acacia dealbata SILVER WATTLE A bushy tree 12–15(30) m with smooth grey bark and silvery-hairy twigs and young leaves. Leaves *2-pinnately divided* with 10–26 pairs of primary divisions; leaflets 2–5 mm long, stipules rudimentary. Flowers *pale, bright yellow* and fragrant, borne in heads 5–6 mm across that form large terminal panicles that exceed the leaves. Fruit to 10 cm, linear-oblong and laterally flattened, bluish-brown. Native to Australia, widely planted as an ornamental and on roadsides.

1. *Acacia farnesiana*
2. *Acacia retinodes*
3. *Acacia saligna* habit; (inset) flower
4. *Acacia cyclops*

(b) Pinnately divided leaves *absent,* reduced to phyllodes (leaf-like blades); flowers borne in *spherical heads.*

Acacia retinodes SWAMP WATTLE A shrub or small tree 8–15(30) m with *spreading or upward-facing branches*. Phyllodes *narrow*, lance-shaped, wavy or straight, light green and leathery with a single vein, 60 mm–14 cm long and often *narrow*, 2–15 mm wide. Flowers pale yellow, borne in dense spherical heads 10–12 mm across, borne up to 10 in a raceme. Fruit 40 mm–12 cm, scarcely constricted between the seeds. Native to Australia, widely planted on roadsides throughout; common.
A. saligna (syn. *A. cyanophylla*) BLUE-LEAVED WATTLE is similar but shorter (<8 m) and with slightly broader phyllodes 5–50 mm wide, and flowers dark yellow, borne in *long, drooping branches*; heads 6–8 mm. Fruit distinctly constricted between the seeds, 50 mm–14 cm. Planted to stabilise dunes, as an ornamental, and naturalised. *A. pycnantha* is similar to the previous species but with similar sized but *sickle-shaped*, asymmetrical phyllodes. Planted and occasionally naturalized.

Acacia cyclops A shrub 2–4 m. Phyllodes elliptic to lance-shaped, blue–green and rather short, *broad and straight*, typically 40–90 mm long and >12 mm wide. Flowers yellow, borne in clusters or 30–40, 4–6 mm wide, solitary or in groups of 2–3 in the leaf axils. Fruit a pod 40 mm–10 cm, splitting to reveal large brown–black seeds to 70 mm across with a bright *orange-red* casing. Native to Australia, widely planted and naturalised.

Vachellia

Similar to *Acacia* (and widely still described under that genus) but with flowers always in spherical (capitate) heads and stipules *spine-like*.

***Vachellia tortilis* (syn. *Acacia tortilis*)** A shrub or small tree with a flat crown. Stipules spine-like, some hooked to 5 mm, mixed with long, straight spines to 10 cm. Leaves with 6–19 pairs of leaflets. Flowers cream to white, borne in spherical heads. Fruits 75 mm–15 cm, contorted or spiralled. Deserts in the southeast. IP. **Subsp. *raddiana* (syn. *Acacia raddiana*)**, also considered to be a distinct species, is often larger with an irregularly more rounded crown and distinct trunk; young branches virtually hairless. Forms savannoid desert landscapes. IP.

Leucaena

Acacia-like trees and shrubs with 2-pinnately divided leaves. Flowers borne in spherical heads. Fruiting pods flattened.

Leucaena leucocephala A shrub or small tree to 18 m with greyish bark, superficially similar to *Acacia*. Leaves rather large, to 35 cm long, oval in outline, 2-pinnately divided with 4–9 pairs of 11–22 leaflets 8–16 x 1–2 mm. Flowers *pale whitish-yellow*, numerous, borne in spherical heads 20–50 mm across. Fruits 14–26 cm long, borne in pendent clusters; *seeds clearly visible*, maroon–brown when mature. Seeds 18–22 per fruit, 6–10 mm long. Native to Central America, widely planted in hot, dry areas.

Delonix

Trees with 2-pinnately divided leaves. *Flowers large and showy, borne in terminal corymbs.* Fruiting pod woody and flattened.

Delonix regia A tree 10–15(18) m with a robust trunk. Leaves 20–60 cm long with stout stalks, 2-pinnately divided into 10–25 pairs, in turn divided into 12–40 pairs of leaflets 5–20 mm x 3 mm. Flowers numerous and conspicuous (the whole canopy appearing red when in bloom), borne in numerous corymbs 15–30 cm long; flowers 30 mm–13 cm across, *bright scarlet*; petals 50–65 mm long, the uppermost paler with dark red markings. Fruit flexuous, turning dark brown and *woody and persistent*; *large*, 30–75 cm long, 50–76 mm wide with 30–45 seeds. Planted in the hottest areas in the south and east.

Erythrostemon

Shrubs or trees with alternate, 2-pinnately divided leaves. Flowers typically yellow or red and showy. Fruit a laterally compressed pod. Planted ornamentals native to tropical and subtropical America.

Erythrostemon gilliesii (syn. *Caesalpinia gilliesii*) A shrub 1–4 m with 2-pinnately divided leaves 10–15 cm long, divided into 3–10 pairs, in turn with (6)7–10 pairs of leaflets 5–6 x 2–4 mm. Flowers borne in extended racemes to 20 cm long; flowers showy, with 5 yellow petals 20–35 mm and 10 *long-protruding, red stamens* 70 mm–12.5 cm, at first upward swept, later drooping. Fruit a flattened, sickle-shaped, splitting pod, covered in short, red glandular hairs, with few seeds. Widely planted in gardens in warmer parts.

Ceratonia CAROB

Shrubs and trees. Normally dioeceous. Flowers inconspicuous (greenish), *regular;* stamens 2–8. Fruit a non-splitting, pod-like legume.

Ceratonia siliqua CAROB TREE An evergreen shrub or tree to 10 m with leaves to 24 cm, pinnately divided into 1–5 pairs of dark green, rounded, leathery, untoothed leaflets; terminal leaflet absent. Flowers green or reddish and small with 5 sepals but without petals, borne in lateral racemes *directly from the trunk* in early autumn. Fruit large, 45 mm–20 cm, linear-oblong and laterally flattened, bluish-brown when ripe and pendent. Very common maquis, roadsides and field boundaries; a relic of cultivation.

---

1. *Vachellia tortilis* subsp. *raddiana*
2. *Leucaena leucocephala*
3. *Delonix regia*
4. *Delonix regia* fruits
5. *Erythrostemon gilliesii*

Senna

A mainly tropical genus of herbs, shrubs and tress with pinnately divided leaves. Flowers typically yellow with 5 sepals and petals, petals not fused in a tube; stamens 10 (7 fertile, all free). Fruit pod-like, several-seeded.

Senna didymobotrya (syn. *Cassia didymobotrya*) CASSIA An evergreen, vigorous and robust shrub to 5(9) m with leaves about 50 cm long with *14–18 pairs* of broadly elliptic leaflets. Flowers bright yellow, borne in large, dense, terminal *racemes with numerous blackish-brown buds* at the apex. Native to Tropical America, widely planted.

Senna corymbosa (syn. *Cassia corymbosa*) POPCORN CASSIA An evergreen or semi-evergreen, bushy and leafy shrub to 1(2) m with pinnately divided leaves with *2–3 pairs* of broadly elliptic leaflets, rounded at the base. Flowers numerous, bright yellow, borne in loose terminal clusters, each to 30 mm across with 5 unequal petals; stamens curved, and exceeded by their petals. Native to Tropical America, widely planted.

Cassia

A reduced genus, which formerly included many species now transferred to the similar, and closely related genus *Senna*.

Cassia artemisioides A shrub 1–2 m with *whitish to greyish*-green pinnately divided leaves with (3)4–6(10) pairs of *long, narrow* leaflets 10–40 mm long and just 1 mm wide, on stalks 5–15 mm long (leaves vaguely *Artemisia*-like). Flowers yellow, cup-shaped, borne in racemes of 4–12 in the leaf axils; petals 5, 7–10 mm long; stamens 10, all fertile. Fruit an elongated, scarcely curved, flattened pod 40–80 mm with 2–12 brownish to blackish seeds. Occasionally planted in warmer areas.

SUBFAMILY CERCIDOIDEAE

Trees and shrubs with leaves often 2-pinnately divided (sometimes modified into phyllodes) and alternate. Flowers borne in spikes, panicles or racemes; sepals and petals (3)5(6), free or fused; stamens often reduced to 3, 4 or 5, frequently >100. Fruit often a 1-valved, many-seeded pod.

Cercis

Shrubs and trees with *entire leaves*. Flowers with 10 free stamens. Fruit flattened with a dorsal *wing*.

Cercis siliquastrum JUDAS TREE A deciduous shrub or small tree to 10(12) m. Leaves heart-shaped 10–12 cm long, blunt, long-stalked and hairless. Flowers pink (rarely white), 14–20 mm, borne in clusters arising directly from older branches, among previous blooms' pods, before the leaves appear; calyx bell-shaped. Fruit pendent, linear-oblong, laterally flattened, 50 mm–10 cm, with a narrow wing along 1 edge. Widely planted as an ornamental. Throughout, except the north and many islands (not CR).

1. *Ceratonia siliqua*
2. *Ceratonia siliqua* male flowers
3. *Ceratonia siliqua* co-sexual flowers
4. *Ceratonia siliqua* fruits
5. *Senna didymobotrya*
6. *Senna corymbosa*
7. *Cassia artemisioides*
8. *Cercis siliquastrum*
9. *Cercis siliquastrum* fruits

SUBFAMILY PAPILIONOIDEAE

A diverse subfamily of trees, shrubs and herbs, often with tendrils. Leaves variously pinnately or palmately divided, or with 1 or 3 leaflets. Flowers solitary or in racemes or panicles; sepals (3)5, fused; petals (0)5(6); stamens typically 10. Fruit variable, often pod-like.

Robinia

Trees with pinnately divided leaves. Flowers borne in pendent racemes; stamens 10, inserted (9 inferior, united into a tube, 1 free). Fruit a 2-valved, splitting, pod-like legume.

Robinia pseudoacacia FALSE ACACIA A deciduous tree to 25(52) m with irregularly fissured, grey-brown bark, and spine-like stipules on younger branches. Leaves 10–25 cm long, pinnately divided with 9–19 leaflets, more or less hairless. *Flowers white*, scented, with a standard yellowish at the base, borne in *pendent* racemes. Fruit linear-oval, 50 mm–10 cm long, flattened. Native to North America, commonly planted as an ornamental. Throughout; frequently naturalised. **R. viscosa** is similar but with pale pink to mauve flowers. Planted.

---

1. *Robinia pseudoacacia*
2. *Anagyris foetida*
3. *Calicotome villosa*
4. *Calicotome villosa* in fruit
5. *Cytisus hirsutus* subsp. *polytrichus*
6. *Spartium junceum*

Anagyris

Large shrubs with trifoliate leaves. Yellow flowers borne in short racemes; stamens 10. Fruit a pod-like legume.

Anagyris foetida BEAN TREFOIL A poisonous, deciduous, unpleasant-smelling shrub 2–4 m. Leaves trifoliate with narrowly elliptic leaflets 6–40 mm long, silvery-hairy below and with papery stipules 5–10 mm. Flowers yellow, borne in short clusters; calyx bell-shaped, bluish. Fruit 60 mm–20 cm long. Dry, rocky places. Throughout (rare in the north).

Calicotome

Spiny shrubs with simple, inconspicuous leaves. Flowers with tubular calyx with 5 teeth; stamens 10, fused in a tube. Fruit a pod-like legume.

Calicotome villosa HAIRY THORNY BROOM A superficially gorse-like, erect shrub 1.5–3 m with slender spines on the branches. Leaves trifoliate, silver-hairy below, leaflets 4.5–15 mm. Flowers yellow and large, to 18 mm, borne in umbrella-shaped clusters of 3–5; calyx tubular with 5 teeth, the upper part falling in flower to leave a cup-shaped structure. *Fruit woolly-hairy* (hairs to 2 mm), 21–41 mm long. Locally common on maquis and garrigue. Throughout (absent from CY).

Cytisus BROOM

Spineless shrubs with leaves with 1 or 3 leaflets. Flowers yellow (sometimes white), borne in racemes; all 10 stamens in a tube. Fruit several–many-seeded.

Cytisus hirsutus A very variable (with many forms described), spreading to erect, softly to *woolly-hairy* perennial with branches 20 cm–1(2) m. Leaflets (4)6–20 x (2)4 x 10(18) mm long, oblong to elliptic. Flowers yellow or pinkish yellow, the standard sometimes brown-spotted. Fruit 25–40 mm, linear and variably hairy. GR, TR. Subsp. *polytrichus* has a prostrate habit and densely woolly-hairy fruits. Rocky ledges on mountains and in gorges. GR.

Cytisus scoparius BROOM A spineless, much-branched shrub 1–2.5 m, with long, slender and flexible stems, normally *5-angled and ridged*, hairy or not. Leaves small, trifoliate (1-foliate and stalkless on young branches); oval-elliptic, 11–14 mm *with short stalks* (13 mm). Flowers large, 15–20 mm, golden yellow, *solitary or paired*. Fruit oblong, compressed, 25–50 mm and *hairy along the margins only*. Woods on acid soils. GR (north).

Cytisus villosus A shrub 1.5–3 m similar to *C. scoparius* with *5-angled*, hairy branches, stalked leaves (stalks 10–17 mm), all trifoliate, with elliptic leaflets 20–47 mm. Flowers yellow and large (17–22 mm) *with darker red markings on the base of the standard petal*, borne solitary or *2–3(4) in each leaf axil*, forming lax, leafy clusters. Fruit 35–50 mm. Woods and rocky slopes; local. AA (local), GR, IA (Corfu), PN, TR.

Spartium

Spineless, broom-like shrubs with stiff, rush-like branches; leaves simple, often absent or inconspicuous. Flowers with stamens 10, fused in a tube. Fruit many-seeded.

Spartium junceum SPANISH BROOM A large, spineless, broom-like shrub to 3 m with cylindrical, blue–green, rush-like stems. Leaves sparse, linear-oval and soon-falling, 15–30 mm. Flowers large, 20–28 mm, and bright yellow, solitary but in large numbers; sweetly scented; *calyx spathe-like*. Fruit 40 mm–12 cm, flattened. Common on dry slopes and in woods. Throughout.

Genista GREENWEED

Spiny or non-spiny shrubs with simple or trifoliate leaves and yellow flowers; *calyx tube equalling or exceeding the lips*; stamens 10, fused in a tube. Fruit 1–several seeded.

(a) Spineless shrubs with simple leaves. Flowers borne in *racemes* (terminal or axillary).

Genista tinctoria DYER'S GREENWEED A tufted, *erect or ascending* small shrub 60 cm (1 m) with simple, linear-lance-shaped, stalkless, hairless leaves to 30 mm. Flowers 10–15 mm, yellow, borne in rather lax, *long racemes*, terminal or in the axils; flowers hairless. Fruit oblong, flat and hairless, 15–30 mm. GR.

Genista sakellariadis A *prostrate* shrub with flexuous, ascending branches. Leaves *simple*, 4–12 x 1–4 mm, elliptic, hairy on both sides, stalkless. Flowers yellow, borne singly on stalks 3–5 mm long in racemes in the axil of each bract; corolla 8–12 mm; standard broadly oval. Fruit narrowly oblong and hairy. An endemic of rocky ledges on Mount Olympus. GR.

Genista carinalis A short, much-branched, often tufted or spreading shrub 10–40 cm. Stems sparsely hairy. Leaves simple, narrowly elliptic. Flowers short-stalked, borne in terminal racemes; calyx with teeth longer than its tube; corolla yellow, small (6–9 mm), hairless, the keel much longer than the standard. Fruit *ovoid* and sparsely hairy, 1–2-seeded. Absent from islands (except those adjacent to the mainland). GR, PN, TR.

(b) Spineless shrubs with *winged stems*. Flowers borne in dense terminal *racemes*.

Genista sagittalis A *prostrate* dwarf shrub with woody, mat-forming stems and simple or scarcely branched flowering stems 10–50 cm; stems *distinctly winged*; wings *entire* (not lobed), constricted at the nodes. Leaves 5–10 x 4–7 mm, elliptic and hairy beneath. Flowers yellow, borne in *congested terminal racemes*; corolla 10–12 mm; standard hairless, equalling the wings and keel. Fruit 14–20 mm. GR.

(c) Spineless shrubs with simple leaves. Flowers *solitary* in the axils of leafy bracts (not in distinct racemes).

Genista anatolica A much-branched dwarf shrub just 50 mm–20 cm. Young twigs greyish with dense, adpressed hairs. Leaves simple, linear to narrowly elliptic, tapering to an abrupt point. Flowers yellow, solitary, on short stalks, in the axils of leafy bracts, forming rather indistinct inflorescences; corolla 9–10 mm, with a keel *much exceeding* the standard. Fruit 1-seeded, ovoid. AA (north), TR.

(d) *Spiny* shrubs with *trifoliate* leaves. Flowers *solitary* in the axils of leafy bracts (not in distinct racemes).

Genista acanthoclada A *dense, rigid, often mound-forming* shrub 40 cm (1 m) with twigs forming straight spines. Leaves small, virtually stalkless, with 3 linear leaflets. Flowers solitary in the axils; calyx with lobes about as long as their tube; corolla 8–10 mm, bright yellow, the standard slightly shorter than the keel. Fruit ovoid, with adpressed hairs, 1-seeded. Common to subdominant in the Aegean. AA, CR, PN. *G. fasselata* is similar but with stouter twigs with more robust spines with blackish tips and hairless calyx and corolla; calyx lobes 1/3 the length of their tube. Strongly eastern, where it often replaces the previous species. AA (east; common on KP island group), CY, IP.

LEGUMINOSAE (FABACEAE)

1. *Genista tinctoria*
2. *Genista sakellariadis*
3. *Genista sagittalis*
4. *Genista acanthoclada* habit
5. *Genista acanthoclada*
6. *Genista fasselata*
7. *Retama raetam*

Retama

Rush-like, branched, spineless shrubs with slender, alternate branches and simple leaves. Flowers 10 stamens, fused in a tube. Fruit 1–2-seeded, (normally) non-splitting.

Retama raetam WHITE BROOM An erect, much-branched, spineless shrub to 2 m with pendent lateral branches, silvery when young. Leaves sparse, linear and silvery-hairy, soon-falling. Flowers *borne abundantly along drooping branches, white*, to 17 mm, borne in dense racemes; sweetly scented; calyx soon-falling. Fruit club-shaped, to 20 mm long and beaked. Coastal sands, deserts and maquis. IP. **R. monosperma** has white flowers borne in lax, lateral clusters with a standard petal 9–10.5 mm long and fruits rough, 10–22 mm. Mainly near the coast. AA, GR, PN.

Adenocarpus

Spineless shrubs with trifoliate leaves. Flowers orange–yellow; stamens 10, fused in a tube. Fruit many-seeded, splitting.

Adenocarpus complicatus ADENOCARPUS An erect, sparsely hairy, spineless shrub to 3 m, rather broom-like but leafy, with leaves trifoliate with narrow elliptic, silkily-hairy leaflets 6–13 mm, and flowers *orange–yellow*, the standard 10–15 mm, borne in *long, lax terminal clusters* of >7; calyx tubular, with a bilobed upper lip, and trilobed lower lip. Fruit oblong-lance-shaped, to 30 mm. AA (north), GR (north).

Lupinus LUPIN

Herbs with palmate leaves. Flowers borne in long, conspicuous terminal racemes; stamens 10, fused in a tube. Fruit with 2–many seeds, splitting.

(a) Flowers yellow.

Lupinus luteus YELLOW LUPIN A robust, hairy annual to 1 m. Leaves palmately lobed with 5–9 oblong leaflets, flowers bright *yellow*, 15–18 mm, borne in long whorls along the raceme, scented. Fruit 40–60 mm, densely hairy, black when ripe. Sandy places, and cultivated for fodder. Infrequent in the region. GR (northeast), PN.

(b) Flowers bright blue or bicoloured.

Lupinus pilosus A soft-hairy annual with brownish hairs, especially below, and leaflets *broadly oblong to lance-shaped*. Flowers borne in whorled racemes; corolla bright *blue* to bicoloured, 15–20 mm; calyx with a 2-lobed upper lip. Fruit short and broad, to 50 mm long. Seeds few and large, 9–12 mm. AA, CR, IP. **L. angustifolius** has sparse, sub-adpressed hairs, *narrowly linear* leaflets and smaller flowers 11–13 mm; calyx deeply 2-lobed. Fruit to 40 mm; seeds small (6 mm), pale grey with brown variegation. Throughout. **L. gussoneanus** has brownish hairs throughout, and leaflets broadly oblong to lance-shaped. Flowers 10–14 mm; calyx deeply 2-lobed. Fruit to 50 mm; seeds to 6 mm, creamy-brown with dark spots. Throughout.

(c) Flowers whitish or cream to pale blue upon opening.

Lupinus albus WHITE LUPIN An erect, sparsely hairy annual to 50 cm–(2) m with palmate leaves with 5–7(9) leaflets, hairless above. Flowers *alternate*, white to pale blue, *with a dark blue-tipped keel*, 18–20 mm long, borne in lax, terminal racemes; corolla 12–16 mm; upper lip of calyx just *shallowly* toothed. Fruits long-hairy, 60 mm–13 cm. **Subsp. albus** has *white* flowers. Seeds pale brown. Widely cultivated. Throughout (not CY). **Subsp. graecus** has *pale blue* flowers and *dark variegated* seeds. Throughout.

Lupinus hispanicus **(syn. *L. gredensis*)** A woolly-hairy annual 25–80 cm. Leaflets 40–60 mm, egg-shaped to oblong, hairless except along the margins. Flowers *yellowish-cream*, ageing pink–purple, borne in regular whorls in racemes 50 mm–16 cm; calyx *deeply 2-lobed*. Fruit 40–50 mm, short-hairy, yellow, ageing black. Seeds 4.5–6 mm, reddish-brown, variegated and warted. Hillsides and river beds. Very rare and local; distribution disjunct across the Mediterranean (more common in the west). AA (Lesbos, Samos), GR (northeast – rare), IP (Golan Heights – rare), TR (west).

Lupinus palaestinus A sparsely hairy annual to 1.2 m. Leaflets egg-shaped to linear. Flowers *pale yellow, tinged with blue*, borne in *remote whorls* in long racemes to 26 cm; bracts linear, soon-falling; flower-stalks scarcely shorter than the calyx. Fruit 70 mm–16 cm, broad, very hairy, with 3–4(6) seeds. Local in sandy habitats in the southeast. IP (Sharon and Philistine Plains and the coastal Galilee).

Colutea BLADDER SENNA

Deciduous shrubs. Flowers yellow or orange, borne in axillary clusters with a broad standard and blunt keel; stamens 10, 1 free. Fruit *strongly inflated* and later papery.

Colutea arborescens BLADDER SENNA A lax, much-branched deciduous shrub 2–3(5) m with pinnately divided leaves with 3–6 pairs of oval to elliptic, virtually hairless leaflets 17–23 mm. Flowers yellow, borne in lax axillary racemes of 3–7, each 16–20 mm long; calyx 6.5–8 mm long, the lower teeth to 0.7–2 mm long; *ovary hairless*; wings flat, nearly as long as the keel. Fruits very distinctive: 52–55 mm, *strongly inflated* and pink to red-flushed, later brown and papery; hairless except for along the keel. Widespread in light woodland and maquis. GR (+ adjacent islands), IA, PN. ***C. insularis*** is very similar but with leaves *with adpressed hairs* on both sides and longer flowers to 25 mm. AA (Rhodes). ***C. cilicica*** is an eastern form with folds of the standard petal almost obsolete, and the rolled (not flat) wings just *exceeding* the keel. IP (Mount Hermon – rare), LB, SY, TR.

1. *Lupinus pilosus*
2. *Lupinus hispanicus*
3. *Lupinus palaestinus*
4. *Colutea arborescens*

Astragalus

Annual or perennial herbs with pinnately divided leaves terminating in a single leaflet. Flowers borne in lateral clusters; keel blunt (not toothed or pointed); stamens 10, 1 free. Fruit variable (sometimes inflated). A very large genus of which only the most widespread or conspicuous are described here.

(a) Typically woody-based perennials with vegetative shoots during flowering; *flowers borne on long stalks* arising more or less from the stock.

Astragalus monspessulanus MONTPELLIER MILK-VETCH A perennial herb to 20 cm with more or less prostrate *rosettes of leaves*, with 7–21 oval leaflets 3–11 mm, adpressed-hairy beneath. Flowers 17–25 mm, mauve, borne in dense, ovoid clusters on long, slender, spreading stalks. Fruit long and cylindrical, 20–30(50) mm. Rocky habitats, especially in mountains in the north. GR (+ adjacent islands: common), IA (Corfu), PN. *A. spruneri* is very similar but with *fewer*, 6–10(12) pairs of leaflets, silvery when young. Fruit smaller, 14–20 mm long, narrowly ellipsoidal. AA, GR, TR.

Astragalus cyprius A tufted perennial with a woody stock, short stems (<30 mm) and mostly basal, crowded leaves divided into 5–12 pairs of *silvery* leaflets to 20 mm. Flowers rather similar to *A. suberosus*, often few (3–12) borne on stalks 30 mm–15 cm, in congested clusters; corolla 15–20 mm, *dirty pink or yellowish*; calyx cylindrical, 8–10 mm. Fruit erect or spreading, *narrowly cylindrical*, 20–30 mm long and straight or upcurved, scarcely compressed. CY.

Astragalus cretaceus A tufted, *virtually stemless* perennial with a woody stock. Leaves to 14(30) cm with (15)20–35 pairs of oblong leaflets 5–8(10) mm with adpressed hairs. Flowers *crimson–purple to brick red*, 16–18 mm, borne in ovoid, dense heads of 15–25; calyx 12–14 mm, sparsely hairy, with linear teeth. Fruit 8–12 mm, oval-lance-shaped, flattened and densely woolly-hairy, with a short, straight beak. Stony mountain slopes and deserts. IP (central – rare), SY, TR.

Astragalus suberosus A short-lived perennial (sometimes annual) with a woody stock and stems to 30(40) cm with simple black or white spreading hairs. Leaves divided into 7–14 pairs of leaflets. Flowers (5)8–12(16), borne in congested clusters, which elongate in fruit, on stalks to 10 cm, equalling or exceeding their subtending leaves; calyx cylindrical, 10–15(16) mm, with adpressed, black hairs; corolla typically *pink* (cream to mauve), 17–26(27) mm. Fruit (15)17–30(38) mm long, narrowly ovoid to cylindrical, upcurved at the tip. AA (rare), CY, GR (+ adjacent islands), IA, PN, TR.

(b) Sprawling annuals. Flowers *yellowish or whitish*. Fruits long, slender, and strongly *curved*.

Astragalus hamosus A spreading annual, branched from the base with stems to 40 cm. Leaves divided into 6–12 pairs of widely spaced oblong to broadly lance-shaped leaflets. Flowers borne in racemes of 2–15(12); calyx 5–6(7) mm with white and dark adpressed hairs; corolla inconspicuous, pale yellow, 7–11 mm. Fruit 30–40 mm, *cylindrical, strongly sickle-shaped* and long-persistent, densely adpressed-hairy. Dry, bare habitats; common in the south. Throughout (except the far north).

Astragalus intercedens A spreading to prostrate desert annual to 20 cm. Leaves 10–30 mm, with 4–6(8) pairs of short-hairy, narrowly oval, notched leaflets. Flowers yellowish-white; calyx 6–8 mm; corolla 10–12(15) mm. Fruits 25–40 mm, erect, linear, compressed, evenly curved, with white, *long* hairs, hooked at the tips. Very rare and local, in deserts. IP (Judean Desert, Dead Sea valley, Negev hills and Eilat).

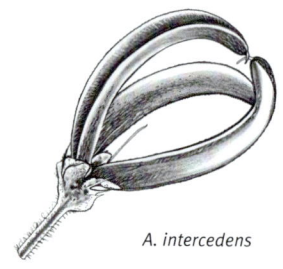

A. intercedens

LEGUMINOSAE (FABACEAE) 227

(c) *Erect* annuals or perennials. Flowers white–cream. Fruit cylindrical or oblong.

Astragalus odoratus LESSER MILK-VETCH An erect to ascending perennial to 30 cm. Leaves pinnately divided with numerous (19–29) leaflets. Flowers 9–12 mm, white–cream to yellow (sometimes flushed with dull violet in bud), borne in rather dense racemes. Fruit 8–9(10) mm, cylindrical, laterally compressed and sparsely short-hairy; beak 2 mm. Grassy and aquatic habitats. GR (north-central), TR.

Astragalus boeticus A hairy, erect annual with pinnately divided leaves with 6–13 pairs of oval leaflets, notched at the tip, hairless above, and slightly hairy below. Flowers 2–5, *pale sulphur yellow–white*, 8–11(14) mm, borne on dense racemes on stalks as long as the leaves or slightly shorter; wings longer than keel. Fruit 20–35(60) mm, oblong, triangular in section, shortly hairy, borne in dense clusters erect at first, then drooping. Sandy places and cliff-tops. AA, CR, CY, GR (rare), IA (rare), IP, PN.

1. *Astragalus monspessulanus*
2. *Astragalus cyprius*
3. *Astragalus cretaceus*
4. *Astragalus hamosus* flowers
5. *Astragalus hamosus* fruits
6. *Astragalus boeticus*
7. *Astragalus intercedens*

(d) Sprawling annuals. Flowers (typically) *pinkish, reddish or purple.* **Fruits borne in spreading,** *star-like* **clusters.**

Astragalus stella A short, sprawling, densely hairy annual with long-spreading to ascending stems to 35 cm. Leaves with 5–11 pairs of oval leaflets, hairy on both surfaces. Flowers 6–14 yellowish, or reddish-violet, (5.5)8–11 mm, borne on dense racemes on rather straight stalks ½ to the same length as the leaves; calyx 5–7 mm. Fruit 10–15 mm, more or less erect, almost straight, laterally compressed, *borne in star-shaped clusters*, each swollen at the base. Very local, on disturbed ground. GR (+ Euboea), TR. ***A. epiglottis*** is similar but with just 3–7(10) leaflets, *very small* flowers just 3–4 mm, with *5 stamens, not 10. Fruit 5–9 mm, distinctly triangular-heart-shaped*, borne in spiky clusters. Local. AA, CR (rare), CY, GR (uncommon), IP, LB, PN, SY, TR.

Astragalus sinaicus A *small*, sprawling annual branched from the base with stems to 12(20) cm, similar to *A. stella* but sparsely hairy and with fewer flowers. Leaves divided into 6–10 pairs of leaflets. Flowers 3–7 borne on virtually *stalkless* in the leaf axil (rarely stalked to 40 mm) in the leaf axils; calyx 4–6 mm with black and white hairs; corolla pale mauve or violet, 6.5–8.8 mm. Fruits 8–12(16) mm, spreading to form a star, *with long, spreading hairs*. Mainly southern and eastern. AA, CR (rare), CY, GR, PN, TR. ***A. asterias*** is similar but with fruits *without* long-spreading hairs. CY, IP.

Astragalus tribuloides A spreading, adpressed silky-hairy annual to 15 cm. Leaves to 10 cm, with 6–10(17) pairs of rather crowded, elliptic-oblong leaflets 5–10 mm. Flowers 6–8 mm, whitish-violet, borne in heads of (2)3–6, almost stalkless in the axils; calyx 4–5 mm, tubular with awl-shaped teeth. Fruits 5–10 mm, 2–3(6) per head, diverging (star-like), with adpressed hairs, oblong-triangular, slightly arched and pointed. Mainly sandy deserts. IP (south), SY. ***A. schimperi*** is similar but with stalked flower-heads and fruits with spreading hairs. Deserts. IP, SY.

Astragalus sesameus A sprawling annual to 36 cm, with hairy stems and rather hairy leaves, similar in most respects to *A. stella* but *with racemes scarcely stalked* and *more erect* and flowers 7–9 mm, bluish-violet (not reddish-violet). Fruit 9–17 mm. IA (records further east probably refer to *A. sinaicus*).

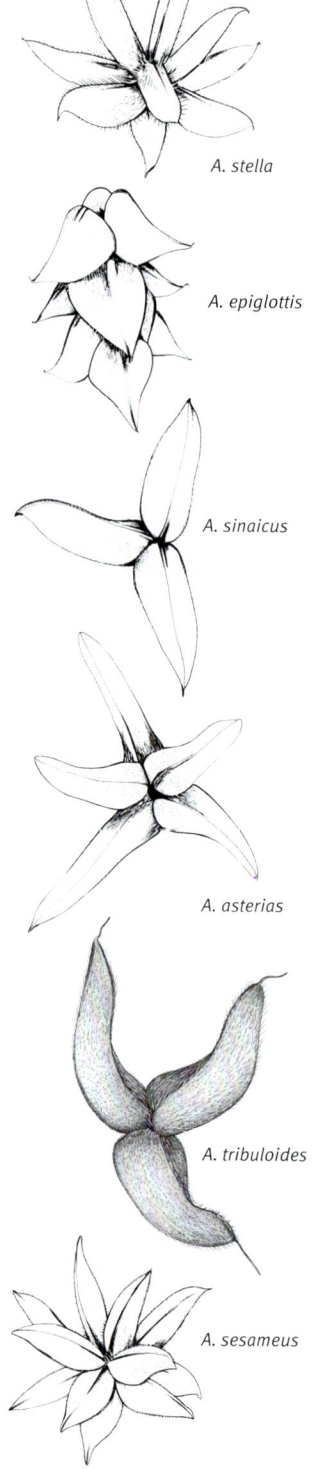

A. stella
A. epiglottis
A. sinaicus
A. asterias
A. tribuloides
A. sesameus

1. *Astragalus epiglottis*
2. *Astragalus tribuloides*
3. *Astragalus schimperi*
4. *Astragalus echinatus*
5. *Astragalus pelecinus* flowers
6. *Astragalus pelecinus* fruits
7. *Astragalus sanctus*
8. *Astragalus palaestinus*
9. *Astragalus callichrous*

(e) Sprawling annuals. Flowers (typically) *pinkish, reddish or purple*. **Fruits with** *numerous thick, flattened, spreading* **hairs.**

Astragalus echinatus A somewhat hairy, sprawling annual to 20(45) cm, with leaflets in 5–9(12) pairs, notched at the tip, hairy beneath and hairless above, 4–11 mm; stipules free, green. Flowers 10–16(20), purplish, 8 mm, borne in very dense racemes with stalks equalling or longer than the leaves; calyx 4–6 mm. Fruit 7–9 mm, oval-triangular, flattened, with *dense, spreading hairs*. AA, CR (rare), PN, SY, TR.

(f) Sprawling annuals. Flowers (typically) *pinkish, reddish or purple*. **Fruits long, strap-shaped, with conspicuous** *saw-like* **margins.**

Astragalus pelecinus (syn. *Biserrula pelecinus*) A branched, sprawling to ascending annual to 25(40) cm, sparsely white-hairy. Leaves with 7–15 elliptic leaflets. Flowers 2–5(11), borne on stalks rather shorter than their subtending leaves; calyx 3.5–4.5 mm; corolla *small*, just exceeding the calyx, blue–mauve. Fruit 15–40 mm, strongly flattened with conspicuous *saw-like margins*, hairless or sparsely hairy. Throughout.

(g) Sprawling to ascending annuals. Flowers (typically) *pinkish, reddish or purple*. **Fruits long and slender.**

Astragalus sanctus An adpressed, silky-hairy, slightly *woody*-based perennial 20–30 cm, with flexuous, brittle stems. Leaves 50 mm–10 cm, with (4)5–7 pairs of oblong to linear-oblong or elliptic-oblong leaflets 8–16 mm. Flowerheads borne on stalks as long as the leaves; flowers lax, 5–15(16) per head; calyx with black and white hairs; corolla 25 mm, yellow, flushed dull purple, almost 2 × as long as the calyx. Fruit 60–75 mm, deflexed, linear, somewhat compressed, *almost semicircular* with adpressed black and white hairs. Desert scrub. IP (mainly south-central).

A. sanctus

1. *Astragalus macrocarpus*
2. *Astragalus aleppicus*
3. *Astragalus angustifolius* subsp. *balcanicus*

Astragalus palaestinus An annual or perennial, branched from the base, to 50 cm. Flower-heads borne on stems longer than the leaf stalks; flowers pinkish, *large 18–20 mm*, with a *strap-shaped* (ligulate) standard petal; calyx 10–12 mm, reddish, with prominent *black and white hairs*; teeth shorter than their tube. Fruit 18–30 mm, *blackish*, inflated, curved and strongly grooved, sparsely spreading-hairy to hairless (not fleecy). Maquis and garrigue. IP (mainly northern), SY.

Astragalus callichrous A soft-hairy, spreading annual to 20(30) cm. Leaves to 10 cm, with 5–8(10) pairs of oblong-elliptic to *linear*, blunt leaflets 5–10(15) mm. Flowers *bright violet–purple*, rather large, 14–20(24) mm, borne in heads of 5–8(15); standard gradually tapered; calyx 6–9 mm. Fruits 25–55 mm, curved, grooved, cylindrical, with spreading hairs, hooked at the tips. A Saharo-Arabian species of fields, bare, dry lowland habitats, and deserts. IP, SY.

(h) Erect, leafy perennials with flowers borne loosely *around the main stem*. **Fruits ovoid to egg-shaped, inflated.**

Astragalus macrocarpus A leafy, adpressed-hairy, clump-forming perennial 30–60 cm (1 m). Leaves with 12–15(22) pairs of oblong-elliptic leaflets 10–20 mm. Flowers yellow, borne in lax racemes around the main stem; calyx sparsely white-hairy with lance-shaped lobes ½ as long as the tube; corolla to 35 mm, with an oblong to spatula-shaped, notched standard, exceeding the straight wings; keel shorter than the wings. Fruit 50–60 mm, ovoid, *inflated*, woody, with a sharp, straight beak to 10 mm. Fields and roadsides. CY (very rare, as **subsp.** *lefkarensis*), IP (frequent in the north, rare elsewhere), LB, TR.

Astragalus aleppicus A leafy, white-hairy, clump-forming perennial, similar to *A. macrocarpus* in general appearance. Leaves with 15–23(30) pairs of elliptic to circular leaflets, sometimes notched, 5–10 mm. Flowers borne 1–4(5) in stalkless racemes around the main stem; calyx 10–13 mm, white-hairy with slender, lance-shaped teeth 1/3 as long as their tube; corolla 25–30 mm, whitish with yellow and green markings; keel somewhat longer than the wings. Fruit 25–30 mm, deflexed, white-hairy, compressed ovoid, with netted veins and a rather long, straight beak (10–15 mm). Mainly dry fields and desert scrub. IP (most common in the northern Negev), LB, SY, TR.

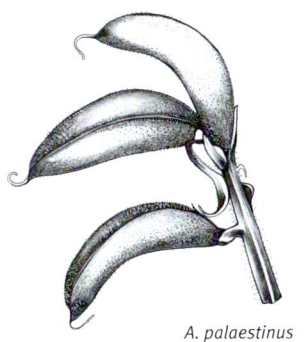

A. palaestinus

A. callichrous

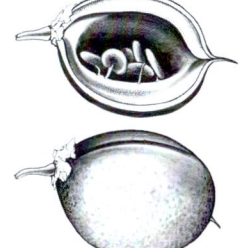

A. macrocarpus

A. aleppicus

(i) Densely spiny, cushion-like shrubs.

Astragalus angustifolius A variable, *compact*, woody, very spiny shrub to 40 cm (1 m) across. Leaves with 6–10 pairs of leaflets 3–5 mm long, narrowly elliptic, covered in short adpressed hairs; leaves soon-falling to leave naked spines. Flowers borne in racemes of (1)3–12; calyx with short black and white hairs; *corolla white or cream*, the keel sometimes purplish, 17–22 mm. Fruit 8–12 mm, exceeding the calyx, broadly cylindrical with dense white hairs. Rocky mountain habitats, sometimes dominant, forming large swathes of spiny thicket vegetation. **Subsp. *angustifolius*** has rounded leaves and racemes of not more than 3–6 flowers. TR. **Subsp. *balcanicus*** has abruptly pointed leaflets and racemes of up to 8 flowers. GR (north). Other regional subspecies have been described, the distinction of which requires further work: **Subsp. *aegeicus*** AA (Lesbos, Samos and Chios). **Subsp. *odonianus*** AA (Thassos). **Subsp. *postianus*** LB, SY. **Subsp. *echinoides*** CR. **Subsp. *erinaceus*** IA (Cephalonia), GR (central), PN.

Astracantha

Spiny shrubs similar to (and often still included in) *Astragalus* but with simple hairs, leaves ending in a sharp spine. Flowers in dense stalkless racemes on the middle of the stems; *calyx densely woolly*, not inflated in fruit.

Astracantha cretica A dwarf shrub forming *dense*, spiny mounds to 30 cm. Leaves with (4)5–7 pairs of leaflets 3–7(9) mm long, narrowly elliptic, variably hairy. Flowers borne in dense clusters; calyx 5–5.8 mm, the teeth *concealed by dense white, woolly hairs to 3 mm*; corolla *pale pink, ageing cream*, 10–15 mm; style *hairy*. Fruit *small*, 5 mm, narrowly ovoid and single-seeded. Several regional subspecies described. AA (Samos), CR, GR, TR. **A. *dolinicola*** is similar but with calyx 10 mm long, hairless below, densely hairy at the apex (but the teeth not obscured); corolla *pink with purplish veins*; style hairless. A rare island endemic. CR (Psiloritis range).

Astracantha thracica A *laxly*-tufted, spiny shrub 20–30 cm. Leaves with 6–8 pairs of elliptic, virtually hairless leaflets. Flowers 2–4(7) per axil, forming dense ovoid clusters of up to 40 flowers; calyx (10)15–22 mm, densely white-woolly; corolla *purplish-pink, 14–23 mm*. Fruit ovoid, densely hairy. **A. *parnassi*** is a similar shrub, forming *large, dense cushion-like mounds* with a *shorter* calyx, just 10–12 mm, the teeth about as long as the tube; corolla 14–17 mm, purplish-pink (rarely yellowish).

Erophaca

Woody perennials similar to *Astragalus* (not recognised as distinct by all authors) but leaflets with *stipels* (stipule-like outgrowths) and *pods greatly inflated*.

Erophaca baetica subsp. *orientalis* (syn. *Astragalus lusitanicus*) An erect, robust perennial to 50 cm (1 m) with grey–green leaves with 7–12 elliptic leaflets, with adpressed hairs. Flowers 25–31 mm, cream or greenish, borne in lateral, long-stalked, dense and many-flowered racemes; calyx teeth longer than their tube. *Fruit oblong, large and inflated*, to 70 x 25 mm, borne in pendulous clusters. Rocky slopes in hill forests and mountains. Only the eastern subspecies (**subsp. *orientalis***) occurs in the region. CY, GR, LB, TR.

Ebenus

Perennial herbs similar to *Astragalus* but with flowers in umbels with *papery bracts*.

Ebenus cretica A much-branched shrublet to 1 m with leaves divided into 3–5 leaflets; leaflets elliptic-oblong, 10–30 x 2–6 mm, grey-silkily-hairy on both sides. Flowers borne in dense, broadly cylindrical racemes; calyx tubular-bell-shaped with 5 narrow teeth; corolla 10–16 mm, *bright pink*;

standard just *exceeding* the keel. Fruit compressed, not exceeding the calyx. CR. **E. sibthorpii** is more slender, woody only at the base with spreading branches and leaves with 2–5 leaflets. Flowers reddish-pink with standard just *shorter* than the keel. AA (east + Rhodes), PN.

Glycyrrhiza

Similar to *Astragalus* but with short, linear or curved fruits and one-celled, not splitting. Leaves pinnately divided with a terminal leaflet; stipules very small.

Glycyrrhiza glabra LIQUORICE An erect, robust, hairless perennial to 1 m with pinnately divided leaves with 4–8(17) elliptic leaflets, sticky beneath, and with *minute or absent stipules*, and *short-stalked* clusters of bluish flowers, 10–13(15.5) mm long. Fruit 10–25 mm long. Scattered across the region, absent from small islands. **G. echinata** SPINY-FRUITED LIQUORICE is fairly similar but with *fruits covered in long, brown-red spines*. Absent from most islands. GR, IP, PN, LB, SY, TR.

Glycyrrhiza flavescens An erect perennial herb 30–85 cm, with leaves with 5–8 pairs of elliptic leaflets 25–40 mm. Flowers golden yellow, borne in dense racemes 80 mm–20 cm; corolla 12–18 mm; calyx with teeth to 3 mm. Fruit 40 x 10 mm, compressed, hairless, ageing dark brown, with several seeds. Fields, maquis and garrigue. SY, TR.

1. *Erophaca baetica* subsp. *orientalis*
2. *Ebenus cretica* PHOTO: NICK TURLAND
3. *Glycyrrhiza glabra*
4. *Glycyrrhiza echinata*
5. *Glycyrrhiza flavescens* PHOTO: ZUHEIR SHATER

Bituminaria

Lax, spineless herbs *smelling of tar* when crushed, with trifoliate leaves. Flowers bluish, borne in dense, spherical heads; stamens 10, fused in a tube. Fruit 1-seeded, non-splitting.

Bituminaria bituminosa (syn. *Psoralea bituminosa*) An erect, branched perennial 20 cm–1 m. Leaves long-stalked, trifoliate with narrow to broad untoothed leaflets 12–90 mm long. Flowers 11–18 mm long, blue–violet borne on long-stalked clover-like clusters. Fruit ovoid, 13–19 mm, flattened with a sickle-like beak. Common on waste ground and disturbed habitats. Throughout.

Vicia VETCH

Climbing or scrambling annual or perennial herbs with unwinged stems, often with tendrils; leaves pinnately divided. Flowers with 10 stamens, 9 in a tube, 1 free. Fruit flattened and splitting.

(a) *Flowers small*, 2–10(15) mm, borne in clusters of 1–9(20) on stalks equalling or longer than the leaves.

Vicia hirsuta HAIRY TARE A slender, hairy to nearly hairless annual to 80 cm with leaves with 4–10 pairs of linear to oblong leaflets, each to 12 mm, notched at the tip and with a fine point; tendrils branched. Flowers white, tinged purple and small, 3–5 mm long in 1–9-flowered racemes; *calyx teeth equal*. Fruit hairy, 6–11 mm, black when ripe; seeds mostly 2. Grassy habitats. Throughout (rare and patchy in the south). The following species are similar: *V. parviflora* (syn. *V. tenuissima*) has leaves with *2–4(5) pairs of* leaflets and 1–4(5)-flowered racemes that greatly exceed the leaves; flowers 6–9 mm with *limb of standard longer than its claw*. Fruit hairless, 12–17 mm; seeds 4–6(8). Throughout. *V. ervilia* has 10–16 pairs of linear leaflets with a short terminal point in place of a tendril. Flowers 6–8 mm, whitish or pale pink with darker veins. *Fruit hairless and constricted between the seeds*; 14–22 mm; seeds 2–4. Local. AA, CR, CY, GR, IP, TR.

Vicia tetrasperma A more or less hairless annual with 3–6(8) pairs of linear leaflets and entire stipules. Racemes with *1 or 2(6) flowers, about equalling (not exceeding) the leaves*; corolla pale purple, 12–15 mm, the calyx teeth unequal. Fruit hairless, 20–35 mm, *with 3–10 seeds*. Rare. GR, TR. *V. pubescens* is similar but *slightly hairy throughout*, with up to 6-flowered racemes and hairy fruits. Throughout, except the far north.

(b) Flowers 8(10)–20(24) mm, borne in clusters of 6–30 on stalks longer than the subtending leaves.

Vicia cracca TUFTED VETCH A variable, often hairy, vigorous, scrambling perennial to 2 m with leaves with 5–15 pairs of leaflets and branched tendrils, and entire, ½-arrow-shaped stipules. Flowers 8–12(13) mm, *borne in dense, long narrow racemes of 10–40 blue–violet flowers* on stout stalks, drooping; calyx teeth very unequal (the upper minute). Fruit hairless, 10–25 mm with 2–6(8) seeds. Grassy habitats on higher ground. GR, IA, PN.

Vicia villosa FODDER VETCH A variable annual herb to 2 m, usually scrambling or climbing, similar to *V. cracca*. Leaves pinnately divided with 4–12 pairs of linear-elliptic leaflets, and *unlobed*, hairy stipules and branched tendrils. Flowers 10–12(20) mm, violet to purple, often with creamy wings borne in conspicuous racemes of 10–30 flowers; stalk shorter than the subtending leaf; *calyx swollen at the base; standard petal with a basal stalk-like part 2 x the length of the blade*. Fruit brown and hairless, 20–40 mm with 2–8 seeds. Throughout. **Subsp.** *microphylla* has leaflets 5–15 mm long, 2–6-flowered racemes; wings purple, violet or white and fruit sparsely hairy when young. Virtually throughout the range of the species.

Vicia benghalensis A short to tall, hairy annual to 80 cm with leaves with 5–9 pairs of linear to elliptic leaflets; tendrils branched; stipules toothed or not. Flowers 2–20, *reddish to burgundy-coloured with blackish tips*, 15–18 mm long, with calyx convex at the base, borne in racemes longer than the leaves. Fruit hairy, 21–40 mm, with 2–5 seeds. Garrigue, pathsides. Virtually throughout (not CY).

Vicia onobrychioides FALSE SAINFOIN A hairy or hairless perennial to 60 cm. Leaflets in 4–9 pairs, linear to oblong and *narrow*, just 1–4(5) mm wide, with tendrils present; stipules toothed. Flowers large, 14–24 mm, borne in rather 1-sided, long-stalked racemes of 4–12, bright *blue–violet with a paler, whitish keel*. Fruit 27–35 mm, reddish and hairless with 2–8 seeds. Disturbed and sandy waste places. GR.

1. *Bituminaria bituminosa*
2. *Vicia hirsuta*
3. *Vicia cracca*
4. *Vicia villosa* subsp. *microphylla*
5. *Vicia benghalensis*
6. *Vicia lutea* (pink-flowered form)

(c) Flowers *yellow or white*, <6 per cluster, on stalks *much shorter* than the subtending leaves.

Vicia lutea YELLOW VETCH A tufted, prostrate, hairless to hairy annual to 50 cm with leaves with 3–9 pairs of linear to oblong, bristle-tipped leaflets; tendrils simple or branched. Flowers 17–25 mm, hairless, borne in groups of 1–3, dull *yellowish-white* (sometimes flushed purple) with a hairless standard; calyx teeth unequal. Fruit yellow–brown and densely hairy, 25–43 mm, with 2–4 seeds. Coastal, sandy habitats. Throughout. *V. hybrida* is similar but with *solitary* flowers 18–25 mm with a *hairy standard*, borne on short stalks, and unequal calyx teeth. Fruit 27–32 mm with 2–5 seeds. Damp habitats. Throughout. *V. grandiflora* LARGE YELLOW VETCH is similar to *V. lutea* but with *large flowers* 22–35 mm long, yellowish or white, sometimes purple-tinged. Mainly in the north. GR, PN (west), TR.

Vicia faba BROADBEAN An erect, hairless, square-stemmed annual to 80 cm (1 m) with 2–3 pairs of large, oval, grey–green leaflets without tendrils. Flowers white with blackish-purple blotches, borne in virtually stalkless clusters of 1–6. Fruit long, to 30 cm with 4–8 seeds. Native to Asia, naturalised.

1. *Vicia hybrida*
2. *Vicia grandiflora*
3. *Vicia lunata*
4. *Vicia narbonensis*
5. *Lathyrus annuus*

(d) Flowers *bi-* or *tricoloured*, <6 per cluster, on stalks *shorter* than to just exceeding their subtending leaves.

Vicia lunata A slender, early-flowering, trailing annual with distinctly angled, hairless stems to 20 cm. Leaves oblong, 10–40 mm long with 6–10 shortly oblong leaflets 0.8–2 mm across; tendrils branched or unbranched. Flowers 2–5 in lax racemes on stalks to 50 mm typically equalling or just exceeding their subtending leaves; flowers 9–12 mm, *bicoloured*, the standard lavender *blue* with darker veins, the wings bright *yellow*; calyx 3–5 mm with virtually equal teeth. Fruit crescent-shaped, flattened, 10–20 mm. Rocky igneous mountainsides on garrigue or under pines. CY, SY, TR.

Vicia melanops BLACK VETCH A sprawling, hairy annual to 50 cm with leaves with 4–8 pairs of leaflets. Flowers *tricoloured,* 17–21 mm long, solitary or in clusters of 2–3 on stalks *much shorter* than the leaves; standard *yellow–green,* the keel and wings *dark black–purple*; calyx hairy. Fruit hairy on the margins only, to 32 mm with 2–4 seeds. Garrigue. GR, PN, TR.

(e) Flowers a shade of *purple*, ≤6 per cluster, on stalks *much shorter* than the subtending leaves.

Vicia narbonensis PURPLE BROADBEAN A downy, erect, square-stemmed annual to 60 cm with *large, grey–green leaves with 1–3 pairs of leaflets 11–39 mm wide* (rather like a broadbean). Flowers 10–30 mm, borne in clusters of 1–3(6), *dark purple*. Fruit short-hairy, 30–50(70) mm with 4–7 seeds. Damp fields and ditches. AA (east), CY, GR, IP, TR.

Vicia sepium BUSH VETCH A scrambling, spreading perennial to 1 m, hairless or hairy with leaves with 3–9 pairs of almost rounded leaflets ending in bristle-tips; stipules ½-arrow-shaped and spotted. Flowers 12–15 mm, *dull bluish-purple with red–purple calyces* with unequal teeth, borne in short-stalked clusters of 2–6. Fruit black and hairless, 20–35 mm with 3–10 seeds. Grassy places on higher ground; absent from hot, dry areas and the far south. GR, TR. *V. pannonica* is similar, with 4–19 leaflet pairs, clusters of 1–4 dull purple (sometimes cream) flowers 20–35 mm borne on short stalks; *the standard petal hairy on the back*. Fruit yellowish and hairy, 20–35 mm, with 2–8 seeds. AA, CY, GR, TR.

Vicia sativa COMMON VETCH A variable, vigorous, sprawling, hairy annual to 1.5 m. Leaves with 3–8 pairs of linear to heart-shaped leaflets, *tendrils branched*; *stipules toothed*. Flowers purplish-red, *solitary or paired* (up to 4), and often large, *20–30 mm long*; calyx teeth equal. Fruit yellowish or blackish, long, 35–68 mm, with 4–9 seeds. Frequent in grassy and disturbed habitats. Throughout. *V. lathyroides* SPRING VETCH is similar but small and prostrate to 20 cm with leaves with 2–4 pairs of bristle-tipped leaflets, *unbranched or undeveloped tendrils*, and *small, solitary flowers just 6–9 mm long*. Garrigue and bare ground. Throughout.

Lathyrus PEAS

Similar to *Vicia* but typically with angled or *winged stems*, and often fewer, parallel-veined leaflets, sometimes reduced to a simple blade or tendril. Flowers with 10 stamens, 9 in a tube, 1 free. Fruit a pod-like legume.

(a) Flowers yellow.

Lathyrus annuus ANNUAL YELLOW VETCHLING A climbing or scrambling annual to 1 m with leaves with 1 pair of lance-shaped leaflets, arrow-shaped stipules and branched tendrils. Flowers 12–18 mm, borne in erect, long-stalked racemes of 1–3, corolla *yellow to orange* and red-veined; calyx teeth equal. Fruit pale brown and glandular when young, 40–80 mm. Disturbed habitats. Throughout.

LEGUMINOSAE (FABACEAE)

Lathyrus aphaca YELLOW VETCHLING A hairless, waxy grey–green scrambling annual to 40 cm (1 m) with unwinged (angled) stems. Mature plants with *leaves reduced to tendrils, but stipules large and leafy, broadly triangular and paired*, 6–50 mm. Flowers borne solitary on erect stalks, yellow, 10–13 mm long. Fruit hairless, 17–35 mm long. Grassy and disturbed habitats. Throughout.

Lathyrus ochrus WINGED VETCHLING A hairless, pale greysh annual to 1.5 m with *very broadly winged stems*, and leaves borne on *leaf-like stalks*; lower leaves oval-oblong, upper leaves with 2–3(4) pairs of leaflets. Flowers *pale yellow*, usually solitary, 13–21 mm long. *Fruit 39–65 mm with 2 wings* along the upper edge. Garrigue. Throughout.

(b) Flowers reddish, purple, pink or white. Leaves simple or with just 2 leaflets; stems with wings <1 mm or wingless.

Lathyrus nissolia GRASS VETCHLING A delicate, ascending, hairless to downy annual with *unwinged stems* to 90 cm. *Leaves grass-like*, lacking leaflets and tendrils, and with very small, narrow stipules to 2 mm. Flowers solitary or paired, long-stalked and crimson, 8–18 mm long. Fruit pale brown, 22–60 mm. Grassy habitats, often near the coast. Throughout, except most islands (present on Corfu).

Lathyrus cicera RED VETCHLING A slender, hairless annual to 80 cm (1 m) with *narrowly winged stems*. Leaves with 1 (rarely 2) pair(s) of lance-shaped to oval leaflets, with simple or branched tendrils on the upper leaves. Flowers 12–18 mm, solitary on stalks to 30 mm, *red, orange or brownish*; *calyx teeth equal*. Fruit hairless, 25–50 mm, and with 2 keels on the upper edge. Garrigue and grassland; common. Throughout. ***L. setifolius*** NARROW-LEAVED RED VETCHLING is similar but with scarcely winged stems and smaller orange–red flowers 9–13 mm with slightly *unequal calyx teeth*. Fruit 19–33 mm. Garrigue; rather local. Throughout.

1. *Lathyrus setifolius*
2. *Lathyrus sylvestris*
3. *Lathyrus clymenum*

Lathyrus gorgoni is similar to *L. cicera* but with *large, pale orange flowers* with spreading sepals. **L. hirsutus** HAIRY VETCHLING is similar to the previous species in form, but sparsely hairy and with *pale violet–blue flowers with pink wings*, 7–15 mm. Fruit 15–45 mm. GR (north).

Lathyrus tuberosus FYFIELD PEA A scrambling, hairless perennial with *angled but mostly unwinged stems* to 1.2 m. Leaves with a single pair of elliptic leaflets and simple or branched tendrils and narrowly ½-arrow-shaped tendrils. *Flowers bright reddish-pink*, borne on long stalks in groups of 2–7, 12–20 mm long; calyx teeth triangular. Fruit brown and hairless, 16–28 mm long. Local in grassy and waste habitats. GR (north).

(c) Flowers reddish, purple, pink or white. Leaves simple or with just 2 leaflets; *stems broadly winged.*

Lathyrus sylvestris NARROW-LEAVED EVERLASTING PEA A climbing, hairless or downy perennial herb with *broadly winged stems* to 2 m. Leaves with a single pair of narrowly lance-shaped, 3-veined leaflets 2–19 mm wide with branched tendrils; *stipules lance-shaped, ‹½ the width of the stem.* Flowers 13–19 mm, borne on rather long stalks (30 mm–21 cm), pinkish-purple flushed yellow; calyx teeth shorter than their tube. Fruit 50–70 mm. GR. **L. latifolius** BROAD-LEAVED EVERLASTING PEA is similar, to 3 m, with *more broadly oval leaflets 2–40 mm across, and stipules ›½ the width of the stem*; flowers 3–12 and larger, 15–30 mm and magenta–pink. GR, IA.

(d) Flowers red, purple, pink or white. Leaves with 4–12 leaflets; *stems broadly winged.*

Lathyrus clymenum A scrambling, hairless annual to 1 m with *broadly winged stems and leaves with winged, leaf-like stalks*; leaves with 4–8 pairs of narrower, *linear-elliptic* leaves 0.5–14 mm wide. Flowers variably pink, dull yellow or crimson, 12–25 mm, borne in racemes of 1–2(3). Fruit *grooved* on the upper side. Roadsides, hedges and tracks. Throughout, though rarer in the east (not Rhodes; rare on small islands). Some authors recognise forms with *very narrow leaflets* (‹5 mm), bicoloured flowers with white or pink wings and ungrooved pods as *L. articulatus.*

Pisum PEA

Herbs with unwinged stems, *large, leafy stipules* (larger than the leaflets) and branched tendrils. Flowers with 10 stamens, 9 in a tube, 1 free. Fruit a pod-like legume.

Pisum sativum WILD PEA A variable, clambering, hairless annual to 2 m (often less), stems *not winged*. Leaves with 1–3 pairs rounded to elliptic leaflets, more or less heart-shaped at the base, tendrils branched; stipules large and leafy, often blotched. Flowers 15–35 mm, borne in 1–3-flowered racemes, wing *petals fused to the keel*; standard lilac–purple, wings maroon. Fruit hairless, net-veined, 30 mm–12 cm. Throughout. *P. fulvum* is more slender with flowers just 15–17 mm, uniformly dull orange with a paler keel. AA (rare), CR (rare), CY, IP, SY, TR.

Ononis RESTHARROW

Herbs or subshrubs with simple or trifoliate leaves (often both), the veins of the leaflets ending in marginal teeth; tendrils absent. Flowers with calyx deeply toothed with nearly equal teeth; all 10 stamens forming a tube. Fruit straight, 1–many-seeded, splitting. Many closely related species occur in the region, of which just a few are decribed here.

(a) Flowers white, pink or purple.

Ononis reclinata A low, slender, *spreading* annual with *shaggy-hairy* stems ‹15 cm; middle leaves trifoliate, the lowermost and uppermost simple. Flowers *small*, just 5–10 mm, pink and solitary forming loose, leafy inflorescences, *abruptly recurved* at the apex; corolla *equalling* the calyx; calyx with *entire teeth*. Fruit 8–14 mm with numerous, *10–15(20) seeds*. Widespread in coastal habitats. Throughout. ***O. verae*** is similar but with all leaves simple, the upper broadly *linear*. Flowers borne on stalks scarcely recurved at the apex; corolla distinctly *exceeding* the calyx. Fruits with *fewer* (about 8) seeds. A rare endemic of western Crete and associated southwest islets. CR.

Ononis spinosa SPINY RESTHARROW A *very spiny* dwarf-shrub to 70(80) cm, with 2 opposite rows of hairs on young stems. Leaves trifoliate, or simple above. Flowers large, bright pink, 10–20 mm, usually borne singly at each node, forming a *narrow, lax* inflorescence. Fruit hairy, 6–10 mm, usually with a single (2–4), warty seed(s). Dry, rocky and waste places and grassland. Probably throughout.

Ononis diffusa A *very sticky, glandular-hairy* annual to 40(60) cm. Leaves trifoliate, with *toothed* margins. Flowers small, 8–14 mm, the corolla exceeding the calyx; pink with a whitish keel, borne singly at each node in a *dense, oblong spike*; calyx 7–8 mm with narrow teeth, longer than their tube. Fruit 4.8–5.5 mm with 3–4 seeds to 2 mm across, slightly curved. Coastal habitats; local. AA, CR (rare), CY. ***O. serrata*** is very similar but shorter, to 30 cm, with shorter, narrower leaflets and smaller corolla, just 6–8 mm long, not exceeding its calyx. Seeds 2–5, 1.5 mm across. AA (southeast), CR (offshore islets), CY, IP. ***O. mitissima*** is similar to *O. diffusa* but a rather taller and more erect, scarcely hairy annual, to 60 cm, with trifoliate bracts with conspicuous white stipules. Flowers pink, exceeding their calyces, 10–12 mm; calyx teeth broad, equalling their tube; flowers borne in *short, dense* inflorescences. Fruit 5–6 mm with 2–3 seeds. Scattered, virtually throughout; absent from the north.

(b) Flowers yellow.

Ononis natrix LARGE YELLOW RESTHARROW A much-branched, green, *sticky* subshrub to 40(60) cm, with stems woody below, densely leafy. Leaves trifoliate with oval, toothed leaflets. Flowers yellow, the standard petal veined with red externally, (6)11–25 mm, *solitary* but borne in loose, leafy clusters. Fruit 11–25 mm, pendulous and hairy, with 2–27 seeds. Common and widespread across the region, especially near the coast. Recorded throughout but locally split by some authors; for example, the widespread ***O. ramosissima*** – also treated as a subspecies – has *short glandular stem hairs*, corolla 9–16(18) mm, and calyx teeth *just* exceeding their tube; also ***U. talaverae*** (CR).

Ononis viscosa A variable species, similar to *O. natrix* but a *taller*, softly hairy annual to 1.3 m with flowers 5–16 mm with a pinkish, glandular-hairy standard, and *3-veined* calyx teeth; flower-stalks 10–40 mm. Fruit 10–25 mm with 3–10 seeds. Throughout. ***O. sicula*** (also treated as ***O. viscosa*** subsp. ***sicula***) is short, to 15(25) cm with *narrowly* oblong-*linear*, sharply toothed, rather erect leaflets, glandular-hairy flower-stalks *without* long spreading hairs; corolla 5–10 mm, exceeded by the calyx lobes. Fruit 9–14 mm, hairy, with 10–20 seeds. Dry habitats and cultivated land. CY, IP (common). ***O. pubescens*** is a *short* annual to 75 cm, similar to *O. viscosa*, also extremely sticky, often with some leaves 1-foliate. Flowers larger, 12–22 mm, yellow, the standard red-veined, the calyx teeth *5-veined with broad, oblong to lance-shaped teeth*. Fruit ovoid to ellipsoid, 8–12 mm with just 2–3 seeds. Virtually throughout.

Ononis ornithopodioides BIRD'S-FOOT RESTHARROW A short, erect, much-branched annual with glandular-hairy stems, to 30 cm. Leaves trifoliate with oval leaflets, broadest above the middle. Flowers yellow (occasionally with pink veins), 7–8.5 mm, borne in sparse, leafy panicles, which elongate in fruit. Fruit linear, 12–20 mm, *constricted between the seeds*; seeds 7–10, 2 mm across. Garrigue and pine woods. Throughout, except the north.

Ononis pusilla A short-hairy, sometimes rather straggling perennial with stems woody below, to 35(55) cm. Leaves trifoliate with elliptic to rounded leaflets, often notched. Flowers yellow, 6–12 mm long, borne in lax spikes with leafy bracts; *calyx lobes hairy, long, equalling the corolla, and* *later opening to become star-like.* Fruit small, 4.5–10.5 mm long with 3–10 seeds. Local in rocky habitats, scree and in pine woods; local and absent from many islands, much of the west and northeast. CY, GR, IA, IP, PN.

Ononis variegata A *low, spreading, mat-forming annual* to 40 cm, branched nearly from the base; densely hairy with *bluish-green*, toothed leaves with distinct veins. Flowers yellow, 10–15 mm long, borne in lax, terminal racemes on stalks to 40 mm; standard hairy on the back. Fruit oblong and slightly hairy, 8–11 mm. Local on coastal dunes. AA (local), CY, IA, IP, PN (west), SY.

1. *Ononis reclinata*
2. *Ononis verae*
3. *Ononis spinosa*
4. *Ononis mitissima*
5. *Ononis sicula*
6. *Ononis pubescens*
7. *Ononis pusilla*

Melilotus MELILOT

Annuals or biennials with trifoliate leaves often with toothed leaflets. Flowers small and sweet-smelling, borne in elongated racemes; stamens 10, 9 in a tube, 1 free. Fruit straight, 1–2-seeded, (normally) non-splitting.

(a) Stipules of middle leaves usually *distinctly toothed*. Flowers yellow.

Melilotus indicus SMALL MELILOT An erect, branched or simple annual to 40 cm with trifoliate leaves with toothed leaflets and stipules with *few, papery teeth* at the base. *Flowers tiny*, 2–3.5 mm, pale yellow to whitish and borne in *dense, many-flowered spikes* (10–40); wings and keel equal, and shorter than the standard. Fruit more or less spherical, pale brown when ripe, 1.5–3 mm and hairless. Common on disturbed ground and on damp, sandy soils. Throughout. **M. italicus** is similar to *M. indicus* but to 80 cm, with 20–40 larger flowers (6)7–8(9) mm long. Fruit 4–5 mm, strongly *wrinkled* and with small depressions when ripe. Dry open habitats. Throughout.

Melilotus segetalis An annual with racemes *equalling or exceeding* their leaves, with *15–100 larger* flowers (4)5–7.5 mm; standard shorter than the keel. Fruit *yellow, egg-shaped* and with concentric grooves, 4–5 mm. Damp habitats. AA (local), CY, GR, IA, PN. **M. sulcatus** FURROWED MELILOT is very similar to *M. segetalis* but with small flowers, 2.8–4.5 mm, borne on stalks *shorter than* or equalling the subtending leaf. Fruit spherical, 3–4.5 mm. Drier habitats than the previous species. Throughout.

Melilotus messanensis (syn. **M. siculus**) An annual to 60 cm similar to *M. segetalis,* with yellow flowers, 4–5 mm, with more or less equal standard and keel that exceed the wings, borne on erect to sub-erect stalks, in 4–10(14)-flowered racemes much *shorter* than the subtending leaves. Fruit yellowish, dark brown when ripe, with concentric grooves, 7–8(9) mm, pointed. Cultivated and damp maritime habitats. Throughout.

(b) Stipules of middle leaves mostly *entire* (or minutely toothed). Flowers yellow.

Melilotus altissimus TALL MELILOT A tall, branched biennial to 1.5 m with oblong leaflets, the uppermost almost parallel-sided, toothed and with bristle-like stipules. Flowers yellow, 50–70 mm long, *the standard and keel equal*. Fruit net-veined and black, 5–7 mm, *hairy*. Various disturbed habitats. GR, IA.

Melilotus neapolitanus (syn. **M. spicatus**) A short annual 15–40(80) cm, hairy above with yellow flowers 4–5 mm with equal wings and keel, borne in *short, lax racemes* of (6)8–16(20) flowers. Fruit small, 3–4 mm, spherical, not flattened, with a short beak, and *netted surface*. Dry, sandy habitats. AA, CR, GR, IA (rare), PN, TR. **M. elegans** is a similar annual to 2 m with stems hairy above, and *numerous* small flowers borne in racemes of 20–35; flowers *larger*, 4.5–5 mm with the standard and wing petals equal, both *shorter* than the keel. Fruit oval, compressed and with very prominent *transverse veins*, 3–4.5 mm. AA (rare), CY, IP.

(c) Stipules of middle leaves mostly *entire* (or minutely toothed). Flowers white.

Melilotus albus WHITE MELILOT A rather vigorous annual to 1.5 m with slender branches and entire stipules. Flowers borne in rather lax racemes, corolla *white*, 4–5 mm. Fruit 3–5 mm, ridged, brown and hairless. Dry, disturbed habitats. Throughout, except IA; rare in the Aegean.

Trigonella FENUGREEK

Annuals with trifoliate, toothed leaves. Flowers borne solitary in leaf axils or in lateral racemes; stamens 10, 9 in a tube and 1 free. Fruit straight or curved with 1–many seeds in 2 rows; eventually splitting.

(a) Flowers *yellow*, borne numerously (to 15) in stalked racemes. Fruit shape variable.

Trigonella stellata A prostrate, sprawling annual. Leaves long-stalked; leaflets narrowed at the base, toothed and notched at the tips. Flowers borne in dense clusters of 3–15, nearly *stalkless, congested at the base* of the plant and in the leaf *axils* along the stems. Fruit 5–8 mm long, cylindrical, divergently spreading in a star, with a netted surface. Bare, arid and desert fringe habitats. IP.

Trigonella graeca (syn. *Melilotus graecus*) An ascending annual 10–30 cm with rounded, rather fleshy leaflets 10–15 x 7–12 mm. Flowers borne in dense, many-flowered heads on stalks 40–60 mm; calyx bell-shaped, 3 mm, the teeth as long as tube; corolla 7–10 mm, yellow. Fruit 12–20 x 10–15 mm, round to oval, flat, membranous, with transverse veins and a *membranous wing* on the upper edge. Seeds 2–3, 4 mm, ovoid, brown, warty. Stony places. AA (Levitha), CR, GR, IA, PN.

Trigonella corniculata SICKLE-FRUITED FENUGREEK An *erect or spreading, more or less hairless* annual to 50 cm, superficially similar to *Melilotus*. Leaves with 3 oval leaflets. Flowers yellow, 5–7 mm borne in *long-stalked cylindrical racemes* of 8–15; *calyx teeth unequal*. Fruit *linear and pendent, slightly curved*, 10–18 mm (without beak). Various dry, grassy habitats. Throughout except IP.

1. *Melilotus indicus*
2. *Trigonella stellata*
3. *Trigonella graeca*
4. *Trigonella arabica*
5. *Trigonella schlumbergeri*

Trigonella spicata A spreading to erect, hairless or adpressed-hairy annual 50 mm–30 cm with toothed, narrowly oblong leaflets. Flowers rather large, borne on stalks longer than their subtending leaves, in dense heads; calyx 4 mm with unequal teeth; corolla bright yellow, 6–7 mm. Fruiting heads ellipsoidal to cylindrical; fruits nodding with a *strongly curved beak*, the seed-bearing part narrowly ovoid, 7 x 4 mm, hairy. Seeds 1(2), with small warts. AA, GR, PN (northeast), SY, TR. ***T. cephalotes*** is similar but with smaller flowers, 3–4 mm and rounded fruiting heads; fruits with seed-bearing part only 4.5 x 2.5 mm with a beak *scarcely curved*. Rare. AA (Fourni, Samos), GR (southeast – rare), TR.

Trigonella spruneriana A sparsely hairy, ascending annual 50 mm–20 cm with oblong leaflets, toothed towards the tips. Flowers borne 5–12 in short racemes; calyx bell-shaped with triangular teeth shorter than their tube; corolla 5–7 mm, pale yellow. Fruit 20 x 2 mm, *hairy*, linear, sickle-shaped in the upper half, tapering into a short beak; fruits not constricted between the seeds. AA, GR (southeast), CY, IP, LB, TR. ***T. smyrnaea*** is similar but with *smaller* hairy fruits 6–10 x 2–3 mm, *constricted* between the seeds. Local. AA (northeast), TR.

(b) Flowers *blue*, borne numerously (to 15) in stalked racemes.

Trigonella rotundifolia A hairy, spreading to ascending annual 50 mm–25 cm with *broadly oval, coarsely toothed* leaflets. Flowers borne in dense heads of 12–15; calyx tubular with more or less equal teeth; corolla *blue*. Fruit 6–12 mm, cylindrical or slightly compressed, slightly curved and short-beaked. Seeds 4–6. AA, GR (southeast), TR. ***T. caerulea*** has *narrowly elliptic* leaflets and fruits to 3.5 mm with a slender beak. AA (Thasos – rare), PN (northeast), TR.

(c) Flowers 1–2(6) borne virtually stalkless in the leaf axils; white, cream, yellow or bluish. Fruit linear.

Trigonella foenum-graecum CLASSICAL FENUGREEK A rather hairy annual to 50 cm. Leaves trifoliate, oblong, toothed near the tip. Flowers cream, flushed with purple at the base, 12–18 mm long, solitary or paired in the axils of upper leaves. Fruit linear and erect, 50 mm–10 cm with a long beak 10–35 mm, hairless, with longitudinal veins. Cultivated and sometimes naturalised. GR, IA, TR.

Trigonella gladiata An ascending to erect annual 50 mm–25 cm with hairy stems and oblong to triangular leaflets, toothed towards the tips. Flowers borne stalkless in the leaf axils, solitary or occasionally paired; calyx 3 mm; corolla 7–9 mm, cream, with violet markings on the standard. Fruit with seed-bearing part broadly linear, compressed, 30 x 5 mm, slightly curved, with a *long beak*, 15–25 mm. Seeds 2–6. Throughout, except the southeast (not CY or IP). ***T. cariensis*** is similar but with corolla *larger*, 17–20 mm, *pale blue*. Seed-bearing part of fruit abruptly contracted into a long beak 25–40 mm. AA, GR (southeast), TR.

Trigonella spinosa A sparsely hairy annual 50 mm–20 cm with angular stems and short-stalked leaves with narrowly oblong-triangular, toothed leaflets. Flowers virtually stalkless, borne in clusters of 2–6 in the leaf axils; calyx bell-shaped with teeth shorter than the tube; corolla yellow, 4 mm. Fruit linear, 30–50 mm long, slightly compressed, curved (sometimes in a circle), hairless, without a beak. Seeds numerous. AA (south), CR, CY, IP, PN (Kithira).

(d) Flowers in umbels of 2–8, *whitish*. Fruits *expanded, flattened* and papery when ripe. East mainland only.

Trigonella arabica A virtually hairless, ascending annual. Leaflets wedge- to heart-shaped, coarsely toothed at the base. Flowers borne in umbels of 3–8 on stems somewhat longer than the leaves; corolla 8–10 mm, *white* (rarely yellowish); calyx 2–4 mm. Fruit 15–30 mm, hairless, broadly curved-*oblong* (5–8 mm wide), *flattened,* with transverse veins and *spiny-ciliate margins* on both sides.

A Saharo-Arabian annual of sand flats and deserts. IP (mainly east and central). *T. schlumbergeri* is a similar annual with minute, wedge- to heart-shaped leaflets toothed towards the tips. Flowers borne in umbels of 2–4 on stems longer than the leaf stalks; corolla 6 mm, whitish. Fruit 10–20 mm, broadly *half oval*, flattened with transverse veins, papery when ripe with a *torn-toothed wing*. Sandy habitats and deserts; local. IP (east and central).

Medicago MEDICK

Annual or perennial herbs typically with trifoliate leaves. Flowers yellow (sometimes purple) borne in small lateral clusters; stamens 10, 9 in a tube and 1 free. Fruit highly variable but important for identification; often curved, spiralled or spiny, 1–many seeded. Only common and widespread species are described here.

(a) Fruit sickle- or bean-shaped.

Medicago lupulina BLACK MEDICK A spreading, hairy annual to 80 cm with leaves with 3 round to diamond-shaped leaflets, often notched at the tips. Stipules toothed or entire. Flowers small, 2–3 mm, yellow, and borne in ball-like, dense clusters of 20(50) flowers. Fruit 1.5–3 mm across, coiled, black when ripe and net-veined. Common in waste places. Throughout.

M. lupulina

Medicago monspeliaca (syn. *Trigonella monspeliaca*) STAR-FRUITED FENUGREEK A prostrate, spreading, finely hairy annual to 20(40) cm with trifoliate leaves with oval, toothed or untoothed leaflets. Flowers small and yellow, borne 4–12(15) in stalkless clusters. Fruits 8–13 mm, borne in *spreading, star-like clusters* (similar to *Trigonella* with which it was previously grouped). Throughout.

(b) Fruit spineless, with a spiral of 1 or more turns.

Medicago sativa LUCERNE A variable (with many forms described), hairy perennial to 90 cm. Leaflets oblong to linear, toothed at the apex. Stipules toothed at the base. Flowers 5–12 mm, violet or blue, borne in rounded, dense racemes of 10–30(50) flowers. Fruit 5–9 mm across, spiralled with 2–3(4) turns and a hole through the centre, not spiny. Common in waste places and on roadsides. Throughout.

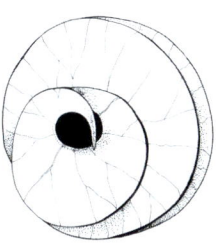

M. sativa

Medicago arborea TREE MEDICK A silvery-grey leafy *shrub to 2 m* with silky-white younger branches. Leaflets narrowed at the base, slightly toothed at the apex. Stipules untoothed. Flowers yellow, 9–13 mm, borne in dense racemes of 8–20. Fruit 9–15 mm across, slender and coiled with 1 turn, hairless and net-veined. In disturbed habitats and gardens. Centre of region; naturalised elsewhere. Throughout.

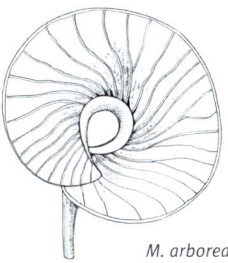

M. arborea

Medicago orbicularis LARGE DISK MEDICK A short, hairless or slightly hairy annual 30–60 cm. Leaflets oval-wedge-shaped, toothed at the apex. Stipules deeply toothed. Flowers 5 mm, yellow, in racemes of 2–4(5). Fruit *large and disk-like*, 10–20 mm across, smooth and spiralled anticlockwise with 4–7 turns, without a central hole. Frequent in waste places. Thoughout.

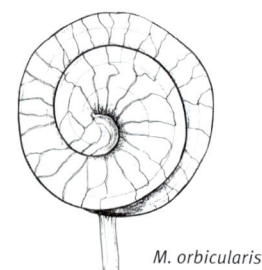

M. orbicularis

Medicago scutellata An ascending annual 25–50 cm, *densely* covered in *glandular* and simple hairs. Leaflets 10–25, to 12 mm, with toothed margins, almost *hairless above*; stipules typically with 6 pointed teeth. Flowers borne 1–2, 6–9 mm; calyx teeth as long as the tube. Fruit protruding sideways from the calyx when young, 7–15 mm, *spineless, glandular,* with 5–7 *cup-shaped* coils, *all concave* (in the same direction), each enclosing the next. Fallow fields, roadsides and red clay soils. Throughout except northern and central Greece. **M. rugosa** is similar, with fruits with flattish coils with *distinct transverse ridges* and a dorsal suture elevated in the middle of the *thick* coil edge. Virtually throughout.

M. scutellata

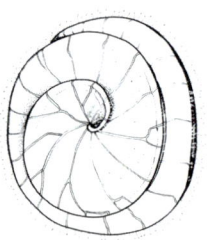

M. rugosa

Medicago blancheana A vigorous, spreading, slightly hairy annual 20–30 cm with deeply toothed leaflets somewhat *hairy* above, 10–15 mm with a small terminal tooth. Flowers yellow, 6.5(8) mm, borne on long stalks. Fruit 9–12 mm across, rounded, with 4–6 spirals, usually without spines (or with slight protrusions to 1 mm protruding at 30°–70°), with netted veins; coils convex and decreasing in size at both ends, and typically with at least some coils *alternately concave and convex*. Cultivated land. Strongly southeastern. AA (east), CY, CY, IP, IR.

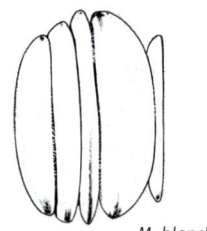

M. blancheana

(c) Fruit normally with spines or projections, with a spiral of 1 or more turns (many very similar closely related and hybridising species; close attention to the fruits required for accurate identification).

Medicago radiata A trailing annual 15–50 cm. Leaflets 15–25 mm; stipules deeply cut. Corolla yellow; standard almost as wide as long, wings longer than the keel. Fruits 18–28 mm across with *one coil*, with a *paper-thin* edge; spines in a *single row*, forming an extension to the fringe of the dorsal suture (like fruits of *Trigonella* spp.). Dry, bare places and deserts. IP (mainly central), LB, SY, TR.

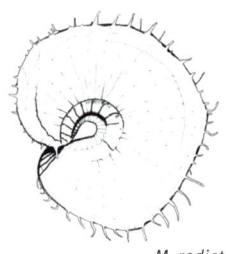

M. radiata

1. *Medicago lupulina*
2. *Medicago monspeliaca*
3. *Medicago sativa*
4. *Medicago arborea*
5. *Medicago orbicularis*
6. *Medicago scutellata*

Medicago arabica SPOTTED MEDICK A low, prostrate and spreading, more or less hairless annual to 60 cm with trifoliate leaves. Leaflets notched and with a *conspicuous dark spot*. Flowers yellow, 4–6 mm borne in racemes of 1–5. Fruit cylindrical, 4.5–5 mm across, hairless and with 3–5(6) anticlockwise turns with stout *spreading spines* 2–3.5 mm and a margin with 3 grooves. Common on fallow land and in woods and pine forests. Thoughout.

Medicago intertexta (syn. ***M. ciliaris***) A short annual 15–70 cm with spreading stems. Leaflets sometimes with a dark spot and finely toothed. Stipules toothed. Flowers dull yellow, 5–8 mm, borne in clusters of 1–4(7). Fruit large, 10–17 mm across, *glandular-hairy to hairless, spiny*, with 5–8(10) anticlockwise spirals. Damp waste places. (*M. ciliaris*, which has hairy fruits, is generally considered to be indistinct). Throughout.

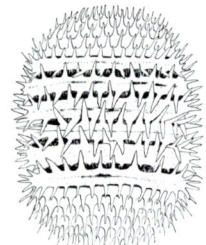

M. intertexta

Medicago marina SEA MEDICK A creeping, *prostrate, white-downy* perennial to 50 cm. Leaflets oval, toothed at the apex; stipules toothed or not. Flowers bright yellow, 7–9 mm in short clusters of 5–12(15). Fruit 4–6(7) mm across with 2–4 anticlockwise spirals and a hole in the centre, white-downy with 2 rows of spines to 1.5 mm (sometimes reduced to warts). Common on coastal sands and dunes. Throughout.

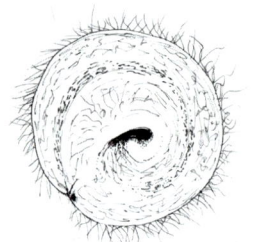

M. marina

Medicago littoralis A short, spreading, hairy annual with purplish stems to 80 cm. Leaflets oval to heart-shaped, toothed at the apex and *hairy on both surfaces*. Stipules toothed. Flowers yellow to orange, 5–7 mm long in clusters of 2–5(8). Fruit 3.5–5(6) mm long, cylindrical with 3–7 turns, shiny, spiny or not, and hairless, with a groove along the margin. Almost always on maritime sands; common. Throughout. The following are similar: **M. rigidula** has larger, spiny fruits 7–8(8) mm long with 4–7 tight anticlockwise spirals, spines 1.5–3 mm, with a netted glandular-hairy surface throughout. Fallow land. Throughout. **M. doliata** (syn. **M. aculeata**) has solitary, spiny (rarely spineless), tightly spiralled fruits 7–12 mm across with *hairs along the margins*, hairless otherwise. Throughout, except smaller islands. **M. constricta** has few stems to just 15 cm, smaller flowers 4–5 mm and broadly cylindrical, *hairless*, somewhat longer than wide with *dense, hardened coils*, with or without a narrow, veinless border. Throughout.

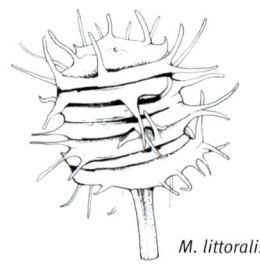

M. littoralis

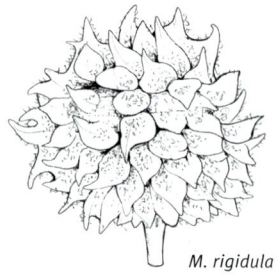

M. rigidula

M. doliata

Medicago polymorpha HAIRY MEDICK A variable, prostrate to ascending, hairy or hairless annual with purplish stems to 50 cm and trifoliate leaves, *toothed at the apex*. Stipules *deeply toothed with slender teeth*. Flowers yellow, 6–6.5 mm long, borne in clusters of 2–9. Fruit 6–7 mm across, variable, typically flattened with 4–6 anticlockwise turns, with 2 rows of *gooved spines* 0.5–0.8 mm, the surface netted (*not shiny*) and the apical spiral larger than the rest. Common on roadsides, grassy places and cultivated fallow land. Throughout. **M. italica (syn. M. tornata)** is similar to *M. polymorpha* but with fruits 5–8 mm with 2 (1–6) spirals and *spines 2 mm, not grooved*; margins of the spirals sharply defined. Throughout. ***M. truncatula*** has fruits 7–10 mm, with *much-thickened spines, swollen at the base*, 1.3–5 mm, and margins of the spirals thickened. Fields and grassy habitats. Throughout.

Medicago rotata An ascending annual to 50 cm. Leaflets 11–24 mm; stipules toothed with a long terminal tooth. Flowers 1–5, yellow with wings just shorter than keel. Fruits 8–9 mm across, cylindrical, with *short, blunt spines* 0.5–2 mm, inserted in the margins of the coil edges almost at right angles to the face of the *loose coils*; apical coil concave. CY, IP, LB, SY TR.

Medicago laciniata is similar to *M. polymorpha* but with leaflets *deeply toothed or lobed*, and flower-stalks rather longer than their subtending leaves' stalks. Fruit 6–8 mm, spiny, hairless, with 4–6 turns. Deserts and semi-deserts. IP (common in the east). **M. praecox** has less distinctly toothed leaflets and flower-stalks much *shorter* than their leaves' stalks. Fruit 4–5 mm, spiny, with 3 turns. Local, but widespread. Throughout, except the southeast.

Medicago minima SMALL MEDICK A prostrate to ascending, densely hairy annual, similar to the *M. polymorpha* group, but with *stipules entire or slightly toothed, the teeth if present short*. Flower-stalks as long or a little longer than their leaves' stalks; flowers 2.5–4.5 mm. Fruits borne in clusters of 4–6, ball-like, spiny, 2.5–3.5(5) mm across (rather large relative to the plant) with 3–5 lax, sharp-edged spirals and spines to 1 mm, weakly hooked; not distinctly flattened, and often rather shiny (less netted) on the surface. Throughout.

M. polymorpha

M. truncatula

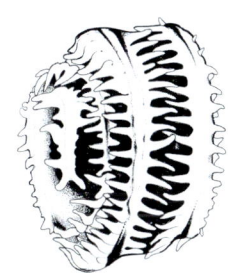
M. rotata

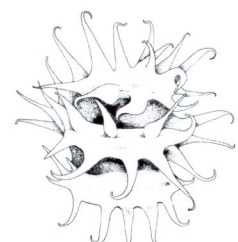
M. praecox

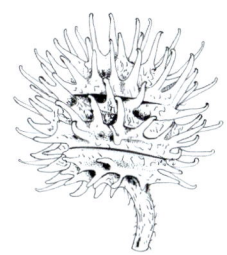
M. minima

Medicago coronata A slender, sparsely hairy annual with ascending stems 10–30 cm and small, toothed leaflets. Flowers 6–12, borne on *long, slender stalks*, much-exceeding the leaves; corolla small, 3 mm, yellow. Fruits initially protruding laterally from the calyx, borne in dense clusters, *small*, just 2.5 mm across when mature, slightly hairy, with few, broad-margined coils and short, stout spines pointing in opposing directions. Throughout.

Medicago disciformis A sparsely hairy annual with several ascending stems 15–30 cm and broadly oval leaflets hairy on both surfaces, toothed towards the tips. Flowers *few*, 1–3, borne on long stalks exceeding the leaves; corolla yellow, 5–6 mm. Fruit often solitary, 6–8 mm across, hairless; coils with broad, *veinless* margins and slender spines 2–3 mm, the *ultimate coils without spines*. Throughout, except the far southeast.

(d) Fruit with spines or projections, with a spiral of 1 or more, *greatly thickened* turns, appearing turban-like.

Medicago turbinata A robust, prostrate or ascending, *densely hairy* annual to 50 cm with reddish stems and toothed, trifoliate leaves. Flowers yellow–orange, 6–6.5 mm long, borne in clusters of 2–9. Fruit barrel-shaped, 6–7.5 mm with 4–6 clockwise or anticlockwise turns, a single groove along the margin, and *short, broad spines* just <1 mm long. Various, often disturbed habitats. Throughout except the far southeast. **M. murex** is rather similar but hairless (except sometimes leaf undersides), with fruits with 5–9 anticlockwise turns, with 3 grooves along the margin, and *spines thick, thorn-like and pointed, 1–8 mm*. Similar habitats. Throughout.

M. turbinata

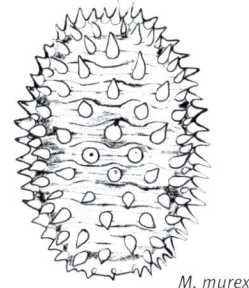

M. murex

(e) Fruit not coiled or sickle-shaped, borne at the end of *long, elongated stalks*.

Medicago hypogaea (syn. *Factorovskya aschersoniana*) A prostrate, mat-forming annual 20 mm–15 cm, often purplish with hairless stems. Leaves with 3 wedge-shaped leaflets 3–5 mm with few, large teeth, hairy beneath. Flowers yellow, very small; corolla 2–3 mm, solitary. Fruit 5–6 mm, hairy, 2-lobed, borne on *elongated stalks* 30–60 mm. Dry fields, pastures, and bare, open habitats at sea level. CY, IP (north), SY, TR.

---

1. *Medicago marina*
2. *Medicago littoralis*
3. *Medicago constricta*
4. *Medicago polymorpha*
5. *Medicago truncatula*
6. *Medicago laciniata*
7. *Medicago minima*
8. *Medicago coronata*
9. *Medicago disciformis*

Lotus BIRD'S-FOOT TREFOIL

Herbs with leaves with 5 leaflets (the lower 2 resembling stipules). Flowers often yellow, borne in flat-topped clusters; stamens 10, 1 free. Fruit several-seeded, splitting, smooth, or keeled and square in cross section.

(a) Plant a spreading, often woody-based perennial. Fruit elongated, not winged or keeled.

Lotus halophilus A slender, spreading annual to 20 cm, clothed in short, slightly adpressed hairs, with small leaflets 3–7 mm, oval below, elliptic above. Flowers yellow, 1–2(4) borne on stems 6–15 mm long; calyx narrowly bell-shaped with teeth shorter than or equalling the tube; corolla yellow, 6–8 mm, scarcely exceeding the calyx. Fruit 15–30 mm long, linear-cylindrical. Throughout. **L. peregrinus** is similar but with larger leaflets 10–15 mm long and flowers 8–12 mm. Fruit 25–30 mm, slightly compressed. Throughout.

Lotus corniculatus COMMON BIRD'S-FOOT TREFOIL A very variable (with many forms described), sprawling or creeping, almost hairless perennial to 50 cm with a woody stock and solid stems, and pinnately divided leaves that appear trifoliate because the lowermost pair is at the base of the stalk; leaflets oval, lance-shaped or round. Flower-heads with 2–8 yellow–orange flowers 10–16 mm (red in bud) on long stalks to 80 mm long. Fruit 15–30 mm long. Grassy places. GR, PN. Throughout. **L. pedunculatus** (syn. **L. uliginosus**) is similar but a vigorous, leafy perennial to 1 m, with *hollow and creeping, rooting stems*. Flowers 10–18 mm. Fruit 15–35 mm. Wet, marshy habitats. GR, IA, PN.

(b) Plant a *silvery-white hairy* perennial. Fruit elongated, not winged or keeled.

Lotus creticus SOUTHERN BIRD'S FOOT TREFOIL A variable (with many forms described), low, spreading, *grey–silver-hairy perennial* with stems to 1.5 m and leaves with oblong leaflets. Flowers 12–18 mm long, yellow with a purple-tipped keel, borne in long-stalked racemes; calyx distinctly 2-lipped. Fruit straight or curved, many-seeded and 20–40 mm. Rare in the area covered and strongly eastern; sandy maritime habitats. AA (Kos + Nisiros), IP, SY. **L. cytisoides** is similar but *less silvery* and with *small* flowers 8–14 mm long, with a short *curved keel* and notched standard petal. Mainly maritime habitats. Throughout, except the north.

(c) Plant a herbaceous, hairy to hairless (*not* silvery-white) annual. Fruit elongated, not winged or keeled.

Lotus edulis EDIBLE LOTUS A short, spreading, slightly hairy annual to 50 cm with oval leaflets. Flowers yellow with a purple-tipped keel, 10–16 mm long; calyx bell-shaped. Fruit oblong and curved, *very inflated*, with a groove along the back, 15–35 mm long. Garrigue and roadsides. Throughout, except the north.

Lotus conimbricensis A more or less hairless annual (or with sparse, long, white hairs) to 30 cm with oval to diamond-shaped leaflets. Flower-stalks shorter than the leaves, with solitary flowers; corolla 4.5–7 mm, *white to pink* with a violet keel. Fruit slender, 25–40(70) mm and curved upwards. Dry grassy places and dunes. Throughout, except the north.

1. *Lotus peregrinus*
2. *Lotus creticus*
3. *Lotus cytisoides*
4. *Lotus edulis*
5. *Lotus ornithopodioides*
6. *Lotus glinoides*
7. *Lotus tetragonolobus*
8. *Lotus tetragonolobus* fruit
9. *Lotus palaestinus*

Lotus ornithopodioides A short, spreading, hairy annual to 50 cm with stalked leaves with oval to diamond-shaped leaflets, the lower 2 triangular and heart-shaped at the base. Flowers yellow, 6–8 mm long, borne in *stalked, terminal heads of 2–5 on stalks slightly longer than the leaves* (at least in fruit); calyx 2-lipped; *teeth very unequal*. Fruit 20–40 mm long, *constricted* between the seeds. Damp habitats near water. Throughout. ***L. angustissimus*** SLENDER BIRD'S-FOOT TREFOIL is similar but with flowers in heads of 1–2(3) on stalks as long, or longer than the leaves (10–40 mm) and *calyx teeth nearly equal*. Fruit 15–25 mm, *much longer than calyx*. Throughout, except for some islands. ***L. halophilus*** is a low, whitish, prostrate annual similar to *L ornithopodioides* but with *scarcely stalked or unstalked leaves*. Maritime sands. Virtually throughout.

Lotus glinoides A prostrate, spreading annual with rather thick, blue–grey, trifoliate leaves with sparse to dense white hairs; leaflets broadest above the middle. Calyx with long, sparse, white hairs; flowers pink. Fruit narrowly cylindrical, reddish and shiny, often borne in divergently spreading pairs. Rare, in deserts. IP (south only).

(d) Plant a herbaceous annual or perennial. Fruit *square* in cross section, with a *keel or wing along the angles*.

Lotus tetragonolobus (syn. *Tetragonolobus purpureus*) ASPARAGUS PEA A low, sprawling to ascending, hairy annual to 60 cm. Leaflets 10–40 mm, oval to diamond-shaped and broadest above the middle. Flowers *dark red*, 18–23 mm long, solitary or paired, on stalks about equalling the leaves; calyx lobes *lance-shaped*. Fruit 35–75 mm, hairless, with broad wings along the angles. Roadsides, fallow land and garrigue. AA, CR, CY, GR, PN. ***L. palaestinus*** (syn. *Tetragonolobus palaestinus*) is similar but with flowers often *solitary* (less often 2) on stalks *shorter* than the subtending leaves and with *long, tapering, linear* calyx lobes; corolla *just* exceeding the calyx (not by x 1.5). Fields and hillsides. IP (north – common). ***L. conjugatus*** (syn. *Tetragonolobus conjugatus*) is similar but with smaller flowers, which are usually paired, 10–14 mm. Fruits *without wings* but with 2 thick ribs. Rare but distribution poorly known owing to confusion with the previous species. Rare. AA (east).

Cytisopsis

Related to *Lotus*. Spineless shrubs, often prostrate. Leaves simple or digitate; stipules absent. Flowers stalkless; stamens in 2 groups. Fruit a woody capsule.

Cytisopsis pseudocytisus A small, often prostrate shrub to 30 cm. Leaves with 5–7 linear to spatula-shaped, silky-hairy leaflets 7–12 mm. Flowers borne 1–3 in umbels; calyx hairy, often reddish; corolla dark yellow to yellowish-orange, 14–30 mm, silky-hairy; wings and standard exceeding the keel; standard oval. Fruit 15–20 mm, reddish, woody when ripe, persisting. Stony, rocky hillsides or maquis. IP (Mount Carmel – very rare elsewhere), TR.

Tripodion

Prostrate hairy annuals. *Calyx distinctly swollen* with sub-equal teeth; stamens 10, 1 free. Fruit non-splitting.

Tripodion tetraphyllum (syn. *Anthyllis tetraphylla*) BLADDER VETCH A low, spreading, hairy annual to 40 cm. Leaves normally with 3–5 leaflets, the terminal lobe largest and oval. Flowers to 25 mm, yellow with darker wings and a red-tipped keel, borne in dense lateral clusters of 1–7; calyx with silvery hairs, becoming very inflated and bladder-like in fruit, pale green tipped with red. Fruit 8–10 mm. Very common in a range of dry and disturbed habitats.

Ornithopus BIRD'S FOOT

Annual, erect or prostrate herbs with pinnately divided leaves with numerous leaflets. Flowers with 10 stamens, 9 in a tube, 1 free. Fruits *slender, often curved* (falcate).

Ornithopus pinnatus ORANGE BIRD'S FOOT A spreading, prostrate, slightly hairy annual. Leaves *pinnately lobed* with (3)6–7(14) leaflets. Flowers *orange–yellow*, 4.5–5.5 mm, borne in heads of 2–5(7) *without a leafy bract* below (bract absent or papery). Fruit linear, 16–33 mm long, flattened and markedly constricted between the seeds. Grassy and bare, stony habitats. Throughout except CY and much of central mainland GR.

Ornithopus compressus A low, prostrate to ascending, very hairy annual to 75 cm with pinnately divided leaves with *(2)8–15(18) pairs* of leaflets. Flowers 5–7 mm, yellow, borne in heads of 1–5, *subtended by a leafy bract*. Fruit slender, flattened and curved, slightly constricted between the seeds, 18–42 mm. Locally common in grassy places and on sand dunes. Throughout.

Galega

Herbs with simple, pinnately divided leaves and prominent stipules. Flowers borne in erect clusters; stamens 10, 1 fused to the other 9. Fruit splitting, not inflated.

Galega officinalis GOAT'S RUE A vigorous, erect, hairless or virtually hairless perennial to 1.5 m with pinnately divided leaves with 9–17 narrowly oval leaflets. Flowers white, sometimes flushed blue–mauve, borne in erect, long-stalked, cylindrical racemes; calyx teeth bristle-like. Fruit cylindrical, not inflated, to 30 mm long. Ditches and damp habitats, rather local. AA (rare), GR, IA.

1. *Cytisopsis pseudocytisus*
2. *Tripodion tetraphyllum*
3. *Ornithopus compressus*

Trifolium CLOVER

Annuals or herbaceous perennials with trifoliate leaves. Flowers numerous, short-stalked, typically clustered into rounded, congested heads; stamens 10, 9 in a tube and 1 free. Fruit 1–9-seeded, often enclosed in calyx. Numerous species occur in the region; just a small subset are described. Calyx shape and vein number are important.

(a) Calyx 5-veined; flowers typically *yellow* with a *boat-shaped* standard (similar species, difficult to distinguish).

Trifolium campestre HOP TREFOIL An erect or spreading *hairy* annual 50 mm–20(30) cm. Leaves alternate, the terminal leaflet longer-stalked than the laterals. Flowers, *yellow*, corolla (4)5–7 mm long, borne in *rounded heads of >20, 8–15 mm across*. Common in a range of habitats. Throughout. *T. dubium* LESSER HOP TREFOIL is similar but smaller, virtually hairless with flower-heads of 5–20, 5–9 mm across; corolla just 3–4.5 mm. AA (Thasos), GR, PN (north). *T. micranthum* is similar to the previous species but with stalkless leaflets, flower-stalks shorter or slightly exceeding their subtending leaves and flowers 3–8 per head, *minute*, 1.5–3 mm. Uncommon. AA, CR, GR, PN (northwest). *T. phitosianum* is similar to all the previous species but with long-stalked, lax heads of 2–5(7) flowers with very short calyx tubes (<1 mm) with very unequal teeth to 2 mm; corolla 4 mm, pale *lilac*. CR.

(b) Calyx with *>5 veins*; flowers typically whitish or pink (or red), borne in *elongated heads*.

Trifolium angustifolium NARROW-LEAVED CRIMSON CLOVER A somewhat hairy annual to 50 cm. Leaves with *linear-lance-shaped*, pointed leaflets. Flowers *pink*, 10–13 mm, borne in stalked, *cylindrical* heads, opening from the top downwards; corolla not (or scarcely) exceeding the calyx; calyx lobes unequal (the lowermost longest) and very slender and sharp, hairless at the tips. Disturbed or sandy habitats. Throughout. *T. infamia-ponertii* is similar but with cream to pinkish flowers and calyx teeth less sharply tipped, with *spreading hairs right to the tips*. Much-confused with the previous species. Virtually throughout except the far southeast and central mainland Greece.

Trifolium purpureum PURPLE CLOVER A robust, variable annual to 40 cm, rather similar to *T. angustifolium,* with *reddish-purple* flowers 15–25 mm opening from the base upwards, the lowermost withered when the uppermost open; calyx teeth *unequal*. AA, GR, IA (rare), IP, PN (rare) (probably elsewhere). *T. pamphylicum* is similar (also treated as a form of the previous species) but very slender, with narrowly elliptic leaflets and calyx tube short (<2.5 mm) with *virtually equal* teeth; corolla 12 mm, just exceeding the calyx, pinkish, ageing darker purple. AA (rare), CY, SY, TR.

Trifolium prophetarum An adpressed-silky annual 10–15 cm, similar to *T. pamphylicum* but with oblong-elliptic leaflets 10–20 mm, opposite uppermost leaves and calyx 6(7) mm with *very narrow, thread-like* calyx lobes, which are almost equal, and 2 x the length of their tube; corolla exserted from calyx for 1/3 its length; standard much longer than the wings. Seeds solitary, brown, smooth 1.2 mm. A rare endemic of desert and maquis scrub. IP. (Kinnroth Valley and Samarian Desert; very rare elsewhere).

Trifolium arvense HARE'S-FOOT CLOVER A softly hairy annual to 40 cm with narrow leaflets, scarcely toothed. Flowers pale pink or white, *borne in dense, elongated ovoid or cylindrical, stalked heads* to 25 mm long, the flowers usually *exceeded* by the calyx teeth; corolla just 3–6 mm, with hairless wings; calyx lobes more or less equal. Common in sandy or disturbed areas, sometimes in large numbers. Throughout. *T. affine* is similar but with more densely woolly flower-heads and corolla equalling or longer than the calyx, with hairy wings. AA (east), GR (northeast), TR.

Trifolium incarnatum CRIMSON CLOVER A robust, hairy annual to 65 cm with branched, erect or ascending stems. Leaflets oval to rounded, toothed towards the apex; stipules oval. Flowers 9–16 mm, very conspicuous, *bright blood-red*, to 12 mm long, borne in dense cylindrical heads; calyx 10-veined. Grassy habitats. GR, IA.

(c) Calyx with >5 veins; flowers typically whitish or pinkish, borne in *stalkless, rounded heads* in the leaf axils, often on spreading or prostrate stems.

Trifolium suffocatum SUFFOCATED CLOVER A distinctive *prostrate*, tufted, low annual <50 mm. Leaves with oval leaflets overtopping the *unstalked* rounded flower-heads, which are congested at the base of the plant. Flower-heads terminal and densely crowded; corolla white, 3–4 mm and shorter than the calyx; calyx tube nearly cylindrical. Local on coastal sands and bare ground. Virtually throughout.

1. *Trifolium campestre*
2. *Trifolium angustifolium*
3. *Trifolium purpureum*
4. *Trifolium pamphylicum*
5. *Trifolium prophetarum*
6. *Trifolium arvense*

Trifolium scabrum ROUGH CLOVER A spreading, often *prostrate* and *rather downy* annual to 20 cm with *leaves with prominent paler veins*, which are thickened at the leaf margins. Flowers 4–7 mm, rather inconspicuous, cream–white, turning pink with age, borne in small, unstalked heads to 10 mm across; calyx soon becomes stiff after flowering, with *starry, spreading, recurved, spiny teeth*, equalling or exceeding the corolla. Various disturbed habitats, towns and gardens. Throughout. ***T. lucanicum*** is similar but with ovoid-cylindrical flower-heads with rounded subtending leafy bracts, which conceal the lowermost flowers; corolla just *exceeding* the calyx, pink. Damp habitats. More common in the west. AA (rare), CR, GR, IA, PN.

Trifolium cherleri A spreading, *densely long-hairy* annual. Flowers white or pink, the corolla *equalling or shorter* than the calyx; *flower-heads densely woolly-hairy*. Throughout.

Trifolium tomentosum WOOLLY TREFOIL A slender, creeping annual to 15 cm, rather similar to *T. resupinatum* (see p. 260), but with leaflets 4–12 mm and smaller flowers 3–6 mm long and *inflated, spherical fruiting heads clothed in soft, white hairs*, short-stalked or stalkless (calyx teeth obscured). Throughout.

(d) Calyx with >5 veins; flowers typically whitish or pink (or yellowish), borne in *distinctly stalked*, often *rounded* heads terminally or in the leaf axils.

Trifolium clypeatum A hairy, sparingly-branched annual 20–40 cm with alternate lower leaves and opposite upper leaves; leaflets wedge-shaped, shallowly toothed towards the tips. Flower-heads ovoid; calyx tube 10-ribbed, the lower teeth longer than the others; corolla *long*, 20–30 mm, greatly exceeding the calyx, white to pale pink. Fruiting heads 25–40 mm with persistent, *spreading calyx teeth, the lower teeth 2 x as long as the others*. Restricted to the south and east. AA, CR (very rare), CY (common), IP, PN (northeast), SY, TR.

Trifolium lappaceum A slender, ascending to erect, virtually hairless annual 10–30 cm. Leaves with shallowly toothed, egg-shaped leaflets. Flowers borne on stalks much longer than their subtending leaves, in heads 12–16 mm across; calyx bell-shaped, 20-veined, *hairless*, with teeth slightly longer than their tube, with a broad, 3-veined base; corolla just exceeding the calyx, white or pink. Throughout.

Trifolium repens WHITE CLOVER A variable, creeping perennial to 50 cm with stems rooting at the nodes, often hairless. Leaves trifoliate with green, oval to elliptic leaflets, paler along the veins. Flowers normally white, sometimes pink or reddish, 7–12 mm long, borne in dense, long-stalked spherical heads; calyx tube longer than wide, the lobes triangular-lance-shaped; flowers scented. Fruit linear and compressed between the seeds. Cultivated for fodder, common in grassy places. Throughout. ***T. nigrescens*** is superficially similar, short, usually hairless and with oval to heart-shaped leaves, notched at the tips; stipules triangular. Flowers 5–9 mm, whitish (or pinkish), turning *grey* (or grey–brown), borne in rounded heads on stalks exceeding the leaves. Throughout. ***T. petrisavii*** is similar but with pink flowers becoming *dark* brown and firm in fruit, the fruit strongly constricted between the seeds. AA (east), GR (north – rare), SY, TR.

Trifolium pallidum A slender, ascending, much-branched annual 10–40 cm similar to the previous species, with egg-shaped to elliptic leaflets, the uppermost narrower and finely and sharply toothed; stems with adpressed hairs above. Flower-heads 15–25 mm across; calyx conspicuously 10-veined with very slender teeth arising from a 3-veined base; corolla 13 mm, usually just exceeding the calyx, cream to pale pink, the standard exceeding the wings and keel. Virtually throughout.

1. *Trifolium cherleri*
2. *Trifolium tomentosum*
3. *Trifolium clypeatum*
4. *Trifolium clypeatum* in fruit
5. *Trifolium lappaceum*
6. *Trifolium repens*
7. *Trifolium pallidum*

Trifolium stellatum STAR CLOVER A short, erect, hairy annual to 20(30) cm with stems simple or branched from the base. Leaflets oval and slightly toothed with oval, *toothed* stipules with bright green veins. Flowers pink or yellowish, 12–18 mm, borne in solitary, large *spherical* heads to 25 mm across in fruit; calyx equalling the corolla, densely white-hairy with slender, reddish, *spreading lobes* (star-like). Common in disturbed and grassy habitats. Throughout.

Trifolium echinatum SPINY CLOVER A variable, short, erect or sprawling, hairy annual. Leaflets oval to oblong and broadest above the middle; the *uppermost leaves opposite*. Flowers whitish or cream, flushed pale pink towards the apex, to 12 mm long, borne in rounded heads on rather long, slender stalks. *Fruiting heads rather spiny, with spreading, pointed calyx lobes*; teasel-like. Grassy and damp habitats. Local. AA (rare), CY, GR, IP, PN, SY, TR. *T. leucanthum* is a similar, densely hairy annual to 45 cm distinguished by its wedge-shaped leaflets and white or pink flowers, 7–9 mm, with corolla *equalling* (not exceeding) the calyx; calyx 10-veined with *equal*, lance-shaped, 3-veined teeth. Flowers borne in more or less spherical heads, often in pairs, on stalks to 12 cm. Dry places and pine woods. AA (rare), CY, GR, IP, PN, SY.

Trifolium resupinatum REVERSED CLOVER A spreading or sprawling more or less hairless annual to 30 cm. Leaves trifoliate with wedge-shaped leaflets 10–20(25) mm long. Flowers borne *upside down* in circular, flattened heads (superficially similar to *T. tomentosum*); pink to purple, 5–8 mm long, on rather short stalks; calyx conspicuously inflated and papery in fruit, rather hairy. Frequent in moist, sandy or grassy habitats. Throughout.

Trifolium fragiferum STRAWBERRY CLOVER A creeping perennial with stems to 30 cm, rooting at the nodes. Leaflets oval, without whitish marks. Flowers pink or purplish, 5–7 mm long, borne in rounded heads to 15 mm across; calyx *greatly swollen in fruit to give the appearance of a grey–pink berry*. Damp and grassy habitats and marshes. Throughout. *T. physodes* is similar but with stems not rooting at the nodes, and *larger flowers to 14 mm long*. Garrigue and pine forests. Throughout, except the north.

(e) Calyx with >5 veins; flowers typically whitish or pink (or yellowish), borne in terminal, short-stalked, rounded heads with 2 leafy bracts close below.

Trifolium pratense RED CLOVER A perennial to 60 cm, initially rosette forming, later spreading; rather hairy. Leaflets circular to oval, more hairy below than above. Flower-heads large, *to 30 mm across*, spherical or ovoid, solitary or paired, and usually *stalkless*; calyx tube 10-veined with triangular teeth that are bristle-pointed at the tips, the lowest tooth *much longer than the others*; corolla 12–18 mm, usually pink, rarely cream or white. Mainly in cool, grassy or mountain habitats. CY, GR (common), IA (rare), PN. *T. ochroleucon* SULPHUR CLOVER is similar in most respects but with *yellow–white* flowers, sometimes flushed pink, 15–18 mm long borne in virtually stalkless heads and unmarked leaves. Grassy habitats, usually above sea level. AA (north), GR, IA (Corfu), PN, TR. *T. caudatum* is similar to *T. ochroleucon* but with shorter, pink–purple flowers and calyx tube with long, sparse, spreading hairs. AA (Samos), TR.

Trifolium spumosum A spreading to erect annual to 30(70) cm, rather similar to *T. resupinatum*. Leaves trifoliate with wedge-shaped, toothed leaflets that are not strongly veined. Flower-heads more or less spherical, borne on short stalks; corolla 12–16 mm, pink, slightly exceeding the calyx; calyx *hairless* and inflated in fruit, forming somewhat spiky fruiting heads. Throughout.

Trifolium squamosum An erect or ascending more or less hairly perennial to 40 cm with elliptic leaflets; *stipules long and spreading*. Terminal flower-heads becoming ovoid, short-stalked and subtended by a pair of leaves; flowers 7–9 mm, pink–white; calyx teeth unequal and each with 3 veins >½ their length, exceeded by the corolla; calyx tube broadly cylindrical. Fruiting heads *resemble miniature teasels* for their spreading calyx teeth, broadly *cylindrical,* about 12 mm wide.

Widespread though local, in coastal grassland. Throughout (except CY and IP). ***T. squarrosum*** is similar but larger with lax flower-heads, hairy calyces *constricted* at the tips and with hairs with pimpled bases, and *large, ovoid* fruiting heads 20 mm across. Rare and local. AA (very rare), IA, PN (northwest).

Trifolium hirtum **HAIRY TREFOIL** A short, spreading, *hairy*, branched annual to 35 cm. Leaflets wedge-shaped and finely toothed towards the tip. Stipules *long and straight* with a hairy tip. Flowers pink to purple, 12–17 mm, borne in large, *densely hairy* heads to 20 mm across with a pair of leaves immediately below; corolla greatly *exceeding* the calyx; calyx 20-veined and with long hairs. Dry, stony habitats. Throughout except CR and IA.

1. *Trifolium stellatum*
2. *Trifolium resupinatum*
3. *Trifolium pratense*
4. *Trifolium eriosphaerum*
5. *Trifolium uniflorum*

(f) Flowers borne in *lax*, short-stalked heads with 2 leafy bracts close below, terminally and in the leaf axils. Fruiting heads conspicuously white-hairy.

Trifolium eriosphaerum An adpressed-hairy, prostrate annual 10–40 cm. Lower leaves long-stalked, upper leaves stalkless; leaflets 6–10(20) mm, egg-shaped with teeth near the blunt tips. Flowers borne in rows of 4–6 in small, congested heads; corolla 12–18 mm (much exceeding the calyx), pink–purple; calyx lobes thread-like and *white-woolly*. Fruiting heads *white-woolly*, borne on *downward-spreading* stalks in rows along the trailing stems. Seeds solitary, 1.5 mm. Grassy habitats, paths and fields. IP (common in the north), LB, SY.

(g) Flowers borne solitary or in groups of 2–3, *not in dense heads*.

Trifolium uniflorum A low, prostrate or *mat-forming* perennial to 30 cm. Leaflets rounded to diamond-shaped, strongly veined and hairy beneath; stipules broadly triangular. Flowers *solitary to few*, 1–3(5) in the leaf axils, white, cream or pink, 15–22(30) mm long; flowers rather large relative to the plant; standard petal strongly recurved; calyx with 10 veins. Fruit beaked with 3–5 seeds. Rocky habitats and tracks. Common in the Aegean, rare elsewhere. AA, CR, GR, PN, TR (western Anatolia).

Dorycnium

Herbs with 5-foliate leaves. Flowers white or pink; stamens 10, 9 fused and 1 free. Fruits swollen, exceeding the calyces; 1–many-seeded, splitting.

(a) Flowers >10 mm.

Dorycnium hirsutum DORYCNIUM A sprawling perennial to 50 cm (1.5 m), *densely woolly hairy*. Leaves pinnately divided with 5 leaflets, the lowermost pair stipule-like (true stipules minute). Flowers with a corolla *11–18 mm*, pale pink or white, with a dark red or blackish keel, borne in a compact, stalked raceme of 5–11. Fruit oblong, 6–10(13) mm. Common on sandy garrigue and dunes. Throughout (absent from CY).

(b) Flowers small, just 3.5–5 mm.

Dorycnium rectum A perennial with pinnately divided leaves, slightly hairy, with *numerous small flowers*; corolla 3.5–5 mm, white with a purplish keel. Damp habitats. Throughout.

Anthyllis

Perennials and shrubs with simple, trifoliate or pinnately divided leaves and stipules small or absent. Flowers numerous, borne in dense heads or interrupted racemes; stamens 10, 1 variably fused to the other 9. Fruit non-splitting.

(a) *Spiny* shrubs with leaves with 1–3(5) leaflets. Main inflorescence axis forming a terminal spine.

Anthyllis hermanniae An intricately spiny, leafy shub to 50 cm, with simple or more often trifoliate leaves. Flowers yellow, solitary or in groups of 2–5 forming interrupted racemes; the *main inflorescence axis forms a terminal spine*; calyx 3–5 mm with triangular teeth, shorter than their tube. Fruits 1-seeded. Garrigue; common throughout most of the Greek territories. AA, CR, GR, IA, PN, TR.

(b) Plant herbaceous, or woody only at the base.

Anthyllis vulneraria MEDITERRANEAN KIDNEY VETCH A variable, tufted, hairy perennial. Lower leaves with a single leaflet, upper leaves with 1–9(15) elliptic leaflets. Flowers red or purple, 8–20 mm long borne in long-stalked heads with a pair of *leaf-like bracts* beneath; calyx shiny with silky hairs, purple-tipped. Fruit 3.5–7 mm. Common in a range of dry habitats, often coastal. Throughout except the southeast (absent from CY and IP). **Subsp. *rubriflora*** is the most frequent and widespread form in the region (many other local forms described).

Hymenocarpos

Annuals similar to *Anthyllis*, with flattened, multi-seeded fruits with winged, membranous margins.

Hymenocarpos circinnatus An annual to 30 cm with dense, spreading hairs. Lower leaves simple, upper leaves with 2–3 pairs of leaflets, the terminal much larger than the laterals. Flowers deep yellow, 5–7 mm; calyx with teeth longer than their tube. Fruit 10–15 mm across, disk-like, flattened, with a slightly toothed margin. Garrigue and maquis. Throughout.

1. *Dorycnium hirsutum*
2. *Anthyllis hermanniae*
3. *Anthyllis vulneraria* subsp. *rubriflora*
4. *Hymenocarpos circinnatus*

LEGUMINOSAE (FABACEAE)

Coronilla

Hairless shrubs with pinnately divided (rarely trifoliate) leaves. Flowers with 10 stamens, 9 in a tube, 1 free. Fruit a cylindrical, pod-like legume.

(a) Plant a spreading shrub typically taller than 1 m.

Coronilla glauca (syn. *C. valentina*) SCORPION SENNA *A blue-green dwarf shrub* to 1.5 m, with pinnately divided leaves with (normally) 2–3 pairs of notched, elliptic, not fleshy, *short-stalked leaflets*. Flowers 8.5–12(14) mm, borne in lateral clusters of 4–8(13), yellow and *strongly scented*; calyx bell-shaped. Fruit 9–40(65) mm long, with 1–4(10) swollen segments. Garrgiue and maquis. Common on the northern Greek mainland; rare elsewhere. AA, CR, GR, IA.

Coronilla juncea RUSH-LIKE SCORPION VETCH A hairless shrub to 2 m with *rush-like stems* with long internodes and few leaves with 2–3 pairs of round to elliptic *fleshy leaflets* that are soon-falling. Flowers 8–12 mm, yellow, borne in very long-stalked clusters of 5–11(24). Maquis and hill slopes; a primarily western Mediterranean species. IA (Corfu).

1. *Coronilla juncea*
2. *Coronilla scorpioides*
3. *Securigera securidaca*
4. *Securigera parviflora*
5. *Securigera varia*

(b) Plant largely herbaceous, often lax, and usually <50 cm.

Coronilla scorpioides ANNUAL SCORPION VETCH A prostate to sub-erect, short, hairless, blue-green, *lax annual* to 40(55) cm. Lower leaves simple or trifoliate, the terminal leaflets elliptic to rounded, *much larger* than the other leaflets; *upper leaves simple or trifoliate*. Flowers yellow, often with brownish veins, 3–8 mm long borne in stalked heads of 2–5. Fruit *long and cylindrical*, ridged, 18–75 mm, curved, with 2–11 jointed segments. Locally common in dry, open habitats. Throughout. *C. repanda* is similar to *C. scorpioides* but with *upper leaves pinnately divided (2–4 pairs of leaflets) with a single terminal leaflet*, flowers becoming reddish on drying and a *curved*, segmented pod. Coastal habitats. PN.

Securigera

Herbaceous perennials with furrowed or ridged stems. Flowers with 10 stamens, 9 in a tube, 1 free. Fruit straight or slightly curved, cylindrical with 3–8(12) segments, scarcely constricted or with constrictions absent.

(a) Annuals with inflorescences of 3–9 flowers.

Securigera securidaca An annual to 40 cm, branched from the base. Leaves with 4–7 pairs of oblong, somewhat bluish leaflets. Flowers 4–8, short-stalked, borne in *spherical heads*; corolla 10–15 mm, *yellow*, with reddish veins. Fruit linear, compressed, to 10 cm long, with thickened margins. Throughout (rarer in the north).

Securigera parviflora A virtually hairless annual with weak stems 15–60 cm, branched form the base. Leaves with 3–6 pairs of oblong-elliptic, blunt leaflets. Flowers in heads of 5–9, borne on long, straight stalks; corolla 7–11 mm, *pink or yellow* (often co-occurring). Fruit linear, cylindrical, *curved* and jointed with 4–8 segments. Scattered but uncommon in mainland Greece. AA, CR, CY, GR, SY, TR. *S. carinata* is similar but with conspicuous lateral keels on the fruit. Rare. AA (Rhodes), TR (southwest). *S. cretica* has fewer (3–6), smaller (4–7 mm), white or pink flowers and *straight, erect* fruits. AA, CR, CY, GR, SY, TR.

(b) Perennials with inflorescences of 10–40 flowers.

Securigera varia (syn. *Coronilla varia*) CROWN VETCH A straggling to ascending, often patch-forming, leafy perennial to 1.2 m. Leaves pinnately divided with 11–25 pairs of oblong to elliptic leaflets with narrow, membranous margins. Flowers 10–20 *pink* (rarely white or purple), 8–15 mm long, borne in long-stalked, spherical heads. Fruit 20–60(80) mm long with 5–8(12) segments. Grassy habitats in the north (over-recorded elsewhere). GR (north). *S. globosa* is similar but bushy, much-branched, *woody* at the base. Flowers 15–40, usually white. Fruit with just 3–5 segments. An island endemic chasmophyte of cliffs and gorges. CR.

Hippocrepis

Herbaceous perennials with pinnately divided leaves. Flowers borne in axillary clusters; stamens 10, 9 fused and 1 free or partially so. Fruit divided into distinct, *horseshoe-like* 1-seeded sections, strongly compressed.

(a) Plant an annual herb.

Hippocrepis ciliata A slender annual herb to 25(35) cm with pinnately divided leaves with 3–6 pairs of linear to oval leaflets. Flowers borne in clusters of 2–6 on stalks *equalling* the leaves; corolla yellow, 2–4.5 mm long. Fruit 15–30(40) mm, flattened and curved upwards (sometimes into an almost complete circle), with very pronounced segments, and hairs on 1 side; *swellings on the interior side of the curve*. Local on coastal sands, dry slopes and pine forests. Throughout, except the southeast (not SY, IP; rare in TR).

Hippocrepis biflora A spreading, virtually hairless annual to 25 cm with pinnately divided leaves with (3)4–6 pairs of oblong, notched leaflets. Flowers 1–2(3), small, almost *stalkless in the leaf axils* (not in erect inflorescences), the short stalks *hairless*; flowers yellow, 5–8 mm; blade of the standard *gradually tapering* into the claw. Fruit very distinctive: broadly linear, 10–35(40) mm long, slightly curved and with (4)7–8(10) horseshoe-shaped constrictions. Dry, stony habitats especially in the east. Virtually throughout. **H. unisiliquosa** is similar to *H. biflora*, but with flower-stalks *hairy*, the blade of the standard *abruptly* distinct from the claw. AA (mainly east), CR (Koufinisi), CY, GR (southeast – rare), IP, PN (northeast – rare). **Subsp. *bisiliqua*** has flowers in groups of (1)2–3 and fruit sections 2–3 mm wide and 6–10 seeds. CY, IP. **Subsp. *unisiliquosa*** has solitary (sometimes paired) flowers and fruit sections 3.5–6 mm wide and 10 or more seeds. AA.

Hippocrepis multisiliquosa A variable annual similar to *H. biflora*, though variable. Stipules with 2 gland-like dots. Flowers larger, 4.5–7 mm long, solitary or in clusters of 3–6, each with hairless stalks, the cluster borne on a distinct stalk (peduncle). Fruit *curved downwards* (sometimes into an almost complete circle); hairless, 30–35 mm; *swellings on the exterior (convex) side of the curve*. Similar habitats. AA, CY, GR (southeast), IP, PN (northeast, Elafonisos, Kithira), SY.

(b) Plant a perennial or deciduous shrub.

Hippocrepis comosa HORSESHOE VETCH A spreading, almost hairless perennial to 30(50) cm with pinnately divided leaves with 7–25(31) leaflets to 8 mm long and yellow flowers 5–10(14) mm borne in dense, terminal, rounded, long-stalked clusters of 4–8(12); petal claw exceeding the hairy calyx. Fruit spreading, 10–30 mm long with 3–6(7) horseshoe-shaped constrictions, smooth or with minute swellings. Dry pastures and grassy habitats. GR, PN. **H. emerus (syn. *Coronilla emerus*; *Emerus major*)** is a *woody shrub to 2 m* with pinnately divided leaves with 2–4 pairs of leaflets. Flowers yellow, borne in clusters of 2–4(5), 14–21 mm long. Fruit with 3–12 segments, 50 mm–11 cm long. Woods and thickets. **Subsp. *emeroides*** is the form in the region. Virtually throughout (except CR and smaller islands).

Scorpiurus

Annual herbs with simple, alternate leaves. Flowers yellow; stamens 10, 9 fused and 1 free. Fruits elongated, strongly curved, with swellings or spines. Some authors recognise numerous taxa, most of which probably belong to the two species described below.

Scorpiurus muricatus A low, hairy or hairless sprawling annual to 60 cm with simple, elliptic, entire leaves with 3 prominent veins; the upper leaves short-stalked or unstalked. Flowers yellow, sometimes flushed with red, 9–13 mm long, borne in long-stalked clusters of 1–5; *calyx tube shorter than its teeth* (teeth 2–2.5 mm). Fruit to 50 mm long, coiled and twisted (so appearing short), variably covered in short, robust hairs. Common in dry, rocky and sandy habitats. **S. vermiculatus** is similar but with *solitary* flowers (rarely 2–3) 11–14 mm long, and fruits swollen, with *warts not hairs*, 5–7 mm across. Similar habitats.

1. *Hippocrepis ciliata*
2. *Hippocrepis biflora*
3. *Hippocrepis unisiliquosa* subsp. *bisiliqua*
4. *Hippocrepis unisiliquosa* subsp. *unisiliquosa*
5. *Hippocrepis emerus* subsp. *emeroides*
6. *Scorpiurus muricatus*
7. *Scorpiurus muricatus* flowers
8. *Scorpiurus vermiculatus*

Hedysarum SAINFOIN

Annual herbs with pinnately divided leaves. Flowers borne in dense axillary clusters; stamens 10, 9 fused and 1 free. Fruit broad, flattened and constricted between the seeds.

Hedysarum coronarium ITALIAN SAINFOIN A robust, rather hairy perennial to 1 m. Leaves pinnately divided with 3–5 pairs of elliptic to rounded leaflets, hairy beneath, untoothed. Flowers *magenta*, 15–18(20) mm long, borne in conspicuous, dense and long-stalked racemes of up to 30(50). Fruit with up to 4 spiny segments; hairless, 4–6 mm. Cultivated for fodder and naturalised on disturbed and fallow ground.

Hedysarum spinosissimum An ascending or spreading annual with stems 50 mm–35 cm and leaves with 5–7(9) pairs of narrowly elliptic to oblong leaflets 4–12 mm long, hairy below; stipules free, papery. Flowers borne on stalks exceeding the leaves in dense heads of 3–8; calyx 5–6 mm with virtually equal teeth much longer than their rube; corolla pale pink to pinkish-orange, 8–10(12) mm. Fruits with (1)2–3(4) *round segments covered in hooked spines*. Rocky slopes and garrigue, often near the sea; strongly southern and eastern. Virtually throughout.

Onobrychis SAINFOIN

Herbs with pinnately divided leaves. Flowers with 10 stamens, 9 in a tube and 1 free. Fruit hard, often rough-surface and jointed, non-splitting.

Onobrychis venosa A tufted perennial with short to absent vegetative stems. Leaves oblong, 40 mm–15 cm long with 5–9(11) oval to almost circular leaflets with *conspicuous bronze to purple veins*, with adpressed silky hairs below. Flowers 10–11 mm, borne in inflorescences arising from the leaf axils in dense clusters on long, spreading to ascending stalks 10–17 cm; corolla cream to lemon yellow with conspicuous red to purple veins. Fruit flattened, virtually circular, upcurved, 10–15 mm across with radiating veins. Garrigue or coastal sands; an island endemic. CY.

Onobrychis viciifolia PURPLE SAINFOIN An erect perennial 60(80) cm with pinnately divided leaves with 13–29 linear leaflets, adpressed hairy beneath. Flowers bright pink with darker veins, borne in dense, elongated heads of up to 50. Fruits 5–8 mm, pitted and with short spines. Widely cultivated and naturalised. GR (north).

Onobrychis caput-galli COCKSCOMB SAINFOIN A slender, grey-hairy, sprawling annual to 60 cm with leaves with 4–9 pairs of leaflets. Flowers 4–5 mm, rather inconspicuous, pink, borne on stems as long as or longer than the leaves, in clusters of 2–8; *calyx hairless*. Fruit 6.5–10 mm, distinctive in shape, hard, *compressed, with deep pits and slender spines*. Various habitats. Throughout. *O. crista-galli* is similar but with a *hairy calyx tube*. Fruits spiny and with rather *spiralled lobes* (like those of *Medicago*). AA (east), CY, IP. *O. aequidentata* is similar to the previous species but with fruits with *cockerel-crest-like lobes* (not with numerous spines). Throughout, except the far southeast.

Wisteria

Deciduous, woody vines with pinnately divided leaves. Flowers borne in long racemes; stamens 10, 9 fused, 1 free. Fruit a flattened, pod-like legume.

Wisteria sinensis CHINESE WISTERIA A vigorous, deciduous climber to 15 m with twining stems. Leaves alternate and pinnately divided with 9–13, untoothed leaflets 20–60 mm. Flowers pale lilac and pea-like, borne in long, drooping racemes 15–20(30) cm. Fruit oblong and velvety. Native to China, widely planted. Throughout. *W. floribunda* is similar but with leaves with up to 19 leaflets and *long racemes to >1 m in length*. Native to Japan, widely planted.

1. *Hedysarum coronarium*
2. *Hedysarum spinosissimum*
3. *Onobrychis venosa*
4. *Onobrychis viciifolia*
5. *Onobrychis caput-galli*
6. *Onobrychis crista-galli*
7. *Onobrychis aequidentata*
8. *Wisteria sinensis*

POLYGALACEAE | MILKWORT FAMILY

Herbs and small shrubs with simple, opposite or alternate leaves. Flowers cosexual, zygomorphic, in slender terminal racemes or spikes; sepals 5, corolla with 3(5) fused petals, stamens 8(9–10 in non-natives). Fruit a 2-seeded capsule.

Polygala

Annual herbs, perennials and shrubs. Flowers with 5 sepals and 3 (rarely 5) petals; stamens 8; ovary 2-parted.

(a) Plant an erect, slender annual with hairless stems.

Polygala monspeliaca A slender annual to 30 cm, *erect*, *simple*, or with *few branches*. Leaves linear-lance-shaped and broadest above the middle, blunt or finely pointed. Flowers borne in lax, terminal spikes; sepals rather large and white, 7.5–8 mm long with conspicuous veins, much exceeding the petals; bracts shorter than the flower-stalks; *corolla tube shorter than the wings*. Fruit winged. Open woods and maquis; common and widespread. Throughout.

(b) Plant a perennial *without* leaves in a basal rosette; leaves not distinctly stalked.

Polygala vulgaris COMMON MILKWORT A variable, short perennial with much-branched and erect or spreading stems to 30 cm, with scattered, linear-lance-shaped, all alternate leaves. Flowers borne in dense racemes, *small, 4–7 mm*, blue (frequently white or pink); outer sepals greenish, inner sepals slightly shorted than the corolla and with branched veins; bracts *shorter* than the flower buds and mature flower-stalks. GR, PN, TR.

Polygala nicaeensis NICE MILKWORT A variable perennial similar to *P. vulgaris* but with *larger flowers, 7.5–11 mm long*, borne in long, lax racemes of 10–40 flowers with bracts exceeding the flower-stalks when mature; sepals with netted veins; corolla tube <2/3 as long as the wings; keel *not* exserted. AA, GR, IA, PN. The following subspecies are widely recognised but their level of distinction is unclear: **Subsp. *mediterranea*** is a widespread mainland subspecies with linear to linear-lance-shaped leaves and scarcely hairy stems. AA, GR. **Subsp. *tomentella*** is similar but with densely hairy stems. AA, GR.

Polygala venulosa A small, woody-based perennial to 25 cm, similar to *P. nicaeensis*. Leaves alternate, stalkless, narrowly egg-shaped below, elliptic to lance-shaped above. Flowers borne in dense, terminal racemes; flower-stalks 1–1.5 mm; corolla pale pink or mauve, the *tube long, at least 2/3 as long as the wings*; wings whitish to lilac with greenish veins; keel *exserted*. AA, CR (common), CY, GR (southeast), PN (east), TR.

Polygala major LARGE MILKWORT A woody-based perennial to 30 cm with erect, dense, terminal spikes of *large*, rosy-purple (rarely blue or white) flowers, *2 x the length of the bracts and the flower-stalks*; inner sepals to 12 mm long with 3–7 veins netted at the margins and shorter than the petals. Fruit long-stalked. Grassy habitats above sea level. GR.

(c) Plant a shrub with woody branches.

Polygala myrtifolia An erect shrub to 2 m (often less) with oblong leaves, which are blunt and broadest above the middle. *Flowers large and showy: bright pink–purple, to 20 mm long*, borne in short racemes; inner sepals slightly longer than the petals. A native of South Africa, widely planted (throughout); scarcely naturalised.

ROSACEAE | ROSE FAMILY

Annual or perennial herbs, shrubs and trees. Leaves alternate, simple or compound, and with stipules (often soon-falling). Flowers with 4(5)(–10) sepals and (0)4–16 petals; stamens usually 2–4 x as many; carpels 1–many. Fruit highly variable, for example an achene, drupe follicle or capsule.

Agrimonia

Rhizomatous perennials with erect, hairy stems. Flowers yellow, in spike-like racemes; petals 5; stamens 5–20; carpels 2. Fruit with hooked bristles

Agrimonia eupatoria A hairy perennial 40–90 cm with leaves pinnately divided into 3–5 elliptic, toothed pairs of leaflets, alternating with much smaller leaflets. Flowers yellow, borne in long, very narrow spike-like racemes; petals 3.5–5.5 mm. Fruit nodding, grooved, with forward-directed, hooked bristles; achenes 1–2. Grassy habitats, pastures and olive groves. Throughout.

1. *Polygala monspeliaca*
2. *Polygala vulgaris*
3. *Polygala nicaeensis* subsp. *mediterranea*
4. *Polygala venulosa*
5. *Polygala myrtifolia*
6. *Agrimonia eupatoria*

Rosa ROSE

Prickly shrubs, generally deciduous with pinnately divided, toothed leaves and stipules. Flowers terminal, usually with 5 sepals and petals; stamens numerous; styles separate or fused into a column. Fruit a nut, enclosed in a fleshy (often edible) structure, called a *hip*. Not all species are described; many absent from hot, dry areas.

(a) Styles fused in a column; outer sepals usually entire.

Rosa sempervirens A vigorous, trailing or scrambling evergreen shrub to 5(10) m with stems with sparse, curved spines. Leaves leathery, *hairless, dark green and shiny*; leaflets 5–7, oval and sharply toothed (18–29 teeth). Flowers 3–10, borne in loose corymbs, white, 25–60 mm across with *entire*, soon-falling sepals; *styles hairless and united into a column*. Hip rounded to egg-shaped, to 10 mm, red. Scrub and open woods. Throughout, except the far southeast (not CY or IP).

Rosa arvensis A weak, trailing perennial with few, curved to straight, *slender* spines. Leaves with 5–7 elliptic, short-toothed leaflets, grey-hairy below, *sparsely hairy to hairless above*. Flowers to 32 mm across, borne (1)2–6 in loose corymbs; sepals *entire* (or occasionally with few lateral lobes); petals white. GR, PN.

Rosa heckeliana A small rhizomatous shrub 15–60 cm with scarcely curved to straight spines. Leaves with 5–7 small, oval to rounded, *small* leaflets 5–12 mm, distinctly *velvety-hairy on both sides*. Flowers short-stalked and solitary, 20–32 mm across, pale pink or white; sepals entire (or occasionally with few lateral lobes). CR (mountains in the west), GR, PN.

(b) Styles free; outer sepals usually lobed.

Rosa canina DOG-ROSE A lax shrub to 3(4) m with arching stems and *stout, curved spines*, longer than the width of the base. Leaves with 5–7 leaflets to 40 mm long, hairless, green. Flowers borne in *small* clusters (usually up to 4), 30–50 mm across and pink or white, with *hairless stalks*; sepals with narrow, usually entire, projecting side lobes. Hip red and hairless, without sepals when ripe. Grassy places, generally above sea level. Throughout. **R. phoenicia** is similar but with leaflets with large, broad teeth and inflorescences of *10–20 flowers*. AA (east), CY, IP, TR. **R. dumalis** is similar to R. canina but with *grey-green* leaves, often with a waxy bloom and long-persistent sepals. CR, GR.

Rosa × damascena DAMASK ROSE A suckering shrub with unequal prickles. Leaves with 3–5(7) leaflets. Flowers pink, showy and scented, borne 1–3, without bracts; sepals deflexed and pinnately lobed, falling before the fruit ripens; styles free; petals double. A hybrid of ancient origin, cultivated for rose water and rose oil; formerly an important cash crop in Syria. SY, TR.

Rosa spinosissima (syn. *R. pimpinellifolia*) BURNET ROSE A small, patch-forming shrub to just 50 cm (1 m), suckering freely. Stems with *numerous straight spines* along the main stems (shorter and less numerous on flowering branches). Leaves with 7–9(11) oval leaflets up to 15 mm. Flowers *solitary*, creamy-white, 30–40 mm across, with erect sepals and without a bract. Hip small and *purple–black* when ripe. Scrub and thickets. GR (north).

Rosa pulverulenta A variable, *short,* much-branched shrub to just 15–50 cm (1 m) tall. Spines straight to hooked, often mixed with bristles. Leaflets 5–7, small, rounded, hairy on both sides. Flowers solitary or few; sepals lobed, almost erect and persistent in fruit; petals pink to white, 24–44 mm across. AA (local), CR, GR, IP, TR.

Poterium BURNET

Herbs with pinnately divided leaves. Flowers inconspicuous, borne in dense, spherical, terminal heads; unisexual and cosexual mixed; petals absent; stamens 0–50. Fruit 1–3 achenes. DNA-based analyses support the distinction of this genus from *Sanguisorba* under which it was previously described.

Poterium sanguisorba (syn. *Sanguisorba minor*) SALAD BURNET A variable (with many forms described), greyish, bushy perennial with densely hairy stems 20–70(80) cm. Leaves forming a basal rosette, pinnately divided, with 9–12(25) pairs of elliptic, toothed leaflets. Flowers tiny, borne in egg-shaped heads 6–15 mm, the upper flowers female with reddish styles, the lower male with yellow anthers; sepals bright green; petals absent. Fruit enclosed in a ridged receptacle. Common in grassy places and fallow land. Throughout. Co-occurring forms include: **subsp.** *sanguisorba* (syn. *Sanguisorba minor* subsp. *minor*), the common form with unwinged fruit receptacles, and *P. verrucosum* (syn. *Sanguisorba minor* subsp. *verrucosa*), which is now generally recognised as a distinct species, differing in the fruit being enclosed by a sub-spherical receptacle with very prominent, lobed ridges. Probably throughout.

Poterium creticum (syn. *Sanguisorba cretica*) A *woody-based chasmophyte, forming suspended clumps* on rocky ledges with persistent dead leaves. Basal leaves numerous, long-stalked with 2–5 pairs of leaflets 12–20 mm long, broadly oval to rounded with 16–28 teeth. Flowers borne on stems exceeding the basal leaves; stem-leaves reduced. Flowering heads few, spherical, 15 mm across; lower flowers cosexual, upper flowers female; sepals greenish. Fruit 4-angled. CR (gorges in the southwest).

Sarcopoterium

Like *Poterium* but an intricately *spiny shrub*.

Sarcopoterium spinosum THORNY BURNET A low, *rounded, intricately spiny shrub with bare, grey, zig-zagging branches*. Leaves pinnately divided with 4–7 pairs of leaflets, soon-falling in summer. Flowers unisexual, borne in dense clusters. Fruit red and berry-like. Very common to subdominant in the south; rarer in the north and west. Throughout.

1. *Rosa sempervirens*
2. *Rosa canina*
3. *Poterium sanguisorba*

ROSACEAE

Potentilla CINQUEFOIL

Herbs or shrubs with lobed or pinnately divided leaves. Flowers with (4)5(6) petals; styles feathery in fruit; stamens (5)10. Fruit a head of achenes. A large genus but few species are common in the Mediterranean.

Potentilla micrantha A rhizomatous perennial with leafy rosettes, without stems rooting at the nodes. Leaves trifoliate with leaflets 10–40 mm with 12–22 teeth, sparsely hairy above, grey beneath. Flowers borne on short stems arising from the basal leaf axils; petals white to pink, equalling the sepals, which are visible between the petals (distinguishing it from the superficially similar wild strawberry, *Fragaria*). Mountain habitats. AA (local; north), GR, IA, PN.

Potentilla reptans CREEPING CINQUEFOIL A creeping perennial with a central rosette and stems *rooting at the nodes* with long-stalked hairless leaves with 5(7) leaflets; stipules leafy and entire. Flowers solitary in the leaf axils, yellow, 15–25 mm across with 5 notched petals; petals 2 x the length of the calyx. Damp, grassy habitats above sea level more or less throughout but absent from hot, dry areas.

Rubus BRAMBLE

A complex genus of typically woody, scrambling vines. Flowers with 5(–8) sepals and 5(–10) petals; stamens numerous. Fruit a head of 1–many 1-seeded drupes. Many closely related species, which are difficult to identify in the field; only the most common species are described here.

Rubus sanctus A spreading perennial bramble with robust, thorny, arching or prostrate stems, angled and grooved; prickles broad-based. Leaves with 3–5 leaflets and an abundance of *star-like hairs on the mid-vein* and terminal leaflets rounded with an abrupt end-point. Flowers with 5 pale pink, crumpled petals 10 mm long, borne in terminal, narrowly pyramidal inflorescences. Fruit black and shiny (a blackberry). Thickets and scrub. Throughout. The following species are similar: *R. canescens* has slender as well as broad-based prickles. GR, IA, PN, LB, SY, TR. *R. hirtus* also has slender as well as broad-based prickles, but leaves *green on both sides*. Mainly northern. GR, TR.

Crataegus HAWTHORNS

Spiny trees and shrubs with lobed or toothed leaves. Flowers borne in flat-topped clusters stamens (5)10–20; carpels 1–5. Fruit a berry.

Crataegus monogyna HAWTHORN A thorny shrub or small tree to 10(15) m with a dense crown above. Spines 10–25 mm, leaves oval to wedge-shaped, with (3)5–7 lobes, darker above, *with basal lobes with entire margins* (rarely with 1–2 teeth). Flowers white, with petals 5–8 mm and borne in clusters; *1 style*. Fruit a bright red berry 8–10(13) mm. Common in thickets, hedges and on roadsides. Throughout.

Crataegus azarolus MEDITERRANEAN MEDLAR A deciduous tree to 10 m with a dense crown; young shoots white-cottony, later blackish and with few spines. Leaves hairless or virtually so, oval with 1–2 pairs of lateral lobes, the lowermost cut to 50–60% the depth of the leaf blade. Flowers white,

---

1. *Sarcopoterium spinosum* habit; (inset) fruit
2. *Potentilla micrantha*
3. *Rubus sanctus*
4. *Crataegus monogyna*
5. *Crataegus monogyna* fruit
6. *Crataegus azarolus*
7. *Crataegus azarolus* var. *aronia*
8. *Crataegus rhipidophylla*

to 20 mm across, borne in dense clusters on *hairy flower-stalks*; *styles 2–3*. Fruit sub-spherical, to 25 mm long and orange–red or yellow. Various scrubby and wooded habitats. AA (Rhodes, Limnos, Simi), CR, CY, IP, TR. **C. azarolus var. aronia (syn. C. aronia)** is the common variety which has grey–green leaves, hairy on both sides and *yellow* fruits 10–12(15) mm across. **Var. *azarolus*** has green leaves, sparingly hairy above or hairless on both sides, and *red* fruits 13 mm across.

Crataegus rhipidophylla A thorny shrub or small tree to 7 m with hairless to sparsely hairy twigs. Leaves dark green and variably hairy above, grey–green below; wedge-shaped at the base, with 2–4 pairs of pointed lobes, each with 6–25 fine-toothed margins; leaves 20–65 mm long. Flowers 5–15(20) borne in lax corymbs; petals 5–7 mm; styles (1)2. Fruit 6–8(15) mm, spherical to cylindrical, bright red. GR.

Crataegus heldreichii A deciduous shrub or small tree to 6 m with densely downy twigs, the previous year's *pale* grey- to straw-coloured. Leaves slightly leathery, *wedge-shaped at the base* with 2–4 pairs of lobes; densely downy beneath. Flowers 5–20, borne in compact clusters; sepals broadly triangular and *deflexed*; petals 4–6 mm; styles 3–5. Fruit 6–8 mm, red with 3–5 seeds. AA (rare), GR, IA, PN.

Crataegus orientalis A shrub or small tree to 6 m with densely downy twigs, the previous year's *dark* reddish-black or brown. Leaves *grey-woolly* on both surfaces, with parallel-sided lobes, toothed at the apex, *gradually* tapering into the stalk. Flowers 4–18, borne in compact clusters; petals 4–9 mm; styles 2–5. Fruit large, 10–18 mm, red. AA (north), GR, IA (Corfu), PN, TR.

Pyracantha FIRE THORN

Spiny trees and shrubs. Flowers numerous, borne in flat-topped clusters; petals 5; stamens ~20; carpels 5. Fruit a berry.

Pyracantha coccinea FIRE THORN A small, very densely branched, spiny evergreen shrub to 2(6) m. Leaves 20–70 mm, elliptic, deep green and blunt-toothed, more or less hairless or hairy beneath when young, and with hairy stalks. Flowers creamy white, 8 mm across; stamens 20; styles 5. Berry orange or red with small persistent sepals. Cultivated and common as a hedge plant. Throughout.

Prunus CHERRY

Spineless or spiny deciduous or evergreen trees or shrubs with large, simple leaves. Flowers solitary or in racemes; sepals and petals 5; stamens 15–30(–numerous); carpel 1. Fruit a drupe (often edible).

(a) Flowers in clusters in elongated racemes borne on leafless stalls arising from the leaf axils.

Prunus laurocerasus An evergreen tree to 8 m with glossy, hairless, rigid leaves. Flowers white, borne in elongated racemes 80 mm–13 cm long, scarcely exceeding the subtending leaf. Fruit 12 mm, black. Widely planted.

(b) Flowers in clusters of 1–3(5); fruit characteristically plum-like, with a waxy bloom.

Prunus avium A deciduous *tree to 30 m* with smooth purplish-brown bark on the young branches, with prominent horizontal scar-like lenticels. Leaves alternate, simple, with a pair of glands on the upper part of the stalk; leaves 40 mm–12 cm, elliptic-oval and toothed; hairy beneath. Flowers white, borne 2–5 in umbels. Fruit spherical, 9–15 mm across, dark red to blackish, borne on *stalks at least 2 x its length*. Throughout (except CR and most of AA).

Prunus spinosa A deciduous shrub to 3 m with dull, dark brown, thorny twigs. Leaves 15–35 mm, narrowly oval-elliptic with toothed margins, dark green and hairless above, grey–green and hairy beneath; stalks 2–10 mm. Flowers white, petals 5–8 mm. Fruit 10–15 mm *blue–black* with a paler bloom. Frequent on the western and northern mainland. AA (rare), GR, IA, PN, TR. Most populations in the region belong to the poorly defined *subsp. dasyphylla,* which has small, spherical to conical fruits.

Prunus cerasifera CHERRY PLUM A shrub or tree to 8(10) m with *glossy, hairless* leaves 25–55 mm, markedly wavy-margined or toothed with forward-pointing lobes; *young twigs green, shiny and hairless*. Flowers white (or pink), borne mostly solitary, with or earlier than the leaves. Fruit round, red or yellow, 20–30 mm across. Woods, ravines and planted. Throughout except smaller islands, the south and southeast.

Prunus domestica PLUM A shrub or tree to 8(12) m with dull brown bark, and spiny branches in naturalised populations; *young twigs dull brown or grey*. Leaves 30–80 mm, *dull green and hairy, at least when young*. Flowers white, appearing with the leaves; petals 7–12 mm. Fruit a pendent round blue–black, red, green or yellow drupe 20–60 mm across (a plum). Widely planted throughout. **Subsp. *domestica*** is spineless with *greenish* white flowers. **Subsp. *insititia*** is the semi-wild, *spiny* form with pure white flowers. Scattered across the mainland.

Prunus prostrata PROSTRATE CHERRY A rigid, *low and spreading shrub* growing prostrate on the rock, with elliptic-egg-shaped leaves (4)7–12 mm, distinctly toothed. Flowers bright pink, mostly solitary; petals 6–7 mm. Fruit small, bright red, to 8 mm across. AA (local), CR, GR, PN.

1. *Prunus laurocerasus*
2. *Prunus avium*
3. *Prunus spinosa*
4. *Prunus spinosa* subsp. *dasyphylla*
5. *Prunus cerasifera*

1. *Prunus persica*
2. *Prunus dulcis*
3. *Prunus dulcis* fruit
4. *Prunus webbii*
5. *Cydonia oblonga*
6. *Pyrus spinosa*
7. *Pyrus spinosa* fruit
8. *Pyrus communis*
9. *Pyrus syriaca*

(c) Flowers in clusters of 1–3; fruit characteristically peach- or almond-like and densely downy.

Prunus persica PEACH A small tree to 6 m with *velvety* fruits (hairless in var. *nucipersica*, the nectarine); fruits yellow, orange or red, *large* rounded. Flowers solitary, remaining *mid-dark pink*; petals 10–20 mm. Cultivated for peaches. Throughout. *P. armeniaca* has broader, almost round leaves on stalks to 40 mm long, flowers always pale *pink–white*; cultivated for apricots. Throughout.

Prunus dulcis ALMOND A deciduous shrub or small tree to 8 m, densely branched and spiny. Leaves oblong to lance-shaped, hairless and toothed, 40 mm–12 cm. Flowers appearing before the leaves, pink in bud, fading to white, borne in pairs; petals 15–20 mm. Fruit oval and laterally *strongly compressed*, grey–green to yellow and conspicuously *velvety*, becoming dry and splitting, 35–46(50) mm. Maquis and in cultivation. Throughout.

Prunus webbii CRETAN WILD ALMOND A thorny, intricately branched shrub to 4 m, similar to *P. dulcis* but with more distinctly spreading branches, with twigs hairless with young; leaves smaller, 15–35 mm, both surfaces hairless. Flowers pale pink; petals 9–13 mm. Fruits *slightly compressed*, 12–25 mm, slightly grey-downy. AA (local), CR (+ KP), GR, IA, PN, TR. *P. graeca* is an intricately branched, thorny shrub to 2 m, similar to *P. webbii* but with twigs *white-downy* when young (later hairless and blackish-brown). Leaves 8–15 mm, narrowly egg-shaped to rounded, bright green and hairless above, densely white-downy beneath; stalks 1–2 mm. Flowers white, initially tinged red. Fruit compressed, grey-downy, 12–18 mm. AA (east), TR.

Cydonia QUINCE

Small, spineless, deciduous trees with simple leaves. Flowers solitary; stamens 15–25; carpels 5. Fruit hard and pear-like.

Cydonia oblonga QUINCE A shrub or small tree to 3(6) m with shoots hairy when young, later hairless. Leaves oval and entire, grey-hairy below. Flowers to 40–50 mm across, borne on short, hairy stalks; petals pink, exceeding the sepals. Fruit to 12 cm, many-seeded, yellow, pear-like and fragrant (a quince). Cultivated widely and persistent on derelict land. Throughout.

Pyrus PEAR

Small, spiny-branched deciduous trees with simple leaves. Flowers in clusters; stamens 20–30; carpels 2–5. Fruits large, fleshy and characteristically pear-like.

Pyrus spinosa A large, thorny, deciduous shrub or tree 3–7 m with downy twigs. Leaves 20–50 mm, much longer than wide, lance-shaped to narrowly egg-shaped, dark green and hairless above, downy beneath when young. Flowers white, borne in small umbels; petals 8–15 mm; anthers red to purple and conspicuous. Fruit (9)12–24(27) mm, yellowish-brown. Common in pastures throughout the area. *P. elaeagrifolia* is similar but with broader leaves, grey–green above and *persistently* grey-downy beneath. GR (northeast).

Pyrus communis PEAR A tree to 10 m with an open crown. Leaves *hairy-margined, untoothed*. Flowers with white petals 13–15 mm. Fruit the cultivated pear, *large*, 50–60 mm across. Cultivated widely in the region. Throughout.

Pyrus syriaca A small, usually spiny tree to 12 m with hairless twigs. Leaves cobweb-hairy when young, narrowly lance-shaped to oval-oblong, broadest in the lower half, 40–80 mm, with wavy-toothed margins; slightly leathery, dull beneath, with a rounded to wedge-shaped base. Flowers borne numerously in corymbs, white, 30 mm across; flower-stalks thick, 25 mm (50 mm in fruit). Fruit spherical to pear-shaped, 20–25 mm long with persistent sepals. Stony slopes and open woods. IP (north-central), SY, TR.

ROSACEAE

Malus APPLE

Trees with simple, toothed, deciduous leaves. Flowers borne in corymbs; stamens 15–50; carpels 3–5. Fruit large, fleshy and characteristically apple-like.

Malus pumila APPLE A tree to 10(20) m with leaves to 15 cm, *downy* on the undersurface, with stalks shorter than the blades. Flowers white (or pinkish), borne in clusters of 4–8; petals 13–29 mm long. Fruit to 12 cm across, variously coloured. Cultivated throughout; possibly naturalised.

Malus sylvestris A more or less spiny tree or shrub 2–10 m with leaves 40 mm–11 cm, *hairless* on both surfaces. Flowers large (30–40 mm across), white or pink; sepals 3–7 mm, downy within. Fruit yellowish-green, *small*, 25–30 mm across. Cultivated widely in the region. Throughout.

Eriobotrya LOQUAT

Small evergreen trees with large, leathery, entire leaves. Flowers borne in branched terminal clusters; petals 5; stamens 15–25; carpels 5. Fruits plum-like.

Eriobotrya japonica JAPANESE LOQUAT A dense, robust, small tree to 10 m. Young stems covered in red–brown felted hairs. Leaves elliptic, large, 10–30 cm long, dark green and shiny above, felted below; strongly veined. Flowers white, borne in terminal panicles 70 mm–17 cm long; petals 6–11 mm. Fruit ovoid, 30–60 mm, yellow when ripe with 2–3(7) large seeds, 10–20 mm. Widely cultivated as an ornamental and for its edible fruits in gardens. Throughout (especially hot, dry areas).

1. *Malus sylvestris*
2. *Eriobotrya japonica*
3. *Elaeagnus angustifolia*
4. *Paliurus spina-christi*
5. *Paliurus spina-christi* fruits

ELAEAGNACEAE

A family of trees and shrubs with scale-like hairs and alternate leaves. Flowers unisexual or cosexual; tepals and stamens 4; fruit drupe-like.

Elaeagnus

A genus of about 70 species of shrubs and trees, characterised by a covering of minute silvery scales, 4-parted calyces and no petals; stamens 4. Fruit fleshy and drupe-like.

Elaeagnus angustifolia OLEASTER A variably spiny small tree or shrub to 7(10) m with silvery young branches covered in minute scales (easily rubbed off). Leaves oblong-lance-shaped, green above and silver-scaly below, 15 mm–10 cm. Flowers butter-yellow and very fragrant, about 8–10 mm across, borne in axillary clusters. Fruit olive-shaped, 9–16 mm. Native to Asia, planted for ornament.

RHAMNACEAE | BUCKTHORN

Trees or shrubs with simple leaves with stipules. Flowers with 4–5 sepals and petals; petals small or absent, often hooded over the stamens; stamens 4–5. Fruit a fleshy black berry.

Paliurus

Spiny deciduous or evergreen shrubs or trees with alternate, simple leaves. Flowers stalked, cosexual, 5(6)-parted. Fruit dry, non-splitting, disk-like or hemispherical.

Paliurus spina-christi A spiny, deciduous shrub 2–3(5) m with alternate, arching branches with paired sharp, curved spines. Leaves 1–23(30) mm, oval. Flowers yellow, or whitish-yellow, 5(6)-parted, borne numerously in short cymes in the axils petals 1–1.6 mm. Fruit a brown, *papery disk* 18–26(30) mm across with a broad undulate margin. Common to subdominant in Greek mainlaind deforested vegetation. Throughout except CR and most of AA.

Rhamnus

Evergreen or deciduous shrubs with alternate, or almost opposite, toothed leaves and winter buds with scales.

(a) Leaves not, or scarcely leathery; always *deciduous*.

Rhamnus pumila A *spineless, spreading shrub to just 20(40) cm* with hairy young branches. Leaves not distinctly leathery, alternate, oblong to elliptic, 10–35(70) mm, with sparse, small teeth, hairless but with downy stalks 5–28(40) mm, *deciduous*. Flowers 3–5 mm across, borne in clusters; flower-stalks to 4 mm; calyx lobes 4, triangular and hairless. Fruits green and hairless, 3–6 mm. Usually in mountains. GR.

Rhamnus saxatilis ROCK BUCKTHORN A shrub similar to *R. pumila* but often *taller, to 1.5(2) m*, with prominent spines; leaf stalks short, (1)2–5(8) mm; leaves 10–25 mm. Flowers solitary, or in small fascicles, greenish-yellow. Fruit 5–8 mm, 2–3-lobed, black. Limestone rocks and garrigue. A variable species with regional forms described. AA (local), CR (mountains), GR, TR.

Rhamnus cathartica BUCKTHORN *A deciduous shrub or small tree to 6(8) m with scattered thorns and buds with dark scales. Leaves oval to elliptic, 40–90 mm, hairless and finely toothed with 2–4(5) conspicuous side veins. Flowers borne in small clusters; sepals and petals 4, male and female flowers separate. Fruit a blackish berry, 6–10 mm with 3–4 seeds. Local in limestone woodland. GR.*

(b) Leaves thick and leathery, evergreen or deciduous

Rhamnus alaternus MEDITERRANEAN BUCKTHORN An erect, superficially holly-like evergreen shrub to 5 m, not spiny and more or less hairless. Leaves 10–60 x 10–35 mm, alternate, *leathery*, oval and shiny dark green, toothed or sparsely toothed, with 3–6 side veins. Flowers borne in dense, cylindrical racemes, yellowish, borne in small clusters in the leaf axils, with 5 sepals and lacking petals. Fruit a berry, red ripening black, 4–6 mm with 2–3 seeds. Very common on maquis and garrigue. Throughout.

Rhamnus lycioides A rigid, intricately branched shrub, similar to *R. alaternus* but generally with narrower leaves, (4)6–40 x 0.5–8 mm; *spines present* and generally *4-parted* (not 5-parted) flowers. Fruit 4–6 mm, hairless. Throughout. The following co-occurring subspecies are considered by some authors to be distinct: **Subsp. *oleoides*** is evergreen with leaves 10–40 mm, leathery with conspicuous veins. **Subsp. *graeca*** is deciduous with rather thin, *smaller* leaves 6–18 mm, without conspicuous veins. **R. *pichleri*** is similar to *R. lycioides* but growing *prostrate* upon rock, and with fruits 3–4 mm, minutely hairy.

ULMACEAE | ELM FAMILY

Trees with simple, toothed, deciduous leaves; leaves usually assymetrical at the base. Flowers borne in small axillary clusters, cosexual or male, regular and inconspicuous; perianth bell-shaped, 4-5-lobed; stamens 4-5; styles 2. Fruit a 2-winged achene, notched at the tip.

Ulmus ELM

Distinctive for having leaves assymetrical at the base and unmistakable flowers and fruits. A complicated genus, mainly in temperate parts.

Ulmus canescens MEDITERRANEAN ELM A deciduous tree to 20 m with greyish bark and slender twigs, *white-downy* in their first year. Leaves alternate, oval-elliptic, bluntly toothed and *grey-hairy below*, with 12-16(18) pairs of veins, with scallop-toothed margins. Flowers borne in clusters before the emergence of the leaves; stamens purplish. Fruit a brown, broadly winged nut 7-15(17) mm, notched at the tip. Rocky gorges and ravines. Throughout. *U. minor* is similar (the two taxa are sometimes treated as subspecies), but with *hairless* twigs and leaves; leaves toothed, with 7-12 pairs of lateral veins. Fruit a broadly winged nut, the seed positioned above the middle. Throughout.

CANNABACEAE | HEMP FAMILY

A small family of annual or perennial herbs with lobed or palmate leaves. Flowers usually unisexual, the male with 5 stamens, the female with 2 styles. Fruit an achene.

Cannabis

Aromatic herbs with alternate upper leaves (sometimes opposite below), often dioecious. Male flowers borne in panicles; female flowers borne in compact racemes.

Cannabis sativa HEMP An erect, rather lax, strong-smelling annual to 2.5 m with alternate leaves, palmately lobed to the base into 3-9 narrow, toothed segments to 15 cm. Male flowers with tepals 3.5 mm, borne in branched clusters, female flowers borne 5-8 in compact racemes in the axils. Naturalised in sandy and disturbed habitats. A native of Asia, now a cosmopolitan weed. Throughout.

Celtis

Deciduous trees. Flowers male or cosexual; perianth 5-parted; stamens 5. Fruit a drupe.

Celtis australis EUROPEAN NETTLE TREE A deciduous tree to 25(30) m with smooth, grey bark and simple, alternate, rough, long-pointed, regularly sharp-toothed leaves 40 mm-15 cm long and small, green, solitary (sometimes 2-3) cosexual flowers that lack petals. Fruit a small, purple, berry-like drupe 8.5-12 mm, borne in clusters. Rocky ravines and planted. Virtually throughout. *C. tournefortii* is similar but shorter, to 6 m with pointed, oval to diamond-shaped, blunt-toothed leaves, *hairy beneath*. Rare and local. AA (local), CR (local), CY, GR, PN, TR.

---

1. *Rhamnus alaternus*
2. *Rhamnus alaternus* fruits
3. *Rhamnus lycioides* fruits
4. *Rhamnus lycioides* subsp. *oleoides* flowers
5. *Ulmus canescens*
6. *Ulmus minor*
7. *Cannabis sativa*
8. *Celtis australis*

MORACEAE | FIG FAMILY

A large, mainly tropical family of trees including the fig, mulberry and breadfruit. Flowers small and inconspicuous, crowded; male flowers with 4–5 stamens; styles 1–2. Fruit a mass of drupes surrounded by a succulent perianth.

Morus

Spineless trees with toothed to lobed leaves. Flowers unisexual, 4-parted, borne in short, dense spikes. Fruit fleshy.

Morus alba A smooth, slender tree. Leaves 60 mm–18 cm, oval, rounded or heart-shaped at the base, virtually hairless. Fruit compound and fleshy (a syncarp), 10–25 mm across, white, pink or purplish. Native to China, widely planted and naturalised. *M. nigra* has leaves hairy beneath and *dark purple* fruits. Widely planted.

Ficus FIG

Trees with (normally) simple leaves. Flowers unisexual; male flowers with 3-parted perianth; female flowers with 5-parted perianth. Fruit succulent (a fig).

Ficus carica FIG A deciduous tree 4–5(10) m with greyish branches and large, palmately lobed leaves to 35 x 28 cm. Flowers minute, borne within a *syconium*. The aggregate fruit is a fig; pear-shaped and purplish when mature and edible, 50–80 mm. Widely planted and naturalised on rocky slopes and near streams in the region; an important constituent of the Mediterranean maquis. Throughout.

Ficus elastica RUBBER PLANT A tall tree 30–35 m with large, thick, leathery green or yellow–green leaves 10–35 cm long. Native to the tropics; commonly planted as an exotic ornamental in gardens, parks and on roadsides in the region. Throughout.

Ficus microcarpa INDIAN LAUREL A tree with a grey trunk, oval, pointed, dark green, leathery leaves, and small yellow–brown fruits. Native to the tropics; occasionally planted as an ornamental, in the very warmest parts of the region.

URTICACEAE | NETTLE FAMILY

Annual or perennial herbs with opposite or alternate, simple leaves. Flowers typically unisexual, greenish with perianth with 1 whorl of 4, without petals; male flowers with 4 stamens; female flowers with 1 style. Fruit an achene.

Urtica NETTLES

Annual or perennial herbs with stinging hairs. Monoecious or dioecious; flowers usually inconspicuous and borne in dense clusters in the axils, forming elongated inflorescences; perianth of free tepals; stamens 4.

(a) Plant a *perennial* with markedly elongated inflorescences.

Urtica dioica STINGING NETTLE A familiar, robust rhizomatous, little-branched perennial to 1.5 m, with many leafy stems and with stinging bristles; plant rather *dull green*. Leaves 25–80 mm across, oval, heart-shaped at the base and toothed. Racemes to 10 cm, elongated, and exceeding the stalks. Damp, disturbed habitats, grassy places and woods; absent from hot, dry areas but very widespread. AA (rare), CY (rare), GR, IA, PN, TR.

(b) Plant an *annual* with markedly elongated inflorescences.

Urtica membranacea (syn. *U. dubia*) MEMBRANOUS NETTLE A bright green annual to 1.5 m. Leaves 20 mm–12 cm x 15–80 mm, with stalk of similar length to the blade. *Racemes long and conspicuous*, 15–90 mm long, often arching; lower racemes female, shorter than their stalks, upper racemes male and longer than their stalks, often coiling. Common in disturbed and grassy places with nutrient-rich soils, often coastal. Throughout.

(c) Plant an annual or perennial with inflorescences spherical or in clusters *not markedly elongated*.

Urtica urens SMALL NETTLE A short annual 10–60 cm, with abundant stinging hairs but otherwise virtually hairless, similar to *U. membranacea* but with *very short, spreading clusters of flowers to 20 mm or less*, not in elongated racemes, the flowers upward-swept in fruit; male flowers few; female tepals hairy-margined and unequal (outer segments smallest). Waste and agricultural land. Throughout.

Urtica pilulifera ROMAN NETTLE A hairy annual to 1 m with oval to lance-shaped leaves, toothed or untoothed, 50–80 mm across. Male flowers borne in interrupted racemes 40–70 mm long, female flowers borne on the same plant in *spherical heads, hanging from the upper leaf axils*, 5–10 mm. Waste places and near buildings. Widespread but local. Throughout.

1. *Morus alba*
2. *Ficus carica*
3. *Urtica urens*
4. *Urtica pilulifera*

Parietaria PELLITORY-OF-THE-WALL

Annuals or perennials with alternate leaves. Flowers borne in clusters in the leaf axils, mostly unisexual; perianth of equal tepals, eventually enclosing the fruit; stamens 3–4.

Parietaria judaica PELLITORY-OF-THE-WALL A short, tufted and *spreading, densely hairy perennial* with much-branched, reddish stems to 80 cm. Leaves oval, pointed, 10–50(70) mm. Plants monoecious; flowers borne in clusters of ≥3; *perianth tubular*, 3–3.5 mm. Walls and damp rocks. Common in towns. Throughout. **P. cretica** is similar but an annual with *small* leaves 15(20) mm long and just 3 flowers in each cluster. AA (common), CR (common), CY, GR, IA, PN, TR.

Parietaria officinalis An erect to ascending perennial similar to *P. judaica*, but much taller and *erect* with scarcely branched stems and bell-shaped, red perianth. Distribution unclear but probably local. GR, IA (Corfu), TR. **P. lusitanica** has abruptly tapering leaves with rather long stalks and flowers borne in clusters of 3–7 with a much smaller perianth to just 1.5–1.7 mm long (in fruit), scarcely exceeding its bracts. Throughout.

FAGACEAE | OAK FAMILY

Evergreen or deciduous trees or shrubs with alternate, simple leaves. Monoecious; male flowers borne in catkins, stamens 4–20; female flowers 1–few with 3–9 styles. Fruit 1–3(6) nuts enclosed in cup-like cupule of fused scales.

Castanea SWEET CHESTNUT

Deciduous trees with erect catkins; male flowers with 10–20 stamens; female flowers 3 at the base of long catkins. Fruits 1–3 enclosed in spiny, splitting cupule.

Castanea sativa SWEET CHESTNUT A large deciduous tree to 35 m, trunk with grey–brown bark, often with spiralled fissures. Leaves 10–30 cm long, oblong-lance-shaped, with pronounced veins and sharply toothed. Male flowers yellowish in long catkins (to 18 cm); female flowers few, on lower branches. Fruit a chestnut in a spiny, splitting husk. Native stands sometimes difficult to distinguish from naturalised specimens; also planted. Scattered across most of the region except many islands (not CY, IP and much of the east).

Quercus OAK

Evergreen or deciduous trees or shrubs. Male flowers with 4–12 stamens; female flowers 1–3. Fruit a distinctive nut (acorn) within a cupule.

(a) Tree evergreen or semi-evergreen.

Quercus infectoria A semi-evergreen shrub or small tree 2–5(10) m with virtually *hairless* twigs. Leaves rigid, *dull* above, slightly leathery, elliptic to broadly oblong, 30 mm–10 cm with 5–10 pairs of lateral veins; margins entire, spiny-toothed or shallowly lobed, both surfaces bright to grey–green with scattered hairs; virtually hairless below when mature. Fruit ripening in its first year with a cupule 12–22 mm across with *adpressed*, grey scales; acorn protruding for about 2/3 its length. Throughout (in its wider treatment). The following taxa are widely recognised but only differ subtly: **Subsp.** *infectoria* has small leaves with short stalks and smallish acorns. AA (Rhodes, Samos, Chios, Lesbos). *Q. boissieri* is a rather *tall tree* to 20 m with rather large, oblong to oblong-lance-shaped leaves 50 mm–10 cm, undulate, regularly sharp-toothed (8–11 teeth on each side), densely hairy beneath when young; later hairless or nearly so. Fruit 30–40 mm. IP, SY, TR. **Subsp.** *veneris* has widely oval leaves with longer stalks and large acorns. AA (Rhodes, Thasos), CY. *Q. petraea* is similar but with herbaceous (not leathery) leaves, clearly lobed, and with *grooved* leaf stalks. GR (north), TR.

Quercus coccifera KERMES OAK A dense, evergreen shrub to 2(15) m with scaly, grey bark. Leaves 15–50 mm, oval-oblong with spiny teeth, short-stalked (1–5 mm), leathery, and *hairless when mature*, with veins prominent above but not beneath; midrib straight. Fruit 10–20 mm, ripening in the second year with conspicuous scales on the cupule; borne solitary or in pairs. A red dye is derived from a scale insect that commonly infects this species. An important component of garrigue and maquis ecosystems across the region. Throughout. The following subspecies are widely

---

1. *Parietaria judaica*
2. *Parietaria officinalis*
3. *Castanea sativa*
4. *Quercus infectoria* subsp. *veneris*
5. *Quercus boissieri*

recognised but their distinction is often unclear: **Subsp.** *coccifera* is the typical low shrubby, small-fruited form in the north and west. **Subsp.** *calliprinos* (also treated as a true species) has a more robust habit, its large, flat leaves 40–50 mm long and *larger* fruits with erect, longer, appressed scales on the cupule, which is 20–30 mm across. CY, IP, SY, TR (south). *Q. aucheri* is very similar but with *small* leaves just 15–35 mm, blue–grey and waxy beneath, with dense and persistent hairs. AA (east), TR (southwest).

Quercus alnifolia GOLDEN OAK An evergreen shrub or much-branched, wide-crowned small tree to 10(14) m (typically shorter). Leaves 15–60 mm long, dark, shiny green above, golden-downy below, rigid and leathery, the margins variably lobed, or not. Fruits maturing in their first year, solitary or in clusters; cupule 8–15 mm across with linear, strongly recurved scales; acorn 20–25 mm. Rare, only on the igneous rock formations of the Troodos Mountains in Cyprus in altitudes between 400 m and 1800 m among *Pinus brutia* trees. CY.

Quercus ilex HOLM OAK A large evergreen tree to 27 m with downy young branches and grey bark. Leaves 20–80(90) mm, leathery, oblong to lance-shaped, untoothed or sometimes spiny-toothed, *downy*, with 7–15 pairs of *prominent veins beneath* when mature; stalks 3–10 mm. Fruit 15–35 mm, bitter-tasting, with dense scales on the cupule. Common in ravines and gorges, sometimes in great stands; often planted. Throughout except the southeast (not CY or IP).

(b) Tree deciduous.

Quercus pubescens (syn. *Q. humilis*) DOWNY OAK A deciduous shrub or tree to 25 m with densely downy young branches. Bark grey and fissured, finely cracking into scales. Leaves grey–green, oblong to lance-shaped and bluntly-lobed, to 12 cm long, *densely downy beneath*, particularly when young; short-stalked (5–12 mm), the stalks not grooved above. Cup with narrow, closely appressed scales. Woods and rocky slopes; the most common species in much of Greece. Throughout, except the southeast.

Quercus ithaburensis A deciduous tree to 15 m with a stout trunk and *broad, semi-circular crown*; twigs densely downy. Leaves 40 mm–13 cm, oval-oblong with 4–7 pairs of short, triangular, rather pointed lobes. Fruit ripening in the second year, conspicuously *large,* the cupule 40–80(90) mm across with linear to oblong scales; acorn protruding for about ½ its length. **Subsp.** *ithaburensis* has an *oblong* acorn + cupule outline, with 28 mm of the acorn protruding. IP (possibly SY). **Subsp.** *macrolepis* has a *spherical* acorn + cupule outline, with just 12 mm of the acorn protruding. Forms extensive forests in the west. AA, CR, GR, IA, PN, TR.

Quercus cerris A deciduous tree to 25 m. Leaves 50 mm–13 cm, oblong-lance-shaped to elliptic with 4–9 pairs of short, triangular, rather *pointed* lobes, the upper surface rough. Fruit ripening in its second year; cupule (20)30(35) mm across, borne on stout stalks 1–7 mm, hemispherical with narrow spreading to deflexed scales; acorn protruding for ½ its length. Mostly in the east. AA (rare), GR, IP, LB, SY, TR.

Quercus libani A deciduous shrub or small tree with hairless branches and leaf stalks. Leaves hairless on both sides when mature, glossy above, rounded to wedge-shaped at the base and stalked; blade lance-shaped with coarse, upward-pointing, triangular teeth. Fruits virtually stalkless, solitary; cupule with greyish, *addressed, diamond-oval scales,* the uppermost longer; acorn slightly protruding. Woods. IP, LB, SY, TR.

1. *Quercus coccifera* fruits
2. *Quercus coccifera* subsp. *calliprinos*
3. *Quercus alnifolia*
4. *Quercus ilex*
5. *Quercus ilex* fruits
6. *Quercus pubescens*
7. *Quercus ithaburensis* subsp. *macrolepis*

JUGLANDACEAE

Trees with alternate, pinnately divided leaves. Flowers unisexual; the male in drooping catkins with 3–many stamens, the female in terminal spikes, each flower with 2 styles. Fruit a drupe or winged nut.

Juglans

Trees with pinnately divided leaves with 3–9 leaflets. Female flowers borne in racemes, erect in fruit. Fruit an unwinged nut.

Juglans regia WALNUT A large, deciduous tree to 24 m with pale, smooth bark, becoming fissured. Leaves alternate, pinnately divided with 7–9 elliptic, untoothed lobes to 15 cm. Male flowers borne in drooping catkins to 15 cm, female flowers in short, erect spikes. Fruit a large, edible nut >30 mm. Throughout, except for parts of the far southeast.

CASUARINACEAE

Superficially conifer-like trees and shrubs native to Australia, Southeast Asia, Malesia, Papuasia, and the Pacific Islands, characterised by drooping horsetail-like twigs and cone-like fruiting structures.

Casuarina

Evergreen trees native to Australia. Leaves much reduced, borne on drooping greyish twigs. Flowers unisexual or cosexual, borne in catkin-like inflorescences.

Casuarina equisetifolia A large, evergreen, *conifer-like* tree to 3 m with a straight trunk with light grey–brown bark and much-branched crown. Twigs needle-like, grey green, 23–38 mm x 0.5–1 mm, jointed, with reduced, minute, tooth-like leaves in whorls of 2–8 per node. Flowers unisexual; male flowers borne in elongated spikes 7–40 mm; female flowers borne on short, cone-shaped spikes 10–24 mm. Fruit solitary, grey–brown, nut-like, borne in a woody, cone-like structure. Native to Australia, planted in hot, coastal areas.

BETULACEAE

Deciduous trees or shrubs with simple, alternate leaves. Flowers unisexual, inconspicuous; stamens 2–14; styles 2. Fruit a nut, winged or not.

Alnus ALDER

Trees with entire leaves. Male flowers 3 per bract, with minute tepals and 4 stamens; female flowers borne in small, erect stalked groups, 2 per bract. Fruit narrowly winged.

Alnus glutinosa ALDER A tree to 29 m with dark brown, fissured bark and hairless twigs. Buds purplish, short-stalked. Leaves oval with a cut-off or notched tip and *abrupt to wedge-shaped at the base,* often with double-toothed margins, and hairless. Male catkins cylindrical and pendent, to 50 mm long; female catkins oval, 8–28 mm, with female flowers 3–8 per stalk, and woody when mature; all catkins appear before the leaves. Local in wet habitats (absent from hot, dry areas and most islands). AA (local), GR (common), PN (northwest), TR.

Carpinus HORNBEAM

Trees with entire leaves. Male flowers 1 per bract, without a perianth; stamens ~10; female flowers borne in pendent catkins, 2 per bract. Fruit winged or not, each nut with an enlarged, usually 3-lobed bract.

Carpinus orientalis EASTERN HORNBEAM A deciduous shrub or small tree to 15 m with smooth, purplish-grey bark. Leaves 30–50 mm, oval to elliptic, pointed and finely toothed, slightly hairy. Male catkins pendent, female catkins borne on the same tree with large, leafy, *unlobed* bracts (unusual in the genus), to 20 mm. Fruit a small winged nut, 3 mm. Mainly northern, absent from most islands. GR, PN, SY (north), TR.

CUCURBITACEAE | CUCUMBER FAMILY

Herbs, often climbing or trailing with tendrils, and alternate leaves. Flowers unisexual, sepals and petals 5; stamens usually 3; stigmas 2–3 on 1 style. Fruit succulent and berry- or pod-like.

Cucumis

A mostly tropical African genus of climbing or trailing, mostly monoecious, bristly herbs with *simple* tendrils. Female flowers solitary; male flowers in clusters; calyx 5-lobed, corolla 5-lobed, yellow; stamens 3, free. Fruit smooth, bristly or spiny; seeds smooth and compressed.

Cucumis prophetarum A rough, grey, prostrate perennial with numerous angle-grooved stems with bent joints. Leaves firm, oval with heart-shaped bases to palmately 3–5-lobed, with crimped margins; tendrils short. Flowers small, unisexual, the male in groups of 2–3, the female solitary; calyx tube 3–5 mm with spreading, linear lobes; corolla 3–5 mm. Fruit ovoid to spherical, softly bristly, striped green and white, yellow when ripe, 30–40 mm. Mainly desert scrub. IP (eastern), SY.

Citrullus

Prostrate herbs with *branched* tendrils. Flowers solitary; corolla 5-parted beyond the middle, bell-shaped. Fruit fleshy or dry.

Citrullus colocynthis DESERT GOURD A bristly, grey, spreading perennial with branched tendrils. Leaves triangular in outline, deeply 3–7-lobed, the lobes again incised. Flowers solitary, bright yellow. Fruit *melon-like*, pale green with dark green mottled stripes, later all yellow; about the size of a grapefruit. Seeds smooth, without markings. Flat sandy places, where water collects seasonally, in otherwise arid habitats. CY, IP, SY, TR.

Ecballium

Herbaceous, bristly monoecious perennials without tendrils. Corolla yellowish; stamens 5. Fruit an explosive pod.

Ecballium elaterium SQUIRTING CUCUMBER A spreading, very bristly perennial to 1.5 m without tendrils, with a tuberous rootstock. Leaves rough, with bristles, more or less triangular and long-stalked (to 13 cm). Flowers small, 20–50 mm across and pale yellow; male and female flowers borne separately. Fruit bristly and pod-like, 30–45 mm borne on long stalks, *exploding violently* from the point of attachment when ripe. Sandy waste ground. Common throughout the Mediterranean region.

Bryonia BRYONY

Climbing dioecious perennials with long, spiralling, unbranched tendrils. Flowers greenish white; stamens 5. Fruit a red berry.

Bryonia dioica (syn. *B. cretica* subsp. *dioica*) WHITE BRYONY A dioecious, climbing, hairy perennial to 4 m with simple, *coiled tendrils*. Leaves deeply palmately 5-lobed and plain green. Flowers greenish-white with darker veins, 5–12 mm across, borne several together. Fruit a berry 5–9 mm, green with white markings then red when ripe. Woods, thickets and maquis. Throughout (rarer on the mainland).

CELASTRACEAE | SPINDLE FAMILY

Shrubs or trees (or woody climbers) with simple leaves. Flowers cosexual or unisexual with a nectar disk, with 4–5 sepals, petals and stamens; style 1. Fruit variable, often a succulent, 3–5-angled capsule with seeds with a bright orange–red aril.

Euonymus SPINDLE

Shrubs or small trees with deciduous or persistent, opposite leaves. Flowers cosexual (rarely unisexual) with 4–5 fused sepals and 4–5 stamens. Seeds with a conspicuous aril.

Euonymus europaeus SPINDLE A bushy, deciduous shrub or small tree to 2–3(8) m with green, 4-angled twigs. Leaves elliptic to narrowly lance-shaped, 30–80 mm (11 cm) long, opposite, pointed and finely toothed. Flowers borne in branched inflorescences in the leaf axils, greenish-white, 4-parted with lobes 8–15 mm across. *Fruit 4-lobed and coral-pink, opening to expose orange seeds*. Woods and scrub, often on higher ground. GR, IA (Corfu), PN, TR.

OXALIDACEAE | OXALIS FAMILY

Perennial herbs, often with a bulbous stock or rhizomes, with clover-like leaves. Flowers with 5 petals and sepals; stamens 10; styles 5. Fruit a capsule.

Oxalis

Perennial herbs, often on disturbed ground, with clover-like leaves with 3 leaflets (ternate). Flowers regular with 5 petals and sepals. Root morphology an important diagnostic.

(a) Flowers yellow.

Oxalis pes-caprae BERMUDA BUTTERCUP A low, tufted perennial with numerous leaves arising from a bulbous stock; far-spreading; leaves withering soon after flowering. Leaves trifoliate and clover-like. *Flowers bright yellow, tubular* with petals 13–26 mm, borne in loose umbels on long stalks to 30 cm. An abundant and highly invasive weed in some coastal regions and disturbed land. Throughout. *O. corniculata* PROCUMBENT YELLOW SORREL has yellow flowers but with *stems rooting at the nodes* and (often) dull-purple leaves. Petals 5–9 mm long. Cultivated land and coasts. Throughout.

1. *Juglans regia*
2. *Casuarina equisetifolia*
3. *Cucumis prophetarum*
4. *Citrullus colocynthis*
5. *Ecballium elaterium*
6. *Bryonia dioica*
7. *Oxalis pes-caprae*
8. *Oxalis corniculata*
9. *Oxalis articulata*

(b) Flowers bright pink (species virtually identical and observation of the roots necessary).

Oxalis articulata PINK OXALIS A tufted perennial with *brown scaly rhizomes* and long-stalked, all basal, clover-like leaves with heart-shaped leaflets, often covered with orange or brown dots. *Flowers pink,* borne in broad, umbel-like clusters; on stalks to 35 cm; petals slightly hairy, 12–19 mm long. Waste ground and gardens, virtually throughout. *O. debilis* is scarcely distuinguishable, also with leaves with orange dots and pink flowers with petals 10–18 mm long, differing mainly in producing *underground bulbils*. Distribution not known owing to confusion with the previous species.

HYPERICACEAE

Shrubs or herbs, often with numerous glands and simple, opposite or whorled leaves. Flowers regular, yellow, with 5 free petals and sepals; stamens numerous; styles 3–5. Fruit a capsule, or succulent and berry-like.

Hypericum

Herbs and shrubs easily recognised by their opposite leaves and flowers with 5 yellow petals and many stamens. Numerous species in the area, just a few of which are described here.

(a) Plant a herbaceous perennial.

Hypericum perforatum PERFOLIATE ST JOHN'S-WORT An erect herb to 50(80) cm with a cylindrical stem with *2 lines running down its length*. Leaves oval to linear, to 30 mm long, *scarcely stalked*, hairless and blunt with numerous transparent glands. Flowers bright yellow, with petals 9–15 mm, often with black dots along the edges; sepal margins entire. One of the most common species in a range of habitats and altitudes. Throughout. *H. triquetrifolium* is similar but shorter, <50 cm, bushier, often with much-branched, spreading stems and *small leaves clasping the stem and with markedly wavy margins*. Various habitats including garrigue, car parks and path margins. Throughout except the north.

Hypericum perfoliatum An erect to spreading, normally hairless perennial to 70(80) cm with stems with 2 lines. Leaves blue–green, opposite, lance-shaped, *clasping the stem at the base*, without wavy margins, 8–50 mm long. Flowers yellow with petals 8–12 mm and blunt sepals with black markings. Fruit 8–10 mm with raised orange warts. Various habitats. Throughout, except the far east, southeast and north.

Hypericum tetrapterum SQUARE-STALKED ST JOHN'S-WORT An erect, herbaceous perennial to 60 cm (1.2 m) with *square stems with 4 winged angles* (0.25–0.5 mm). Leaves small and oval and unstalked, 9–35 mm. Flowers pale yellow with petals 5–6.5(7) mm long; sepals narrow and pointed, without black dots. Waterlogged habitats and woods. AA (rare), GR, IA (Corfu), PN.

(b) Plant a woody-based perennial or shrub.

Hypericum olympicum An erect to spreading, hairless, woody-based perennial 10–50(70) cm with 2-lined stems. Leaves 5–28(36) mm, narrowly oblong to narrowly elliptic or lance-shaped, grey–green. Flowers yellow, few, *large* (20–60 mm across) with unequal, leafy, hairless sepals, sometimes with black glands; petals persistent, rarely with marginal black glands. Seeds pitted. Dry stony habitats. GR, TR.

Hypericum hircinum STINKING TUTSAN An erect, semi-evergreen shrub to 1.5(3) m with (2)4-angled stems, *smelling of goats when crushed*. Leaves narrowly lance-shaped to broadly oval and unstalked, 12–50 mm long. Flowers yellow with petals 10–8(20) mm long borne in few-flowered, branched terminal inflorescences; *sepals shorter than the petals* (4–7 mm); *stamens longer than the petals*. Fruit ellipsoidal, 6–13(16) mm long. Damp and shady habitats. Virtually throughout.

Hypericum aegypticum EGYPTIAN ST JOHN'S-WORT A low, spreading shrub to 30 cm. Leaves small, green and rather crowded, narrowly oblong and leathery and slightly concave. *Flowers borne singly* (but many to a stem), in long spikes (unless stunted), bright yellow, rather flax-like. Coastal cliffs and scree; rare and local. CR (near Sitia in the east; possibly Akrotiri), IA, PN (west).

Hypericum empetrifolium An erect, tufted shrub to 50 cm. Leaves 2–12 mm, borne in *whorls of 3*, hairless. Flowers borne in elongated panicles; sepals with stalkless, black glands along the margins; petals and stamens *soon-falling*. Capsule rough. Rocky habitats. AA, CR, CY, GR, IA, PN, TR.

ELATINACEAE | WATERWORT FAMILY

Small aquatic annuals with simple, opposite leaves and leafy stipules. Flowers minute, solitary or 2–5 in the leaf axils, cosexual and regular; sepals and petals 3–4; stamens 2 x as many as petals; styles 3–4, very short. Fruit a capsule.

Elatine

Aquatic or water-margin (sometimes terrestrial) annuals, superficially similar to Portulacaceae or Caryophyllaceae but with distinctive floral characteristics and leafy (not papery) stipules, unlike the latter family.

Elatine alsinastrum A short, hairless annual to 50 cm with leaves in *whorls* of 6–18, linear to lance-shaped or oval, to 25 mm if submerged, 5–13 mm if exposed. Flowers solitary, minute, reddish or greenish, unstalked; stamens 8. Fruit a tiny 4-parted capsule. Wet, muddy and aquatic habitats; rare and local. Throughout, but absent from many islands (e.g. CY). **E. macropoda** is a similar, smaller annual 20 mm–10 cm, often patch-forming, with *opposite leaves* and stalked flowers. Saline damp habitats. AA (local), CY, IA (Corfu), IP, PN (northwest), TR (possibly extinct).

VIOLACEAE | VIOLET FAMILY

Herbs and shrubs. Flowers zygomorphic, solitary; sepals 5, separate; petals 5, the lowermost forming a lip, and extended behind into a spur; stamens 5; style 1. Fruit a 3-valved capsule.

Viola VIOLETS

Herbs or shrubs with alternate stalked leaves with stipules at the base. Flowers with 5 sepals and 5 petals; stamens 5; carpels 3. Fruit a capsule. Early spring-flowering. Numerous species occur in the region, many not covered here.

(a) A woody-based shrublet. Flowers bright yellow.

Viola scorpiuroides A woody, ascending, grey-hairy *small shrub*. Leaves broadly egg-shaped and pointed. Stipules linear, as long as the leaves. Flowers borne on stalks with minute bracts, bright *yellow*, 10–15 mm, with 2 brownish spots at the base of the lower petal. Spur 4 mm. Rocky garrigue. AA (Antikythera), CR (mainly west and east), GR (+ Kithira).

(b) Annual or perennial herbs. Stipules fringed with narrow teeth. Flowers typically mostly white or blue–violet, with *downward-pointing* lateral petals.

Viola odorata SWEET VIOLET A small perennial 80 mm–20 cm with *creeping, rooting stems above ground*, rounded, blunt leaves, and stipules broad and short-fringed. Flowers usually dark violet–purple (rarely white) and fragrant; spur 4–5(7) mm. Throughout. *V. hirta* HAIRY VIOLET lacks creeping stems and has narrower, triangular-oval leaves hairy on both sides. *Flowers not fragrant*; spur 3–4(5) mm. GR (north).

Viola alba MEDITERRANEAN WHITE VIOLET A variable, small, slightly hairy perennial 50 mm–15(20) cm, similar to *V. odorata* but with *short, non-rooting* runners. Leaves long-stalked, borne in basal tufts, slightly hairy or hairless, oval-triangular, pointed, dark green. *Stipules long, linear-lance-shaped and deeply fringed.* Flowers *fragrant,* white or blue–violet, to 20 mm across, with an upturned spur 3.5–6 mm. Many forms described. Throughout, except the far southeast.

Viola reichenbachiana A rosette-forming perennial with side shoots, to 15 cm. Leaves 20–40 mm, 1–1.5 x as long as wide, heart-shaped at the base. Stipules narrowly lance-shaped and fringed with *long, slender* teeth, exceeding the stipule width. Flowers 12–18 mm across; sepals pointed; petals narrow, *violet*, darker at the base; *stigma hairless*. Spur slender, straight and deep *violet*, 3–6 mm. Shady woods. AA (rare), CR (central mountains), GR, IA, PN, TR. *V. sieheana* is similar but with stipules with short teeth, the *uppermost becoming large and leafy*. Flowers larger with very wide, *pale blue* to whitish petals, the lowermost deeply cupped; stigma with a tuft of hairs. Spur very stout, whitish. Among boulders in woods. AA (rare), CY, GR, PN, TR. *V. riviniana* is similar to *V. reichenbachiana* but with stipules with equal or shorter teeth. Sepal appendages *conspicuous*, 2–3 mm, concealing the top of the flower-stalk; spur stout, blunt and whitish. Woods. GR, PN. *V. canina* lacks a basal leaf rosette, has leaves longer than wide and upper stipules long, and toothed. Rare. AA (north), GR (north).

(c) Annual or perennial herbs. Stipules often leafy and divided but *not fringed*. Flowers white, yellow, violet or multi-coloured, with *spreading to upward-pointing* lateral petals.

Viola arvensis An erect, branched annual to 20(40) cm with short, deflexed hairs. Leaves 20–50 mm, oblong-spatula-shaped with wavy margins; stipules as long as the leaves and coarsely lobed with a terminal leaf-like segment. Flowers 10–15 mm across, the lower petal cream to yellow, the others cream to bluish-violet; sepals lance-shaped, equalling or exceeding the petals. Spur equalling calyx. AA (Samothraki, Thasos), GR, TR. *V. kitaibeliana* (syn. *V. hymettia*) is a similar, but smaller, bristly-hairy annual to 25 cm, with tufted leaves, oval below, narrower above, with *wavy-toothed margins*. Stipules pinnately divided with a large terminal lobe. Flowers violet or white with a yellowish centre and with darker veins; spur 1.5–3 mm, *exceeding* the calyx. Grassy habitats. Throughout (rare or absent on CR). *V. parvula* is similar to the above species but white *woolly-hairy* and with cream flowers, the petals equalling the calyx and the spur almost concealed within it. Mountains. AA (Samothraki, Thasos), CY, GR, IP, LB, PN, SY, TR (south). *V. phitosiana* is very similar to the previous species, distinguished by its bushy habit and *long spur 3–5 mm*. GR, PN.

Viola rausii A small, clump-forming perennial. Leaves sparsely to densely hairy, 10–28 mm, oval to elliptic. Stipules shorter than the leaves, entire to 3(5)-divided. Flowers large, borne on smooth, erect stalks 50 mm–15 cm; flowers not scented, pale violet with a yellow centre; sepals 7–15 mm, pointed and smooth; upper petals *broadly rounded*, *large* (12.5–20 x 8–15.5 mm); lateral petals narrowly egg-shaped to elliptic with 3 long, dark purple veins; lower petal triangular to semicircular

---

1. *Hypericum triquetrifolium*
2. *Hypericum olympicum*
3. *Hypericum hircinum*
4. *Viola scorpiuroides*
5. *Viola odorata*
6. *Viola reichenbachiana*
7. *Viola sieheana*
8. *Viola arvensis*
9. *Viola rausii*

with 5 darker veins. Spur slender, *long*, 8–16 mm. Seeds brown. A rare endemic of rocky meadows around the summits of Mount Pilion and Ossa in eastern Thessaly. GR. The following rare, narrow endemics are closely related and similar: **V. euboea** on Euboea, **V. athois** on Mount Athos, and **V. samothracica** on Samothraki.

Viola fragrans A hairless to slightly short-hairy, woody-based perennial to 15 cm. Leaves elliptic to lance-shaped, gradually narrowed into stalks of similar length to the blades. Stipules similar to the leaves but smaller. Flowers borne on long stalks; corolla longer than wide, white, yellow or mixed, often with violet veins; lateral petals bearded at the base. Spur short and blunt, exceeding the calyx. CR (high mountains).

Viola heldreichiana A virtually hairless, *very small* annual, just 20–80 mm. Leaves long-stalked and broadly elliptic. *Stipules reduced to lobes or teeth*. Flowers small, to 10 mm across, cream tinged violet; sepals 3–5 mm. AA (east), CY, TR.

PASSIFLORACEAE | PASSION FLOWER FAMILY

A pantropical family of vines and shrubs. Flowers regular, unisexual or cosexual, with 3–5 sepals, petals and stamens. Fruit a capsule or berry.

Passiflora

Mostly vines, with tendrils borne in the leaf axils, and spirally arranged leaves. Flowers complex, with a tubular calyx, 5 petals and 5 sepals and a corona of thread-like elements, as well as 5 conspicuous stamens on a column with the ovary and 3 stigmas.

Passiflora caerulea COMMON PASSION FLOWER A vigorous, tropical-looking, climbing vine with coiled tendrils. Leaves alternate, palmately lobed with 5–7 oblong lobes, untoothed and dark green above, paler below. Flowers very distinctive, consisting of a tubular calyx, 5 greenish-white petals and 5 sepals and a corona of thread-like elements banded purple, white and blue, and 5 conspicuous stamens on a column, which carries the ovary and 3 stigmas. Fruits ovoid, orange, and rather large, to 80 mm long. Very commonly cultivated in towns and gardens; occasionally in ruderal habitats.

SALICACEAE | WILLOW FAMILY

Deciduous trees and shrubs usually with alternate, simple, toothed leaves. Flowers reduced, borne in catkins; male flowers with 1–many stamens; female flowers with 1(–2) short styles. Fruit a 2-valved capsule.

Populus

Deciduous trees with buds with unequal scales and oval or triangular, entire, toothed to lobed leaves. Flowers appearing before the leaves in stalked catkins, each flower with a stalked, cup-shaped disc and subtended by a bract; stamens 4 to many. Capsule 2–4-valved. Seeds numerous. Many non-native species cultivated.

Populus nigra A tree to 30 m with *trunk with prominent bosses* and a broad, uneven crown and cylindrical yellowish twigs, greyish when mature. Leaves 50 mm–10 cm, oval-diamond-shaped and long pointed, with minutely wavy-toothed margins; leaves on the shorter shoots smaller and broader. Stamens 20–30. Fruiting catkins 10–15 cm long. Capsule 2-valved. Scattered across the regions but absent from many islands.

Salix

Trees or shrubs with winter buds with 1 outer scale. Flowers borne before or after the leaves. Stamens 1–5(12). Few species in the Mediterranean. A complex group: hybrids are common and it is impossible to see all traits at any one time of year (many more species in cooler parts, and non-native cultivated taxa are not included here).

(a) Some or all leaves virtually opposite.

Salix amplexicaulis A shrub to 3(5) m with slender, flexible, hairless twigs; previous years' twigs yellowish to reddish-brown. Leaves *opposite* or virtually so, with short stalks (0.5–3 mm), 20–50 mm long, oblong to lance-shaped, dark green above, hairless on both sides. Catkins opposite, 15–45 mm long, dense, cylindrical. River banks. GR, PN, TR. **S. purpurea** is similar but with some *alternate,* longer leaves to 12 cm on stalks to 8 mm. Less common than the previous species. GR, PN.

(b) Leaves alternate.

Salix triandra A shrub or small tree, 4–10 m, with smooth, *flaking* bark and hairless, greenish to reddish-brown twigs. Leaves 50 mm–10 cm, 3–7 x as long as wide, oblong to lance-shaped and short-pointed; leaves hairless (even when young), dark green above, much paler beneath; *stipules persistent*. Catkins appearing with the leaves, erect, cylindrical, 30–50 mm (the female stouter than the male); stamens 3. River banks, absent from hot, dry areas. GR, IA, PN.

Salix pedicellata MEDITERRANEAN WILLOW A deciduous shrub or tree to 8 m with flaking bark and grey-downy twigs. Leaves 50 mm–10(12) cm, broadly oblong to lance-shaped, toothed or scarcely toothed; stipules large and heart-shaped, soon-falling; leaves *virtually hairless on both sides*. Catkins 30–60(90) mm long. Streams, rivers and ravines; rare or absent on most islands. AA (Ikaria), CR (west), LB, TR. **S. alba** is similar but with leaves *densely and persistently hairy* beneath. Catkins oblong, dense, 40–50 mm long. Throughout, except small islands.

1. *Passiflora caerulea*
2. *Populus nigra*
3. *Salix triandra*

EUPHORBIACEAE | EUPHORBIA FAMILY

A large family of herbs, shrubs, trees and lianas, often with a white latex. Monoecious or dioecious; perianth absent or 3(5)-lobed; male flowers with 1–many stamens; female flowers with 2–3 styles. Fruit a capsule.

Euphorbia SPURGE

Leaves normally alternate and entire, with a sticky, milky latex. Flowers in small groups surrounded by a cup-shaped structure with 4–5 round or crescent-shaped glands, a solitary female flower and male flowers with 1–several stamens, the whole structure forming a cyathium. Often difficult to identify in the field; many species occur in the region, so mainly widespread and common species are described.

(a) Mature plant more or less *leafless and succulent* (cactus-like).

Euphorbia ingens CANDELABRA TREE A large, imposing, *cactus-like* perennial to 9 m. Stems bright green and 4-angled with irregular, spiny margins. Native to South Africa, occasionally planted in landscaped areas or cultivated in parks and gardens.

(b) Plant a rigid intricately branched and spiny shrub, typically <50 cm.

Euphorbia acanthothamnos A mounded, *densely and intricately branched* shrub 10–30(50) cm with paired spines formed from old forked twigs. Leaves elliptic to oval, entire. Ray leaves similar to the stem-leaves; raylet leaves wedge-shaped, yellow; rays 3(4). Capsule 3–4 mm with short, conical tubercles. Seeds smooth, 2 mm. Exposed, rocky limestone garrigue; sometimes a subdominant component of the Aegean flora. AA, CR, GR (eastern coasts + islets), IA, PN (eastern coasts + islets), TR (southwest – rare).

(c) Plant a robust *shrub* (>50 cm), often woody at the base, and spineless.

Euphorbia dendroides TREE SPURGE A stout, shrubby and woody shrub to 2(3) m. Leaves oblong to lance-shaped, 35–70 mm, borne in the autumn and winter and falling in late spring to leave bare branches; leaves turning red with age. Umbel with 4–8(10) rays; glands almost circular and irregularly lobed; bracts yellowsh and broader than the leaves. Capsule smooth. Seeds laterally compressed. Rocky coastal slopes and dunes; sometimes subdominant. Throughout, except CY.

Euphorbia characias LARGE MEDITERANNEAN SPURGE A *large*, robust, *hairy* perennial with very thick, *unbranched* stems, to 1.5 m. Leaves bluish or greyish green above, whitish beneath, linear-lance-shaped, untoothed and crowded towards the upper stems, 25–90 mm (13 cm) long. Umbels with 9–20 short rays forming dense, rounded or oblong apical heads. *Bracts fused in pairs around the flowers*, glands dark red–brown, notched or with short horns. Capsule smooth, softly hairy; seeds greyish. Exposed or shady, dry places. Throughout, except the southeast (absent from CY and IP).

---

1. *Euphorbia ingens*
2. *Euphorbia acanthothamnos*
3. *Euphorbia dendroides*
4. *Euphorbia characias*
5. *Euphorbia hierosolymitana*
6. *Euphorbia helioscopia*
7. *Euphorbia peplus*
8. *Euphorbia amygdaloides*
9. *Euphorbia deflexa*

Euphorbia hierosolymitana A hairless, much branched, compact shrub 20–50 cm (1 m). Leaves yellowish green, 14–50 mm. Ray leaves elliptic to egg-shaped, 5–15 mm, yellowish. Umbels yellow with 5 main rays, 2–3 x diverging. Seeds ovoid, 2.3–3 mm, uniformly yellowish or light to mid-brown. Maquis and grassy or rocky hillsides on limestone. AA (east), IP (north), LB, SY (west), TR. *E. lemesiana* has recently been described as distinct based on its more lax habit, deep green leaves, purplish to reddish flower-heads (cyathia) and seeds *marbled dark brown*. Igneous rocks. CY (Lemesos Forest and Stavrovouni area – rare).

Euphorbia sultan-hassei A virtually hairless, non-spiny, mound-forming shrub to 1.5 m across with lax branches, leafy above. Leaves lance-shaped to elliptic, green, paler beneath, 30–50 mm long. Rays 3–5; raylet leaves narrowly egg-shaped, yellowish-green; glands kidney-shaped, orange–yellow. Capsule 3.8–4.8 mm, covered in tubercles. Seeds rough, brown. Rocky gorges. CR (southwest – rare).

(d) Plant herbaceous with oval to rounded glands in cyathium *without horn-like projections*.

Euphorbia helioscopia SUN SPURGE A short, erect hairless annual, normally with a single stem, to 30(50) cm. Leaves oval or spoon-shaped, broadest above the middle and *toothed* in the upper ½, 4–35(60) mm long. *Umbel 5-rayed* (small plants with 2–3 rays), *with 5 distinctive bracts* at the base, yellowish and similar in shape to the leaves. Glands oval and untoothed. Capsule smooth and unwinged. Seeds brown and wrinkled. A common ephemeral on waste ground and bare soil. Throughout.

(e) Herbaceous, sparingly-branched annuals with glands in cyathium sickle-shaped or (often) with horn-like projections.

Euphorbia peplus PETTY SPURGE An erect, hairless annual, 20 mm–20(40) cm, branched at the base. Leaves green, oval to rounded, untoothed and short-stalked, 4–30 mm. Umbels with *3 main rays* (rarely 2–5), with 3 triangular-oval to spoon-shaped, *green* and unstalked bracts. Glands kidney-shaped with long, slender horns. Capsule smooth but with *2 winged keels*; seeds pale grey and *pitted*. Common on disturbed and waste ground. Throughout. Forms with poorly developed umbels and scarcely pitted seeds have traditionally been described as *E. peploides*. *E. falcata* SICKLE SPURGE is similar to *E. peplus* but with few, lance-shaped, 3-veined unstalked leaves 2–25(40) mm long with a waxy bloom and an *unwinged* capsule with *unpitted* seeds; seeds with 4–5(7) *grooves*. Similar habitats. Virtually throughout.

Euphorbia exigua DWARF SPURGE A very small, hairless, grey–green annual, 20 mm–20(30) cm, branched from the base. Leaves *very narrowly* lance-shaped, untoothed and unstalked, 2–25 mm long. Rays 2–5(7), branched x 1–3(5), with narrowly triangular bracts. Glands crescent-shaped with 2 horns. Capsule shallowly grooved, 1.5 mm. Seeds wrinkled and grey. Very common in cultivated, grassy and fallow habitats. Throughout.

(f) Herbaceous, *many-stemmed,* thin-leaved perennials with glands in cyathium sickle-shaped or with horn-like projections.

Euphorbia amygdaloides A hairy, tufted perennial to 90 cm, sometimes slightly woody at the base. Leaves (10)25–70 mm long, oblong-lance-shaped to spatula-shaped and entire, not glossy, slightly hairy. Ray leaves broadly oval; raylet leaves *cone-like* (forming a cup 15–20 mm across); rays (3)5–11, up to 4 x branched; glands with 2 horns. Capsule 3–4 mm. Seeds ovoid, blackish, 2–2.5 mm. Damp, shady woods; not at sea level. AA (Thasos), GR, IA, PN, TR.

Euphorbia deflexa A hairless, blue–grey, clumped perennial to 35 cm, branched from the base. Leaves 3–15 mm, rounded to egg-shaped or oblong, entire and short-stalked. Ray leaves oval to diamond-shaped; raylet leaves diamond-shaped; rays (3)5(9), up to 3 x branched; glands with *2 long, slender horns*. Capsule 3 mm, rough along the keels. Seeds 2–5 mm, ovoid to cylindrical, pale grey, *deeply pitted*. Mountain rocks and screes. AA (Samothraki), CR (eastern mountains – rare), GR, PN (Mount Killini – rare or extinct).

Euphorbia nicaeensis A hairless, blue–grey, often red-tinged perennial to 80 cm. Leaves 10–75 mm, lance-shaped to oblong, leathery, 3(7)-veined. Ray leaves oval-elliptic to round; raylet leaves yellowish; rays (3)5–18, 1–2 x branched; glands often with 2 short horns. Capsule 3–4.5 mm, hairless or hairy. Seeds ovoid, nearly smooth, pale grey. Open, rocky habitats at high altitude. TR.

Euphorbia cyparissias CYPRESS SPURGE A hairless, rhizomatous perennial with erect stems to 50 cm, usually unbranched at the base and branched above to form numerous non-flowering shoots. *Leaves crowded, long and narrow, 0.5–2(3) mm wide*, stalkless, turning yellowish with age (not greyish or bluish), thin. Umbels with *many rays* (11–18); glands horned. Capsule 3–3.5 mm. Inland habitats. GR, TR.

Euphorbia segetalis (syn. *E. pinea*) A hairless annual or perennial, simple or with some branches at the base, to 80 cm (1 m). Leaves alternate and *narrow* (2–4 mm), linear to linear-lance-shaped, 27–40 mm long. Rays 4–6(8), *bracts diamond-shaped* and yellow-green, glands with 2–4 horns. Capsule rough and glandular; seeds pale grey. Open, maritime habitats. Perennial forms were formerly described as *E. pinea*. GR (north).

(g) Herbaceous, many-stemmed, *fleshy-leaved perennials* with glands in cyathium sickle-shaped or with horn-like projections.

Euphorbia myrsinites BROAD-LEAVED GLAUCOUS SPURGE A short, *prostrate* perennial with unbranched, fleshy, densely leafy stems radiating from the centre 20–40 cm long; not, or scarcely, woody. *Leaves succulent, grey, borne in a distinctly regular arrangement around the stem*; oval to rounded and pointed at the tips. Rays 5–12, bracts bright yellow–green and heart- or kidney-shaped; glands weakly horned. Capsule 5–7 mm. Seeds smooth to rough and grey–brown. Rocky slopes, often in mountains. GR (common), IA, PN, TR.

1. *Euphorbia nicaeensis*
2. *Euphorbia cyparissias*
3. *Euphorbia myrsinites* habit; (inset) flower
4. *Euphorbia veneris*

Euphorbia rigida NARROW-LEAVED GLAUCOUS SPURGE A hairless and distinctly bluish perennial, similar to *E. myrsinites* but taller, with numerous stout, erect to ascending stems (30)40–50(80) cm arising from a *woody* stock. Leaves fleshy, lance-shaped, 4 x as long as broad (>35 mm long), and pointed, often flushed with purple. Rays 6–12, 1–2 x forked; ray leaves oblong. Glands minutely *horned*. Capsule strongly *3-sided*. Seeds smooth and whitish. Local on garrigue and maquis on higher ground. GR, IA, PN, SY, TR. *E. veneris* is a very similar island endemic, which is much *less robust*, to 35 cm, with *smaller leaves*, 10–25(30) mm long, and perfectly smooth, greyish seeds. Locally common in Troodos mountains, up to the snow line. CY.

Euphorbia paralias SEA SPURGE A clump-forming, hairless, stiffly erect, fleshy perennial to 60(80) cm. Leaves grey–green, regularly and closely set around the stem, *overlapping*, oval, *broadest towards the base* and concave above; *midrib obscure below*, 8–20 mm long. Umbels with 3–6 rays, bracts oval and concave. Glands kidney-shaped with long horns. Capsule rough along the back, 4–5.5 mm. Seeds pale grey and *smooth*. Common in rocky and sandy maritime habitats and dunes. Throughout.

Euphorbia terracina A hairless, *succulent* perennial, 40 mm–90 cm, with erect to ascending stems and non-flowering lateral branches. Leaves oblong to linear-lance-shaped, minutely toothed, regularly and closely set around the stem and overlapping but *flat*, 4–60 mm long. Umbels with 2–5(6) rays with as many oblong to diamond-shaped, green bracts. Glands with 2, long, slender horns. Capsule smooth; seeds pale grey and smooth with a boat-shaped, fleshy structure (caruncle) attached. Open habitats. Throughout.

(h) *Chamaesyce* group: small-flowered, often low, *prostrate* annuals; closely related, very similar species, now reclassified in the genus *Euphorbia*.

Euphorbia peplis (syn. *Chamaesyce peplis*) PURPLE SPURGE A *prostrate*, hairless annual with 4 (sometimes 3 or 5) main branches at the base. Stems red or purple, *leaves fleshy*, grey–green and small to 11 mm, opposite, oblong, with a single rounded lobe at the base. Flowers tiny, greenish with semicircular red–brown glands, borne laterally or in clusters but not in umbels. Capsule nearly smooth and purplish. On sandy and shingly shores. Throughout. *E. serpens* (syn. *Chamaesyce serpens*) is very similar (often confused with *E. peplis*) but with up to 16 branches and leaves symmetrical rather than with rounded lobes at the base. Native to Tropical America; hot, dry areas. AA, CR, CY, IP (probably elsewhere).

Euphorbia prostrata (syn. *Chamaesyce prostrata*) A spreading herb, similar to *E. peplis* but often hairy, and with *up to 10 branches at the base* and stems (which are hairy above), *leaves slightly serrated on the margins*, and *capsules hairy on the keels*. A North American weed naturalised locally in sandy places; absent from most small islands. AA (Rhodes), CR, CY, GR, IA, IP, PN. *E. maculata* (syn. *Chamaesyce maculata*) is similar, with up to 8 branches and *leaves often with a dark central spot* and with a capsule either virtually hairless, or more often *entirely covered with closely adpressed hairs*. A North American weed naturalised in ruderal places. Probably throughout. *E. chamaesyce* (syn. *Chamaesyce canescens*) is similar but with up to 25 branches, leaves often without a dark spot, and with capsules with *spreading* (not adpressed) hairs. Probably throughout.

---

1. *Euphorbia rigida* habit; (inset) flowers
2. *Euphorbia paralias*
3. *Euphorbia peplis*
4. *Euphorbia serpens*
5. *Euphorbia maculata*
6. *Euphorbia chamaesyce*

Chrozophora

Monoecious, hairy annual or perennial herbs or shrubs with simple, alternate leaves. Male flowers with 5–15 stamens; female flowers with 3 stigmas. Fruit a capsule.

Chrozophora tinctoria A grey-hairy annual, 10–40 cm, with simple, wavy-margined leaves, 20–60(80) mm. Flowers unisexual, the male with yellow triangular corolla lobes 1.5–2 mm and 10 stamens; female flowers with greenish, linear corolla lobes, 3–3.5 mm. Fruit sub-spherical, 5–8 mm across. Local in disturbed habitats. Throughout. *C. obliqua* is densely white-hairy with star-shaped hairs throughout, and with flowers with just 4–7 stamens. AA, CR (rare), CY, GR (mainly southeast), IP, PN.

Mercurialis MERCURY

Annual or perennial herbs (or shrubs) with opposite leaves. Flowers unisexual, green and inconspicuous; sepals 3; stamens 8–25; carpels 2(3–4). Fruit a 2(3–4)-parted capsule.

Mercurialis annua ANNUAL MERCURY A dioecious, branched, erect *annual*, 30–70 cm, more or less hairless. Leaves 20–70 mm long, opposite, oval to elliptic, toothed and long-stalked, shiny *light* green, with minute hairs along the margins (<0.4 mm). Male flowers borne on dense, long, erect greenish spikes, female flowers few, borne in lateral clusters. Fruit 2.4–2.6 mm, 2-lobed and bristly. A common weed on disturbed ground. Throughout.

Mercurialis ovata A dioecious, patch-forming, *rhizomatous perennial*, 15–30 cm. Leaves opposite, broadly oval-elliptic, *dark* green. Male flowers borne in long-stalked spikes in the axils; female flowers solitary, stalked. Fruit 2(3)-parted. Seeds 3 mm. AA (local), GR, IA (Corfu).

Ricinus CASTOR OIL PLANT

Shrubs without a milky latex, with *palmately lobed leaves*. Flowers unisexual; perianth with 3–5 tepals; *stamens conspicuous in fascicles* (branched); ovary 3-parted. Fruit a capsule.

Ricinus communis CASTOR OIL PLANT A robust annual or shrub to 5(7) m, flushed red, bronze or purple. Leaves shiny, large, 10–36(60) cm across and palmate, with 5–9 coarsely toothed lobes. Flowers borne in large terminal panicles with the male below with yellowish stamens, and the female above and with bright red stigmas. Fruit a 3-parted, spiny capsule 18–20 mm across; seeds bean-like, 10–15 mm long. Native to tropical Africa, widely planted in towns and naturalised in warmer parts.

LINACEAE | FLAX FAMILY

Annuals or perennials with simple, opposite or alternate leaves. Flowers in branched inflorescences; sepals and petals 4–5, free; stamens 4; styles 4–5. Fruit a 8–10-valved capsule.

Linum FLAX

Hairless annual or perennial herbs or shrubs. Flowers 5-parted with white or blue petals. Capsule 10-valved, often short-beaked.

(a) Annuals with yellow flowers.

Linum nodiflorum A hairless annual to 40 cm with solitary or few, branched, winged stems. Lower leaves spatula-shaped, upper leaves linear and 1(3)-veined; all leaves with very finely toothed margins. Flowers yellow, borne in very lax inflorescences; petals 20 mm, long-clawed; sepals 8–13 mm. Capsule 5–6 mm. Garrigue and maquis; local. Throughout.

Linum strictum UPRIGHT YELLOW FLAX A short, erect annual 10–45 cm with narrowly lance-shaped leaves with inrolled, very rough margins. Flowers small with yellow petals 6–12 mm long, which exceed the long-pointed sepals, borne in branched, spreading clusters or short lateral clusters in a rigid, open, corymb-like inflorescence. Coastal sands and other dry places. Throughout. **Subsp. *spicatum*** is considered by some authors to be distinct, characterised by its adpressed, *spike-like* inflorescence. Throughout the range. **L. *corymbulosum*** is a similar, erect, slender annual to 40 cm. Leaves 10–20 mm, linear, with rough margins. Flowers borne in *lax* inflorescences, on long, very *slender* stalks; sepals oval; petals 6–10 mm, yellow. Capsule 2.5–3 mm. Throughout, except smaller islands.

(b) Woody-based perennials (or shrubs) with yellow (to orange) flowers.

Linum arboreum A hairless *shrub* to 1 m with thick, persistent, 1-veined, spatula-shaped leaves, often crowded. Infloresecence rather few-flowered and compact; flowers yellow with petals 12–18 mm; sepals 5–8 mm. Capsule 6–8 mm, beaked, more or less equalling the sepals. A rare chasmophyte of limestone cliffs and gorges. AA (southeast: Astipalea, Chalki, Rhodes, Simi), CR (+ KP), TR.

Linum mucronatum A virtually hairless, greyish, *shrubby* perennial to 30 cm, woody at the base. Leaves 10–30 mm, typically 1-veined (some 3–7-veined). Flowers *yellow, apricot or orange*, borne in 3–many flowered cymes, often with widely spreading branches; sepals linear to lance-shaped, 6–13 mm, often keeled, with papery, glandular margins; petals 15–28 mm, broadly egg-shaped. Capsule 5–6 mm with a beak to 1 mm. Fields and rocky habitats in the east; rare and local. IP (mainly central), LB, SY, TR.

Linum maritimum A slightly hairy or hairless perennial *with a woody stock* and long, erect or ascending stems. *Leaves small* and upward-pointing, greyish, lance-shaped to narrowly elliptic, the *lower leaves opposite*, 3-veined, the upper leaves alternate and 1-veined. *Flowers yellow*, borne in lax clusters, the petals to 13 mm long. Damp, saline soils. GR (southeast), IP, PN, TR.

---

1. *Chrozophora tinctoria*

2. *Mercurialis annua*

3. *Ricinus communis*

Linum flavum YELLOW FLAX A variable, often tall, hairless, erect, branched perennial to 60 cm with a woody stock and *rather large, dark green, 3(5)-veined leaves 20–35 mm long and 3–12 mm wide*, the uppermost lance-shaped, those beneath spatula-shaped. Flowers *large and bright yellow*; petals 20 mm long. Dry, grassy habitats. GR, PN.

(c) Flowers blue; sepals without glandular hairs.

Linum bienne PALE FLAX An annual or perennial herb with slender, erect to spreading stems, often branched below (not in small specimens), 10–60 cm. Leaves alternate, linear and long-pointed, mostly 3-veined, 0.5–1.5 mm wide. Flowers *pale blue*, borne on slender stalks in loose clusters or singly, petals 8–12 mm, exceeding the oval, long-pointed, papery-margined sepals (4–6 mm long). Capsule 4–6 mm. Common on garrigue. Throughout. ***L. usitatissimum*** CULTIVATED FLAX is probably derived from *L. bienne* and similar, but *usually an unbranched annual* to 85 cm with *larger, darker blue or white flowers* with petals 12–20 mm long. Capsule 6–9 mm. Disturbed habitats. GR, IA, PN.

1. *Linum nodiflorum*
2. *Linum strictum*
3. *Linum strictum* subsp. *spicatum*
4. *Linum arboreum* habit
5. *Linum arboreum*
6. *Linum mucronatum*
7. *Linum pubescens* subsp. *pubescens*

(d) Flowers white.

Linum tenuifolium A stiffly branched, shrubby hairless perennial to 50 cm with numerous stiff, linear leaves to 1 mm across. Flowers conspicuous, white with a purplish centre (yellow in bud), with petals 15–30 mm long, much-exceeding the 1-veined sepals; stamens usually markedly protruding the corolla tube; sepals 3-veined. CR (west – very rare), GR, IA, PN, TR.

(e) Flowers pink.

Linum pubescens A slender annual 70 mm–10(25) cm. Leaves alternate, stalkless, 12–23 mm with 3–5 veins. Flowers conspicuous, borne in terminal inflorescences of 3–5; sepals 5, free, 1-veined, 9–12 mm; petals oblong-oval, *bright pink* with darker veins and often bluish at the base. Hillsides, maquis and fallow fields. AA, CR, CY, GR, IP (north), PN, SY. Two regional forms occur: **Subsp. *pubescens*** AA, CY, IP (north), SY (west), TR. **Subsp. *sibthorpianum*** CR, GR, PN.

Radiola

Annuals with opposite leaves. Flowers 4-parted with white petals. Capsule 8-valved.

Radiola linoides A much-branched, *extremely slender* and small annual to 10 cm with 1-veined leaves. Flowers tiny with 4 sepals and petals to 1 mm. Capsule 0.7–1 mm. Open, sandy ground. AA, CR, GR (uncommon).

GERANIACEAE | GERANIUM FAMILY

Herbs with alternate palmately or pinnately lobed leaves. Flowers borne in cymes, umbels, or solitary, usually more or less regular with 5 sepals and petals; stamens (3)5 or 10; style 1. Fruit with 5 1-seeded portions united into a prominent beak.

Geranium

Annuals or perennials with simple, palmately lobed leaves. Flowers regular with 10 stamens; style with 5 branches. Fruit beaked.

(a) Perennials with large flowers; petals long, 7–20 mm, *notched*.

Geranium sanguineum A rhizomatous perennial with diffuse, branched, erect to ascending stems with long, white spreading hairs. Leaves mostly on the stems, 30–50(80) mm wide, divided for 85% to the base into 5–7 lobes, each with 1–3 linear-oblong, pointed segments on each side. Flowers large, borne on stems 70 mm–15 cm, usually *solitary*, *bright reddish-purple* (sometimes pink); petals 15–20 mm; sepals 8–13 mm. Fruit hairy, without ridges. Rocky ledges, gorges and mountains; mainly northern. GR, IA, TR.

Geranium peloponesiacum An erect perennial branched from the base. Leaves 80 mm–12 cm wide, divided for 65% to the base into 3–5 oval-diamond-shaped, irregular lobes with blunt segments. Flowers blue–violet, borne in corymbs; flower-stalks densely hairy; petals 15 mm; sepals 10 mm. Fruits hairy. Shady habitats. PN.

Geranium pyrenaicum An erect, hairy perennial to 60 cm with palmately lobed leaves to 50 mm across, rather deeply divided into 5–7(9) lobes (2/3 to the base) with straight, entire margins, wavy towards the tip. Flowers large, borne in pairs, and *pink–purple or lilac* with *deeply notched* petals 7–10 mm long, all 10 stamens with anthers. Local in dry, grassy habitats. GR, PN, TR.

(b) Annuals with small flowers; petals usually <7(10) mm; *sepals spreading or ascending*.

Geranium molle DOVE'S-FOOT CRANE'S-BILL A short, sprawling annual to 40 cm with stems branched from the base, *grey–green and softly hairy*. Basal leaves long-stalked, rounded or kidney-shaped, divided into 5–7(9) wedge-shaped, 3-lobed segments ($1/2$–$2/3$ to the base); upper leaves more deeply divided and short-stalked or unstalked. Flowers pink–purple; petals 4–6 mm, *deeply notched*, borne in lax clusters; outer stamens lacking anthers; flower-stalks with short and long hairs. Field and grassy habitats; common throughout. Less hairy, taller forms have traditionally been described as *G. brutium*. **G. pusillum** SMALL-LEAVED CRANE'S-BILL is similar to but with flower-stalks with hairs all short (not some long); flowers small with petals 2.5–4 mm. Garrigue. Widespread but local and scattered. AA (rare), CY, GR, PN. **G. rotundifolium** is similar but with leaves shallowly 5–9-lobed (<$1/2$ to the base) and bright pink flowers with *unnotched or slightly notched* petals, rounded at the tips, 5–7 mm long. Similar habitats. Throughout.

Geranium dissectum CUT-LEAVED CRANE'S-BILL A spreading, hairy annual to 60 cm with ascending flowering stems. Leaves circular in outline but *deeply dissected into 5–7 lobes almost to the base*, with sub-lobes. Flowers bright pink with shallowly notched petals, 4.5–6 mm; *flower-stalks <15 mm long*; sepals spreading and with pointed tips. Fruit ridged and hairy. Common in a range of habitats, especially damp, disturbed, grassy places. Throughout. *G. columbinum* LONG-STALKED CRANE'S-BILL is similar but with larger flowers on *long stalks, 25–60 mm*; petals not notched, 7–10 mm long. Throughout, except small islands.

(c) Annuals with (normally) small flowers; petals 5–10(14) mm; *sepals erect and curved at the tips.*

Geranium robertianum HERB ROBERT A hairy, *very aromatic* annual or biennial to 50 cm; usually *strongly flushed with red* or purple. Leaves palmate, the lower leaves with 3–5 pinnately lobed segments. Flowers pink (sometimes white); petals slightly notched or rounded, 8–10(14) mm long; pollen orange. Fruit hairy and ridged. Cool, damp shady places and woods, mainly on the northern and western mainland; rare or absent on islands. AA (Thasos), GR, IA (rare), IP, PN, TR. *G. purpureum* LITTLE ROBIN is similar but less flushed with red, flowers purplish-pink and smaller; petals 5–9 mm; pollen yellow. Similar habitats. Throughout, except the far southeast. *G. lucidum* SHINY CRANE'S-BILL is similar to *G. robertianum* but readily distinguished by its *hairless* (or sparsely hairy), *shiny leaves*, which are circular in outline, divided into 5(7) lobes. Petals deep pink, with rounded tips, 8–10 mm; sepals keeled on the back. Throughout.

Erodium STORK'S-BILL

Annuals or perennials like *Geranium* but generally with pinnately lobed leaves and often slightly zygomorphic flowers with 2 petals larger than the other 3; stamens 5. Fruit beaked. Difficult to distinguish in semi-arid habitats in the east where several similar taxa occur.

(a) Some (or all) leaves *shallowly lobed* (in at least some leaves to <½ the width of the blade).

Erodium gruinum An annual or perennial 15–50 cm with spreading or deflexed hairs. Leaves to 10 cm, oval to lance-shaped, lobed, sometimes with a pair of free leaflets at the base; lobes and leaflets irregularly toothed. Flowers violet, borne in umbels of 2–6; bracts hairless, whitish; petals 20–25 mm; sepals 15–20 mm. Fruit with ascending, whitish hairs and a beak 60 mm–11 cm. Throughout except the north.

Erodium chium THREE-LOBED STORK'S-BILL A robust, hairy perennial or biennial to 40 cm. Leaves oval, those below divided into 3 toothed, blunt lobes. Flowers pink–purple, 10–18 mm across, borne in 2–8 flowered clusters on non-glandular flower-stalks; 2 petals slightly larger than the remaining 3; sepals with hairs *not glandular*. Fruit with short white hairs, the beak 20–40 mm long. Open, rocky places and roadsides. AA, CR, GR (far southeast), PN (east).

Erodium malacoides MALLOW-LEAVED STORK'S-BILL An erect to sprawling, glandular-hairy biennial to 40 cm. Leaves oblong, heart-shaped at the base, those below toothed, sometimes 3–several-lobed, covered in shiny glands. Flowers purplish pink, 11–18 mm across, borne in 3–7-flowered clusters with at least 3 bracts at the base, borne on glandular-hairy stalks; *sepals with glandular hairs*. Fruit beak 20–35 mm long. Very common on bare or cultivated ground. Throughout. *E. laciniatum* is similar to *E. malacoides* but with leaves more deeply cut and not markedly glandular-hairy beneath, and flowers in clusters of 4–9 with just 2 bracts at the base. Fruit beak longer, to 90 mm. Coastal sands. Throughout except the north and central Greece.

---

1. *Geranium sanguineum*
2. *Geranium molle*
3. *Geranium dissectum*
4. *Geranium robertianum*
5. *Geranium lucidum*
6. *Erodium gruinum*

(b) All leaves 1–2-*pinnately divided* into discrete leaflets.

Erodium cicutarium COMMON STORK'S-BILL A very variable, erect or prostrate, hairy (sometimes sticky), annual to 60 cm. Leaves *deeply pinnately divided* without smaller lobes between the larger ones. *Stipules pointed* and whitish. Flowers purplish, pink or white, (7)10–18 mm across with 3–7(12) in a cluster; petals 4–9 mm, the upper 2 petals normally larger and with a blackish patch; sepals with inconspicuous netted veins and a bristly point. Bracts brownish. Fruit hairy, with a beak 15–40 mm. Very common in a range of open and disturbed habitats. Many variants have been described. Throughout. *E. touchyanum* is a similar, densely hairy desert annual, flowering after rain. Sepals with *conspicuously* netted veins, the points *without* bristles; petals 5–7 mm. IP (mainly eastern). *E. acaule* STEMLESS STORK'S-BILL is similar to *E. cicutarium* but *stemless* (though some populations of *E. cicutatium* appear stemless in exposed habitats). Leaves with stalks shorter than the blades. Flowers lilac without darker patches, to 22 mm across. Fruit with white hairs and a beak to 50 mm. Dry habitats (less frequent than *E. cicutarium*). GR, IA, IP, PN, TR (probably LB, SY).

1. *Erodium laciniatum*
2. *Erodium cicutarium* fruits
3. *Erodium touchyanum*
4. *Erodium moschatum*
5. *Erodium crassifolium*
6. *Lythrum salicaria*

Erodium moschatum MUSK STORK'S-BILL A spreading annual to 60 cm, similar to *E. cicutarium* but always stickily-hairy and *smelling of musk*, leaflets only shallowly lobed (<½ to the midrib). *Stipules blunt*. Flowers larger, to 28 mm across, violet or pinkish purple. Fruit with beak 20–45 mm long. Cultivated and waste ground. Throughout (rarely far from coasts).

Erodium crassifolium A perennial with *tuberous roots*, rather woody at the base, to 30 cm. Leaves deeply divided with slightly blunt lobes. Flowers *bright pink*, borne in umbels of 3–5; sepals 6 mm; petals 8–10 mm. Fruit body with *short, stiff* bristles; beak 5–9 mm, not twisted and with long, pale brown hairs on the inner surface. Local, and strongly southern and eastern; deserts and stony limestone habitats. CR (southeast), CY (south), IP (south), SY.

Erodium botrys MEDITERRANEAN STORK'S-BILL A short, hairy annual to 50 cm with an obvious stem above ground. Leaves bristly, to 50 mm across, oval and deeply pinnately lobed and toothed, at least on the upper leaves. Flowers to 30 mm across, bluish with darker veins, borne in clusters of up to 4; bracts brown. Fruit with a long beak, 60 mm–(9)11 cm. Dry rocky habitats and roadsides. Throughout (but uncommon).

LYTHRACEAE | LOOSESTRIFE FAMILY

Annual or perennial herbs with leaves simple and opposite or in whorls of 3. Flowers cosexual, regular, usually with (4)6 sepals and 6 petals (0–5) often pink or purple; stamens (2)6–12; style 1. Fruit a 2-valved capsule.

Lythrum LOOSESTRIFE

Herbs with tubular or bell-shaped calyx with 4–6 teeth and 4–6 petals <8 mm long.

(a) Erect plants with small, normally *solitary* (or 2) pink–purple flowers.

Lythrum junceum A hairless perennial to 70 cm with much-branched, sparse stems. Leaves mostly alternate, elliptic and stalkless. *Flowers small*, borne 1(2) in each leaf axil, purple, rarely white, solitary; petals 6, 5–6 mm long; *stamens 12, some or all protruding*. Damp habitats. Virtually throughout.

Lythrum hyssopifolia GRASS POLY An erect, hairless annual to 25 cm, similar to *L. junceum*, with linear-lance-shaped, rough-margined leaves. Flowers pink, borne 1(2) in each leaf axil with *4–6 stamens, not protruding*; petals 2–3 mm. Seasonally flooded areas and damp places. Throughout.
L. tribracteatum is similar to *L. hyssopifolia* but with *broadly* linear to oval leaves, and with appendages on the inner sepals *longer (not shorter) than the sepals themselves*; petals purple, just 2–3 mm long. Scattered in distribution. Virtually throughout (absent from many islands).
L. thymifolia is similar to *L. hyssopifolia* but with *tiny, narrow leaves 0.75–1(2.5) mm wide,* and flowers with 4 petals 1.5–3 mm long and 2–3 stamens. AA (Lesbos), GR, IA, IP, PN (Flafonisos), TR.

(b) Erect plants with conspicuous pink–purple flowers clustered in whorls; petals >8 mm long.

Lythrum salicaria PURPLE LOOSESTRIFE An erect, more or less hairless or shortly grey-hairy perennial to 1.5 m; stems sparingly branched with stalkless leaves in whorls of 3 or opposite; lance-shaped. *Flowers clustered in whorls, large, conspicuous, and pink–purple*, borne in long, dense terminal spikes; petals 8–10 mm; stamens 12, some or all protruding. Damp places and on riverbanks. AA (Lesbos, Thasos), GR (common), IA (uncommon), PN, TR.

(c) Creeping annuals with prostrate or ascending (rarely erect) stems.

Lythrum portula WATER PURSLANE A low to *prostrate creeping, hairless annual with stems rooting at the internodes*, to 25 cm. Leaves opposite, fleshy, often reddish and tapered into a short stalk, oval-spatula-shaped. Flowers purple, very small, borne solitary in the leaf bases; petals 6 or absent, 1 mm long; stamens normally 6. Seasonally flooded and waterlogged muddy ground. GR, PN (Elafonisos). *L. borysthenicum* is similar but *bristly*, at least when young, and with *stems not rooting at the internodes*. Local. AA, CR, TR.

Punica POMEGRANATE

Fruit-bearing deciduous shrubs or small trees, best known for the pomegranate. Flowers with 4–6(9) sepals and petals and as many or 2 x as many stamens. Fruit a capsule or berry. Recently established as part of the family Lythraceae.

Punica granatum POMEGRANATE A deciduous shrub or small tree to 5 m with spiny, 4-angled young stems. Leaves opposite, shiny, bright green, roughly oblong, untoothed and virtually unstalked. Flowers to 40 mm across, with 5–9 scarlet, crumpled petals and a fleshy calyx (hypanthium), borne in clusters of 1–3 near the ends of the branches. Fruit spherical, to 90 mm across. Widely cultivated in the region, and naturalised.

ONAGRACEAE WILLOWHERB FAMILY

Annual or perennial herbs with simple, alternate or opposite leaves. Flowers mostly regular, with 2–4 free sepals and petals and 2, 4 or 8(10–12) stamens; style 1. Fruit a capsule or berry splitting lengthways with distinctive cottony seeds, or a 1–2-seeded nut.

Oenothera

Annual or perennial herbs with alternate leaves. Flowers regular, large, borne in leafy spikes; sepals and petals 4; stamens 8, in 2 whorls; stigma deeply 4-lobed. Fruit elongated; seeds small and numerous. An American genus cultivated for ornament in the region.

Oenothera speciosa An erect, simple to branched annual or perennial herb 10–50 cm. Leaves 10–80 mm long, oval to lance-shaped, the lowermost pinnately divided. Flowers rather large and wide open, *whitish-pink* with fine veins, greenish-yellow in the very centre; petals broad and notched, 25–40 mm; sepals 15–30 mm; stigma *conspicuously 4-lobed*. Fruit 15–20 mm. Native to North and Central America; a casual escape of cultivation. GR, IA.

Epilobium WILLOWHERB

Perennial herbs with opposite lower leaves opposite. Flowers pink or purple with 4 sepals and petals; stamens 8; ovary 4-parted. *Fruit a linear capsule splitting into 4 valves to reveal seeds with long plumes of hairs.*

(a) Stigma with 4 lobes in a cross.

Epilobium montanum A perennial 10–80 cm. Leaves 25 mm–14 cm, oval, short-stalked and usually toothed. Buds pointed; petals 6–10 mm, purplish-pink. Seeds 1 mm. Mountains. GR.

1. *Punica granatum*
2. *Oenothera speciosa*
3. *Epilobium montanum*
4. *Myrtus communis* subsp. *communis*; (inset) *Myrtus communis* detail
5. *Eucalyptus camaldulensis*
6. *Melaleuca citrina*
7. *Biebersteinia orphanidis*
   PHOTO: ARNE STRID

Epilobium hirsutum GREAT WILLOWHERB A robust, densely and softly hairy perennial to 1.8 m with spreading non-glandular and glandular hairs. Leaves opposite, lance-shaped, unstalked and *partially clasping the stem below*; markedly toothed. Flowers with bright pink, notched petals 10–16(18) mm long, borne in a leafy raceme; stigma 4-lobed. Common in damp places and near rivers. Throughout. *E. parviflorum* SMALL-FLOWERED HAIRY WILLOWHERB is similar to (and hybridising with) the previous species but *smaller* in all parts (to 75 cm), with leaves not clasping the stem and *small, pale pink flowers with petals 5–9 mm long*. Local in damp places. Throughout (except small islands).

(b) Stigma club-shaped (clavate).

Epilobium palustre MARSH WILLOWHERB A perennial to 60 cm with sparse, adpressed, non-glandular hairs (some glandular hairs above) *rounded stems without lines or ridges, and with untoothed, virtually stalkless leaves* (leaf stalks short, <4 mm). Flowers pale pink to white, borne in lax, coarsely hairy racemes; petals 4–7 mm; *stigmas club-shaped*. Marshes and fens. GR (north). *E. roseum* SMALL-FLOWERED WILLOWHERB is similar but with *distinctly stalked leaves* (4–15 mm long) *with toothed blades; stem with 2 raised lines*. Petals 4–7 mm. Damp habitats on higher ground. GR (north).

Epilobium tetragonum SQUARE-STALKED WILLOWHERB An erect perennial to 75 cm, similar to the previous species but with inflorescences with dense, white, adpressed hairs (no glandular hairs), distinguished by its *clearly 4-ridged, often winged stems*. Petals pink–purple, 5–7 mm. Fruits 65–80 mm (10 cm) long. Throughout. *E. obscurum* SHORT-FRUITED WILLOWHERB is similar to *E. tetragonum*, also has has 4-ridged stems, petals 4–7 mm, and *shorter fruits, (30)40–60(56) mm long*. Throughout (except small islands).

MYRTACEAE | MYRTLE FAMILY

A mainly tropical family of shrubs and trees with normally opposite, simple leaves. Flowers with 4–5 sepals and petals and numerous stamens; style 1. Fruit a many-seeded capsule or berry.

Myrtus MYRTLE

Trees or shrubs with opposite leaves. Flowers with 5 free sepals and petals; stamens numerous. Fruit a berry.

Myrtus communis COMMON MYRTLE An erect, much-branched evergreen shrub to 1.5 m, glandular-hairy when young. Leaves opposite, shiny deep green, lance-shaped and pointed, *aromatic* when crushed, 20–50 mm long. Flowers white, to 30 mm with rounded petals and numerous, conspicuous protruding stamens. Berry blue–mauve then bluish-black when ripe, 7–10 mm long. Common on maquis and coastal garrigue throughout, also planted. Subsp. *communis* is the most widespread, common form.

Melaleuca BOTTLEBRUSHES

Shrubs native to Australia. Leaves alternate. Flowers borne in dense inflorescences, petals 5; stamens numerous. Fruit a capsule.

Melaleuca citrina (syn. *Callistemon citrinus*) BOTTLEBRUSH An evergreen shrub to 2 m with stiff, arching stems. Leaves leathery green, narrowly elliptic, entire, and lemon-scented when crushed. Flowers crimson, borne in brush-like heads with numerous prominent stamens. Fruit a capsule. Native to Australia; widely planted.

Eucalyptus

Evergreen trees native to Australasia with peeling bark. Leaves of saplings broad and erect; leaves on mature trees narrow and pendent. Flowers borne in clusters, with >100, prominent stamens. Fruit a woody capsule concealed by a fleshy cap (transformed sepals) when in bud. Widely planted and important for timber production in the region.

Eucalyptus camaldulensis RIVER RED GUM A tree to 15(50) m with smooth, white, peeling bark. Juvenile leaves oval-lance-shaped, grey–green, mature leaves much narrower, 80 mm–25(30) cm long. Flowers yellowish-white, borne in clusters of 5–12, and *distinctly stalked; stalks 6–15(20) mm long*. Fruit *hemispherical*; longer than wide, to 6 mm and with a broad, raised rim. Commonly planted for timber.

NITRARIACEAE

A small family of annuals, perennials and shrubs with alternate, often fleshy leaves. Flowers with 5 persistent sepals and 5 petals; stamens 10–15; style 1. Fruit a drupe or capsule.

Peganum

Perennials with alternate, irregularly lobed leaves, often stalkless. Flowers borne solitary in the leaf axils opposite the leaves, with persistent sepals; sepals and petals 5; stamens 12–15. Fruit 3(4)-parted. Previously classified in the family Zygophyllaceae.

Peganum harmala An erect, hairless perennial with 5–13 branches, 40–45(70) cm. Leaves alternate, rather fleshy, with minute, soon-falling stipules, and palmately, *irregularly* divided into 3–5 linear, lance-shaped or elliptic lobes 30–50(60) mm long and 1.5–3 mm wide. Flowers greenish- or yellowish-white, borne on stout stalks opposite the leaves; calyx lobes entire or slightly toothed; petals 10–13(19) mm. Capsule (8)9–13 mm, strongly 3(4)-lobed, with 35–47 dark brown seeds. Derelict land and dry, semi-arid scrub. Scattered from mainland Greece south and eastwards.

BIEBERSTEINIACEAE

A family with just one genus containing 5 species, which occur from the eastern Mediterranean to Central Asia. Leaves compound. Flowers with ovaries with a single ovule per chamber.

Biebersteinia

The only European representative of the genus. Perennial herbs with pinnately divided leaves. Flowers in spikes; style arising from the inner side of the carpel; stigmas united. Fruits without beaks.

Biebersteinia orphanidis A distinctive, rather *Acanthus*-like, glandular-hairy perennial with a tuberous, woody stock, 35–50 cm. Leaves oblong-lance-shaped in outline, short-stalked, shiny green, to 35 cm, pinnately divided into deeply, irregularly toothed to lobed leaflets; leaves mostly basal. Flowers small and pink, borne in compact, spike-like clusters 50–10 cm long; petals exceeded by the sepals. Fruit 6 x 4 mm, rough, brown. Very rare; long presumed extinct in Greece but discovered relatively recently in clearings in *Abies* forests. PN (north), (also TR, east).

ANACARDIACEAE

A large, mostly tropical family of shrubs, trees and lianas. Leaves alternate, simple or pinnately divided. Flowers with 5 sepals and petals and 5–10 stamens; styles 3. Fruit a small, 1-seeded drupe.

Pistacia MASTIC

Shrubs and trees (mostly) with alternate leaves. Dioecious; flowers unisexual; stamens 3–5. Fruit a drupe.

(a) Leaves pinnately divided, without a terminal leaflet; stalks *winged*.

Pistacia lentiscus MASTIC A small, evergreen, dark green tree or shrub 6–8 m. Leaves dark green, pinnately divided *without* an end leaflet; leaflets 10–50 mm, borne 4–14, oval, leathery and untoothed, on winged stalks. Individual flowers rather inconspicuous, borne in dense *spike-like* clusters, the male with dark red anthers, the female greenish. Fruit 3.5–5 mm, spherical, red then black and shiny. One of the most abundant plants in the region. Throughout. Mastic (the resin) was formerly used as a chewing gum. *P.* × *saportae* is of hybrid origin between *P. lentiscus* and *P. terebinthus*; vegetative parts similar to *P. lentiscus* but with female flowers like *P. terebinthus*; fruit whitish. Widely reported but distribution improperly known.

(b) Leaves pinnately divided, *with* a terminal leaflet; stalks *winged*.

Pistacia atlantica A deciduous tree to 7 m with a rounded crown. Leaves with 2–5 pairs of oblong to lance-shaped leaflets 25–80 mm long, dark green above, paler beneath, with narrowly winged stalks. Fruits red, borne in panicles, 5–8 mm. Dry hillsides and roadsides in the east. AA (east), CY, IP, SY, TR.

(c) Leaves pinnately divided, *with* a terminal leaflet; stalks *unwinged or narrowly winged*.

Pistacia terebinthus TURPENTINE TREE A hairless, aromatic *small, deciduous tree* 8–10 m with unwinged, pinnately divided leaves; leaflets 3–11 mm, and *with an end leaflet present*. Flowers brownish-red borne in lax, long-branched panicles. Fruit 5–9(12) mm. Maquis, dry slopes, rocks and pine woods; common, sometimes abundant. Throughout.

Pistacia vera PISTACIO NUT A tree to 10 m with pinnately divided leaves with 3–5 rather broad, pointed leaflets 50 mm–12 cm, minutely hairy when young. Fruit conspicuous: a pale brown nut 15–30 mm long, borne numerously in lax panicles. Cultivated for its edible fruit.

Schinus

Resinous trees native to South America with pinnately divided leaves with stalkless leaflets. Flowers with 4–5 sepals and petals and 8–10 stamens. Fruit berry-like.

Schinus molle CALIFORNIAN PEPPER TREE A small evergreen tree to 15(25) m with slender, pendent branches. Leaves pinnately divided with 11–47 linear-lance-shaped, toothed leaflets 15–60 mm, hairy when young. Flowers 3–5 mm, yellow–white, borne in small, much-branched, pendent inflorescences; sepals and petals 5. Fruit 6–8 mm, pink, spherical. Widely planted as an ornamental; doubtfully naturalised. *S. terebinthifolia* has leaves with 5–15 broader leaflets, 30–60 mm and numerous red fruits, 4–5 mm; similar in general appearance to *Pistacia terebinthus* but with greenish-yellow (not red–purple) flowers with 10 stamens (not 3–5). Often planted in hot, dry areas.

1. *Pistacia lentiscus*
2. *Pistacia lentiscus* fruits
3. *Pistacia atlantica*
4. *Pistacia terebinthus* flowers
5. *Pistacia terebinthus* fruits
6. *Schinus molle* flowers
7. *Schinus molle* fruits
8. *Schinus terebinthifolia*

Rhus SUMACH

Shrubs or small trees with pinnately divided leaves and thick twigs. Flowers often tiny and densely clustered, with 5 sepals and petals and stamens.

Rhus coriaria SUMACH A softly hairy shrub or tree 1–4(5) m with more or less evergreen leaves and densely downy shoots. Leaves pinnately divided with 7–21(25), toothed, green leaflets, *the stalk between the leaflets slightly winged*, at least towards the end; leaf stalk 20–30 mm. Flowers small and whitish, borne in very dense hairy panicles to 17–25 cm long. Fruit 4–6 mm, spherical and woolly, ageing brown–purple. Scattered throughout.

Cotinus

Deciduous shrubs with simple leaves and slender twigs. Flowers borne in diffuse, plume-like panicles. Fruit 1-seeded.

Cotinus coggygria SMOKE TREE A rounded, deciduous, hairless shrub to 5 m. Leaves 30–80 mm, purple, simple, oval, untoothed and long-stalked. Flowers borne in large, lax panicles 15–20 cm long with many sterile branches bearing spreading hairs giving the impression of a *haze of smoke*. Fruit kidney-shaped, 3 mm long. Mountain slopes and woods. CR (rare or extinct), GR, IA, PN (rare).

SAPINDACEAE

A large family of perennials, lianas and trees (genera very distinct from each other). Leaves pinnately or palmately lobed or divided, stalked; stipules absent. Flowers unisexual or cosexual with 4–5 sepals and (0)4–5 petals; stamens 5–9; styles 1–2. Fruit variably dry or fleshy.

Aesculus HORSECHESTNUT

Deciduous trees opposite leaves. Flowers with 5 sepals forming a tubular calyx; stamens 5–9; style 1. Fruit a large capsule with 3 valves and 1–3 large seeds.

Aesculus hippocastanum HORSECHESTNUT A large, deciduous tree to 39 m with opposite, palmately divided leaves with 5–7 leaflets 10–25 cm long, the whole leaf to 60 cm across. Flowers white marked with pink–red, borne in erect panicles 15–30 cm in spring. Fruit a horsechestnut: a brown, shiny seed (conker) borne in a spiky green capsule 50–80 mm in the autumn. Native, locally, to mountain forests in the Balkans but planted throughout.

Acer MAPLE

Deciduous trees with opposite leaves. Flowers with 4–5 sepals and 0 or 4–5 petals; stamens 8; styles 2. Fruits with 2 parts, each 1-seeded with a long wing.

(a) Trees with leaves 3–5(7)-lobed and a *latex*. Inflorescence a corymb; stamens inserted on the middle of the disk.

Acer campestre A shrub or small tree up 20(25) m. Leaves 40–70 mm across, bluntly 3–5-lobed, rather thick in texture, with *minutely fringed margins*. Flowers few, greenish, borne in erect, hairy corymbs, opening alongside the leaves. Fruit usually hairy (sometimes hairless), with *horizontal wings*. Strongly western. GR, IA, PN.

1. *Rhus coriaria*
2. *Rhus coriaria* in fruit
3. *Cotinus coggygria*
4. *Aesculus hippocastanum*
5. *Acer campestre* foliage
6. *Acer obtusifolium*
7. *Acer negundo*
8. *Acer sempervirens*
9. *Acer sempervirens* foliage

(b) Trees with leaves 3–5(7) lobed, without a latex. Inflorescence a panicle or corymb; stamens inserted on the inner margin of the disk.

Acer pseudoplatanus A tree to 30 m. Leaves 10–15 cm across, 5-lobed to about ½ the diameter; lobes pointed, coarsely toothed. Flowers numerous, greenish, borne in narrow, pendent panicles, usually appearing alongside the leaves. Fruit hairless, with pointed wings usually diverging at a right angle. GR, PN.

Acer heldreichii A tree to 25 m. Leaves 50 mm–14 cm (typically <70 mm), *deeply* 5-lobed with the middle lobe free *nearly to the base*; lobes pointed, with 2–3 large teeth on each side. Flowers rather few, virtually hairless, yellowish, borne in rather erect panicles, opening alongside leaves. Fruit hairless, with curved wings diverging at an obtuse angle. GR, PN.

(c) Trees with leaves 3–5(7)-lobed, rather leathery and sometimes evergreen, without a latex. Inflorescence a corymb; stamens inserted on the inner margin of the disk.

Acer obtusifolium SYRIAN MAPLE A small evergreen tree to 10 m. Leaves leathery, dark green, obscurely to distinctly 3-lobed, (20)40–50(65) mm long, hairless on both sides. Flowers borne in lax corymbs, the male and cosexual flowers in the same inflorescences; male flowers with 4–5 sepals to 5 mm; petals 4–10, similar to the sepals; stamens inserted towards the centre of a conspicuous disk. Fruit with 2 spreading to erect, blunt wings. Rocky pine forests, often by streams. CY, IP (upper Gallillee – rare), LB, SY.

Acer negundo A tree to 20 m. Leaves 50 mm–10 cm across with *3–5(7) oval-pointed leaflets*. Flowers without petals, greenish, opening before the leaves; male inflorescence a corymb, the female a lax, pendent raceme. Fruit hairless, with wings diverging at an acute angle. Widely planted as an ornamental and occasionally naturalised. GR (mainly the northeast).

Acer sempervirens An *evergreen* shrub to 5(12) m. Leaves 20–50 mm across, *undivided to 3-lobed*, leathery, *green and hairless beneath*, short-stalked. Flowers few, greenish-yellow, borne in erect, hairless corymbs. Fruit with wings more or less parallel or diverging at an acute angle. AA, CR (common), GR, PN, TR (southwest).

Acer hyrcanum BALKAN MAPLE A tree to 15 m. Leaves 40 mm–10 cm, with (3)5 long, narrow, parallel-sided lobes, the lower surface hairless or slightly hairy. GR, PN, SY, TR.

Acer monspessulanum MONTPELIER MAPLE A shrub or small tree 8–12(15) m with greyish, finely cracked bark and leathery, *3-lobed* leaves 15–45 mm with blunt, *untoothed* lobes, shiny above and greyish below. Flowers yellow–green, borne in lax, erect clusters to 50 mm, which appear before the leaves. Fruit 20–30 mm. GR, PN, TR.

Cardiospermum

Herbaceous annuals or perennials, sometimes woody below, often with lobed leaves and small, soon-falling stipules. Flowers unisexual and zygomorphic; sepals 4(5); petals 4; stamens 8; style 1. Fruit a capsule.

Cardiospermum halicacabum BALLOON-VINE A tall, rather hairy annual 1–3 m, climbing with tendrils. Leaves trifoliate with deeply lobed and tooted leaflets 20–55 mm long. Flowers small and greenish-white with 4 sepals and 4 petals 2.5–5.2 mm long; stamens 8, of 2 sizes. *Fruit inflated and bladder-like*, 13–35 mm long, pendent; seeds black. Widespread as a native in tropical regions; naturalised widely. Probably local throughout.

RUTACEAE

A large and widely distributed family of aromatic herbs shrubs and trees. Leaves opposite or alternate with translucent glands. Flowers with 4–5 sepals and petals, free; stamens 2 x as many; style 1. Fruit a berry, capsule or drupe.

Haplophyllum

Perennials and shrubs. Sepals and petals 5; stamens 10; ovary 3–5-parted. Fruit a capsule. Many rare and local species to the east of the region covered (eastern Turkey). The following species are still frequently classified under *Ruta*.

Haplophyllum buxbaumii (syn. *Ruta buxbaumii*) A non-woody, rather robust, hairy perennial 15–48(50) cm. Leaves alternate, broadest above the middle, sometimes lobed with linear-lance-shaped segments. Flowers borne in broad, spreading inflorescences; sepals oval; *petals 5*, bright yellow, oblong to elliptic, 5–7(9) mm. Fruit a hairless capsule. Scrub, fields and disturbed ground. AA (Lesbos + east Rhodes), CY, IP, LB, SY, TR.

Haplophyllum suaveolens (syn. *Ruta suaveolens*) A short perennial 10–25 cm, similar to *H. buxbaumii* but with simple (rarely lobed) leaves, the uppermost forming a whorl below the flowers; hairless or sparsely hairy. Flowers yellow, borne in compact clusters; sepals lance-shaped, dark blackish-green when dry and white-hairy; *petals 5*, oval, 6–7 mm. Capsule 5 mm, warty. Dry, rocky slopes. TR.

1. *Cardiospermum halicacabum*
2. *Haplophyllum buxbaumii*
3. *Ruta chalepensis*
4. *Ruta chalepensis* subsp. *fumariifolia*

Ruta RUE

Strong-smelling perennials and shrubs. Flowers yellowish with 4–5 sepals and petals, often lobed; stamens (8)10; ovary (4)5-parted. Fruit a capsule.

Ruta chalepensis FRINGED RUE A hairless shrub 20–60 cm. Leaves rather long-stalked, divided into narrowly oblong-lance-shaped segments 1.5–6 mm wide. Flowers borne in lax inflorescences; petals oblong, *fringed with cilia*, the cilia not as long as the petal width; sepals hairless, triangular-oval. Capsule hairless, with pointed segments. Throughout. **Subsp. *fumariifolia*** is a dwarf form that occurs in hot, dry habitats in the southern Aegean that is considered to be distinct by some authors. AA (southwest), CR (+ KP).

Ruta graveolens COMMON RUE A distinctly grey–blue shrub, with dense branches 14–45 cm, superficially similar to *R. chalepensis*. Leaves broad, with oval-lance-shaped leaf segments. Flowers with *toothed* petals. Fruits 3.5–9.1 mm. Garrigue and maquis. AA (local), CR (southwest – rare), GR, PN. *R. montana* is similar but with flowers with *erect, unfringed petals* 3–6.5 mm, exceeded by the narrow, linear lobed leaves, giving the inflorescence a leafy appearance; leaf segments narrow (just 0.5–1.4 mm wide). Fruit 2.1–4.3 mm. GR (rare).

Citrus

Small evergreen trees with glossy leaves. Flowers solitary to few, often fragrant; petals (4)5(–8); stamens numerous (16–20 to 100). Fruit large and edible. Relationships are complicated by a long history of cultivation and interbreeding.

Citrus medica CITRON An evergreen shrub or small tree 2–5 m with reddish, spiny young branches. Leaves slightly toothed and scented when crushed, with scarcely winged stalks. Fruits yellow, *very large*, often misshapen, to 25 cm. Locally cultivated throughout. The following cultivars are frequently cultivated throughout the region: *C. × limon* LEMON An evergreen, tree to 4 m with a rounded crown, flowering and fruiting throughout the year. Leaves elliptic-lance-shaped and shallowly toothed. Flowers male or cosexual, with >4 x as many stamens as petals, petals white, often streaked purple. Fruit a lemon, yellow and warty when ripe. *C. × aurantium* SEVILLE ORANGE A small tree to 10 m with flexible spines. Leaves with a rounded base and *broadly winged stalks*. Fruit an orange. *C. × sinensis* SWEET ORANGE is similar but with narrowly *winged leaf stalks*. *C. × paradisi* GRAPEFRUIT is similar to *C. sinensis* but with a distinct, broadly winged stalk, and large, yellow fruits. *C. × aurantiifolia* LIME has leaves resembling those of an orange tree and small, *spherical* fruits 25–50 mm across, harvested green (yellowish when mature).

Dictamnus BURNING BUSH

Perennials with pinnately divided leaves with a *winged axis*. Flowers slightly zygomorphic, borne in spike-like inflorescences; stamens 10; style 1. Fruit 5-lobed.

Dictamnus albus BURNING BUSH A strong-smelling, erect perennial to 90 cm with glandular hairy, pinnately divided leaves with 7–9 oval, finely toothed, leathery leaflets 25–75 mm. Flowers large, zygomorphic, white or pink, veined violet, borne in long, lax, leafless spikes; upper petals erect, 24–29 mm, the lower deflexed; stamens long-projecting, 25–31 mm. Fruits deeply 5-lobed. Open woods and maquis, mainly on the mainland. GR, IA, PN, TR.

SIMAROUBACEAE

A predominantly tropical family of plants with alternate, pinnately divided leaves. Flowers small, usually unisexual with (3)5(7) sepals and petals and 2 x as many stamens; ovary superior, surrounded by a disk with 2–5 fused or free carpels. Fruit an aggregate of winged achenes.

Ailanthus

Trees with flowers with 5–6 sepals fused to the middle, the male flowers with 10 stamens; ovary superior surrounded by a disk. Fruit winged.

Ailanthus altissima TREE OF HEAVEN A rapidly growing tree to 20(30) m with smooth bark and large, pinnately divided leaves with 5–12 pairs of oval-lance-shaped leaflets 40 mm–17 cm long and terminal panicles of small, strong-smelling greenish-yellow, 5(6)-parted flowers, unisexual; petals 2.2–4.5 mm. Fruit in clusters of 3-winged carpels, 25–50 mm, reddish brown. Native to China but very commonly planted as an ornamental and naturalised. Throughout.

---

1. *Citrus medica*
2. *Citrus × limon*
3. *Citrus × aurantium*
4. *Citrus × aurantiifolia*
5. *Dictamnus albus*
6. *Ailanthus altissima*

MELIACEAE

A family of mostly trees and shrubs characterised by alternate, usually pinnately divided leaves without stipules. Flowers cosexual (or cryptically unisexual) borne in panicles, cymes, spikes, or clusters; stamens 3–10–numerous. Fruit a berry.

Melia

Trees with 2-pinnately divided leaves, usually toothed. Flowers with 5 sepals and petals; stamens 10–12, in 1–2 whorls. Fruits berry-like.

Melia azedarach INDIAN BEAD TREE A deciduous tree to 15 m with furrowed bark. Leaves 20–40(60) cm, alternate and 2-pinnately divided; leaflets elliptic and toothed or lobed, 50–70 mm. Flowers lilac and scented, borne in panicles; petals 5, each 8–12 mm. Fruit a berry 8–15(25) mm across, yellow and *long-persisting* (even when not in leaf). Native to India and China, widely planted along roadsides.

CYTINACEAE

A small family of obligate root-parasites of other shrubs. Flowers unisexual and ant-pollinated. Seeds numerous, in a pulp, dust-like when dry and windborne.

Cytinus

Parasitic plants without chlorophyll or obvious leaves or stems. Perianth 4-lobed; stamens 8–10. Fruit a capsule; seeds minute. Host-specific races are probably in the process of forming cryptic species.

(a) Flowers bright yellow.

Cytinus hypocistis CYTINUS An unmistakable, bright yellow parasite of various Cistaceae species, with underground stems 30 mm–16 cm (stemless above ground), and yellow, orange or red oval-oblong scale leaves. *Flowers bright yellow*, in dense clusters of 4–14(20), subtended by 2 bracteoles the same colour as the scale leaves. Garrigue, maquis and dunes. Throughout (although records in the southeast seem to correspond better with the next species).

(b) Flowers pale pink.

Cytinus ruber Similar to the above species, but with about 20 flowers with *crimson* (not red–orange) scale leaves and bracts contrasting a *white–pale pink* (not yellow) perianth 10–23 mm that slightly exceeds the bracteoles. Parasitic on pink-flowered *Cistus* spp. Garrigue and maquis, rather local. Often described as a form of the previous species but shown to be genetically distinct. Throughout except the far north.

1. *Melia azedarach*
2. *Cytinus hypocistis*
3. *Cytinus ruber*
4. *Gossypium herbaceum*
5. *Abutilon fruticosum*

MALVACEAE | MALLOW FAMILY

A large family of herbs or shrubs with star-shaped (stellate) hairs, and alternate, often palmately lobed leaves with stipules. Flowers often conspicuous, usually with both a calyx and epicalyx (an important character); sepals and petals 5; stamens numerous; styles 1 or 5–many. Fruit a nut with 1–3 seeds, or capsule splitting into nutlets.

Gossypium COTTON

Herbs and shrubs (to small trees). Leaves usually 3–9-palmately lobed (sometimes entire). Flowers solitary, borne in the leaf axils; epicalyx segments 3, large, leafy; corolla usually yellow with a crimson centre. Fruit a capsule; seeds covered in a mass of cotton.

Gossypium herbaceum A shrubby, rather woody annual to 1.5 m, usually hairy. Leaves 20–50 mm with hairy margins, palmately (3)5(7)-lobed; lobes oblong, elliptic or oval. Flowers erect, borne on stalks 7–15 mm; epicalyx segments 10–20 mm with heart-shaped bases; calyx cup-shaped, black-dotted; petals 25–35 mm, broad, yellow with a crimson centre. Fruit a capsule 25–30 mm long, beaked, 3–5-parted, woody and widely splitting when mature, revealing *cottony mass* concealing the 5–7 seeds per chamber. Native to sub-Saharan Africa and Arabia; planted in hot, dry parts; an important crop in Syria. **G. hirsutum** is similar, with long-toothed epicalyx lobes, cream flowers and large capsules 40–60 mm. Native to Central America. Planted as an ornamental in the east.

Abutilon

Mainly tropical and subtropical perennial herbs or shrubs. Flowers often yellow; calyx 5-lobed. Fruit with 5–many mericarps.

Abutilon fruticosum A shrubby perennial to 1(2) m. Leaves to 60 mm (10 cm), heart-shaped at the base, toothed. Flowers solitary; calyx 4–10 mm, divided to the middle; corolla yellow-orange; petals 7–10(17) mm. Mericarps 3-seeded. Deserts. IP (south and east).

Malope

Similar to *Malva* (see below) but epicalyx with 3-heart-shaped lobes (wider than the sepals) and fruit strawberry-like.

Malope malacoides A woody-based perennial with erect to ascending, bristly to almost hairless stems. Leaves oblong-lance-shaped, with double-scalloped margins, bristly along the veins beneath. Flowers pink, borne singly in the leaf axils; epicalyx segments *broadly triangular–heart-shaped*; sepals lance-shaped, 11–16 mm; petals 22–40 mm. Fruits hairy, ridged. Rocky slopes and fallow land; rather rare and local. AA (local), GR, IA, PN, TR.

Malva MALLOW

Annuals or perennials with epicalyx of 3(–10) segments, free or fused, and white–pink petals; carpels numerous; fruit splitting into nutlets. The traditional distinction between *Lavatera* and *Malva* is based on fusion or non-fusion of the epicalyx, but this character is now established to be artificial therefore the genera are now merged. A difficult genus; observation of mature fruits often necessary.

(a) Annuals. *Epicalyx segments 6–10.*

Malva setigera (syn. *Althaea hirsuta*) ROUGH MARSH MALLOW A slender annual to 60(70) cm with simple and star-like hairs with swollen bases. Lower leaves rounded and kidney-shaped, 10–40 mm across, long-stalked and toothed or shallowly lobed, more deeply palmately lobed further up the stem. Flowers pink, solitary, long-stalked and small, cup-shaped, the petals scarcely exceeding the sepals; petals 12–16 mm. Epicalyx segments 6–9(10), and nearly as long as the sepals. Fruit hairy and ridged. Dry fallow land and maritime waste places. Throughout (except the far southeast).

(b) Annuals with leaves often lobed but not deeply divided. Epicalyx segments 3; *petals large, 25–45 mm.*

Malva trimestris (syn. *Lavatera trimestris*) ANNUAL LAVATERA An erect, somewhat *bristly* annual to 1.2 m, with long-stalked, rounded to heart-shaped, toothed leaves; the upper shallowly 3–7-lobed, deep shiny green. Flowers large, the petals 25–45 mm long, pink (sometimes white), solitary. Epicalyx segments shorter than the sepals; sepals fused over most of their length, and *enlarging in fruit*. Fruits hairless, ridged, and *covered by a disk-like expansion* projecting from the central axis. On cultivated, fallow and waste ground. AA (east), IP.

1. *Malva parviflora*
2. *Malva moschata*
3. *Malva alcea*
4. *Malva punctata*
5. *Malva sylvestris*
6. *Malva multiflora*
7. *Malva arborea*

(c) Annuals with leaves often lobed but not deeply divided. Epicalyx segments 3; *petals small, <5 mm.*

Malva parviflora A hairy or hairless annual to 50 cm. Leaves rounded to heart-shaped with 3–7, shallow, rounded and toothed lobes. Flowers borne in clusters of 2–4, pale mauve or lilac, *small – the petals <5 mm long, short-stalked* with linear-lance-shaped epicalyx lobes; *sepals hairless at margins. Fruit with enlarged, spreading, papery sepals*, fruit strongly netted and hairy or not; with stalks <10 mm. Throughout.

(d) Annuals or perennials usually <1 m with *upper leaves deeply divided*. Epicalyx segments 3; petals >(10)16 mm.

Malva moschata MUSK-MALLOW An erect, branched perennial herb to 80 cm with scattered *simple hairs*. Basal leaves kidney-shaped, 3-lobed, to 80 mm long and long-stalked; stem-leaves *deeply palmately cut* into narrow, further divided segments. Flowers pale to bright rose pink, borne singly or in pairs in the leaf axils in loose terminal clusters; petals >16 mm; epicalyx segments linear lance-shaped and *>3 x as long as wide*. Nutlets smooth with *long hairs*. Grassy habitats, generally above sea level. GR, PN. *M. alcea* GREATER MUSK-MALLOW is a similar perennial to 1.2 m with *star-like* (not simple) on all vegetative parts. Epicalyx segments 7–8 mm, <3 x as long as wide. Nutlets hairless or hairy. Dry and waste habitats. (north).

Malva cretica An erect annual to 40 cm with bristly, long, spreading hairs. Lower leaves more or less rounded, wavy-lobed; *upper leaves deeply divided into 3(5) oblong, toothed lobes*. Flowers borne in the axils on long stalks; epicalyx segments 4, linear to narrowly triangular; sepals 7–10 mm, *narrowly triangular, much longer than wide*; petals 10–35 mm. Fruits hairless with small ridges, the angles slightly winged. Throughout. **Subsp. *cretica*** is the form in the region that has star-shaped and simple hairs on the flower-stalks and petals 1–1.5 x as long as the sepals. *M. punctata* has lower leaves shortly 5-lobed, and 3-lobed upper leaves with a *long central lobe*. Petals 15–30 mm. Virtually throughout.

(e) Annuals or perennials, often to 1 m with leaves often lobed but not deeply divided. Epicalyx segments 3; *petals >9 mm*.

Malva multiflora (syn. *M. pseudolavatera, Lavatera cretica*) SMALL TREE MALLOW An erect or ascending, *non-woody* annual or biennial to 1.5 m, with lower leaves circular to heart-shaped, shallowly 3–7-lobed, the upper leaves more deeply 5-lobed. Flowers *lilac*, borne in the axils of the leaves in clusters of 2–8, on unequal stalks shorter than the leaves. Petals 10–20 mm long, epicalyx segments to 6 mm long, and free *almost* to the base (but fused there), shorter than the long-pointed sepals; *sepals just shorter or as long as sepals in fruit*. Fruit smooth or slightly ridged. Common on waste ground throughout. Similar to *M. sylvestris*, which has narrow epicalyx segments that are *free* completely to the base.

Malva sylvestris COMMON MALLOW A variable, erect to spreading annual or perennial herb to 1 m. Leaves roughly kidney-shaped or heart-shaped with 3–7 toothed lobes. Flowers borne in clusters of 2 or more, *bright pink or purple with darker veins*, the petals 12–30 mm long and bearded; petals about *>2(4) x the length* of the downy sepals. Epicalyx segments *free* to the base. Fruit sharply angled, hairy or not, and netted. Common on disturbed ground; frequently confused with *M. multiflora*, which has epicalyx segments *fused* at the base, and less strongly marked flowers. Throughout.

Malva neglecta DWARF MALLOW An erect or ascending annual similar to *M. sylvestris* but smaller, to 60 cm, with leaves less deeply lobed, petals *pale lilac* and *2–3 x the length* of the sepals (9–13 mm); epicalyx segments linear-lance-shaped. *Fruit smooth* and hairless; *fruit stalks remaining erect*. Waste places. Throughout. *M. nicaeensis* is similar to *M. sylvestris* but with broader epicalyx segments, leaves not distinctly heart-shaped at the base, pale mauve petals 10–12 mm. Fruit *netted; fruit stalks recurved*. Similar habitats. Throughout.

(f) *Robust, woody perennial* with leaves often lobed but not deeply divided. Epicalyx segments 3; petals normally >(10)15 mm.

Malva arborea (syn. *Lavatera arborea*) TREE MALLOW A robust, *woody* biennial or perennial to 3 m, downy above with star-shaped hairs. Leaves large, circular palmately lobed with 5–7 lobes to 20(22) cm long, velvety. Flowers borne in the leaf axils in clusters of 2–7, forming a long, terminal inflorescence; petals 14–20 mm, *pink–purple with darker veins*. Epicalyx segments to 10 mm long, *exceeding the sepals* and greatly enlarged in fruit and fused at the base. Fruit hairy or not, and sharply angled. Maritime rocks. Throughout (not CY or some smaller islands).

Althaea

Erect biennials or perennials with flowers with an epicalyx with 6–10 lobes fused at the base and shorter than their calyx. Fruit splitting into many disk-like, 1-seeded nutlets.

Althaea officinalis MARSH MALLOW A tall, velvety downy perennial to 1.5 m with all hairs star-shaped. Leaves triangular-oval, entire or scarcely 3–5-lobed, often folded and fan-like, to 10 cm across. Flowers solitary or clustered in the leaf axils, pale lilac–pink, borne on stalks shorter than their adjacent leaves; petals 15–20 mm. Fruits hairy, sepals curved outwards in fruit. Damp, saline coastal habitats; rare or absent on most islands. AA (north: Lesbos, Samothraki, Thasos), CR (west – very rare), GR, IA (uncommon), IP, PN (north), TR.

Althaea cannabina A tall, erect, hairy perennial to 1.8 m with green, deeply 5-lobed leaves with oblong, double-toothed lobes. Flowers large and pale rose–pink, the petals >2 x as long as the calyx (12–30 mm). GR, TR.

Alcea HOLLYHOCK

Tall, erect biennials or perennials with *very large* flowers borne in tall, wand-like inflorescences; epicalyx with 6–7 lobes fused at the base and shorter than their calyx. Fruit splitting into 1-seeded nutlets.

Alcea rosea HOLLYHOCK *A tall perennial to 3 m with stiff, erect, unbranched stems*, and large, flat palmate, bluntly toothed leaves. Flowers large and showy, usually pink but variable in colour, particularly in cultivated forms; petals 25–50 mm. Nutlets winged. Possibly of hybrid origin. Mediterranean; widely naturalised.

Hibiscus HIBISCUS

Epicalyx with 3(10–13) segments; sepals 5, often fused and persisting in fruit; carpels 5, fused. Fruit a splitting capsule.

(a) Woody-based shrubs.

Hibiscus syriacus COMMON HIBISCUS An erect shrub to 3 m with diamond-shaped, toothed leaves, which are somewhat 3-lobed and short-stalked. Flowers variable in colour from pink to white with a purplish centre, borne solitary or in pairs in the upper leaf axils. Capsule yellow-hairy. Widely planted as an ornamental. Throughout. *H. rosa-sinensis* has shiny, oval, sometimes slightly irregularly toothed leaves, and conspicuous and showy bright red flowers. Very commonly planted in towns and gardens. Throughout.

(b) Non-woody annuals or perennials.

Hibiscus trionum BLADDER HIBISCUS A bristly, short, normally erect annual to 50 cm. Leaves deeply divided into 3(5) pinnately lobed segments. Flowers solitary and long-stalked, cream-coloured with a *deep violet centre*, petals 15–25 mm long; epicalyx with 10–13 linear lobes; *calyx inflated, with purply and bristly veins*. Fruit a capsule consealed within the inflated calyx. Very locally naturalised on plains and scrub. AA (Lesbos, Rhodes), CR (rare), GR (rare), IA (rare), PN (rare), IP, TR.

Dombeya

Trees or shrubs native to Africa and Madagascar. Flowers borne in pendent clusters; *petals persistent*; typically with 15–25 stamens + 5 conspicuous sterile stamens (staminodes); carpels 3–5. Fruit a capsule.

Dombeya × *cayeuxii* An exotic-looking tree, probably of hybrid origin (*D. burgessiae* × *D. wallichii*), to 5 m with large, drooping leaves and pendent inflorescences of pink flowers, which have long-persisting petals; epicalyx not persistent. Sometimes planted in towns in the warmest parts.

Ceiba

Trees from the tropics with straight, swollen trunks, alternate, often palmately lobed leaves and regular flowers with 5 free petals and filaments fused in a column.

Ceiba insignis WHITE FLOSS SILK TREE A distinctive, deciduous tree with a grey, *markedly swollen, bottle-shaped trunk* when mature, sparsely *covered in stout spines*. Leaves dark green and palmately lobed to the base; lobes 5–7, lance-shaped and pointed, often shallowly toothed. Flowers borne in late summer, large and showy with 5 free petals to 12 cm long, white, flushed with cream to golden yellow, often with reddish markings; stamens fused into a column. Fruit *large and pear-shaped* splitting to reveal *cotton-like seeds*, which are conspicuous in the absence of leaves in winter. Planted in towns and parks. *C. speciosa* is similar but with *pink* flowers. Occasionally planted.

Brachychiton

Trees and shrubs native to Australia. Monoecious; the unisexual flowers with a bell-shaped, lobed perianth; stamens typically 10–30 plus the same number of sterile stamens (staminodes); carpels 5.

Brachychiton populneus A tree to 18 m (often much less) with a distended trunk, and pointed, simple or broad-lobed leaves. Flowers *bell-shaped*, variably greenish, yellowish or pinkish, with small red markings or entirely red within, and with usually 5 or 6 unequal, pointed lobes. Fruit a splitting pod. Planted in hot, dry areas.

NEURADACEAE

A small family of hairy, prostrate annuals. Flowers solitary; sepals and petals 5; stamens and styles 10. Fruiting carpel dry, disk-like, convex.

Neurada

Unmistakable for their white-hairy, prostrate habit and disk-like fruiting heads.

Neurada procumbens A prostrate, white-hairy annual to 14(32) cm, much-branched. Leaves 6–25 mm, oval, irregularly lobed. Flowers greenish, borne solitary in the leaf axils; petals 5, 2–4.3 mm long; stamens and styles 10. Fruiting carpel distinctly disk-like, 8–15 mm across, spiny above and 10-valved. Sandy deserts and coastal plains. CY (south), IP (common on coasts).

1. *Althaea officinalis*
2. *Alcea rosea*
3. *Hibiscus rosa-sinensis*
4. *Hibiscus trionum*
5. *Ceiba speciosa*
6. *Brachychiton populneus* fruits
7. *Neurada procumbens*

THYMELAEACEAE

Hairless shrubs with simple, untoothed, alternate leaves. Flowers in clusters or racemes, regular and cosexual; calyx a tube, petal-like; true petals absent; stamens 8, fused to the surface of the calyx tubes; style solitary. Fruit a drupe, berry or nut.

Daphne

Shrubs with simple leaves. Flowers with 4 petal-like lobes; stamens 8. Fruit fleshy, enclosed in a persistent calyx.

Daphne gnidium An erect, lax, virtually hairless shrub to 2 m with branches bare beneath; superficially rather *Euphorbia*-like when not in flower, but without a white latex when cut. Leaves 20–30 mm, pale green, leathery, linear and pointed. Flowers cream-white, 5–6.5 mm long, borne in dense panicles in later winter to spring. Berry deep red, ageing black, 7–8 mm. Maquis, garrigue and open woods. GR. ***D. gnidioides*** is similar but with flowers borne in *dense* terminal heads, in *autumn*. AA (east), CR (south – rare), TR (south, southwest).

Daphne oleoides A small shrub rather similar to *D. gnidium*, but shorter and more compact with ascending to almost prostrate stems to 40(60) cm with leaves completely hairless when mature, and white to cream flowers borne in groups of 2–6 *with longer, pointed lobes*; 9–14 mm long. Fruit 10 mm. Rocky habitats at high altitude. CR, CY, GR, IA, PN, TR. ***D. euboica*** is similar, often taller (to 70 cm) and with flowers borne in clusters of 3. Rocky woods. A rare endemic of Euboea + adjacent mainland.

Daphne laureola SPURGE LAUREL An erect, hairless, evergreen shrub to 1.5 m with leathery, *broadly lance-shaped*, alternate, *dark green and shiny* leaves, 30 mm–12 cm. Flowers borne in short, tight clusters of 2–10; *yellowish-green*, 12–14 mm long, slightly nodding, with 4 petal-like lobes; fragrant. Fruit spherical and black when ripe, 8–10(13) mm. Mountain forests. GR.

Daphne sericea A dense, rounded shrub to 70 cm with hairy young shoots. Leaves rather crowded, dark green and leathery, broadest above the middle and narrowed into short stalks; hairy beneath. Flowers pinkish (sometimes cream), borne in dense terminal heads, fragrant; 8–12 mm long. Fruit a red–brown berry. Maquis and open woodland. CR, GR, LB, TR.

Thymelaea

Dwarf evergreen shrubs or annual herbs with small, unstalked leaves. Flowers normally yellowish, borne singly or in clusters in the leaf axils. Fruit nut-like.

(a) Plant a perennial shrub or subshrub.

Thymelaea hirsuta A dwarf evergreen shrub to 2 m with white-downy stems; superficially similar to shrubby Amaranthaceae, with small, scale-like succulent leaves 2.5–6 mm overlapping along the stem, shiny green and *hairless outside, white-downy inside*. Flowers 3–5 mm, unisexual or cosexual; yellowish, borne in inconspicuous clusters of 6–12. Fruit 3–5 mm. Common in bare, dry and sandy places, often on sea cliffs. Throughout, except the north.

1. *Daphne gnidium*
2. *Daphne oleoides*
3. *Daphne laureola*
4. *Daphne sericea*
5. *Thymelaea hirsuta*

Thymelaea tartonraira A low, *dense, almost prostrate and often cushion-like shrub* to 50 cm. Leaves broadest towards the tips and *greyish*, 5–15 mm. Flowers 3.5–6 mm, pale yellow, borne in clusters of 2–5 at the base of the leaves. Fruit 2.5–3.5 mm. Coastal garrigue and scrub; absent from most islands. Throughout.

(b) Plant an annual.

Thymelaea passerina A very slender, sparingly branched, late-flowering, more or less *hairless annual* with stiffly erect stems to 50(80) cm. Leaves linear and flax-like, alternate and pointed, 4–8 mm. Flowers 2–3 mm, greenish, borne in clusters of 3–7, in the leaf axils in elongated, lax inflorescences. Fruit 2–3 mm. Various dry habitats; widespread but rarely common. AA (Rhodes, Samothraki, Thasos), CY, GR, IP, PN, TR.

CISTACEAE | ROCK ROSE FAMILY

An important family of herbs and shrubs in the Mediterranean. Leaves normally opposite, stipules often present. Flowers often showy, cosexual, solitary or in lax, terminal clusters; sepals 3–5, petals 5, free, stamens numerous. Fruit a capsule with 3–5 valves.

Cistus ROCK ROSE

Shrubs with opposite leaves without stipules. Flowers solitary, often showy; stamens 50–150; carpels 5(6–12). Fruit a capsule.

(a) Sepals 5; flowers pink.

Cistus creticus A variably compact bush to 1.4 m with sticky young branches. *Leaves short-stalked* (3–10 mm), oval to elliptic, 15–45 mm long, green to grey–green. Flowers pink–purple; petals 17–20 mm; sepals long-pointed. Capsules 7–10 mm. Stony pastures and field boundaries. Throughout, except the far north and southeast. Two subspecies are generally recognised, the distinction and distribution of which is rather unclear. **Subsp. *creticus*** has small, undulate-crimped edged, sticky-*glandular* leaves to 25 mm. Most common in the south. **Subsp. *eriocephalus* (syn. *C. villosus*)** has larger, flatter, scarcely glandular leaves, and whitish downy flower-stalks and calyx. Most common further north.

Cistus parviflorus is a more *compact* shrub than the previous species, often hemispherical, with 3-veined leaves, smaller, pale pink flowers and very short style (<0.5 mm, rather than 2.5–4 mm). AA, CR, CY, GR (Sterea Ellas), IA (Zakynthos – rare), TR (south).

(b) Sepals 5; flowers white.

Cistus monspeliensis NARROW-LEAVED ROCK ROSE A slightly sticky bush to 1.8 m, lax below, compact above. Leaves mid- green, *narrow*, linear to lance-shaped, scarcely tapered at the base and unstalked, 15–45(70) mm. Flowers small, white; petals 9–14 mm; sepals 5, the outer 2 wedge-shaped at the base. Capsule 4 mm. Pine woods, maquis and cliff-top garrigue. AA (rare), CR, CY, GR (west and east coasts; rarer in the north), IA, PN, TR.

Cistus salviifolius SAGE-LEAVED ROCK ROSE A low, spreading, almost prostrate shrub 20–90 cm. Leaves *sage-like*; *short-stalked*, mid-green, oval-elliptic with a rounded or wedge-shaped base, 8–18(45) mm long, 3-veined and hairy on both surfaces. Flowers white, borne solitary or in clusters of up to 4; petals 14–20 mm; sepals 5, the outer 2 with a heart-shaped base. Capsule 5–7 mm. Common on garrigue and dunes. Throughout.

(c) Sepals 3; flowers white.

Cistus ladanifer GUM ROCK ROSE An aromatic, *extremely sticky*, lax, erect shrub to 2(4) m. Leaves linear-lance-shaped, dark green (paler beneath), 40–80 mm long, 3-veined in the lower 1/3, and scarcely stalked. Flowers solitary, *large* (50–80 mm across), white, often with a crimson blotch at the base of each petal; petals 30–55 mm; *sepals 3*. Capsule 10–15 mm. Native to the western Mediterranean, naturalised in Cyprus. CY (north and west).

1. *Cistus creticus* subsp. *creticus*; (inset) *C. creticus* subsp. *eriocephalus*
2. *Cistus parviflorus*
3. *Cistus monspeliensis*
4. *Cistus salviifolius*
5. *Tuberaria guttata*

Tuberaria (syn. Xolantha)

Annuals or perennials with basal rosettes and erect flowering stems. Flowers yellow, sepals 5, the outer 2 smaller. Capsule 3-valved. Most floras classify under *Tuberaria* rather than *Xolantha*.

Tuberaria guttata (syn. *Xolantha guttata*) ANNUAL ROCK ROSE A very variable, hairy, low annual to 42 cm with a basal leaf rosette, dying when mature, and a normally unbranched flowering stem. Leaves elliptic to oval, often with down-turned margins, 16–73 mm. Flowers yellow, with petals with or without a dark brown or purple spot at the base, 3–9 mm; flower-stalks longer than the sepals at the point of flowering. Maquis and garrigue. Throughout.

Helianthemum

Dwarf shrubs or herbs with opposite leaves. Flowers borne in 1-sided clusters; sepals 5, the outer 2 smaller; stamens numerous, style long and S-shaped. Fruit a 3-valved capsule.

(a) Plant an annual; flowers yellow.

Helianthemum salicifolium WILLOW-LEAVED ROCK ROSE A low, hairy, branched, erect or spreading *annual* to 30 cm. Leaves 5–25 mm, oval-lance-shaped, flat and short-stalked. Flowers yellow, borne in lax clusters; petals 2–7 mm long and narrow, shorter to slightly longer than the sepals; bracts large and leafy; *flower-stalks spreading in fruit,* upturned at the apex. Sandy and rocky slopes; widespread and common. Throughout. ***H. ledifolium*** is taller, to 45(60) cm, *hairier*, with *petals 6–8 mm, shorter than the sepals*, and flower-stalks *erect* in fruit. AA (local), GR, IP, TR. ***H. sanguineum*** is very small – usually <10 cm, *sticky*, and with flower-stalks *strongly bent downwards* in fruit. Rare and local. AA (Milos – rare), CY.

1. *Helianthemum obtusifolium*
2. *Helianthemum apenninum*
3. *Helianthemum vesicarium*
4. *Fumana thymifolia*
5. *Fumana arabica*

(b) Plant a woody-based perennial (or subshrub), often (not always) with stipules; flowers yellow.

Helianthemum obtusifolium A sprawling to sub-erect shrub to 25 cm, densely white-downy, at least in the upper half. Leaves lance-shaped to elliptic, 15–25 mm long with recurved margins, narrowed at the base into a short but distinct stalk. Flowers creamy yellow, borne in *simple* (or sparingly branched) inflorescences; petals 15 mm long; 3 inner sepals boat-shaped, 6–9 mm, the outer 2 much narrower, 4–5 mm, with hairy margins. Garrigue; an island endemic. CY.

Helianthemum nummularium COMMON ROCK ROSE A variable, prostrate to ascending dwarf shrub 10–35 cm. Leaves oblong-lance-shaped with *flattish margins*, hairy or hairless above, and grey, or white-hairy beneath, 10–25 mm. Stipules small and *leaf-like*. Flowers yellow, solitary or in clusters of up to 15(18); petals 8–12 mm; sepals hairless. Dry grassland and rocky outcrops. Mainly northern and inland; absent from most islands. GR, IA, LB, PN, SY, TR.

Helianthemum syriacum (syn. *H. lavandulifolium*) A grey-felted, low, *woody, stout* shrub to 50(85) cm. Leaves lance- to linear-lance-shaped, grey–green with recurved margins and white-downy beneath, 10–50 mm. Flowers bright yellow, numerous, borne in *distinctive forked (compound) inflorescences, at first coiled*, later spreading, and bearing numerous drooping fruits; petals 5–10 mm. Pine woods and garrigue. Throughout, except the north.

(c) Plant a subshrub or shrub. Flowers white.

Helianthemum apenninum WHITE ROCK ROSE A variable (many subspecies described), lax, spreading shrub to 40 cm with linear to oblong, grey-hairy leaves 4–20 mm. Flowers white with a yellow claw to each petal, in clusters of 2–10(15); petals 8–13 mm. Grassy and rocky habitats; rare and local. AA, GR (+ adjacent islands), PN.

(d) Plant a subshrub. Flowers typically pink.

Helianthemum vesicarium A small shrub 15–30 cm with adpressed, fine white hairs. Leaves 10–20 mm, oblong to linear with down-turned margins, with long, linear stipules (½ as long as the leaves). Flowers borne 3–8 in simple, 1-sided racemes, typically bright *pink* (or white, reddish or purplish); inner sepals 4-veined with sparse bristles. Fruiting calyx papery and inflated, exceeding the fruit. Deserts and scrub. IP (mainly east-central).

Fumana

Dwarf shrubs similar to *Helianthemum* with narrow, lance-shaped to linear, usually *alternate* leaves. Outer stamens sterile.

(a) Stipules present.

Fumana thymifolia THYME-LEAVED FUMANA A small, much-branched somewhat thyme-like (not aromatic) dwarf shrub to 20 cm with erect or ascending branches. Leaves 4–12 mm, *opposite* at least below, unequal on the stem and reduced above; *oval to oval-lance-shaped*, often *stickily hairy* and with *down-turned margins*; *stipules present*, as well as short shoots in the leaf axils. Flowers yellow, borne in 4–8-flowered clusters; the inner sepals much larger than the outer 2, membranous and green-veined; flower-stalks 8–12 mm. Common in a range of habitats. Throughout. *F. laevipes* is similar but with alternate, linear, *bristle-like* leaves 8–10 mm. Flowers 5–10 per cluster; flower-stalks 8–12 mm. Coastal habitats. Rare and local. AA, CR (west + KP), IA (Corfu), PN (Elafonisos).

Fumana arabica ARABIAN FUMANA A small, much-branched shrub to 25 cm with *leaves all alternate*, equally spaced along the stems; stipules present. Flowers yellow and rather large, 19–20 mm across, borne up to 7 in a raceme; the 2 outer sepals much smaller than the inner 3. Maquis and garrigue. Throughout, except the north.

(b) Stipules absent.

Fumana procumbens PROCUMBENT FUMANA A small, finely hairy-stemmed shrub to 35 cm similar to *F. thymifolia* but more prostrate and with linear leaves 12–18 mm, *all alternate; stipules absent. All flowers solitary* (or few), arising laterally; petals yellow with a dark golden spot at the base, *borne on short, thick stalks* (0.8 mm across), curving in fruit. North of the region. GR, IA, PN. *F. scoparia* is similar but with many erect flowering stems 40–80 mm, curved in bud with 2–3 flowers; inflorescence very sticky-glandular-hairy. Rare and local but widespread. GR, PN.

TROPAEOLACEAE | NASTURTIUM FAMILY

Herbaceous annuals and perennials, often with a rather succulent habit and hairless, alternate leaves. Flowers showy, with 5 sepals and petals and 8 stamens; stigmas 3. Fruit 3-parted, breaking into succulent segments.

Tropaeolum

Climbing or scrambling annuals with peltate leaves. Flowers zygomorphic with 5 free petals and sepals; stamens 8. Fruit 3-parted.

Tropaeolum majus NASTURTIUM A vigorous, creeping, hairless annual to 2 m with circular, blue-green, shallowly 5–6-lobed, upward-facing, *peltate leaves* (stalk attached to the centre of the blade). Flowers large, solitary, long-spurred yellow or orange–red 25–60 mm across borne in the leaf axils; spur 25–40 mm. Native to South America but often escaping. Throughout.

RESEDACEAE | MIGNONETTE FAMILY

Annual or perennial herbs with alternate, simple or pinnately divided leaves. Flowers borne in long spikes; flowers with 4–6(8) free sepals and the same number of often yellow, distinctively shaped petals; stamens 7–numerous (25); style 0. Fruit a capsule.

Reseda

Herbs with flowers with 4–8 sepals and petals, 10–25 stamens and *bottle-shaped fruits*.

(a) Flower spikes whitish; leaves clearly divided or lobed; capsule with 4 terminal teeth.

Reseda alba WHITE MIGNONETTE An erect to ascending, hairless perennial to 75 cm with stems branched above. Leaves pinnately lobed with 10 or more narrow, untoothed lobes. *Flowers white*, with 5–6 sepals and petals, the petals all lobed to 1/3 or more of their length; *stamens 10–12*; flowers borne in long spikes. Fruit 6–15 mm, *4-angled* (4 carpels) and elliptic, constricted apically and erect when ripe. Common in disturbed areas and sandy waste places. Throughout. *Reseda decursiva* has a petal claw not or scarcely winged and absent or poorly developed disk. IP, SY.

(b) Flower spikes whitish; leaves divided or not; capsule with 3 terminal teeth.

Reseda phyteuma A rather lax, branched annual to 30(50) cm with leaves all linear-spatula-shaped (sometimes with a single pair of lateral lobes). Flowers white with 6 sepals and petals with deeply cut lobes, and prominent stamens. Fruit 11–25 mm, 3-lobed and distinctly *nodding* in fruit, with conspicuous and persistent *sepals, enlarged in fruit*. Cultivated ground and roadsides; local and strongly western. AA (Thasos), GR, IA, PN (northwest; rare in northeast).

Reseda alopecuros A woody-based, densely leafy annual with robust stems bearing dense, *conspicuous, white hairs to 2 mm*. Lower leaves in rosettes, 40–60 mm with undulate margins; stem-leaves 6–80 mm with larger terminal lobes; leaves sparsely hairy. Flowers whitish, borne in dense racemes to 45 cm; flower-stalks ribbed and hairy; petals 5–6 mm; upper petal with *linear appendages with densely ciliate margins*; stamens 18–20, almost exceeding the petals, filaments persistent. Capsule 10–15 mm. Maquis and fields. IP (north), SY.

(c) Flower spikes *yellowish*; leaves divided or not; capsule with 3 terminal teeth.

Reseda lutea WILD MIGNONETTE An ascending, leafy perennial to 75 cm, bushy and sometimes woody at the base. Leaves stalked and pinnately lobed with 1–4 pairs of leaflets. Flowers pale *greenish-yellow*, borne on short flower-stalks (<6 mm), each with 6 petals and 5–6 sepals, borne in many long narrow spikes; petals entire to 3-lobed; stamens 15–20. Fruit 7–20 mm, oblong and erect, 3-parted. Seeds smooth. Very common in disturbed areas and sandy waste places. Throughout.

1. *Reseda alba*
2. *Reseda phyteuma*
3. *Reseda alopecuros*
4. *Reseda lutea*
5. *Reseda minoica*; (inset) fruits
6. *Reseda decursiva*

RESEDACEAE

Reseda minoica An annual to short-lived perennial 10–70 cm with stems branched from the base. Leaves entire or 3(5)-lobed, bristly on the veins and margins. Flowers borne in lax racemes with erect to spreading stalks, 3–5 mm, elongating to 12 mm in fruit; sepals and petals 5, the upper petals divided into 3 lobes, further divided into many *linear-spatula-shaped segments*. Fruit 7–12 mm with short teeth. Seeds ridged. AA (Anafi, Salamis), CR (central, south + Gavdos), CY, TR.

Reseda luteola WELD A tall, erect biennial to 1.3 m. All *leaves unlobed*, lance-shaped and with wavy margins. Flowers greenish-yellow with 4 sepals and petals, borne in long, slender spikes. Fruit 3–6 mm, rounded with 3 pointed lobes. Local in disturbed areas and sandy waste places. Throughout.

Ochradenus

Dioecious, hairless desert shrubs with spine-like branches. Leaves stalkless, solitary or in clusters. Flowers borne in spikes or racemes; sepals 5–6; petals sometimes absent; stamens 10–22, inserted on a disk. Fruit a capsule or berry.

Ochradenus baccatus A straggling desert shrub with slender, yellowish-green branches. Leaves 10–40 mm, linear, pointed. Flowers *bright greenish yellow*, borne in dense, rigid racemes 50 mm–15 cm; bracts 2.5 mm; flowers cosexual or female: cosexual with 5(6) oval sepals, 1 mm; petals 0 or 2–3, <1 mm; stamens 10–12; female flowers with petals absent or reduced. Berry *white*, 3–4 mm across. Seeds 1.5 mm. IP (south – especially the Southern Negev and Arava Valley).

1. Ochradenus baccatus
2. Capparis spinosa
3. Thlaspi perfoliatum
4. Lunaria annua

CAPPARACEAE | CAPER FAMILY

A small family of herbs and shrubs. Flowers with 4 petals, 4 sepals and 6–numerous stamens. Fruit a berry with numerous seeds.

Capparis CAPER

Shrubs with spiny stipules and simple leaves. Flowers with petals much shorter or longer than the sepals. Fruit a succulent berry (caper).

Capparis spinosa CAPER A spreading or pendent shrub, often sprouting directly out of old walls. Leaves alternate, fleshy, grey–green and oval to circular, blunt or slightly notched at the tip with curved spines at the base of the stalks, *very sparsely hairy*; stipules *weak or virtually absent*. Flowers conspicuous, with white petals 20–35 mm, and numerous fine, violet stamens. Fruit a large fleshy berry 20–30 mm. Throughout (locally replaced by the next species). *C. zoharyi* is very similar but *erect*, with *distinct stipules* running along the stems. Throughout except the west (distribution unclear owing to confusion with the previous species).

BRASSICACEAE | CABBAGE FAMILY

A large family of annual and perennial herbs (or shrubs). Leaves alternate and simple or pinnately divided. Flowers often in racemes, characteristically with 4 sepals and petals forming a cross; stamens usually 6 (2 shorter). Fruit dry, sometimes non-splitting; shape of the ripe fruit important for identification, classified as a *silicula* (broad and variously shaped), or a *siliqua* (long and thin).

Thlaspi

Annual or perennial herbs with stalkless, more or less clasping stem-leaves. Flowers borne in racemes without bracts; sepals erect; petals usually white or purplish. Fruit a silicula, with or without apical notch, the valves keeled, often winged; stigma somewhat 2-lobed. Seeds 1–8 in each chamber.

Thlaspi perfoliatum (syn. *Microthlaspi perfoliatum*) An erect, hairless *annual* 40 mm–30(40) cm, without a crowded rosette of basal leaves; basal leaves entire to toothed, stalked and oval-shaped; stem-leaves stalkless, with heart-shaped, clasping bases. Flowers white; petals 1.5–2.5 mm; sepals 1–1.5(2) mm. Siliqua 4–6(7) mm. Seeds 3–4(5) per chamber, smooth. Throughout.

Lunaria

Biennial to perennial herbs, virtually hairless with oval- to heart-shaped, toothed leaves, short-stalked above. Flowers white or violet. Fruit a *large, persistent silicula*. Seeds few.

Lunaria annua A biennial (rarely annual) to 1 m. Leaves irregularly and coarsely toothed, the uppermost short-stalked to virtually stalkless. Flowers unscented, pink–purple (rarely white); petals 14–25 mm; sepals violet, 5.5–10 mm. Fruit *large, disk-like and papery* when mature, 30–65(70) mm across, the septum persistent. Frequently cultivated for ornament, common in woods. Throughout, except much of the east and southeast.

Erophila

Annual herbs with leaves all basal. Flowers inconspicuous, in racemes without bracts; stamens 6. Fruit a siliqua. Seeds 20–50.

Erophila verna (syn. *Draba verna*) A small annual 20 mm–20 cm, hairy below, often flushed purple. Leaves in basal rosettes, linear-lance-shaped to egg-shaped, entire or slightly toothed. Flowers small, white or pinkish, borne 5–10 in corymb-shaped racemes; petals 1.5–6 mm; sepals 1–2.5 mm. Fruit a siliqua, 5–12 mm. Bare ground after winter rain. Probably throughout, except for hot, dry areas.

Aubretia

Perennials, often cushion or mat-forming, with simple leaves. Flowers with pink to violet, long-clawed petals; filaments of the outer stamens with a toothed appendage. Fruit a siliqua (rarely a silicula) with seeds in 2 rows in each chamber; close observation of hairs on the fruits necessary for identification.

Aubretia deltoidea A variable, tufted to straggling perennial. Leaves variably linear, spatula- or diamond-shaped, entire or with 1–3 pairs of teeth. Flowers reddish-purple or violet (rarely white); petals 12–28 mm; sepals 6–10 mm. Fruit a siliqua 6–16(22) mm, 2–5(7) x as long as wide, with both *long, unbranched, forked hairs and much shorter adpressed, star-shaped hairs*; style 4–8 mm. Widespread, more or less throughout except the southeast (with numerous local forms described). The following rare or local species are all similar, and close attention to hairs on the fruits is necessary: *A. thessala* has fruits ›12 mm long with mostly small, star-like hairs and occasional forked hairs; *never simple* hairs. Mount Olympus. GR. *A. erubescens* has smaller fruits (6.5)7–12(14) mm. Mount Athos. GR. *A. glabrescens* has *large* fruits 20–35 mm with *only* star-shaped hairs (or hairless). Mount Smolikas. GR. *A. gracilis* has linear to lance-shaped leaves and fruits 20–25 mm, 6 x as long as wide. GR, PN.

Clypeola

Small, ephemeral annuals with entire, simple leaves. Flowers pale yellow, borne in bractless racemes; sepals erect-spreading; stamens 6. Fruit a rounded silicula.

Clypeola jonthlaspi An erect annual 30 mm–28 cm with white hairs. Leaves 3–18 mm, lance- to spatula-shaped. Flowers pale yellow, borne on recurved stalks 2–4 mm; sepals 1–1.5 mm, much-exceeding the petals. Fruit a *circular to elliptic silicula* 5 mm across, with short rigid hairs, or hairless. Throughout.

Alyssum

Annual or perennial early-flowering herbs with branched or star-like hairs (rarely mixed with unbranched hairs). Flowers yellow; sepals erect-spreading; filaments of 2 kinds: long and winged, or short with an appendage. Fruit a silicula. Seeds 1–2(6) in each chamber, often winged. A taxonomically difficult group to identify in the field. Many local endemics exist, which it is beyond the scope of this book to describe, so just a handful are included here.

(a) Low subshrubs or woody-based perennials.

Alyssum akamasicum A weak, straggling, silvery-downy perennial with erect to almost prostrate stems, 80 mm–15 cm. Leaves elliptic to egg- or spatula-shaped, 10 mm long. Flowers initially crowded, bright yellow; petals 3.3 mm; sepals erect, boat-shaped, 2–3 mm. Fruit an almost circular, *hairy* silicula to 6 mm long. A rare island endemic of garrigue at sea level. CY (Akamas Peninsula).

Alyssum troodi A much-branched, silvery-downy subshrub 10–25 cm with a woody rootstock with numerous vegetative shoots. Leaves broadly egg- to spatula-shaped, 5–10 mm long, gradually narrowing into the stalks; those of flowering shoots more distant. Flowers golden yellow; petals 3 mm; sepals 2 mm, erect. Fruit an *elliptic* silicula, narrowed at both ends, *hairless*, 7–8 mm. A rare island endemic on rocky mountainsides. CY. ***A. cypricum*** is similar but with narrow leaves and *sparsely hairy* fruits just 4 mm long. Rocky mountainsides. CY, TR.

(b) Annual herbs with circular fruits.

Alyssum simplex A small annual with few prostrate to acending stems 50 mm–20 cm. Leaves elliptic or broadest above the middle, often withered below, covered in adpressed, 6–8-rayed white hairs. Flowers small, borne in crowded heads, eventually elongating in fruit; sepals *soon-falling*; petals 2.5 mm, often slightly notched; fruits circular, with dense 8-rayed hairs. Seeds narrowly winged. Common in bare habitats and dunes after winter rain. Throughout. ***A. strigosum*** is virtually indistinguishable but with deeply notched petals and fruits with both multi-rayed and forked (2-rayed) hairs. Throughout, though rare in the west (possibly absent from IA).

1. *Erophila verna*
2. *Aubretia deltoidea*
3. *Aubretia thessala*
4. *Clypeola jonthlaspi*
5. *Alyssum akamasicum*
6. *Alyssum simplex*
7. *Alyssum strigosum*
8. *Alyssum damascenum*

1. *Aethionema saxatile* subsp. *graecum* habit; (inset) flowers
2. *Aethionema saxatile* subsp. *creticum*
3. *Arabis turrita*
4. *Arabis verna* habit; (inset) flowers
5. *Arabis collina* habit; (inset) flowers
6. *Arabis purpurea*; (inset) fruits
7. *Arabis cypria*

Alyssum damascenum A small, grey, adpressed-hairy, *desert* annual, branched from the base, to 12 cm. Leaves all egg-shaped to oblong, to 20 mm. Flowers yellow, *minute*, borne in dense racemes; petals divided. Fruits 3–5 mm across, rough, circular, the style as long as or exceeded by the pod; seeds wingless. Rocky deserts. IP (Moav and Samarian deserts; rare elsewhere), SY.

Aethionema

Hairless annual or perennial herbs, with entire, stalkless leaves. Flowers borne in raceme; sepals erect; petals entire, of various colours; the 4 inner stamens with winged and bent filaments, the wing sometimes ending in a tooth above. Fruit a flattened, winged silicula flattened with 1–4 seeds in each chamber.

Aethionema saxatile A small, ascending or erect woody-based perennial (rarely annual) 30 mm–13(35) cm. Lower leaves >5 mm, usually bluntly oval; the upper narrower and more pointed. Flowers white or purplish; sepals 1.5–3 mm; petals 2.5–8 mm. Fruit a silicula 5–9 mm, rounded, sometimes broader than long, with up to 8 seeds. Mainly in mountains; absent from the southeast. AA, CR, GR, IA, PN, TR. Subsp. *graecum* has flowers with petals 4–6.5(8) mm and sepals 2–3 mm. GR, PN. Subsp. *creticum* is generally very *short*, to just 10 cm, with *smaller* flowers with petals just 2.5–4(5) mm and sepals 1.5–2.3 mm. AA, CR, PN.

Arabis

Annual or perennial herbs with unbranched, branched or star-shaped hairs. Leaves simple. Flowers usually white, pink or purple. Fruit a siliqua with seeds usually in 1 row in each chamber. Seeds often winged.

(a) Flowers small, white, whitish-violet or yellow, with petals <8 mm long.

Arabis turrita A hairy biennial or perennial 20–80 cm with long-stalked, egg-shaped, toothed basal leaves; stem-leaves stalkless, with a heart-shaped, clasping base. Flowers pale yellow; petals 6–8 mm. Fruiting head elongated, 1-sided; fruit a siliqua 10–14 cm long, with thickened margins, at first erect, curved when mature. Cliffs, gorges and mountains. AA (Thasos), GR, IA, PN.

Arabis verna A small, hairy annual 50 mm–40 cm with several ascending flowering stems arising from a rosette of stalked, oval, entire leaves; stem-leaves few (often 1–2), heart-shaped at the base, toothed. Flowers very few (<10) borne on stalks <2 mm; petals 5–8 mm, white to pale violet. Fruit a siliqua 45–60 mm long on thick stalks. Seeds narrowly winged. Shady cliffs and rocks. Throughout.

(b) Flowers white, pink pr purple, with petals >8 mm long.

Arabis collina A perennial 10–30 cm with long-stalked leaves crowded at the base, stem-leaves broadly oval below, lance-shaped above. Flowers white to purplish-pink, borne 10–18 in rather crowded terminal racemes; petals 8–10 mm. Fruit a siliqua 60–90 mm long, curved, on stalks 8–12 mm. Crevices on rock faces, gorges and cliffs. AA (Thasos), GR, IA, PN.

Arabis purpurea An erect, tufted perennial 60 mm–25 cm with a tough, woody, creeping rootstock. Leaves densely downy with short, soft, greyish hairs, 8–20(40) mm long, egg- to spatula-shaped, borne in loose rosettes. Flowers pink or purplish (rarely white); petals 10–15 mm long. Fruit a siliqua, spreading to deflexed, 25–60 mm long. Shaded woods and cliffs on igneous rocks; an island endemic, locally common in the Troodos. CY. *A. cypria* is similar but with leaves 20–80 mm with long hairs and *large,* pale pink or white flowers with petals 10–18 mm long. CY (north).

Isatis WOAD

Erect annuals or perennials with simple leaves. Flowers small and yellow. Fruit winged, 1-seeded, pendent and non-splitting.

Isatis lusitanica An annual 10–70 cm with sparse, slender stems. Basal leaves borne in rosettes, pinnately lobed to almost entire, egg-shaped, with conspicuous central midribs, soon wilting; stem-leaves thin, oblong, with semi-clasping bases. Flowers yellow, 3–5 mm. Fruits pendent, broadly linear-wedge-shaped, 12–22 mm long, flattened, tapered at the base, purple when ripe. Maquis, fields, desert fringe habitats. AA (KP + Rhodes), IP (common in the north), LB, SY, TR.

Isatis tinctoria WOAD A large, erect, much-branched biennial to 1.5 m with grey–green leaves; leaves arrow-shaped and clasping the stem above. Flowers yellow, borne in dense and much-branched racemes. Fruit a *pendent*, oblong-elliptic silicula 10–25 mm. Used by the Romans as a medicinal plant, cultivated in the Middle Ages as a source of the blue dye *indigotine*. Waste places and on disturbed ground on the mainland. GR, PN, TR. *I. tomentella* is similar but with spreading hairs and narrower, *short-hairy* fruits. AA (Chios – rare), GR, IA (Corfu), PN, TR.

Capsella SHEPHERD'S PURSE

Annuals or perennials with simple basal leaves. Petals normally white. Fruits triangular- to heart-shaped.

Capsella bursa-pastoris SHEPHERD'S PURSE A distinctive, sparsely hairy annual to 40 cm with variable leaf shape, and scentless white flowers borne in a long raceme; petals 2–3 mm. Fruit 5–9 mm and *heart-shaped* with straight to slightly convex sides. A very common weed in dry waste places throughout.

Ionopsidium (syn. *Jonopsidium*) DIAMOND FLOWER

Hairless, slender annuals with 4-parted, unequal flowers.

Ionopsidium albiflorum An ephemeral annual with stems simple or branched from the base. Basal leaves entire or tooth-lobed, with stalks equalling the blade. Flowers white, the petals all equal, 1.5 x the length of the calyx, borne in inflorescences with leafy bracts. Fruit an oblong silicula. Seeds 5–6 in each chamber. AA (Skiros).

Sisymbrium ROCKET

Annuals or perennials with simple, entire to deeply lobed leaves, the lowermost often withering in bloom. Petals yellow. Fruit a beakless siliqua, much longer than broad; seeds in 1 row under each valve.

(a) Fruits adpressed to the stems.

Sisymbrium officinale HEDGE MUSTARD A bushy, more or less hairless annual or biennial to 1 m with grey–green dense, oval, deeply lobed leaves with a large terminal lobe. Flowers yellow, borne on slender, branched inflorescences *without bracts*, small; petals 3–4.2 mm. Fruits 10–20 mm, cylindrical, straight, *closely and densely adpressed along slender stems*. Common on roadsides and near buildings. Throughout.

(b) Fruits erect to spreading.

Sisymbrium erysimoides FRENCH ROCKET An annual with erect stems to 60 cm and variously entire to (often deeply) divided leaves. Flowers yellow, rather crowded, with *small* petals ≤3.5 mm, not exceeding the sepals or stamens. Fruits long, spreading to erect-spreading, 20–50 mm with *robust stalks about 1 mm wide* (>1/3 the width of the fruit). Dry, semi-arid habitats. IP (common in the east).

Sisymbrium irio LONDON ROCKET An erect, much-branched annual to 60 cm with variously (some deeply) pinnately lobed leaves, the end lobe pointed. Flowers pale yellow, petals 2.5–3.5 mm, equal to or longer than the sepals. *Fruits 30–60 mm long, overtopping the open flowers*, erect to spreading, with *slender stalks* (0.3–0.6 mm wide, <1/3 the width of the fruit). Throughout.

Sisymbrium altissimum TALL ROCKET An erect annual to 1 m, hairy below, hairless above. Leaves deeply cut into narrow, toothed, triangular lobes, *the stem-leaves unstalked and with long, narrow, linear lobes*. Flowers yellow, petals 7–8 mm, 2 x the length of the sepals. Fruit 50–90 mm long. Native to southeast Europe, widely naturalised. Throughout. *S. orientale* EASTERN ROCKET is similar, grey-hairy, and with short-stalked uppermost leaves with few (0–2) lateral lobes, the middle lobe linear-lance-shaped. Flowers with petals >6 mm (much longer than sepals and stamens). Fruits 40 mm–10(12) cm, slender, and erect to spreading. Throughout.

Bunias

Biennials to perennials with simple basal leaves. Petals typically yellow. Fruit an irregularly ovoid silicula with warty or wing-like crests, not splitting.

Bunias erucago SOUTHERN WARTY CABBAGE A bristly hairy annual or biennial 15–70 cm (1 m). Lower leaves pinnately lobed or wavy-margined, upper leaves unlobed, toothed or not. Flowers borne in racemes of 10–30, yellow, with notched petals 9–10(12) mm long. Fruit 8–12 mm, square in section, with *toothed wings along the angles*. Waste places and cultivated land. Throughout, except much of the southeast (absent from CY and IP).

1. *Isatis lusitanica*
2. *Capsella bursa-pastoris*
3. *Sisymbrium erysimoides*

Chorispora

Variably hairy annual or perennial herbs. Leaves lobed to entire, usually stalkless above. Flowers often showy; sepals erect, much-exceeded by the petals; racemes elongating in fruit; stamens 6. Fruit a linear, segmented siliqua; seeds not winged.

Chorispora purpurascens A short-hairy annual to 30 cm. Leaves oblong-elliptic, toothed to lobed with oblong to triangular segments; upper leaves less divided. Flowers violet–pink, with dark veins, borne in bractless racemes; sepals 10 mm, reddish; petals 15–18 mm. Fruit 40–60 mm, beaked. Fields, garrigue and deserts. IP (mainly near Gilead, Jordan – rare), SY, TR (rare).

Erysimum WALLFLOWER

Annuals, herbaceous perennials or shrubs with erect, often slightly winged stems and leaves entire, covered in branched hairs. Petals yellow or red. Fruit a flattened to 4-angled siliqua. A large number of similar species occur in the region.

(a) Single-stemmed annuals without vegetative shoots or previous years' leaf scars. Various habitats.

Erysimum repandum An annual with a *simple, erect stem* 20–40 cm. Leaves linear to narrowly elliptic, toothed. Flowers lax, borne in racemes elongating in fruit; petals 6–8 mm, pale to bright yellow; flower-stalks short and robust. Fruit square in cross section. AA (Lesbos), GR (north and northeast).

(b) Biennials to perennials, often with vegetative shoots or previous years' leaf scars. Various habitats.

Erysimum × cheiri WALLFLOWER A perennial to 50(80) cm with woody, erect stems covered in branched hairs. Leaves narrowly elliptic and untoothed. Flowers usually dark *orange–yellow*, borne in racemes; petals 20–35 mm; sepals erect and ½ the length of the petals. Fruit flattened, 25–80 mm long and almost erect. Widely grown as an ornamental and naturalised in disturbed habitats and near coasts.

Erysimum mutabile A more or less *mat-forming* perennial with basal leaves in a rosette and very *slender* spreading stems to 25 cm; leaves spatula-shaped, entire; stem-leaves progressively smaller. Flowers 3–6 in lax inflorescences; petals bright yellow, 7–10 mm, sometimes purple-tinged; sepals reddish. Fruit sub-cylindrical, 10–24 mm. Mountains in Crete only. CR.

Erysimum creticum A biennial with few (1–3) stems, 15–50 cm. Leaves withered during flowering, oblong-lance-shaped with distant, simple teeth. Flowers bright yellow; petals 9–13 mm. Fruits square in cross section, with 3–4-rayed hairs. CR (east). *E. krendlii* is similar but with 6-rayed hairs on the fruits. An island endemic. AA (Samothraki). *E. horizontale* is similar but with *double-toothed* leaves. AA (Lesbos, KP). *E. smyrnaeum* has double-toothed leaves and stems with conspicuous *ridges*. AA (Kos), CR (west – rare), LB, TR.

(c) Perennials with a woody stock, growing *on cliffs* (chasmophytes).

Erysimum candicum A woody-based, chasmophytic shrub 25–50 cm. Leaves clustered at branch tips, 20 mm–12 cm long, virtually stalkless, elliptic-lance-shaped, (usually) entire and hairy. Flowers 5–12, borne in racemes or in lateral clusters; petals 19–22 mm, pale to bright yellow; sepals erect. Fruit erect-spreading, 40–65 mm, cylindrical to flattened. Seeds winged, dark brown. Cliffs and gorges. Two similar regional subspecies exist: **subsp. *candicum*** AA (Anafi), CR.

1. *Chorispora purpurascens*
2. *Erysimum* × *cheiri*
3. *Erysimum candicum* subsp. *candicum*
4. *Moricandia arvensis*
5. *Malcolmia maritima*
6. *Malcolmia flexuosa* subsp. *naxensis*
7. *Malcolmia africana*

subsp. *carpathum*. KP. The following regional taxa are rather similar: ***E. corinthium*** has fruits square in cross section. GR (southeast), PN. ***E. naxense*** has a *prominent leafy rosette* and terminal, many-flowered inflorescences; plants typically dying after flowering. AA (Naxos – common). ***E. rhodium*** is tall with *toothed* leaves, and fruits borne in rather 1-sided infructescences. AA (Rhodes, Simi), TR (Marmaris Peninsula). ***E. senoneri*** has distinctly *stalked* leaves. AA.

Moricandia

Hairless plants with simple, fleshy leaves. Petals *violet*. Fruit linear, 4-angled.

Moricandia arvensis VIOLET CABBAGE A hairless, blue–grey annual or perennial to 65 cm with slightly fleshy leaves, shallow-lobed below, clasping the stem with heart-shaped bases above. Flowers violet–purple borne in racemes of 10–20; petals 21–29 mm. Fruit linear, 30–60 mm and 4-angled. A mainly western and North African species, rare in the region. IA (Cephalonia).

Malcolmia

Annuals with branched hairs. Petals normally *pink or violet*. Styles absent, stigmas 2-lobed. Fruits linear with seeds in 1 row under each valve, not winged.

(a) Flowers borne in racemes *without* bracts.

Malcolmia maritima An annual 10–35 cm, with 3–4-forked hairs. Leaves dark green, egg-shaped to oblong, wedge-shaped at the base, entire or toothed. Flowers pink to violet, borne in inflorescences without bracts; fruiting stalks 0.5–1 mm across when mature (*thinner* than the fruit itself); sepals 6–10 mm; petals 12–25 mm. Fruit a siliqua 35–80 mm. Western only. IA, PN. ***M. flexuosa*** is similar but with light green leaves, fruiting stalks 1–2.5 mm in diameter at the base, *about as wide as the fruit*; siliqua up to 3 mm across. AA, CR, CY, GR, PN, TR. Two widely distributed races are recognised that are scarcely distinct: **subsp. *flexuosa*** is erect, sparingly branched, to 15(20) cm with few flowers 5–12(15) mm across. AA (southeast), CR (rare), CY, TR (southwest). **Subsp. *naxensis*** is more variable; typically spreading, much branched, to 40(60) cm with larger flowers 9–10(25) mm across. More or less throughout the species range.

Malcolmia graeca An annual 50 mm–20 cm, with 3–4-forked hairs. Leaves oval-oblong, wedge-shaped at the base, entire to lyre-shaped, lobed. Flowers purple to violet, borne in inflorescences without bracts; fruiting stalks 4–10 mm across when mature; petals 8–17 mm. Fruit a siliqua 25–75 mm, erect to spreading. GR, PN (common).

Malcolmia chia An annual to 20 cm, with 3–4-forked hairs. Leaves oval-oblong, the basal wedge-shaped, entire or toothed. Flowers *small, pale pink* to violet, borne in inflorescences without bracts; fruiting stalks 4–10 mm when mature; sepals with 2-forked hairs; petals just 6–10 mm. Fruit a siliqua (25)30–70 mm, erect to *spreading*. Coastal sands. AA, CR, CY, GR (rare), PN (rare), TR.

Malcolmia bicolor An annual 70 mm–35 cm, with 3–4-forked hairs. Leaves oval to oblong-lance-shaped, usually more or less entire. Flowers pink, the petals with a yellow claw; fruiting stalks 3–5(8) mm when mature (about *as wide as the fruit*); sepals with mainly hairs attached at the middle; petals 4.5–11 mm. Fruit a *short* siliqua, 10–30(40) mm, erect to spreading but *straight*, and rigid. GR.

(b) Flowers borne in racemes *with* bracts.

Malcolmia africana A variable, typically *desert* annual 20–30(40) cm, branched, erect, rough-hairy, with leafy stems. Leaves stalked below, oblong to almost elliptic, stalkless above, entire to wavy-toothed. Flowers borne in lax racemes of 10–20, 5–7 mm across, pale lilac or pink (sometimes whitish); fruiting stalks (1)2–4 mm when mature and as thick as the fruit; sepals (2.5)4–5 mm, often persistent in fruit; petals 4–8 mm. Fruit a siliqua 40–70 mm long, 1.5–1.8 mm wide, rather straight and spreading, variably hairy with almost adpressed, short or mixed *short and long bristles*. Seeds numerous, oblong, 1 mm. Rocky deserts and maritime sands; rather uncommon and strongly southeastern. AA (local), CR, GR (southeast), IP (southern and eastern deserts), PN (Kithira), SY, TR.

Ricotia

Grey annuals and perennials. Leaves with 3 leaflets or pinnately divided. Sepals pouch-like. Fruit a compressed siliqua.

Ricotia lunaria An annual 15–40 cm with spreading stems. Leaves to 10 mm, 2–3-divided into stalked divisions with oval, lobed segments. Flowers large and few; petals heart-shaped (deeply notched), 10–18 mm with a long, whitish claw and pink limb. Fruit 15–30 mm, *flattened and elliptic*, with 1–7 round seeds in single rows. Rocky habitats in the east. IP (north), SY, LB.

Ricotia cretica An annual 10–25 cm. Leaves 2–3 x dissected into narrowly oval to elliptic lobes. Flowers pale pink; petals 10–12 mm, notched and rather long-clawed. Fruit a siliqua 30–50 x 8–9 mm (5–8 x as long as wide), not beaked or winged. Seeds up to 10. Screes and rocky ledges in gorges; an island endemic. CR. **R. isatoides** is similar but with *smaller* flowers with petals to just 5 mm, with violet veins, and *shorter* fruits about 2 x as long as wide with a marginal wing. Seeds normally 1. An island endemic. KP.

Matthiola STOCK

Grey-leaved plants with simple basal leaves. Flowers with deeply 2-lobed stigmas; petals white or purple. Fruit linear with seeds in 1 row under each valve, broadly winged.

(a) Fruit with 3 virtually equal triangular horns at the apex all at least 2 mm long.

Matthiola tricuspidata An annual to 30 cm, with dense, short hairs, especially on the stem. Leaves spatula-shaped to oblong-elliptic, blunt and markedly wavy-lobed; lower leaves to 70 mm, those above to 35 mm. Flowers purple; sepals 8–11 mm; petals 6.5–11 mm. Fruit 30–55 mm, spreading to deflexed, *the apex with 3 virtually equal horns to at least 3 mm*. Coastal sands and sandy waste places. Probably throughout.

1. Ricotia lunaria
2. Ricotia cretica
3. Matthiola tricuspidata

(b) Fruit *narrowly cylindrical* with apical horns normally >1.5 mm.

Matthiola fruticulosa   A fleshy, simple or branched perennial to 40 cm with dense hairs and a *long, woody stock*. Leaves to 50(60) mm, narrowly oblong-lance-shaped to linear, with markedly wavy-toothed, often inrolled margins. Flowers variably violet, brownish, whitish or yellowish; sepals 8–12 mm; petals 6–13 mm. Fruit 30–80 mm (12 cm), circular in cross section, with blunt, *short* horns (usually <2 mm). Maquis and garrigue, sometimes in gorges and mountains. GR, PN (possibly elsewhere). **Subsp.** *fruticulosa* has a branched habit and stem-leaves. GR (southeast), PN. **Subsp.** *valesiaca* has a simple habit and most or all leaves in a basal rosette. GR (north). *M. longipetala* is similar but an annual, rarely with leafy rosettes at the base. Flowers with *long*, linear to oblong, pinkish purplish or yellowish petals, 15–25 mm. Fruit with spreading, recurved or upward pointing, rather *longer* horns >2 mm. Mainly in the east. AA (Rhodes), CY, GR (southeast), IP, PN (northeast), TR.

1. *Matthiola fruticulosa* subsp. *valesiaca*
2. *Matthiola longipetala*
3. *Matthiola incana*
4. *Matthiola sinuata*
5. *Lobularia maritima*
6. *Lobularia libyca*
7. *Fibigia clypeata*
8. *Iberis sempervirens*

Matthiola parviflora An annual to 20 cm with deeply pinnately lobed lower leaves. Flowers small, with *brownish-purple petals 6–10 mm (the limb just 2.5–4 mm)*; sepals 4–6 mm. Fruit narrowly *cylindrical*, 50 mm–11 cm long, *slightly constricted at intervals*; horns 1–1.5 mm, straight and pointed. Mainly deserts. IP (Dead Sea Valley and adjacent areas).

(c) Fruit *compressed*, usually *without conspicuous horns* (or with only 2 horns >2 mm).

Matthiola incana HOARY STOCK A stout, bushy perennial to 80 cm with a woody stock and numerous non-flowering shoots, stems and leaves grey–green and hairy. Leaves lance-shaped, usually untoothed and unlobed. Flowers pink, purple or white, the petals 20–30 mm long. Fruit long and thin, 45 mm–15 cm, laterally compressed; hairy but not glandular. Local on sea cliffs and other coastal habitats. Throughout. *M. sinuata* is similar to *M. incana* but with lower leaves deeply pinnately lobed and wavy-margined, the upper leaves entire; plant with conspicuous large glands on the upper parts. Flowers pale purple; petals 15–25 mm. *Fruit glandular and sticky* (with large, yellowish or blackish glands when mature), 50 mm–15 cm. Throughout except the north and southeast (not CY or IP).

Lobularia SWEET ALISON

Annuals or perennials with narrow, untoothed leaves. Petals white or purplish. *Fruit a disk-like siliqua.*

Lobularia maritima SWEET ALISON A greyish, downy, spreading woody-based *perennial* with stems to 30 cm, often prostrate. Leaves narrow, untoothed, linear-lance-shaped and pointed. Flowers in dense rounded racemes that lengthen in fruit, white; sweetly scented; petals 2.5–4.5 mm long. Silicula flattened, oval to circular, often hairy, 2–3.5 mm, usually with 2 seeds, which are winged over at least some of their perimeter. Rare and local in the area. AA (local), GR, PN. *L. libyca* is a similar *annual* with smaller flowers (petals 1.5–2 mm long) and seeds winged along the entire perimeter. Rare and local, in dry habitats. AA (Santorini), IP.

Fibigia

Perennials or subshrubs with entire leaves. Petals yellow. Fruit strongly compressed; seeds in 2 rows, flattened and *winged*.

Fibigia clypeata An erect, densely grey-felted, tufted perennial 15–75 cm with entire, elliptic to spatula-shaped leaves to 10 cm long. Flowers small and yellow, with petals 7–13 mm, borne in often branched racemes. Fruits distinctive: conspicuously flattened, bat-like and grey-felted, 21–28 mm long, borne on elongated stems 10–20 cm long. Rocky habitats (occasionally on walls). GR, PN, IP (fairly common).

Iberis CANDYTUFT

Herbs with simple, entire to lobed leaves. Flowers borne in flat-topped clusters, the outermost pair of petals of each flower much longer than the inner. Fruit compressed with 1 seed under each valve. Many closely related species, which has led to confused distributions.

Iberis sempervirens EVERGREEN CANDYTUFT A branched, hairless, evergreen, woody-based shrublet to 25 cm with ascending flower shoots and flat, narrowly spatula-shaped, thick leaves. Flowers *pure* white, borne in flat-topped *racemes*. Fruit rounded and broadly winged, especially at the tips, 4–7 mm. Mountains on the mainland (or an escape from cultivation). CR, GR, PN.

Iberis carnosa An annual or perennial to 30 cm, sometimes purplish. Leaves undivided, fleshy, spatula-shaped; stem-leaves linear to oblong. Flowers borne in *dense corymbs*, white to purplish-pink with *much larger* outer petals. Fruit 5–7 mm, compressed with 2 triangular, blunt wings, the style forming a notch. Cliffs and screes; rather local. AA (local), GR, PN, TR.

Biscutella BUCKLER MUSTARD

Perennials with simple leaves. Petals yellow. Fruit distinctive, with the appearance of 2 1-seeded disks fused edge to edge. A rather difficult group with many species besides the one described here.

Biscutella didyma An erect, simple or branched, bristly-hairy annual 40 mm–25 cm. Basal leaves stalkless in loose rosettes, elliptic to lance-shaped, 10–60 mm long, often with slightly toothed margins; stem-leaves few and reduced, semi-clasping. Flowers borne in sparsely branched inflorescences; petals pale yellow, 1–2 mm long; sepals elliptic, 1.5–2 mm. Fruit a hairless silicula with valves 4–5 mm across. Stony pastures and rocky slopes. Throughout. Several forms described, for example **subsp. *dunensis*,** an island endemic subspecies. CY.

Lepidium PEPPERWORT

Leaves simple to 2–3-pinnately divided. Flowers white or reddish. Fruits flattened and strongly keeled or winged.

(a) Inflorescence a dense, terminal raceme (several similar species occur).

Lepidium spinosum An annual or biennial with simple, hairless stems. Basal and lower stem-leaves pinnately divided, the linear lobes, lobed again at the base; middle and upper stem-leaves linear or linear-oblong. Flowers with sepals 1–1.5 mm, white-margined; petals 1.5 x as long as the sepals; flower-stalks 1–2 mm in fruit. Fruit a silicula 5–6 x 3 mm, egg-shaped, deeply notched at the apex, hairless, the wings very long above. Frequent in the Aegean; rare elsewhere. AA, CR, GR (east), IA (Cephalonia), PN (south and east), SY, TR.

(b) Inflorescence a corymb-shaped panicle.

Lepidium draba (syn. *Cardaria draba*) HOARY CRESS An erect, hairless or slightly downy, greyish perennial to 90 cm. Leaves oblong, pointed, toothed and long-stalked at the base, clasping the stem above. Flowers borne in dense, umbel-like clusters, white, petals 3–4 mm. Fruit 2.5–4 mm, heart-shaped with a protruding style (0.5–1.2 mm), inflated, not splitting. Common in disturbed, grassy habitats and roadsides. Throughout.

(c) Inflorescences mostly opposite the leaves (species previously described under *Coronopus*).

Lepidium didymum (syn. *Coronopus didymus*) LESSER SWINECRESS A small, spreading or ascending biennial to 40 cm, strong-smelling when crushed. Leaves divided, feathery, at first in a rosette, later on spreading stems. Flowers inconspicuous, petals 0.5 mm, shorter than the sepals, or absent. Racemes elongated in fruit, flower-stalks longer than the fruit. *Fruit dumb-bell shaped*, 1.2–1.7 mm, veined and with a notch at the apex, style absent. Sandy waste places. AA (rare), CR (west), GR (southeast), IA, PN (rare). **L. coronopus** (syn. *Coronopus squamatus*) is similar, but with flowers with white petals to 2.5 mm across, and *ridged, kidney* (not dumb-bell)-*shaped* fruits, 2–3 mm. Probably scattered throughout.

1. *Biscutella didyma*
2. *Biscutella didyma* subsp. *dunensis*
3. *Lepidium spinosum*
4. *Lepidium didymum*
5. *Diplotaxis acris*
6. *Zilla spinosa*

Diplotaxis WALL ROCKET

Annuals or perennials with (normally) deeply lobed leaves. Petals yellow. Fruits linear and flattened with seeds in 2 rows.

Diplotaxis acris An erect annual 50 mm–50 cm, almost hairless, weakly branched from the base. Leaves mostly basal, rather fleshy, egg-shaped to oblong, toothed with stalks 1/3–1/2 as long as the blade. Flowers purplish-pink, borne in rather flat-topped terminal clusters; petals 12–20 mm; flower-stalks ascending, about equalling the flowers. Fruit a short-stalked (7–15 mm), erect to ascending siliqua 20–50 mm. Sandy deserts. IP (south).

Diplotaxis erucoides An annual 20–80 cm with *white flowers with violet veins*; petals 8–10(13) mm. Fruit 25–35(48) mm, erect to spreading. Local but widespread (not CY). **D. muralis** WALL ROCKET is similar but a sparsely hairy annual to 60 cm with yellow flowers; petals 6–8(10) mm. Fruit 15–40 mm. Local but widespread. Throughout (except IP and much of the east).

Zilla

Spiny desert shrubs with leaves in rosettes, later leafless. Flowers pink. Fruit a silicula.

Zilla spinosa A *robustly thorny*, hairless, grey–green shrub to 1 m with divergent branches, virtually *leafless* when mature, initially with leaves in basal rosettes. Lower leaves fleshy, oblong, weakly lobed, tapering into their stalks, soon withering; upper leaves alternate, oblong to linear, irregularly ribbed or pitted, to 30 mm. Flowers borne few, in terminal racemes, pale violet–pink with yellowish centres; petals 15–17 mm, entire, with long claws. Fruit 5–10 mm across, spherical with dark green markings and a robust beak. Deserts in the southeast. IP (especially Arava Valley and Edom).

Brassica CABBAGE

Annuals or perennials with leaves entire or pinnately lobed. Petals yellow or white. Fruit a beaked siliqua with 1 row of seeds under each valve; valves rounded on the back with a single prominent vein.

(a) Stem-leaves not distinctly clasping the base of the stem: either stalked or narrowed to the base.

Brassica nigra BLACK MUSTARD An annual to 1.2 m with stems branched from the middle or from near the base. Lower leaves round to lyre-shaped, and pinnatisect (lobed to the midrib), with 1–3 pairs of lateral lobes and a much larger terminal lobe; *bristly on both surfaces*; upper leaves linear-oblong, usually entire and without bristles. All leaves *stalked*. Sepals erect-spreading; petals yellow, 7.5–12 mm. Fruit 8–25 mm long and slender – more so, gradually towards the seedless beak, on short stalks *adpressed to the stem*. Local on maritime grassy and sandy places. Probably throughout.

Brassica cretica A shrubby perennial 50 cm–1.5 m. Leaves stalked or virtually stalkless, hairless and rather leathery. Flowers borne in simple or branched racemes without bracts; petals 15–25 mm, white, cream or pale yellow. Fruit an erect to spreading, beaked siliqua 40 mm–10 cm long, borne on slender stalks; seeds many, in a single row. AA, CR, GR, IA, IP (Mount Carmel only), PN, TR. **Subsp.** *aegaea* has short-stalked to stalkless leaves and yellow flowers. AA (mostly common; not Rhodes), GR, IA, PN, TR. **Subsp.** *cretica* has long-stalked lower leaves and white or cream flowers. CR, PN (Korinthos). **Subsp.** *laconica* has short-stalked lower leaves and yellow or cream flowers. PN.

1. *Brassica cretica*
2. *Brassica oleracea*
3. *Brassica hilarionis*
4. *Sinapis alba*
5. *Cakile maritima*

(b) Stem-leaves distinctly *clasping* the base of the stem.

Brassica oleracea WILD CABBAGE A robust, hairless, greyish biennial to perennial to 1.5(2) m with a thick woody trunk with leaf scars. Basal leaves large and fleshy, to 30 cm long, undulate, lobed and with winged stalks; upper leaves unlobed and *clasping at the base*. Flowers pale yellow, borne in *elongated racemes*; petals 12–30 mm; *sepals erect*. Fruit an ascending beaked siliqua with a single row of seeds, 50 mm–10 cm long. GR.

Brassica hilarionis An erect to spreading, hairless perennial 50–80 cm with stout, often ridged stems. Leaves thick and fleshy, grey-bluish, the lowermost in a basal rosette, 50 mm–20 cm long with bluntly toothed margins; upper leaves typically clasping or with rounded lobes. Flowers *white*; sepals 8–10 mm; petals 1.5–2.5 mm. Fruit a broadly linear siliqua 40–75 mm long, slightly constricted between the seeds; beak 7–9 mm. An island endemic of rocky cliffs and fissues. CY.

Brassica rapa TURNIP An annual or biennial to 1.5 m with a swollen taproot. Leaves bright green, often hairy below, those above with a grey–blue sheen, not fleshy and bristly hairy. Flowers pale yellow, *overtopping the buds*; petals 6.5–12 mm; *sepals erect-spreading*. *Fruits long and linear*, 50 mm–10 cm long, narrowed into a slender beak to 30 mm long. Cultivated and naturalised throughout.

Sinapis MUSTARD

Annuals with crimped to lobed leaves. Flowers with spreading sepals and yellow sepals. Fruit a siliqua, splitting lengthways, with a distinct beak; valves with 3(7) veins.

Sinapis alba WHITE MUSTARD A tall, normally bristly annual (sometimes hairless) to 70 cm. Leaves *all stalked* and pinnately lobed. Flower pale yellow. Fruit 20–40 mm long, the beak flattened and sword-like, as long as, or exceeding the valves (10–30 mm). Common in disturbed, cultivated and waste areas. The source of white mustard. Throughout. **S. arvensis** CHARLOCK is similar but with lance-shaped *unstalked* upper leaves, and a larger fruit 25–45 mm long, with a *conical beak shorter than the 3–7 valves* (7–16 mm). Similar habitats. Throughout.

Hirschfeldia HOARY MUSTARD

Annuals or weak perennials with lobed to divided leaves. Petals yellow; sepals erect. Fruit a siliqua with a short, swollen, club-shaped beak; valves with 1–3 strong veins.

Hirschfeldia incana HOARY MUSTARD A tall, lax, erect annual to 1.2 m. Lower leaves stalked and pinnately divided with an oblong end-lobe and up to 9 pairs of lateral lobes; the uppermost unlobed. Flowers pale yellow, borne in crowded terminal racemes; petals 6–9 mm. Fruit 6–17 mm, closely adpressed to the stem, peg-shaped with a swollen, 1-seeded upper segment, and flattened, 2–6-seeded lower segment. Disturbed, sandy ground; frequent. Throughout.

Cakile SEA ROCKET

Hairless annuals with succulent and blue–grey leaves. Petals white, pink or violet. Fruit a 2-parted siliqua.

Cakile maritima SEA ROCKET A variable short, rather succulent, spreading, hairless annual to 50 cm. Leaves grey–green, irregularly pinnately lobed, the lobes narrow and untoothed, or undivided. Flowers lilac to white, borne in racemes that elongate significantly in fruit; petals 4–10 mm long. Fruit 7–25 mm, brown and succulent; bipartite, *the lower segment with an arrow-shaped base*, the upper oval and 4-angled. Common on maritime sands. Throughout.

Rapistrum

Annuals or perennials with toothed to lobes leaves. Petals yellow. Fruit a 2-parted siliqua: the lower slender with 0–1(3) seeds, the upper spherical, with 1 seed, wrinkled and with a persistent style.

Rapistrum rugosum BASTARD CABBAGE An annual to 80 cm, bristly hairy below and hairless above. Lower leaves pinnately divided, often toothed and stalked. Petals *lemon yellow*, 6–8 mm long. Fruit 4–10 mm, the upper segment ovoid-spherical *and abruptly contracted into the beak*; the lower segment cylindrical or swollen. Common in ruderal maritime habitats. Throughout. **R. perenne** is similar but a perennial with *bright yellow flowers*. East of range only.

Crambe SEAKALE

Large, bushy perennials with lobed leaves. Flowers with white petals and sepals erect to spreading. Fruit a 2-parted siliqua: the lower section stalk-like and seedless, the upper spherical and 1-seeded.

Crambe hispanica SEAKALE A slender, bristly hairy annual to 75 cm (1.2 m). Lower leaves rather shiny, round-lyre-shaped, divided and with a single large, kidney-shaped terminal lobe, and 0–2 pairs of inconspicuous lateral lobes. *Petals white*, 2–4 mm, sometimes purple at the base, borne in a *much-branched panicle*. Fruit with a short lower part (1 mm) and a *spherical* upper part (3–4.5 mm) *containing a single seed*. Arable land. AA (local), GR (mainly western), IP (north), PN (mainly western), SY.

Raphanus RADISH

Annuals or perennials with a peppery smell; leaves shallowly lobed. Petals white, mauve or yellow; sepals erect. Fruit a cylindrical siliqua, elongated into a seedless beak, constricted between the seeds.

Raphanus raphanistrum WILD RADISH A variable short to tall, bristly annual to 80 cm, erect and branched. Flowers white to pale yellow or mauve, *with lilac or reddish veins*, borne in branched racemes; petals 10–25 mm. Fruit 20–60 mm long, *jointed and beaded*. Very common on arable land, sea cliffs, sandy and waste places. Throughout.

Eruca

Annuals with deeply lobed leaves. Flowers with erect sepals; petals white or yellow with purple veins. Fruits with seeds in 2 rows under each 1-veined valves.

Eruca vesicaria (syn. *E. sativa*) A bristly annual to 1 m with stalked, lobed leaves with a large terminal lobe. Flowers with erect, purple sepals and white or pale yellow, purple-veined petals 15–20 mm. Fruit a small, *unbeaded siliqua 10–18 mm long with a flattened, sword-shaped beak*. Disturbed habitats. Probably throughout.

Erucastrum HAIRY ROCKET

Annuals or perennials with deeply lobed leaves. Petals yellow. Fruits linear, constricted between the seeds; seeds in 1 row under each 1-veined valve.

Erucastrum gallicum HAIRY ROCKET A rough-hairy annual to 60 cm with leaves deeply cut into distant, parallel-sided toothed lobes. Flowers pale yellow with erect sepals; petals 7–8 mm; flower-stalks with *pinnately lobed bracts*. Fruit curving upwards, 20–45 mm. Disturbed, sandy habitats. GR (north).

Erucaria

Annuals branched from the base, hairless or with simple hairs. Leaves lobed. Flowers white or violet; sepals erect; stamens 6. Fruit cylindrical, 2-parted beaked.

Erucaria pinnata A virtually hairless, *desert* annual, branched from the base, 15–50 cm. Leaves 90 mm (12 cm), pinnately lobed; lobes linear, entire or toothed. Flowers 5–8 mm, whitish-violet; petals egg-shaped with a narrow claw; sepals erect, greenish, about ½ as long as the petals. Fruit 15–28 mm narrowly cylindrical, obscurely 2-parted, the *upper part much longer* than the lower, tapering and often *strongly curved to hooked*. Seeds small and pale brown. Sandy deserts. IP (mainly western Negev).

Teesdalia SHEPHERD'S CRESS

Annuals with lobed or divided leaves. Petals white. Fruit a compressed siliqua, keeled to narrowly winged round the edges.

Teesdalia coronopifolia A small annual 80 mm (45 cm) tall, often with ascending basal branches. Basal leaves borne in small, crowded rosettes, 20–50 mm, stalked, narrowly lance-shaped with short pointed side lobes and a 3-lobed terminal segment; stem-leaves (if present) with fewer lobes or entire. Flowers white, borne in racemes on rather long ascending, slender stems; petals more or less equal and as long as the sepals; style absent. Fruit a rounded silicula <3 mm, winged towards the upper part and with upward-curved margins. Bare ground. Throughout (very rare in parts of the east and southeast).

1. *Rapistrum rugosum*
2. *Raphanus raphanistrum*
3. *Eruca vesicaria*
4. *Erucaria pinnata*

Carrichtera

A monotypic genus of annuals with pinnately divided leaves. Flowers pale yellow; filaments free. Fruit a 2-parted, pendent siliqua.

Carrichtera annua A small, rigid, bristly annual to 40 cm with linear 2–3-pinnately divided leaves. Flowers borne singly in elongated, leaf-opposed spike-like inflorescences; petals pale yellow, exceeding the hairy sepals. Fruit with 2 transverse segments: the lower ellipsoid with 2 chambers and 2 boat-shaped, 3-veined valves; the upper strongly flattened and sterile; seeds 3–4 per chamber. Common in deserts in the southeast, extending locally to dry areas further west. AA (rare and local), CR (local), CY, GR (southeast), IP (common), PN.

Nasturtiopsis

Desert annuals with *yellow* flowers; stigma simple. Fruits oblong and stiff; seeds in 1–2 rows.

Nasturtiopsis coronopifolia An ascending annual 10–25 cm with spreading bristles (the stem almost hairless). Basal leaves in rosettes, with toothed lobes, hairy, rather resembling those of *Plantago coronopus*; stem-leaves small and stalkless. Flowers borne on stalks 12–15 mm, bright yellow; petals almost circular; stigmas almost stalkless. Fruits linear, erect or incurved, as long as their stalks, hairless; seeds in single rows. Dry, sandy habitats and deserts. IP (mainly the Negev Highlands and Dead Sea Valley).

Notoceras

Annual herbs with entire leaves. Flowers yellow, borne in racemes. Fruit a 4-sided, short siliqua.

Notoceras bicorne A small, spreading to ascending annual to 30 cm with entire, narrowly lance-shaped to linear leaves to 25(30) mm with white adpressed hairs. Flowers small and yellow, borne in compact racemes, elongating in fruit; petals 2 mm; sepals 1.2–1.6 mm. Fruit erect, adpressed to the stem, to 10(12) mm long. Seeds oblong, rough, 1.2 mm. Deserts and other dry habitats. IP (east – common).

1. *Carrichtera annua*
2. *Nasturtiopsis coronopifolia*
3. *Notoceras bicorne*

SANTALACEAE

Woody or herbaceous perennials, hemi-parasitic on the roots of surrounding vegetation. Flowers small, cosexual or unisexual; stamens 3–5; style 1. Fruit a berry or nut.

Thesium

Perennial hemi-parasitic herbs with simple or branched stems. Leaves alternate, stalkless, linear and entire. Flowers borne in racemes or panicles; perianth 5-lobed; stamens 5. Fruit a nut.

Thesium humile A slender, ascending to erect, branched annual 70 mm–20 cm. Leaves linear, 10–35 mm. Flowers inconspicuous, greenish-white or yellowish, 1.25–1.5 mm; tepals triangular, 0.5 mm. Fruit 2.5–3 mm with an oblong, erect, net-veined nut, and persistent perianth ¼ x as long; fruiting stalk slightly swollen. Grassy hillsides and maquis. Probably throughout, except for cooler areas inland.

Osyris

Dioecious shrubs with angled stems and entire leaves. Flowers with 3–4 tepals and 3–4 stamens or 1 style. Fruit a berry.

Osyris alba A superficially broom-like, yellowish shrub 50 cm–1.5(2) m. Leaves small, 15–20(40) mm long, and narrow, just 2–3(4) mm wide, alternate, linear and entire, with a single mid-vein. Bracts green, leaf-like and persistent in fruit. Flowers sweetly scented, yellow, with 3(4) tepals, the male in small clusters, the female solitary, and borne on separate shrubs; stamens with a cluster of hairs. Fruit a red berry 6–7 mm long. Woods and scrub; common. Throughout.

Viscum MISTLETOE

Hemi-parasitic herbs with opposite leaves. Flowers 3–5, virtually stalkless; tepals 4; stamens 4; style absent. Fruit a sticky 1-seeded berry.

Viscum album MISTLETOE A hemi-parasitic shrub with rounded clumps to 1(2) m across on tree branches. Leaves 20–80 mm long, leathery, borne in pairs, widest towards the tip, with 3–5 veins. Flowers yellowish and inconspicuous. Fruit a sticky, white berry 6–10 mm. CR, GR, PN, TR. **Subsp.** *album* is parasitic on deciduous trees, extending locally into the north of the region. GR, PN (west). **Subsp.** *abietis* has leaves <4 x as long as broad and yellowish berries. On *Abies*. GR, PN. **Subsp.** *austriacum* has leaves >4 x as long as broad and white berries. On *Pinus nigra*. **Subsp.** *creticum* has a hemispherical habit and short, broad leaves. On *Pinus brutia*. CR. *V. cruciatum* is similar to *V. album* but with *red fruits*. IP (rare and local).

Arceuthobium

A mainly North American genus of hemi-parasites. Leaves scale-like, in fused pairs, sheathing the stem. Flowers with 2–5 tepals. Fruit a dry, green berry.

Arceuthobium oxycedri A small, yellowish, tufted hemi-parasitic perennial on the branches of *Juniperus* spp., to 12 cm. Leaves scale-like, 0.5–1 mm in pairs. Flowers inconspicuous, unisexual. Fruit a green berry 2–3 mm. AA (north), GR, LB, PN, SY, TR.

SANTALACEAE–FRANKENIACEAE–TAMARICACEAE

1. *Thesium humile*
2. *Osyris alba*
3. *Viscum album* subsp. *austriacum*
4. *Frankenia pulverulenta*
5. *Tamarix smyrnensis*
6. *Tamarix nilotica*

FRANKENIACEAE | SEA HEATH FAMILY

Dwarf shrubs (sometimes annual herbs) with opposite, entire leaves and no stipules. Flowers usually cosexual with 5 partly fused sepals and 5 free petals; stamens usually 6; style 1, divided. Fruit a small capsule.

Frankenia

Woody-based subshrubs, easily distinguished by their distinctive flower structure and small, *Erica*-like leaves.

(a) Plant an *annual* (scarcely woody at the base).

Frankenia pulverulenta ANNUAL SEA HEATH An *annual* with numerous *prostrate branches* 50 mm–17(30) cm, not particularly woody, often spreading in a circle, with oval leaves 5–6(8) mm long, hairless above, crispy-hairy below, often reddish. Flowers borne in the axils of the branches and upper leaves; stalkless, pink, the petals notched. Dry and saline habitats, mainly in the south and east. AA, CR, CY, GR (local), IA, IP, PN (local); probably elsewhere in the east.

(b) Plant a perennial, with *flowers arranged in dense terminal clusters or spikes*.

Frankenia hirsuta HAIRY SEA HEATH A low, prostrate, mat-forming *perennial* with branches 10–20(40) cm long, *white-bristly-hairy* at least above, not strongly white-encrusted, but often white powdery above, with pale to mid-pink flowers borne at the ends of the branches (not in lateral clusters in the axils). Saline habitats. Throughout.

TAMARICACEAE | TAMARISK FAMILY

Deciduous shrubs or trees with alternate, small (often scale-like), stalkless leaves. Flowers typically in catkin-like racemes, with (4)5 sepals and petals; stamens 5; styles 3–4. Fruit a capsule; seeds with hairy tufts.

Tamarix TAMARISK

Shrubs or small trees a distinctive habit and simple, alternate, small, scale-like leaves. Flowers tiny, borne in catkin-like spikes. Fruit a capsule with hairy, wind-borne seeds. Species often very similar and difficult to distinguish in the field; confused and conflicting accounts exist in the literature for the region.

(a) Flowers typically with 5 sepals, petals and stamens.

Tamarix smyrnensis A lax shrub or small tree to 5 m. Leaves 1–2 mm with a narrow and *semi-clasping* base. Flowers with 5 sepals, petals and stamens; petals 1–1.5 mm, *keeled*, pinkish or whitish with a yellowish-brown central part, borne in short racemes 10–25 mm long arising from *current year's shoots*; styles 3. CR, CY, GR, (IP), (LB), (SY), TR (easternmost localities in parenthesis may actually correspond to *T. hohenackeri*, which produces racemes 15–90 mm long and petals *incurved* and touching).

Tamarix nilotica A hairless, greyish or green shrub to 6 m. Leaves to 3 mm, scale-like, broad-based and half to fully *clasping*, oval to triangular. Flowers white (or pinkish), borne in rather lax racemes to 10 cm, petals 5, flat, elliptic, soon-falling; bracts exceeding the flower-stalks; stamens inserted into notches of the 10-lobed disk. Fruit a capsule 4–5 mm. Sandy habitats and deserts; a mostly Saharo-Arabian species; local outside the far southeast. AA (Samos, Patmos, Rhodes), CR, IA, IP, LB.

(b) Flowers typically with 4 sepals, petals and stamens.

Tamarix parviflora A shrub or tree to 5 m with brown–purple bark, hairless or minutely hairy. Leaves pointed, 3–5 mm long with membranous margins. Flowers white or pink, typically with *4 petals, sepals and stamens*, borne in racemes to 30 mm long and up to 5 mm wide on the *previous year's shoots*; sepals finely toothed; bracts tiny; both *bracts and sepals purple at the tips*. Riverbanks and roadsides; local. AA, CR (populations here often described under *T. cretica*), GR, IA, PN, TR. ***T. dalmatica*** is similar, also hairless, and with *thick (not slender) racemes to 12 mm wide*; flowers pale pink and typically 4-parted. Coastal marshes. CY.

PLUMBAGINACEAE | THRIFT FAMILY

Perennial herbs with basal, untoothed, simple leaves without stipules. Flowers regular, borne in lax or tight clusters, 5-parted with papery calyx lobes that persist in fruit; petals fused in the lower part to form a tube; stamens and styles 5. Fruit a 1-seeded capsule.

Acantholimon

Spiny, cushion-forming shrubs. Leaves alternate, simple. Inflorescence made up of 1-flowered spikelets grouped into panicles; calyx funnel-shaped with a slender tube, 10-ribbed at base, 5-toothed at the apex; stamens free; styles 5. Fruit cylindrical, not splitting when ripe. Conflicting accounts exist in regional floras.

Acantholimon aegaeum A small, rather dense, spiny shrub with alternate, needle-like leaves to 1 mm with small, round, limey secretions. Flowers borne in more or less stalkless inflorescences; calyx 10–15 mm, funnel-shaped, with purple veins; petals longer, pink. Rare on bare limestone slopes. AA (Chios, Samos), TR (west). ***A. androsaceum*** is similar but with leaves *without* limey secretions and inflorescences of just 2–3 flowers. Rare, on rocky slopes. CR (west, central).

Plumbago

Herbaceous perennials and shrubs with clasping leaves. Inflorescence a dense, terminal spike; corolla tubular; stamens 5, free; style 1 with 5 stigmas. Fruit dry, 1-seeded and 5-valved.

Plumbago europaea A tall, erect, much-branched perennial to 1.2 m with alternate, oval to oblong leaves to 10 cm long, clasping and heart-shaped at the base, *glandular along the margins*. Flowers violet or lilac–pink, the lobes darker *20 mm long*, borne in a spike-like inflorescence; lobes oval. Fruit a 5-parted capsule. Coastal sands, disturbed habitats and hedgerows. Throughout. ***P. auriculata*** is a scrambling shrub to 6 m with leaves *without* glandular margins. Flowers sky-blue, *40 mm long*. Native to South Africa, planted, sometimes naturalised.

Armeria THRIFTS

Tufted perennials with basal leaves. *Flower-heads dense and hemispherical* with papery bracts, borne on long, leafless stalks; stamens and styles 5. Rather few species in the region.

Armeria canescens A variable perennial. Leaves 20–75 mm, *dissimilar* and varying in dimension, the outermost linear and broadest above the middle, the inner longer and linear to linear-lance-shaped, flat or with curled edges, and with membranous margins. Flowers pinkish or whitish, borne in heads to 25 mm across on scapes to 15–40(70) cm long; bracts 6–9.5 mm, pale brown and with papery margins. Rocky habitats, generally on higher ground. AA (Limnos), GR, IA (Cephalonia – rare), PN. ***A. icarica*** has linear leaves all similar 14–45 mm long, calyx 6–6.5 mm, pink flowers, and shorter scapes 10–20 cm long. AA (Ikaria). ***A. johnsenii*** is very similar but with a *short* calyx just 4–4.5 mm, *white* flowers and scapes with *yellowish glands* beneath. AA (Euboea).

Limonium SEA LAVENDER

Perennial herbs with simple, leathery leaves forming basal rosettes and flowers in *branching cymes*; flowers persisting and papery in fruit; stamens and styles 5. A highly complex genus with numerous species, including the highest number of endemics for Greece (c. 80); many are of uncertain status, however, and few are described here.

(a) Perennials with *lobed* leaves. Flowering stems conspicuously winged or angled.

Limonium sinuatum WINGED SEA LAVENDER   A bristly perennial 10–40 cm. Leaves wavy-lobed, 40 mm–15 cm long and 8–30 mm wide, borne in a basal rosette. Stems with 3–4 undulate wings. Flowers cream; calyx 11.5–14 mm, *conspicuous, blue–purple and with a papery margin*. Saline coastal areas. Throughout. ***L. lobatum*** is similar but an *annual* with *angular* (not winged) stems and *yellowish* flowers. GR, IP.

1. *Acantholimon aegaeum*   PHOTO: ARNE STRID
2. *Plumbago europaea*
3. *Plumbago auriculata*

1. *Limonium sinuatum*
2. *Limonium antipaxorum*
3. *Limonium elaphonisicum*
4. *Limonium narbonense*
5. *Limoniastrum monopetalum*

(b) Herbaceous *annuals* with leaves in flat rosettes.

Limonium echioides A slender annual to 45 cm with *small rosettes of flat, bristly, leaves*, which age red–purple; leaves broadest above the middle, gradually narrowing at the base, 7–55 x 3–16 mm. Flowers pink–white and short-lived, borne on widely branching, slender, red, 1-sided branches; spikes 20 mm–18 cm with 1–2 spikelets per cm; calyces persistent; calyx lobes with up to 10 curved spines. Fixed dunes, widespread. AA, CR (+ KP), CY, GR (southeast), PN (east).

(c) Perennials with a woody stock. Leaves <80 mm long.

Limonium bellidifolium A woody-based, hairless perennial 90 mm–30(40) cm. Leaves 14–40 mm, spatula-shaped and pointed (1)3(5)-veined, usually *withered* or absent during flowering. Inflorescence with numerous non-flowering branches; spikes dense, spreading; spikelets 1–3-flowered; inner bract 1.6–3.9 mm, transparent for about 1/3 of its length; outer bract transparent except on the keel; calyx 2.6–3.8(4) mm; corolla 4.5–5 mm, pale violet. Coastal saltmarshes; scattered across Greece. AA, GR, IA, PN.

Limonium antipaxorum A small, tufted perennial with a sparingly branched woody stock and green (ageing reddish), spatula-shaped leaves, tapering into the stalk, forming several lax rosettes; leaves with a single prominent mid-vein and slightly down-turned margins. Stems green, spreading and reticulated (zigzagging), the branches mostly *diverging at an angle* of 90°–180°; flowers violet, rather few. IA (Paxos and Antipaxos). Many closely related and similar Greek endemics, which there is not space to cover in detail here, are also distributed on the Ionian Islands and western coast of the Peloponnese. They include: ***L. arcuatum*** IA (Othonoi, Corfu), CR; ***L. cephalonicum*** IA (Cephalonia); ***L. ithacense*** IA (Ithaka, Cephalonia); ***L. saracinatum*** IA (Lefkada, Meganisi, Ithaka, Echinades, Cephalonia); ***L. coronense*** PN; ***L. zacynthium*** IA (Zakynthos); ***L. phitosianum*** IA (Zakynthos, Strofades); ***L. damboldtianum*** IA (Lefkada, Cephalonia); ***L. kardamylii*** PN; and ***L. messeniacum*** PN.

Limonium elaphonisicum A hairless, tufted perennial with a stout woody stock, similar in general appearance to the *L. virgatum* group (below). Basal leaves almost erect, 15–30(40) mm long, broadest above the middle, v-shaped in cross section, gradually tapering to the base. Stems erect, 20–30 cm, sparsely branched from the base, with *constricted nodes* and *numerous tiny raised glands* throughout; sterile branches few. Flowers borne few in spikelets; inner bracts 7–8 mm, 3-coloured (green below, banded reddish and then papery at the apex); calyx 8–8.5 mm; corolla bluish-violet. A very rare island endemic of rocky coastal sands. CR (Elafonisi Island and adjacent mainland coast + Gavdos).

Limonium virgatum A tufted, hairless, cushion-forming perennial with a stout taproot and branched woody stock. Leaves crowded, 15–30 mm, narrowly spatula-shaped, single-veined. Flowers borne on several slender stems 10–30(50) cm, branched from the middle; some branches sterile; spikes dense, 1-sided with 2–3-flowerd spikelets; bracts rust-coloured; calyx 5–6 mm; corolla 8 mm, violet. Scattered throughout. ***L. aegaeum*** is similar but with larger, often slightly sickle-shaped leaves 20–90 mm long and stems branched from near the base with numerous sterile branches; spikelets 2–7-flowered, fan-shaped. AA, CR, GR (southeast). ***L. roridum*** is similar to *L. virgatum* but the whole plant with conspicuous raised *glands*, the basal leaves broad and thick; stem with *swollen internodes*; the nodes constricted and fragile; branches often with small rosettes of leaves in the axils. AA, CR, GR (southeast), PN (south), TR (west). ***L. proliferum*** (AA, CR) and ***L. xerocamposicum*** (CR – east) may be local forms of the previous species.

Limonium sieberi A tufted perennial with a stout woody stock. Leaves 20–45 mm, spatula-shaped, thick and rigid with a single prominent vein, sometimes with 2 weak lateral veins; leaves fresh green (but described as grey–blue in some floras). Stems few, 15–30 cm, erect, slender with few, long branches; spikes lax; spikelets 8–10 mm, 1–2-flowered; bracts rust-coloured; calyx 5.5–8 mm;

corolla pale violet. AA, CR, GR (southeast), PN (south and east), TR (southwest). *L. graecum* is similar to *L. sieberi* but with grey-bluish (not fresh green), spatula- to oblong-lance-shaped leaves 10–40 mm and flexuous stems 5–12 cm. AA. *L. palmare* is possibly indistinct from the previous species. AA. *L. galilaeum* is similar to *L. sieberi* but with abundant (not few) non-flowering branches, scale beneath the first inflorescence branch 1.5 (not 2.5–4) mm, and flowers per spikelet 1–2 per cm (not 3–4). Calcareous sandstone and on limestone outcrops. IP (northwest). *L. sartorianum* is similar to *L. sieberi*, but with smooth, gland-dotted leaves and broader inner bracts. AA (Andros).

Limonium sitiacum A hairless, rather loosely tufted perennial with a laxly branched woody stock. Leaves more or less erect, in loose rosettes, spatula-shaped, sometimes folded, tapering into their stalks; leaves leathery, greyish and with conspicuous glands. Stems 10–30 cm, with few branches, rough; sterile branches absent or few; spikes long and lax; spikelets 8.8–10 mm with 1–3 flowers; calyx 6.3–7.3 mm; corolla 9 mm, lilac. AA (southeast – rare), CR (east + nearby islets). *L. vanandense* (KP), *L. samium* (Samos) and *L. monolithicum* (Rhodes, Simi + adjacent islets) are all very similar to, and may all be local forms, of *L. sitiacum*.

(d) Perennials with a woody stock. Leaves >80 mm long.

Limonium narbonense A hairless perennial with a stout woody stock. Basal leaves *large*, 80 mm–18 cm, rather erect, oblong to broadly spatula-shaped, thick, with a single mid-vein and numerous weak lateral veins. Stems 30–70 mm, repeatedly branched above; sterile branches few to absent; spikes numerous, 10–20 mm, dense; calyx 4–5 mm; corolla 6–8 mm, blue–violet. Scattered across the region (not CR). *L. hirsuticalyx* apparently differs only in its short spikelets and short calyx with hairy veins. AA (east-central). *L. compactum* apparently differs only in its very densely crowded spikelets. AA, GR (east coasts), PN (east coasts; local).

Goniolimon

Very similar to *Limonium* but typically perennial herbs with compressed stems in dry stony habitats (not exclusively sea cliffs). Leaves few, leathery, in basal rosettes. Flowers with club-shaped stigmas.

Goniolimon dalmaticum A perennial 70 mm–30 cm with stout, angular to compressed stems. Leaves 50–80 mm long, lance to spatula-shaped, sharp-pointed, *covered in white dots*. Flowers borne in *dense* spikes 15–30 mm long with 8 spikelets per cm; calyx 7–8 mm with a sparsely hairy tube; corolla reddish-purple. GR (northeast, north-central). *G. sartorii* is similar but with leaves 3–4 x as long as wide, abruptly contracted into the stalks, pale pink flowers with hairless calyx, and grows on maritime rocks. AA, GR (southeast). *G. heldreichii* has whitish flowers borne in lax spikes (3 spikelets per cm; spikelets 1-flowered). GR (northeast).

Limoniastrum

Similar to *Limonium* but *large, spreading shrubs* with leaves alternate along the stems and with chalky glands. Flowers with 5(6) stamens.

Limoniastrum monopetalum LIMONIASTRUM A fleshy, spreading shrub 50 cm (2 m) with silvery-green, fleshy spoon-shaped leaves 20–60(90) mm long covered in white scales, sheathing the stem at the base. Flowers bright pink and conspicuous, later violet, to 16 mm borne in loosely branched spikes; corolla with 5 spreading, oval petals. Rare and local in maritime sands and saltmarshes (occasionally also planted in coastal areas). AA (Kithnos, Limnos), CR (mainly offshore islets), GR (southeast), PN (southeast).

POLYGONACEAE | DOCK FAMILY

Herbs or small shrubs without latex, alternate leaves and stipules that form a membranous sheath around the stem. Flowers cosexual or unisexual, often small and greenish or reddish; tepals 3–6; stamens (3)6–9; stigmas 2–3; sessile or with styles. Fruit an achene.

Coccoloba

A neotropical genus of about 120 trees, shrubs and climbers. Flowers borne in panicles; tepals 5; stamens 8. Fruit a nut surrounded by a fleshy, enlarged perianth.

Coccoloba uvifera SEA GRAPE An evergreen, branched shrub or small tree (to 10 m) with a stout trunk with greyish bark. Leaves to 20 cm, broadly oval to circular, heart-shaped at the base, leathery, glossy and with red veins; short-stalked. Flowers fragrant, borne rather numerously in racemes 10–20 cm long, drooping in fruit. Fruit 15–20 mm, spherical to egg-shaped, purplish when mature. Occasionally grown in hot, coastal areas.

Polygonum

Annuals or perennials with tap-roots. Flowers single or few (<6) in the leaf axils, exceeded by the leaves; stamens 8; stigmas 3, virtually sessile. Achene with 3 rounded angles.

Polygonum maritimum SEA KNOTGRASS A prostrate or ascending, branched perennial with stems 60–80 cm with a woody stock. Leaves narrowly elliptic, blue–grey and sessile, with down-turned margins, 15–25 mm long and 6–9(16) mm wide. Stipules silvery, reddish at the base with *8–12 conspicuous branched veins*. Flowers white or pink, 5-lobed and 3–4 mm across, solitary or 2–3 in the nodes. Common on maritime sands and shingle. Throughout.

Persicaria

Annuals or perennials with rhizomes. Flowers borne numerously in terminal or axillary, leafless, spike-like clusters; stamens 8; style 1. Achene 3-angled or winged.

Persicaria maculosa (syn. *Polygonum persicaria*) REDSHANK An erect or sprawling *hairless annual* to 80 cm with branched stems reddish below and swollen at the leaf nodes. Leaves lance-shaped and tapered at the base, *often with a dark central spot. Flowers pink*, borne in dense terminal (or in axillary) leafless, cylindrical spikes. Widespread and common in damp waste areas. Throughout except most smaller islands and the far southeast. *P. lapathifolia* (syn. *Polygonum lapathifolium*) PALE PERSICARIA is similar but slightly hairy with greenish stems, the flower-stalks with yellow glands, and *greenish-white flowers*. Similar habitats. Throughout (except much of AA).

Rumex DOCK

Perennials with terminal or axillary racemes, or panicles with whorled flowers; stamens 6; styles 3. Achene with 3 acute angles.

(a) Plant a herbaceous annual with basal leaves *heart-shaped or rounded, ending abruptly or gradually into the stalk* (not or scarcely lobed at the base).

Rumex cyprius An erect or sprawling, hairless annual to 30 cm with fleshy, ribbed, reddish stems. Leaves 10–60 mm long, triangular, narrowing abruptly at the base into stalks 10–60 mm long (without distinct lobes). Flowers borne in racemes to 10 cm, cosexual or unisexual. Fruit valves rather large and conspicuous, 10–20 mm across with reddish-pink veins. Dry stony slopes, fields or near coasts. CY, IP.

Rumex bucephalophorus HORNED DOCK A variable, reddish, erect annual to 30(50) cm, branched or not. Leaves normally small, 6–35(65) mm long, lance-shaped, oval or spoon-shaped, stalked, and greyish-green. Flowers borne on variously sized flower-stalks, very small, red and in clusters of 2–3 in the leaf axils, forming a long, dense spike. *Valves triangular-oval or narrow, with 3–4 teeth.* Common on waste ground, maritime sands and sea cliffs. Throughout.

Rumex vesicarius BLADDER DOCK An annual with oval-triangular to rounded leaves. Flower-stalks each with 2 flowers, one of which is smaller and concealed by the *conspicuous*, inflated, rounded valves 12–18(23) mm long, *flushed bright pink or crimson* with darker netted veins when in fruit. Mainly arid and desert fringe habitats. AA (rare), CR (northwest), GR (southeast), IP (south – common), PN (rare – northeast).

1. *Coccoloba uvifera*
2. *Polygonum maritimum*
3. *Rumex cyprius*
4. *Rumex bucephalophorus*
5. *Rumex vesicarius*

(b) Plant a herbaceous perennial with basal leaves triangular or arrow-shaped (generally *lobed* at the base).

Rumex acetosella SHEEP'S SORREL A spreading to erect, *slender* perennial to 45(80) cm with *small leaves*; leaves *arrow-shaped with small, forward-directed lower lobes and a large, oval-lance-shaped central lobe* 6–60 mm long. Flowers greenish or reddish, borne in simple or branched racemes, unisexual, the male and female flowers borne on separate plants. Fruit valves equalling the achenes. Fallow land and roadsides. Throughout except the far southeast.

Rumex thyrsiflorus (syn. *R. acetosa*) COMPACT DOCK A little-branched perennial to 1 m (often less) with basal arrow-shaped leaves, the middle lobe long and lance-shaped, 20–90 mm (12 cm) long, and the *lateral lobes backward-pointing*; stem-leaves becoming progressively narrower, and eventually very narrowly linear and clasping the stem. Flowers greenish or red, and borne in *very dense panicles*. Locally common on dunes or bare ground. GR (north).

Rumex tuberosus A *tuberous* perennial with a slightly woody stock and erect to ascending, ridged stems 10–60 cm. Leaves long-stalked with a arrow- to heart-shaped blade 15–50 mm long. Flowers borne in branched racemes. Fruits with valves 3–8 mm, net-veined, heart- to circular-shaped, with small warts; tepals deflexed in fruit. AA, CR, GR, IA, PN, TR.

(c) Plant a much-branched, *woody-based* perennial with basal leaves triangular or arrow-shaped (lobed at the base).

Rumex scutatus A woody-based, much-branched perennial to 40(65) cm with leaves 10–23(45) x 5–26(40) mm, variably triangular-heart-shaped with *diverging basal lobes*, rather thick and blue–grey. Flowers unisexual, reddish, borne in branched clusters. Fruits with oval to rounded valves, heart-shaped at the base, *as long as wide* (4.5–6.5 mm). Mainly in mountains. AA (Samothraki), GR, PN, TR.

Emex

Annuals superficially similar to *Rumex*. Flowers unisexual; stamens 4–6; styles 3. Fruit a *spiny nut*.

Emex spinosa EMEX A hairless, short, somewhat fleshy annual with sprawling stems to 50(60) cm. Leaves oval, heart-shaped at the base, 12–14 x 8(10) cm, long-stalked (to 25 cm). Male flowers stalked in terminal clusters, female flowers sessile at the base. *Fruit a spiny nut*, 3 mm. Locally common on maritime sands and disturbed ground, especially in the south and east (rare in the west). Probably throughout, except the north.

Fallopia

A small genus of annuals or herbaceous perennials and woody vines, previously included in the genus *Polygonum*. Flowers with 8 stamens; styles 3. Achene 3-angled.

Fallopia convolvulus An annual, clockwise-twining, climbing or prostrate vine to 1 m with angular stems. Leaves heart- or arrow-shaped, pointed and mealy beneath, 30–70 mm long. Flowers greenish- or yellowish-white, borne in loose clusters in the leaf axils. Fruit a triangular nut borne on a short stalk 1–3 mm long. Ruderal and bare places. Throughout. *F. dumetorum* is similar but with more rounded stems and *fruits borne on stalks 4–8 mm long,* often deflexed. Fruit black, smooth and glossy. Hedges, scrub and degraded woodland. GR, PN.

CARYOPHYLLACEAE | PINK FAMILY

Herbs with opposite leaves. Flowers cosexual, regular, with 4–5 free or fused sepals and 4–5 petals (absent in some species), typically 8–10 stamens and 2–5 styles. Fruit a many-seeded capsule, sometimes a berry or 1-seeded achene.

Gymnocarpos

Erect, woody undershrubs with opposite, stalkless, *fleshy* leaves. Flowers stalkless; sepals and petals 5, free; stamens 5; stigmas 3. Fruit dry, non-splitting.

Gymnocarpos decandrus A woody-based shrub, often growing in bare rock. Leaves greyish, cylindrical, fleshy, stalkless, hairless and pointed, 8–16 mm. Flowers stalkless, yellowish-green to purplish, rather small and inconspicuous; sepals 2–3 mm long, with whitish margins; petals pointed, 1–1.5 mm long; stamens opposite the sepals and slightly exceeding the petals. Fruit membranous, 1-seeded and enclosed by persistent sepals. Dry maquis and rocky deserts. Common in southern Israel, probably absent elsewhere. IP (south).

Arenaria SANDWORT

Leaves opposite and stipules present. Flowers with 10 stamens and 3 styles. Capsule with 6(–10) teeth. Many mountain-dwelling species occur besides those described here.

Arenaria serpyllifolia THYME-LEAVED SANDWORT A variable, downy, grey–green prostrate or ascending annual with unstalked, oval leaves. Flowers small, to 8 mm across, white, with unnotched petals, *exceeded by their sepals*. Fields and dry habitats. Throughout (in some places in the Aegean and south and east, replaced by the next species). **A. leptoclados** is similar (also widely treated as a subspecies) with conical, straight-sided capsules; sepals 2–3 mm long. Throughout.

Minuartia SANDWORTS

Small, slender annuals or perennials. Flowers with 0 or 5 petals, 10 or fewer stamens and 3(–5) styles. Capsule with 3(–5) teeth.

Minuartia picta A small, tufted annual (50)60 mm–10(12) cm with many spreading stems, branched at the base, glandular-hairy. Leaves 10–20 mm, linear to bristle-like. Flowers borne up to 30 in lax inflorescences with slender, spreading branches; petals pink in bud, opening white to pale pink; sepals (1.5)2–2.5 mm, hairless, oval with wide papery margins; petals 1.5–2 x as long as the sepals. Capsule ovoid, 1.5–2 x as long as the calyx. Plains, fields and semi-deserts. CY, IP (mainly southern and eastern), SY, TR.

Minuartia mediterranea MEDITERRANEAN SANDWORT A low, erect or sprawling, hairless annual 20–60 mm (12 cm) high with linear-lance-shaped, flat, pointed leaves. *Flowers borne in dense clusters*, with purple-flushed sepals; *flower-stalks shorter than the sepals*; *petals not notched, ½ the length of the sepals*. Fields and pastures. Throughout. **M. hybrida** FINE-LEAVED SANDWORT is a similar, slender annual 30 mm–20 cm without non-flowering shoots, and very *lax inflorescences*; petals absent or 1/2–1/3 of the length of the sepals. Throughout.

1. *Emex spinosa*
2. *Fallopia convolvulus*
3. *Gymnocarpos decandrus*
4. *Minuartia picta*
5. *Cerastium glomeratum*
6. *Paronychia argentea*
7. *Paronychia capitata*
8. *Illecebrum verticillatum*

Cerastium MOUSE-EAR

Annuals or perennials, often hairy. Flowers with 4–5 sepals and petals (or absent), 3–5(6) styles and 4, 5 or 10 stamens. Fruit a capsule with 2 x as many teeth as styles. Numerous similar species in the area, just one widespread taxon described here.

Cerastium glomeratum STICKY MOUSE-EAR A short, erect or ascending annual to 45 cm, covered in sticky glandular hairs. Leaves oval-elliptic, to 20 mm long, hairy. White flowers borne in *dense clusters*, the *sepals hairy to the tips*, petals more or less equalling the length of the sepals. Very common on bare ground and in dry, grassy places. Throughout.

Paronychia

Perennial herbs with small opposite leaves and *conspicuous silvery stipules*. Flowers small, borne in dense clusters *surrounded by silvery bracts*; stamens 5; styles 1–2.

(a) Sepals with papery margins and *bristle-tips*.

Paronychia argentea A branched, *mat-forming perennial* with stems 50 mm–50 cm long. Leaves oval-lance-shaped, greyish, in opposite pairs, bristle-tipped and almost hairless. Stipules membranous, and shorter than the leaves. Flowers borne in lateral and terminal clusters 10–15(25) mm across; with prominent *membranous, silvery bracts* (most conspicuous in mature or fruiting heads), concealing the flowers; sepals with bristle-tips 0.4–0.65 mm. Common in maritime sandy places. AA, CR, TR. *P. echinulata* is a similar, spreading *annual* with stems to 20 cm and flowers with bracts much *shorter* (the flowers not concealed by them). Absent from most of the Greek mainland, but present on Euboea. AA, CR, CY, IP, LB, PN (rare) TR.

(b) Sepals with green margins, *without bristle-tips*.

Paronychia capitata A grey–green, hairy, tufted or mat-forming, short-lived perennial with stems to 12(15) cm. Leaves stiff, oblong to linear-lance-shaped, 3–6 mm, hairy, equalling the length of the stipules; stipules 4 per node. Flower-heads rather dense with bracts completely concealing the flowers (as in *P. argentea*) and *markedly unequal*, somewhat fleshy *sepals inwardly curved at the tips*. Much more frequent in the western Mediterranean. GR, PN (east). *P. cephalotes* is similar but with the sepals *equal* (or virtually so) and not inwardly curved when in fruit. GR, TR. *P. macrosepala* (syn. *P. kapela* subsp. *insularum*) has rather fleshy, silvery leaves and bracts scarcely concealing the flowers. AA (common), CR, GR (east), PN, TR. *P. chionaea* (syn. *P. kapela* subsp. *chionaea*) has just *2 stipules* at each node. AA (east – rare).

Illecebrum

Similar to *Paronychia* but with white, spongy sepals, which persist in fruit. Stamens 5; stigmas 2. Fruit a 1-seeded, 5-valved capsule.

Illecebrum verticillatum CORAL NECKLACE A distinctive, low, hairless annual superficially similar to *Paronychia*. Stems creeping, rooting at the base, reddish, to 20 cm. Leaves 2–5 mm, opposite and with small stipules, Flowers borne in 2 whorl-like clusters at each node, 4–6-flowered with silvery bracts. Damp, sandy and gravelly places; rare and local. AA, IA (Corfu), PN (southeast).

Pteranthus

Small, fleshy annuals with minute, inconspicuous flowers. Calyx with hooded lobes, tipped with spiny appendages; petals absent; stamens 4.

Pteranthus dichotomus A spreading to ascending annual 10–15 cm with stems 1–3 x divergently branching, jointed at the nodes. Leaves linear, fleshy, 8–18 mm. Flowers borne in leafy, corymbose panicles with an *expanded, leaf-like* peduncle to 9 x 6 mm; calyx closed, 4-parted, with a *spiny appendage*; petals absent; stamens 4, opposite the calyx lobes; stigma 2-parted. Fruit 1-seeded; seeds compressed, 1–2 mm. Sandy habitats and deserts. IP (south and east).

Herniaria RUPTURE-WORT

Similar to *Paronychia* but often with tiny leaves and flowers, and *inconspicuous bracts*. Stamens 5; stigmas 2. Fruit an achene.

Herniaria hirsuta HAIRY RUPTURE-WORT A variable annual (with many forms, sometimes treated as species) with bright green to grey–green leaves 4–8(11) mm long, which are clothed in *dense, straight, white, spreading hairs*, and flowers 1.3–1.6 mm with 2–5 stamens, borne in roundish clusters of 7–12 in the leaf axils; *calyx hairy*. Bare habitats. Throughout. **Subsp.** *hirsuta* has sepals with short, straight hairs and usually 3–5 stamens. Throughout the range. **Subsp.** *cinerea* has sepals with long, stout hairs and 2 stamens. Possibly throughout. **H. glabra** is similar but with leaves without hairs except for the margins, and *calyx hairless;* flowers 1.3–1.5 mm. Sandy and fallow ground. Throughout the mainland; absent from most islands.

Polycarpon ALLSEED

Small herbs with forking stems and opposite or whorled leaves with papery stipules. Flowers with *keeled and hooded sepals*; stamens (1)3–5; stigmas 3. Fruit a 3-valved capsule.

Polycarpon tetraphyllum FOUR-LEAVED ALLSEED A small (sometimes minute) hairless annual without a woody stock, 20 mm (35 cm) high. Leaves mostly in *whorls of 4*, oval, green or purple. Flowers in branched clusters, white, tiny to 2 mm across with notched petals shorter than the sepals; stamens 3–4. Seeds brownish, with protuberances. Common on dunes and bare, sandy places inland. Throughout.

1. *Pteranthus dichotomus*
2. *Herniaria hirsuta*
3. *Herniaria hirsuta* subsp. *cinerea*

Sagina PEARLWORT

Small, often tufted, moss-like herbs. Flowers minute with 4–5 sepals and petals or petals absent; stamens 4, 5, 8 or 10; styles 4–5. Capsule 4–5-valved.

Sagina apetala ANNUAL PEARLWORT A small, annual herb with very slender, *sub-erect stems to 15 cm, all producing flowers*. Leaves linear, and tapered at the tip. Flowers solitary, small with 4 oval sepals, often hooded; petals minute and falling early. Common in sandy places. Throughout.
S. procumbens PROCUMBENT PEARLWORT is a similar, moss-like, bright green *mat-forming perennial* with a short, non-flowering main stem bearing a central, dense leaf rosette and numerous lateral stems to 20 cm ascending from rooting bases. Sepals and stamens 4(5). Common in sandy and urban waste places. AA (local), GR, IA, PN (local).

Spergula

Slender annuals with opposite leaves appearing whorled. Petals almost as long as sepals or slightly longer, white, entire; stamens 5–10; stigmas 5. Fruit a 5-valved capsule.

Spergula fallax A slender, spreading to ascending annual with hairless stems 12–30 cm, branched from the base. Leaves *channelled beneath*, borne in whorls at the nodes, 10–25 mm, linear, hairless, often with sterile shoots. Flowers white, solitary or few, borne on long, slender stalks; sepals 3–4.5 mm with white, papery margins; petals *shorter* than the sepals. Seeds blackish-brown and *smooth with a shiny white membranous wing*, almost as broad as the seed. Dry and semi-arid areas. IP (mainly eastern).

Spergula arvensis A slender annual, branched from the base 10–30(60) cm, sparsely glandular-hairy throughout. Leaves in interrupted whorls of 6, linear, slightly fleshy, *not* channelled beneath. Flowers white, borne on long, slender stems; sepals 3–5 mm, slightly *exceeded* by the petals; stamens 5–10. Capsule exceeding the sepals. Seeds *black, scarcely* compressed, *keeled* (or only narrowly winged, <1/10 the seed width). Throughout, except many islands (absent from CY).
S. pentandra is similar but with 5 stamens and *strongly* compressed, *winged* seeds. Absent from the south, east and most islands. AA (west, north), GR, PN, TR.

Spergularia SEA-SPURREY

Annuals or perennials with narrow, opposite (sometimes seemingly whorled) leaves and leafy tufts at each node; leaves fleshy or not. Petals purple, pink or white; stamens 5–10; stigmas 5. Capsule 5-valved.

(a) Petals normally *exceeding* the sepals; seeds winged or partially winged.

Spergularia media GREATER SEA-SPURREY A short almost hairless perennial to 40 cm with fleshy leaves ending in an abrupt point, rounded beneath. Stipules broadly triangular. Flowers 10–12(13) mm across with white or pink *petals equalling or exceeding the sepals*; sepals 4–6 mm; stamens 0–10. Capsule 7–9 mm, greatly exceeding the calyx; seeds dark brown, all (or mostly) winged. Sandy shores and in saltmarshes. AA (rare), GR (mainly northeast), PN (rare).

(b) Petals *equalling or shorter* than the sepals; seeds winged, not winged or both.

Spergularia marina LESSER SEA SPURREY An annual to 35 cm with very fleshy leaves, normally a slender (not woody) stock, and short stipules that form a sheath. Inflorescence sparingly branched; flowers 5–8 mm across, petals pink above and white below, *not exceeding the sepals*;

sepals 2.5–4 mm; stamens 2–7(10). Capsule 3–6 mm, exceeding the calyx; seeds light brown, unwinged, winged or mixed. Common on sandy shores and in saltmarshes. Throughout.

(c) Petals equalling or shorter than the sepals; seeds not winged.

Spergularia bocconei A slender annual or biennial to 20 cm, with *densely hairy inflorescences*. Leaves scattered (not in dense clusters); stipules triangular. Petals pink with a white base, equalling or shorter than the sepals, which are 2–4 mm. Capsule 2(–4) mm, shorter than the calyx; seeds grey–brown, not winged. Ruderal habitats and fields. Throughout.

Spergularia rubra SAND SPURREY A *sticky-hairy* annual or perennial with spreading stems to 25 cm. Leaves ending in an abrupt point, *not fleshy*, grey–green and seemingly *whorled*. Stipules silvery and lance-shaped. *Petals equalling or shorter than the sepals*, uniformly pink; sepals 3–4(5) mm. Capsule 3.5–5 mm; seeds not winged. Sandy (not saline) soils. Throughout, except CY.

1. *Spergula fallax*
2. *Spergularia marina*
3. *Spergularia bocconei*
4. *Agrostemma githago*
5. *Saponaria cypria*
6. *Saponaria officinalis*

Agrostemma

Annuals with sepals fused into a tubular calyx with long, green leaf-like teeth; stamens 10, styles 5. Fruit a capsule with 5 teeth.

Agrostemma githago CORNCOCKLE An annual to 70 cm (1 m) covered in adpressed, greyish hairs. Leaves narrowly lance-shaped and pointed, 5–7 mm across. Flowers borne on long individual stalks; petals 20–35 mm, pale reddish-pink, shallowly notched and shorter than the long-pointed, linear sepals; sepals 15–18 mm with teeth 20–35 mm. A rather local but widespread arable weed. Throughout.

Saponaria SOAPWORTS

Annuals or perennials with a smooth, tubular calyx, without an epicalyx; styles 2. Fruit a capsule with 4 teeth.

Saponaria cypria An erect to spreading perennial 80 mm–20(35) cm with a thick, woody stock. Basal leaves loosely clustered, virtually hairless, egg- to spatula-shaped, 25–50 mm long. Flowers solitary or in lax cymes, opening in the evening, closing by the middle of the following day; calyx 25 mm, densely hairy; petals pink, 30–40 mm. Capsule cylindrical, 20 mm long. Rocky mountainsides and screes among pines or by streams; an island endemic. CY (common in the Troodos).

Saponaria officinalis SOAPWORT A *tall, robust*, hairless perennial to 90 cm with oval, strongly veined leaves, the largest ›50 mm. Flowers borne in rather compact clusters, 25 mm across, pale pink with unnotched petals; calyx flushed red and inflating in fruit. Damp woods, especially in the north. AA (local), GR, PN, TR.

Saponaria calabrica CALABRIAN SOAPWORT A *short*, reddish, stiffly spreading-ascending perennial, similar to *Silene* in general appearance. Leaves spoon-shaped to oval. Flowers pink–purple, to 10 mm across, borne in lax clusters; petals not notched; calyx long-tubed, purple and hairy. Garrigue, scrub and woods (usually inland). AA (local), GR (mainly south), IA (Corfu, Kalamos, Lefkada), CR, PN, TR (northwest).

Silene CAMPION

Herbs with flowers in branched inflorescences. Sepals fused into a tube, often with 5 teeth; petals separate; styles 3 or 5, protruding. Fruit a capsule. Numerous species occur in the region, but only a subset are described here.

(a) *Dioecious* biennials or perennials with 5 styles (3 in other taxa described).

Silene latifolia (syn. *S. alba*). WHITE CAMPION A variable, densely *stickily hairy*, rather tall perennial to 1 m with oval leaves, the lowermost stalked. Flowers unisexual and more numerous on male plants: large and white with deeply notched petals; calyx red-veined, *not inflated*, 15–22 mm and 10-veined (male) or 20–30 mm and 20-veined (female). Woods and scrub. GR, PN (north-central).

(b) Annuals. Calyx hairless.

Silene cretica CRETAN CATCHFLY A *slender*, erect annual 20–50 cm, hairy at the base, *hairless above*, and sticky. Basal leaves oval and broadest above the middle and stalked; upper leaves pointed and unstalked. Flowers bright pink, borne in lax, branched clusters with notched petals; *calyx distinctly ridged and hairless*, 11–16 mm. Cultivated and disturbed ground. Throughout, except most of mainland Greece.

(c) Annuals. Calyx hairy with (15)30–60 veins, variably inflated in fruit. Many closely related taxa.

Silene conica An erect, hairy annual 10–25 cm with broadly linear leaves. Flowers few, borne on *short, stout* stalks 5–10 mm; calyx 30-veined, toothed, 13–18 mm; petals claw not exserted, the limb (3)4–5(6) mm, pink to magenta. Capsule 7–12 mm, pear-shaped, included within the calyx. Seeds grey, minutely warty. AA (Limnos, Thasos), CR (east), GR, IA, PN. *S. subconica* is similar, glandular-hairy above with *long, slender* flower-stalks (6–25 mm); fruiting calyx very inflated, and petals with claw just exserted and pale pink limb 7–10 mm. Absent from the west. AA (east), GR, PN, TR. The following more local species are all similar to *S. conica*: **S. grisebachii** has a calyx 15–18 mm, *not* inflated in fruit and pink petal limb 7–10 mm. GR (north + adjacent islands). **S. lydia** has *long* calyx teeth, equalling the tube, and bright magenta petal limb 5–7 mm. AA (Samos), GR (north-central + northeast), TR. **S. macrodonta** has calyx with *60 very close-set* veins, cylindrical in flower, broader but not conspicuously inflated in fruit. AA (southeast), CY, IP, SY. **S. sartorii** has calyx 10–13 mm, pear-shaped but scarcely inflated in fruit and almost *smooth* seeds. AA. CR (central, south), GR (far southeast), PN (northeast).

1. Silene subconica
2. Silene colorata
3. Silene palaestina
4. Silene damascena
5. Silene dichotoma

(d) Annuals. Calyx hairy with 10 veins, not usually inflated in fruit.

Silene colorata A variable, erect or spreading, short annual. Leaves oval-linear, the lower leaves stalked. Flowers with *bright pink* (rarely white) *deeply notched petals*, borne in lax clusters; calyx club-shaped, 10–15 mm long, variably hairy. Seeds ›1 mm. Very common on coastal waste land and dunes. Throughout (except the north and KP). *S. discolor* has petals pink above, yellowish below and minute seeds (‹1 mm). Replacing *S. colorata* in eastern Aegean Islands. AA (east).

Silene palaestina A sticky, sparingly bristly, erect annual, simple or divergently branched below, 30–50 cm. Leaves oblong-linear to lance-shaped. Flowers 12–20 mm, bright *purplish-pink*, borne in stiff racemes; calyx 10–15 mm, membranous, *red-veined*, tapering-cylindrical from the base, club-shaped in fruit with blunt, oblong teeth and with *long, dense bristles*; flower-stalks *shorter* than their calyx (uppermost very short); petal limb split to the base. Capsule oblong, 7–10 mm. Sandy habitats. IP (mainly coastal plains), LB. *S. damascena* is similar but with lower flower-stalks *longer* than their calyx; calyx reddish and sparingly glandular with *shorter* bristles; petal limb paler pink to whitish, split to the *middle*; corolla with a *dark red central ring*. Dry fields and deserts. IP (mainly north and east), LB.

Silene dichotoma A variable, short-hairy annual to 80 cm with sparse, *stiffly divergent branches*. Leaves linear to spatula-shaped. Flowers *white* (rarely pink), borne on very short stalks (to almost stalkless); calyx 7–15 mm, oblong-cylindrical, bristly hairy along the conspicuous 10 green (slightly winged) veins; petals divided beyond the middle into oval-oblong lobes. Capsule oblong, exceeded by the calyx. Fields, rocky habitats and deserts. AA (mainly east), CR, CY, GR (northeast) IP, TR.

Silene bellidifolia is similar in form to *S. colorata*; a sparsely bristly annual 25–70 cm with simple or branched stems. Leaves bristly hairy, lance-shaped to spatula-shaped below and oval-lance-shaped above. *Flowers more or less stalkless* in closely spaced inflorescences; petals pink and deeply notched; *calyx long, 14–18 mm*. IA, CR (west), GR (west; Egina) PN (Mani).

Silene apetala A small, erect annual 10–35 cm with linear to narrowly oval leaves. Lower flower-stalks 3 x as long as the calyx; calyx (6)7–10 mm, becoming bell-shaped in fruit; petals *pale pink*, sometimes *absent* (or included in calyx). Capsule 4–6(7.5) mm. Seeds dull blackish-brown. AA (Rhodes), CR (northeast islets + KP), GR. *S. alexandrina* is similar but has calyx with red veins with long, white hairs, and rose-lilac corolla. CR, CY, IP.

Silene sedoides A *small to minute*, often tufted, *densely hairy*, purplish annual 40 mm–10(15) cm. Leaves fleshy, spatula-elliptic and ‹10 mm long. Flowers pinkish or whitish; calyx 5–6.5(7) mm, purple, long, contracted at the base, and hairy; petal limb often with a darker spot at the base. Coastal rocks. Throughout. **Subsp.** *sedoides* is the common form. **Subsp.** *runemarkii* occurs in the Peloponnese. PN (Malea Peninsula). *S. aegaea* is similar but with calyx 7–9 mm, petal limb *without* a dark spot at the base, seeds to almost 1 mm. Very rare. AA (Ikaria, Tinos). *S. pentelica* is similar to *S. sedoides* but with petal limb without a dark spot at the base and sharply *keeled* seeds. Rare and local. AA (central), GR (southeast).

Silene gallica SMALL-FLOWERED CATCHFLY A downy, erect annual with simple or branched stems 80 mm–45(60) cm, sticky above. Leaves oval and stalked beneath, narrower and unstalked further up the stem. Flowers small, yellowish-white or pinkish, borne in a more or less *1-sided inflorescence with alternating, short-stalked flowers*; calyx small, 6.5–10(12) mm, cylindrical to ovoid with long teeth, ¼ the length of the tube, *sticky-hairy and 10-veined*. Very common in waste places. Throughout. *S. sclerocarpa* (syn. *S. cerastoides*) is similar but ascending and *woolly hairy* (especially below), the flower-stalks short, to 24 mm (not exceeding the calyx); calyx 9–11 mm long, *strongly contracted at both ends*, with dark red, somewhat netted veins with forward-pointing, stiff bristles; corolla pink. Bare, sandy habitats and dunes; rare and local. AA (Sterea Ellas), CR (west and east + KP), GR (southeast).

Silene noctiflora NIGHT-FLOWERING CATCHFLY An erect annual to 50 cm with simple or branched stems, which are hairy below and sticky-hairy above. Leaves narrowly oval and unstalked except at the base. Flowers whitish-yellow with deeply notched petals, rolling inwards (exposing yellow) during the day and opening fully at night, and fragrant; calyx 20–30 mm. GR (northeast).

(e) Biennials and perennials. Calyx hairless.

Silene vulgaris BLADDER CAMPION A variable perennial to 80 cm with a branching woody stock and several erect or ascending shoots, all flowering, usually (not always) hairless. Leaves elliptic to oval, the upper stalkless. Flowers white, with *strongly inflated, 20-veined calyx* 18–20 mm, persisting in fruit. Capsule with erect-spreading teeth. Very common and widespread in a range of habitats. Throughout. *S. behen* also has flowers with an inflated, hairless calyx, similar to the previous species, but an *annual* with broad, bluish leaves. Throughout, except the north.

Silene multicaulis A variable, weak-stemmed, often tufted perennial to 40 cm with 0–few vegetative shoots. Leaves linear to linear-lance-shaped. Flowers few; calyx (12)14–18(22) mm, pale green or wine-coloured; petals purplish or green, the claw long-exserted. Capsule equalling or to 2 mm longer than the calyx. CR (eastern mountains), GR (+ adjacent islands), PN.

1. *Silene apetala*
2. *Silene alexandrina*
3. *Silene sclerocarpa*
4. *Silene vulgaris*
5. *Silene sedoides* subsp. *sedoides*

(f) Biennials and perennials. Calyx hairy.

Silene succulenta A *fleshy, very sticky* (often with sand adhered) perennial with a woody stock and numerous trailing stems. Leaves thick, egg- to spatula-shaped. Flowers large, solitary or paired (rarely 3) in the axils of leafy bracts; calyx 15–20(22) mm with greenish or reddish veins; petals white to pale pink, the claw long-exserted, the limb 6–8 mm. Capsule 10 mm, included in the calyx. Rare and local, on dunes. CR (mostly on offshore islets), IP (coasts).

Silene italica ITALIAN CATCHFLY An erect, branched perennial to 70 cm with a woody base, sticky above. Leaves oval-elliptic, pointed and sticky-hairy, downy below. *Flowers erect on branched stems with >1 flower*, with deeply bilobed, *white petals*, yellowish above and greenish or reddish below (often inrolled during the day); calyx 14–21 mm, purplish, cylindrical and *hairy*. Common in dry, grassy places. Throughout, except southern islands (not CR, KP or Rhodes). **S. sieberi** is similar but tufted with a woody stock and short vegetative shoots, and petals *yellowish* above. CR (common). **S. spinescens** is similar but with densely grey, short-hairy leaves and cream to greenish-yellow flowers borne on 1–3-flowered branches diverging from the main axis almost at right angles. GR (southeast + adjacent islands), PN (northeast).

(g) Fruit a *shiny berry* loosely surrounded by a bell-shaped calyx.

Silene baccifera (syn. *Cucubalus baccifer*) BERRY CATCHFLY A downy, brittle-stemmed perennial to 1 m, often straggling around surrounding vegetation; not a typical member of its family. Leaves oval, pointed and short-stalked. Flowers rather large and conspicuous, drooping with green–white petals; calyx 8–15 mm. Fruit a spherical green, later black berry 6–8 mm. Grassy habitats and woods, mainly in the north. AA (west), GR.

Dianthus PINK

Annual or perennial herbs typically with 5 pink petals and calyx base surrounded by an epicalyx of paired bracts; stamens 10; styles 2. Capsule with 4 teeth. A highly complex genus with numerous species of which just a handful are described here.

(a) Bushy shrubs growing directly from rocky cliffs (chasmophytes) with leathery to succulent leaves.

Dianthus fruticosus A short, mound-forming, hairless shrub to 30 cm high and 20 cm–2 m across with numerous, densely leafy shoots. Leaves linear-lance-shaped to egg-shaped, blunt to pointed, leathery or succulent. Flowers pale pink to purplish-red with darker markings, borne in dense inflorescences of 3–8; epicalyx scales numerous (6–20); calyx 12–25 mm. A chasmophyte of coastal cliffs, mainly in the Aegean area, split into many geographical subspecies: **subsp. *carpathus*** KP; **subsp. *fruticosus*** AA (Cyclades); **subsp. *creticus*** CR; **subsp. *amorginus*** AA (Cyclades), CR (northeastern islets); **subsp. *karavius*** AA (east); **subsp. *occidentalis*** IA (south), CR (west + nearby islets), PN; **subsp. *sitiacus*** CR; **subsp. *rhodius*** AA (southeast).

Dianthus juniperinus A short, mound-forming, hairless shrub with linear to linear-lance-shaped, *pointed*, leathery leaves 10–40 mm. Similar to *D. fruticosus* but with calyx 7–23 mm with *few* (4–8) scales. An endemic chasmophyte of Crete where 7 similar subspecies have been described across the length of the island, varying subtly in leaf, calyx and petal characteristics: **subsp. *bauhinorum*,** subsp. *juniperinus*, subsp. *heldreichii,* subsp. *pulviniformis,* subsp. *idaeus*, subsp. *kavusicus* and subsp. *aciphyllus*.

(b) Annuals or perennials (sometimes woody beneath).

Dianthus crinitus A hairless, tufted perennial with a slightly woody stock and sparingly branched stems 10–40 cm. Leaves rather grey–green. Flowers solitary; calyx 24–32 mm, dull purple; petal with limb 12–17 mm, white to pale pink, *deeply divided into numerous long, fine segments*. AA (Rhodes).

Dianthus armeria DEPTFORD PINK An erect, hairy *annual or biennial* to 60 cm with dark green oblong basal leaves and linear stem-leaves. Flowers open in sunlight, *bright red-pink* across, subtended by *long, leafy green bracts*; *petals rather pointed and toothed*; calyx 15–20 mm. Widespread in grassy habitats. GR, IA.

---

1. *Silene succulenta*
2. *Dianthus fruticosus* subsp. *carpathus*   PHOTO: ELEFTHERIOS DARIOTIS
3. *Dianthus armeria*
4. *Dianthus juniperinus* subsp. *bauhinorum*   PHOTO: NICK TURLAND
5. *Dianthus crinitus*   PHOTO: ARNE STRID

CARYOPHYLLACEAE

Dianthus gracilis A laxly tufted, hairless perennial, woody below with linear leaves with rough margins. Flowers solitary or in clusters of 1–4; calyx 9–19 mm with 4–6 epicalyx scales; petals bearded, pinkish-purple above, greenish or cream beneath, the limb 5–10 mm. Several regional subspecies described. AA (north), GR.

Dianthus pinifolius A tufted perennial, slightly woody below with rigid, *bristle-like* leaves. Flowers borne in dense heads; calyx 8–12(15) mm with epicalyx scales rather spine-pointed; petals rose-pink. Several regional subspecies described. AA (north), GR, PN, TR.

Dianthus viscidus A weak perennial with spreading to erect flowering stems 10–25(40) cm with linear leaves. Flowers borne in dense clusters of 2–7 (rarely solitary); calyx 10–18 mm with 4 epicalyx scales reaching the teeth of the calyx; petals with limb 6–8 mm, sparsely bearded, pink with white spots, irregularly toothed. GR, IA (Corfu), PN.

Dianthus sylvestris WOOD PINK A variable, erect, tufted perennial with flower stems to 60 cm. Basal leaves slender and wiry, <1 mm wide, rough-margined and often recurved. Flowers generally mid-pink, *scentless, with hairless petals*; epicalyx blunt and leathery; calyx 12–29 mm long. Rocks, cliffs, dry wooded slopes. GR, IA.

Petrorhagia

Similar to *Dianthus* with epicalyx of papery scales at the base of a single flower, or with several bracts at the base of a dense flower-head; stamens 10; styles 2. Capsule with 4 teeth.

(a) Flowers solitary, not clustered, borne in a lax inflorescence.

Petrorhagia saxifraga (syn. *Tunica velutina*) TUNIC FLOWER A hairless, *mat-forming* perennial 50 mm–45(50) cm with spreading, then ascending flowering shoots. Leaves linear, pointed and rough-edged. *Flowers solitary*, pale pink to white, borne on *long stalks*; petals notched; *bracts much shorter than the calyx; calyx small, just 3–6(7) mm. Seeds minutely netted, not warted.* Rocks and stony or sandy habitats in the west. GR, IA, PN. *P. illyrica* is similar in habit, with linear, 3-veined leaves and white to pink or yellow flowers, often veined or spotted with pink in the centre, borne in spreading to erect panicles; petals oblong and *not notched*. Rocky habitats. CR, GR, PN.

(b) Flowers 1–few, clustered, borne in a dense flower-heads.

Petrorhagia prolifera (syn. *Kohlrauschia prolifera*) PROLIFEROUS PINK A lax, *hairless annual* 60 mm–50(70) cm. Leaves linear, greyish and rough-edged, fused at the base into *leaf sheaths that are about as long as wide*. Flowers small and pale to mid-pink, borne in a dense cluster but opening one at a time, *surrounded brown, papery bracts*; bracts equalling the calyx; calyx 10–13 mm. *Seeds minutely netted, not warted.* Maquis in the north and west. AA (Samothraki), GR, IA, PN (north). ***P. dubia*** (syn. *Kohlrauschia velutina*) KOHLRAUSCHIA is very similar to the previous species but with calyx smaller, 8–14 mm, *all* inflorescence bracts short bristle-tipped (mucronate) and *seeds covered in sparse, cone-like warts*. Maquis. Throughout.

1. *Petrorhagia dubia*
2. *Anabasis articulata*
3. *Bosea cypria*

AMARANTHACEAE | AMARANTH FAMILY

Herbs and shrubs with alternate, simple leaves. Flowers unisexual, greenish with 2–5 tepals (female) and 3–5 bracteoles (male and female); stamens 1–5; styles 2–3. Fruit an achene or 1-seeded capsule. Often in maritime or waste habitats; many introduced species, particularly in eastern semi-arid areas. Many genera were traditionally described under the family Chenopodiaceae.

Anabasis

Articulated, succulent small shrubs with opposite leaves forming segments. Flowers cosexual or female, 1 or several in the upper leaf axes; tepals and stamens 5; stigmas 2.

Anabasis articulata An articulated, fleshy, cushion-forming shrub 10–30(50) cm, woody below; branches short and much divided, greyish or reddish, the segments fragile, to 8 mm. Leaves opposite, clasping, 1–2 mm. Flowers to 5 mm across with 5 winged, rounded, straw-coloured to reddish, papery tepals 4 x 7 mm, similar in fruit. Very common in semi-arid to arid habitats in the east; absent elsewhere. IP (south), SY.

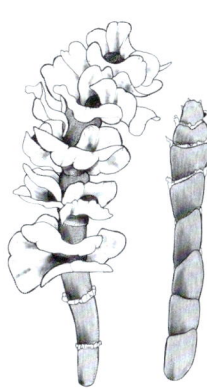

A. articulata

Bosea

Erect or scrambling shrubs, often intricately branched. Leaves alternate, entire. Flowers inconspicuous, often in panicles; tepals and stamens 5; stigmas 2–3. Fruit a large-seeded berry.

Bosea cypria A bushy, erect to arching, intricately branched shrub 1–1.5 m. Leaves hairless, broadly lance-shaped to elliptic, 25–60 mm long with a pale midrib. Flowers borne in leafless terminal panicles and in spikes in the axils along the branches, stalkless, cosexual with tepals to 2.5 mm, greenish-brown or with darker streaks. Berry 4–5 mm, red, with a large seed to 4 mm across. Dry, rocky ground and cliffs; an island endemic, fairly common on Cyprus. CY.

1   2   3

Einadia

Mealy perennials with stalked, flat, normally opposite leaves. Flowers borne in panicles; flowers cosexual or unisexual; tepals (4)5; stamens 0–3. Fruit *berry*-like.

Einadia nutans A herbaceous perennial to 60 cm, sometimes woody at the base. Leaves 5–20(25) mm, lance- to arrow-shaped, abruptly contracted into the stalks, mealy when young. Flowers tiny, borne in panicles; tepals 0.5–1.5(2) mm, fused below, mealy to hairlesss; stamens 0–2(3). *Fruit 3–5 mm, red, berry-like spherical*. Native to Australia, naturalised in bare places. IP.

Amaranthus

Annuals of waste land and disturbed habitats. Flowers inconspicuous with a brownish, papery perianth; tepals (2)3–5 (or absent). Fruit 1-seeded. A hand lens is required to examine the tepals. Most species are naturalised in dry or arid areas from Central and South America. Distributions are provisional; most are widespread, ephemeral weeds.

(a) Fruit non-splitting.

Amaranthus deflexus A slender, much-branched, spreading annual to 40 cm. Leaves 30–50 mm, diamond- to oval-shaped, *ending in an acute point*, the margins finely wavy. Inflorescence a brownish, dense, terminal, spike-like, interrupted and leafy below. Bracteoles as long as the tepals, oval and pointed; tepals 2–3, 1.2–1.5 mm. Fruits *inflated*, larger than the seeds, non-splitting. A cosmopolitan weed. Throughout.

Amaranthus muricatus A hairless, spreading perennial to 60 cm. Leaves 20–50 mm, *narrow*, linear to lance-shaped or narrowly oval (>3 x as long as wide), long-stalked. Flowers borne in long panicles, branched below; bracteoles as long as or *shorter* than the tepals and not spine-pointed but rather papery with green mid-veins; tepals 5, 2 mm long. Fruit non-splitting. Very local. AA (rare), GR (rare), IP (Esdraelon Plain).

Amaranthus viridis An annual to 80 cm, spreading to erect, *hairless*. Leaves diamond-shaped to almost circular (<3 x as long as wide), with light or dark spots on the upper surface, often with wavy margins. Flowers borne in clusters in the axils, forming dense, more or less leafless and *weak* spikes towards the apex; lateral inflorescence branches spreading; bracteoles as long as the tepals; tepals 3(4–5), 1–1.5 mm, whitish with green mid-veins. Fruit contorted and non-splitting. Common in coastal areas. Throughout.

Amaranthus blitum An ascending to erect annual to 80 cm with diamond to oval-shaped leaves, often reddish and speckled. Flowers borne in clusters in the axils and forming a short, dense spike above; tepals 3, much longer than the bracteoles, unequal. Fruit compressed, non-splitting. Local throughout. **A. emarginatus** is similar but prostrate to mat-forming and rather fleshy with leaves indented at the tips, unmarked. Dry, seasonally flooded ground; casual. GR (north, northeast).

(b) Fruit splitting transversely. Flowers borne in terminal, virtually leafless spikes or panicles.

Amaranthus retroflexus PIGWEED An erect, robust, hairy annual to 1 m with few side branches, green (rarely red). Leaves oval to lance- or diamond-shaped and stalked. Flowers small and inconspicuous, greenish-*white* with 5 *linear, tapering tepals 2–3 mm with mid-veins ending below the tips*; borne in compact greenish white spikes that are leafless towards the top, densely *hairy* below. Fruit capsule splitting when ripe. A North American native widely established on wasteground. Throughout.

AMARANTHACEAE | 389

Amaranthus hybridus An erect annual, 20 cm–1 m, similar to *A. retroflexus* with broadly lance- to diamond-shaped, long-stalked leaves. Inflorescences rather sparse, compound; *tepals 3(5), tapering to a long point*, as long as the fruit; bracteoles in female flowers 4–6 mm, exceeding the tepals x 1.5. Fruit splitting transversely. Common and widespread in dry but seasonally wet habitats. Throughout. **A. cruentus** is similar but often reddish or purplish, with dense, *longer* terminal inflorescences, often nodding above and with short branches at the base; bracteoles 2–4 mm (equalling the tepals) with a short spine-point (mucro); most branches of the inflorescence laxly spreading. An ephemeral escape. Throughout, except the north. **A. hypochondriacus** is similar but with *broad, erect* panicles to 20 cm wide. An ornamental escape. Local, throughout. **A. palmeri** is similar to *A. retroflexus* but taller and dioeceous with leaves with prominent white veins beneath. Flowers borne in erect, dense, narrow inflorescences to 40 cm long. CR (rare), PN, IP. **A. powellii** is also similar to *A. retroflexus* but often >1 m tall with leafless, bright green inflorescences, much elongated above. Scattered probably throughout, except the far west.

1. Einadia nutans
2. Amaranthus deflexus
3. Amaranthus muricatus
4. Amaranthus viridis
5. Amaranthus retroflexus
6. Amaranthus cruentus

(c) Fruit splitting transversely. Flowers all borne in clusters in the leaf axils.

Amaranthus albus An erect annual 10–60 cm with rigid, whitish stems. Flowers borne in small clusters in the axils, exceeded by the leaves; bracteoles 2 x as long as the tepals, rigid and spine-like; tepals 3. Fruit splitting transversely. Widely naturalised. Throughout.

Amaranthus blitoides A prostrate to ascending, often mat-forming annual. Leaves with white, narrow, whitish margins. Flowers borne in clusters in the axils, green or purplish, overtopped by the leaves; bracteoles exceeded by the tepals; tepals 4(5), unequal. Fruits splitting transversely. Dry habitats and farmland. Throughout. **A. *graecizans*** is similar but ascending to erect to 70 cm with leaves not or scarcely margined and 3 tepals becoming papery when mature. Possibly native, now a cosmopolitan weed. Throughout.

Haloxylon

Desert-dwelling shrubs or small trees with thick basal trunks and leaves reduced to scales. Flowers male, or cosexual with 2 short stigmas. Fruit a winged achene.

Haloxylon salicornicum WHITE SAXAOUL A rather diffuse to intricately branched shrub to 1 m with virtually leafless, slightly succulent, jointed, woody stems. Leaves minute, short-triangular and scale-like, with membranous margins, woolly within. Flowers borne on short spikes at the end of lateral young shoots, each with 2 oval-concave bracts, woolly at the base; stigmas 2. Fruiting body (perianth) distinctly winged, to 8 mm across. Dry, semi-arid areas. IP.

Atriplex ORACHE

Herbs and shrubs, often greyish, with flat leaves and inconspicuous flowers. Flowers unisexual, the male flowers with 5 tepals, female flowers with no tepals but 2 bracteoles, enlarged in fruit.

(a) Shrubby or woody-based perennials with greyish leaves, in maritime or desert habitats.

Atriplex semibaccata A short, greyish, woody-based perennial to 40 cm with spreading to prostrate, herbaceous branches. Leaves 8–15 mm, oblong to oval-lance-shaped, wedge-shaped at the base, wavy-toothed. Female flowers grouped around the male, in clusters in the axils, sometimes spike-like; fruiting bracts stalkless, diamond to broadly oval-shaped, fused in the lower half, *fleshy, reddish and berry-like* when mature. Native to Australia; naturalised, especially in the northern Negev. IP.

Atriplex halimus SHRUBBY ORACHE An erect to ascending, woody *shrub* to 2.5 m. Leaves small, to 30 mm long, alternate, *silvery-white*, narrowly oval, slightly diamond-shaped, or angled leaves. Flowers yellowish, borne in leafless branched terminal spikes; *bracteoles 1.5–3 mm, fused only at the base*. Sandy shores, cliffs and estuaries. Throughout (rare in the west). **A. *glauca*** is a similar whitish, shrubby perennial to 50 cm with *prostrate to ascending stems*. Leaves stalkless, 10 x 7 mm. Bracteoles *oval-diamond-shaped* and with protuberances on the surface. Dry, saline habitats. IP.

Atriplex portulacoides (syn. *Halimione portulacoides*) SEA PURSLANE A low shrub, similar in form to *A. halimus*, with spreading and often rooting branches, more or less mat-forming. Leaves upward-pointing, silvery, the *lowermost opposite*, narrowly oval and untoothed. Flowers small and green or reddish, borne in more or less leafless panicles; *bracteoles 2.5–5 mm, fused to just $>1/2$ their length*. Common to abundant in saltmarshes. Throughout.

1. *Atriplex semibaccata*
2. *Atriplex halimus*
3. *Atriplex halimus* fruits
4. *Atriplex glauca*
5. *Atriplex portulacoides*
6. *Atriplex holocarpa*
7. *Atriplex suberecta*
8. *Atriplex prostrata*

(b) Leafy annuals with green or mealy leaves, in various (often disturbed) habitats.

Atriplex holocarpa An erect, greenish to grey-mealy annual to short-lived perennial to 30(40) cm. Leaves broadly oval to circular, to 70 mm long. Flowers unisexual, inconspicuous, borne in clusters in the axils. Fruits with fused, *strongly inflated* bracts, 8–12 mm, so as to appear *bobble-like, spongy* in consistency. Seeds broadly elliptic. Native to Australia, naturalised in disturbed hot, dry areas. IP (south).

Atriplex suberecta A sparsely spreading, mealy, green to ash-coloured annual (or short-lived perennial) to 1 m with leaves 5–11 mm, oval to diamond-elliptic in outline, wedge-shaped at the base, with toothed margins, without basal lobes. Flowers 2–3 mm, clustered in the axils, dispersed all the way up the stem; fruiting bracts diamond to broadly oval, fused in the lower half, with toothed margins. Native to Australia; naturalised, especially in the Negev. IP (mainly central).

Atriplex prostrata (syn. *A. hastata*) SPEAR-LEAVED ORACHE A variable, tall, hairless and branched annual to 2 m, mealy when young. Leaves green–grey with spreading triangular basal lobes, the *largest lower leaves with straight-edged lower lobes at right angles to the leaf stalk*. Flowers reddish, borne in dense clusters in leafy panicles; bracteoles small, entire and triangular, 2–6 mm. Common on sandy shores and disturbed ground. Throughout.

Atriplex patula COMMON ORACHE A very variable, erect or prostrate annual to 1 m with ridged, often reddish stems. Leaves mealy, usually diamond- or arrow-shaped and coarsely toothed below, with *forward-pointing lobes*, the upper leaves linear and slightly toothed. Flowers greenish; *bracteoles 3–7(20) mm, diamond-shaped and scarcely toothed or entire*. Arable and coastal habitats. Throughout, except smaller islands.

(c) Leafy annuals with slivery-white leaves.

Atriplex rosea A *silvery-white* erect or ascending, much-branched annual to 1.5 m with oval to diamond-shaped leaves with wavy-toothed margins, and mainly *lateral leafy flower spikes* all along the stem; bracteoles 2–3(4) mm, pinkish or whitish, diamond-shaped and toothed. Cultivated ground and waste places. Throughout, except smaller islands.

Bassia

Small desert annuals and perennials with adpressed to spreading hairs (often concealing flowers and fruits). Leaves alternate, linear, often succulent. Flowers cosexual or unisexual; tepals 5; stamens 3–5, exserted. Fruit a capsule.

Bassia arabica A yellowish-green perennial subshrub, densely branched, 50–70 cm. Leaves oblong to narrowly oval with blunt tips, 18–25(32) mm. Flowers borne in spike-like inflorescences with solitary stalkless flowers in the axils of leafy bracts; tepals 2.5–3 mm. Fruiting perianth slightly enlarged with papery lobes. Seeds broad, oval, pale to dark brown. Deserts in the southeast. IP (common in the Judean and Negev Deserts; rare elsewhere), SY (Syrian Desert).

Chenopodium GOOSEFOOTS

Herbs or small shrubs with alternate, often *mealy* leaves and inconspicuous cosexual or female flowers; bracteoles absent, tepals 4–5; *stamens 5*. Leaves and seeds are important diagnostics. A complex genus; species described here are the most widespread and easily identifiable. The genera *Oxybasis*, *Dysphania* and *Blitum* were previously grouped within this genus and are very similar but now established to be genetically distinct.

(a) Leaves *heart-shaped* at the base.

Chenopodium hybridum MAPLE-LEAVED GOOSEFOOT A tall and erect hairless annual to 1 m with leaves scarcely mealy and *large, to 15 cm long, oval to triangular with few, long, acute lobes and slightly heart-shaped at the base*. Tepals rounded. Seeds with deep oval pits on the surface. Absent from islands. GR, PN.

(b) Leaves almost *entire*, not heart-shaped at the base; *perianth mealy*.

Chenopodium vulvaria STINKING GOOSEFOOT An erect annual to 40 cm, *not mealy* (except for perianth) and *unpleasant-smelling* when crushed. Leaves small (<25 mm), diamond-shaped with basal lobes but more or less entire. Tepals rounded. Seed *with faint furrows and zygomorphic thickenings*. Saline waste places. Probably throughout.

(c) *Leaves toothed or lobed*, not heart-shaped at the base; perianth mealy.

Chenopodium album FAT HEN A tall, erect, *deep green* but *grey-mealy* plant to 1.5 m. Stems stiff, often marked with red. Leaves oval to diamond-shaped and markedly longer than broad (to 80 mm) and *entire to coarsely and bluntly toothed*. Inflorescence usually leafless at the very top; tepals slightly keeled. Seeds smooth or faintly ridged. Very common on disturbed ground. Throughout. ***C. murale*** is similar and also mealy, occasionally reddish with leaves with coarse, *irregular, forward-pointing teeth* (nettle-shaped). Flowers mealy, with blunt keels on the back of the *toothed tepals*; inflorescence leafy to the top. Seeds minutely pitted. Sandy, disturbed and maritime waste habitats. Throughout.

1. *Bassia arabica*
2. *Chenopodium murale*
3. *Dysphania ambrosioides*
4. *Beta maritima*
5. *Salicornia procumbens*

Dysphania

Very similar to *Chenopodium* but markedly *glandular*-hairy, aromatic and sticky.

Dysphania ambrosioides An annual (or short-lived perennial) to 2 m, hairy, glandular and aromatic. Leaves lance-shaped, entire or toothed. Flowers borne in panicles with stalkless upper branches, usually with bracts; sepals free almost to ½ their length, rounded on the back. Seeds 0.5–0.8 mm across. Coastal. Probably throughout.

Dysphania botrys (syn. *Chenopodium botrys*) STICKY GOOSEFOOT An erect, long stemmed, *sticky* glandular-hairy aromatic annual with alternate, oblong, *pinately divided, rather oak-like leaves*. Flowers small and green, in long, narrow clusters. Mainly in the north. AA (rare), GR, PN (local).

Blitum

Sparsely mealy perennials. Flowers arranged in terminal inflorescences; stamens 4–5.

Blitum bonus-henricus (syn. *Chenopodium bonus-henricus*) GOOD KING-HENRY An erect, branched, rhizomatous *perennial* to 50 cm with stiff stems that are often streaked red. Leaves up to 10 cm long and *broadly triangular* with basal lobes, otherwise entire or shallowly lobed. Inflorescence a terminal, narrowly pyramidal raceme. Damp or fertilised waste places and other disturbed habitats. GR, PN.

Beta BEET

Erect or ascending annuals or perennials with swollen roots and with alternate, not mealy leaves and cosexual flowers with 5 tepals.

Beta vulgaris SEA BEET A variable, hairless, *reddish* and rather fleshy annual or perennial to 1.5 m, but often prostrate. Leaves oval-lance-shaped, dark green, leathery and shiny, more or less untoothed, to 20 cm. Flowers small and green or purplish, borne at the top of dense, long spike; stigmas mostly 2; lower bracts 2–20 mm; bracts absent above. Common on maritime sands, cliffs and shingle. Throughout. *B. maritima* (syn. *B. vulgaris* subsp. *maritima*) is treated by some authors as a distinct species, differing mainly in its sprawling (sometimes erect) habit, and uppermost bracts 10–20(35) mm, or absent. Throughout the range of the species.

Salicornia GLASSWORT

Saline succulent herbs with slender, finger-like branches and minute leaves and flowers. Flowers in groups of 1–3, the central extending to make the group triangular. A difficult and poorly understood genus with weak morphological distinction between species, and the same species given different names across regions; therefore the true number of taxa in the region not known.

Salicornia europaea COMMON GLASSWORT An erect, much-branched perennial to 35 cm with fleshy, articulated, sub-cylindrical stems, clear green, becoming yellowish, suffused with pink or red with age; branches straight, the terminal spike (5)10–30(40) mm; lower fertile segments 1.9–3.5 mm, 2–4 mm wide at the narrowest point. Flowers with 1(2) stamens, the central flower larger than the 2 laterals; fertile segments with convex sides. Saltmarshes. Apparently widespread but the true distribution not known (in its strictest sense *S. europaea* may be an Atlantic species). *S. procumbens* is very similar but with all 3 flowers the same size; fertile segments with straight (to slightly convex or concave) sides. GR (northeast). *S. perennans* is difficult to distinguish from the previous species, and is possibly the common form in much of Greece and Turkey.

Sarcocornia (including *Arthrocnemum*)

Similar to *Salicornia* but *woody-based, shrubby* perennials; flowers minute, in groups of 3 in a row.

Sarcocornia perennis (syn. *Arthrocnemum perenne*) A small, very succulent rather *short, spreading shrub to 30(70) cm with creeping, subterranean stems, at first dark green*, ageing red–brown or yellowish. Leaves opposite, scale-like and fused in pairs, appearing as segments. Flowers tiny, borne groups of 3, each with 2 stamens; flowers falling to leave 3 scars in the segment. Maritime sands. AA, CR, CY, GR (southeast + adjacent islands), IA, IP (Acco plain). ***S. fruticosa*** (syn. *Arthrocnemum fruticosum*) is very similar but with stems *erect to 1 m*, stout and non-rooting, and *blue–green*. Saltmarshes and coasts; absent from many islands. AA, CY, GR, IA, IP (Acco plain and Dead Sea Valley).

Suaeda SEABLITE

Herbs or small shrubs in saline habitats with fleshy, alternate, linear leaves and minute flowers with 5 tepals and 2–3 bracteoles; flowers cosexual and female.

(a) Plant a woody-based shrub.

Suaeda vera SHRUBBY SEABLITE A succulent, woody shrub to 1.2 m, leaves *densely crowded, blunt and fleshy*, alternate, small, 5–18 mm, semi-cylindrical, sessile and blue–grey, becoming reddish or purplish. Flowers small, greenish, borne 1–3 in the axils of the upper leaves; stigmas 3. Maritime slopes and saltmarshes. Throughout, except eastern AA.

1. *Suaeda vera*
2. *Salsola kali*
3. *Salsola vermiculata* flowers
4. *Salsola vermiculata* in fruit
5. *Salsola tetrandra*

(b) Plant a succulent, annual herb.

Suaeda maritima ANNUAL SEABLITE A succulent annual to 30(75) cm with prostrate or ascending stems and distinct, *pointed, alternate, fleshy leaves* 3–25 mm long, ranging from blue–green to reddish or purple; *leaves slightly concave above in cross section*. Flowers minute, green and borne in the leaf axils stigmas 2. Seeds ›1.5 mm long. Muddy coastal habitats. GR (north and west + adjacent islands), IA, PN.

Salsola SALTWORT

Herbs or shrubs with fleshy, cylindrical leaves, and small cosexual flowers borne in the leaf axils. Tepals 5, often developing a transverse wing on the back in fruit.

(a) Plant herbaceous, with leaves and bracts sharp-pointed or spiny.

Salsola kali PRICKLY SALTWORT A variable, hairy annual with a prostrate, spreading or erect habit to 50 cm (1 m). Leaves linear and narrow, 10–40(70) mm long, blue–green, *spine-tipped* and broadest at the base; opposite below. Flowers small and 5-lobed, solitary in the leaf axils with a pair of stiff, spiny bracts, which exceed the flowers. Coastal environments. Throughout (if treated in its wider form). ***S. soda*** is similar but hairless, with ½-cylindrical, clasping, *not spine-tipped* leaves 15–75 mm long, and flower bracts equalling the flowers. Similar habitats. Scattered throughout except many islands (absent from CR, LB, SY).

(b) Plant a shrub or subshrub, with leaves and bracts *not* spiny.

Salsola vermiculata A succulent *shrub* to 1 m with *very small, alternate, crowded leaves* 5–12(25) mm long, semi-cylindrical, oval at the base and semi-clasping the stem, *often pubescent*. Flowers borne in inflorescences with regularly arranged primary branches 20–40 cm long; flowers small and green or pink-tinged; fruiting perianth to 12 mm across, stigmas shorter than the style; sepals with a dorsal wing in fruit. Dry, saline habitats. IP. ***S. tetrandra*** is similar in form, with minutely white-woolly, blue–grey, *opposite* leaves. Sepals without a dorsal wing in fruit. Semi-desert and saline habitats. IP.

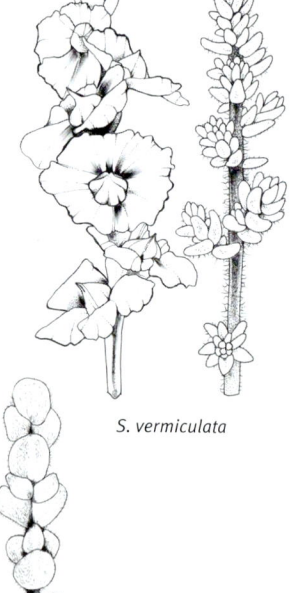

S. vermiculata

S. tetrandra

Salsola aegaea (syn. *Caroxylon aegaeum*) is similar to the previous species but to ‹30(60) cm with much shorter branches, lower leaves to 5(10) mm, virtually hairless. Inflorescence with *short* primary branches to just 10 cm, irregularly branched. Island sea cliffs. AA, CR.

Maireana

Woody, succulent shrubs native to Australia. Fruiting perianth with horizontal wings (superifically like that of *Salsola*).

Maireana brevifolia COTTON BUSH An erect, succulent, virtually hairless shrub 20 cm–1 m with woody, ridged stems. Leaves alternate, fleshy, cylindrical to ovoid, 2–5 mm, narrowing into short stalks, grey–green or purplish. Flowers borne solitary in the leaf axils, green, cosexual; fruit to 3 mm, *prominent*, surrounded by a fruiting perianth, 5 spreading, often overlapping winged bracts, pale green, pinkish when mature, *small*, 6–8 mm across. Native to Australia, naturalised. Rare, in semi-arid habitats; perhaps under-recorded. IP (centre).

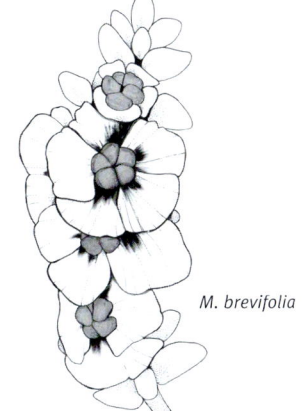

M. brevifolia

AIZOACEAE

Herbaceous annuals or perennials with simple, opposite or alternate, succulent leaves. Flowers regular with numerous linear 'petals' (derived from staminodes, described as petals below); sepals 4–5(6); stamens 3–numerous. Fruit a capsule with (1)3–numerous seeds.

Aizoon

Annual or perennial herbs with a crystalline (*papillose*) surface. Leaves (mostly) alternate. Flowers with 4–5 tepals, often fused below into a short tube; ovary superior. Fruit a capsule with numerous seeds.

Aizoon canariense A short, slightly succulent annual to 30 cm, sprawling when mature. Leaves 40 mm wide, opposite below, alternate above, oblong-lance-shaped, blunt and minutely hairy. Flowers solitary and unstalked with 5 yellow tepals 1–3 mm long and yellow stamens. Semi-desert habitats. IP (mainly Dead Sea Valley). *A. hispanicum* has leaves just 10 mm across and tepals yellow or whitish, 7–15 mm long. Rare in the area; bare sand and deserts. CR (central, south + Gavdos), IP (south and east).

Carpobrotus HOTTENTOT FIG

Fleshy perennials with leaves triangular in cross section. Flowers with numerous petals and 8–20 stigmas. Fruit succulent, the seeds embedded in mucilage.

Carpobrotus edulis HOTTENTOT FIG A succulent, trailing, mat-forming perennial with woody stems to 2 m long. Leaves opposite, 3-angled and upwardly curving, finely toothed along the edge and tapered towards the apex, 40 mm–9(13) cm long. Flowers large, 8–10 cm across, solitary, virtually stalkless, bright pink or yellow with yellow stamens. Fleshy, edible fruit. Native of South Africa, naturalised and invasive in some coastal areas. Throughout.

Malephora

Succulent perennials native to Africa. Leaves opposite and 3-angled. Flowers showy with numerous (65) petals and (150) stamens. Fruit an 8–12-parted capsule.

Malephora purpureocrocea A prostrate, spreading succulent perennial with stems with long internodes and *a white–grey bloom*. Leaves up to 50(60) mm long, opposite, cylindrical and bluntly 3-angled. Flowers typically 50 mm across, with numerous (to 65) bright pink petals and yellow–orange stamens. Occasionally planted in coastal areas.

1. *Maireana brevifolia*
2. *Aizoon canariense*
3. *Aizoon hispanicum*
4. *Malephora purpureocrocea*
5. *Mesembryanthemum crystallinum*
6. *Mesembryanthemum nodiflorum*
7. *Phytolacca acinosa*
8. *Boerhavia helenae*

Mesembryanthemum

Succulent annuals or biennials *densely covered in crystal-like vesicles*. Flowers with numerous stamens. Capsule with 4(5) valves.

Mesembryanthemum crystallinum ICE PLANT A spreading annual covered in glistening, frost-like *crystalline hairs*. Leaves stalked, flat, fleshy, oval, alternate and untoothed, to 11.5 cm long. Flowers solitary, 20–30 mm across, virtually stalkless, yellowish or whitish. In sandy, rocky and saline environments. AA, CR, GR (southeast), IP. *M. nodiflorum* is similar but less crystalline and with *cylindrical-linear leaves* 10–25(30) mm long, red-tinged; flowers smaller, to 15 mm across. Similar habitats. AA, CR, GR, IP, PN.

Mesembryanthemum cordifolium (syn. *Aptenia cordifolia*) A mat-forming perennial to 60 cm high with stems to 3 m, with slightly fleshy, bright green, oval leaves, which are pointed at the apex and heart-shaped at the base, 13–56 mm long. Flowers bright magenta, 10–18 mm across. An ornamental native to South Africa, widely planted. Throughout.

PHYTOLACCACEAE

Herbaceous plants, shrubs and trees with alternate, entire leaves. Flowers borne in leaf-opposed racemes; perianth 5(4–9)-parted and persistent in fruit; stamens 5–25(30); ovary composed of 5–15 1-seeded carpels. Fruit succulent, berry-like and lobed.

Phytolacca

Herbs or shrubs. Flowers unisexual or cosexual; stamens 5–25(30); carpels 5–15. Fruit a berry.

Phytolacca acinosa POKEWEED A tall, leafy, herbaceous perennial to 1.5(3) m with thick, ribbed stems, often reddish. Leaves large and oval, 12–20(25) cm long. Flowers green or pink, borne in erect, cylindrical spikes; stamens and styles 7–10, tepals more or less equal. Fruits borne in clusters of red, later black berries 4 mm long, borne in dense cylindrical spikes. Native to North America, widely naturalised. (*P. americana* is often recorded, possibly in error; it has narrower leaves and long, arching racemes.)

NYCTAGINACEAE

Trees, shrubs and woody climbers native to South America with opposite leaves. Flowers with 5 showy bracts, cosexual; stamens 5; style 1. Fruit surrounded by the perianth tube, forming a false fruit.

Boerhavia

Succulent perennial herbs or shrubs with opposite, entire leaves. Flowers cosexual with 5 lobes; stamens 2–5. Fruit a 10-ribbed, club-shaped anthocarp (calyx + enclosed seed). Some authors treat *Commicarpus* (which has fruits with large, wart-like glands) as distinct.

Boerhavia helenae (syn. *Commicarpus helenae*) A straggling, branched perennial with hairless, whitish stems. Leaves oval to rounded with undulating margins, and stalks to 15 mm. Flowers borne 3–5 per node in very lax, rigid inflorescences, small, 6–6.5 mm long, pinkish red; flower-stalks short and stout (1–4 mm). Fruit an antocarp 5–7 mm, with a *ring of large wart-like glands* around the apex. Desert and semi-desert habitats. IP (Dead Sea Valley).

Bougainvillea

Vigorous climbing South American shrubs with flowers with showy bracts; stamens 8. Fruit a 5-lobed achene.

Bougainvillea glabra BOUGAINVILLEA A vigorous woody, virtually hairless climber to 10 m. Leaves opposite or alternate, to 60 mm long, untoothed and stalked. Flowers in groups of 3, inconspicuous, whitish-yellow and funnel-shaped, surrounded by *large, flower-like crimson bracts*, 1 to each flower. Very commonly planted. *B. spectabilis* is very similar but with leaves with a densely hairy underside, and flowers with straight hairs and a purplish, not greenish outer corolla. Bracts red, orange or violet. Planted in the region.

Mirabilis

South American tuberous perennials typically with fragrant, deep-throated flowers; stamens 3–5. Fruit a 0–5-angled achene.

Mirabilis jalapa MARVEL OF PERU A hairless or short-hairy perennial to 1.5 m with large, oval, untoothed, leaves narrowing to a point; rather wrinkled, stalked below and short-stalked to sessile above. Flowers borne in terminal clusters, variably white, yellow, pink, red or variegated; tube 25–35 mm long, 5-lobed. Native to tropical America, widely planted and naturalised.

PORTULACACEAE | BLINKS FAMILY

Annual or perennial herbs, often fleshy, with cosexual flowers with only 2 opposite sepals and 4–6 petals; stamens 3–14; styles 1–3(6). Fruit a capsule.

Montia

Stem-leaves in several pairs, opposite or alternate. Flowers in terminal or axillary groups, white or pink; stamens 3–5; style 1, ovary superior. Fruit normally a 3-seeded capsule.

Montia fontana BLINKS A small, bright green, dense, patch-forming annual or perennial 10 mm–20(50) cm with oval, opposite, spatula-shaped stem-leaves (basal leaves absent). Flowers white, tiny, borne in small loose clusters with stems that lengthen in fruit; petals <2 mm, scarcely longer than the 2 sepals. Local, in wet pastures. AA (north-central), GR (north).

Portulaca PURSLANE

Herbaceous annuals with 1–few, stalkless flowers; stamens numerous and ovary ½ inferior. Fruit a many-seeded capsule.

Portulaca oleracea PURSLANE A fleshy, prostrate or ascending, patch-forming, leafy annual with stems to 30(50) cm, Leaves to 30 mm long, spoon-shaped and more or less opposite and densely crowded beneath the flowers. Flowers yellow with 4–5(6) petals 4–8 mm long, which soon fall, revealing blunt sepals. Very common in bare or exposed habitats. Throughout.

1. *Bougainvillea spectabilis*
2. *Hylocereus undatus*
3. *Opuntia maxima* fruits
4. *Opuntia maxima*
5. *Opuntia dillenii*
6. *Cornus mas*

CACTACEAE | CACTUS FAMILY

A family of (typically) spiny succulents from the Americas, often with showy flowers. Cultivated and naturalised in the region.

Hylocereus

Succulent plants from Tropical America. Stems succulent, 3-angled. Flowers opening at night, white with numerous tepals; stamens numerous. Fruit fleshy, many-seeded.

Hylocereus undatus DRAGON FRUIT A robust cactus-like perennial to 2 m with yellowish-green stems with spiny, undulate margins. Flowers conspicuous, about 25–30 cm across, white, with greenish outer tepals. Fruit reddish with a white pulp and >200 seeds. Planted in hot dry areas, sometimes as a hedge.

Opuntia PRICKLY PEAR

Woody cacti with jointed stems and soon-falling leaves. A confused genus with conflicting names and descriptions in the literature (morphological traits are often continuous, or not clear or useful in the field). Native to North and South America. Further work is required to accurately resolve the genus in the Mediterranean.

(a) Stem segments usually *without stout spines*. Flowers yellow.

Opuntia maxima (syn. *O. ficus-indica*) PRICKLY PEAR A robust, blue–green cactus 5–6 m; woody and trunk-like at the base (rather tree-like) with *large*, flattened jointed stem segments 30–50 cm long and soon-falling, inconspicuous leaves. Bristles hooked and yellowish, *straight spines usually absent*. Flowers bright yellow. Fruit egg-shaped, 50–60(90) mm long, yellow or reddish when ripe; edible. Widely planted, the most common naturalised species in the region. Throughout (often described under the apparently misapplied name, *O. ficus-indica*).

(b) Stem segments generally with conspicuous, *stout spines*. Flowers yellow.

Opuntia dillenii A shrub to 1.5 m with blue–green stem segments 70 mm–40 cm long, *with 1–10 conspicuous, stout, yellow spines of various length*. Flowers yellow. *Fruits 50–75 mm, purple–red*. Planted in hot, dry areas. *O. stricta* has a *spreading habit* with erect branches, *shorter* than the previous species, to 1 m high; stem joints 20–30 cm; spines normally present. Fruits reddish. Planted in hot, dry areas.

CORNACEAE | CORNUS FAMILY

Trees or shrubs with simple, opposite leaves. Flowers borne in axillary or terminal inflorescences, cosexual; sepals, petals and stamens 4; ovary inferior, formed of 2 carpels. Fruit a berry or drupe.

Cornus DOGWOOD

Shrubs with opposite leaves, cosexual flowers in clusters and succulent fruits (drupes).

Cornus mas CORNELIAN CHERRY A deciduous shrub or tree to 8 m with oval to elliptic, pointed, untoothed leaves 40 mm–10 cm. Flowers bright *yellow*, small, 4–5 mm across, borne in congested clusters of 10–25 *before the emergence of the leaves*. Fruit ovoid, 12–15 mm long, shiny red when ripe. Mountain woods. GR (north), IA (Corfu – rare), PN (north), TR.

Cornus sanguinea DOGWOOD A deciduous shrub 1.5–5 m with erect, *dark red twigs* and elliptic to oval, pale green, hairy, untoothed or toothed leaves. Flowers dull *white*, 7–11 mm across, borne in dense, umbel-like clusters. Fruit rounded, 5–8 mm across, *purplish-black* when ripe. Woods, scrub and ditches. Distribution as above.

BALSAMINACEAE | BALSAM FAMILY

Herbaceous hairless annuals with simple leaves and translucent, fleshy stems. Flowers strongly zygomorphic; sepals 3, petal-like; stamens 5; style 1. Fruit a capsule that violently expels seeds when ripe.

Impatiens

Herbaceous annuals with succulent stems swollen at the nodes and opposite or alternate leaves. Flowers zygomorphic. Fruit an explosive capsule.

Impatiens balfourii An annual herb to 1.2 m with hairless, reddish-translucent, fleshy stems. Leaves alternate, oval-lance-shaped, toothed, stalked, to 40 mm long. Flowers borne in racemes, to 20 mm long, pink and white, the lower white sepal forming a long, slender spur >10 mm long. Native to the Himalayas but naturalised in towns. GR.

PRIMULACEAE | PRIMULA FAMILY

Herbaceous annuals or perennials. Flowers regular with free to fused sepals; petals 5(–9), fused (sometimes only just, seemingly free); stamens usually 5; ovary with a single style. Fruit a capsule.

Lysimachia

Erect or spreading perennials, or dwarf shrubs. Leaves flat, opposite (rarely alternate). Flowers often borne in *panicles or racemes* or solitary in the leaf axils; calyx with 5 teeth; corolla with short tube and 5 lobes; stamens 5. Fruit a capsule with 5(7) valves; seeds numerous.

(a) Flowers pink or purple, borne in long, spike-like racemes.

Lysimachia atropurpurea An erect annual 20–65 cm, densely glandular-hairy. Leaves alternate, stalkless, 30 mm–11 cm long, spatula- or linear-lance-shaped, dotted with reddish glands. Flowers stalkless, borne in terminal spikes; bracts 2.5–5.5 mm; calyx lobes linear-lance-shaped, glandular-hairy and gland-dotted; corolla *dark purple*, hairless, with lobes 5.5–7.5 mm and with *long-protruding stamens*; style spine-like on ageing. Capsule 4–5 mm, hairess, gland-dotted. Maquis, fields, roadsides. AA (northeast: Lesbos, Limnos, Samos, Samothraki), GR, PN, TR. **L. dubia** is similar, taller (to 1.1 m), often hairless, or sparsely hairy, with *pink flowers* with lobes 5–7 mm; style not spine-like when mature. Capsule 2–3 mm without dotted glands. Wet habitats and pine forests. AA (Rhodes + Lesbos), GR, SY, TR.

(b) Flowers yellow, borne solitary in the leaf axils.

Lysimachia serpyllifolia A dwarf shrub with spreading to ascending stems. Leaves opposite, small, oval, 6–11 mm, scarcely stalked. Flowers bright yellow and *pimpernel-like*, 6–8.5 mm across, long-stalked (at least 2 x as long as subtending leaves), borne solitary in the leaf axils. Mountain scrub and stony ground. CR, GR (Sterea Ellas), PN.

Anagallis PIMPERNEL

Hairless annuals or perennials with opposite or alternate, simple leaves. Flower with 5 corolla lobes exceeding their tube. Fruit a splitting capsule. DNA data indicate the species below should be classed with *Lysimachia*, but they are still described under *Anagallis* in most floras.

Anagallis arvensis (syn. *Lysimachia arvensis*) SCARLET PIMPERNEL A low, weedy annual with prostrate or ascending 4-angled stems to 40 cm. Leaves often >10 mm, oval, opposite and gland-dotted; unstalked. Flowers *blue, red or orange*, 4–7(10) mm, long-stalked (35 mm) and becoming curved in fruit; *petals with minutely hairy margins*, sometimes toothed at the tip. Very common in towns, waste places and near the coast. Throughout. *A. foemina* (syn. *Lysimachia foemina*) is often treated as a subspecies, but is apparently genetically distinct, and has flowers always blue, with narrower petals, *without hairy margins*; corolla 5–8 mm. Probably throughout.

PRIMULACEAE

Primula PRIMROSE

Perennial herbs with leaves in basal rosettes. Flowers borne in umbels (rarely solitary); calyx bell-shaped or cylindrical, with 5 teeth; corolla with cylindrical tube and 5 teeth; stamens not protruding. Capsule with valves; seeds numerous.

Primula veris COWSLIP A short-hairy, early-flowering perennial with leaves in a lax rosette, 50 mm–20 cm (enlarging in fruit), oval to oblong, wider at the base, and abruptly contracted into the stalks; stalks sometimes winged. Flowers borne in umbels on hairy stalks 10–30 cm; calyx 8–20 mm with triangular lobes; corolla 8–28 mm across, bright to deep brownish-yellow with orange spots at the base of the lobes. Common on hillsides and in pastures in the north; rare or absent elsewhere. AA (Thasos), GR, PN, TR.

Primula vulgaris PRIMROSE An early-flowering shaggy-hairy perennial with rosettes of irregularly toothed, egg-shaped leaves, hairless above. Flowers borne singly on stalks to 15 cm (rarely in an umbel); corolla pale yellow, 20–40 mm across; calyx 10–20 mm, tubular. Moist rocky hillsides, often in or near seeps and streams. Throughout, except arid areas in the east and most islands. **Subsp.** *vulgaris* has leaves that gradually narrow into the short, winged stalk and pale yellow flowers. Throughout the range. **Subsp.** *rubra* has leaves more wedge-shaped and abruptly contracted into their stalks at the base, and *pale pink to red–pink* flowers. Often near meltwater. AA (Andros, Ikaria), GR (north, east).

Samolus

Hairless perennials with leaves in basal rosettes. Flowers white with 5 sepals and corolla lobes. Capsule with 5 teeth.

Samolus valerandi BROOKWEED A hairless perennial herb with a rosette of leaves 10–50 mm long, and erect flowering stems 50 mm–70 cm. Leaves rather shiny, spoon-shaped and scarcely stalked below; stalkless above. Flowers small and white, cup-shaped, to just 3 mm across and with 5 petals. Fruit 3 mm. Damp rocks and seeps, often coastal; local. Throughout.

Coris

Flowers reddish-violet, corolla distinctly 2-lipped with 3 longer upper lobes and 2 shorter lower lobes; calyx with 10 or more lobes.

Coris monspeliensis A distinctive, short annual or biennial with ascending to erect stems 10–25(40) cm, woody at the base. Leaves 4–9 mm, very narrowly linear (almost needle-like), densely arranged, unstalked and alternate up the stem. Flowers borne in dense terminal heads; corolla 9–16 mm, pinkish blue and 2-lipped and divided into unequal lobes; stamens 5. Fruit 1–2.5 mm. Dry, rocky habitats in the west; rare and local. IA (Zakynthos), GR, PN.

---

1. *Anagallis arvensis*
2. *Primula veris*
3. *Primula vulgaris*
4. *Primula vulgaris* subsp. *rubra*
5. *Samolus valerandi*
6. *Coris monspeliensis*

Cyclamen CYCLAMEN

Perennials with simple leaves and solitary flowers with 5 *strongly deflexed petal lobes*. Capsule 5-valved. Previously classified in the family Myrsinaceae, but most recent DNA analysis confirms that the genus is in the Primulaceae.

(a) Autumn (to winter)-flowering. Fruit stalks coiling from the top downwards.

Cyclamen cilicium An autumn-flowering perennial with small, smooth tubers to 50 mm across. Leaves oval with heart-shaped bases, unlobed and scarcely toothed, 14–56 mm across, marbled grey or cream. Flowers fragrant, pink (or white) to crimson; petal lobes 14–19 mm long, strongly twisted. Fruit stalks coiling from the top downwards. Scree among conifers. TR (south). *C. mirabile* is similar, but with tubers with a corky surface, leaves marbled with red when young, and the petal-lobes toothed towards the apex. Pine forest and maquis. TR (southwest).

Cyclamen cyprium An autumn- to winter-flowering perennial with greyish tubers to 50 mm rooting from one side. Leaves broadly heart-shaped, toothed or not, 40 mm–14 cm across, with pale marbling. Flowers appearing before, or with immature leaves, fragrant, white to pinkish with a dark magenta spot at the base of each lobe; lobes twisted. Fruit stalk coiling from the top downwards. Rare, in rocky, shady habitats among trees and srubs. CY (west and north).

Cyclamen hederifolium IVY-LEAVED CYCLAMEN An autumn-flowering perennial with tubers rooting from the upper surface. Leaves normally *conspicuously angled or lobed*, slightly toothed, pointed, grey or cream-marbled 25–80 mm long, produced during or shortly after flowering; leaves and flowers extending outwards in bud. Flowers pale pink or white with purple in the throat; lobes (12)14–25 mm long, with crimson–purple at the base; style scarcely protruding. Fruit stalks coiling from the top downwards. Deciduous woods and stony garrigue. AA, GR, IA, PN, TR. *C. confusum* (considered by some authors to be indistinct) has wide-mouthed flowers and leaves with fleshy, shallow-lobed margins but is otherwise similar. CR (rare – west).

(b) Spring-flowering. Fruit stalks coiling from the middle or base.

Cyclamen graecum An autumn- to winter-flowering perennial similar to *C. hederifolium* with corky tubers 30 mm–10 cm across, rooting only from the base. Leaves appearing after flowering, 30–70 mm (10 cm) across, with heart-shaped bases and minutely toothed margins, scarcely or *not angled or lobed*. Flowers white or pale pink with dark purple throat markings; lobes 20 mm. Fruiting stalk coiled from the base or middle. AA, CR, CY (north), GR, PN, TR.

(c) Spring-flowering. Fruit stalks coiling from the top downwards.

Cyclamen repandum A spring-flowering perennial with smooth, later corky, tubers to 45 mm across. Leaves with heart-shaped bases, often lobed, 43 mm–11 cm across with shallow to coarsely toothed margins, marbled or speckled with white or grey–green. Flowers fragrant, appearing with the leaves, pale to dark pink, sometimes with darker markings around the mouth; lobes broadly elliptic, 17–31 mm. Fruit stalk coiling from the top downwards. **Subsp. *repandum*** has dark green leaves with broad, irregular grey markings and deep pink flowers with lobes 17–24 mm. Deciduous

1. *Cyclamen cilicium*
2. *Cyclamen mirabile*
3. *Cyclamen cyprium*
4. *Cyclamen hederifolium*
5. *Cyclamen graecum*
6. *Cyclamen repandum* subsp. *peloponnesiacum*
   PHOTO: ELEFTHERIOS DARIOTIS
7. *Cyclamen repandum* subsp. *rhodense*
   PHOTO: ELEFTHERIOS DARIOTIS
8. *Cyclamen creticum*
9. *Cyclamen libanoticum*

PRIMULACEAE | 407

scrub. GR, IA. **Subsp. *peloponnesiacum*** has deep green, shiny leaves with paler spots and *large*, pale to dark pink flowers with lobes 20–30 mm long. Woods. PN. **Subsp. *rhodense* (syn. *C. rhodense*)** has greyish leaves with paler markings and *white* to pale pink flowers with rose–pink throats. Pine woods. AA (Rhodes + Kos). *C. creticum* is similar but with *smaller* (normally) *white* flowers with lobes 15–26 mm and leaves 35–90 mm with rather sharply toothed margins. Shady habitats, among boulders, tree roots and by old walls. CR (+ KP).

Cyclamen libanoticum A spring-flowering perennial with a velvety (later corky) tuber to 45 mm across. Leaves with heart-shaped bases, 40–80 mm across with paler marbling, and angled lobes but untoothed margins. Flowers appearing with mature leaves, fragrant, white turning pink; lobes broadly oval, scarcely twisted. Fruit stalk coiled from the top downwards. LB (mountains northeast of Beirut).

Cyclamen coum A spring-flowering perennial with a tuber 20–35 mm across. Leaves 30–70 mm across, with or without paler marbling, kidney-shaped to circular with entire to scarcely undulate toothed margins. Flowers deep pink with a *darker, pale-centred blotch* at the base of each lobe; lobes 7–15 mm. Fruit stalk coiled from the top downwards. Deciduous and evergreen woodland. IP (north – rare), LB, TR.

(d) Spring-flowering. Fruit stalks curved, not coiled.

Cyclamen persicum PERSIAN CYCLAMEN A winter- to spring-flowering tuberous perennial with heart-shaped, finely and bluntly toothed (not angled) green or marbled leaves to 14 cm across, purplish beneath. Flowers scented, white, pink or mauve, with lobes (20)25–37(45) mm long, *markedly twisted*; flowers appear with mature leaves. Fruit stalks curved but *not coiled*. Rocky garrigue in east and southeast. AA (southeast), CR (northeast – rare), CY (common), IP, SY, TR (south).

STYRACACEAE | STORAX FAMILY

Mainly shrubs and trees with spirally arranged simple leaves without stipules. Flowers regular, perianth with 4–5(7) lobes; stamens (5)8–10(20); styles and stigmas 1. Fruit a dry capsule, sometimes winged, or drupe.

Styrax

Shrubs or small trees with flowers in short racemes; calyx bell-shaped, entire or lobed; stamens 8–16. Fruit a dry drupe with 1–2 large seeds.

Styrax officinalis STORAX A deciduous shrub or tree 2–5(7) m with white-hairy twigs. Leaves 50 mm–10 cm, alternate, oval to oblong, untoothed and short-stalked. Flowers *white*, bell-shaped and pendent, 18–22 mm, borne in clusters of 3–6, fragrant; petals 5–6, fused at the base; stamens yellow and prominent. Fruit 1-seeded, ovoid, white-woolly with a persistent calyx. Ditches, woods and gorges; widespread but strongly southern and eastern. AA (patchy), CR, CY, GR (rare), IP, LB, PN (rare), SY, TR.

ERICACEAE | HEATHER FAMILY

Evergreen trees and shrubs with alternate, opposite or whorled leaves, or herbaceous perennials. Flowers often borne in clusters (rarely solitary); sepals and petals (3)4–5, petals fused; stamens 2 x as many as petals; style 1. Fruit a capsule, drupe or berry.

Monotropa

Mycoheterotrophic perennials devoid of chlorophyll. Leaves reduced to scales. Flowers in terminal racemes.

Monotropa hypopitys YELLOW BIRD'S NEST A short, yellowish perennial to 30 cm. Stems *nodding* at the apex in bud, erect in fruit, with numerous scale-leaves to 13 mm below. Flowers pale, dull yellow; petals (6)8–13 mm, recurved at the tips; stamens exceeded by the petals. Fruits more or less spherical. Frequent in mountain deciduous woods in the northern Balkan Peninsula; rare or absent elsewhere. GR, IA, PN.

---

1. *Cyclamen coum*
2. *Cyclamen persicum* pink form; (inset) white form
3. *Styrax officinalis*
4. *Monotropa hypopitys*; (inset) fruits

Arbutus

Evergreen shrubs with alternate leaves. Flowers with 5 petals, borne in terminal clusters; stamens 10. Fruit a warty berry.

Arbutus unedo **STRAWBERRY TREE** An evergreen shrub or small tree 4–5(11) m. Bark dull brown and fissured; young twigs at least partially *downy*. Leaves 80 mm long, oblong-lance-shaped, short-stalked and somewhat toothed, more or less hairless. Flowers scented; white, tinged green or pink, bell-shaped with recurved petal lobes, 7–8(11) mm, borne in drooping panicles in *autumn*. Fruit a spherical berry 7–15(20) mm, ripening deep crimson; strawberry-like. Common on maquis and acidic woodland. Throughout (rarer in the east). *A. andrachne* is similar but with smooth, papery, *peeling* and *orange–red* bark; young shoots hairless. Leaves <2 x as long as wide, often entire. Flowers borne in *spring*. Evergreen maquis and lower mountain slopes. Throughout.

1. *Arbutus unedo*
2. *Arbutus andrachne* habit
3. *Arbutus andrachne*
4. *Erica arborea*
5. *Erica manipuliflora*
6. *Erica multiflora*
7. *Rhododendron luteum*

Erica HEATHER

Shrubs with whorled, narrow to linear leaves. Flowers bell-shaped to spherical, borne in spikes or panicles, generally 4-lobed; stamens 8. Fruit a dry capsule.

(a) Stamens included in the corolla; shrub >1 m tall.

Erica arborea TREE HEATH A tall shrub or small tree to 4(7) m with densely hairy young twigs. Leaves in groups of 3–4, to 4–9 mm long. Flowers pure *white*, broadly bell-shaped, 2–4 mm long with erect lobes, borne in dense terminal panicles; sepals 1.2–2 mm. Fruit capsule 2 mm. Common to abundant in a range of habitats, often above sea level. Throughout, except the far southeast.

(b) Stamens longer than the corolla with anthers projecting.

Erica manipuliflora A spreading to ascending shrub to 50(75) cm with flexuous stems. Leaves 4–8 mm in whorls of 3–4. Flowers pink; sepals 1 mm; corolla 3–3.5 mm, broadly bell-shaped with erect lobes; anthers with divergent lobes. Evergreen scrub and maquis. Scattered more or less throughout (not IP).

Erica multiflora An erect, branched shrub to 1(3) m with leaves in whorls of 4–5, linear, 6–11 mm long, and dense, rounded, terminal clusters of pale pink flowers, 3.5–5.5(7) mm, *borne on long stalks, up to 3 x as long as the calyx* (sepals 1.5–2 mm); anthers completely projecting. Capsule 2–2.5 mm. Garrigue and maquis. IA (Cephalonia, Lefkada, Zakynthos).

Rhododendron

Evergreen shrubs with alternate leaves. Flowers showy, with 5 petals fused to form a bell-shaped corolla; stamens 5 or 10. Fruit a capsule.

Rhododendron luteum YELLOW RHODODENDRON A deciduous shrub 1–3 m. Leaves short-stalked, lance-shaped to elliptic with ciliate margins, sparsely hairy beneath when young. Flowers large, funnel-shaped, 30–50 mm across, fragrant and dark yellow. Fruit a woody capsule to 20 mm. Damp coniferous woods; rare and local in the region as a native. AA (Lesbos), TR.

RUBIACEAE | MADDER FAMILY

A large family of herbs with opposite or distinctly whorled leaves; stipules present between each pair of leaves, often leaf-like. Flowers typically small, funnel-shaped and tubular, borne in dense heads, branched cymes or panicles; sepals 0 or minute, petals 4–5; stamens 4–5; styles 1–2 (if 1, branched); ovary inferior. Fruit fleshy, dry or berry-like, 1–2-seeded.

Rubia

Scrambling perennials or shrubs with leaves in whorls of 4, equal and stalkless. Flowers with 4-lobed, yellow corolla; the terminal cosexual, the laterals male. Fruit a pair of smooth nutlets.

Rubia peregrina WILD MADDER A trailing or scrambling hairless, evergreen perennial to 7 m with a creeping rootstock. Stems square and rough with down-turned bristles. Leaves in whorls of 4–6(8); oval to elliptic, tough, leathery and dark shiny green, 1-veined, *4–27 mm wide*, margins rough. Flowers pale yellow–green, 3.5–8 mm, 5-lobed, forming dense, leafy panicles. Fruit 3–7(9) mm, black and fleshy when ripe. Coastal maquis. AA, CR, GR, PN, TR (northeast). *R. tinctorum* MADDER is similar to *R. peregrina* but with leaves in whorls of 4–6, which are *paler green, not notably leathery, and with prominent lateral veins*. Flowers pale yellow, 3.5–6 mm. Fruit red–brown or blackish, 2.5–6.5 mm. Scattered throughout. ***R. tenuifolia*** is similar to *R. peregrina* but woody, except for the end branches and the leaves that are less rough (to *smooth*) along the margins. Flowers borne in *dense* inflorescences in the leaf axils only (not terminal). Throughout, though rare in the west, interior mainland, and north.

Plocama

Herbs or shrubs, strong-smelling when crushed, with opposite leaves (sometimes seemingly whorled). Corolla funnel-shaped with 4 spreading lobes; stamens 4, inserted to projecting. Fruit a drupe or splitting into 1-seeded mericarps.

Plocama calabrica (syn. *Putoria calabrica*) PUTORIA A *strong-smelling*, dwarf, much-branched, prostrate to spreading, almost thyme-like shrublet to 1 m with opposite, elliptic to oval, leathery leaves 10–28 mm with down-turned, minutely hairy margins; stipules small and inconspicuous. Flowers 13–19 mm, borne in dense terminal heads; pink and funnel-shaped with a long tube and 4 spreading, linear-lance-shaped lobes 2.8–4.7 mm; stamens projecting. Fruit 4–5 mm, 2-lobed and red or blackish when ripe. Coastal garrigue and gorges; strongly western. AA (south), CR, CY, GR, IA (common), PN, TR.

Crucianella

Annual or perennial herbs with whorled leaves and yellowish, stalkless or virtually stalkless flowers borne in 2- or 4-ranked *compact spikes*; corolla fused below into a tube.

(a) Succulent perennials with *whitish stems*.

Crucianella maritima CRUCIANELLA A prostrate to spreading low shrub to 50 cm with whitish, hairless stems. Leaves 3–11 mm, borne in groups of 4, forming regular, symmetrical, close-set whorls; grey, leathery, spine-tipped with a whitish margin. Flowers yellow, 10–12.6 mm long, 5-lobed. Coastal sand dunes; absent in most of the area. IA.

(b) Annual herbs; greenish or reddish.

Crucianella latifolia A rather lax, inconspicuous annual to 40 cm with leaves 3–28 mm borne in whorls of 6–8, linear-lance-shaped and broader below, not spine-tipped. Flowers *4-parted*, somewhat exceeding the bracts; corolla 4.5–6.2 mm, borne in rather flattened, *cylindrical inflorescences*, cream or purplish; bracts fused. Bare places. Throughout. ***C. angustifolia*** is similar but with slender, roughly *4-dimensional inflorescences* with flowers in whorls and bracts free; corolla 4–4.7 mm. Bare and dry habitats. Virtually throughout (not CY, IP).

Crucianella graeca A slender, erect annual to 35 cm. Lower leaves broadly elliptic with rough margins; upper leaves in whorls of 6(8), 10–20 mm, often with down-turned margins. Flowers borne in dense terminal spikes to 70 mm long, *5-parted,* clearly *exceeding* the subtending bracts; corolla 7–8 mm long, beige, tinged purple–brown. GR, PN. ***C. bithynica*** is similar but with leaves in whorls of 4. AA (Lesbos), GR (northeast), TR.

1. *Rubia peregrina*
2. *Rubia tenuifolia*
3. *Crucianella latifolia*
4. *Cruciata articulata*
5. *Plocama calabrica*

Cruciata

Herbs with 3-veined leaves borne in whorls of 4–10. Flowers yellow, cosexual or male. Fruit 2 hairless nutlets.

Cruciata articulata A small annual to 15(17) cm with erect to ascending branches and sharply angled stems with distant internodes below, shorter above, the uppermost *concealed by large bracts*. Leaves in whorls, 4–12 mm, egg-shaped to oblong, short-stalked and hairless, those above yellowish. Flowers borne in cymes of 5–7; corolla pale yellow to greenish, 1.25–2.6 mm across with oval lobes. Fruits usually solitary, kidney-shaped, 2–3 mm. Bare slopes and fields or maquis. IP (north), LB, SY, TR.

Cruciata laevipes CROSSWORT A softly hairy, weak-stemmed, light green perennial 15–75 cm with weakly 3-veined leaves 10–27 mm, borne in whorls of 4 and almost stalkless clusters of 5–11 yellow flowers forming interrupted spikes; flower clusters exceeded by the leaves. Fruit 2.5–2.8 mm, smooth and hairless, borne on recurved stalks. Grassy habitats in mountains. GR, PN (north), TR.

Asperula

Annual or perennial herbs or subshrubs with square stems and linear leaves in whorls of 4 or more. Flowers with 4-lobed corolla. Fruit a pair of nutlets. Many rare, local species are not covered here.

(a) Plant an annual; flowers blue.

Asperula arvensis BLUE WOODRUFF A short, slender, hairless annual to 50 cm with linear to linear-lance-shaped, equal leaves in whorls of 6–8. Flowers blue or blue–violet, with slender tubes 5–6.5 mm long, much longer than the lobes, borne in clusters surrounded by leafy bracts. Fruit smooth and brown. Disturbed ground. Virtually throughout.

(b) Plant a perennial or subshrub; flowers white, pink or reddish.

Asperula arcadiensis A slender, woody-based, *tufted* perennial 80 mm–15(18) cm. Leaves (4)8–10(12) mm, broadly lance-shaped, densely grey-hairy with weakly down-turned margins. Flowers pink, borne in stalkless, terminal, few-flowered heads; corolla tubular-bell-shaped, hairless, with tube 8–12 mm and incurved lobes 2–3 mm. Fruit 2 mm, hairless. Mountain rocks; a rare Greek endemic. PN (south).

Asperula aristata A subshrub, woody at the base with green or greyish, slightly hairy, 4-angled stems to 95 cm. Leaves 3–8.5 x 0.3–1.1 mm, in *whorls only of 4*, lance-shaped to linear, colourless at the tips, the margins often down-turned. Flowers 5–8(9) mm, *pale pink* (sometimes greenish or yellowish), funnel-shaped and with 4 petal lobes. Garrigue and maquis. GR, IA, PN.

Asperula rigida A dwarf shrub with angled stems to 40 cm. Leaves in whorls of 4(6), linear with down-turned margins. Flowers *small*, borne in very lax inflorescences; corolla 3 mm, pink. CR (common). ***A. brevifolia*** is very similar but with yellowish-green or brownish, larger flowers, 4–5.5 mm long. AA (Rhodes + adjacent islets).

Asperula laevigata A delicate perennial with long stems to 84 cm with *broadly elliptic to oval leaves* 2.4–8 x 1.7–5 mm, borne in whorls of 4 with a prominent mid-vein and netted veins either side. *Flowers white*, 1.3–1.9 mm long, narrowly funnel-shaped with 4 lobes. GR, IA, PN.

(c) Flowers yellow (or brownish to purplish-yellow).

Asperula lutea A variable, woody-based perennial to subshrub with stems to 25 cm. Leaves in whorls of 4, 6–12 mm, flat or with slightly down-turned margins. Flowers almost stalkless, 4–6 mm, *yellow* or brownish to purplish-yellow. GR (southeast), PN.

Galium BEDSTRAW

Similar to *Asperula* but with rounded stems, and leaves in whorls of 4–12. Flowers with white to yellow 4-lobed corolla. Fruit a pair of bristly nutlets with hooked bristles. Numerous species with fluctuating taxonomy; just a few are described here.

(a) Leaves 3-veined, in whorls of 4. Flowers white to green, ≤4 mm across.

Galium rotundifolium ROUND-LEAVED BEDSTRAW A short, leafy, sparsely hairy perennial with *broad, almost circular, fine-pointed leaves* in whorls of 4, each with *3 parallel veins*. Flowers 3–4 mm, white or greenish, borne in lax clusters. Fruit with hooked bristles. Virtually throughout, except smaller islands (not CY).

(b) Leaves 1-veined in whorls of 4 or more. Flowers white (or pinkish, greenish or yellowish), ≤5 mm across.

Galium aparine GOOSE-GRASS An annual with spreading stems to 1(3) m, often stouter and more hairy at the nodes, with strongly recurved prickles on the stems. Leaves 3–5 mm, narrowly oblong, in whorls of 6–8, rough-margined. *Corolla short-tubed, 1.5–2(3) mm across, whitish.* Nutlets 1.1–4.1 mm with hooked bristles. A common weed on disturbed ground. Throughout. **G. spurium** is very similar but with greenish corolla just 0.8–1.2 mm. Virtually throughout (not CY). **G. verrucosum** is similar to *G. aparine*, but small and slender, to 25 cm with weak, recurved bristles on the angles and leaves 4–15 mm. Flowers borne in cymes of 1–3 shorter than their subtending leaves. Fruits borne on recurved stalks, spherical, 4–6 mm, coarsely warty. Virtually throughout. **G. tricornutum** is similar to *G. aparine* but with very rough leaves and stems and flowers borne in cymes of 2–5, shorter than, or equalling, their subtending leaves. Fruits minutely warted. Throughout.

Galium ionicum (syn. *G. mixtum*) An erect to ascending, hairless or short-hairy plant with stems 25–70 cm and stolons often present; non-flowering branches short and few. Leaves in whorls of 6–10(12), 10–30 mm, linear-lance-shaped with rough, somewhat down-turned margins. Flowers 3–5 mm, white (rarely yellowish or greenish), borne in dense, ovoid inflorescences with spreading branches. Fruit dark brown. Stony slopes and crevices. IA.

1. *Asperula arcadiensis*
2. *Asperula aristata*
3. *Galium aparine*
4. *Galium ionicum*
5. *Galium intricatum*

Galium capitatum A slender, much-branched annual to 25 cm with virtually hairless stems. Leaves in whorls of 6–9, shorter than the internodes, 6–10 mm long; stems with minute bristles on the angles (or almost hairless). Flowers in heads of 8–15; corolla 1–1.5 mm with reddish lobes. Fruit 1 mm, kidney-shaped, sparsely hairy or warted. AA (local), GR (southeast), PN. ***G. incrassatum*** is similar but with fruiting stalks thickened and rigid. CR (west).

Galium debile A rhizomatous perennial to 80 cm with 4-sided stems, often somewhat bristly along the angles. Leaves in whorls of (4)6, linear-lance-shaped with rough margins. Flowers borne in lax panicles; corolla 2–3 mm, white or pinkish. Fruits 2–3 mm, warty. AA (local), CR, GR, PN, TR.

(c) Leaves 1-veined in whorls of 4 or more. Flowers inconspicuous, often greenish, typically <1(2) mm.

Galium intricatum A slender, intricately branched annual 35 cm, similar to *G. floribundum*. Stems slightly hairy above. Leaves 4–12 mm, rather broadly lance-shaped, *scarcely* bristle-tipped. Flowers with rather broad, sharp-pointed to bristle-tipped, reddish-yellow corolla lobes, borne in ovoid inflorescences. Fruit <1 mm, with hooked hairs, hairless, or minutely warty. AA, GR, IA, PN.

Galium divaricatum A very slender, intricately branched annual 10–25 cm with hairless or minutely hairy stems. Leaves narrow to almost linear, borne in whorls of 6(9), 4–10 mm, minutely toothed (rough) along the margins and *bristle-tipped*. Inflorescence dominating the plant, open; flowers minute, <1 mm across, cup-shaped, yellow to reddish; stalks elongating and deflexing somewhat in fruit. Fruit <1 mm, broadly kidney-shaped and minutely warty. Throughout. ***G. floribundum*** is similar but with leaves in whorls of up to 12 and brownish-purple flowers. AA (east), TR.

Galium murale A short, sprawling annual with stems to 25(30) cm, much-branched from the base and sparsely hairy. Leaves 1.7–6 mm long, narrowly elliptic with a *short spine-tip*, in whorls of 4–6. Flowers inconspicuous, yellowish, minute, just 0.4–0.65 mm, borne in lax, few-flowered inflorescences (appearing paired in leaf nodes); corolla lobes pointed, erect. Nutlets 1.1–1.7 mm, rather cylindrical with spreading lobes and unevenly hairy. Common on waste and fallow land. Throughout. ***G. verticillatum*** is similar but with whitish (or reddish) flowers 0.5–1.75 mm across and *markedly deflexed leaves* 1.7–5 mm long. Nutlets 0.7–1.2 mm, evenly hairy. Probably throughout.

(d) Leaves 1-veined in whorls of 4 or more, stems with smooth angles or no angles; flowers *yellow* and conspicuous, ≤3.5 mm across.

Galium verum LADY'S BEDSTRAW A creeping or scrambling perennial to 1 m with faintly 4-angled stems and erect flowering stems. Leaves linear and spine-tipped, 5–21(30) mm long, and *narrow*, to 0.8(1.4) mm wide, dark green above and downy below with inrolled margins, 8–12 in a whorl. Flowers *bright yellow* (most species in the genus have white flowers), 2–3.5 mm across, borne in erect, terminal inflorescences. Nutlets 0.8–1.8 mm, smooth and black. Grassy habitats, woods and woodland clearnings, generally above sea level. AA (Lesbos), GR, PN (centre), TR.

Sherardia

Annuals with leaves in whorls of 4–6. Flowers lilac, borne in dense terminal and lateral clusters surrounded by 8–10 leafy bracts. Fruit a pair of nutlets with a persistent calyx.

Sherardia arvensis A small, slender, hairy or hairless annual with spreading stems to 40 cm or less. Leaves 5–18 mm, in whorls of 4–6, soon withering below, pale green. Flowers borne in clusters of 4–10; corolla pale lilac, 4–5 mm, with tube longer than the lobes. Very common on fallow land and in grassy places. Throughout.

Valantia

Small annuals with leaves in whorls of 4. Flowers with 3–4 lobes, whitish or yellowish, borne in clusters of 3 in the leaf axils. Fruit 2 hairless nutlets.

Valantia muralis A prostrate or ascending annual with fleshy, softly *hairy* stems to 23 cm. Leaves in whorls of 4, oval and broadest above the middle, 1–5(8) mm long, downward-pointing. Flowers greenish yellow, 1–1.5 mm, borne in clusters at the base of the leaves. Fruits borne on stalks with *conspicuous horn-like appendages*; nutlets usually solitary, and smooth. On rocks, cliffs and old walls; frequent. Throughout. *V. hispida* is very similar in form, but with stems that are very bristly towards the apex (not soft-hairy), and fruits without horn-like appendages, *densely bristly*; nutlets usually 2, brown or blackish and rough. Throughout. *V. aprica* is a rather similar, mat-forming perennial, *not* conspicuously bristly, and always with *solitary* (not 1–2) fruits. CR (mountains), GR (southeast – rare), IA (Corfu), PN.

Theligonum

Virtually hairless annuals with opposite to alternate leaves. Monoecious; male flowers borne 2–3 with 2–5 lobes; female flowers 1–3 with 2–4 poorly defined lobes. Fruit nut-like.

Theligonum cynocrambe A small, prostrate or ascending annual 80 mm–40(50) cm, unpleasant smelling. Leaves opposite below, alternate above, 10–20(40) mm, untoothed and slightly succulent. *Flowers inconspicuous*, 2–3 mm, solitary or in small lateral clusters of 2–3; stamens much-exserted. Fruit 1.6–2.3 mm, slightly fleshy. Rocky, damp and shady habitats. Probably throughout.

1. *Sherardia arvensis*
2. *Valantia muralis*
3. *Valantia hispida*

GENTIANACEAE | GENTIAN FAMILY

Hairless annuals or perennials with opposite, untoothed leaves. Flowers with 4–5(8) petals fused into a tube; calyx deeply lobed; stamens as many as petals; styles 1–2 (if 1, branched). Fruit a 2-parted, splitting capsule.

Centaurium CENTAURY

Hairless annuals or perennials, often small. Flowers pink (rarely white) with 4(5) calyx and corolla lobes; stamens 5; style 1, divided.

(a) Flowers pink.

Centaurium erythraea COMMON CENTAURY A short biennial to 50 cm with a solitary stem, branched above. Leaves 20–60 mm, elliptic to oval, 3–7 veined, the lowermost in a *distinct rosette*; stem-leaves smaller. Flowers pink, purplish or sometimes white, virtually stalkless, borne in *flat-topped* clusters; corolla lobes 4.5–6 mm. Common in dry grassy places and maritime environments. Throughout. **Subsp.** *erythraea* has stems branched only above, dense inflorescences and corolla lobes 4–6 mm (<½ the length of the tube). Throughout. **Subsp.** *rhodense* has stems branched from the base or middle, rather laxer inflorescences and corolla lobes 5–9 mm (>½ the length of the tube). AA, CR, CY, IA (the common form). *C. pulchellum* LESSER CENTAURY is similar but smaller, to 20 cm (often <60 mm), *lacking a distinct basal leaf rosette*, with the stem usually forked in the lower part with *wide-spreading branches*, and with fewer flowers borne in *lax, open clusters*; corolla lobes 2–4 mm. Damp, grassy places. Throughout. *C. tenuiflorum* SLENDER CENTAURY is similar to *C. pulchellum,* also lacking a basal rosette, and with *stems erectly branched above*; flowers borne in *dense, narrow* clusters; corolla lobes 2–4 mm. Damp, marshy, coastal places. Throughout. *C. serpentinicola* is similar but with flowers 12–16 mm, distinctly exceeding the calyx teeth. AA (Lesbos, Rhodes), CY, TR (south, southwest).

1. *Centaurium erythraea* subsp. *erythraea*
2. *Centaurium erythraea* subsp. *rhodense*
3. *Schenkia spicata*

(b) Flowers yellow.

Centaurium maritimum YELLOW CENTAURY A short annual to biennial to 30 cm with a solitary stem, branched above. Leaves 3–10 mm, fleshy, elliptic to oval, the lower 2 small, *not* forming a distinct rosette, the upper leaves much longer. *Flowers pale yellow*, borne in loosely branched clusters, sometimes solitary; petals lobes 5, elliptic, 4–7 mm. Sandy and grassy maritime habitats. Throughout.

Schenkia

Small annuals with straight, erect stems. Similar to, and still widely still included in the genus *Centaurium*.

Schenkia spicata (syn. *Centaurium spicatum*) SPIKED CENTAURY An erect annual to 50 cm, similar to *Centaurium* but easily distinguished by its *spike-like inflorescences, 60 mm–25 cm, with erect-pointed, pink flowers*; flowers 10–16 mm, short-stalked (1.5 mm); corolla lobes 3.5–5 mm. Damp coastal habitats and saltmarshes. Throughout.

Blackstonia YELLOW-WORT

Herbs with pairs of stem-leaves fused at the base. Flowers yellow, short-tubed, with 6–8-numerous spreading lobes; style 2-lobed.

Blackstonia perfoliata YELLOW-WORT An erect, hairless, bluish plant to 40(50) cm with stems branching from the base or from the middle, from a basal rosette of broadly oval leaves, which soon wither; stem-leaves opposite, oval-triangular, pointed and narrowed at the base and joined there, *almost encircling the stem*. Flowers yellow, with a short corolla tube and 6–8 spreading lobes 4.2–10 mm; calyx divided into 12 *narrowly linear segments to 1 mm wide, free almost to the base*. Widespread and common. Throughout. *B. acuminata* is similar but with stem-leaves *scarcely or not fused at the base*, and with *broader, linear to almost lance-shaped calyx segments to 3 mm wide*, fused at the base; corolla lobes 6–9 and 5–7.5(10) mm. Throughout except the north and southeast.

APOCYNACEAE | PERIWINKLE FAMILY

Trees, shrubs, climbers and herbs with opposite, untoothed leaves, often with a milky latex when cut. Corolla with a distinct tube and 5 lobes; stamens 5, inserted in the tube; style 1. Fruit a pair of follicles; seeds often with hairy tufts. A large family with several subfamilies, subject to many recent changes in taxonomy; now including species formerly classified in the family Asclepiadaceae.

SUBFAMILY APOCYNOIDEAE

Apocynum

Rhizomatous perennials with opposite, deciduous leaves. Flowers borne in terminal inflorescences; *corolla bell-shaped*.

Apocynum venetum A rhizomatous perennial to 50 cm. Leaves 20–40(70) mm long, narrowly oblong with close teeth (rough-margined), hairless. Flowers pale to bright *pink*, borne in *red-stemmed, branched clusters*; calyx lobes 15–20 mm long, lance-shaped with papery margins; corolla 4–5 mm, whitish with lobes as long as the tube. Follicles to 15 cm, pendent. Coastal sands; rare and local. CY, TR.

Nerium

Shrubs with opposite leaves. Large pink or white flowers in terminal clusters; stamens with woolly filaments. Fruit a pair of follicles.

Nerium oleander OLEANDER A robust, evergreen, upright shrub to 4(6) m with long, erect branches. Leaves opposite, leathery, linear-lance-shaped, 16 mm–19 cm. Flowers pink, red or white, sweetly scented, borne in showy terminal clusters; corolla lobes 13–26 mm. Follicles large, 40 mm–16 cm long; seeds with brown hairs. Extremely poisonous. Seasonally drying river banks and gullies; also commonly planted on roadsides. Throughout, except the far north.

SUBFAMILY RAUVOLFIOIDEAE

Amsonia

Erect perennial herbs. Leaves alternate, deciduous. Corolla with slender tube. Carpels surrounded by a disk.

Amsonia orientalis A conspicuous, erect perennial with unbranched stems to 50 cm. Leaves to 40 mm long, lance-shaped, narrowed or rounded at the base, thin, with hairy margins. Flowers *blue-violet* (rather *Plumbago*-like); calyx lobes 2–3 mm with hairy margins; corolla 10 mm across with tube 10–15 mm. Follicles 50–80 mm, erect. Wet habitats near the sea. GR (north), TR.

Vinca PERIWINKLE

Perennials with trailing stems and pairs of leathery leaves. Flowers characteristically periwinkle-like; solitary in the upper leaf axils; corolla propellar-shaped. Fruit a pair of follicles.

Vinca major GREATER PERIWINKLE A spreading evergreen shrub with long trailing stems to 1.5 m, often rooting down. Leaves *large*, 25–90 mm, shiny dark green, oval with *minutely hairy margins*. Flowers *bluish-violet*; corolla 30–50 mm across (tube 12–15 mm; calyx lobes 7–17 mm), hairy on the margin. Follicles to 50 mm; often not produced. Cultivated and widely naturalised, probably throughout. *V. herbacea* is similar but *not evergreen* and with smaller flowers, 20–35 mm across. Woods and pastures. AA (rare), GR, IA, PN.

Cascabela

Shrubs or trees exuding a milky latex. Leaves arranged spirally and short-stalked. Flowers with 5 sepals and corolla lobes. Fruit a drupe. Cultivated in the region.

Cascabela thevetia A shrub 2–3 m with a milky sap. Leaves linear-lance-shaped, (50)70 mm–16 cm, virtually stalkless (rarely with stalks to 40 mm). Flowers conspicuous, *yellow to orange*; calyx lobes 8 mm; corolla tube 10–35(40) mm, the lobes to 30 mm; anthers attached to, and included within, the corolla; style 13–15 mm. Fruit pear-shaped, wider than long, to 35(50) mm across, with persistent calyx. Seeds normally 2(3–4). Native to Tropical America, widely planted in warmer areas.

1. *Nerium oleander*
2. *Nerium oleander* fruits
3. *Vinca major*
4. *Cascabela thevetia*
5. *Caralluma europaea*
6. *Caralluma sinaica* habit; (inset) flowers

SUBFAMILY ASCLEPIADOIDEAE (previously classified in the family Asclepiadaceae)

Caralluma

Succulents with erect stems and soon-falling leaves leaving scars (leaves often absent). Fruit paired, erect or spreading follicles.

Caralluma europaea CARALLUMA A short *succulent* to 20 cm with purplish or greyish, square stems to 30 mm wide; leaves soon-falling and inconspicuous. Flowers distinctive: starfish-like, 10–20 mm across, yellow with maroon stripes and a darker centre, fringed with short hairs; lobes 4.7–5 mm wide; flowers borne in crowded terminal clusters. Follicles 80 mm–13 cm. Rocky habitats in the far southeast. IP.

Caralluma tuberculata A short succulent perennial to 15 cm with angular stems 8–13 mm wide. Flowers borne in terminal cymes, dark purple, deeply divided, hairless. Follicles 80 mm–10.5 cm, hairless, gradually tapering. Arid habitats. IP (very rare – Edom area).

Caralluma sinaica A clump-forming perennial to 30 cm with slender, greyish, branched stems to 10 mm across at the base, narrower above (2–5 mm); stems spotted with dark maroon. Flowers in clusters of (1)3–5, *scattered along the upper stems* (*not* in congested terminal clusters); flowers with slender, recurved stalks 3–4 mm; flowers sand-coloured with maroon markings, with long, spreading hairs. Follicles greyish, spotted purple, to 60(80) mm. Arid, rocky habitats. IP.

Pergularia

Twining, bristly or prickly herbaceous perennials. Leaves opposite, downy on both sides. Flowers borne in the axils or in clusters; calyx lobes 5; corolla yellowish or greenish-white; corona in 2 series, the outer hairless and 5-lobed, the inner with 5 fleshy lobes, each with 2 spurs, one directed outward and downward and the other upward with the tip. Fruit a lance-shaped follicle.

Pergularia tomentosa A twining shrub with milky sap and branches covered with dense, ash-coloured hairs. Leaves 15–35 mm long, heart-shaped at the base, more or less hairy above and velvety below; leaf stalks 4–12 mm. Flowers borne in umbels arising from the leaf nodes; sepals 2.5–3.5 mm, hairy; corolla dull white, tinged pink, 7.5–10 mm with lobes 7 mm long. Follicles 50–75 mm, tapering, *softly spiny*. Common in deserts in southern Israel, rare or absent elsewhere. IP (south).

Leptadenia

Virtually leafless desert shrubs with a broom-like habit. Flowers with anthers obtuse at the tip, not appendaged or in a disk-like structure. Fruit a spineless follicle.

Leptadenia pyrotechnica A stiff, leafless shrub with wand-like, cylindrical and tapering branches to 2.5 m. Leaves linear to linear-lance-shaped, virtually stalkless, 25–60 mm. Individual flowers virtually stalkless, borne in clusters on short, recurved stalks; calyx 1.5 mm; corolla 3–3.5 mm long, 5-parted, whitish-grey with *woolly* lobes with recurved margins. Fruit with pendent, virtually hairless follicles 75 mm–11(13) cm. Hot, dry deserts. IP (Dead Sea Valley – very rare).

Pentatropis

Twining perennials with oblong to elliptic leaves. Flowers with a crown-like structure, the longer lobes of which alternate with the anthers.

Pentatropis nivalis A slender, herbaceous, twining perennial 1(5) m; stems eventually corky. Leaves elliptic to oval, 15–45 mm long, blunt at the apex and rounded at the base, sparsely hairy to hairless. Flowers borne in stalkless (rarely short-stalked) inflorescences with 2–3(8) flowers open at any one time; calyx lobes 1 mm, triangular, hairless; corolla with cream or green, often brown-tinged, triangular lobes 4–6 mm, densely hairy at least in part; corona lobes fleshy. Fruiting follicles narrowly lance-shaped, 45–60(80) mm long, strongly beaked, smooth and hairless. Scrambling in thickets, or desert saltmarshes. IP (south).

Solenostemma

Stiffly branched, erect desert shrubs with oval-oblong leaves. Flowers white, borne in *umbels* in the leaf axils. Fruit an ovoid, smooth follicle.

Solenostemma arghel A desert shrub with stiff, leafy branches 60 cm–1 m. Leaves 20–40 mm, virtually stalkless, fleshy, bluish and velvety, oval-oblong to elliptic. Flowers 7–8 mm, white, borne in *umbels* in the leaf axils, short-stalked. Fruit a solitary oblong-ovoid follicle to 50 mm. Rare, in deserts. IP (Jordan Valley and Dead Sea Valley).

1. *Pergularia tomentosa*
2. *Araujia sericifera*
3. *Calotropis procera*

Araujia

Climbing, hairless perennial vines with opposite leaves. Sepals large and leaf-like; corolla erect; stamens 5. Fruit pendent.

Araujia sericifera CRUEL VINE A climbing vine to 7(10) m with a milky latex. Leaves 45–70 mm, opposite, dark green, shiny and leathery, oval-triangular. Flowers 21–28 mm across, 5-lobed with a tube 11–16 mm, white-tinged with yellow or violet and generally moth-pollinated. Fruit 85 mm–13 cm, grey–green, pear-like with a smooth but clefted surface, containing numerous seeds, each with a silky parachute attached for wind-dispersal. Native to South America; locally naturalised. GR (northeast), IP (north-central – rare); probably locally elsewhere.

Calotropis

Small desert trees with large oblong-round leaves. Fruit a large, inflated, smooth follicle, spongy within.

Calotropis procera A small, erect tree to 3(5) m with *corky, white* bark and milky sap. Leaves large, 50 mm–15(20) cm, oblong-round, stalkless and mealy-velvety, especially below. Flowers 25 mm across with 5 oval, spreading lobes, whitish flushed with dark pink–mauve. Fruit a pair of *strongly inflated, soft*, ovoid to spherical follicles 65–95 mm (12 cm). Mainly desert plains, sometimes cultivated as an ornamental. IP (south), LB.

Gomphocarpus SILKWEED

Erect perennials. Flowers with a corona of 5 horns *without smaller projections*. Fruit 1(2) inflated, bristly, erect-spreading follicles.

Gomphocarpus fruticosus BRISTLE-FRUITED SILKWEED An erect shrub to 1.2(2) m with linear-lance-shaped, hairless leaves 39 mm–12 cm long. Flowers *white*, borne in hairy, stalked umbels; corolla lobes 6.2–8.3 mm, hairy along the margins. Fruit conspicuous; follicles green, *inflated, pointed*, hairless between the prominent bristles, *25–45(60) mm across*, with 120 cottony seeds. Native to South Africa but planted as an ornamental in developed areas, and naturalised locally throughout.

1. *Gomphocarpus physocarpus*
2. *Gomphocarpus sinaicus* PHOTO: ORON PERI
3. *Vincetoxicum hirundinaria*
4. *Periploca laevigata*
5. *Periploca laevigata* in fruit
6. *Periploca graeca*
7. *Periploca aphylla*
8. *Periploca aphylla* (detail)

G. physocarpus is very similar but with larger, *rounded (not pointed) pods*, 45–60 mm across, with 100 seeds. Native to Asia. CR (probably elsewhere).

Gomphocarpus sinaicus A *desert* shrub, to the previous species, to 1 m. Leaves 40–80 mm. Flowers *yellowish, hairless*, 15–25 mm across. Fruit a *long-pointed* follicle, mealy between the prominent *red*, soft bristles. Stony deserts. IP (south).

Vincetoxicum

Short, erect herbaceous perennials with oval to heart-shaped leaves. Flowers with 5-lobed corolla and a corona of rounded scales. Fruit 1–2 follicles.

Vincetoxicum hirundinaria SALLOWWORT A variable, erect (rarely climbing), hairless, short to tall perennial with stalked, pointed, heart-shaped or oval, opposite leaves. *Flowers greenish-white or yellowish*, with lobes 1.8–5 mm, borne in clusters of 5–12 at the bases of the uppermost leaves. Follicles to 60 mm long, often paired, with silky seeds. Rocky garrigue, woods and pastures; local. GR.

Vincetoxicum canescens A soft-hairy perennial with *spreading* stems 20–50 cm. Leaves almost stalkless, grey–green, oval. Flowers 8 mm across, borne in short-stalked clusters of 5–12 in the leaf axils, *dull yellow*. Follicles 2–3, 30–50 mm long, ovoid. AA (eastern), TR (southwest). *V. creticum* is similar but almost hairless and with narrower (lance-shaped) follicles. CR.

Vincetoxicum fuscatum is similar to the previous species but with *dark purplish-brown* flowers with lobes 4 mm. GR (north and central), IA (Cephalonia – rare), TR. *V. speciosum* has blackish-purple flowers with white hairs, with lobes 5–6 mm. GR (north), TR.

Cynanchum

Climbing herbaceous perennials or shrubs with heart-shaped leaves. Flowers with corona of 5 long appendages. Fruit 1(2) pendent, spineless follicles.

Cynanchum acutum STRANGLEWORT A hairless, twining, blue–green, woody climbing perennial to 3(4) m with slender stems. Leaves 22–78 mm (10 cm), *arrow-head-shaped*, deeply heart-shaped at the base and stalked. Flowers tubular, scented, borne in lateral or terminal umbels, with a corolla of 5 triangular, projecting lobes 4.9–7.2 mm. *Fruits long and spindle-like*, pointed, 17 mm–18 cm long. Scrub and field margins, ditches; often coastal. Local, but scattered throughout.

SUBFAMILY PERIPLOCOIDEAE (preciously classified in the family Asclepiadaceae)

Periploca

Climbing perennials with oval-lance-shaped leaves and greenish-brown flowers; corolla lobes hairy and fleshy; corona lobes without conspicuous veins, and with 5 slender appendages; filaments free. Fruit 1–2 follicles.

Periploca laevigata (syn. *P. angustifolia*) A *branched shrubby perennial* to 3 m with only the upper branches weakly climbing, often with few leaves. Flowers borne in few-flowered clusters, with *yellow–green* lobes, 6–7 mm, with a brown–purple centre, with whitish appendages; hairless except for woolly spots on the lobes. Fruit with follicles spreading horizontally. Very rare in dry coastal garrigue; a predominantly North African species. CR (a Libyan element on offshore islets), SY.

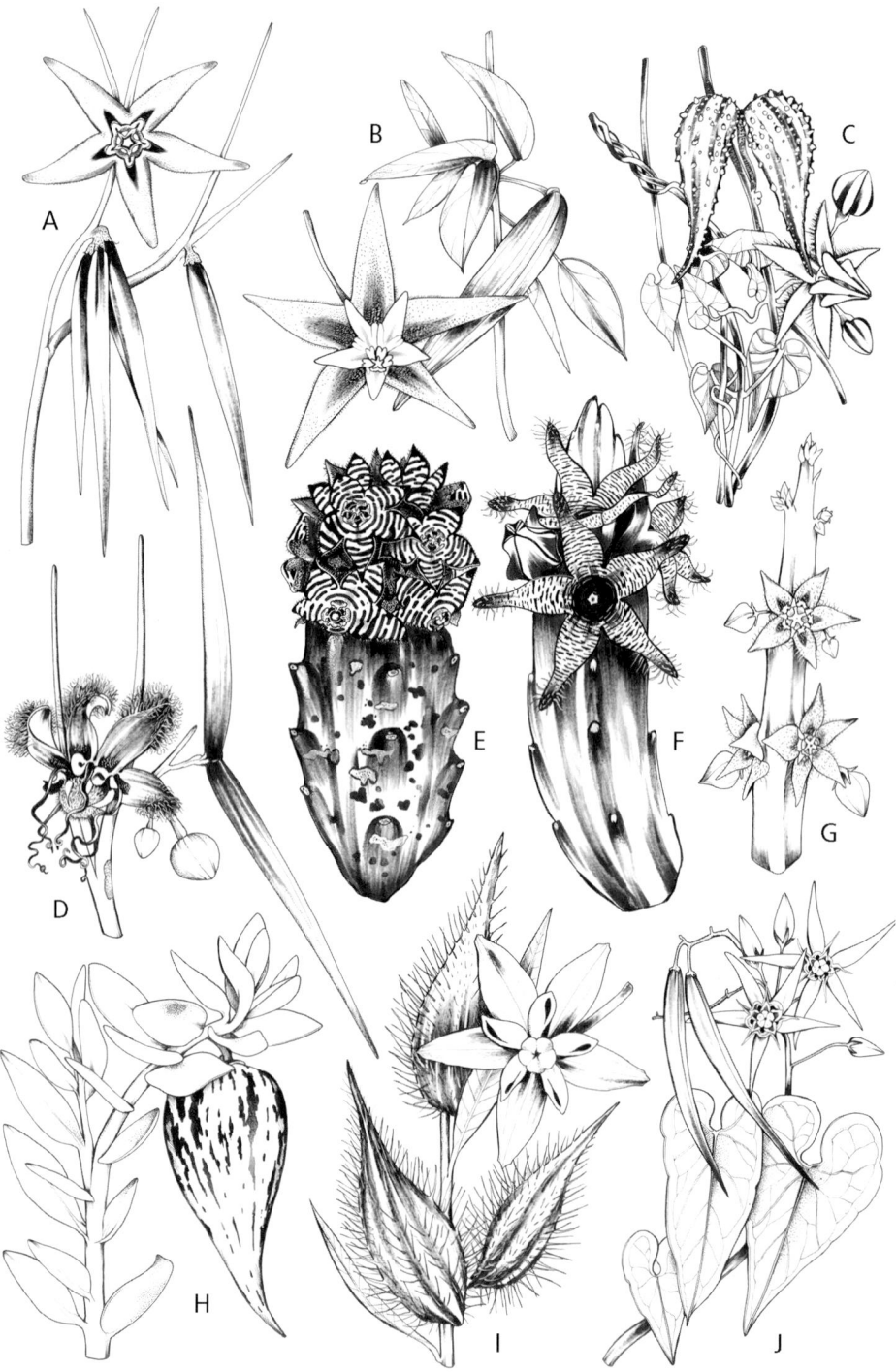

Apocynaceae species: **A.** *Leptadenia pyrotechnica*; **B.** *Pentatropis nivalis*; **C.** *Pergularia tomentosa*; **D.** *Periploca aphylla*; **E.** *Caralluma europaea*; **F.** *Caralluma tuberculata*; **G.** *Caralluma sinaica*; **H.** *Solenostemma arghel*; **I.** *Gomphocarpus sinaicus*; **J.** *Cynanchum acutum*. Illustrations not to scale.

Periploca graeca SILK-VINE A vigorous *twining, deciduous, woody perennial* to 12(15) m with dark green, stalked, shiny, opposite leaves 30 mm–12 cm. Flowers *brownish-green*, star-like with hairy, spreading-deflexed lobes 9.5–10 mm. Fruit 10–15 cm, follicles spreadingly divergently. Woods and hedges near streams. Widespread. AA, GR, IP, PN (northeast), TR.

Periploca aphylla A rigid, *virtually leafless* shrub 1–2(3) m with long, naked, arching branches rather like those of *Spartium junceum*. Leaves few or absent, virtually stalkless and sharp-tipped, 2–6 mm. Flowers brown–purple with lobes *bearded* on the the upper half of the margin; tube short; lobes 2–6 mm; corona lobes to 6 mm. Fruits velvety when young, the follicles spreading horizontally, rigid, woody when mature, 50 mm–10 cm long. Semi-desert scrub. IP.

Cyprinia

Similar to *Periploca* but with thin, *hairless*, yellow to white corolla lobes and ribbon-like corona lobes with visible veins.

Cyprinia gracilis A woody climbing or scrambling perennial with hairless young shoots. Leaves lance-shaped to elliptic, 40–50 mm long, green and shiny, with short stalks to 5 mm. Flowers 12–14 mm across with hairless, deeply dissected calyx; corolla lobes white to yellowish-green, 6 mm long; corona lobes ribbon-like, 4 mm and forked. Fruits hairless, cylindrical, uniformly thick and slightly arched, to 80 mm. Maquis scrub and forests. CY, TR (south – rare).

BORAGINACEAE | BORAGE FAMILY

Annual or perennial herbs or shrubs often with bristly stems and leaves; bristles often with swollen bases. Leaves simple and alternate. Flowers in spiralled clusters, short-stalked, blue in many species; stamens 5, borne on the corolla tube; ovary deeply 4-lobed; style 1 (sometimes split). Fruit 4 nutlets, often concealed in a persistent calyx.

Alkanna

Perennial (or annual) herbs with flowers in cymes with bracts, identified by hairiness, flower colour and their warty fruits (nutlets). A taxonomically difficult group in the region.

(a) Flowers mostly white, cream or pinkish upon opening.

Alkanna sieberi A woody-based perennial with spreading stems to 15(20) cm. Leaves oblong to linear, 30–40 mm long, with tubercle-based hairs and non-glandular hairs; stem-leaves shorter. Flowers with corolla tube to 2 x as long as the calyx, 6–8 mm across, *whitish*-pink to cream, ageing blue–lilac, with a darker throat. Fruits to 2 mm, strongly curved. Dry, sandy habitats. CR.

Alkanna oreodoxa A tufted, white-downy perennial to 25 cm with short crisped hairs mixed with weak hairs, 1–2 mm. Basal leaves elliptic or broadest above the middle, 10–60 mm; stem-leaves smaller. Flowers borne in cymes extending to 20 cm in fruit; calyx 5–7 mm; corolla 14–17 mm, the tube pink, yellowish-brown or blue, the lobes paler (to white). Fruits 2.2 mm, densely warty. A rare endemic of limestone rocky crevices. TR (Antalya).

Alkanna sartoriana An ascending, bristly perennial with glandular and non-glandular hairs, to 40 cm. Basal leaves 50–70 mm long. Flowers white (or slightly pink or yellow-flushed), with a darker throat, (10)12–13 mm across. Fruits 2 mm, warty. Dry, sandy habitats, rare. PN (northeast).
A. chrysanthiana is a rare, close relative with cream flowers. PN (Mani).

(b) Flowers mostly blue upon opening.

Alkanna sfikasiana A tufted, bristly perennial 15–30(35) cm. Basal leaves in rosettes, lance- to broadly elliptic-lance-shaped, 10–18(22) cm with pimple-based bristles; stem-leaves stalkless, oval-oblong, smaller. Flowers rather crowded in terminal clusters; corolla 11–14 mm, deep *cobalt blue with an orange throat*; tube brownish-cream; stamens included. Fruits 3.5–4.2 mm, warty. A rare endemic of maquis and conifer forests on limestone. PN (Parnonas massif).

Alkanna tinctoria A grey–green, often mound-forming perennial to 50 cm with stiff, tubercle-based hairs and short, non-glandular hairs. Basal leaves broader above the middle; stem-leaves oblong. Flowers borne in long, many-flowered racemes; corolla *bright blue*. Fruits 2 mm, recurved. Rocky garrigue. Throughout.

(c) Flowers yellow upon opening.

Alkanna graeca A perennial to 40(80) cm with bristly (non-glandular) hairs. Basal leaves to 10 cm, with entire, flat margins; stem-leaves shorter and oval-lance-shaped. Flowers *yellow to brownish-orange*, 8–10 mm wide. Fruits 2.5–3.5 mm. Mountain rocks. GR, IA, PN.

Alkanna orientalis A *glandular*-hairy perennial herb with ascending to erect stems to 50(80) cm. Basal leaves 10–15 cm, broadest above the middle, with smooth margins. Flowers bright yellow, 8–12 mm wide. Fruits 3 mm, slightly *recurved*. Rocky habitats in the east. AA (east – local), IP, SY, TR. *A. hellenica* is similar but with undulate leaf margins. GR (Sterea Ellas), PN (north and central).

Arnebia

Erect or prostrate, annual or perennial bristly herbs. Leaves stalkless, alternate. Flowers usually borne in terminal inflorescences; calyx 5-lobed to the base, exceeded by the corolla; corolla throat hairless. Nutlets (1–)4, rough and warty, *enveloped by stiff calyx bristles*.

Arnebia decumbens An erect, bristly annual to 17(30) cm, branching from the base, with long, yellowish hairs to 1.4 mm. Lower leaves linear-oblong, 11–40 mm, upper leaves linear to lance-shaped, pointed. Flowers yellow; corolla tube 10–11 mm (1–2 x as long as the calyx); stigma 2–4-parted. Fruiting raceme *elongated and lax*. Nutlets irregularly warty, 2–2.8 mm long, beaked, grey–brown. Mainly on desert sands. IP (especially the Negev).

Buglossoides

Bristly annuals or perennials. Flowers often borne in branched panicles; stamens 5, included within the (typically) whitish or blue corolla. *Nutlets minutely warty*.

Buglossoides arvensis (syn. ***Lithospermum arvense***) A (normally) *erect* annual 10–35 cm with adpressed hairs. Leaves oblong-lance-shaped. Flowers 5–9 mm with a cylindrical tube and without scales in the throat, *white* (rarely pale blue); calyx teeth linear-lance-shaped with stiff white hairs, *long* (exceeding corolla). Nutlets of fruit 3–4 mm, pale brown, without prominent swellings; fruit stalks not thickened. Open grassy habitats. Throughout. *B. incrassata* (syn. *Lithospermum incrassatum*) is similar but with short, *spreading* stems, blue (rarely white) flowers with short-lobed calyx, and *strongly thickened* fruiting stalks. Virtually throughout (absent from smaller islands).

Buglossoides tenuiflora (syn. ***Lithospermum tenuiflorum***) is similar to *B. arvensis* but smaller, 10–18 cm, with inflorescences less elongated in fruit; corolla *minute*, with *limb just 1–2 mm across*. Nutlets small (2 mm), easily detached, and with *lateral swellings* and distinctly incurved beak. Rare and local on rocky slopes and in deserts; strongly eastern. CY, GR (southeast), IP (mainly southern and eastern), SY (Syrian Desert), TR.

Buglossoides purpurocaerulea (syn. ***Lithospermum purpurocaeruleum***) A rhizomatous perennial with spreading, non-flowering stems rooting at the apex, and more or less erect flowering stems to 50(60) cm with adpressed to spreading hairs. Leaves lance-shaped to narrowly elliptic. Flowers *purplish*, turning blue, rather large, 10–16(18) mm, borne in dense, later elongating, cymes. Nutlets bluntly keeled, white and shiny. GR, PN.

---

1. *Alkanna sieberi*
2. *Alkanna oreodoxa*
3. *Alkanna sfikasiana* PHOTO: ELEFTHERIOS DARIOTIS
4. *Arnebia decumbens*
5. *Buglossoides arvensis*
6. *Buglossoides tenuiflora*

Rindera

Perennials with flowers borne in bractless, terminal cymes; corolla usually purple. Nutlets large, flattened, smooth and with *broad wings*.

Rindera graeca A woody-based perennial to 30 cm, with stems not leafy above. Leaves in basal rosettes, linear to elliptic-lance-shaped, with greyish, silvery hairs. Flowers borne in nodding cymes of up to 16; corolla tubular-cylindrical, 9–15 mm, dark maroon. Fruit a *winged* nutlet, 11–14 mm. Mountain rocks. GR (+ Euboea), PN.

Onosma

Bristly perennials with *pendulous* flowers; corolla often yellow, typically cylindrical to club-shaped. Many species in the region and adjacent areas, particularly in Turkey (c. 95).

(a) Corolla typically ≤20 mm, yellow upon opening.

Onosma frutescens A mound-forming subshrub with non-woody flowering stems 10–20 cm long. Leaves oblong, often broadest above the middle, with coarse, pimple-based hairs, among shorter hairs. Flowers borne in dense inflorescences with nodding flowers; calyx purplish, white-hairy, inflated in fruit; corolla 17–20(22) mm, yellow, often bright orange–red distally when mature; stamens slightly protruding. Rocks on cliffs; often chasmophytic. AA (local), GR (Sterea Ellas), PN (common), SY, TR.

Onosma graeca A robust biennial to 40 cm, branched above, with coarse, spreading bristles to 8 mm mixed with short, fine hairs. Leaves to 15 cm, lance-shaped to broadly linear, borne in rosettes; stem-leaves much shorter. Flowers in lax inflorescneces; calyx hairy; corolla 15–18 mm, *narrow*, hairless, yellow, turning orange–brown. AA, CR, PN, TR (rare).

(b) Corolla *large,* ≥20 mm, cream or yellow upon opening (white in *Onosma stridii*).

Onosma echinata A *conspicuously* densely and long-bristly biennial to perennial, 30–60 cm. Stems stiff, branching into panicles. Leaves oblong-linear to linear-lance-shaped, the lowermost tapering into their stalks. Flowers borne in racemes nodding in bud, loose and erect in fruit; corolla *pale yellow to whitish-cream,* about ¼ longer than the calyx; calyx c. 20 mm, with linear lobes; anther tips exserted. Nutlets to 5 mm, warty. Rare and local, in dry habitats and desert scrub. IP (mainly Negev and Judean Deserts + Lower Jordan Valley), SY.

Onosma taurica A tufted perennial with stems to 25(30) cm, not, or scarcely branched, with rather long, white pimple-based hairs. Leaves whitish, 25–60 mm, gradually tapering into long stalks, the uppermost linear to oblong and stalkless. Flowers borne in 1–2 cymes, which elongate and straighten in fruit; calyx 12–13 mm; corolla white, cream or yellow, 22–25 mm, bell-shaped. TR. **Var. *brevifolia*** grows to 20 cm and has linear-spatula-shaped, blunt leaves. **Var. *taurica*** is shorter, 12–25(30) cm with spatula to linear-lance-shaped, *pointed* leaves.

Onosma heterophylla A woody-based perennial to shrub to 50 cm with 0–few non-flowering rosettes during flowering. Basal leaves egg- to lance-shaped with variably sized star-shaped hairs with up to 15 branches; lower stem-leaves similar in size to the basal leaves. Flowers borne in nodding cymes; calyx lobes green, corolla *lemon yellow* (17)20–30 mm, rather broadly cylindrical-bell shaped. Rocky woods and scrub on the mainland. GR, PN, TR. *O. aucheriana* is similar but with lower stem-leaves *much smaller* than the basal leaves. AA (northeast), GR (northeast).

Onosma stridii A mound-forming perennial with sterile shoots and rather adpressed, pimple-based bristles. Flowers *large and white*; corolla hairless, 24–27 mm; calyx 10–14 mm. Very rare, on serpentine gravel on Mount Kallidromo and Mount Chlomo only. GR (central).

(c) Corolla ≤20 mm, white, cream or yellow upon opening.

Onosma mersinana A perennial with slender, creeping branches to 25 cm. Leaves linear to narrowly lance-shaped with *down-turned margins*, the lowermost *persistent* throughout flowering; bristles stout, adpressed and with pimple-like bases. Flowers borne in solitary cymes, cream; corolla 17 mm; calyx 12–15 mm and white-bristly. Rare. TR (Mersin). **O. lineariloba** is similar, with broader leaves, withered during flowering. TR. **O. lycaonica** has distinctly *adpressed* bristles, short calyx and longer corolla, white turning pink then bluish. TR. **O. subulifolia** has club-shaped, white flowers. TR.

1. *Onosma frutescens*
2. *Onosma echinata*
3. *Onosma stridii* PHOTO: ARNE STRID
4. *Onosma alborosea*
5. *Onosma orientalis*

Onosma juliae A woody-based, tufted perennial to 15(20) cm. Leaves blue–green, *broadly* oval-spatula-shaped, borne in rosettes; bristles adpressed and *very dense*, easily detached. Flowers creamy-yellow; corolla to 20 mm; calyx 14–16 mm. A rare endemic. Rocky habitats. TR (Eastern Taurus range in Karaman province).

Onosma gigantea A very *robust*, clumped biennial to 60(95) cm branched below the middle, bristly. Leaves lance-shaped to oblong, to 30 cm long, with stalks to 70 mm. Flowers borne in numerous cymes; calyx 13–16 mm with linear lobes; corolla golden yellow, 17–20 mm. Dry hillsides in the east. IP, SY, TR.

(d) Corolla whitish-pink upon opening.

Onosma alborosea An erect to ascending, usually simple perennial to 20(25) cm with spreading to adpressed hairs. Leaves 25–60 mm, broadest above the middle with very dense, adpressed hairs. Flowers congested; calyx 14–16 mm with densely hairy linear-lance-shaped lobes; corolla 18–30 mm, whitish-pink becoming pink or purple, eventually blue or purple. Variable, with regional subspecies. SY, TR.

(e) Corolla whitish-blue upon opening.

Onosma orientalis (syn. ***Podonosma orientalis***) A soft-hairy perennial with tubercle-based bristles on the leaves, to 16 cm. Leaves 8–30 mm, oval to lance-shaped, stalkless. Flowers borne in dense inflorescences, becoming lax and elongated; calyx 10–12 mm with broadly lance-shaped lobes; corolla cylindrical, *pale to mid-blue*, 10 mm, exserted, with deflexed, often yellowish lobes; anthers exserted. Nutlets minute, 2.5 mm, with beak laterally incurved. Limestone rocks and cliffs in the east. IP, LB, SY, TR.

Phacelia

Hairy annuals with pinnately divided leaves. Flowers borne in spiralled, terminal and lateral cymes; calyx divided more or less to the base; corolla bell-shaped; stamens protruding. Fruit a 2-valved capsule.

Phacelia tanacetifolia A hairy, erect to ascending annual to 70 cm. Leaves pinnately divided. Flowers numerous, blue to mauve; calyx deeply divided; corolla 6–10 mm. Native to Mexico; naturalised and locally invasive in Cyprus (possibly casual elsewhere). CY (west-central).

Paracaryum

Loosely branched, hairy perennials. Fruits conspicuous; the nutlets flattened, winged and sub-circular.

Paracaryum lithospermifolium A rather small, woody-based perennial with few, more or less erect stems to 12(25) cm. Leaves to 40 mm, linear or broadest above the middle, forming rosettes; stem-leaves oblong and smaller. Flowers borne in few-flowered, forked, slender cymes; corolla 3–5 mm with lobes equalling the tube, wine-red to purplish-blue. Fruit comprising nutlets 5–6(9) mm across with prominent, spreading, toothed wings. Limestone and igneous slopes; scattered across the region. **Subsp.** *lithospermifolium* has adpressed stem hairs. IP, TR. **Subsp.** *cariense* has spreading, bristly stem hairs. AA (Samos), CR (western mountains), CY, TR. *P. aucheri* is similar but with larger flowers, 7–9 mm, the corolla with short lobes and style 5–6 (not 1–2) mm. AA (rare – Samos), TR. *P. rugulosum* is a desert species with nutlets with *wings incurved*. IP (south).

Heliotropium

Small, bristly annuals or perennials. Flowers white or violet, borne on the upper side of outwardly coiled branches; stamens included in the corolla (or virtually so); style terminal (not basal).

Heliotropium europaeum **HELIOTROPE** A short, greyish, erect or spreading, much-branched, softly hairy annual to 50(60) cm. Leaves 30–70 mm, oval to elliptic, grey and stalked. Flowers white with a yellow throat, small, 2–4 mm across, borne in distinctly 1-sided, spiralled spikes; *calyx divided almost to the base*. Fruit splitting into 4 nutlets. Bare, cultivated and waste ground; widespread and common. Throughout. The following species are similar: **H. lasiocarpum** has dense cymes and *densely hairy* nutlets. AA, TR. **H. supinum** has spreading, *almost prostrate* stems and leaves green above, and a *flask-like calyx with short lobes, divided by >1/3 its length*, which encircles the fruit. Scattered throughout. **H. dolosum** has calyx teeth curving upwards in fruit, *scented flowers* and a hairy stigma. Nutlets hairless. Widespread but local, absent from the west. **H. bacciferum** is a *woody-based perennial*, eventually with a stout stock to 15 mm thick. Leaves 5–55 mm with down-turned margins. Flowers borne in somewhat reduced inflorescences; corolla white, 2.5–3 mm. Arid habitats. IP.

1. *Phacelia tanacetifolia*
2. *Paracaryum lithospermifolium* subsp. *cariensis* PHOTO: NICK TURLAND
3. *Paracaryum rugulosum*
4. *Heliotropium europaeum*
5. *Neatostema apulum*

Heliotropium curassavicum A *greyish, fleshy perennial* with stems to 70 cm. Leaves narrow and *hairless*, blunt. Flowers 3 mm across, white with a greenish-yellow centre. Nutlets 4. Native to America; locally naturalised. AA (local), IA (Cephalonia), GR (northeast + southeast), PN (northeast).

Neatostema

Small, bristly annuals. Flowers *yellow*, borne in branched or spike-like inflorescences; stamens included in the corolla.

Neatostema apulum YELLOW GROMWELL A short annual to 25(30) cm with erect, bristly stems, branched above. Leaves 45(70) mm long, narrowly oblong, and bristly along the margins, those on the stems *erect*. Flowers yellow, borne in dense inflorescences to 90 mm long; corolla 2–3.5 mm across. Nutlets pale brown and warted. Fairly common in bare and stony habitats. Throughout.

Lithodora

Dwarf shrubs with entire leaves. Corolla tubular-bell-shaped; stamens included in the corolla. Fruiting heads dense.

Lithodora hispidula A much-branched, dense, rigid shrub to 35 cm (1 m). Leaves oblong to narrowly egg-shaped, 15(20) mm long leathery, dark green and bristly, especially beneath. Flowers borne 1–4 in cymes; calyx 7 mm; corolla bluish-violet, twice as long as the calyx, 10 mm across with short, rounded lobes. Nutlets minutely warty. **Subsp.** *hispidula* AA (east), CR (+ KP). TR (west). **Subsp.** *versicolor* CY, SY (north),TR (south). *L. zahnii* (*L. hispidula* subsp. *zahnii*) is considered by some authors to be a subspecies of the former and has larger flowers, with calyx 8–11 mm; corolla pale blue or white, 13–15 mm across with spreading, blunt lobes. Nutlets smooth. Limestone cliffs and ravines; a rare endemic. PN (western Mani peninsula, southeast of Kalamata).

Cerinthe HONEYWORT

Virtually hairless, blue–grey annuals or perennials. Flowers with *cylindrical, tubular corolla*; stamens included.

Cerinthe major HONEYWORT A rather fleshy, grey–green, almost hairless annual 30–50(70) cm. Lower leaves 85 mm long, oblong, clasping the stem with a heart-shaped base, leaves covered with conspicuous *white swellings*. Bracts oval to heart-shaped, 12.5–20 mm. *Flowers dark purple throughout*, 17–22 mm long, the short petal lobes 3–4 mm, recurved; flowers borne in a *drooping* inflorescence. Common on coastal sands and garrigue, mainly in the west. AA, CR, GR (local), IA, PN. *C. minor* LESSER HONEYWORT is a similar annual with *small yellow flowers, 11–12 mm long*, sometimes spotted violet at the base, with *straight lobes, 3.5–5 mm*. Woods. Local and absent from islands. GR, PN, TR.

Cerinthe palaestina An annual 50 cm–1 m. Leaves blunt, the lower egg- to spatula-shaped, tapering into the stalks, the uppermost stalkless. Flowers borne in dense inflorescences; bracts often purplish, equalling or exceeding the calyx; calyx lobes unequal; corolla 20–25 mm, cream-coloured with minute violet spots between the lobes; lobes broader than long. Nutlets marbled, 4–5 mm. A rare and local endemic. IP (north).

1. *Lithodora hispidula*
2. *Lithodora zahnii*
3. *Cerinthe major*
4. *Cerinthe minor*
5. *Cerinthe palaestina*
6. *Nonea philistaea*

Nonea

Bristly herbs. Flowers borne in clusters with many bracts; corolla yellow, brown, purple or white with a tuft of hairs in the throat; stamens included or slightly protruding.

Nonea philistaea A prostrate, sticky-hairy, rough annual 15–35 cm. Leaves lance-shaped to oblong, tapered at the base; upper (floral) leaves triangular, tapered at the tips and much exeeding the flowers. Flowers 8–12 mm, small, creamish-white, flushed pink to pale orange, borne in leafy racemes; calyx split for 1/3 its length, becoming inflated in fruit. Nutlets 3–4 mm, net-veined with a lateral beak directed inwards. Dry habitats, fallow fields and desert scrub. IP (especially Kinnroth Valley and Negev).

Nonea echioides A bristly annual to 50 cm, branched below. Leaves oblong, lance-shaped or linear, 20–70 mm long; stem-leaves semi-clasping at the base. Flowers white, 5–6 mm long; corolla funnel-shaped; calyx 4–5 mm (8–11 mm in fruit), divided 1/3 to the base. Nutlets kidney-shaped, 1–1.5 mm. Roadsides, dry fields. AA, CY, GR, IA, PN, SY, TR.

Echium BUGLOSS

Bristly herbs or shrubs. Flowers with zygomorphic, funnel-shaped corolla, borne in spiralled cymes making up dense or lax panicles; stamens unequal and normally at least some exserted (an important trait).

(a) Stamens not projecting from the corolla tube; corolla short, 7.5–13 mm long.

Echium parviflorum SMALL-FLOWERED BUGLOSS A rough, bristly-hairy annual with ascending stems to 40(65) cm, with oblong leaves, stalked below and stalkless above, with adpressed hairs. *Flowers small* with corolla 7.5–13 mm long, pale to dark blue with white throats, borne in lax, leafy clusters, with *stamens not protruding*. Coastal sandy and rocky habitats. Probably throughout, although rare in the west. *E. arenarium* is similar but often prostrate, with violet flowers with corolla just 6–10 mm long. Virtually throughout (absent in the north).

1. *Echium italicum*
2. *Echium angustifolium*
3. *Echium vulgare*
4. *Echium plantagineum*
5. *Symphytum brachycalyx*
6. *Borago officinalis*

(b) Stamens projecting from the corolla tube; corolla 9–16(18) mm long.

Echium italicum PALE BUGLOSS A large, erect biennial to 1 m with dense, spreading, white or yellowish bristles. Basal leaves lance-shaped, to 20(25) cm long and 20 mm wide, narrowed into a basal stalk; stem-leaves stalkless. Inflorescence spike like or branched at the base, therefore pyramidal (depending on size of plant). *Corolla pale in colour*: whitish, pinkish or pale blue, 9–12 mm long and narrowly funnel-shaped; finely hairy outside; stamens 4–5 and markedly protruding. Pastures and fallow land. Virtually throughout, except the far southeast.

(c) Stamens projecting from the corolla tube; corolla (10)14–35(40) mm.

Echium angustifolium A grey-bristly perennial 25–50 cm. Leaves in rosettes with short, white, spreading bristles, *linear*, or broadest above the middle; upper leaves narrowly oblong-lance-shaped. Flowers borne in several dense cymes; corolla 16–22 mm, *remaining red* when mature (sometimes fading purplish); stamens protruding. Throughout, except the north.

Echium vulgare VIPER'S BUGLOSS A bristly hairy biennial to 1 m with 1–several stems. Leaves covered in bristle-like hairs, elliptic and stalked at the base, narrowly lance-shaped and stalkless along the stem. Inflorescence spike-like; flowers *blue* to blue–purple *with 4–5 long-protruding stamens*; corolla hairy all over, 10–21(25) mm. Nutlets ridged. Dry open habitats and fallow land. GR (north). *E. plantagineum* is similar to *E. vulgare* but *softly hairy* (rather than coarsely bristly) with reddish bristles. Leaves with prominent midribs and lateral veins. Flowers borne in branched, panicle-like inflorescences, dark blue–purple, ageing red, with *fewer*, 2(4) protruding, and reddish stamens; corolla (15)18–30 mm *hairy on the veins and margins only*. Common in sandy maritime habitats. Throughout (except some parts of the north).

Echium candicans PRIDE OF MADEIRA A bristly hairy, *large, shubby perennial* to 2 m, with many ascending flowering stems, woody at the bases. Leaves dark green, rough, elliptic in shape and pointed at the tips. Flowers bright blue, borne in dense inflorescences on stems leafy below. Native to Madeira; occasionally planted in towns.

Symphytum

Bristly perennials with flowers borne in short, bractless terminal cymes; calyx lobed almost to the base; corolla variously coloured; throat with 5 long scales; style exserted. Nutlets ovoid, often with a thickened collar.

Symphytum brachycalyx A hairy perennial 20–60 cm. Leaves linear-oblong to oval, or narrowly oval to linear-lance-shaped, stalked below, stalkless above. Flowers 9–15, white, 12–15 mm; calyx 7–9 mm (15 mm in fruit), divided 1/4 or 1/3 to the base with linear-lance-shaped lobes; corolla with linear scales equalling or exceeding the stamens. Nutlets 2.5–3 mm, constricted at the base, warty. Woods and rocky, shady habitats, often near water. IP (Judean Mountains and the north), SY, TR.

Symphytum creticum (syn. *Procopiania cretica*) A perennial 10–50 cm with short and long, hooked and straight bristles. Leaves oval, stalked, the uppermost stalkless with bases running down the stem (decurrent). Flowers borne numerously in cymes; calyx 4.5–7(8) mm with pointed lobes; corolla *blue–violet* (rarely white) with lobes 2–4 x as long as the short tube; throat scales with projections (papillae). CR, IA (Zakynthos – possibly), PN. *S. insulare* (syn. *Procopiania insularis*) is similar (treated as the same species by some authors) but with blunt calyx lobes and corolla with shorter lobes, just 2–3 x as long as the tube; throat with scales with slender projections (3 x as long as wide); base of filaments surrounded by a cup-shaped scale. AA, KP.

Symphytum ottomanum A slender perennial 30–80 cm. Leaves oval to lance-shaped, stalked, the lowermost narrowed into winged stalks, the uppermost wedge-shaped, stalkless with bases slightly running down the stem (decurrent). Flowers small, borne numerously (c. 20) in cymes; calyx 3–5 mm with pointed lobes; corolla *whitish to pale yellow*, 5–7 mm; throat scales exserted, 2–5.5 mm, exceeding the stamens. Wooded habitats on the mainland. Nutlets 2 mm. IA (Cephalonia – rare), GR (north), TR.

Borago BORAGE

Bristly annuals or perennials. Flowers blue (rarely white), short-tubed with widely spreading lobes, and stamens equal, forward-pointing in a cone (included to exserted).

Borago officinalis BORAGE A bristly annual with robust, generally branched stems to 60 cm. Basal leaves oval and light green, stalked; stem-leaves smaller and unstalked. Flowers borne in loosely branched cymes, bright blue and star-like with a white centre, deflexed lobes 7–10 mm, and an exposed *cone of blackish anthers*; calyx 7–15 mm (20 mm in fruit). Locally common on cultivated, sandy and waste ground. AA, CR (common), CY (naturalised), GR (local), IA, PN.

1. *Anchusa undulata* subsp. *hybrida*
2. *Anchusa azurea*
3. *Anchusa cespitosa* PHOTO: ARNE STRID
4. *Anchusa strigosa*
5. *Anchusa aegyptiaca*
6. *Anchusa cretica*
7. *Anchusa variegata*

Anchusa ALKANET

Bristly annual or perennial herbs. Flowers with blue, purple, white, funnel-shaped corolla tube; stamens equal and not exserted.

(a) Flowers blue (occasionally whitish).

Anchusa undulata A hairy biennial 40 cm–1 m with elliptic leaves to 20 cm forming a bushy basal rosette; *leaves with undulate margins*, and some hairs with white, pimple-like bases, among dense, short hairs. Corolla *deep violet–blue*, funnel-shaped, corolla 5–10 mm across; bracts oval-lance-shaped and *shorter* than, or scarcely exceeding the calyx; calyx divided for about 2/3 its length. Coastal sands. Throughout. Several subspecies have been described, the most widespread being **subsp. hybrida.** *A. officinalis* is a similar perennial with rather uniform bristles – scarcely pimple-based, *flat-margined leaves*, and bright, clear-blue flowers. Virtually throughout, except for CR, CY and the southeast.

Anchusa azurea LARGE BLUE ALKANET A robust species to 1.8 m with dense bristles, often with white, pimple-like bases. Leaves to 25(40) cm long and 50 mm wide; lance-shaped. Flowers large, the corolla 8–15(20) mm across, deep blue or purple with a whitish centre (sometimes all white or cream); flowers borne in a large, loose and much-branched inflorescence; calyx divided almost to the base into linear lobes, erect in fruit. Frequent on bare ground. Throughout.

Anchusa cespitosa A tufted, *woody-based* perennial forming flat *mats*, with prostrate stems, to 50 cm across. Leaves all basal, linear-oblong, often with undulate margins, with pimple-based bristles. Flowers borne in virtually stalkless cymes of 1–3(5); calyx divided almost to the base; corolla 10–15 mm, bright blue. A local island endemic. CR (west).

Anchusa strigosa A rather *robust, erect biennial* to perennial to 80 cm with sparse, pimple-based bristles. Leaves elliptic, rather leathery. Flowers bright blue (or whitish), borne in panicles of lax cymes; corolla 9–15 mm; calyx 6–8 mm, shorter than the corolla tube and divided to at least 2/3 its length into rather *broad, spreading lobes* in fruit. Eastern only. AA (Rhodes), CY, IP, LB, SY, TR.

(b) Flowers pale yellow or white.

Anchusa aegyptiaca EASTERN ALKANET A short, very bristly annual with dense, short hairs, prostrate stems 15–35(40) cm and oval leaves with undulate margins to 10(15) cm; bristles with conspicuous white bases. *Flowers pale whitish-yellow*, the corolla 4.5–6.5 mm across, borne in lax, unbranched inflorescences; calyx 5 mm. Disturbed habitats and coastal garrigue in the south and east (rare or absent elsewhere). AA, CR, CY (common), IP, SY.

(c) Flowers white, blue or pink. *Only 2 fertile stamens. Described as a separate genus Anchusella* by some authors.

Anchusa cretica (syn. *Anchusella cretica*) An annual to 30 cm with prostrate to ascending stems. Leaves broadest above the middle, the uppermost lance-shaped, irregularly toothed and with conspicuous, erect, pimple-based bristles. Flowers *bright blue*, often with purple margins; corolla zygomorphic with rounded lobes; calyx divided almost to the base into linear-lance-shaped segments, *maroon and hairy*. GR, IA, PN. **A. variegata (syn. Anchusella variegata)** is similar but with whitish to pinkish flowers with corolla lobes each with a large purple spot. Fairly common across southern Greece. AA (local), CR (common), GR (south), IA, PN.

Hormuzakia

Similar to *Anchusa* and but with a distinctive *congested, aggregate* inflorescence and helmet-shaped nutlets (mericarpids).

Hormuzakia aggregata A densely bristly annual, divergently branched from the base and above, 50 mm–40 cm. Leaves oblong to spatula-shaped or linear, slightly undulating, stalked below, stalkless above. Flowers borne in *dense, terminal, corymb-like* clusters (unlike related genera); flowers virtually *stalkless*; corolla blue and exserted, 10 mm across; calyx 5–6 mm. Nutlets 3–5 mm, short-beaked, with an inflated ring at the base. Mainly maritime sands. AA, CY, IP (common on coastal plains), LB, TR.

Omphalodes

Small annuals. Flowers with corolla very short with a throat closed by 5, blunt, forward-pointing lobes; stamens included.

Omphalodes luciliae A small, tufted perennial to 20(40) cm. Basal leaves blunt, long-stalked, oval to elliptic; stem-leaves smaller and virtually stalkless. Flowers about 7 mm across; calyx hairless; corolla pale sky-blue with yellowish central folds. Rocky crevices in mountains. GR, PN, TR.

Cynoglossum HOUND'S TONGUE

Hairy biennial to perennial herbs. Flowers borne in branched cymes without bracts; corolla with a short tube and 5 spreading petal lobes with scales closing the throat; stamens not protruding. Nutlets egg-shaped and with barbed spines.

Cynoglossum creticum BLUE HOUND'S TONGUE A short, robust, softly hairy biennial 30–70(90) cm with erect, angular stems branched above. Leaves to 30 cm, lance-shaped, untoothed and densely hairy to felted on *both* surfaces, sometimes clasping the stem. Flowers purplish in bud, opening pale blue–violet with fine, *darker blue venation*, the corolla (5)7–9 mm across, borne in branched cymes that elongate in fruit. Nutlets without a distinct border, and with dense hooked spines, 6 mm. A range of habitats, often on disturbed ground. Throughout. *C. columnae* has deep, dull blue or purple flowers without conspicuous veins, just 5–6 mm long. Nutlets with a wide border. Throughout, except CY, IP and much of the east.

Cynoglossum sphacioticum A small *woody-based perennial* to 15 cm. Leaves lance-shaped, *woolly-hairy*. Flowers borne in cymes; corolla 5 mm, dark purplish-blue. Nutlets to 7 mm, reddish, ovoid, convex, densely covered in blunt bristles and borderless. CR (western mountains). *C. troodi* is a similar island endemic. CY (Troodos Mountains).

Pardoglossum

Hairy biennial to perennial herbs closely related and morphologically similar to *Cynoglossum* but established to be a genetically distinct group.

Pardoglossum cheirifolium (syn. *Cynoglossum cheirifolium*) A biennial 25–40(65) cm with leaves to 18 cm, conspicuously *white-felted* on both surfaces; somewhat undulate. Flowers with corolla 3–6.5 mm across, *reddish-purple* or bluish, borne in crowded racemes, elongating in fruit. Nutlets with a distinctly *thickened border*, and with dense hooked spines, or smooth. Dry, stony places and waste ground. IA.

Myosotis FORGET-ME-NOT

Hairy annuals or perennials. Flowers with corolla tube short and the throat enclosed by 5 short scales; stamens equal and included. Numerous species occur in Northern Europe and in cooler parts of the region covered.

(a) Calyx with almost adpressed, *straight* hairs only.

Myosotis laxa TUFTED FORGET-ME-NOT An erect to ascending annual to 40(70) cm, not conspicuously hairy, but with adpressed hairs on the stems and leaves; leaves to 55 mm. Flowers small, 4–5(6) mm across, pale blue with rounded petal lobes; flowers borne sparsely along the stem, densely at the apex; calyx lobed >½ to the base with triangular lobes, and with straight hairs. Local, in wet habitats. GR, PN (north).

(b) Calyx tube with spreading to deflexed *hooked* hairs.

Myosotis ramosissima EARLY FORGET-ME-NOT A low, *bristly*, erect to spreading, short-lived annual to 25 cm. Flowers pale sky-blue, just (0.8)2(4) mm across, with corolla tube shorter than the calyx; calyx teeth spreading, flower-stalks equalling calyx, and inflorescence longer than the leafy part of the stem when in fruit. Nutlets borderless. Wet, sandy and grassy habitats; probably the most common species. Throughout. ***M. congesta*** is similar with *dark blue* flowers c. 1 mm and bordered nutlets. AA (local), CR, GR.

1. *Hormuzakia aggregata*
2. *Cynoglossum creticum*
3. *Cynoglossum columnae*
4. *Myosotis ramosissima*

Myosotis arvensis An annual or biennial to 40 cm. Lower leaves spatula-shaped; stem-leaves oblong, all with soft, straight, spreading hairs. Flowers borne in long racemes, rather congested above; calyx tube with hooked or deflexed hairs; corolla 3 mm, bright blue with a yellow centre. Nutlets narrowly bordered. AA (local), GR, PN.

Myosotis sylvatica A biennial to short-lived perennial to 45 cm with spreading hairs below and adpressed hairs above. Basal leaves spatula-shaped *without* distinct stalks. Flowers with hooked hairs on the calyx and straight hairs on the lobes; corolla 5–8 mm, pinkish in bud, bright blue with a yellow centre when open. Nutlets narrowly bordered. Mountain habitats. Absent from most islands. GR, PN, TR.

Myosotis refracta A slender annual to 15(25) cm with soft, spreading hairs, and hooked hairs at the base of the stem. Flowers borne in *long, slender, 1-sided racemes*; calyx deflexed on short, *curved stalks* in fruit; corolla 1.5–2 mm, rather cupped, blue. Nutlets narrowly bordered. Rocky, limestone habitats. AA (local), CR (mountains), GR, PN (rare).

CONVOLVULACEAE | CONVOLVULUS FAMILY

Typically climbing annuals or perennials (sometimes shrubs) with alternate leaves (leafless parasites in *Cuscuta*). Flowers 1–few in leaf axils, large and showy in some genera; sepals and petals 4–5, fused at the base or to form a funnel-shaped corolla tube; stamens 4–5; style(s) 1–2. Fruit a capsule.

Cuscuta DODDER

Leafless, parasitic twining herbs virtually without chlorophyll. Flowers small with lobed 4–5-lobed corolla; styles 2. Fruit a 2–4-valved capsule. A number of species have become globally naturalised, but a lack of characters that are useful for identification in herbarium specimens has led to conflicting information in the literature.

(a) Style solitary. Flowers borne in lax, spike-like clusters.

Cuscuta monogyna A sparingly branched, robust twining annual. Flowers borne in *lax, spike-like* clusters; corolla lobes almost erect and half as long as the broadly cylindrical tube; *style solitary*. Capsule 4–6 mm. Parasitic on *trees and shrubs*. Rather rare and local. CY, GR, IP, PN, TR.

(b) Styles 2. Flowers normally 4-parted, stems reddish, yellowish or whitish.

Cuscuta palaestina A twining annual with slender, often reddish stems. Flowers borne in spherical clusters 4–6 mm across; flowers 4-parted, 1.5–2 mm; calyx teeth fleshy; corolla lobes almost *erect*, as long as the tube; styles + stigma *longer* than the ovary. Capsule rounded, 1.5–2 mm. Parasitic on dwarf shrubs; mostly in the south and east – there one of the most common species. Throughout, except the north. ***C. rausii*** is similar but with flowers in *distinctly stalked, lax clusters* of 2–19; calyx exceeding the corolla; style and stigma together longer than the ovary. On shrubs such as *Genista fasselata*. KP, IP.

1. *Cuscuta palaestina*
2. *Cuscuta epithymum*
3. *Cuscuta brevistyla*
4. *Cuscuta planiflora*
5. *Cuscuta approximata*

Cuscuta europaea A twining annual with rather *thick, purplish or reddish stems* 1.6 mm wide and *4-parted* flowers 2–3.5 mm borne in clusters of 5–20; filaments very *thin*; styles + stigma *shorter* than the ovary. Parasitic on various herbs, particularly *Urtica* spp. High-altitude habitats in woods and hills. GR (north), IA (local), TR.

(c) Styles 2. Flowers normally 5-parted, stems reddish or purplish.

Cuscuta epithymum COMMON DODDER A spreading annual with *reddish*, thread-like stems 0.8 mm wide. Flowers 1.5–4.9 mm, pale pink with a reddish calyx, scented and 5-parted with spreading petals borne in ball-like clusters of 5–30; *styles + stigma longer than the ovary*. Parasitic on various herbs and shrubs; one of the commonest species. Throughout. ***C. suaveolens*** has grey–purple stems, 5-parted, and *stalked* white flowers with greenish sepals, borne in *lax* clusters; stigma club-like. A sporadic South American weed parasitic on cultivated legumes. AA (Thasos – casual).

Cuscuta atrans A twining annual with very slender *purplish* stems. Flowers 5-parted, virtually stalkless, 2–2.5 mm; calyx enclosing the corolla tube, pale purple, fleshy; lobes pointed; styles somewhat shorter than the ovary. A poorly known species, parasitic on *Verbascum spinosum*, *Astracantha cretica* and other thorny bushes in mountains. CR. ***C. brevistyla*** also has slender reddish to purplish stems (to blackish or yellowish in places) and 5-parted, *white* (or partially translucent) flowers, often in rather large clusters; calyx teeth triangular-oval; styles *very short*, together with the stigma, shorter than the ovary. Rather uncommon. AA (rare), GR, IP (probably elsewhere).

(d) Styles 2. Flowers normally 5-parted, stems often yellowish (to red-tinted).

Cuscuta planiflora A twining annual with very *slender* yellowish stems (reddish in places and in juveniles) 0.5 mm wide. Flowers in *compact, stalkless clusters* of 5–20(30), (4)5-parted, *whitish*, 2.5–3.5 mm; *calyx lobes with crystalline margins*. Parasitic on various herbs and shrubs (especially legumes). Locally throughout. ***C. approximata*** is similar but with calyx lobes with smooth (not crystalline) edges and often bright golden yellow, fleshy and shiny tips. Parasitic mainly on dwarf shrubs. Most common in hot, arid areas on the mainland. Throughout, except for most islands.

Cuscuta campestris FIELD DODDER A twining annual with rather *thick, yellow–orange* stems 1.2 mm wide and small 5-parted flowers 1.5–3 mm across borne in lax clusters 10–12 mm across; *stigma capitate (ending in a conspicuous knob)*; calyx lobes *triangular and pointed*; corolla urn-shaped and persistent (the *ovary becoming enlarged and conspicuous* within the perianth). Capsule 2–3 mm. Native to North America but very widely naturalised and parasitic on various herbs. Fields and garrigue. Throughout. *C. scandens* is similar, with flowers in clusters just 5–9 mm across; calyx lobes *oval and blunt*. Capsule large, 3.5–4 mm. AA (local), GR.

Calystegia BINDWEED

Perennial herbs with twining or prostrate stems, with a white latex, and triangular leaves. Flowers typically large and solitary, with scarcely divided, broadly funnel-shaped corolla; style 1. Fruit not lobed.

(a) Prostrate perennials.

Calystegia soldanella SEA BINDWEED A low, hairless, prostrate, spreading perennial to 50 cm. Leaves 40 mm–12 cm, kidney-shaped, deep green and veined, rather fleshy and long-stalked (30–90 mm). *Flowers pale pink* with white stripes, 40–45 mm; solitary. Very common on coastal dunes. Throughout.

(b) Climbing perennials.

Calystegia sepium HEDGE BINDWEED A vigorous climbing perennial with twisted stems to 4 m, which clamber on other plants for support. Leaves light green, matt (not glossy), arrow-shaped, 40 mm–13 cm long. *Flowers white*, 40–55 mm long with 2 epicalyx bracts, which exceed the sepals and surround them, *not, or scarcely, overlapping at the base*. Capsule 10–16 mm. Various waste and damp habitats. Throughout. *C. silvatica* GREAT BINDWEED is very similar, differing in having larger flowers, 47–65(75) mm across, and the 2 epicalyx bracts strongly inflated and markedly *overlapping* at the base, *concealing* the sepals. Capsule 10–12 mm. Waste habitats and thickets. Throughout.

Convolvulus BINDWEED

Shrubby, trailing or climbing herbs with entire or variously divided and lobed leaves. Flowers 1–few; calyx 5-parted, corolla funnel-shaped, opening fully in daylight; style 1. Fruit unlobed.

(a) Flowers normally at least partly blue.

Convolvulus tricolor DWARF CONVOLVULUS A hairy, short, spreading annual to 40 cm. Leaves 13–40 mm, oval to elliptic, untoothed and unstalked. Flowers 20–30 mm, conspicuously *with 3 bands of colour*: yellow in the throat, then white, and bright blue on the perimeter, solitary or paired on short stalks. Capsule hairy, 5–6 mm. Local, on bare and sandy ground and waste places near the sea. GR (northeast), IP (north). (Probably elsewhere.) *C. pentapetaloides* is similar but smaller still; the flower just 8–9 mm and distinctly 5-lobed. Throughout. *C. siculus* is similar to *C. pentapetaloides* but with leaves 12–60 mm, lance-shaped to oval, untoothed and *stalked* (0.5–35 mm). Flowers blue with a yellowish centre, 7–12 mm across, distinctly 5-lobed and solitary or paired on short stalks. Capsule 4–6 mm. Dry, bare places. Throughout except the north and smaller islands.

1. *Calystegia soldanella*
2. *Calystegia silvatica*

(b) Flowers pink, white or yellowish. Leaves linear to lance-shaped. Plant tufted or mat-forming (not climbing).

Convolvulus cantabrica PINK CONVOLVULUS A spreading to ascending or prostrate, tufted perennial to 75 cm with a woody stock. Leaves 12–93 mm long, linear and often broadest above the middle, *covered in long, silky, spreading hairs*. Flowers pale pink, 17–23 mm across, borne in lax clusters on stems exceeding their adjacent leaves; ovary hairy. Capsule 6–8 mm. Various dry and bare habitats; common and widespread. Throughout, except most islands (present on IA).

Convolvulus libanoticus A small, woody-based, mat-forming perennial. Leaves linear or broadest above the middle, only the midrib distinct, hairless above. Flowers 10–15 mm, reddish-pink, hairy outside, with 5 dark purplish blotches just above the throat. Rare, in rocky mountain habitats. CR (possibly), GR (Mount Giona), PN (Mount Killini), LB, TR (south).

Convolvulus lineatus A small, woody-based, mat-forming perennial with non-woody, spreading stems to 25 cm. Leaves lance-shaped, silver silkily hairy, with pointed tips. Flowers borne in small clusters; sepals 6–8 mm with adpressed hairs; corolla 15–25 mm, whitish to rose-pink. Capsule enclosed within the calyx. AA, GR, IA.

(c) Flowers pink, white or yellowish. Leaves more or less heart-shaped. Plant generally climbing (or trailing).

Convolvulus arvensis BINDWEED A creeping or twining, more or less hairless perennial to 2 m with arrow to oblong-shaped, stalked leaves 10–75 mm. Flowers pale pink to white with paler stripes, 15–20(25) mm across, solitary or paired on stalks shorter than their leaves; scarcely scented; *ovary and sepals hairless*. Capsule 6–8 mm. Frequent on fallow land and waste ground. Throughout.
C. betonicifolius is similar but more *densely hairy* with larger, darker pink flowers 21–30(45) mm in clusters of 1–2(3); sepals more or less blunt and hairy; ovary *hairy*. AA (east), CY, GR (north and east + adjacent islands), IA (Corfu), IP, TR.

Convolvulus althaeoides MALLOW-LEAVED BINDWEED A trailing or twining, hairy perennial to 2 m with a slender, creeping rootstock. Leaves 30 mm–12 cm, mallow-like; greyish, heart-shaped below and stalked, deeply lobed above with linear divisions; all hairy with somewhat wavy margins. Flowers pale pink with a *darker throat*, 27–37(40) mm across, solitary or paired, long-stalked. Capsule 8–10 mm. Garrigue and exposed maquis; common. Throughout. ***C. elegantissimus*** is similar but more slender, silver silkily hairy and flowers *uniformly pink*. Throughout, except the far north.

Convolvulus stachydifolius A trailing perennial to 1 m, similar in form to *C. althaeoides* but with spreading, crisped to *backward-directed* hairs. Leaves 15–60 mm, rough-hairy, oval-kidney-shaped with scalloped to toothed margins, the uppermost lobed. Flowers borne 1–3(5) on stems *much exceeding* the leaves (30–90 mm); at a divergent angle from the main stem; sepals 6–8 mm, broadly elliptic, papery-margined and pointed; corolla pink–purple, 25–35 mm; ovary hairless. Deserts, plains and bare roadsides. IP (fairly frequent), LB, SY.

Convolvulus coelesyriacus A scarcely hairy *annual* with spreading to ascending stems 15–40 cm. Lower leaves stalked, 15–25 mm, *broadly oval, blunt and heart-shaped at the base*; upper leaves toothed to lobed, the uppermost often linear with 2 lower lobes. Flowers solitary; sepals rounded; corolla 12–25 mm, bright pink; the stalks curved in fruit. Capsule hairless. Only in the far south and east. AA (island of Megisti), CY, IP, LB, TR (south).

---

1. *Convolvulus tricolor*
2. *Convolvulus pentapetaloides*
3. *Convolvulus cantabrica*
4. *Convolvulus arvensis*
5. *Convolvulus althaeoides*

Convolvulus scammonia A trailing, hairless perennial to 80 cm. Leaves stalked, arrow-shaped, sometimes toothed. Flowers borne in the axils, long-stalked, solitary or few; outer sepals shorter than the inner; corolla broadly funnel-shaped and large (rather like *Calystegia* spp.), 30–45 mm across, *cream or pale yellow* with 5 pale maroon stripes outside. Capsule hairless. AA (east), SY, TR.

(d) Plant a shrub.

Convolvulus oleifolius A shrub with *very narrow, linear leaves*. Flowers pink, to 30 mm across, borne in lax clusters. Garrigue and rocky habitats. Throughout, except the far west (absent from IA). *C. dorycnium* is similar but a woodier, more *intricately branched* shrub, densely hairy but not silvery. Flowers pink, 20–25 mm across, borne in branched inflorescences. Dry garrigue in the south and east. AA, CR, IP, SY, TR. *C. argyrothamnos* is similar to *C. oleifolius* but more robust, branched, and woody almost throughout, with long, silvery, silky-hairy leaves. A rare island endemic chasmophyte in gorges. CR.

Ipomoea

Climbing annuals or tuberous perennials with oval to heart-shaped leaves. Flowers with funnel-shaped, unlobed, often showy corolla; style 1. Fruit not lobed.

(a) Flowers white or yellowish.

Ipomoea imperati (syn. *I. stolonifera*) FIDDLE-LEAF MORNING GLORY A fleshy, *creeping* perennial to 1 m with alternate, entire to deeply 3-lobed leaves 40–60 mm, heart-shaped at the base. *Flowers large and white* or pale yellow, sometimes with a purple centre, 35–50 mm across. Coastal sands. Native to the Americas but locally naturalised in the south and east. AA, CR, CY, IP, LB, TR.

(b) Flowers blue or purplish.

Ipomoea cairica A twining or trailing perennial herb with slender, angular stems to 5 m. Leaves 40–50 mm, palmately lobed to the base and stalked (stalks 20–80 mm); basal lobes again lobed or toothed. Flowers 1–few, borne on stalks 20–80 mm; sepals unequal, the outer 4–6.5 mm, the inner 5–9 mm, hairless; corolla pinkish-purple with a dark centre (rarely white); stigma 2-lobed. Capsule spherical, 10 mm across. Of uncertain origin; occasionally naturalised in hot, dry areas. IP (mainly Sharon Plain + Lower Gallilee and adjacent areas); possibly elsewhere.

Ipomoea indica (syn. *I. acuminata*) MORNING GLORY A perennial herb with long trailing stems to 15 m. Leaves to 18 cm, oval and *entire* to deeply 3-lobed, heart-shaped at the base. Flowers large and conspicuous: sky-blue or purplish (rarely white), fading pink, 60–86 mm. Naturalised in thickets, usually near towns. IA, PN. *I. purpurea* is similar but with always entire, oval leaves 80 mm–16 cm and *deep purple flowers* 40–50(60) mm. A native of tropical America, commonly naturalised. Scattered throughout. *I. nil* (syn. *I. hederacea*) is an exotic ornamental native to tropical America with *3-lobed* leaves and light purple flowers 20–40 mm. Planted in developed areas and occasionally naturalised.

---

1. *Convolvulus stachydifolius*
2. *Convolvulus stachydifolius* (left, basal leaf; right, upper leaf)
3. *Convolvulus oleifolius*
4. *Ipomoea imperati* PHOTO: NICK TURLAND
5. *Ipomoea cairica*
6. *Ipomoea nil*

SOLANACEAE | POTATO FAMILY

Herbs or shrubs with simple, entire or pinnately divided, alternate leaves. Flowers with a star- or bell-shaped corolla, the 5 petals fused below; stamens 5, attached to the corolla tube; style 1; ovary superior with 2 compartments. Fruit a berry or 2–4-valved capsule.

Lycianthes

Herbs, shrubs, small trees and climbers with simple, often unequally paired leaves. Calyx bell-shaped; corolla 5-lobed, white, blue or purple. Fruit a berry.

Lycianthes rantonnei BLUE POTATO BUSH An arching shrub 2–3 m, more or less hairless with angular stems. Leaves elliptic, to 10 cm long, entire and wedge-shaped at the base; leaf stalks to 15 mm, winged. Flowers borne in umbels of 2–5, each with stalks 15–25 mm; calyx 4–5 mm with linear lobes; corolla *bright dark blue*, 20–30 mm across. Fruit a yellow berry to 30 mm across. Native to South America, widely planted in urban areas.

Atropa

Herbaceous, almost hairless perennials with solitary or paired entire leaves. Corolla bell-shaped and shortly 5-lobed, stamens not projecting. Fruit a berry.

Atropa belladonna DEADLY NIGHTSHADE A faintly unpleasant-smelling and poisonous, stout, erect, branched, short-lived perennial to 1(2) m, glandular but not distinctly hairy. Leaves oval-pointed, rather large 80 mm–20 cm long. Flowers *bell-shaped and liver-coloured* (rarely green–yellow), stalked; corolla 24–30 mm long. *Fruit a black, shiny, fleshy berry* 15–20 mm across framed by the 5-lobed persistent calyx. Mountain limestone woods and gorges. GR, PN (rare), TR.

Nicandra

Herbaceous, hairless annuals with toothed to lobed leaves. Corolla broadly bell-shaped and shallowly lobed; calyx much-inflated. Fruit a dry berry.

Nicandra physalodes APPLE OF PERU A hairless, unpleasant-smelling, vigorous, much-branched annual to 80 cm with stalked, oval and toothed to lobed, large leaves. Flowers borne singly in the leaf axils; corolla 12–20 mm, blue to violet with a white centre, soon closing; calyx much enlarging in fruit (25–35 mm). A native of Peru, widely naturalised.

Lycium

Woody, almost hairless shrubs with simple, deciduous leaves. Flowers dull purple, rather deeply lobed, often with stamens protruding. Fruit a berry.

Lycium schweinfurthii A deciduous, rigid shrub 3–5 m with intricately branched, arched stems; many of the lateral branches comprising robust spines; shoots hairy when young. Leaves rather few, borne in clusters, broadest above the middle, 10–50 mm long, blue–green and hairless. Flowers borne 1–3 on short stalks (<4 mm); calyx 1.5 mm, cup-shaped; corolla mauve, *long and slender, 15–20 mm*, greatly exceeding the (small) calyx, with rounded, spreading lobes 2.5–3 mm across. Fruit 4 mm across; fruiting calyx not enlarged. Coastal habitats. AA, CR, CY, IP (coastal plains + western Negev).

SOLANACEAE | 451

Lycium europaeum TEA TREE A deciduous, robustly spiny shrub to 3(4) m. Leaves elliptic, broadest above the middle, 20–73 mm long. Flowers borne in clusters of 2–5, with a pink or white corolla, narrowly funnel-shaped, 10–17 mm long (less slender than *L. schweinfurthii*); lobes 3–5 mm; stamens usually protruding; calyx 2–3.5 mm. Fruit a spherical, reddish berry, 5–6 mm. Hedges and thickets. Throughout (not CR, CY or much of AA).

Lycium barbarum DUKE OF ARGYLL'S TEAPLANT A deciduous shrub to 2.5 m with arched, grey-white branches with few, slender spines. Leaves alternate or in clusters, narrowly elliptic and broadest at the middle; untoothed, 20 mm–10 cm long. Flowers in small clusters of 1–3(7); corolla red-purple, turning brown with age, trumpet-shaped, *small*, 7–12 mm; lobes 4–5 mm; stamens long-protruding; calyx 3.5–5.5 mm. Fruit an orange-red, ovoid (slightly elongated) berry, 10–20 mm. Native to China, widely naturalised. GR, IA. TR. *L. chinense* is very similar but with a *shorter calyx*, just 1.8–3.3 mm; corolla 10–15 mm; lobes 5–8 mm. *Fruit smaller*, more rounded, 5–10 mm long. GR, TR.

1. *Lycianthes rantonnei*
2. *Atropa belladonna*
3. *Lycium schweinfurthii*
4. *Lycium chinense*

Hyoscyamus HENBANE

Glandular-hairy annuals to biennials with simple, toothed to lobed leaves. Flowers borne numerously in rows along outwardly coiled, leafy stems; corolla funnel-shaped. Fruit a splitting capsule.

(a) Flowers whitish to yellow, often with a dark purple throat.

Hyoscyamus aureus A rather woody-based, glandular-hairy perennial with spreading stems, or suspended on cliffs, to 80 cm. Leaves stalked, oval, sharply and irregularly toothed. Flowers borne in long cymes; calyx 15–20 mm with broadly triangular teeth; corolla 30–45 mm, *strongly uneven (zygomorphic)*, *golden yellow* with a dark brownish-purple throat; filaments violet and long-protruding with white to violet anthers. Old walls and limestone cliffs. AA, CR, CY, IP, LB, SY, TR.

Hyoscyamus albus WHITE HENBANE A sticky annnual or short-lived perennial with long, sparsely branched stems to 80 cm, often woody below. Leaves 60 mm–20 cm long, broadly oblong and wedge-shaped or heart-shaped at the base, with wide teeth along the margin. *Flowers stalkless*, at least above, borne in long, dense, 1-sided spikes; calyx densely hairy and swollen below, ending in short, triangular teeth; corolla 20–30 mm, greenish or *yellowish-white* with a green or purple throat; *stamens not, or scarcely protruding*. Capsule 10 mm. Common on waste ground and near buildings. Throughout. *H. desertorum* is a similar *desert* annual, *conspicuously glandular*, with oval to diamond-shaped leaves. IP, SY.

Hyoscyamus pusillus An annual to 35(60) cm, woody below, glandular-hairy. Leaves mostly basal, oval to lance-shaped, 30 mm–10 cm. Flowers borne solitary in the leaf axils, virtually stalkless; calyx tubular (8)11–14 mm (to 27 mm in fruit); corolla yellow with a dark purple throat, just exceeding the calyx, with unequal lobes; stamens not protruding. Capsule 4.5–6(7) mm; seeds 1.2 mm, compressed, greyish. Rare and local, in deserts. IP (southeast).

(b) Flowers with conspicuous dark, netted veins.

Hyoscyamus niger HENBANE A sticky, unpleasant-smelling annual or biennial with long, sparsely branched stems to 80 cm, often woody below. Leaves 60 mm–20 cm long, oval and coarsely toothed, forming a basal rosette; stalked below but stalkless above and somewhat *clasping the stem*. Flowers virtually stalkless, borne in long, dense, 1-sided spikes; calyx densely hairy and swollen below (especially in fruit), ending in short, triangular teeth; corolla to 20–30 mm, slightly zygomorphic, *dull pale yellow with purple veins*; stamens slightly protruding. Capsule 10 mm. Disturbed habitats. GR, PN (rare), TR.

Hyoscyamus reticulatus A stout, robust biennial (to perennial), branched from the middle or above, 40–50 cm, almost hairless to cobweb-hairy. Leaves lance-shaped and *deeply pinnately lobed*; stalkless but not clasping above, the uppermost (floral) leaves entire, exceeding the flowers. Corolla 20–35 mm, nearly *regular*, greenish to purplish with *netted violet veins*; lobes rounded, ¼ the length of their tube; fruiting calyx 20–30 mm, with recurved, sharp-pointed teeth ½ as long as their tube. Field margins, wadis and roadside scrub near deserts. IP (frequent in northern Negev and Judean Mountains; rare elsewhere), SY.

Physalis

Annuals or perennials with simple, often toothed leaves. Flowers with broadly funnel-shaped corolla with 5 spreading lobes. Fruit a berry, often concealed by a markedly inflated calyx.

Physalis philadelphica An erect, branched annual to 60 cm with ridged stems. Leaves alternate, rounded with wavy-toothed margins, to 10(13) cm long. Flowers borne in leafy cymes, nodding, 22–30 mm across; calyx 5–6-lobed, 6–7.5(10) mm long, expanding in fruit; corolla 5–6-lobed, yellow, to 30 mm across; anthers 1.25–4 mm. Fruit a large yellow, reddish or purple fleshy berry to 60(85) mm, enclosed in the expanded calyx tube. Native to the Americas; locally naturalised in Greece, possibly elsewhere. GR (Sterea Ellas). *P. angulata* is similar, with anthers to just 2 mm, and *smaller* yellowish-green berries 10–12 mm. A casual native to the Neotropics. GR, IA, PN.

Physalis ixocarpa A *sparsely hairy, branched, erect annual* 40–60 cm. Leaves lance-shaped, entire or with coarse teeth, and rounded bases; stalked. Flowers borne solitary in the leaf axils; calyx oval, green and purple-veined, with 5 teeth; corolla bright yellow with brownish markings in the centre; filaments purple. Fruit borne in a very inflated, greenish, purple-veined calyx 30–50 mm enclosing a berry 13–40 mm. Native to Mexico; naturalised. GR (west), PN.

1. *Hyoscyamus aureus*
2. *Hyoscyamus albus*; (inset) spotted form
3. *Hyoscyamus desertorum*
4. *Hyoscyamus pusillus*
5. *Hyoscyamus niger*
6. *Hyoscyamus reticulatus*

SOLANACEAE

1. *Physalis philadelphica*
2. *Withania somnifera*
3. *Solanum dulcamara*
4. *Solanum nigrum*
5. *Solanum luteum*
6. *Solanum luteum* fruits

Withania

Shrubs with opposite leaves and bell-shaped flowers. Fruit a berry *surrounded by an inflated calyx.*

Withania somnifera ASHWAGANDHA A woody-based perennial to 1.5 m with sparingly branched stems. Leaves 40 mm–10 cm, wedge-shaped at the base. Flowers borne 4–16 in crowded clusters arising from the axils; calyx 5 mm, densely downy; corolla small, yellowish-green with spreading lobes. Fruit a red berry 5–8 mm. Dry habitats and deserts. AA (rare), CR, CY, GR, IP, TR.

Solanum NIGHTSHADE

Herbs or shrubs with simple leaves. Flowers borne in 1-sided cymes or umbels; leaf-opposed; corolla star-shaped with spreading petal lobes; stamens protruding, forming a cone around the stigma. Fruit a berry.

(a) Plant a spiny shrub.

Solanum linnaeanum APPLE OF SODOM A very *spiny shrub* to 2 m with much-branched, stout stems covered in yellow spines; slightly hairy. Leaves 50 mm–13 cm, oval and pinnately, bluntly lobed; prickly, and stalked. Flowers with a violet, 5-lobed corolla 20–30 mm across, borne solitary or in clusters. *Berry large and spherical, 20–40(50) mm across*, marbled green and white, later yellow or brown and shiny, tomato-like. Native to Africa, occasionally naturalised in waste places and near buildings. GR.

(b) Plant not spiny and flowers purple.

Solanum dulcamara WOODY NIGHTSHADE A scrambling, woody-based, shrubby perennial to 3(7) m with oval-lance-shaped leaves to 80 mm long, simple, or the lower often lobed at the base. *Flowers with purple petals with contrasting yellow anthers*, borne in lax clusters. Fruit *a red berry 8–12 mm long*. Woods and damp habitats, often on higher ground. AA (local), GR, IP, PN, TR.

(c) Plant not (or scarcely) spiny and flowers white. A complex group with similar features, and hybridising; consequently with conflicting regional accounts.

Solanum nigrum BLACK NIGHTSHADE A variable, hairless to hairy *annual* to 70 cm with stems spreading and blackish. Leaves oval-lance-shaped, toothed, lobed or entire; stalked. Flowers with white petals, star-like with a yellow cone of anthers borne in clusters of 5–10; corolla 5–6 mm. Berry 6–10 mm, round and green ripening matt-black, borne on erect (to spreading) stalks. A common weed on cultivated land. Throughout. *S. nigrum* has a number of subspecies and is one of a complex group of taxa: *S. americanum* (under which *S. nigrum* is also treated as a synonym) has erect fruiting stalks but is otherwise similar. *S. luteum* (syn. *S. villosum*) is *densely long-hairy* with more deeply lobed leaves, clusters of 3–6 flowers, and fruits *reddish or yellowish* (not black), 6–10 mm, often longer than wide. Common in dry, bare places. Throughout. *S. sinaicum* has *blackish-violet* striping on the outer corolla. IP. *S. elaeagnifolium* has leaves white-downy beneath, and with small spines. IP. *S. chenopodioides* is a *softy short-hairy perennial* with purplish-blackish fruits 6–8 mm borne on *deflexed stalks*. Native to South America; widely naturalised.

(d) Plant an ornamental, evergreen climber. Flowers white.

Solanum laxum (syn. *S. jasminoides*) POTATO VINE An ornamental *evergreen, climbing perennial* to 5 m. Leaves 25–75 mm, bright green, leathery and shiny; pointed and lance-shaped with heart-shaped base; stalked. Flowers with a white corolla 15–18 mm across, star-like, with a yellow cones of anthers, borne in lax, many-flowered, showy cymes. Fruit 4–5 mm. Widely planted in developed areas. Throughout.

Mandragora MANDRAKE

Leaves basal. Flowers short-stalked and arising directly from the root stock. Fruit a berry. The roots were formerly used in the Mediterranean to relieve pain and induce sleep.

Mandragora officinarum MANDRAKE A ground-hugging, stemless, hairless or hairy, autumn, winter or spring-flowering perennial with a robust rootstock. Leaves to 30(45) cm, forming a large, flat *rosette*, oval, bright green and shiny, stalked and distinctly wavy-margined. Flowers with a greenish-white to blue–violet corolla, 25–50 mm across with 5 triangular lobes 20–34 mm, borne in the centre of the rosette. Fruit an orange or yellow, egg-shaped berry 16–25(40) mm across, held in a large, persistent calyx that equals or exceeds the fruit. Fallow land and garrigue. Strongly southern. AA, CR, CY, GR (rare), IP, PN (rare), TR (possibly elsewhere). Autumn-flowering forms referred to as **M. autumnalis** are no longer widely considered to be distinct.

Datura (including *Dutra*)

Erect annuals with simple leaves. Flowers regular, showy, trumpet-shaped, upward-pointing; calyx with 5 teeth. Fruit a large, spiny capsule.

Datura stramonium A stout and vigorous, unpleasant-smelling, normally *hairless* annual to 1.5 m with stout, spreading stems. Leaves oval to elliptic, lobed with jagged teeth, 50 mm–18 cm. Flowers with a white corolla, sometimes flushed with purple, trumpet-like, 50 mm–10 cm long, borne solitary in the leaf axils of the upper leaves; calyx large, to ½ the length of the corolla (30–50 mm), sharply angled. Fruit an erect, large, spiny capsule 35–70 mm long. Frequent in disturbed habitats. Throughout. **D. ferox** is very similar but with smaller flowers; calyx 25–40 mm and corolla 40–60 mm. *Fruit with sparse, stout, long spines 10–30 mm*. Distribution unclear owing to confusion with the previous and next species.

Datura innoxia (syn. *Dutra innoxia*) A perennial (now classified by some authors under the separate genus *Dutra*), *softly hairy*, to 50 cm (2 m), with *large flowers* with corolla 14–16.5 cm long, hairless outside, and capsules 30–50 mm, *nodding* when mature. Disturbed habitats. Casual throughout.

Brugmansia

Evergreen shrubs and trees with entire leaves and large, showy, drooping, fragrant flowers. Native to South America but very commonly planted. Fruit large, smooth and pod-like. A highly complex genus in which numerous hybrids and cultivars exist, which are difficult to distinguish; below is a simplified and approximate guide only.

Brugmansia versicolor ANGEL'S TRUMPET An ornamental shrub to 3(5) m with thick, stiff stems. Leaves oval, stalked and untoothed. Flowers showy, pendent, white, flushed with yellow or pink with age, trumpet-like, *very large, 30–45(50) cm long* with a slender tube, and fragrant; borne solitary in the upper leaf axils; calyx inflated, slit along 1 side. Fruit large and egg-shaped. Widely cultivated. **B. arborea** is a similar small tree to 7 m with white, pale greenish or white, nodding flowers; *corolla smaller, 12–17 cm long*. Fruit soft-walled when ripe; seeds angular. Commonly planted in towns. **B. sanguinea** is an ornamental tree to 10 m similar in form to the previous species but with *tubular red, orange or yellow flowers*; calyx ribbed. Fruit smooth. Commonly planted in resorts and gardens.

1. *Mandragora officinarum*; (inset) immature fruits
2. *Datura stramonium*
3. *Datura innoxia*; (inset) fruit
4. *Datura innoxia* habit
5. *Brugmansia arborea*

Cestrum

Spineless, hairless shrubs. Flowers with tubular corolla, borne in clusters. Fruit a berry. Native to tropical America.

Cestrum nocturnum NIGHT JASMINE A shrub to 4 m with alternate, simple, lance-shaped, pointed leaves. *Flowers white*, fragrant, abundant and showy, 20–25 mm long with 5 stamens, not protruding. Fruit a 2-parted berry to 10 mm. Planted in the region, possibly naturalised. *C. parqui* is a similar shrub to 4 m with *yellow flowers* 19–27 mm. Often planted.

Nicotiana TOBACCO

Shrubs or perennials, sticky, with simple, entire leaves. Flowers with corolla elongated and funnel-shaped, borne in branched, leafless clusters. Fruit a 2-valved capsule. Native to South America, widely planted and naturalised.

Nicotiana glauca SHRUB TOBACCO A hairless, lax shrub *or small tree to 6 m* with long grey branches and sparse, *grey–green*, stalked, elliptic-lance-shaped, pointed leaves 21 mm–12 cm. Flowers greenish-yellow, borne in lax panicles; corolla tubular, 27–45 mm long. Fruit an egg-shaped capsule 8.5–15 mm, exceeding the persistent, papery calyx. Widely naturalised in waste places, derelict sites and on walls. Throughout.

1. *Nicotiana glauca*
2. *Solandra maxima*
3. *Syringa pubescens*
4. *Fraxinus angustifolia*
5. *Fraxinus ornus*

Solandra

Vigorous shrubby vines native to Central and South America with shiny foliage. Flowers very large, showy. Fruit a large capsule.

Solandra maxima HAWAIIAN LILY A shrubby climber with oval, dark green leaves and *large, showy, trumpet-shaped flowers to 20 cm long*; yellow–cream with purple veins and prominent stamens. Fruit a berry. A striking, exotic ornamental widely planted in towns, gardens and resorts in the region.

OLEACEAE | OLIVE FAMILY

Trees and shrubs, usually with opposite, simple leaves. Flowers in cymes or panicles; calyx with 4 small teeth; corolla with 4(–6) free or fused petals; stamens usually 2; style 1. Fruit a 2-valved capsule, 2–4-seeded-berry or winged nut or achene.

Syringa

Deciduous shrubs or small trees with opposite, entire to toothed leaves. Flowers borne in panicles, appearing after the leaves; corolla lilac (rarely white) with tube 2 x as long as lobes. Fruit a capsule.

Syringa pubescens A shrub 1–4 m with 4-angled to cylindrical branches. Leaves oval to elliptic, 15–80 mm long. Flowers lilac, borne in erect, lateral panicles to 16 cm long; calyx 1.5–2 mm; corolla to 18 mm with a long tube and oval, spreading lobes; anthers usually purplish-black. Fruit a capsule to 20 mm. Native to Korea and China; naturalised locally in dunes in the northeast; possibly elsewhere. GR (northeast).

Jasminum JASMINE

Woody climbers and shrubs with alternate compound leaves (rarely simply and opposite). Flowers yellow or white with 4–6 petals united into a tube. Fruit (usually) a 2-lobed black berry.

(a) Flowers yellow.

Jasminum fruticans WILD JASMINE An evergreen shrub to 2 m with slender, 4-angled stems. Leaves alternate, usually trifoliate, shiny, with oval-lance-shaped leaflets 7–25 mm. *Flowers yellow*, 9–18.5 mm long, unscented, borne in clusters of 1–5; lobes 4–10.5 mm. Fruit a black, shiny berry 7.8–13.2 mm. Garrigue and maquis. GR, TR.

Jasminum mesnyi An arching, evergreen shrub with angular stems and opposite, deep green, trifoliate leaves with elliptic leaflets. Flowers 15–22 mm long, borne singly and laterally; bright yellow, often with an orange-tinged centre, to 40 mm with 6–8 overlapping, oval lobes 6–8 mm. Native to China, widely cultivated.

(b) Flowers white.

Jasminum officinale WHITE JASMINE A vigorous twining deciduous climber with pinnately divided, *opposite leaves* with 5–7 leaflets. Flowers *white*, 22 mm long, and strongly fragrant, borne in umbels; corolla lobes 15–20 mm. Fruit a blackish berry. Commonly cultivated in gardens and hedges.

Fraxinus

Deciduous trees with compound leaves. Flowers with 4 petals or petals absent; calyx 4-lobed or absent; stamens 2. Fruit a 1-seeded winged samara.

Fraxinus excelsior ASH A tall, erect tree to 30(45) m with smooth silvery bark and large compound leaves with 9–13 pointed, oval, *stalkless lateral leaflets* 22 mm–11 cm. Flowers unisexual or cosexual, brownish-purple, borne on the previous year's twigs before the leaves in *paniculate axillary clusters*; calyx and corolla absent. Fruit a winged samara 28–48 mm long. Widespread in open woods and river banks on the northern mainland. GR, PN, TR. *F. angustifolia* NARROW-LEAVED ASH is similar but with *pale brown (not blackish) winter buds and flowers borne in racemes*. Mixed woods and riversides. AA (Lesbos + Samothraki), GR, IA (Corfu), IP (north), PN, TR. *F. ornus* FLOWERING ASH has at least some short-stalked leaflets (stalks 24–88 mm) and conspicuous, whitish, cosexual or male flowers borne in sweetly scented, *terminal pyramidal clusters at the same time as the leaves*. Throughout, except the southeast (not CY or IP).

Olea OLIVE

Small evergreen trees with opposite, entire leaves and flowers with 4 corolla lobes and sepals borne in axillary clusters. Fruit a berry (the olive).

Olea europaea OLIVE A familiar, much-branched tree to 15 m with a grey trunk. Leaves 7–60 mm, opposite; grey–green, silvery beneath, minutely scaly, lance-shaped, untoothed and short-stalked. Flowers small, 6–8.5 mm, whitish, borne in erect clusters. Fruit an olive, 6–15(20) mm. Common in abandoned pastures and extensively cultivated. Throughout. The wild variety (var. *sylvestris*) differs from the cultivar (var. *europaea*) in having spiny lower branches.

Phillyrea

Evergreen shrubs with simple leaves. Flowers greenish or yellowish with 4 corolla lobes and projecting stamens. Fruit a berry.

Phillyrea latifolia An evergreen shrub to 2(4) m with upright branches with finely hairy twigs and leaves 10–33 mm of *2 types*: juvenile leaves oval and heart-shaped at the base, adult leaves elliptic; both leathery dark green, toothed or not, and with 7–10 pairs of *distinct* lateral veins. Flowers 3–7.8 mm, whitish. Fruit small, 4–6.6 mm, fleshy with a point, blue–black when ripe. Common on garrigue, maquis and in rocky, open woods. Throughout.

Ligustrum PRIVET

Deciduous shrubs with simple leaves. Flowers white, densely clustered; petals 4, united into a tube. Fruit a black berry.

Ligustrum lucidum CHINESE PRIVET A small, *evergreen* tree to 10(15) m with elliptic, opposite, glossy, dark green leaves 55 mm–15 cm long, and *hairless young branches*. Flowers typically privet-like: white, small and borne in dense panicles to 20 cm long; corolla 3.8–5.2 mm across, tubular with 4 spreading lobes. Berry black when ripe, 4.5–10 mm long. Native to China and widely planted. *L. vulgare* COMMON PRIVET is similar but a *deciduous* shrub with hairy young branches and *smaller leaves*, 12–64 mm long. Damp habitats and woods; local, and rare in the far south. *L. ovalifolium* is similar to *L. vulgare* but with *oval leaves* 17–59 mm, hairless young branches, and leaves remaining for most of the winter. Native to Japan, widely planted.

GESNERIACEAE

Perennial herbs with basal rosettes of leaves. Flowers solitary or in small umbels; cosexual and zygomorphic to almost regular; corolla tubular and rather 2-lipped; stamens in equal number to corolla lobes. Fruit a capsule. Representatives in the region are Tertiary relicts: geographically isolated lineages with their nearest neighbours in Asia.

Haberlea

Flower stems (scapes) with bracts. Calyx with 5 lobes, equalling the tube; corolla with unequally 2-lipped mouth and tube exceeding the lobes; stamens 4. Capsule equalling the calyx.

Haberlea rhodopensis A perennial with rosettes of coarsely toothed leaves 30–80 mm long, softly hairy, tapering into the stalks. Flowers borne on stems 60 mm–10 cm with 1–5 flowers; corolla 15–25 mm, blue–violet, hairy within. Very rare, in rocky crevices in mountains. GR (northeast).

1. *Olea europaea*
2. *Phillyrea latifolia*
3. *Ligustrum lucidum*
4. *Haberlea rhodopensis*

GESNERIACEAE

Jankaea

Perennials with leaves densely white-woolly above. Flowers with 4-lobed, bell-shaped corolla with tube as long as lobes; stamens 4.

Jankaea heldreichii A perennial with rosettes of white-hairy egg-shaped leaves 20–40 mm long. Flowers borne 1–2 on stalks to about 40 mm; calyx 5-lobed; corolla 4(5)-lobed, pale lilac; anthers yellow. Capsule 7 mm. Shady crevices, boulders and rocky overhangs; a rare point endemic. GR (Mount Olympus).

PLANTAGINACEAE | PLANTAIN FAMILY

Annual or perennial herbs or shrubs with opposite or whorled, simple or compound leaves. Flowers variable, but usually zygomorphic and 2-lipped; sepals and petals 2–4; stamens 4; style 1. Fruit a capsule or 1-seeded nut. The family includes numerous genera traditionally in the Scrophulariaceae (though the revised classification and taxonomy are not universally accepted).

Plantago PLANTAIN

Small annual or perennial herbs with a basal rosette of leaves, opposite or alternate along the stem. Flowers small and inconspicuous, borne in dense heads or spikes; 4-parted; corolla papery; stamens protruding. Fruit a splitting capsule.

(a) *Stems leafy and branched* (spikes borne in axils opposite the leaves).

Plantago phaeostoma A small, hairy, *divergently branched* annual 40–70 mm. Leaves linear, 10–25 mm long. Flowers borne in spikes on stalks as along as the leaves or shorter; spikes oblong, 14–20 mm, dense; corolla with oval, spike-pointed lobes. Rocky or sandy, arid habitats after spring rain. CR (offshore islet of Koufonisi – very rare), IP (Negev Highlands and adjacent areas).

Plantago indica (syn. *P. arenaria*) BRANCHED PLANTAIN A hairy (but *not* markedly sticky) annual to 30(50) cm with *much-branched stems*. Leaves 40–80 mm, *linear* to linear-lance-shaped, opposite or whorled, not fleshy and normally untoothed. Flowers brownish-white, to 4 mm, borne in round or conical spikes 5–15 mm, on spreading stalks; anthers pale yellow; *inner bracts larger than the outer*. Fairly common on coastal garrigue and in dry, sandy places. Throughout. *P. afra* is similar but usually (not always) extremely *sticky* and glandular-hairy above, *bracts all similar*. On waste and fallow ground. Throughout. *P. squarrosa* is spreading (not erect), with lateral branches as long as the main stem, and linear, fleshy leaves. Strongly southern. AA (rare, or recorded in error, in the east), CR (southwest + offshore islets), CY, IP, TR.

(b) *Leaves borne in a rosette*, linear or narrowly lance-shaped, stems not ribbed; spikes borne on leafless stems.

Plantago coronopus BUCK'S HORN PLANTAIN A very variable, low annual or perennial to 20 cm with solitary or clustered leaf rosettes. Leaves 20 mm–20 cm, linear-lance-shaped, usually *pinnately lobed*, though sometimes unlobed, not particularly fleshy; hairless or finely hairy. Flowers yellowish-brown, to 3 mm, borne in spikes 40–70 mm long, terminating from ungrooved, curved stems exceeding the leaves; anthers pale yellow. Very common in disturbed coastal habitats in the region. Throughout.

Plantago bellardii A low, *densely hairy* annual to 80 mm (16 cm) high with 1 or more leaf rosettes with linear-lance-shaped leaves 15–60 mm, scarely toothed or entire, 3-veined and white-hairy. Flowers brownish, borne in spreading spikes 8–20(48) mm, rather large relative to the leaves, *borne on stalks exceeded by the leaves* (to 13 cm); bracts hairy; corolla lobes oval-lance-shaped. Dry, bare ground and garrigue. Throughout. The following co-occurring subspecies are recognised

1. *Jankaea heldreichii*; (inset) flower INSET PHOTO: NICK TURLAND
2. *Plantago afra*
3. *Plantago phaeostoma*
4. *Plantago coronopus*
5. *Plantago bellardii*
6. *Plantago bellardii* subsp. *deflexa*
7. *Plantago ovata*
8. *Plantago cretica*

but are scarcely distinct: **Subsp. *bellardii*** has stalks equalling or exceeding the leaves, ascending to erect in fruit. **Subsp. *deflexa*** has *short* stalks, arching in fruit. *P. ovata* has flower-stalks with *adpressed* (not spreading) hairs and flat, sub-equal sepals. Arid and bare habitats. CY, IP, LB, SY. *P. cretica* has flowering stalks just 10–30 mm, thickened in fruit and *greatly exceeded* by the leaves; corolla lobes rounded. Southeastern. AA, CR, CY, IP, SY, TR (south).

Plantago albicans SILVERY PLANTAIN A short, tufted, *silver-woolly* perennial to 28 cm with a woody stock. Leaves 30 mm–15 cm, linear and often slightly twisted, 3-veined (obscured by hairs), and untoothed. Flowers greenish, borne in small, oblong spikes to 5 mm–11 cm long, on long, spreading or erect stems; stamens not markedly protruding. *Seeds 4–5 mm*. Locally frequent on dry, bare ground, often coastal. Throughout. *P. amplexicaulis* is similar, less hairy and *less silvery*, with elliptic leaves 20–50 mm broadest above the middle, tapered at the base, and faintly veined. Spikes 10–20(30) mm. *Seeds just 2.5 mm*. Various dry habitats. AA (rare), CR (east), CY, GR (southeast), IP, PN (northeast).

Plantago crassifolia (syn. *P. maritima* subsp. *crassifolia*) A *fleshy* perennial with few rosettes. Leaves 50 mm–20 cm long, linear, not rigid, sparsely toothed. Flowering stems numerous, exceeding the leaves, stout, with adpressed hairs; spikes 20–50 mm long, dense; bracts shorter than the calyx, scarcely keeled. Maritime rocks and dunes; scattered across the region, absent from many islands. Possibly a race of the widespread *P. maritima*. CR, GR, IA, IP.

(c) Leaves borne in a rosette, linear or narrowly lance-shaped, *stems grooved or ribbed*; spikes borne on leafless stems.

Plantago lanceolata RIBWORT PLANTAIN A variable hairy or hairless perennial to 50 cm with 1–several leaf rosettes. Leaves 15 mm–20 cm, linear-lance-shaped or lance-shaped, toothed or untoothed, 3–5-veined, *strongly ribbed* and stalked. *Bracts hairless*. Flowers brown, borne in short, blackish spikes 40(80) mm long on grooved stalks that markedly exceed the leaves; anthers pale yellow. A common weed on fallow land and grassy places. Throughout. *P. lagopus* is similar to *P. lanceolata* but smaller to 15(47) cm and more *white-hairy, especially the bracts*. Spikes 10–30 mm. Common in similar habitats to the previous species, especially near the coast. Throughout.

(d) Leaves borne in a rosette, *broadly* oval or elliptic; spikes borne on leafless stems.

Plantago major GREATER PLANTAIN A short, hairy or hairless perennial with broadly oval to elliptic leaves 50 mm–37 cm in a *single* basal rosette, 3–9-veined, narrowing abruptly into broad stalk at the base; stalk equalling the blade. Spikes *long, dense and slender, 30 mm–32 cm*, borne on unfurrowed, hairy stalks to *shorter than the leaves*; corolla whitish, anthers yellowish. Very common in cultivated and grassy places. Throughout. *P. media* Hoary plantain is similar, with 1–few rosettes with elliptic (not oval) leaves 80 mm (28 cm) long, grey-downy and *gradually narrow into short stalks at the base*. Inflorescence greatly exceeding the leaves; anthers purple and white, and *long and prominent*; spikes 15–60 mm (12 cm). GR (northern and central).

Antirrhinum SNAPDRAGON

Dwarf shrubs or woody-based herbs with entire leaves. Flowers zygomorphic and 2-lipped; stamens 4. Capsule opening by 3 apical pores.

Antirrhinum majus SNAPDRAGON A variable (many forms described, accepted by some as separate species), bushy perennial, to 65 cm (1 m) much-branched below, with stems woody at the base. Leaves lance-shaped to linear and wedge-shaped at the base, opposite or alternate. Flowers with corolla 33–45 mm, bright pink–purple (pale yellow in cultivated forms); calyx 6–10 mm. Fruit capsule 12–15 mm. Rocky slopes and fixed dunes. Throughout (in its wider description). The following similar subspecies probably occur throughout the range: **subsp. *tortuosum*** has *opposite upper leaves much longer than wide (to x 12); inflorescence hairless*; **subsp. *majus*** has leaves up to 9 x as long as wide, the uppermost always alternate, and glandular-hairy inflorescences.

Antirrhinum siculum SICILIAN SNAPDRAGON An erect or spreading perennial, more or less hairless below and glandular-hairy above; stems freely branched, with leaf tufts in the axils. Leaves 20–60 mm. Flowers borne in terminal racemes, with corolla 17–25 mm, *pale yellow*, often flushed with darker yellow and with violet veins. Fruit capsule 10–12 mm, glandular-hairy. Native to Sicily; very locally naturalised in rocky habitats and on old walls. GR, IA (Cephalonia), IP (Tel Aviv) PN.

Misopates

Annuals similar in form to *Antirrhinum* with distinctly unequal, rather long, linear calyx lobes; stamens 4.

Misopates orontium LESSER SNAPDRAGON A short, sparingly branched, more or less hairless annual to 30(70) cm. Leaves 10–55 mm, linear to elliptic, untoothed, opposite below and alternate above. Flowers with corolla 10–17 mm, pale pink or whitish, snapdragon-like; *calyx 12–20 mm with long lobes*. Fruit capsule 5–10 mm, glandular-hairy. Common in a range of habitats. Throughout.

1. *Plantago crassifolia*
2. *Plantago major*
3. *Antirrhinum majus* subsp. *tortuosum*
4. *Antirrhinum majus* subsp. *majus*
5. *Antirrhinum siculum*
6. *Misopates orontium*

1. *Linaria pelisseriana*
2. *Linaria micrantha*
3. *Linaria triphylla*
4. *Linaria haelava*
5. *Linaria haelava* (purple form)
6. *Linaria haelava* (Arava Valley form)
7. *Linaria joppensis*

Linaria

Herbs with simple unstalked leaves, opposite or whorled, alternate above. Flowers in spikes or racemes, snapdragon-like but small; calyx unequally 5-lobed and short; stamens 4. Fruit a capsule opening by slits.

(a) Seeds flat, circular and winged.

Linaria pelisseriana JERSEY TOADFLAX An erect, slender and hairless, grey–green annual with unbranched stems 15–50 cm with narrow, strap-like, pointed leaves 5–47 mm. Flowers many (2–35); corolla 15–20 mm, bright violet with a white throat-boss and slender spur 6–9 mm; calyx 4–6 mm, hairless with white-margined lobes. Capsule 3.5–5 mm; seeds flattened with hairy, *irregularly winged margins*. Widespread in sandy and dry habitats. Throughout.

Linaria vulgaris COMMON TOADFLAX A hairless, grey–green perennial *with numerous erect stems* to 1.2 m with dense linear leaves 15–53 mm. Flowers rather numerous (5–85); corolla 19–28 mm, yellow with an orange spot on the lower lip, and slightly scented; spur long, 8–15 mm and curved. Capsule 5–9 mm. Cool grassy habitats above sea level. AA (Thasos), GR.

Linaria micrantha A leafy, slightly blue–grey annual to 45(55) cm, hairless throughout except on the inflorescence. Leaves 5–40 mm, linear-lance-shaped. Racemes dense in fruit; flowers *very small*; corolla just 2.5–5 mm, lilac; calyx 2.5–5 mm; *spur minute*, 0.5–1 mm, straight or curved; borne in rather inconspicuous small terminal clusters of 8–25. Capsule 3.5–6 mm. Widespread in cultivated and waste places. Throughout. *L. simplex* is similar but with pale yellow flowers; corolla 5–9 mm; spur 2–3.5 mm. Throughout.

(b) Seeds angular and *not* winged.

Linaria triphylla THREE-LEAVED TOADFLAX A robust, hairless annual with thick, erect stems 10–65 cm, branched at the base. Leaves 6–36 mm, *broadly oval, 8–25 mm wide*, 3-veined and whorled in groups of 3. Inflorescence cylindrical and rather dense; corolla 20–30 mm, pale yellow–white, often flushed violet, with an orange throat base; spur 4–7 mm; calyx 9–12 mm with oval lobes. Capsule 6–9 mm. Widespread on open ground. Throughout (rare or absent in IP outside Esdraelon Plain).

Linaria haelava An ascending, *short, desert* annual, hairless below, glandular above, 40 mm–22 cm. Leaves 10–50 mm, thread-like to linear-oblong, tapering, alternate below, the lowermost opposite or whorled. Flowers borne 2–12 in dense racemes, which elongate in fruit; corolla 16–31 mm, *yellow, whitish, or blue-purple with a yellow palate*; spur (10)12–22 mm, straight or scarcely curved, exceeding the remainder of the corolla. Capsule 2.5–4 mm, ovoid, hairless or hairy at the apex. Seeds 0.4–0.6 mm, oblong to kidney-shaped. Dry, sandy plains and deserts. IP (mainly south-central), SY. *L. joppensis* is similar but taller and more erect, with violet flowers; spur rather shorter; calyx with prominent white hairs. IP (coastal plains).

Linaria chalepensis WHITE TOADFLAX A hairless, erect and usually unbranched, slender annual to 40(50) cm with linear leaves 15–50 mm, the lowermost in pairs or in groups of 3. Inflorescence long and lax, with small, pure white to cream flowers; corolla 12–22 mm, upper lip deeply 2-lobed, the lower lip 3-lobed; spur slender and curved, 8–12.5 mm long; calyx 3–5.5(10) mm, *with markedly unequal lobes*. Capsule 4–5 mm. Common in grassy and stony habitats. Throughout.

Linaria genistifolia A rather robust, stiffly erect, rhizomatous, hairless perennial to 80 cm, branched above. Leaves numerous, alternate, narrowly lance-shaped, 20–50 mm long. Flowers short-stalked, borne in lax racemes; calyx to 12 mm, with teeth 3–5 mm; corolla 15–25(50) mm, lemon-yellow; spur 4–25 mm. GR, TR.

Linaria tenuis A more or less erect annual to 60 cm, slightly glandular-hairy above. Leaves alternate, linear and slightly fleshy. Flowers borne in lax, elongated racemes of 5–25; corolla 14–16 mm (including spur), pale yellow with a darker palate; spur 6–7 mm. PN (southeast; recorded as *L. hellenica*) + Elafonisos.

Kickxia

Annuals with oval to elliptic or arrow-shaped leaves, entire to sparsely toothed, with pinnate veins. Corolla strongly zygomorphic; stamens 4. Capsule opening by 2 oblique lids.

(a) Flower-stalks *hairless*.

Kickxia commutata A spreading, glandular-hairy perennial with slender stems to 80 cm, sometimes rooting at the nodes. Leaves 10–45 mm, oval to arrow-shaped, long-hairy. Corolla 7–17 mm, whitish or yellowish with a blue or purple to blackish upper lip and spotted palate; spur strongly curved; flower-stalks 5–25 mm, *hairless*. Capsule leathery, 2–4 mm. Cultivated and waste ground. Throughout.

(b) Flower-stalks usually *hairy*.

Kickxia aegyptiaca A (normally) glandular-hairy, much branched, greyish annual 15–40 cm; branches stiff, becoming spine-like. Leaves bluntly oblong below, *triangular-arrow-shaped* above. Flowers borne in leafy racemes; calyx (3)4–5 mm with lance-shaped, pointed lobes; corolla creamy yellow with a faintly purple-spotted palate, 12.5–14(18) mm (including spur). Capsule 2–2.5 mm, velvety; seeds glandular and warty. Various habitats, often in dry places and deserts. IP, SY. ***K. floribunda*** is similar but with flowers *crowded*, arranged along densely leafy branches in spike-like racemes. Sandy deserts. IP.

Kickxia elatine An annual to 60 cm, similar to *K. commutata* with leaves 3–10 mm. Corolla *yellow* with a purple upper lip, 7–15 mm long; *spur straight*; flower-stalks 5–20 mm, *sparsely hairy* and *longer* than their bracts. Capsule 3–5 mm, hairy above. Throughout. **Subsp.** *elatine* is spreading with few branches; flowers 7–10 mm long. Rather scattered across the range. **Subsp. *crinita*** is rather stouter with several secondary flowering branches; flowers to 15 mm. The most common form in the region.

Kickxia lanigera A long-hairy annual to 1 m, similar to the above species but with *densely white-hairy*, ridged stems and *heart-shaped leaves*. Corolla 8–11 mm long, whitish or cream with a blue to violet upper lip and spotted palate; spur curved; flower-stalks hairy and *shorter* than their bracts. Capsule 2–3 mm, sparsely hairy. Cultivated and waste ground. AA, CY, IP. ***K. spuria*** is similar but with spreading stems and corolla with dark brown upper lip and bright yellow lower lip. Throughout.

Cymbalaria IVY-LEAVED TOADFLAX

Trailing herbs with very slender stems and palmately veined leaves. Flowers with zygomorphic, 2-lipped corolla with spur at the base. Species all similar.

Cymbalaria muralis IVY-LEAVED TOADFLAX A trailing, purplish, tufted, hairless perennial to 60 cm. Leaves alternate, circular with 5–9 lobes; long-stalked. Flowers small; corolla 9–15 mm, lilac with a yellowish palate, borne on long, slender stalks; spur 1.5–3 mm long (about equalling the calyx). Capsule hairless. Damp, shady places. Native northwest of the region covered but widely naturalised, possibly throughout. ***C. microcalyx*** is similar but persistently woolly-hairy throughout. Calyx *tiny* (1–1.5 mm); spur *short* and cone-like. Limestone cliffs; local. AA, CR, PN, TR (southwest). ***C. longipes*** is very similar to *C. muralis* but hairless when mature, with calyx 2–2.5 mm much *exceeded* by the spur. Rocky crevices. AA, CR, CY, GR (southeast), PN (east), SY (west), TR (south, southwest).

Digitalis FOXGLOVE

Tall biennial to perennial herbs with alternate leaves. Flowers showy with long, tubular-bell-shaped, 2-lipped corolla; stamens 4. Capsule opening by 2 valves.

Digitalis ferruginea RUSTY FOXGLOVE An erect, almost hairless perennial to 1.2 m with lance-shaped leaves. Flowers borne in long, dense spikes; corolla 15 mm, dull brownish or reddish-yellow with darker veins; *middle lobe of lower corolla lip large, almost as long as the tube*. Mountain woods. GR, PN, TR.

Digitalis cariensis A slender, clumped perennial, with leafy basal shoots, 30–80 cm. Leaves hairless, linear, 70 mm–11 cm. Flowers borne in elongated racemes; calyx glandular-hairy, with lobes 6–10 mm; corolla 10–15 mm, pale yellow–brown with whitish lower lobes. Capsule 9–10 mm. Conifer forests and oak woods on rocky slopes. AA (Ikaria – **subsp.** *ikarica*) TR (south, southwest). *D. trojana* is probably a form of the previous species. TR (northwest). **D. lanata** is similar but with slightly larger corolla, 12–16 mm; calyx not glandular with lobes 5–6 mm. GR. **D. leucophaea** is probably a form of *D. lanata*. GR.

1. *Kickxia commutata*
2. *Kickxia aegyptiaca*
3. *Kickxia floribunda*
4. *Kickxia elatine* subsp. *crinita*
5. *Cymbalaria muralis*
6. *Digitalis ferruginea*

Digitalis viridiflora A perennial with 0–few stems 30–80 cm. Leaves narrowly elliptic, shallowly toothed. Flowers borne in dense, almost 1-sided racemes; calyx teeth 5 mm; corolla 11–17 mm, broadly tubular, *greenish-white* to cream, veined with brown distally. GR, TR.

Digitalis laevigata A hairless perennial 60 cm–1 m. Leaves oblong to lance-shaped, entire or toothed. Flowers borne in long, lax racemes; calyx lobes oval, pointed, with or without papery margins; corolla 15–35 mm, yellowish with purple–brown veins. Woods and scrub. GR, IA, PN.

Veronica SPEEDWELL

Annual or perennial herbs (sometimes shrubs) with opposite leaves. Flowers often blue, short-tubed and with 4 unequal lobes longer than their tube; stamens 2. Capsule opening by 2 valves. Numerous similar species; only the most common and widespread are described.

(a) Flowers in terminal clusters (not borne in the leaf axils).

Veronica serpyllifolia THYME-LEAVED SPEEDWELL A more or less hairless *perennial* herb with creeping, rooting stems to 30 cm as well as *erect or ascending*, flowering stems. Leaves oval, 8–20 mm. Flowers 5–10 mm across, pale blue to white, borne in lax terminal spikes; flower-stalks longer than calyx. Capsule wider than long, with equal style. Scrub and open woods; a cosmopolitan weed. GR, TR (probably elsewhere). *V. acinifolia* FRENCH SPEEDWELL is a *glandular annual* to 15 cm with smaller leaves, 5–14 mm. Bracts and leaves shallowly toothed. Flowers blue, 2–3 mm across. *Fruit longer than broad, and notched*; seeds flat. AA (local), CR, GR, IA, PN, TR. *V. arvensis* WALL SPEEDWELL is an *erect annual* to 30 cm with oval leaves 2–35 mm, many hairs non-glandular. Flowers 2–3 mm across, blue. *Fruits hairy* and as long as *broad and heart-shaped*. Throughout.

Veronica triphyllos FINGERED SPEEDWELL A sub-erect annual to 20 cm with *palmately lobed leaves* 5–18 mm. Lower bracts and upper leaves deeply lobed at the base (3–7 lobes). Flowers 3–4 mm across. Fruit *as long as broad, shorter than the calyx*, with spreading glandular hairs. GR, PN, TR.

(b) Flowers in terminal clusters, or in clusters arising from the leaf axils, 1–2 per node.

Veronica chamaedrys GERMANDER SPEEDWELL A perennial with erect to ascending stems to 50 cm with hairs along 2 opposite lines. Leaves triangular-oval, toothed, stalkless or stalked to 5 mm, hairy. Flowers 8–12 mm across, bright blue, borne in *terminal racemes or 1–2 racemes at the nodes*. Capsule triangular-ovoid. GR, IA, PN, TR.

(c) Flowers in clusters, borne in the lower leaf axils, with a leafy, non-flowering shoot at the tip of the plant.

Veronica anagallis-aquatica BLUE WATER-SPEEDWELL A hairless or slightly hairy, *erect perennial* to 50 cm with branched or unbranched stems. Leaves opposite, oval-lance-shaped and scarcely toothed; stalked below, unstalked and semi-clasping the stem above. Flowers 4–9 mm across, pale blue with darker veins, borne in slender, paired racemes. Capsules 2.5–4 mm, rounded or elliptic; hairless. Frequent in damp, seasonally flooded habitats or by streams. Throughout. *V. catenata* PINK WATER-SPEEDWELL is similar but with all leaves narrow and stalkless. Flowers 3–8 mm across, *pinkish*. Capsule 2–3 mm, wider than long. Damp dune slacks. GR (north).

(d) Flowers solitary in the leaf axils.

Veronica persica COMMON FIELD-SPEEDWELL A prostrate, hairy annual with stems to 50 cm and triangular-oval leaves 15–17 mm, coarsely toothed (>7 teeth), hairy below. Flowers 8–12 mm across, *bright blue with a paler or white lower lip*, borne solitary in the leaf axils; calyx lobes with rounded bases. Capsule with spreading hairs. Bare and cultivated land. Throughout.

V. polita GREY FIELD-SPEEDWELL is a similar, hairy annual with *dull green* leaves 6–17 mm and *flowers bright blue throughout*, 4–8 mm across. Capsule with many short arched hairs (some glandular). Throughout. ***V. hederifolia*** IVY-LEAVED SPEEDWELL has *kidney-shaped leaves* 8–28 mm, *with 3–7 large, shallow teeth near the base*. Flowers 4–9 mm across, pale lilac, the corolla shorter than the calyx; calyx lobes heart-shaped at the base. Capsule hairless. Throughout.

Globularia

Perennial herbs or small shrubs with alternate, undivided leaves, and flowers in dense rounded heads with an involucre of bracts; stamens 4. Fruit a 1-seeded nut.

Globularia stygia A dwarf shrub with slender, *creeping* woody stems, which root at the nodes. Leaves in rosettes, almost circular, blunt, sometimes weakly notched at the tip, and stalked. Flowers borne in stalkless heads with numerous tapering bracts; flowers pale violet; calyx *regular*, with narrowly lance-shaped, narrow-pointed teeth. A rare mountain-dwelling endemic. PN (north).

1. *Veronica serpyllifolia*
2. *Veronica chamaedrys*
3. *Veronica polita*
4. *Veronica hederifolia*
5. *Globularia stygia* PHOTO: ELEFTHERIOS DARIOTIS
6. *Globularia arabica*

Globularia arabica A low shrub 20–40 cm with short branches; young twigs and leaves covered in minute chalky dots; very similar to *G. alypum*. Leaves scattered along the stems, 5–20 mm, egg- to spatula-shaped, with a narrow, pointed apex (sometimes somewhat 3-toothed). Flowers blue, borne in compact heads; calyx hairy, with lobes 2 x as long as the tube. Nutlets enclosed in a persistent calyx. Dry plains and deserts. IP (mainly Sharon Plain and Negev Highlands).

Globularia alypum SHRUBBY SPHERICALIA A low, branched, *erect*, evergreen shrub, 40–60 cm (1 m) high with brittle twigs and alternate, leathery, spine-tipped, short-stalked, sometimes apically 3-toothed leaves 15–25 mm. Stems leafy up to the flower-heads. Flowers lilac, borne in *dense, rounded heads*, 10–25 mm across; fragrant; corolla 7 mm with a single 3-lobed lip; bracts oval. Cliffs, scree and garrigue. AA (rare), CR, IA (Cephalonia, Zykanthos), GR, PN (common), TR. *G. cordifolia* MATTED GLOBULARIA is a *low, mat-forming perennial* with spreading rooting branches and short, erect, almost leafless stems bearing spherical heads of blue flowers 6–15 mm across; corolla 6 mm. Mountains. GR.

SCROPHULARIACEAE | FIGWORT FAMILY

Herbs (rarely shrubs or trees) with opposite or alternate leaves. Flowers zygomorphic, usually in spikes or racemes; calyx 4–5 lobed or 2-lipped; corolla 5-lobed, often 2-lipped; stamens 2 or 4. Fruit usually a 2-parted capsule. Many genera previously classified in the family now transferred to the Plantaginaceae.

Scrophularia FIGWORT

Annuals or perennials, often with square stems and opposite leaves. Flowers yellow or greenish; corolla with 5, small, spreading lobes; calyx with 5 lobes; fertile stamens 4 (and 1 sterile).

(a) Mature leaves pinnately divided into narrow, toothed or lobed segments.

Scrophularia xanthoglossa A virtually hairless perennial 24–60 cm with stout, fleshy stems, often purplish and woody below. Juvenile leaves broadly oval, undivided, sharply toothed; mature basal leaves stalked, the uppermost stalkless, and 2–3-pinnately divided into oval-elliptic to oblong, toothed segments. Bracts lance-shaped to linear, toothed to lobed, the uppermost entire. Flowers borne 5–24 in alternate cymes on stalks 8–33 mm; flower-stalks 0.5–2 mm; calyx lobes hairless, oval to egg-shaped, 2.4–5 mm with white, papery margins; corolla maroon, 5–6 mm with yellowish-white bordered lateral and lower lobes; stamens exserted. Capsule ovoid to spherical, 4–5 mm. Rocky slopes, desert scrub, plains and roadsides. IP (common), LB, SY, TR.

Scrophularia canina FRENCH FIGWORT A hairless, much-branched perennial to 1.25 m, usually slightly woody at the base. Basal leaves *pinnately, narrowly lobed with teeth*; upper leaves elliptic and toothed, usually *alternate*. Bracts small and not leaf-like. Flowers small, 3.5–6 mm long, borne on *stalks exceeding the calyx*; purple and white, with *stamens clearly projecting*, and borne numerously in lax terminal, cylindrical clusters; calyx lobes with a membranous margin. Dry rocky habitats and garrigue. AA (scattered), GR, IA, PN, TR. *S. lucida* is similar but normally with solitary stems and with *flower-stalks much shorter than the calyx* and with stalkless, glandular hairs and *stamens scarcely projecting*. Rocky cliffs. Throughout. *S. floribunda* is similar to *S. canina* but scarcely woody below, shorter, to 45 cm and calyx teeth with *broad, papery margins*. AA (east), TR.

Scrophularia spinulescens A perennial to 40 cm with a woody stock, covered in spreading hairs, those below shaggy. Stems 0–few, cylindrical, with divergent, rather spine-like branches below. Leaves pinnately lobed and toothed. Flowers borne in long, lax inflorescences; calyx teeth without papery margin; corolla 4 mm, brownish with paler margins. An island endemic. AA (Samothraki).

(b) Leaves undivided (toothed or untoothed); calyx lobes *without a papery margin*.

Scrophularia peregrina NETTLE-LEAVED FIGWORT An annual to 75 cm (1 m) with hollow, 4-angled, often reddish stems. *Leaves entire*, 12 mm–12 cm long, light green, paired and nettle-like, oval and pointed, with unevenly toothed margins. Flowers borne in clusters in the leaf axils in groups of 1–3(6); corolla 4.5–7 mm long and *dark brown–purple*; *sepals green without a papery margin*. Old walls and pastures; widespread and quite common. Throughout except the far southeast (not CY or IP).

(c) Leaves undivided (toothed or untoothed); calyx lobes *with a papery margin*.

Scrophularia nodosa COMMON FIGWORT An erect perennial to 1.4 m with square, unwinged stems, hairless beneath. Leaves 50 mm–20 cm, oval, pointed and coarsely toothed and short-stalked. Flowers 5.2–8 mm long with a greenish tube and purplish upper lip; calyx 5-lobed with a *narrow white, papery border*. Damp, cool habitats; not at sea level or in the south and east. GR, IA, PN, TR.

Scrophularia scopolii A rather hairy perennial to 1 m with *double-toothed leaves*; leaves oval-lance-shaped and heart-shaped at the base; bracts generally not leaf-like. Flowers greenish with a purple–brown upper lip, to 12 mm long, borne in clusters of 4–7 with glandular-hairy stalks. Cool, damp woods on higher ground. AA (northeast), GR, TR.

Scrophularia lyrata An erect, almost hairless perennial to 1 m with stout, scarcely branched, *markedly 4-angled, narrowly winged* stems. Leaves large with 2–3 pairs of small basal lobes and a large, toothed terminal lobe. Flowers borne in inflorescences of 5–9; calyx teeth oval with papery margins; corolla 5–8 mm, greenish with a brown upper lip. Wet places at low altitude. CR (mainly the west).

Verbascum

Herbs with large basal rosettes, and often grey-hairy leaves. Flowers usually with yellow (sometimes white or purple) 5-lobed corolla; calyx with 5 equal lobes; stamens 5; style 1. A difficult group with numerous species and hybridising freely. Only a small subset are described here.

(a) *Flowers in clusters of 2 or more* in each node; filaments with purple hairs.

Verbascum sinuatum A stout, erect, *grey or yellow-woolly* biennial to 1.5 m. Basal leaves 15–45 cm, dense, *forming distinctive rosettes*; *conspicuously wavy-pinnately lobed*. Flowers yellow, 13–25 mm across, borne in clusters on a *widely branching inflorescence*; stamens with violet-hairy filaments; bracts 1.5–4 mm. Rocky, grassy and coastal habitats; one of the commonest Mediterranean species. Throughout.

Verbascum nigrum An erect, sparsely hairy (rarely downy) perennial 50 cm–1 m. Basal leaves 15–40 cm long, oval to oblong, heart-shaped at the base and with crimped margins. Flowers borne in elongated racemes, 2–few per node; calyx 2.5–4.5 mm with linear lobes; corolla 18–25 mm across, yellow (rarely white); stamens 5, with kidney-shaped anthers and violet filament hairs. Capsule 4–5 mm. Mainly North European, extending locally into the region covered. GR (north, central), PN.

(b) *Flowers solitary* in each node; filaments with purple hairs.

Verbascum dumulosum A *short, clumped*, woody-based, densely greyish or yellowish-woolly perennial to 30 cm. Leaves lance-shaped to elliptic, 15–50 mm long with stalks 5–35 mm long. Flowers 10–15 mm across, yellow, borne 1 per node in numerous short, cylindrical, glandular inflorescences of 10–35; calyx 3–6 mm with linear lobes; filaments with violet hairs, the anterior pair hairless near the apex. Capsule spherical, 3–4 mm. Crevices of old walls and ruins. TR.

Verbascum arcturus A woody-based perennial with few to several stems 30–70 cm, white felted below. Leaves in loose basal rosettes, typically with 1–2 pairs of small lateral lobes and a large, oblong, shallowly crimped terminal lobe, grey-felted on both sides; stem-leaves few. Flowers each borne on *long, slender, spreading stalks,* in long, lax, glandular-hairy racemes; calyx teeth triangular; corolla 25–30 mm across, bright yellow, sometimes with purple markings; fertile stamens 4, filaments with purple hairs. An endemic chasmophyte of cliffs and gorges. CR.

Verbascum blattaria A slender biennial, hairless below with erect stems to 1.2 m, stickily hairy above. Leaves 9–19 cm with paler mid-veins, rather upward-pointing, and coarsely and irregularly toothed along the margins, shiny, hairless and wrinkled. Flowers *long-stalked*, yellow, 20–35 mm across, borne in long, slender, lax, *hairless spikes*; filaments with reddish or violet hairs; the upper petals with small reddish blotches at the base; bracts 5–11 mm. GR, IA, PN, TR.

Verbascum chaixii NETTLE-LEAVED MULLEIN An erect biennial to 1.2 m, similar to *V. lychnitis* but with leaves 20–35 cm slightly lobed at the base. Flowers yellow or white, 18–22 mm across, borne on long, slender inflorescences with *upward-arching branches*; *filaments with violet hairs*; bracts 5.5–7 mm. Mainly western. GR, IA.

(c) Filaments with white or yellow hairs; *anthers* (at least some) *elongating down the filament* (decurrent).

Verbascum longifolium A yellowish-white-woolly biennial with upper leaves clasping the stem. Flowers flat, to 30 mm across, the longest filaments hairless; flowers borne in dense spikes with flower-stalks of unequal length (either longer or shorter than calyx). GR, PN.

Verbascum phlomoides A grey to white-downy biennial to 1.2 m. Basal leaves to 40 cm, oblong-elliptic, downy on both sides. Flowers borne in congested clusters in dense inflorescences; calyx 6–12 mm; corolla 30–50 mm across; stamens 5, the lower 2 with anthers extending down the hairless (or virually so) filaments; other filaments with whitish hairs. GR, PN (north), TR (north).

(d) Filaments with white or yellow hairs; *anthers attached to centre of filament.*

Verbascum speciosum A densely grey-hairy, erect biennial with entire, lance-shaped leaves. Flowers borne in clusters, 18–30 mm across, yellow, on larger specimens borne in *pyramidal, top-heavy-seeming inflorescences* with many long, ascending branches; stamens with kidney-shaped anthers and filaments with white hairs. GR, TR.

Verbascum spinosum A thinly white-downy, mounded, intricately branched, *spiny shrub* to 1 m across. Leaves to 50 mm, irregularly toothed or lobed; twigs *ending in a spine*. Flowers solitary; calyx 3 mm; corolla 10–18 mm across, light yellow; stamens 5, all with white-woolly filaments. A locally important element of the Cretan garrigue. CR (mainly western).

Verbascum lychnitis WHITE MULLEIN An erect biennial to 1.5 m with angled, *white-mealy stems*. Leaves 16–35 cm, *dark shiny green and hairless above, powdery-white below*. Inflorescences branched, the *branches erect, parallel to the main stem*, giving a slender appearance; flowers 14–18 mm, white or yellow; *all stamens clothed with white hairs*, borne 2–7 per bract; bracts 10–15 mm. IA. *V. pulverulentum* is similar but the whole plant thickly covered in mealy white hairs, which rub off. Flowers 18–25 mm. GR, IA.

1. *Scrophularia xanthoglossa*
2. *Scrophularia peregrina*
3. *Verbascum sinuatum*
4. *Verbascum nigrum*
5. *Verbascum dumulosum*
6. *Verbascum speciosum*
7. *Verbascum spinosum*
8. *Verbascum arcturus*

Verbascum undulatum A biennial to perennial with several ascending stems to 80 cm, thinly downy throughout. Leaves with undulate margins (like *V. sinuatum* but see filaments). Flowers 2–6 per node, the lowermost lax; calyx 6–12 mm; corolla 25–40 mm across; stamens 5, all with white-woolly anthers. AA (close to mainland), GR, IA (Corfu), PN.

Verbascum graecum A stout biennial to 1.5 m, white-downy below, almost hairless above. Basal leaves to 30 cm long, oval to lance-shaped, often with crimped, undulate margins; stalks to 10 cm. Flowers borne in branched, lax inflorescences; calyx 2–5 mm with linear-oblong lobes; corolla 13–30 mm across; stamens 5 with kidney-shaped anthers; filaments with white hairs. GR, IA, PN.

(e) Flowers *purple*.

Verbascum phoeniceum PURPLE MULLEIN A tall perennial with erect stems, glandular-hairy above. Basal leaves oval and toothed or untoothed; upper leaves stalked. *Flowers violet*, to 30 mm across, borne in lax racemes; filaments with purple hairs. Widely cultivated as an ornamental. GR, IA, PN, TR.

1. *Myoporum laetum*
2. *Myoporum tenuifolium*
3. *Acanthus mollis*
4. *Acanthus hirsutus*
5. *Acanthus spinosus*

Myoporum

A genus of trees and shrubs native to Australasia, formally classified in the Myoporaceae. Flowers with 5 corolla lobes; stamens 4; ovary 2–4-parted. Fruit a berry.

Myoporum laetum An evergreen shrub or tree to 9(13) m with sticky shoots. Leaves 45 mm–10(17) cm, alternate, narrowly lance-shaped, with inconspicuous *forward-pointing teeth* towards the tips, hairless and dotted with oil glands. Flowers 10–15 mm across, borne in lateral clusters of 5–6(10), white with purple markings; stamens 4. Fruit a purple berry 7–10 mm. Native to New Zealand, sometimes planted in towns; occasionally naturalised. *M. tenuifolium* is a similar shrub or round-crowned tree to 8 m with leaves 45 mm–10 cm, lance-shaped, *thin and without teeth*. Flowers borne in *dense cymes* of 5–9 (occasionally solitary). Fruit 7–9 mm, very dark purple when mature. Native to New Caledonia; planted and naturalised. IP.

ACANTHACEAE

Herbaceous perennials with simple, often lobed leaves and erect (usually) unbranched stems; bracts conspicuous and spiny. Flowers borne in dense spikes; calyx 4-lobed and 2-lipped; corolla zygomorphic, 1–2-lipped, the lower lip 3-lobed; stamens 4, not protruding; style 1. Fruit a capsule.

Acanthus BEAR'S BREECH

Leafy perennials with flowers borne in long, dense, terminal spikes. Easily distinguished by the robust habit, large pinnately lobed leaves, spiny bracts and zygomorphic flowers.

(a) Basal leaf margins without stout spines.

Acanthus mollis BEAR'S BREECH A robust, bushy perennial to 75 cm (1 m). Leaves large, 20 cm–1 m, shiny dark green, pinnately lobed, and soft; hairless or nearly so, and *long-stalked*; stem-leaves small and few. Bracts purple-tinged and spiny. Flowers white, 35–50 mm long, borne in dense spikes; corolla 3-lobed; calyx purple and hairless. Various habitats, often in woods or in damp, shady places; also a garden escape. AA (Lesbos, Chios, Agios Evstratios), CR (rare), GR, IA, IP, PN, TR. *A. hungaricus* has dull green basal leaves, which are *short-stalked* and divided into several segments narrowed at base. GR (northeast). *A. greuterianus* has pale green, rather thin leaves with weak spines and sterile bracts at the base of the inflorescence. GR (north).

(b) Basal leaves with conspicuously *spiny* margins.

Acanthus spinosus SPINY BEAR'S BREECH A robust perennial, similar to *A. mollis* but with leaves with *markedly spiny lobes*, and the whole plant often hairy; leaves with prominent white veins beneath. Bracts purple (sometimes pale green). The most common species in much of the region. Throughout, except for parts of the southeast.

Acanthus hirsutus (including *A. syriacus*) A robust, bushy perennial with leaves mostly basal. Basal leaves large, oblong-lance-shaped, tapering into *short stalks*, divided into spiny-margined lobes; stem-leaves usually 1 pair, stalkless. Flowers borne in compact, ovoid spikes with broad, leathery *spiny* bracts (4–6 spines on each edge), tapering to a *spiny point* as long as the flower; outer sepals herbaceous, the uppermost exceeding the corolla; corolla whitish with round lobes. Field edges and waste places in the east only. IP (north), LB, SY, TR. *A. dioscoridis* has entire or slightly divided leaves with stalks to 12 cm and mostly *purple to rose-red* flowers. Rocky igneous slopes, cliffs, and fallow fields. TR.

BIGNONIACEAE

A large tropical and subtropical family of trees and climbers with opposite, sometimes pinnately divided leaves. Flowers with large, tubular corolla with 4 stamens in 2 pairs. Fruit a pod-like capsule splitting lengthways into 2, containing numerous, often winged seeds.

Tecoma

Small trees, shrubs or climbers. Leaves simple or pinnately divided with a terminal leaflet. Flowers typically with a tubular corolla; stamens 4. Fruit a hairless, linear capsule. Seeds winged.

Tecoma capensis CAPE HONEYSUCKLE An evergreen, scambling (to climbing) shrub to 3(7) m with pale, eventually furrowed bark. Leaves glossy green, opposite, pinnately lobed, to 13(15) cm long with 2–7(11) pairs of elliptic to circular leaflets to 15 mm, the terminal the largest; leaflets with coarsely toothed margins. Flowers borne in clusters, *bright orange*; calyx 5–7 mm; corolla tubular, 35–50 mm long, to 70 mm wide at the mouth; styles and stamens protruding. Fruit a slender, wrinkled, pod-like capsule to 65 mm (13 cm). Seeds with papery wings. Native to South Africa, planted in warmer areas. *T. stans* is similar but with leaves with 3–9 leaflets, and broadly tubular-bell-shaped *yellow* flowers 30–43 mm long with reddish veins. Widely planted in hot, landscaped areas.

Pyrostegia

Climbers with tendrils and angled branches. Leaves opposite with 2–3 leaflets, with or without a terminal leaflet. Flowers borne in terminal panicles; calyx bell-shaped; corolla tubular. Fruit a linear capsule.

Pyrostegia venusta A climbing, woody perennial. Leaflets to 12 cm, oval to lance-shaped, leathery and rather scaly above; hairless to finely hair beneath. Flowers *bright orange*, borne in clusters; calyx 4–6 mm with 5–10 veins; corolla tubular, 35–70 mm, curved and hairless, with lobes 9–14 mm. Fruit smooth and leathery, 24–30 mm. Native to South America; planted in warmer parts.

Jacaranda

Trees native to the Neotropics, normally with pinnately divided leaves. Flowers conspicuous, borne in panicles. Fruits pendent, flattened.

Jacaranda mimosifolia JACARANDA A small tree to 8(15) m. Leaves to 45 cm, 2-pinnately divided with numerous, opposite, oval to diamond-shaped, downy leaflets. Flowers blue–purple, trumpet-shaped, 50–60 mm long, 2-lipped, borne in showy, terminal panicles, before the leaves have developed. Fruit oval, brown and papery when ripe, 50–80 mm across; flattened. Frequently planted in towns and gardens.

Catalpa

Deciduous trees native to warm-temperate North America with heart-shaped (sometimes 3-lobed) leaves. Flowers with bell-shaped corolla with 2 stamens. Fruits slender.

Catalpa bignonioides INDIAN BEAN TREE A deciduous tree to 15(18) m with a spreading crown. Leaves large and heart-shaped, long-stalked, 20–30 cm long. Flowers bell-shaped, white with purple spots and yellow markings within, 25–35(50) mm long, borne in large, showy panicles. Fruit slender and bean-like, green and later brown, (15)20–40 cm long, persisting after the leaves. Native to North America, widely planted.

BIGNONIACEAE

Podranea

Evergreen shrubs and vines native to Africa with compound leaves. Flowers showy, with bell-shaped corolla, borne in panicles. Fruit linear, flattened.

Podranea ricasoliana PORT ST JOHN'S CREEPER A vigorous climber with long, twining stems and large, pinnately divided leaves. Flowers pale pink with darker linear markings in the throat, tubular-bell shaped and 5-lobed, borne in pendent bunches. Native to South Africa; planted.

Campsis TRUMPET VINE

Deciduous, woody climbers with aerial roots native to North America and China. Flowers with showy, trumpet-shaped corolla. Fruit elongated, bean-like.

Campsis radicans TRUMPET VINE A vigorous, deciduous climber to 10(12) m with robust, woody stems, clinging with aerial roots. *Leaves hairy*, pinnately divided, with 9–11(13) oval, toothed leaflets 20–50 mm. *Flowers bright orange to scarlet, trumpet-shaped* and weakly 2-lipped, 65–95 mm long. Fruit spindle-like, 60 mm–24 cm; seeds linear, hairless, 20–30 mm. Commonly planted in parks and gardens across the region. ***C. grandiflora*** is similar but with twining stems, *hairless leaves* and larger flowers borne in pendulous clusters. Fruit capsule obtuse. Commonly planted, as are hybrids of the above two species.

1. Tecoma capensis
2. Tecoma stans
3. Pyrostegia venusta
4. Jacaranda mimosifolia
5. Podranea ricasoliana
6. Campsis radicans

LENTIBULARIACEAE | BLADDERWORT FAMILY

Carnivorous perennials. Flowers zygomorphic; calyx 2- to 5-lobed, usually 2-lipped; corolla tubular, the upper lip 2-lobed, the lower 3-lobed; stamens 2; style 0 (or short). Fruit a capsule. Rare in the Mediterranean.

Pinguicula BUTTERWORT

Soft, fleshy perennials with leaves in basal rosettes. Flowers solitary; calyx 2-lipped; corolla 2-lipped and spurred with a 2-lobed upper lip and a 3-lobed lower lip. Capsule 2-valved.

Pinguicula crystallina A perennial with leaves in basal rosettes, 20–60 mm long, broadly egg-shaped, blunt, yellowish and sticky (mucilage-covered). Flowers solitary on 1–5 slender, bractless stalks 40 mm–15 cm arising from the centre of each rosette; corolla 2-lipped, spurred, 14–26 mm (including the spur), pale pink to lilac with a pale yellow throat and spur. Fruit a capsule. Wet rocks, seeps and by waterfalls in mountains; widespread (**subsp.** *hirtiflora* in the north and west; **subsp.** *crystallina* in the south and east): CY, GR, PN (north), TR. *P. balcanica* is similar but with blue-violet flowers, growing at higher atltitudes. GR (north, central).

Utricularia BLADDERWORT

Submerged aquatic carnivorous perennials. Flowers with 2-lipped, yellow corolla borne on erect stems above water. Capsule opening irregularly.

(a) Flowers 12–18 mm.

Utricularia vulgaris GREATER BLADDERWORT An aquatic perennial with submerged, free-floating stems, which bear numerous tiny 'bladders' that trap tiny invertebrates. Flowers *deep yellow*, borne few on erect, leafless, reddish stems 10–25 cm; corolla with *lower lip 12–15 x 14, deeply turned backwards at the edges*; *upper lip 11 x 10 mm, equalling the palate*; spur *broadly* cone-shaped,

1. *Pinguicula crystallina* PHOTO: ARNE STRID
2. *Utricularia vulgaris*
3. *Verbena officinalis*

6–8 mm; flower-stalks 6–12(15) mm, recurved but not elongated after flowering. Stagnant water in marshes and ditches. Very local. GR, IA, PN. *U. australis* is very similar to (and often confused with) *U. vulgaris*, but with *lemon yellow flowers* with the *lower lip not strongly deflexed at the edges, upper lip exceeding the palate*; spur *narrowly* cone-shaped 6–8 mm; flower-stalks 8–15 mm when flowering, elongating to 10–30 mm in fruit. Lakes and ponds. Distribution poorly known owing to confusion with the previous species.

(b) Flowers very small: 6–8 mm.

Utricularia minor LESSER BLADDERWORT An aquatic perennial, distinguished by its very slender purple stems to 15(25) cm and *very small, pale yellow flowers with corolla just 6–8 mm long* (upper lip 4 x 3 mm, lower lip 7 x 6 mm); spur 1–2 mm; flower-stalks 4–8 mm. Very local in pools in marshes on higher ground. GR (north – rare).

VERBENACEAE

Herbaceous annuals perennials or shrubs with opposite leaves and square stems. Flowers borne in clusters or heads; corolla a slender tube and flat limb, often 2-lipped; stamens 4, not protruding; style 1. Fruit berry-like or 4 1-seeded nutlets.

Lantana camara LANTANA A small prickly shrub to 1.5(4) m with square and prickly branches that are hairy when young. Leaves 40 mm–13 cm, opposite, oval, toothed and short-stalked. Flowers 10–11 mm, bright yellow or orange, ageing red, congested in tight, often paired heads carried on long stalks; corolla slightly 2-lipped. Fruit a small black berry 4–7 mm. Cultivated in gardens and frequently naturalised.

Verbena

Herbaceous annuals or perennials. Flowers borne in elongated or flat-topped clusters; calyx and corolla 5-lobed. Fruit 4 nutlets.

Verbena officinalis VERVAIN A rough-hairy perennial to 1.8 m with slender, stiffly erect and square stems; superficially mint-like. Leaves 40–80 mm, opposite, lance-shaped to diamond-shaped in outline but deeply pinnately lobed; stalked below, stalkless and often unlobed above. Flowers stalkless, pink, 4.5–6.5 mm, slightly 2-lipped, borne in long, slender, leafless spikes to 30(55) cm long. Fruit separating into 4 nutlets. Bare waste ground. Throughout.

LAMIACEAE | MINT FAMILY

A large and important family of herbs and shrubs; often glandular and aromatic. Leaves opposite and simple, often toothed or lobed. Flowers zygomorphic, often borne in whorls around the stem; calyx 5-lobed, often a tube with teeth; corolla 2-lipped with 3–5 lobes, the upper lip sometimes reduced; stamens 4, 2 long and 2 short; style 1. Fruit comprising 4 1-seeded nutlets concealed within the persistent calyx.

Vitex

Shrubs with palmately compound leaves with 5–7 leaflets. Flowers in whorls forming terminal spikes; calyx 5-lobed; corolla tubular with 5 lobes.

Vitex agnus-castus CHASTE-TREE An aromatic shrub 1–3 m with white-felted branches and leaves palmately divided into lance-shaped leaflets, green above, white-felted below. Inflorescence a long, terminal, interrupted spikes of lilac (rarely pink) flowers; calyx hairy; corolla 6–9 mm, almost 2-lipped, hairy outside; stamens protruding. Fruit fleshy, reddish-black. Typically by streams, or other damp habitats, near the sea. Throughout.

Ajuga BUGLE

Annual or perennial herbs with entire to deeply divided leaves. Flowers with 5-lobed calyx; corolla pink, white, blue or yellow with a very short upper lip and conspicuous 3-lobed lower lip and with a ring of hairs within; stamens 4, shorter than lower lip.

1. *Vitex agnus-castus*
2. *Ajuga reptans*
3. *Ajuga iva*
4. *Ajuga chamaepitys* subsp. *chamaepitys*
5. *Ajuga chamaepitys* subsp. *palaestina*

(a) Inflorescence with *many-flowered whorls*; flowers blue.

Ajuga reptans BUGLE A variable, short perennial herb *with leafy runners above ground* and erect flowering stems 15–40 cm, usually *hairy on 2 faces only*; leaves 25–50 mm, spoon-shaped and toothed or not, scarcely hairy. Flowers blue–violet, 15–18 mm long. Mountain forests. GR, IA, PN (local), TR (north). ***A. genevensis*** is similar but with *aerial runners absent*, stems hairy on all faces (sometimes just on opposite sides) and rather hairy leaf blades. Local, and absent from hot, dry areas. GR, TR. ***A. orientalis*** EASTERN BUGLE has *distinctively symmetrical, dense pyramidal*, very hairy spikes of flowers, exceeded by their leaves. Virtually throughout (rare or absent on some islands).

(b) Inflorescence normally with *2-flowered whorls*, and leaves undivided.

Ajuga iva SOUTHERN BUGLE A tufted or sprawling, short, softly hairy perennial to 15 cm. Leaves 15–40 mm, broadly linear, normally with short lobes, sometimes unlobed. Bracts similar to the leaves and exceeding the flowers. Flowers white or cream to pink-purple, often with small darker spots; corolla 13–24 mm long, the *upper lip entire and highly reduced*. Common on stony garrigue and cliff-tops. Throughout.

(c) Inflorescence normally with 2-flowered whorls, and *leaves deeply 3-lobed*.

Ajuga chamaepitys GROUND PINE A short, hairy, ascending annual to 19 cm. Basal leaves 23–52 mm soon withering, stem-leaves divided into *3 long, linear lobes*, smelling of pine when crushed. Flowers rather sparse, (5)10–16(25) mm long, yellow with purplish markings, borne 2–4 at each node, and greatly exceeded by their bracts. Throughout, with several widespread forms: **Subsp.** *chamaepitys* is an annual with leaf segments 0.5–2 mm wide and corolla (5)7–15 mm. Nutlets net-veined. **Subsp.** *chia* is a short-lived perennial with leaf segments 1.5–3(4) mm wide and corolla 18–25 mm. Nutlets wrinkled. **Subsp.** *palaestina* is a perennial with *woolly* stems and broader leaf lobes. AA (east), CY, IP, SY, TR.

Teucrium GERMANDER

Herbs or shrubs with toothed to lobed leaves. Flowers with a 2-lipped calyx and corolla with a *single, 5-lobed lower lip*; stamens 4, shorter than the lower lip.

(a) Plant growing in wet habitats (other described species occur in dry, rocky habitats).

Teucrium scordium WATER GERMANDER A softly white-hairy perennial to 40(60) cm with creeping rhizomes and leafy runners from which arise erect or ascending stems. Leaves 35 mm long, oblong and coarsely toothed, rounded at the base, grey–green and more or less unstalked; garlic-scented when crushed. Flowers borne in loose terminal inflorescences; corolla pink–purple, 9–10 mm long. Damp or aquatic habitats; widespread. Throughout.

(b) Flowers white, cream or greenish-yellow.

Teucrium capitatum (syn. ***T. polium*** subsp. ***capitatum***) A grey-hairy dwarf shrub 10–25 cm with bunched leaves with down-turned, shallowly scalloped margins towards the tips. Flower-heads compound, borne in branching inflorescences with small, *dense, congested* heads, <10 mm (as wide as long) of cream, white or pink flowers; corolla with *short, rounded or triangular lateral lobes*; calyx 3–3.5 mm, densely and evenly hairy. Throughout (in its wider treatment). **Subsp.** *capitatum* is the form in the region. ***T. cuneifolium*** is somewhat similar, with leaves with *clearly scalloped* margins and lateral lobes of the corolla *much longer than wide* and oblong-spatula-shaped. Flowers cream. CR (southeast). ***T. alpestre*** A short, dense, mat-forming dwarf shrub, similar to

T. capitatum, with short, slender, flowering shoots, small, egg-shaped leaves and flowers with cream corolla 6–8 mm with oblong-spatula-shped lateral lobes. CR. ***T. gracile*** is very similar but with a taller, robust and clumped habit with ascending branches, often with numerous dead surrounding branches, and leaves broadest above the middle. CR (east + KP).

Teucrium flavum YELLOW GERMANDER A shrubby perennial to 35(65) cm with few ascending stems, stalked, leathery, toothed leaves 18–25 mm, and rather dense, leafless terminal, rather 1-sided clusters of dull, pale sulphur- or green–yellow flowers; corolla 12–20 mm, up to 2.5 x the length of the densely hairy calyx. Rocky places and woods; common, especially in the east. CR, GR, IA, PN, TR. **Subsp.** *glaucum* has leaves to 10 mm, almost hairless, blue–grey beneath. Calyx 9–10 mm with teeth as long as the tube. GR. **Subsp.** *gymnocalyx* has slender, often almost hairless stems and hairless leaves 10–20 mm. Calyx 7 mm, with teeth almost as long as the tube. CR, GR. **Subsp.** *hellenicum* has rather velvety-hairy stems and leaves 10–15 mm, velvety beneath. Calyx 7 mm with teeth almost as long as the tube. GR, IA, PN.

Teucrium spinosum A low, hairy dwarf shrub with woody-based, erect stems, *thorny* from the base, with spine-tipped branches. Leaves stalked, soon-dropping, *lobed*; the uppermost stalkless and toothed. Flowers borne in distant whorls of 2–4 with stiff, *spine-like* bracts exceeding the calyces; calyx with a concave, oval upper lip and 4-lobed lower lip; corolla *white*. Dry fields. IP (mainly the Lower Galilee and Esdraelon Plain).

(c) Flowers pink, reddish-purple or lilac.

Teucrium chamaedrys **WALL GERMANDER** A tufted, spreading, green perennial to 20(35) cm, woody at the base with erect or ascending stems. Leaves small, 10–15(20) mm long, oval-oblong, 2 x as long as wide, blunt and short-stalked, shiny green and usually with distinct, rounded teeth. Flowers pink–purple; corolla 2–17 mm long; calyx reddish, hairy and bell-shaped, regular with 5 teeth. Common in various habitats on the mainland. GR, IA, PN, TR. Numerous subspecies have been described. ***T. divaricatum*** is similar but taller (to 50 cm) with much-branched, *woody* flowering branches, leathery, shallowly toothed leaves 10–20 mm, the uppermost entire, small and purplish. Flowers with corolla 10–16 mm, bright reddish-purple. Virtually throughout. ***T. microphyllum*** is similar to the previous species but a dwarf shrub <30 cm with leaves 3–6 mm and rather smaller, pink flowers 8–10 mm long. AA (local), CR (common). ***T. massiliense*** is similar to *T. chamaedrys* but taller and more slender with small flowers; calyx 2-lipped (not regular); corolla 7–9(10) mm. CR.

Prasium

Small shrubs with flowers with 2-lipped corolla, the upper lip arched over the stamens, the lower 3-lobed; stamens 4.

Prasium majus **PRASIUM** A subshrub to 1 m, hairless or slightly hairy. Leaves 16–43 mm, *shiny dark green*; oval, pointed and toothed, heart-shaped at the base; all stalked; bracts similar. Flowers with a white corolla with tube 10 mm long, upper lip 6–8 mm, lower lip 8–11 mm, borne in terminal racemes; calyx 2-lipped with bristle-tipped teeth; corolla with the middle lobe the largest. Nutlets shiny black when ripe. Widespread and common on rocky maquis. Throughout.

Marrubium HOREHOUND

Perennial herbs with toothed leaves. Calyx with 10 teeth, spreading when in fruit; corolla white (cream or purple) with a flat upper lip; stamens 4, none protruding.

Marrubium vulgare **WHITE HOREHOUND** An erect, white-downy, aromatic, rather nettle-like perennial to 85 cm with erect, square, *branched, white-cottony stems*. Leaves 17–65 mm, oval, heart-shaped at the base and wrinkled on the surface; with stalks shorter than their blades. Flowers small, and rather inconspicuous, borne in dense, *many-flowered* whorls in the leaf axils up the stem; corolla 2-lipped and white, the tube 3.5–5 mm, upper lip 2–3.5 mm, lower lip 1.8–3.5 mm; *calyx with 10 or more equally short, hooked teeth*. Common on rocky and stony ground. Throughout. ***M. peregrinum*** **BRANCHED HOREHOUND** is similar but with narrow, toothed leaves and white flowers that are smaller, particularly relative to the leaves, borne in rather few-flowered (4–10), well-spaced whorls. GR, PN, TR.

---

1. *Teucrium polium*
2. *Teucrium capitatum* subsp. *capitatum*
3. *Teucrium flavum*
4. *Teucrium flavum* subsp. *hellenicum*
5. *Teucrium chamaedrys*
6. *Prasium majus*

Galeopsis HEMP-NETTLES

Annuals usually with toothed leaves. Calyx bell-shaped or tubular, 10-veined, with 5 spine-pointed lobes, exceeded by the 2-lipped corolla, which has a flattened upper lip and 3-lobed lower lip with projections at the base; stamens 4, exceeded by upper corolla lip.

Galeopsis ladanum NARROW-LEAVED HEMP-NETTLE An erect, divergently branched, hairy annual to 76 cm with narrow leaves 13–59 mm with forward-pointing teeth and whorls of few (up to 14) pale purple to bright rosy-purple flowers with yellowish markings; corolla 12–24 mm, hairy; calyx densely hairy, the tube much longer than its teeth. Disturbed habitats, shingle and sandy river banks. GR (north).

Sideritis

Erect, aromatic herbs, perennials and shrubs. Flowers with calyx bell-shaped, 5-toothed and 10-veined, corolla usually yellow and 2-lipped with a flat upper lip; stamens 4, not protruding. A difficult genus for which floras give conflicting information. Hybrids exist.

(a) Slender annuals (to biennials).

Sideritis curvidens A slender, scarcely branched annual to 15 cm with spreading, non-glandular hairs. Leaves short-stalked, elliptic to narrowly egg-shaped with shallow teeth. Inflorescence with few, distant whorls of 4–6 flowers over most of the stem; calyx inflated at the base, with 4 equal teeth shorter than the tube, sharp-pointed; corolla equalling or exceeding the calyx, *white* (rarely pinkish). Throughout (rare in the west). **S. purpurea** is similar but with calyx scarcely inflated at the base and *reddish-purple* corolla *exceeding* the calyx. Mainly southwestern. GR, IA, PN.

Sideritis lanata An erect to ascending, soft-hairy, scarcely branched annual 10–35 cm. Basal leaves long-stalked, broadly elliptic with scalloped margins, stem-leaves short-stalked. Flowers borne in several densely and softly hairy whorls of about 6, merged above; calyx straight with about equal teeth; corolla just exceeding the calyx, with whitish tube and *contrasting purplish-black* lips. Fixed dunes, mainly in the Aegean. AA, GR.

Sideritis montana A hairy annual to biennial to 40 cm with leaves broadest above the middle. Flowers borne in several distant whorls of 4–6 over most of the stem; calyx tube straight with *5 almost equal teeth*, shorter than their tube, rather *bristle-like* at the tips; corolla just exceeding the calyx, with a shallowly 3-lobed upper lip, pale yellow ageing brownish, with dark maroon–brown markings. AA (local), GR, IA, PN, TR.

(b) Woody-based perennials (some local species not described).

Sideritis scardica A white-hairy perennial herb, slightly woody below, with few branches, to 30 cm. Leaves 30–60 mm, narrowly elliptic, somewhat woolly-hairy. Flowers borne in crowded (not distant) whorls (or with one remote whorl below); corolla 12–15 mm, 2-lipped, *pale yellow* throughout, the upper lip with 2 short teeth, the lower lip 3-lobed. GR (north). **S. euboea** is similar but *densely white-woolly* throughout, with lower leaves egg-shaped and with *distant* lower whorls of flowers (those above crowded); corolla *small*, 7–10 mm. AA (Euboea). **S. syriaca** is virtually identical to S. euboea (possibly a form of it) but with spatula-shaped lower leaves and *all* whorls of flowers distant. CR. **S. perfoliata** is similar to S. scardica but the upper corolla lip with 2 brown stripes. CY, GR (north – very rare), IP, SY.

---

1. *Sideritis curvidens*
2. *Sideritis lanata*
3. *Phlomis fruticosa*
4. *Phlomis lunariifolia*
5. *Phlomis lanata*
6. *Phlomis platystegia*

Phlomis

Herbs or shrubs with entire leaves. Flowers with calyx with 5 equal teeth; corolla 2-lipped, the upper lip notably hooded and notched at the tip, the lower lip 3-lobed; stamens 4, protruding or not. Many closely related species in the *P. fruticosa* group.

(a) Flowers yellow.

Phlomis fruticosa JERUSALEM SAGE A *robust*, sage-like, grey-felted, erect or spreading, lax evergreen *shrub* to 1.5 m. Leaves elliptic, usually untoothed, wedge (not heart)-shaped at the base, thick, stalked, greyish above and white-felted below, 30–90 mm. *Flowers large, bright golden yellow*, borne in 1–3 dense whorls of 12–30; bracts just shorter than the calyx; calyx 15–20 mm with short, slender (inconspicuous) teeth just 1–3 mm, and both glandular and non-glandular hairs; corolla 25–35 mm. Nutlets hairy at the tips. Rocky ground, maquis and field boundaries. A common to subdominant component of regional garrigue vegetation. AA (west), CR, CY, GR (southern), IA,

PN, TR. Locally replaced by similar, related species: ***P. bourgaei*** has *glandular* hairs, glandular-bristly bracts and *long* calyx teeth, 4–5 mm. AA (southeast), TR. ***P. lycia*** is very similar but with *yellowish-downy* hairs and whorls of few (6–12) flowers; calyx teeth just 1 mm. AA (southeast), TR. ***P. cretica*** has shorter, broader leaves, which are densely white-downy on both sides with starry hairs, bracts equalling the calyx and spreading calyx teeth; corolla dark yellow. AA (east), CR, PN (southeast), TR (west). ***P. floccosa*** is very similar to *P. cretica* but with lower leaves heart-shaped at the base and linear bracts equalling the calyx with long, spreading hairs; calyx 15–19 mm with spreading, broadly triangular teeth; both bracts and calyx teeth *hooked* at the tips. KP, IP. ***P. pichleri*** has *narrowly oval* leaves with heart-shaped bases and no glandular hairs on the calyx. KP. ***P. chrysophylla*** has adpressed-woolly, golden hairs, somewhat stalked flowers, linear, strongly tapered bracts exceeded by the calyces, broad, short, bristle-toothed calyx lobes and corolla 2 x as long as the calyx. Sub-alpine forests. IP, LB. ***P. cypria*** has a *short* calyx tube (7–12 mm) with *rigid, spine-like* teeth 2–2.5 mm long; corolla pale yellow, 26 mm. Nutlets *hairless* (or virtually so) at the apex. A rare island endemic. CY (north).

Phlomis lunariifolia A *slender,* grey-hairy perennial herb to 1 m. Leaves narrowly oblong-lance-shaped to oval, 40 mm–10 cm long, dull green and finely wrinkled above. Bracts leaf-like, 10–15 mm (about as long as the calyx), thinly woolly. Flowers borne in lax inflorescences of 1–3 distant many-flowered whorls; calyx 14 mm, with 10 veins, alternately ending in hairy, *spreading spine-like teeth 2.5–4 mm long*; corolla bright yellow, 32 mm long. Pine woods and dry limestone rocks. Garrigue and rocky valleys. CY, TR.

Phlomis lanata A *small shrub* just 25–60 cm with densely downy stems. Leaves *broadly elliptic-oblong*, 10–25 mm long, rather thick and wrinked with netted veins beneath. Flowers borne in (1)2–4 distant whorls of 3–10; bracts shorter than the calyx and broadly elliptic; calyx 12 mm with *very short*, rather spine-like teeth; corolla rather small, 20–25 mm, golden yellow with a somewhat paler lower lip. Nutlets hairy. Garrgiue and maquis. CR.

Phlomis platystegia A clump-forming, shrubby *desert* perennial 30 cm–1.2 m, woolly when young, later almost hairless. Leaves scalloped, wrinkled, oval-oblong, 20–90 mm; stem-leaves stalked, wedge- to heart-shaped at the base. Flowers golden yellow, borne numerously in remote whorls of 3–6; bracts oval to oblong, pinnately veined, flat, numerous, 2–8(10) mm wide and nearly as long as the calyx; calyx 12–15 mm in flower, tubular-bell-shaped; corolla 24–30 mm. Nutlets hairless and glossy. Rare, in rocky deserts and ravines. IP (especially Judean Desert, Moav and Edom).

Phlomis armeniaca A *slender*, non-glandular perennial herb to 60 cm. Leaves with adpressed, short hairs, often white-downy beneath; oval-oblong to lance-shaped or linear, tapering towards base, 20 mm–10 cm. Bracts linear-lance-shaped. Flowers borne in 2–5 whorls of 4–10, distant below, congested above; calyx 13 mm, with adpressed to spreading hairs and lance-shaped teeth 4–6 mm; corolla yellow, 25–35 mm. Pine woods and dry limestone rocks. TR.

(b) Flowers pink or purple.

Phlomis samia A rhizomatous herb with few, erect, simple branches to 1.2 m. Leaves long-stalked (stalkless above), 60 mm–14 cm, oval, heart-shaped at the base, dark green above, grey-downy beneath. Flowers with calyx 18–25 mm with long teeth; corolla 30–35 mm with a hooded upper lip, the lower lip brownish-purple. AA (Samos – rare), GR, TR.

Phlomis herba-venti A robust, hairy, branched perennial 25–75 cm with large, lance-shaped, toothed, leathery, shiny green leaves 42 mm–22 cm, paler beneath. Flowers purple, borne in 2–6 dense whorls of 8–16, to 20 mm long; *calyx with spine-tipped teeth ˃ 4 mm*. Dry slopes and maquis. GR, PN.

Eremostachys

Similar to *Phlomis*. Calyx tubular-bell-shaped with 5 spiny teeth; corolla tube not exserted; stamens ascending under the hooded upper corolla lip. Nutlets hairy at the apex.

Eremostachys laciniata A *robust*, erect, usually unbranched perennial 30 cm–1 m. Leaves oblong-elliptic, to 30 cm, 2 x pinnately divided into oblong-lance-shaped to linear, toothed segments. Flowers borne numerously in many whorls in thick, robust spikes; corolla 30–40 mm, white, cream or yellow, often with a purplish lower lip; calyx densely *woolly*, with broad, short teeth ending in spiny tips. Various habitats including hillsides, maquis, plains. IP (common), LB, SY, TR.

Nepeta CAT MINT

Perennial herbs with toothed leaves, plants often male-sterile. Flowers with 5-toothed calyx; corolla white, blue or purple, with flat to slightly hooded upper lip; stamens 4, shorter than upper lip of corolla.

Nepeta cataria CAT MINT An erect, hairy rhizomatous perennial with branched stems 34 cm–2 m and grey, toothed leaves 27–84 mm long, grey-woolly beneath. Inflorescence whitish with numerous false whorls; flowers cosexual; corolla 5.7–9 mm, white with *small purple spots*; bracteoles just 0.2–0.5 mm wide; calyx teeth straight. Fields, cultivated land and woods, often above sea level. GR, IA, PN, TR. *N. italica* has creamy white flowers; corolla 11–13 mm. AA (northeast), IP, SY, TR.

Melissa BALM

Perennials with toothed leaves. Flowers borne in distant whorls in the leaf axils, with 2-lipped calyx; calyx usually pale yellow; stamens 4.

Melissa officinalis A *lemon-scented*, rhizomatous perennial to 90 cm (1.5 m) with few branches. Leaves stalked, 30–70 mm, oval, with toothed margins. Flowers borne in few, distant whorls of 4–10; calyx bell-shaped, 13-veined; corolla cream to yellow (or pinkish), (6)8–12(15) mm. Throughout.

1. *Eremostachys laciniata*
2. *Melissa officinalis*
3. *Prunella vulgaris*

Prunella SELF-HEAL

Perennial herbs entire to divided leaves. Flowers with 2-lipped calyx, the upper broad with 3 short teeth, the lower with 2 narrow lobes; corolla yellow, blue, pink or white with strongly hooded upper lip; stamens 4, shorter than upper corolla lip.

(a) Flower-heads with a pair of leaves immediately below; flowers purple.

Prunella vulgaris COMMON SELF-HEAL A *hairy*, tufted perennial with stems ascending from a creeping base, 50 mm–60 cm. Leaves 17–96 mm, *broadly lance-shaped*, toothed or not, stalked. Flowers blue–purple (rarely white), 11–12 mm long, emerging from very dense, small, cylindrical heads of conspicuous purple, spiky calyces, with *both bracts and a pair of leaves immediately below*. Damp and wooded places above sea level; most common in the north. Throughout, except smaller islands and the far southeast.

(b) Flower-heads with a pair of leaves immediately below; flowers white.

Prunella laciniata CUT-LEAVED SELF-HEAL A short, densely hairy, patch-forming perennial with erect flowering stems 10–37 cm. Leaves 35 mm–11 cm, *pinnately lobed*, stalked, with narrow segments. Flowers pale yellow–white (rarely pink or purplish), 14–16 mm long, borne in heads with a pair of leaves immediately below; calyx with broadly rectangular upper lip with 3 short teeth and lower lip divided to about ½ the length. Dry scrub; widespread but local. AA (rare), CR (rare), GR, IA, PN, TR. *P. cretensis* is similar but with broad, rounded entire to undulating upper calyx lip, the lower lip divided almost to the base. CR.

1. *Lamium amplexicaule*
2. *Lamium purpureum*
3. *Lamium bifidum*
4. *Lamium garganicum* subsp. *striatum*

(c) Flower-heads interrupted from the next pair of leaves by a short stalk; flowers purple.

Prunella grandiflora LARGE SELF-HEAL A leafy perennial 10–48 cm, similar to *P. vulgaris* in general appearance but with *larger* flowers, with corolla 18–32(48) mm long; calyx large (10–16 mm) and purplish, and inflorescence with bracts, but *without* leaves immediately below. Woods and shady places. GR (north).

Lamium DEAD-NETTLE

Annual or perennial herbs with crowded whorls of flowers. Flowers with white pink or purple corolla with hooded upper lip and tubular or bell-shaped calyx with 5 fine-pointed lobes; stamens 4, exceeded by upper corolla lip. Small annuals 10–25 cm, with flowers <15 mm long.

(a) Slender annuals <50 cm, with *pink to purple* flowers <20 mm long.

Lamium amplexicaule HENBIT DEAD-NETTLE A short, scarcely branched, hairy annual 10–40 cm. Leaves 9–20 mm, circular or oval, blunt-toothed and stalked below; bracts kidney-shaped and distinctly *clasping* the stem. Flowers *upward-spreading*; pink–purple, (13)15–20 mm long, with a slender, straight tube; calyx hairy with teeth about equalling the tube. A common weed of cultivated ground. Throughout. *L. purpureum* RED DEAD-NETTLE is similar but with stalked, not clasping upper leaves and bracts, and densely leafy inflorescences, flushed purple above; bracts stalked (not clasping). Flowers 8–12 mm; corolla somewhat exceeding the calyx. Disturbed ground. AA (local), CR, GR (common), TR. *L. macrodon* is similar to *L. amplexicaule* but with *oval, scarcely clasping* floral bracts and calyx teeth *exceeding* the tube, narrowly triangular. AA (rare and local), LB, TR.

(b) Slender annuals <50 cm, with *white* flowers 12–25 mm long.

Lamium bifidum A soft annual, similar to the previous species but with *white* flowers 12–25 mm with a distinctly *2-lobed* upper lip. AA (rare), CR (rare), GR (common), IA, PN. *L. moschatum* has white flowers 16–25 mm and entire upper corolla lip, and upper leaves with distinct white or pink markings. Regional subspecies exist. AA, CY, GR (southeast), TR.

(c) Rhizomatous perennials to 50(80) cm, with flowers >20 mm long.

Lamium maculatum SPOTTED DEAD-NETTLE A variable, hairy, aromatic, patch-forming perennial with erect stems to 18–50(80) cm. Leaves 30–65 mm, nettle-like, oval to triangular, pointed, toothed and stalked, often with a pale central blotch. Flowers whitish or purplish-pink with darker markings; corolla tube curved, (18)20–30(35) mm long, borne in 2–5 whorls of 6–8. Woodlands and mountains inland. GR, TR.

Lamium garganicum LARGE RED DEAD-NETTLE A perennial to 50 cm, rather similar to *L. maculatum*. Leaves to 70 mm, oval-heart-shaped, often toothed. Flowers *large*; calyx 7.5–18 mm with teeth shorter or as long as the tube; corolla *25–40 mm long*, tube *straight* and hairless within, greatly exceeding the calyx, and rose–purple, often with darker markings; upper lip 10–15 mm; lower lip 10–15 mm. Mountain rocks and woods. Throughout. Several co-occurring subspecies exist, the most widely recognised being **subsp. *striatum***, which has stems ≤30 cm with kidney-shaped leaves to 25 mm long, flowers in whorls of ≤10; **subsp. *pictum***, which is *short, ≤10 cm* with heart-oval-shaped leaves, found in the mountains of the mainland, and **subsp. *garganicum***, which is rather tall, >30(50) cm, densely hairy and with *large*, heart-oval-shaped, densely hairy leaves 15–50(70) mm, flowers with lobed upper corolla lip, flowers in whorls of ≥10.

Ballota

Perennials with sterile leaf rosettes. Flowers with 2-lipped corolla with concave upper lip; calyx funnel-shaped, 10-veined and usually 5-lobed; stamens 4, exceeded by upper corolla lip. Few species in the region.

Ballota nigra A moderately branched, weak-stemmed perennial to 1.5 m with short, deflexed hairs. Leaves oval and toothed. Flowers borne in distant whorls of 10–20; calyx 7–10 mm, scarcely expanded into 5 *long*, equal, triangular lobes ending in *teeth*; corolla 9–15 mm, pink to reddish-purple with a hooded upper lip. Widespread and variable with many forms described. Woods and grassy habitats. Throughout.

Ballota acetabulosa A clumped, many-stemmed, rhizomatous, woody-based perennial to 70 cm. Leaves 25–50 mm, broadly oval to circular, net-veined, densely grey-downy. Flowers borne in long, slender inflorescences with whorls of 6–12 crowded flowers, the whorls distant below, merging above; calyx expanded into a *conspicuous plate-like limb*, 12–20 mm across with undulate margins; corolla 15–18 mm, pinkish. Throughout except the southeast (not CY or IP). *B. pseudodictamnus* is similar but with *thickly-woolly* whitish leaves and much smaller, lobed calyx 8–10 mm across. CR (+ adjacent islets), TR (southwest – **subsp.** *lycia*).

Stachys

Annual or perennial herbs, often with toothed leaves. Flowers borne in dense, spike-like inflorescences; calyx tubular or bell-shaped with 5 equal teeth; corolla yellow, pink or purple, 2-lipped with a flat or hooded upper lip and 3-lobed lower lip; stamens 4, exceeded by upper corolla lip.

(a) Flowers whitish-yellow to yellow.

Stachys citrina A woody-based, white-downy perennial with basal rosettes and typically unbranched flowering stems to 35 cm. Basal leaves elliptic to oval, 7–50 mm long, slightly scalloped to entire with stalks 15–35 mm. Flowers lemon yellow, borne in crowded whorls of 6–8, forming a dense head (sometimes with few remote whorls below); calyx regular, 11–15 mm, with almost equal teeth; corolla 20–25 mm. Limestone cliffs and scree. TR.

Stachys ocymastrum An erect, pale green, hairy annual to 70 cm (1.1 m). Leaves 16–65 mm long, oblong and pointed, slightly heart-shaped at the base, toothed and wavy along the margin. Flowers borne in dense whorls of 4–6 along the stem, congested above, *laxer below*; calyx densely hairy with 2 upper teeth as long as the tube and 3 shorter lower teeth; *corolla white*, 10–16 mm (shorter than the calyx), the upper lip entire. Garrigue. CR, IA (Corfu – rare), PN (rare). *S. spinulosa* is similar but more robust with a broadly bell-shaped calyx with teeth half as long as the tube; corolla large, 20–24 mm, cream with pinkish spots on the lower lip. AA, CR (west), GR, PN, TR.

Stachys chrysantha A spreading to ascending, white-woolly perennial to 20 cm, woody below. Leaves 10–25 mm, almost circular to oval, white-woolly. Flowers borne 4–6 per whorl, rather congested; calyx 6–9 mm, with teeth about as long as the tube; corolla 12–14 mm, yellow, woolly; upper lip 4–6 mm, lower lip 6–8 mm. Cliffs and screes. PN.

---

1. *Stachys citrina*
2. *Stachys ocymastrum*
3. *Stachys spinulosa*
4. *Stachys chrysantha* PHOTO: ELEFTHERIOS DARIOTIS
5. *Stachys germanica*
6. *Stachys cretica*

(b) Perennials with pink (or whitish-pink) to purplish flowers.

Stachys germanica DOWNY WOUNDWORT A *densely white-felted*, erect perennial to 1.3 m. Leaves 58 mm–17 cm, *oblong and heart-shaped at the base*, grey-woolly below and grey–green and less hairy above; calyx with unequal teeth, the upper 2 <½ as long as the tube, but longer than the lower 3. Flowers with corolla to 20 mm; cream–white or bright pink–purple, borne in congested terminal inflorescences with equal or longer bracts. Dry grassy places, fallow land and roadsides; common. Throughout. The following closely related species are all rather similar. *S. cretica* MEDITERRANEAN WOUNDWORT is also white-felted, with pink–purple flowers, but with *leaves with rounded or wedge-shaped bases* (not heart-shaped). Several forms described. Throughout, except smaller islands. *S. thirkei* is similar to *S. germanica* but with adpressed-downy hairs and *long, spreading to recurved calyx teeth*. AA (rare), TR. *S. tournefortii* is densely white-downy to woolly with large flowers and bracts *shorter* than the whorls. CR (northwest; otherwise only in Libya).

Stachys balansae An erect, grey-hairy perennial with sterile basal rosettes and simple to sparingly branched flowering stems to 1 m. Leaves to 10 cm, oblong-oval with sparse, adpressed, silky hairs. Flowers borne in congested (10–16 flowers), distant whorls, spaced by 10–6 mm, slightly merged above; calyx 8–9(12) mm, rather bell-shaped with upper teeth almost as long as the tube; corolla (13)15–18 mm, pink, densely hairy, with an entire upper lip. Grassy habitats and woods. TR.

Stachys ionica A slightly woody-based, hairy to *woolly* perennial to 25 cm with a dense, *cushion-forming* habit. Leaves oval to oval-elliptic, rounded or abrupt at the base, *without conspicuous net veins*; upper leaves and bracts *stalked*. Flowers whitish with a pink and spotted lower lip; corolla 20 mm. Rocky cliffs. IA. ***S. swainsonii*** is similar but with upper leaves and bracts stalkless and usually conspicuously net-veined. GR (southern Sterea Ellas), PN.

***Stachys officinalis* (syn. *Betonica officinalis*)** An erect, almost hairless to densely hairy perennial to 1 m. Leaves 30 mm–12 cm long, oblong to oval-oblong and heart-shaped at the base, with coarsely scalloped margins. Flowers borne in whorls forming dense spikes, somewhat interrupted below; calyx 5–9(12) mm with teeth as long as the tube; corolla 12–18 mm, bright reddish-purple (rarely white), the tube exceeding the calyx and with an entire upper lip. Grassy habitats. GR, TR.

(c) *Annuals* with white or pale pink flowers.

Stachys arvensis FIELD WOUNDWORT A small, erect, hairy annual to 20(45) cm with sparse, spreading hairs. Leaves 15–50 mm, oval and heart-shaped at the base, hairy, toothed along the margin and wavy-edged. Flowers borne in whorls of (2)4–6, crowded above, distant below; calyx with teeth as long as the tube, purple-tinged; corolla 6–8 mm, white or pale pink and scarcely exceeding the calyx; upper lip entire. Cultivated land and sandy places. Throughout, except smaller islands (not CY).

1. *Stachys ionica* PHOTO: ELEFTHERIOS DARIOTIS
2. *Stachys balansae*
3. *Stachys officinalis*
4. *Stachys arvensis*

Satureja

Shrubs or perennials with entire leaves. Flowers with bell-shaped, 10-veined calyx with 5 nearly equal lobes; corolla straight and 2-lipped; stamens 4, shorter than the upper corolla lip.

(a) Bracts (bracteoles) on flower-heads fairly conspicuous, about equalling the calyx.

Satureja montana WINTER SAVORY A variable, aromatic, hairless to slightly hairy, rather densely spreading *dwarf shrub*, 13–45 cm. Leaves 12–24 mm, linear to oblong and broadest above the middle, hairless but with *short-hairy margins*. Flowers white to purple, 7.5–12 mm long, borne on many-flowered, leafy terminal whorls; lower *bracts longer than the flowers*; calyx with teeth shorter than their tube. Stony pastures and rocky slopes, common in the northwest. GR. **S. parnassica** A low shrub similar to *S. spinosa*, woody below, with stems to 10 cm, *without spines*. Leaves 5–10 mm, egg-shaped and glandular. Flowers borne in crowded whorls with equal bracts; calyx 3.5–4 mm with teeth as long as or just exceeded by the tube; corolla 6–7 mm. Mountains. GR, PN.

(b) Bracts short or absent. A rigid, eventually spiny shrub.

Satureja spinosa An *intricately branched, rigid, mound-forming shrub* to 15(30) cm, with *spiny branches*. Leaves 4–10 mm, thick and rigid. Flowers borne in few, crowded whorls, often just with 2 flowers; calyx 2.5–4 mm, almost regular; corolla 5–8 mm, white to pale lilac. AA (Samos + Chios), CR, TR (southwest – rare).

(c) Bracts short or absent. A woody-based perennial to shrub, *without* spines.

Satureja icarica A dense, spreading, cushion-forming shrub, similar in most respects to *S. montana* but with *small* leaves (6–10 mm) and flowers (calyx 3 mm; corolla 5–7 mm). AA (Ikaria), TR (northwest – rare).

Satureja spicigera A summer-flowering, woody-based, spreading dwarf shrub to 25(60) cm with stems hairy on 2 sides. Leaves densely hairy and stalkless, (8)15–20 mm (smaller above). Flowers borne in elongated terminal racemes; calyx 3–4.5 mm, hairy, bell-shaped with upper teeth 1/3 the length of the tube; corolla (6)8–10 mm, *white* (or whitish-pink); *stamens long-protruding*. TR.

Satureja thymbra SATUREJA An aromatic *thyme-like dwarf shrub* to 35 cm, grey-hairy. Leaves oblong, broadest above the middle, pointed and rather bristly. Flowers pale pink, to 12 mm long, *borne in rather dense, almost spherical, distant whorls* (unlike most *Thymus* spp.); calyx reddish and bristly. Garrigue and roadsides. Virtually throughout (rare in CY and the north).

Clinopodium

Perennial or annual herbs with entire or toothed leaves. Flowers borne in stalked axillary clusters (sometimes reduced to solitary flowers); calyx 5-lobed, tubular with (11)13(15) veins; corolla tube 2-lipped and straight to curved; stamens 4, not protruding. The genus now includes species classified traditionally under *Acinos* and *Calamintha*.

(a) Flowers 6–18 mm long.

Clinopodium nepeta (syn. *Calamintha nepeta*) LESSER CALAMINT A mint-like, greyish, hairy perennial 20–75 cm with creeping rhizomes; stems erect and branched. Leaves 17–70 mm, oval and shallowly toothed to untoothed; stalked. Flowers with corolla 6–17 mm, pale pink–purple with darker markings, borne in leafy whorls; calyx ribbed and purplish, with white *hairs protruding from the mouth*. Dry, fallow land and waste places. Virtually throughout. **Subsp. *glandulosum*** is the predominant form in the region.

Clinopodium vulgare WILD BASIL A mint-like, softly hairy, aromatic perennial 16–95 cm with erect, branched or unbranched stems. Leaves 14–50 mm, oval-lance-shaped, slightly toothed and short-stalked. Flowers with corolla 9–18 mm, pink–purple, borne in distant whorls along an interrupted spike with prominent calyces; calyx green with purple ribs, more or less 2-lipped and prominently toothed. Disturbed, dry, grassy and stony habitats. Virtually throughout.

Clinopodium acinos (syn. *Acinos arvensis*) An erect to ascending, soft-hairy annual to short-lived perennial to 25 cm. Leaves short-stalked, elliptic, slightly toothed towards the tips, 1.5 x as long as broad. Flowers borne in 3–8-flowered clusters shorter than or equalling the leaves; calyx 5–7 mm; corolla purple, 7–10 mm. AA (rare), GR, TR.

(b) Flowers 25–35 mm long.

Clinopodium grandiflorum (syn. *Calamintha grandiflora*) LARGE-FLOWERED CALAMINT A spreading, sparsely hairy perennial with weakly ascending stems 25–35 cm with oval, coarsely toothed leaves 30–51 mm. Flowers pink, protruding and conspicuous, borne in leafy inflorescences; corolla 25–35 mm long. Mountain woods. GR, IA, PN, TR.

Micromeria

Small, thyme-like subshrubs. Flowers with calyx with 13–15 veins and 5 pointed lobes.

(a) Blade of stem-leaves *broadly* elliptic-oval.

Micromeria nervosa An erect subshrub 15–40 cm with oval to triangular leaves 5–15 mm long with distinct lateral veins. Flowers borne in lax, interrupted inflorescences; *calyx covered in sparse, white, long fine hairs*; calyx teeth unequal, ≥½ as long as their tube. Throughout (rarer in the west). *M. myrtifolia* is similar to *M. nervosa* but with leaves with inconspicuous lateral veins. Flowers with calyx teeth 1/3 as long as their tube. Distribution as for the previous species.

Micromeria sinaica An adpressed-hairy, much-branched shrub 15–40 cm; young branches thick, erect. Leaves 4–8 mm, few, remote, almost stalkless, oval to oblong with somewhat down-turned margins. Flowers 6 mm, pinkish-purple, borne numerously in remote whorls; calyx 2–3 mm, with lance-shaped, straight teeth <½ as long as the tube. Rocky deserts in the southeast. IP (especially Negev Highlands).

(b) Blade of stem-leaves *narrowly* elliptic to linear-oblong. Calyx teeth short and narrowly triangular, ≤½ as long as their tube.

Micromeria sphaciotica A woody-based perennial to dwarf shrub to 15 cm with spreading, woolly hairs on the stems. Leaves short-stalked, narrowly elliptic. Flowers pinkish-purple; calyx 3.5 mm, tubular with straight teeth; corolla tube *exceeding* the calyx; stamens *included*. Rare. CR. *M. carpatha* is similar but with short stem hairs, corolla tube scarcely exceeding the calyx and stamens slightly protruding. KP only.

(c) Blade of stem-leaves *narrowly* elliptic to linear-oblong. Calyx teeth short and narrowly triangular, ≥½ as long as their tube.

Micromeria graeca A slender, lax, *dwarf shrub* 13–60 cm; woody at the base and variably hairy. Leaves small, 7–14 mm long, oval below and linear-lance-shaped above, *with down-turned margins*. *Flowers stalked*, borne in *distant* whorls of 4–12, forming long, slender, terminal inflorescences; corolla 5–7(10) mm, pink–purple; calyx 4–5(6) mm, woolly-hairy in the throat with teeth almost as long as their tube and unequal. Dry, scrubby places, often coastal. AA, GR (coastal), IA, LB, PN, TR (south). Variable with many forms described; **subsp. *graeca*** is the form in the region.

1. *Satureja spicigera*
2. *Satureja thymbra* habit
3. *Satureja thymbra*
4. *Clinopodium nepeta* subsp. *glandulosum*
5. *Clinopodium vulgare*
6. *Micromeria nervosa*
7. *Micromeria sinaica*
8. *Micromeria graeca* subsp. *graeca*
9. *Micromeria juliana*

Micromeria juliana A dwarf shrub 12–40 cm with numerous stiffly erect, unbranched and very densely small-leaved stems; leaves 4–8 mm long, and *linear with down-turned margins*. Flowers tiny, borne in *dense, spike-like, continuous stalkless whorls* of 8–20; corolla 2.5–3 mm, purple and hairy; calyx 3.2–3.5 mm, *hairless in the throat*. Common on garrigue nearly throughout the Greek territory. Throughout except CY, IP and smaller islands.

Hyssopus HYSSOP

Aromatic subshrubs with entire leaves. Flowers with tubular-bell-shaped calyx with 15 veins and 5 sub-equal lobes; corolla blue (or white); stamens 4, protruding.

Hyssopus officinalis HYSSOP An aromatic, virtually hairless, much-branched subshrub 15–52 cm with narrow leaves 11–22 mm, clustered at the nodes. Flowers pale pink, blue or violet, borne in rather compact, elongated spikes; corolla 7–8 mm, *open with stamens long-projecting*. Dry slopes and maquis, also planted. GR.

Thymus THYME

Dwarf shrubs, woody at the base, and characteristically aromatic, with entire leaves. Flowers borne in heads; calyx 2-lipped, the upper lip with 3 short teeth, the lower with 2 long teeth; corolla 2-lipped; stamens 4, protruding (in cosexual flowers). Very diverse and taxonomically complex in the region, particularly mainland Greece and Turkey. Just a small subset are described here.

(a) Perennials, typically with spreading non-flowering shoots and scarcely woody below.

Thymus leucotrichus A *mat-forming* perennial with densely short-hairy young stems and numerous non-flowering shoots. Leaves 7–15 mm, linear-oblong, with scarcely down-turned margins or flat, *rather hairy*. Flowers borne in heads with oval bracts, wider than the leaves; calyx 4–5 mm, deeply 2-lipped; corolla pale mauve to purplish. Several regional forms described. CR (western mountains – the only *Thymus* sp.), GR, PN, TR.

Thymus longicaulis A variable, mat-forming perennial with more or less erect flowering stems 30 mm–10 cm with small basal clusters of small leaves; stem-leaves 5–13 mm, narrowly elliptic and hairless, not particularly leathery. Flowers borne in congested heads, pinkish-mauve; corolla 4–6 mm; calyx 2.5–3.5 mm, the tube shorter than the upper lip. AA (local), GR, TR.

Thymus thracicus A mat-forming perennial with non-flowering shoots with clusters of leaves in the axils. Flowering stems 50 mm–15 cm, usually hairy. Leaves 8–12 mm, flat, narrowly elliptic and with numerous reddish glands. Flowers borne in heads with leafy bracts; calyx 4–6 mm with upper teeth 1.4–2 mm; corolla 6–8 mm, bright mauve. AA (Thasos), GR, TR.

Thymus zygioides A mat-forming perennial with long, leafy, non-flowering shoots; scarcely woody below. Leaves 7–10 mm, linear to egg-shaped, hairless and gland-dotted; the lowermost leaves much smaller and broader. Flowers borne in heads with bracts *slightly* wider than the stem-leaves; calyx 3.5–4.5 mm with very short upper teeth; corolla pinkish-purple; anthers protruding. Woody-based forms on Samos have been described as *T. samius*. Local. AA (north), GR, TR.

(b) Woody-based perennials or shrubs *without* (or with few) spreading non-flowering shoots.

Thymus sibthorpii A variable, erect to ascending, branched, short-hairy perennial to 20(30) cm. Leaves 10–15 mm, short-stalked, elliptic to lance-shaped, leathery but thin, glandular with distinct lateral veins. Flowers with bell-shaped, greenish to beige calyx 2.5–3.5(4) mm with *short upper teeth* (0.7–1 mm), the upper lip longer than the tube; corolla pale pink to reddish-mauve;

infloresence *densely hairy*. Capsule *large*, 16–18 mm. Widespread on the Greek mainland. GR (+ adjacent islands), PN. *T. comptus* is similar but with leaves to just 2.5 mm (not 3–5 mm) wide and with *long* upper calyx teeth 1.2–1.5 mm. GR, TR.

Thymus integer A rather early-flowering, spreading, gnarled, dwarf shrub with stems to 10(20) cm, woody and fissured below (sometimes with prostrate and rooting lateral branches). Leaves stalkless, hairy, narrowly lance-shaped to linear, 3–10 mm, with down-turned margins. Flowers borne in terminal clusters of 5–12; calyx bell-shaped, purple, hairy; corolla tubular, pink or white, 10–15 mm with *conspicuously long, straight tube*. Igneous hillsides and rocky garrigue, or pine forest margins; a rare island endemic. CY.

Thymus teucrioides A bushy shrub to 20(30) cm with slightly hairy stems, superficially similar to *Clinopodium*. Leaves distinctly *stalked*, 3–6 mm long, triangular to oval-diamond-shaped, hairless. Flowers borne in *few, distant whorls* of 4–8 with leafy bracts; calyx 5–6 mm, deeply 2-lipped; corolla 10–14 mm, purplish-pink. Several closely related species occur in GR and PN. AA (Euboea).

Thymus atticus A woody-based perennial with numerous ascending flowering shoots to 15 cm; young stems hairy. Leaves to 10 mm, linear, leathery and flat. Flowers borne in *terminal heads*, with *conspicuously veined bracts much broader than the leaves*; calyx 6 mm; corolla 7–8 mm, white to pale pink with darker markings on the lower lip. GR, PN.

Thymus pulegioides LARGE THYME An aromatic, tufted perennial, somewhat woody beneath, to 30 cm. Leaves elliptic, flat, 3.5–16 mm, *with long hairs at the base, otherwise hairless*. Flower stems *strongly 4-angled with hairs only on the angles*; *flowers pink*, borne in rather *distant whorls*; calyx 3–4 mm. Local in grassy and often damp habitats; absent from hot, dry areas. GR, PN.

1. Thymus sibthorpii
2. Thymus integer

Thymbra

Woody-based, erect, aromatic perennials very similar to *Thymus* (and still widely described under this genus). Leaves linear to lance-shaped. Flowers borne in heads of more or less crowded whorls with bracts similar to the leaves; calyx with 2 lower elongated teeth and 3 upper short, triangular teeth; corolla with lobed upper lip and 3 equal lower lobes; stamens 4, protruding.

Thymbra capitata (syn. *Thymus capitatus*) A mid- to late summer-flowering, dense, often cushion-like dwarf shrub 10–40 cm high with erect, woody branches *with axillary leaf clusters*. Leaves fleshy; *somewhat 3-angled*, (4)5–10 mm long, stalkless, linear, more or less hairy and with inconspicuous lateral veins; gland-dotted with a *flat margin*. Flowers with corolla *mid-pink*, 6–10 mm long, borne in *dense, oblong terminal clusters* with red-tinged bracts that overlap, forming cone-like heads; calyx *many-veined* (20–22). Dry, rocky places. Throughout. *T. spicata* is erect, with fewer branches, leaves 10–18 mm, calyx with 13 veins and corolla 12–15 mm. Local. AA (local), GR (not the north), IP, LB, SY, TR. *T. calostachya* has *white flowers*, and calyx 2–3 mm. Rocky slopes and scree. CR (east – rare).

Ziziphora

Rather thyme-like, very aromatic annuals with erect, hairy, branched stems. Leaves simple, often toothed. Flowers in spike-like or congested inflorescences; calyx cylindrical with 13 veins; corolla with 1 superior lobe and 3-lobed lower lip; stamens 2, protruding.

Ziziphora capitata A single-stemed or scarcely branched annual to 15(25) cm. Leaves in few pairs, narrowly elliptic to oblong, almost hairless. Flowers borne in crowded heads with large bracts; calyx 8–11 mm, tubular, with 13 veins; corolla with a narrow tube, exceeding the calyx by 2–3 mm, reddish-pink with a broad, 3-lobed lower lip. GR, IP, PN, TR.

Ziziphora tenuior A *tiny annual* 30 mm–10 cm with slender stems. Leaves in few pairs, small, just 8–15 mm, linear. Flowers in dense, terminal heads. Calyx 6–8 mm, tubular and with conspicuous veins; corolla 8–11 mm, pale violet, the tube gradually widening towards the apex. AA (rare), IP, TR. *Z. taurica* is similar but larger, with leaves to 30 mm and corolla 12–15 mm. AA (Lesbos), TR.

1. *Thymbra capitata* (habit)
2. *Thymbra capitata*

Scutellaria SKULLCAP

Perennials with entire to toothed or lobed leaves. Flowers borne *2 per node* in the leaf axils; calyx 2-lipped with both lips entire; corolla pinkish to blue with hooded upper lip and obscurely 3–4-lobed lower lip; stamens 4.

Scutellaria columnae A rhizomatous perennial with few, almost erect stems to 1 m with short, non-glandular hairs. Leaves simple, toothed. Flowers borne in long, lax spikes of paired flowers; calyx 2-lipped; corolla *long*, 20–26 mm, strongly curved, lilac–purple with a whitish tube. Widespread in damp woodlands in the west and north. GR, IA (Corfu), PN.

Scutellaria sieberi A woody-based perennial with sparingly branched, ascending stems 15–40 cm. Leaves rather long-stalked with blades 20–50 mm, often slightly heart-shaped at the base and coarsely toothed. Flowers borne in long, dense spikes; calyx green or purplish; corolla 12–16 mm, *pale yellowish-cream* and usually unmarked. A widespread island endemic. CR. **S. hirta** is similar but shorter, 50 mm–15 cm, with a short, dense inflorescence; corolla 10–13 mm with a cream tube and lower lip, and *pale pinkish-purple upper lip*. A rare island endemic in rocky mountain habitats. CR.

Scutellaria orientalis A variable, usually woody-based, sparsely mat-forming perennial with rather long, spreading hairs. Leaves stalked, 8–15 mm long, oval, pale green, divided into blunt, narrowly oblong lobes. Flowers *erect, and bright yellow*; corolla 25–35 mm with a slender tube. AA, TR.

Origanum MARJORAM

Woody-based perennials and small shrubs. Flowers clustered in terminal heads; calyx bell-shaped and white-hairy within, with 13 veins and 5 lobes; corolla purplish (or white); stamens 4, protruding in cosexual flowers.

(a) Dwarf shrubs to just 10 cm with small leaves, 2–6 mm. Calyx *2-lipped* (the upper longer than the lower).

Origanum vetteri A tufted dwarf shrub to 10 cm with small leaves <5(6) mm, broadly oval-triangular, heart-shaped at the base, short-stalked and hairy. Flowers few, nodding, in lax panicles; bracts egg-shaped, hairy; calyx 4 mm, with 3-toothed upper lip 2 x as long as the lower; corolla 8 mm, pink, 2 x as long as the calyx tube. KP.

(b) Dwarf shrubs or perennials up to 70 cm (1.3 m). Calyx teeth more or less *equal*.

Origanum vulgare MARJORAM A rather thyme-like, lax, aromatic perennial with erect, purplish stems to 70 cm (1.3 m). Leaves 15–42 mm, oval and scarcely toothed or untoothed; short-stalked or unstalked and with leaf tufts in the axils. Flowers with corolla 4.5–10 mm, pink to reddish purple, darker in bud, borne in *broad, branched, panicle-like clusters*; *calyx bell-shaped with 13 veins and 5 almost equal teeth*. Grassy places; several regional forms recognised. Throughout.

(c) Dwarf shrubs or perennials up to 70 cm (1.3 m). Calyx more or less 1-lipped.

Origanum onites POT MARJORAM A similar species to *O. vulgare* but with a more compact, many-stemmed habit with stems irregularly covered with minute protuberances; *flowers always white*; calyx 1-lipped (not with 5 almost equal teeth). Garrigue (also planted and naturalised). AA (common), CR, CY, GR (southeast coasts), PN, TR. *O. syriacum* is a similar, more erect shrub to 1.3 m, with leaves 5–20 mm long, conspicuously net-veined beneath. Flowers with overlapping, broadly oval, hairy bracts and sheath-like calyx, 1.8 mm long; corolla white, 4 mm long. Steep, rocky, limestone habitats. CY, LB, SY (west), TR (south).

(d) Ascending, woody-based perennials or dwarf shrubs <1 m. Calyx more or less 1-lipped.

Origanum dictamnus A white-woolly dwarf shrub to 20 cm. Leaves 13–25 mm, almost *circular*, entire, with raised veins. Flowers borne in lax panicles; bracts 7–10 mm, conspicuous, purple, exceeding the calyx; calyx with almost entire upper lip and shallowly toothed lower lip; corolla pink, the tube twice as long as the calyx tube. A locally common island endemic. CR. *O. calcaratum* is similar but with *green* (not white-woolly), broadly *oval* leaves. AA (islands southeast of Naxos), CR (east – rare).

Origanum majorana SWEET MARJORAM A subshrub to 80 cm with round to oval or elliptic leaves to 30 mm. Flowers borne in very dense, pyramidal inflorescences with long branches, green bracts and white (or purple) flowers extending down the stems; calyx 2–3.5 mm, with 1 prominent (upper) lip, *split deeply along 1 side* only (not into 5 equal teeth); corolla 3–7 mm. Garrigue. CY, TR (south).

(e) Ascending, woody-based perennials or dwarf shrubs <1 m. Corolla with *conspicuously long tube* (to 40 mm).

Origanum amanum A late-flowering, spreading dwarf shrub to 20 cm. Leaves 10–15 mm, heart-shaped at the base. Flowers conspicuous, in terminal clusters of up to 6; bracts purplish; corolla pink, ascending, with *long tube*, to 40 mm. Rare, on rocky limestone outcrops. TR (Amanus Mountains), SY.

(f) *Desert-dwelling* shrubs.

Origanum dayi A *sticky*, glandular, sparingly hairy, shrubby perennial 30–80 cm; branches numerous, leafy and brittle. Winter leaves 8–20 mm, oval with heart-shaped bases, spreading, prominently veined; summer leaves 3–5 mm. Flowers borne in racemes along leafy branches from the lower 1/3 to the apex; bracts oval-circular, prominently veined, heart- to wedge-shaped at the base, sharply pointed at the apex, just longer than 1/2 the length of the calyx; calyx 5.5–6.5 mm; corolla pinkish-white (lilac in bud); stamens *long exserted*; anthers pink. A rare endemic of rocky deserts. IP (mainly Negev Highlands and Judean Desert).

Mentha MINT

Aromatic perennial herbs with creeping rhizomes. Flowers cosexual or female in dense whorls; calyx regular or weakly 2-lipped, corolla weakly 2-lipped with 4 subequal lobes; stamens 4, protruding (in cosexual flowers). Frequently hybridising.

(a) Inflorescence terminating in a cluster of leaves.

Mentha pulegium PENNY-ROYAL A strong-smelling spreading or ascending perennial 12–78 cm with erect flowering stems; leaves roughly oval, 8.5–30 mm, finely downy and blunt-toothed. Flowers mauve, borne in *rounded, well-separated* heads, *without* a clear terminal head; corolla 4–5 mm; calyx 2.5–3.5 mm. Pondsides; rare in much of the north of the region. Throughout.

(b) Inflorescence terminating in a cluster of flowers; terminal cluster oblong-spherical.

Mentha aquatica WATER MINT A short to tall, vigorous and leafy, hairy to hairless perennial to 1.5 m, strongly smelling of mint when crushed, with angled, purple stems. Leaves 20–50 mm, oval, pointed, toothed and stalked. Flowers purple–white, borne in very *dense, oblong heads* with 1 or 2 whorls of flowers below; corolla 5.5–7 mm; calyx 3.5–4.5 mm, veined and hairy. Always near water. Throughout (except small, dry islands).

(c) Inflorescence terminating in a cluster of flowers; terminal cluster cylindrical.

Mentha spicata SPEAR MINT A variable, green, mint sauce-smelling perennial 43–84 cm with lance-shaped, toothed, *scarcely hairy leaves* 17–88 mm. Flowers pink or white, borne in long, cylindrical, interrupted, *lax spikes*; corolla 2.5–4 mm; calyx 1.5–2.5 mm, the *tube hairless* (teeth sometimes hairy). Throughout. **M. longifolia** HORSE MINT is similar, though greyish and distinctly *white-downy below*, and leaves 15 mm–12 cm, *grey and densely silkily hairy*. Flowers white or lilac, borne in long, terminal, *dense spikes*, which separate as they mature; calyx 2–3 mm; corolla 3–3.5 mm. Wet habitats. Throughout.

Mentha suaveolens APPLE MINT A rather small perennial to 40(87) cm with a sickly sweet scent and stems variably hairy. Leaves 18–52 mm, *bright green,* stalkless or virtually so, circular to oblong and broadest near the base, toothed, hairy above, and grey-hairy beneath. Flowers borne in many congested whorls, forming long, dense inflorescences; often branched; corolla 3–3.8 mm, pale pink or white; calyx 1.2–2.5 mm, hairy with subequal teeth. Damp habitats; absent from most islands. AA (rare), GR, IP, PN, TR.

---

1. *Origanum onites*
2. *Origanum dictamnus*
3. *Origanum amanum*
4. *Origanum dayi*
5. *Mentha aquatica*
6. *Mentha suaveolens*

Lycopus GYPSYWORT

Herbaceous perennials. Flowers with 5-toothed calyx; corolla with 4 sub-equal lobes; stamens 2, protruding.

Lycopus europaeus GYPSYWORT An erect, slightly hairy, light, bright green perennial, often much-branched, with angled stems to 92 cm. Leaves 21 mm–11 cm, oval and very deeply cut into narrow triangular teeth except at the tip. Flowers small and inconspicuous, 3–4 mm across, white, with 4 roughly equal lobes, borne in interrupted clusters. Common on river banks and near ponds at all altitudes. Throughout (not CY).

Lavandula LAVENDER

Aromatic shrubs with narrow leaves and distinct bracts. Flowers borne in crowded, long-stalked terminal spikes; calyx with 5 small teeth, 13-veined; corolla purple, 2-lipped, weakly zygomorphic; stamens 4, not protruding.

(a) Flower spikes topped by conspicuous bracts.

Lavandula stoechas FRENCH LAVENDER A greyish shrub to 1.5 m with erect, much-branched stems. Leaves 6–37 mm, linear and untoothed. Flowers with a deep mauve corolla, 4.5–5 mm, borne in short, dense spines, topped by *conspicuous purple flower-like bracts* 8–36 mm long; inflorescences longer than their spikes. Common on maquis, dry scrub, open woods and fixed dunes. Throughout.

1. *Lycopus europaeus*
2. *Lavandula coronopifolia*
3. *Lavandula stoechas*
4. *Lavandula pubescens*
5. *Rosmarinus officinalis*

(b) Flower spikes *not* topped by conspicuous bracts.

Lavandula pubescens An erect shrub with angled, *hairy* stems, divergently branched, 40–80 cm. Leaves broad, *pinnately divided* into oblong-linear segments. Flowers borne in long, dense, slender spikes, solitary or in panicles; whorls 2-flowered; bracts oval, shorter than the calyx; calyx 5–6 mm with *unequal*, triangular teeth; corolla blue, with a tube about 2 x as long as the calyx. Desert valleys. IP (Mainly Judean Desert and Dead Sea Valley). *L. coronopifolia* has *lax* spikes, bracts *much shorter* than the calyx, and *equal* calyx teeth. Deserts and scrub. IP (south and east).

Rosmarinus ROSEMARY

Evergreen, aromatic shrubs with linear leaves with inrolled margins. Flowers with 2-lipped calyx with a virtually entire upper lip; corolla blue or pink, strongly zygomorphic; stamens 2, protruding.

Rosmarinus officinalis ROSEMARY A familiar evergreen shrub to 1.8 m, characteristically aromatic; branches brown and woody, erect to spreading. Leaves needle-like, 10–41 mm, linear and leathery, mid to dark green, sharply pointed, with down-turned margins. Flowers white, flushed pale purple, 8.5–13.5 mm long, borne in small lateral clusters; corolla with 2 protruding stamens; calyx bell-shaped. Very common on maquis, garrigue, open woods and fixed dunes. Throughout, except the southeast.

Salvia SAGE

Herbs and shrubs with distinct whorls of (often purple) flowers forming a lax inflorescence. Flowers with both calyx and corolla 2-lipped, the upper corolla lip hooded, the lower 3-lobed; stamens 2, hinged in the middle, joined beneath the corolla hood (shorter than upper corolla lip).

(a) Plant a shrub or subshrub.

Salvia aegyptiaca A woody-based, tufted dwarf shrub with leafy, erect to ascending stems to 25 cm with short to long hairs. Leaves oval-oblong to linear-elliptic, 12–25 mm with crimped to toothed margins. Flowers borne in simple racemes (sometimes branched) with distant whorls of 2–6; calyx 5 mm (larger in fruit); corolla violet–blue or white with darker markings in the lip, 6–8 mm with a straight to deflexed upper lip, exceeded by the lower. Nutlets smooth and black. Arid and semi-arid habtitats. IP (mainly south + Lower Jordan Valley).

Salvia officinalis SAGE A strongly aromatic, greyish shrub to 60 cm with erect branches that are woody beneath. Leaves broadly elliptic, greenish above and white-felted below, with a finely toothed margin. Flowers pale violet, blue, pink or white; corolla 15–25(35) mm long; calyx 10–14 mm, flushed purple, the upper lip 3-toothed; bracts oval and hairy. Dry, stony pastures. GR (northwest). *S. tomentosa* is similar, with flowers with pale blue corolla 22–35 mm with straight upper lip. GR.

Salvia fruticosa (syn. *S. triloba*) THREE-LOBED SAGE An aromatic subshrub to 1.5 m with felted stems. Leaves grey-green and wrinkled above, grey-felted below, and stalked; narrowly oval usually with 3(5) *lobes at the base*. Inflorescence spike-like with 2–6 flowered false whorls; corolla blue–purple or pink, 15–30 mm long; calyx 5–11 mm, bell-shaped and indistinctly 2-lipped, with 5 triangular teeth 1.5–3.5 mm long. Garrigue and gorges. Virtually throughout.

Salvia dominica A sparsely hairy, woody-based shub with numerous erect to ascending stems 50 cm–1 m, ending in *bristly* racemes. Leaves entire, with adpressed hairs, wrinkled and with scalloped to lobed margins, triangular-oval with heart-shaped bases, short-stalked below and stalkless above, 28–65(75) mm long. Bracts heart-shaped, shorter than the calyx. Flowers borne in whorls of 4–6, distant below, somewhat congested above; calyx 7–10 mm with about 10 prominent veins and with bristle-like teeth; corolla 15–18 mm, creamy-*white* with an arched upper lip as long as the calyx. Dry hillsides. CY (rare), IP (common), SY.

Salvia pomifera A deciduous shrub to 1 m with erect to ascending branches. Leaves simple, short-stalked, oblong to lance-shaped or broadly elliptic, greyish and densely hairy when young. Flowers large, borne in few-flowered whorls forming a terminal spike with bracts exceeding the calyx; calyx bell-shaped, enlarged in fruit, reddish-purple; corolla large, 25–30 mm, the upper lip pale violet–blue, the lower lip whitish. AA, CR (west), GR (south), PN, TR.

(b) Plant a herbaceous annual, biennial or perennial; bracts beneath flower whorls conspicuous and partially hiding calyx.

Salvia sclarea CLARY A robust, sticky, strong-smelling biennial or perennial to 1.5 m. Leaves hairy and oval to heart-shaped. Flowers 4–5 per whorl, *lilac or pale blue*, rather large, the corolla 18–30 mm long, with a strongly curved hood, and with *prominent oval or heart-shaped, lilac or white bracts that exceed the flowers*; calyx 10–16 mm with spiny teeth. Absent from smaller islands. AA (rare), GR, IA, IP, PN, TR. **S. aethiopis** is similar but with white-felted stems and smaller, *white* flowers 10–15 mm, with *greenish* bracts, *equalling* the calyx. GR, TR.

Salvia argentea SILVER SAGE A robust perennial herb to 50 cm with spreading hairs, similar to S. sclarea. Leaves stalked, oval-oblong, woolly when young, persistently cobweb-hairy and silvery-white. Flowers borne in few whorls of (4)6–8(10); bracts half as long as the calyx, broadly oval and green or white; corolla *large*, (17)25–30 mm, white, with a pale yellow lower lip with darker margins. AA, GR, PN, TR (north).

(c) Plant a herbaceous annual, biennial or perennial; bracts beneath flower whorls *not* conspicuous or hiding calyx. Flowers mostly purplish or bluish.

Salvia pratensis FIELD SAGE A variable, green perennial herb to 1 m, erect and branched; glandular above. Leaves long-stalked, oval and heart-shaped at the base, crinkled along the margins, with spreading hairs above and more or less hairless beneath. Inflorescence lax and shortly branched. Whorls of 4–6 cosexual or female flowers; *violet–blue* and *large, 20–30 mm*. Grassy places and open woods. CR (rare), GR, IA, PN.

Salvia nemorosa A perennial similar to S. pratensis, 30–60 cm with more or less spreading non-glandular hairs. Basal leaves stalked, oblong, heart-shaped to rounded at the base; stem-leaves rather numerous, stalkless. Inflorescence dense; bracts as long, or longer than calyx, violet; corolla *dark blue–purple*, 8–12(15) mm long. GR (north). **S. amplexicaulis** is similar to S. nemorosa but taller, to 80 cm, with short-stalked to stalkless leaves and violet corolla, (8)9–12 mm, with slightly convex upper lip. GR, PN, TR (north).

Salvia verbenaca WILD CLARY A perennial to 80 cm with erect stems, glandular above. Basal leaves forming a rosette, wrinkled, *deeply lobed*, long-stalked below, and short-stalked or stalkless above. Flowers pale blue or violet, 6–16 mm long, borne in lax whorls forming a spike; calyx 5–12 mm, bell-shaped and distinctly veined; enlarged in fruit. Fallow land, grassy places, roadsides garrigue; common. Throughout.

Salvia viridis RED-TOPPED SAGE A short *annual* with stems simple or branched below, to 50 cm. Leaves hairy, pale green, blunt and long-stalked, with finely toothed margins. Inflorescence an elongated spike of small, pink flowers borne in whorls of 4–6(8); corolla 10–14 mm; calyx 7–12 mm; bracts pointed, *the terminal bracts reddish or deep violet, forming a terminal tuft* (particularly conspicuous in eastern forms). Dry habitats on limestone, also planted; local. Throughout.

1. *Salvia aegyptiaca*	4. *Salvia dominica*	7. *Salvia verbenaca*
2. *Salvia officinalis*	5. *Salvia argentea*	8. *Salvia viridis*
3. *Salvia fruticosa*	6. *Salvia nemorosa*	9. *Salvia viridis* (eastern form)

Salvia indica A sparsely hairy perennial, leafy below, leafless above the middle. Leaves oval, heart-shaped at the base, with wavy-lobed margins; those above triangular and stalkless. Bracts triangular-oval, the lowermost exceeding the calyces, the uppermost broader than long and shorter than the calyces. Flowers *large*, borne in distant whorls of 4–6; calyx sticky-hairy and bell-shaped, with *short, broad, spine-tipped teeth*; corolla *4 x as long as calyx*, soon-falling, with a pale blue–violet upper lip and whitish lower lip with brownish margins *and conspicuous dark blue markings*; stamens long-protruding. Shady, rocky habitats; rare and local. IP (north and central), SY, TR.

Salvia lanigera A densely and long-hairy perennial 20–40 cm with a pungent smell. Leaves oblong to triangular in outline, divided into linear, *wrinkled*, blunt and scallop-margined, linear lobes; uppermost (floral) leaves oval to circular. Flowers borne in remote whorls of 6–8; calyx *white-fleecy*; corolla *long*, 15–25 mm, *deep violet* or blue; upper lip sickle-shaped. Sandy habitats and deserts. IP (common in the south), SY.

Salvia verticillata WHORLED CLARY An erect, hairy, rather unpleasant-smelling perennial 30–80 cm with purple-tinged stems and leaves. Leaves stalked and toothed, oval to heart-shaped with basal lobes. *Flowers lilac–blue or purple, borne in numerous tight whorls of (15)20–30*; corolla 8–15 mm long; calyx 5–7 mm. Dry habitats above sea level in the west. GR, IA, PN. **S. napifolia** is similar but with whorls of 6–12 flowers. TR.

(d) Plant a herbaceous annual, biennial or perennial; bracts beneath flower whorls *not* conspicuous or hiding calyx. Flowers whitish.

Salvia samuelssonii A scentless perennial 25–50 cm with erect stems broadly branching from the middle, with numerous long, white, kinked hairs. Leaves densely hairy, oblong to lance-shaped, 10–20 cm, irregularly lobed, the lower segments remote. Flowers borne in distant whorls of 4–6; calyx tubular, pale green, 25–30 mm in fruit; corolla *cream*, dilated in the throat, the upper lip sickle-shaped, >10 mm. Scrub, maquis and seasonally arid plains. IP (mainly east-central).

(e) Plant a herbaceous annual, biennial or perennial; bracts beneath flower whorls *not* conspicuous or hiding calyx. Flowers mostly pale to bright pink.

Salvia hierosolymitana A perennial 40 cm–1 m with stems with backward-pointing hairs along the angles, otherwise hairless. Leaves 10–25 cm, tooth-lobed; stem-leaves in 1–2 pairs, stalkless; bracts *small*, triangular-oval. Corolla 25–38 mm (3 x as long as the calyx), the upper lip strongly sickle-shaped and pink to blackish red, the lower lip paler to whitish; calyx typically reddish. Strongly eastern. CY, IP (north), LB, SY.

Salvia bracteata A slightly woody-based perennial to 50 cm, often purplish, densely glandular-hairy. Leaves stalked, pinnately lobed with an oval to oblong terminal segment 25–70 mm and 1(2) pair(s) of smaller lateral segments; variably toothed or lobed. Flowers borne in distant whorls of 5–8(10) subtended by bracts 15–30 mm (scarcely expanding in fruit but persistent and *boat-shaped*); corolla pink, 20–30 mm (2 x as long as the calyx), with a *long*, straight, brownish upper lip and whitish, spotted palate. Nutlets spherical, 3.5 mm. Rare, on rocky mountain slopes, oak forests and fallow fields. IP (Judean Mountains; extinct but in cultivation), SY, TR.

PAULOWNIACEAE

Deciduous trees native to China, Laos and Vietnam, with large, opposite, heart-shaped leaves. Flowers zygomorphic, borne in large panicles; calyx and corolla 5-lobed; stamens 4; style 1. Fruit a 2-parted capsule.

Paulownia

Deciduous trees with heart-shaped leaves. Flowers with tubular, foxglove-shaped corolla, borne in erect panicles; stamens 4. Fruit a capsule.

Paulownia tomentosa FOXGLOVE TREE A deciduous tree with a rounded, open crown, to 16 m. Leaves opposite, 3–5-lobed and *large*, 12–25(50) cm across, long-stalked and downy, heart-shaped at the base. Flowers 35–60 mm, blue–purple with a paler throat, 2-lipped and funnel-shaped, rather foxglove-like, borne in terminal panicles. Fruit a 2-parted, pointed, egg-shaped and woody, to 50 mm long. Commonly planted. *Paulownia fortunei* has a longer calyx (to 25 mm) and interior corolla *pinkish-cream* with purple spots, and with *inconspicuous folds*. Planted in IP.

---

1. *Salvia indica*   PHOTO: BANAN AL SHEIKH
2. *Salvia lanigera*
3. *Salvia samuelssonii*
4. *Salvia hierosolymitana*
5. *Salvia bracteata*
6. *Paulownia tomentosa*
7. *Paulownia fortunei*

OROBANCHACEAE | BROOMRAPE FAMILY

A distinctive family of root-parasitic (hemiparasitic or holoparasitic) herbs, attaching to the roots of host plants as seedlings. Flowers borne in spikes or racemes; corolla zygomorphic, more or less 2-lipped; stamens 4; style 1. Fruit a 2-parted capsule. Hemiparasitic species were previously classified in the Scrophulariaceae.

Odontites

Small hemiparasitic annuals or perennials with opposite entire to toothed leaves. Calyx tubular or bell-shaped with 4 entire lobes; corolla open-mouthed. Capsules with few seeds.

(a) Flowers pink–purple.

Odontites vernus RED BARTSIA A variable, short, hairy annual to 50 cm. Stems erect, dark reddish-purple and slightly squared. Leaves 10–50 mm, opposite, lance-shaped and normally toothed; bracts similar, exceeding the flowers. Flowers red–pink, 2-lipped, the lower deflexed, the anthers protruding; corolla 7–12 mm. Grassy and sandy habitats in the north and west. GR, IA, PN.

(b) Flowers yellow.

Odontites luteus An erect, much-branched, virtually hairless annual to 60 cm with linear-lance-shaped leaves 10–32 mm with inrolled margins, shallowly toothed. Flowers *bright yellow*, borne in dense, 1-sided spikes with narrow bracts; corolla 6–8 mm long, *finely hairy, with ciliate margins* and anthers and style long-projecting; *filaments hairy*. Mountains. GR (north).

Parentucellia

Hemiparasitic annuals with opposite leaves. Flowers with tubular to bell-shaped, regularly 4-lobed calyx; corolla long-tubed with open mouth. Capsule with numerous seeds.

Parentucellia viscosa YELLOW BARTSIA A short glandular-hairy annual with erect, normally unbranched stems to 35(60) cm. Leaves 17–35 mm, opposite, lance-shaped, deeply toothed and sessile; bracts similar, and decreasing in size along the stem. *Flowers yellow* (rarely white); corolla 18–23 mm, 2-lipped, the upper lip hooded, the lower lip 3-lobed; calyx teeth linear-lance-shaped and long, 4.5–6.5 mm. Damp grassy places; locally common. Throughout.

Parentucellia latifolia SOUTHERN RED BARTSIA A short annual to 20(40) cm with *triangular-lance-shaped and deeply lobed leaves* 6–15 mm. Flowers small, with a corolla 8–10(15) mm long, pale *reddish-purple* (rarely white); calyx teeth 1.4–1.6 mm long. Sandy and stony places, often coastal. Throughout. **Subsp. *flaviflora*** has *yellow* flowers (otherwise indistinguishable). CY, IP.

Bartsia

Hemiparasitic perennials with opposite, toothed leaves. Flowers with tubular to bell-shaped, regularly 4-lobed calyx and long-tubed corolla with an open mouth. Capsules with few, winged seeds.

Bartsia trixago (syn. *Bellardia trixago*) A short glandular-hairy annual with erect, simple stems to 60(70) cm. Leaves 20–40 mm, opposite, linear-lance-shaped, toothed and stalkless; bracts similar, and decreasing in size along the stem. Flowers with corolla white, normally flushed with pink or bright yellow, 17–24 mm, borne in a dense, *4-sided spike*. Ruderal sites, grassy places and garrigue; locally common. Throughout. **Var. *flaviflora*** has yellow (often marked dark red) flowers but is otherwise indistinguishable. Commonly mistaken for *Parentucellia viscosa*. Rare, but possibly scattered throughout the range.

1. *Odontites vernus*
2. *Parentucellia viscosa*
3. *Parentucellia latifolia*
4. *Parentucellia latifolia* subsp. *flaviflora*
5. *Bartsia trixago* (typical form)
6. *Bartsia trixago* var. *flaviflora*

Lathraea TOOTHWORT

Early-flowering rhizomatous perennial parasites that lack green pigment. Flowers with bell-shaped, 4-lobed calyx and 2-lipped corolla.

Lathraea squamaria TOOTHWORT A short, fleshy, often clumped, *whitish-pink* perennial to 15(30) cm with underground rhizomes. Leaves reduced to scales, crowded below. Flowers borne in dense, spike-like racemes, the flowers somewhat drooping; calyx 10 mm; corolla 14–17(20) mm. Capsule 10 mm with rather large seeds, 1–2 mm. Parasitic on the roots of trees and shrubs (especially *Corylus avellana*) in mountain woods inland; widespread but local on the mainland (absent from islands). GR, PN (north), TR.

Lathraea rhodopea A robust, clumped parasitic perennial, similar to *L. squamaria* but with *tall, yellowish* stems to 50(70) cm, and stout, to 20 mm thick below. Flowers borne numerously in dense, cylindrical racemes; calyx 5–7 mm; corolla 10 mm, the upper lip hooded, dull pinkish-purple. Capsule with few, large seeds, 2.5–3 mm. Parasitic on the roots of various trees (including *Corylus avellana*) in mountain foothills; rare and local. AA (Mount Ipsario, Thasos), GR (northeast).

1. *Lathraea squamaria*
2. *Lathraea rhodopea*

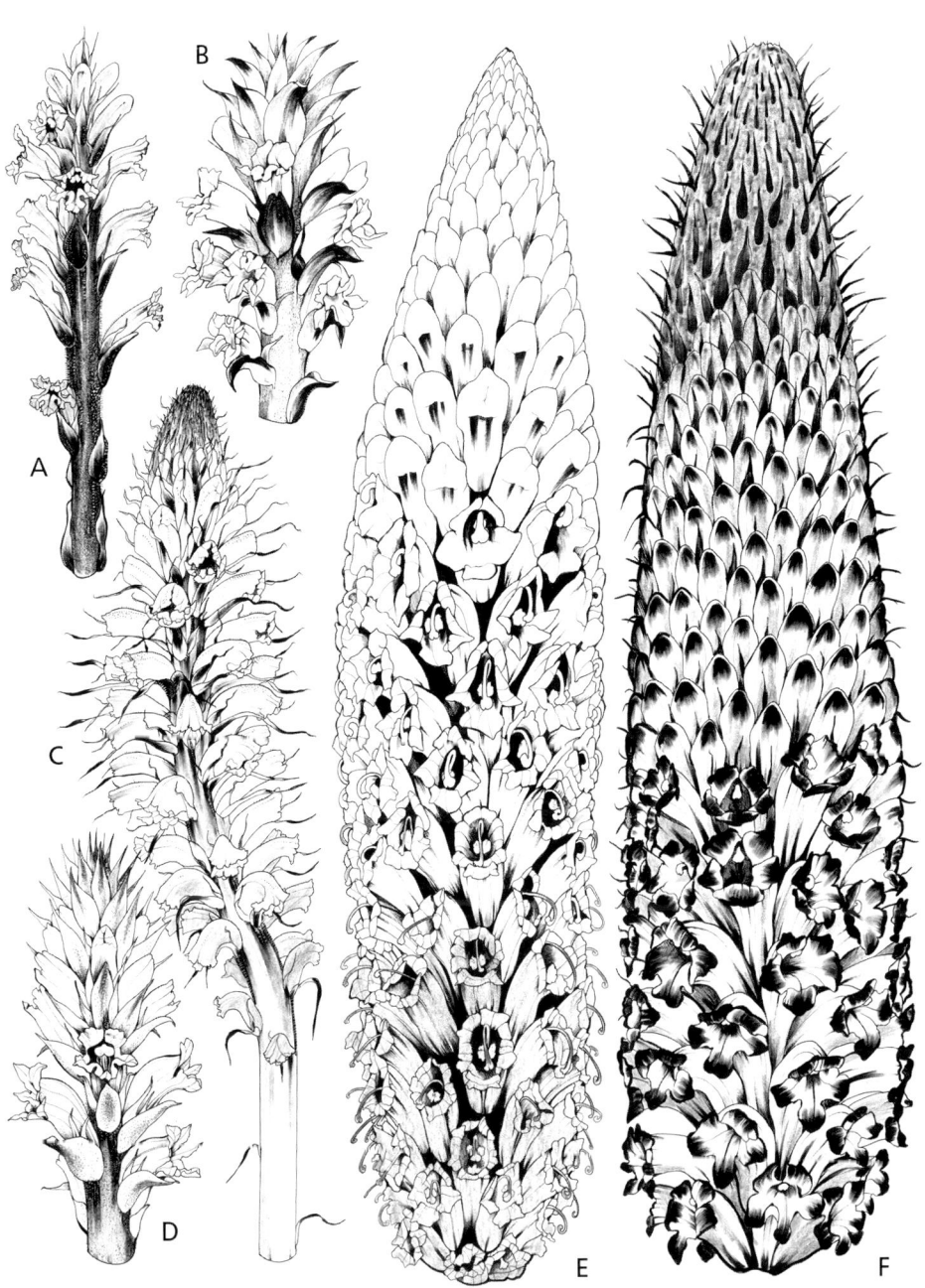

Orobanchaceae species: **A.** *Orobanche cypria*; **B.** *O. hermonis*; **C.** *O. palaestina*; **D.** *O. camptolepis*; **E.** *Cistanche tubulosa*; **F.** *C. violacea*. Illustrations not to scale.

Phelypaea

Simple-stemmed parasites with *solitary* (rarely 2), terminal flowers; corolla more or less 2-lipped with 5 large, spreading lobes. Fruit a capsule.

Phelypaea boissieri A conspicuous, slender parasite with stout, reddish-brown stems to 30(50) cm. Bracts to 20 mm. Flowers 40–60 mm, *bright red*, with equal calyx lobes 20–32 mm long; corolla with distinctly overlapping, large, round lobes, longer than their tube; anthers yellow-white, hairy; stigma large, scarlet. Capsule 10–15 mm, ovoid. A rare parasite of Asteraceae (especially *Centaurea* spp.) in mountain habitats. GR (north-central), TR.

Cistanche

Robust parasites with no green pigment. Flowers dense and numerous; calyx tubular with 5 teeth; corolla with 5 almost equal teeth. Fruit an ovoid capsule. Confused in the region covered; the descriptions below are provisional, in the absence of detailed DNA-based analysis for this area.

(a) Plant hairless. Flowers mainly yellow.

Cistanche tubulosa A *tall*, robust, yellow *desert* herb to 50 cm (1.5 m). Corolla 34–52 mm, bright *yellow* (sometimes tinted purple, especially in bud); bracts grey, *strongly wavy*-margined (to toothed), 16-26(30) mm; calyx lobes subequal (or one shorter); filaments densely hairy below. Parasitic on various desert shrubs, often by roadsides, especially *Atriplex halimus* (also *Zygophyllum, Haloxylon, Calligonum, Prosopis, Pteropyrum* and *Tamarix*). IP, SY. **C. tinctoria** is probably a distinct, mainly *coastal* species. Bracts 14–22 mm with *papery margins;* calyx with equal lobes. Anthers long-pointed. Widely recorded in error as the western Mediterranean species *C. phelypaea*. Usually parasitic on shrubby Amaranthaceae on coastal dunes. AA (islet north of Kasos), CR (eastern islets only), CY (rare – Ayios Island, Cape Gata), SY (requires confirmation).

(b) Plant hairless. Flowers whitish with purple markings.

Cistanche violacea A robust, *purplish* desert herb to 25(30) cm. Corolla 33–34 mm, *long-tubular, whitish,* with *mauve to purple edges;* bracts with strongly wavy-toothed, slightly papery margins, 14–19(28) mm; calyx lobes subequal or with 2 slightly shorter lobes; filaments densely hairy below. Parasitic on various desert shrubs, especially *Anabasis*. Locally common in clayey deserts in IP but frequently recorded in error as *C. salsa* which has hairy bracts; the distinction from purple forms of *C. tubulosa* in the southeast is less clear and intermediates may be hybrids. IP.

(c) Plant with *white-woolly* bracts.

Cistanche salsa A short herb 18–30 (50) cm. Corolla *broadly bell-shaped* with 5 unequal lobes, 25–35 (42) mm long, white to pale lilac with yellow folds; bracts 14–18 (25) mm long, *densely white-woolly*; calyx lobes equal; filaments densely hairy below. Parasitic on various shrubs. Rare in the region; widely recorded in error for *C. violacea*, or purplish forms of *C. tubulosa*. Distribution not fully known. TR (possibly IP, SY). **C. fissa** is similar but with corolla lobes densely woolly on the margins and *inner surface;* hairs *long*. Parasitic on *Haloxylon* and other desert shrubs in bare, rocky deserts. Very rare; previously known from Turkmenistan and Azerbaijan and only discovered in the region during research for this field guide. IP (possibly SY, TR).

1. *Phelypaea boissieri* PHOTO: ARNE STRID
2. *Cistanche tubulosa*
3. *Cistanche violacea*
4. Probable *Cistanche fissa*

Phelipanche BROOMRAPE

Branched or unbranched holoparasites, often bluish. Flowers with *bracteoles* below the calyx; corolla zygomorphic, 2-lipped. Capsule with numerous minute seeds. Host identification useful but difficult in mixed vegetation. Closely related to *Orobanche* (still included in that genus by some authors). Reliable identification of closely related species not always possible in the field. Distributions not properly known.

(a) Filaments hairless to hairy beneath. Often parasitic on Asteraceae and or various cultivated crops.

Phelipanche ramosa BRANCHED BROOMRAPE A variable, blue–purple (sometimes yellow–white), glandular-pubescent annual with, often, *branched stems*, *usually short*, 50 mm–20(30) cm. *Flowers small*, 10–15(22) mm long, pale blue, violet or cream with a white patch at the base; stigmas white or pale blue; calyx teeth 5–15 mm (shorter than the corolla), lower corolla teeth *blunt*. On various hosts, usually *cultivated* (Linaceae, Cannabaceae, Solanaceae and Leguminosae). Sandy and disturbed habitats. Throughout (in its wider treatment). The following species are very similar. *P. aegyptiaca* is taller (to 50 cm), often branched. Lowermost flowers with *stalks 10–20 mm*; corolla 20–35 mm; filaments hairless or slightly hairy below; anthers hairy. Parasitic on various crops (mainly Brassicaceae, Solanaceae and Cucurbitaceae). IP, TR. *P. mutelii* has more erect flowers with distinctly *open* corolla mouth with markedly *divergent*, rounded, broadly elliptic lower teeth; anthers usually hairless (not always). On various *uncultivated* hosts, especially Asteraceae (*Arctotheca*, *Hedypnois*, *Helichrysum*, *Hyoseris*); less often Leguminosae (*Medicago* and *Scorpiurus*). Throughout. *P. nana* is usually *unbranched*, with *pointed* lower corolla lobes,

1. *Phelipanche ramosa* 3. *Phelipanche mutelii*
2. *Phelipanche aegyptiaca* 4. *Phelipanche daninii*

and calyx teeth almost as long as the corolla. Throughout. *P. oxyloba* has long bracts, 5–15 mm (not 6–8 mm) and *acute* (not rounded) lower corolla lobes. Parasitic on *uncultivated* Asteraceae, particularly *Anthemis*. AA (east), CR (rare), GR, IA, TR.

Phelipanche purpurea A (typically) *unbranched* annual, robust, to 30(45) cm with short-hairy, not glandular, bluish stems. Inflorescence dense, with a blue–purplish or greyish appearance. Corolla long, 18–26(30) mm and distinctly constricted at the base; stigma white or pale blue; anthers normally hairless. Parasitic on *uncultivated* Asteraceae, especially *Achillea* (rarely Leguminosae). Rare in the region. GR.

The following are rare and local species similar to the group described above, the taxonomy and relatedness of which remain unclear. Hosts various:

Phelipanche cilicica An annual with short, dense, unbranched and few-flowered spikes. Calyx greyish with lance-shaped teeth shorter than (to equalling) the tube; corolla pale violet, 20–25 mm, rather erect-spreading, hairy. Parasitic on *Phlomis armeniaca* (possibly other species of *Phlomis* and *Stachys*). SY, TR. *P. zosimii* has yellowish stems, very pale (almost white) corolla with almost rectancular, well-spaced lower lobes; stigma whitish. Parasitic on *Zosima absinthiifolia*; a very rare island endemic. CY. *P. cohenii* A robust, thick-stemmed, unbranched annual with stalked lower flowers, calyx teeth equalling or shorter than their tube; corolla 19–22 mm, *tubular*, with hairless stamens inserted 3–4 mm above the base. IP (Mount Hermon). *P. daninii* is similar, with flowers with corolla 16–22 mm and hairless stamens inserted 4–6 mm above the base. On *Moricandia* in deserts. IP. *P. astragali* has flowers with corolla 25–30 mm with hairless stamens inserted 4–6 mm above the base. LB. *P. rechingeri* (syn. *P. nowackiana*) has *yellow* flowers and grows on *Alyssum*. AA.

(b) Filaments hairy beneath. Parasitic on uncultivated Leguminosae or Apiaceae.

Phelipanche lavandulacea A robust annual to 45(59) cm with *numerous, dense* flowers and with a *dark purple or blackish appearance*. Corolla 19–22 mm with *lower lip obtuse*; stigma white, bluish or yellowish; filaments slightly hairy; calyx teeth *not exceeding* the corolla tube. Lower portion of the stem with stalked flowers. Various hosts, particularly *Bituminaria bituminosa*. Local but widespread. Virtually throughout. *P. schultzii* has few or no stalked flowers, *lower corolla lip acute* with divergent lobes, white stigma, with calyx teeth *greatly exceeding the corolla tube, and filiform at the tips*. Stamens inserted 4–5 mm above the corolla base. Parasitic on Apiaceae (especially *Ferula, Thapsia, Elaeoselinum*). Local and rare. AA, GR, PN, TR. *P. heldreichii* is similar to *P. schultzii* but parasitic on *Eryngium campestre*. TR.

Orobanche BROOMRAPE

Unbranched holoparasites, often dull reddish or yellowish, similar to *Phelipanche* but *lacking 2 bracteoles* below the calyx; corolla zygomorphic, 2-lipped. Capsule with numerous minute seeds. Host identification useful but difficult in mixed vegetation; reliable identification closely related species not always possible in the field.

(a) Spikes usually pinkish, purplish or yellowish, slender, often <30 cm; lax to dense; *corolla narrow, tubular and small*, 10–20(25) mm; bracts beneath flowers tapered. Many similar species difficult to distinguish.

Orobanche minor COMMON BROOMRAPE A variable annual with lax to dense spikes to 60 cm. Flowers *small with an evenly arched corolla* 10–18 mm, yellow tinged violet; lower lip evenly lobed without a hairy margin, stigma often pink (but variable); *filaments not densely hairy or woolly*, usually inserted close to corolla base; bracts equalling or exceeding the corolla; calyx lobes variably toothed. Parasitic on various hosts across many families, in disturbed habitats. Possibly throughout except for some islands (distribution unclear owing to confusion with the next species).

Orobanche pubescens An annual to 30(50) cm with *lax* spikes, *densely long-white, hairy*. Corolla 10–20 mm, white to cream with purple veins and evenly curved, with flattened, somewhat *matted* white, crimped hairs; lobes of lower lip equal or the middle lobe slightly larger; filaments *densely hairy below* (inserted 2 mm above the base); stigma lobes yellow, white or mauve. Parasitic on Asteraceae and Apiaceae (including *Galactites*, *Carlina* and *Glebionus*). Common in the region. ***O. grisebachii*** is a poorly known species, allied to *O. pubsecens*, *O. minor* and *O. crenata* (probably much-confused with the previous species), with *whitish* flowers with small, virtually equal lips, the upper tapering to a point, the lower with a larger middle lobe; bracts exceeding the flowers; anthers narrow and *long-pointed*; *stigma yellow*. Parasitic on Apiaceae (hosts poorly known), in bare, sandy habitats. AA, CR, GR, IP, TR.

Orobanche picridis OXTONGUE BROOMRAPE An annual to 60 cm, similar to *O. minor* but *pale throughout*. Corolla 15–22 mm, *whitish, straight along the back*; *stigma lobes dark purple* or orange (sometimes pink); *filaments inserted above the base of the corolla, and densely hairy* at the base; bracts long and filiform. Parasitic on Asteraceae, often *Picris hieracioides*, *Crepis* and *Artemisia* (occasionally *Daucus*). Rare and local. AA (local), CR, GR, IA, PN. ***O. artemisiae-campestris*** has red–orange stems and *dull yellow flowers* 16–22 mm, and pink–red stigma lobes. Parasitic on *Artemisia campestris*. GR, IA.

Orobanche amethystea Like *O. minor* but often more robust and dense, 15–50 cm. Corolla 15–20(25) mm *straight then abruptly arched (geniculate)*, pale yellow tinged and veined with violet (rarely all yellow or orange); bracts exceeding the corolla, often markedly. Parasitic on various hosts, most commonly *Eryngium campestre*. AA, CR, GR, IA, PN, TR. ***O. hederae*** IVY BROOMRAPE is similar, but often *strongly purple with long spikes*; corolla 10–22 mm, cream tinted maroon, inflated at the base, narrow and straight-backed, with notched lobes, the middle lobe square; stigma lobes always *yellow*; filaments scarcely hairy. Parasitic on *Hedera helix* (rarely other species). Rather rare and local. AA, CR, GR, IA, PN, TR.

Orobanche palaestina A hairy annual 30–60 cm with dull orange–pink stems. Flowers *numerous and dense* with corolla rather long, 15–17 mm, scarcely constricted, cream, slightly tinged violet; middle lobe of lower lip exceeding lateral lobes; bracts long and narrow, *conspicuously exceeding the flowers*; calyx lobes usually entire; stigma *dark yellow*. Mainly on *Notobasis syriaca* (also Leguminosae, *Cirsium phyllocephalum* and crops such as *Lactuca sativa* and *Cynara scolymus*). IP, SY, TR (rare).

(b) Spikes slender and *dark red* throughout; corolla small and narrowly tubular, 10–17 mm; bracts beneath flowers tapering to a *slender* point.

Orobanche sanguinea (syn. ***O. crinita***) A slender annual 10–27(45) cm, *dark red throughout*. Corolla *small and narrow,* 12–14(17) mm long with a weakly bilobed upper lip and pointed lower lobes (the middle lower lobe largest); bracts long and narrow, exceeding the flowers; stigma *red*; stamens inserted 1.5–2 mm above the corolla base. Parasitic on *Lotus cytisoides* (sometimes *L. cytisoides* or *Medicago marina*). Widespread but rare, mostly on dunes of small islets. AA, CR (especially islets + KP), GR, IA (widespread), PN.

1. *Orobanche minor*
2. *Orobanche pubescens*
3. *Orobanche picridis*
4. *Orobanche amethystea*
5. *Orobanche hederae*
6. *Orobanche reticulata*
7. *Orobanche sanguinea*
8. *Orobanche caryophyllacea*

Orobanche cypria A short annual 10–22 cm, strongly tinged *dark maroon–red* throughout. Flowers borne in rather compact heads; corolla small, 10–12 mm, slightly curved, pink to red with darker veins; calyx lobes divided, equalling the corolla; bracts shorter than the corolla (8–10 mm); filaments inserted 3 mm above the base, thinly hairy below; stigma pink to purple. Vineyards, hillsides and pine forests, apparently on *Pterocephalus*, *Cistus*, *Salvia* and *Scutellaria*. An island endemic. CY.

(c) Spikes *dark brownish-purple* throughout; corolla 25–27 mm.

Orobanche baumanniorum A *dark brownish-purple*, obscurely glandular-hairy annual to 25 cm, with stems scarcely swollen at the base. Corolla erect-spreading, 25–27 mm long, sometimes with a yellowish lower lip; calyx 10–12(15) mm; stamens inserted 2–2.5 mm above the base of the tube; stigma purple. Parasitic on Dipsacaceae. GR, IA, PN.

1. *Orobanche gracilis*
2. *Orobanche crenata*
3. *Orobanche cernua*

(d) Spikes *short and stout,* yellowish or pinkish; corolla small but not narrowly tubular, 10–17 mm; bracts beneath flowers *broadly triangular.*

Orobanche camptolepis A very *short,* stocky, pink–mauve (or yellowish) *minutely* hairy annual. Flowers with a straight corolla 10–15 mm with an abrupt curve near the apex, much exceeding their *short, broadly triangular, stiffly recurved bracts;* lower lip with long-pointed, spreading lobes, the upper lip deflexed; calyx lobes slender and deeply divided; filaments *hairless;* stigma lobes *pale pink.* Parasitic mainly on *Polygonum* (possibly also *Artemisia* and *Atraphaxis*). Often referred to as a form of *O. cernua* but very distinct. Rare, in mountains. IP (north), LB, SY, TR. ***O. hermonis*** is widely recognised as a species but possibly a highland form of *O. camptolepis;* further investigation is required. IP, LB.

Orobanche sideana A short annual 80 mm–15 cm, similar to *O. camptolepis* (considered by some authors to be a form of it) but *cream to yellowish* throughout. Flowers yellow, corolla (12)15–17 mm long, rather straight along the back with toothed lobes; filaments inserted 3–5 mm from the base, hairy below; bracts 10–12 mm long and *broad* (to 57 mm); calyx lobes linear, nearly divided to the base; bracts and calyces very glandular. Very rare on coastal dunes; parasitic on *Polygonum.* TR.

(e) Corolla *not* narrowly tubular, but broader or more bell-shaped, and rather *large,* (18)20–33 mm long.

Orobanche reticulata THISTLE BROOMRAPE A robust species to 35 cm with stout, purplish or yellowish stems. Flowers dense, few to numerous; corolla 18–22 mm, yellowish with purple markings and *dark glands, abruptly curved along the back,* the lower lobes equal, upper lobes spreading and notched; stamens inserted above the base of the corolla, filaments hairy above; stigma lobes touching and *dark purple.* Parasitic on *Cirsium, Carduus, Scabiosa* and *Knautia.* GR, PN, TR.

Orobanche caryophyllacea CLOVE-SCENTED BROOMRAPE A robust annual to 40 cm with pale, curved stems and darker bracts. Flowers few and *large* relative to plant; 20–32 mm long and *bell-shaped,* weakly *carnation-scented;* pale pink or yellow to violet–brown; *stigma lobes distant and purple;* filaments hairy below; bracts shorter than the corolla; lateral calyx lobes *oval,* toothed and short. Parasitic on *Galium* and *Asperula.* Rather rare and restricted to sub-alpine areas. GR, IA, PN, TR.

Orobanche elatior KNAPWEED BROOMRAPE A stout annual to 75 cm, with *numerous, very dense* flowers 18–25 mm; with corolla lobes toothed and crisped; *stamens inserted well above* (3–5 mm) *the base of the corolla;* stigma lobes *bright yellow* and diverging. Parasitic on *Centaurea scabiosa* and *C. nigra.* Rare, on mountain scree. GR, IA.

Orobanche gracilis SLENDER BROOMRAPE An annual to 40 cm with reddish, lax spikes. Flowers large relative to the stem 18–24(29) mm, broadly campanulate, yellowish veined with red externally and *shiny red inside;* bracts triangular, and shorter than the corolla; filaments short-hairy. Parasitic on legumes (sometimes *Cistus*). Open woods and hillsides. AA (rare), GR, IA, PN.

Orobanche lutea YELLOW BROOMRAPE A yellowish annual with few to many flowers. Corolla 20–33 mm, rather pale, yellow, pink or brown; corolla inflated at the base, then evenly arched, the lobes spreading; stamens inserted up to *7 mm* above the base and *woolly-hairy below;* stigma lobes *bright yellow* and distant. Bracts dark, not exceeding the corolla. Parasitic on legumes (or *Anthemis*). GR (north).

Orobanche alba THYME BROOMRAPE A short, *reddish* (sometimes pale) annual to 30(50) cm. Flowers few, *fragrant;* corolla 15–20(25) mm, pink to dull red, bell-shaped with toothed, *hairy* lobes and *dark hairs* on the surface; upper lip spreading; stamens inserted at least 3 mm *above* the base of the corolla and *hairy; stigma lobes reddish, touching;* bracts short and triangular; calyx lobes *entire* (or weakly toothed). Parasitic on Lamiaceae. Frequent on maquis. AA, CR, CY, GR, IA, PN, TR.

Orobanche crenata BEAN BROOMRAPE An annual to 1 m with densely flowered spikes. Stems purplish, flowers large; corolla (15)20–30 mm, broadly campanulate and *white with violet veins*; *fragrant*; stigma variable in colour. Parasitic on various legumes, particularly on cultivated peas and beans, sometimes in large numbers. Cultivated and fallow ground. Throughout.

(f) Corolla 12–18 mm, *markedly curved* to appear downward-pointing.

Orobanche cumana SUNFLOWER BROOMRAPE A very slender, lax, erect, bluish or brownish plant to 50 cm with *flowers from the base* to the apex of the spike. Corolla 15–18 mm, white or blue, narrow and curved to appear *downward-pointing*; stigma lobes white; bracts and calyx lobes distinctly short and abrupt. A pest on sunflower crops. GR (probably elsewhere). *O. cernua* is similar but with short stems to 23 cm. Corolla 12–18 mm, white to yellow with a *blackish margin*; *downward-pointing*; stigma lobes yellow–white. Parasitic on *Artemisia* and other Asteraceae, often in dry or semi-desert habitats. GR, IP, LB, SY, TR.

1. *Ilex aquifolium*
2. *Michauxia campanuloides* PHOTO: YUVAL SAPIR
3. *Petromarula pinnata* PHOTO: NICK TURLAND

AQUIFOLIACEAE | HOLLY FAMILY

Evergreen trees and shrubs with alternate, often spiny leaves. Flowers dioecious, borne in cymes in the leaf axils, regular, with 4 free sepals and 4 white petals fused at the base; stamens 4; stigma 4-lobed. Fruit a 2(4)-seeded drupe.

Ilex aquifolium HOLLY A shrub or small tree (rarely to 23 m). Leaves glossy, 50 mm–12 cm, oval to elliptic, with *strongly spiny, undulate margins*. Fruit a spherical red berry, 6–10 mm across. Cool, damp, high-altitude habitats in the north of the area. GR (north), TR.

CAMPANULACEAE | BELLFLOWER FAMILY

Annual or perennial herbs, exuding a latex when cut, with alternate leaves. Flowers often large and showy, clustered or solitary; corolla bell-shaped, lobed, often blue or purple; stamens 5, fused or free; style 1. Fruit a capsule.

Michauxia

Tall, robust perennials with *showy white flowers* with 8–10-lobed calyx and corolla; stamens and stigmas 8–10. Capsule nodding.

Michauxia campanuloides A tall, rough-hairy perennial to 1(2) m, branched above, with alternate, toothed to lobed leaves, the uppermost stalkless and somewhat clasping at the base. Flowers borne in lily-like panicles; calyx with 8–10 deflexed lobes 5–10 mm (growing in fruit); corolla *white*, 18–20 mm divided to 1/3 into 8–10 deflexed lobes; anthers yellow; style hairy, to 15 mm. Capsule splitting from the base. Rocky outcrops in the east. IP, LB, SY, TR.

Petromarula

Perennials with flowers in panicles; calyx deeply 5-lobed; corolla funnel-shaped with 5 lobes; stamens 5; ovary 3-parted.

Petromarula pinnata A robust perennial to 80 cm, hairless below, short-hairy above. Leaves large, to 30 cm, pinnately divided into leaflets or deep lobes, the lowermost long-stalked. Flowers pale blue, 10 mm across, borne in small clusters; calyx and corolla deeply divided; style protruding. An island endemic of rocky habitats. CR.

Campanula BELLFLOWER

Annuals or perennials. Flowers with 5-lobed calyx fused to the ovary, and funnel- or bell-shaped corolla. Fruit a capsule or berry. A numerous and complex group, with many mountain dwelling species (particularly in southern Greece), possibly taxonomically over-inflated; just a small subsection are described here.

(a) Short-lived perennials with *flat leaf rosettes*, in *rocky habitats* (chasmophytes). Stigmas (4)5; capsule (4)*5-parted*. Numerous similar species.

Campanula topaliana A patch-forming perennial with numerous, hairy stems 20–40 cm and leaves in rosettes, silky-hairy, lyre-shaped or lobed, *heart-shaped* at the base and toothed with lobed stalks; upper leaves elliptic and toothed. Flowers violet–mauve, borne numerously, terminally or in the leaf axils; calyx teeth oval, about as long as the ovary; corolla 9–12 mm, tubular, hairy, with spreading lobes. Limestone rocks. GR, PN. *C. kamariana* is similar but with strongly deeply divided, sparsely hairy to hairless rosette leaves with long, sparsely lobed stalks. PN (south). *C. andrewsii* is also very similar but with leaves not heart-shaped at the base, with broad stalks and large, rounded, terminal segments; *flowers larger*, corolla 14–20 mm. PN. *C. celsii* is very similar to the previous species but with simple stems (or with short branches) and spatula-shaped to almost lyre-shaped leaves. Comprising local subspecies. AA, GR, PN. *C. euboica* is similar to the above species but *white-downy* throughout. GR (northwest Euboea and nearby mainland).

Campanula tubulosa A biennial or short-lived perennial 15–40 cm, with a rather woody stock. Rosette leaves *long-stalked* with *narrowly elliptical*, scalloped-edged blades. Flowers with triangular calyx teeth up to ½ as long as the corolla, which is 20–30 mm, violet–blue and narrowly bell-shaped. CR (west and central).

Campanula saxatilis A rhizomatous perennial with a woody stock and erect to ascending, fragile stems to 20 cm, branched or not. Leaves borne in rosettes, leathery, hairless to sparsely hairy, toothed or not, the uppermost stalkless. Flowers few, borne in short inflorescences; corolla (10)14–19(23) mm, tubular, covered in short, velvety hairs. A rare chasmophyte of limestone crevices. **Subsp. *saxatilis*** has spatula-shaped leaves and flowers with a narrowly tubular corolla and triangular-lance-shaped calyx lobes. CR. **Subsp. *cytherea*** is larger in all parts, with almost hairless leaves broadest above the middle, more broadly tubular corolla and egg-shaped calyx lobes. PN (Kithira, Andikithira).

(b) Erect to ascending, *slender* plants in various habitats. Stigmas 2–3. Capsule 3-parted, *splitting by apical pores or valves*.

Campanula ramosissima A slender to rather robust annual 20–40 cm, erect, branched or not, with angular, hairy stems. Leaves sparsely hairy, oval-lance to spatula-shaped, scalloped, stalkless above. Flowers numerous, long-stalked; calyx teeth narrowly lance-shaped, *3-veined*, hairy, shorter than to *almost equalling* the corolla; corolla *broad and open*, 10–25(30) mm, violet. Capsule hairy. GR, IA, PN. *C. phrygia* is similar but delicate and with *narrower, 1-veined*, linear-lance-shaped teeth. Local on the northern and eastern mainland. GR, IP, LB, SY, TR.

Campanula sidoniensis An ascending, slender annual with stems sparsely prickly at the angles. Leaves rough or hairless at the margins, the lowermost circular to oval, the uppermost oblong to lance-shaped and stalkless. Flowers *blue*–lilac, funnel-shaped, just exceeding the calyx; calyx *long-bristly*, with linear, 3-veined lobes ½ as long as their tube. Capsule *spherical, bristly*. Thickets and old walls; rather rare. IP (Upper Galilee), LB.

Campanula patula SPREADING BELLFLOWER A hairless or hairy biennial or perennial with erect or ascending, slender stems to 60(70) cm. Leaves gradually narrowed into their stalks. *Inflorescences wide-spreading*, with bell-shaped but open flowers with lobes as long as their tube, pale to purplish-blue (rarely white), 15–25(35) mm long, with erect, slender stalks with a bract in the middle; buds more or less nodding. Capsules 6–11 mm, erect with 10 veins. Grassy places on higher ground. *C. rapunculus* RAMPION is similar to *C. patula* and confused easily with this species, but with *simple, pyramidal inflorescences* with smaller, always pale blue–mauve flowers 10–22 mm, held on ascending to erect short stalks with a bract at the base. Capsule 7–20 mm. Roadsides, meadows, cultivated land and other grassy places, mainly on the mainland. GR (north), IA, IP, LB, SY, TR.

Campanula persicifolia NARROW-LEAVED BELLFLOWER A hairless perennial to 70(80) cm with non-flowering rosettes, and erect, unbranched stems with shiny, oval, blunt-toothed leaves below and toothed, linear-lance-shaped leaves above. Flowers few; calyx teeth 12–20 mm; corolla blue–violet, shallowly divided into broad, triangular lobes, *large and open, (25)30–40(50) mm*. Capsule 11–16 mm. Various habitats, and cultivated for ornament. GR (north), TR.

(c) Plants with flowers mostly in *terminal heads*. Stigmas 2–3. Capsule 3-parted, splitting by basal pores or valves.

Campanula glomerata A very variable bristly or almost hairless perennial 15–80 cm, erect, somewhat branched with stems reddish and angular below. Leaves scalloped, oval-lance-shaped or oblong to elliptic, heart-shaped to rounded at the base. Flowers stalkless, borne numerously in dense heads; calyx teeth lance-shaped, 1/4–1/3 as long as the corolla; corolla hairy or hairless, violet, 12–25(40) mm long with lobes 1/3 as long as the tube. Grassy habitats on higher ground. GR, PN.

(d) Plants with flowers mostly in open racemes or panicles (sometimes solitary). Calyx with *recurved appendages* between the teeth; stigmas 2–3. Capsule 3-parted, splitting by basal pores or valves.

Campanula calaminthifolia A late-flowering tufted, grey-hairy perennial to 20 cm with a woody stock. Basal leaves in rosettes, long-stalked, egg-shaped and blunt; stem-leaves numerous, broadly elliptic, toothed. Flowers short-stalked, borne numerously in rather crowded terminal clusters; calyx teeth narrowly triangular; corolla 12–15 mm, narrowly bell-shaped, blue–violet

1. *Campanula topaliana*   PHOTO: ELEFTHERIOS DARIOTIS
2. *Campanula saxatilis* subsp. *saxatilis*
3. *Campanula ramosissima*
4. *Campanula sidoniensis*
5. *Campanula glomerata*
6. *Campanula calaminthifolia*   PHOTO: ARNE STRID

with some long, thin hairs within. A chasmophyte. AA (Naxos). Numerous, closely related species in the Aegean area, which are difficult to distinguish, besides locality, include: **C. heterophylla**, **C. hierapetrae** and **C. olivieri**.

Campanula incurva A short-lived perennial to 45 cm with large leafy rosettes. Leaves to 80 mm, oval, heart-shaped at the base, slightly fleshy and softly-hairy; stem-leaves almost stalkless. Flowers few but conspicuous; calyx teeth triangular and shorter than the corolla tube; corolla *large*, bell-shaped, 35–50 mm long and *wide*, to 40 mm across, pale blue to purplish; lobes short and recurved. AA (local), GR (north, west and adjacent islands).

(e) *Annuals*, often with *paired* (dichotomous) branches. Calyx *without* recurved appendages between the teeth; stigmas 2–3. Capsule 3-parted, splitting by basal pores or valves.

Campanula drabifolia A *slender*, erect to ascending annual to 20 cm, often branched divergently from the base, slightly bristly throughout, similar in appearance to *C. ramosissima*. Basal leaves deeply toothed to 3-lobed. Flowers terminal or borne in the leaf axils on short stalks to 2 mm; calyx teeth triangular; corolla broadly bell-shaped, 7–10 mm with spreading lobes to 6 mm, violet–blue. GR (south and southeast), IA, PN. Many closely related species occur in the area, including: **C. rhodensis** (AA), **C. simulans** (AA, TR), **C. creutzburgii** (CR), **C. delicatula** (AA, CY, TR), **C. erinus** (widespread) and **C. scutellata** (GR – north).

Campanula erinus ANNUAL BELLFLOWER A slender, rough-haired, spreading, branched annual to 35 cm with oval- to wedge-shaped leaves 10–20 mm, and tiny, pale blue (or reddish or white) flowers with *inconspicuous* corolla, just 2–5 mm, borne in the axils of the branches; flowers very short-stalked. Capsule 2.5–3.6 mm. Throughout.

(f) *Perennials without* paired (dichotomous) branches. Calyx *without* recurved appendages between the teeth; stigmas 2–3. Capsule 3-parted, splitting by basal pores or valves.

Campanula cretica A woody-based perennial to 60 cm. Basal leaves long-stalked, 60 mm–14 cm long, shallowly toothed and heart-shaped at the base; upper leaves reduced. Flowers 3–10, *large*, often almost stalkless, borne in more or less 1-sided racemes; calyx teeth linear-lance-shaped, half as long as the corolla; corolla 35–50 mm, purplish-blue to whitish, sparsely woolly-hairy within. CR (mainly the west). **C. boreosporadum** is similar but small and slender with basal leaves to just 30 mm, stems to 30 cm and flowers *with stalks 2–5 mm*; corolla 18–28 mm. Rare. AA (Sporades). **C. samothracica** is also smaller, rather tufted with leaves to 60 mm and corolla 30–40 mm, lilac–pink to purplish-blue, hairy at the tips. AA (Samothraki).

Asyneuma

Similar to *Campanula*, flowers borne in elongated spikes; *corolla lobes linear*.

Asyneuma limonifolium A slender, erect, variable branched (to unbranched) perennial 15–70(90) cm with mostly basal lance-shaped, long-stalked leaves, toothed or not. Flowers blue–lilac, borne in lax, elongated spikes, 1–3 in the axils of tiny triangular bracts, virtually stalkless; *corolla lobes linear*. Mountains, gorges and garrigue; widespread. AA (local), GR, IA, PN, TR. **A. pichleri** has several stem-leaves (leaves not all basal) and spreading to *deflexed* calyx teeth. CR (western mountains), GR. **A. giganteum** is stout and woody-based with capsules opening from pores around the *middle* (rather than apex). A very rare chasmophyte. AA (Rhodes, Chalki), KP.

Jasione SHEEP'S-BIT

Hairless or hairy, annual or perennial herbs, usually with narrow leaves. Flowers lilac, blue or white, borne in small, compact, terminal, hemispherical heads. Capsule splitting by 2 apical short valves.

Jasione heldreichii An erect biennial to perennial with few branches, 20–40 cm, stiffly-hairy below and hairless above. Leaves linear-lance-shaped, bristly. Flower-heads borne on long, slender stalks, 15–20 mm across with lance-shaped leafy bracts beneath with 1–4 teeth; corolla bright blue, divided almost to the base. GR, TR.

Solenopsis

Delicate annual or perennial herbs with simple, thin leaves. Flowers bell-shaped, zygomorphic, *2-lipped*, blue or lilac. Capsule 2-parted.

Solenopsis laurentia (syn. *Laurentia gasparrinii*) A small, *slender*, more or less *hairless annual* 60 mm–20 cm. Leaves 12–20 mm oblong to spatula-like, entire, often forming a basal rosette; small *stem-leaves* present. Flowers solitary; small, 3–5(6) mm, borne on stems with 1 or 2 bracteoles; calyx 5-parted; corolla blue, lilac or white, 3.5–6 mm. Capsule 2–3.5 mm. Widespread but uncommon; damp habitats. AA, CR (west – rare), GR (west). **S. minuta** is similar but an *annual or perennial* with *leafless* stems. Corolla 6–8 mm. IA (Cephalonia), CR. Plants occurring after spring rain with a small, and strictly *annual habit* are widely referred to as **subsp. annua**. CR.

Legousia

Annual herbs. Flowers borne in panicles or racemes; calyx 5-lobed; corolla flattish, 5-lobed, fused at the base; stamens 5; style 1, stigmas 3. Fruit a cylindrical, 3-valved capsule.

Legousia speculum-veneris LARGE VENUS'S LOOKING GLASS A hairy, ascending to erect annual 10–25 cm, *much-branched from the base*. Leaves to 20 mm, alternate, oblong and slightly wavy, unstalked, or scarcely stalked beneath. Flowers violet with a whitish centre, 15–23 mm across, star-like with 5 blunt lobes 8–9 mm, borne in lax panicles; calyx teeth 6–8 mm (almost as long as petal lobes); *filaments hairless*. Capsule 10–14 mm. Agricultural land and stony habitats. Virtually throughout.

---

1. *Asyneuma limonifolium*
2. *Solenopsis minuta* subsp. *annua*

Legousia hybrida VENUS'S LOOKING GLASS An annual to 30 cm, similar to the above species but with *markedly wavy leaves* to 30 mm, and few, pink, violet or white flowers borne in terminal clusters with petal lobes just 2.5 mm, about ½ the length of the calyx lobes (5–6 mm); *calyx teeth erect in fruit*. Capsule 15–25 mm, constricted above. Fallow land and olive groves. Scattered more or less throughout.

Legousia falcata SPICATE VENUS'S LOOKING GLASS An erect to spreading annual, 15–50 cm, with hairless or bristly stems. Leaves 20–40 mm, oval, slightly wavy-margined and usually short-stalked below. Flowers violet to blue–purple, solitary or paired with lobes 7 mm, borne in a *spike-like inflorescence*; calyx teeth linear and curved, 4–8 mm (equal or ½ length of the tube), spreading, often exceeding the petals. Capsule 10–22 mm, narrowed at the apex. Maquis and stony pastures. Throughout. *L. scabra* is a similar annual to 90 cm with oval to elliptic leaves and straight calyx teeth 8–10 mm (1/2–1/3 length of tube) and corolla lobes 6–7(12) (almost equal). Capsule 10–22 mm. Throughout (except the far north).

MENYANTHACEAE | BOGBEAN FAMILY

Aquatic or marsh-dwelling herbs with creeping stems bearing alternate lower leaves and floating or emergent upper leaves. Flowers with 5 fringed petals; stamens 5; style 1. Fruit a capsule.

Nymphoides

Leaves simple and rounded, floating on the water surface. Flowers yellow with fringed petals, long-stalked. Fruit an irregularly splitting capsule.

Nymphoides peltata FRINGED WATER-LILY An aquatic perennial with creeping stems to 1.5 m, with floating leaves; resembling a water-lily in habit, but with very different flowers; leaves round to kidney-shaped, 30–90 mm across. Flowers solitary or few, bright yellow, 30–40 mm across with 5, distinctly *fringed*, *lobes*. Local in the region and restricted to still and sluggish water. GR.

ASTERACEAE | DAISY FAMILY

The largest family of flowering plants (see also Orchidaceae). Herbs, perennials or shrubs with alternate, opposite or rosette leaves, simple or compound. Flower-heads typically with an involucre of closely overlapping bracts around the base; flowers (*florets*) borne in congested flower-heads (*capitula*); those in the centre (*disk florets*) tubular and often distinct from the edge (*ray florets*); stamens 5, fused around the style; ovary inferior. Fruit an achene, often (but not exclusively) with an appendage of bristles, hairs or scales (a *pappus*).

SUBFAMILY ASTEROIDEAE

Plants without white latex, and stem-leaves spirally alternate (or absent; rarely opposite). Flower-heads usually yellow and radiate (with a central disk of tubular flowers, surrounded by ray flowers) or discoid (with tubular – often orange – flowers only).

Filago CUDWEED

Downy or woolly annuals with alternate, untoothed leaves. Flower-heads inconspicuous, often numerous (8–40) per cluster. Pappus of simple hairs or absent.

(a) Plant erect, with flower-heads (15)20–40 in rounded clusters, *not exceeded by their leaves*.

Filago germanica (syn. *F. vulgaris*) COMMON CUDWEED An erect, white-woolly annual to 30(40)cm, branched or unbranched below, always branched above in 2–3 forks. Leaves erect, lance-shaped, to 20 mm long, wavy-edged and untoothed; widest in the basal ½. Flower-heads (15)20–40, borne in dense *rounded* clusters 10–12(20) mm across, not exceeded by the leaves immediately below; flower bracts 4–4.5 mm with a long, transparent bristle-tip. Common on bare disturbed, sandy ground. Throughout, except the southeast. *F. eriocephala* is very similar but with broadly ovoid (rather than spherical) clusters of flower-heads with bracts to 3 mm without long bristle-tips. Throughout.

(b) Plant erect, with flower-heads (5)10–20(25) per cluster, *exceeded by their leaves*.

Filago desertorum A small, sparsely grey-downy annual with a short primary stem surrounded by a ring of longer, prostrate to ascending branches. Most leaves congested below the flower-heads, oblong-spatula-shaped. Flower-heads 3.5–5 mm long, borne in congested yellowish clusters of 6–12, exceeded by the leaves. Bare and arid habitats, flowering after winter rain in the far southeast. IP (south and east).

Filago pyramidata BROAD-LEAVED CUDWEED An erect, white-woolly annual always branched and spreading *from the base*, often almost prostrate. Leaves oval and bristle-tipped. Flower-heads 5-angled, 10–20 mm, borne in dense clusters of (5)10–20(25) in the branch axils and terminally, *exceeded* by 2–4(5) leaves immediately below; outer flower bracts *with yellow points*, curving inwards in fruit. Open fallow and sandy waste places, and sandy cliff-tops; common. Throughout.

---

1. *Filago germanica*
2. *Filago desertorum*
3. *Ifloga spicata*

(c) Plant often short (to virtually stalkless), with leaves and flower-heads congested, rosette-like (many species previously described under *Evax*).

Filago pygmaea (syn. *Evax pygmaea*) A very small, grey-felted annual to 40 mm, branched at the base. Leaves oblong, *narrow* and blunt, to 15 mm long and 5 mm wide, spreading in a flat rosette; upper leaves 2–3 x longer than flower-heads; all leaves without distinct stalks. Flower-heads borne in very compact clusters to 35 mm across; brownish-yellow. Fairly common on dry, bare and stony places. Probably throughout. *F. contracta* (syn. *Evax contracta*) is similar but with *almost erect* (not spreading, flat) leaves. Strongly southern and eastern. AA, CR, CY, IP, SY, TR. *F. wagenitziana* has *narrow*, linear-lance-shaped leaves white-downy beneath, green above. Bracts centrally green, with contrasting *broad*, translucent margins. CR (west).

Filago cretensis A very small, grey-hairy annual with stems 20–60 mm (12 cm) tall with 0–few branches. Stem-leaves few, broadest above the middle and pointed. Flower-heads cylindrical to egg-shaped, few (5–7), in clusters 5–7 mm across, slightly overtopped by the leaves; bracts slender, with hairless, papery margins, becoming erect in fruit. Regional forms described. AA, CR. *F. aegaea* (syn. *Evax aegaea*) is similar but with clusters of flower-heads scarcely overtopped by the leaves and bracts rigid and inward-curving in fruit. AA, CR, PN (rare).

Logfia CUDWEED

Small, woolly annuals. Previously treated either as distinct or under *Filago*, but recently resurrected based on DNA analysis; characterised by flower-heads solitary or in small clusters and *achenes always hairless*.

Logfia minima (syn. *Filago minima*) SMALL CUDWEED A short, grey silky-hairy annual to 25 cm. Stems very slender, branched above the middle. Leaves linear-lance-shaped, 4–10 mm long. *Flower-heads in clusters of 3–7, with 5 marked angles*, 2–5 mm, ovoid to pyramidal, *not* overtopped by the leaves immediately below; outer flower bracts *woolly at the base*, but yellow and *hairless at the tip*. Fallow and sandy waste places. AA (rare), GR, TR.

Logfia gallica (syn. *Filago gallica*) NARROW-LEAVED CUDWEED An annual similar to *L. minima*, to 33 cm with linear, *thread-like leaves (4)12–20(26) mm long*, the most apical *overtopping* the flower-heads; flower-heads leafy, borne 3–14, 3.5–5 mm; bracts woolly and yellowish at the tip. Coastal scrub, pastures and tracksides. Throughout.

Ifloga

Typically small, desert-dwelling annuals. Female florets with divided styles. *Pappus feathery (plumose) above.*

Ifloga spicata A variable annual 10(50)mm–10(12)cm, branched from the base. Leaves crowded, with adpressed, silky hairs on the upper surface, often with sand grains adhered, *linear*, 6–20 mm long with rather down-turned margins. Flower-heads yellowish or reddish, made up of 26–38 female florets and 19–26 cosexual florets; heads 3–4 mm long, borne terminally congested in leafy racemes, greatly exceeded by the leaves; bracts translucent and whitish to golden. Coastal and desert sands in the far southeast. IP (south + western plains).

1. *Helichrysum sanguineum*
2. *Helichrysum sibthorpii* PHOTO: ARNE STRID
3. *Helichrysum sibthorpii*

Gnaphalium CUDWEED

Inconspicuous annuals and perennials, often white woolly-hairy. Flower-heads small, yellow–brown and bell-shaped with papery bracts, borne in clusters of (3)5–40. Pappus of simple hairs.

Gnaphalium sylvaticum (syn. *Omalotheca sylvatica*) HEATH CUDWEED A grey-hairy perennial to 60 cm with non-flowering shoots and densely leafy stems. Leaves lance-shaped to linear, *decreasing in size along the stem*, green above, woolly beneath, 1- or indistinctly 3-veined. *Inflorescence elongated*, lax and interrupted below, with numerous yellowish and small, narrowly bell-shaped flower-heads; florets compact; involucral *bracts dark blackish-brown*. Heaths and scrub on the mainland. GR, TR.

Gnaphalium uliginosum (syn. *Filaginella uliginosa*) MARSH CUDWEED A short (often tiny) annual to 20 cm with *stems thickly white-woolly and branched throughout*, with short lateral branches. Flower-heads in clusters of 3–10, *overtopped by the surrounding leaves*; bracts outward-spreading, flat and star-like after the seeds are shed. Disturbed habitats. GR (north).

Helichrysum CURRY PLANT

Dwarf greyish aromatic shrubs with alternate, untoothed leaves. Flowers borne in dense clusters with papery bracts. Pappus absent.

(a) Plant an erect *annual*. Flower-heads *pale yellow*.

Helichrysum luteoalbum (syn. *Gnaphalium luteoalbum*, *Laphangium luteoalbum*) JERSEY CUDWEED A *slender annual* (more similar to the group above, in which it is often still included), to 50 cm, *white-woolly*, unbranched or with branches spreading from the base, then erect. Leaves broadly lance-shaped, blunt and running down the stem, *woolly on both sides*. Flower-heads terminal and densely ovoid, and *not overtopped* by their surrounding leaves (at least when mature); bracts elliptic, *shiny, straw-yellow*, only the outermost woolly below, and not bristle-tipped; florets pale yellow, with red stigmas. Disturbed habitats. Throughout.

(b) Plant an erect *perennial*. Flower-heads persistently *red*.

Helichrysum sanguineum An erect, soft-woolly to cobweb-hairy perennial, simple or branched from above, 25–75 cm. Leaves oblong to spatula-shaped, those above tapering at the base and running down the stem (decurrent). Flower-heads 4–7 mm, borne in dense corymbs; bracts 30–35, *deep crimson*, small, hooded and blunt, the outermost almost circular, the innermost spatula-shaped; florets yellow. Achenes minute. Hill scrub and maquis. IP (north), LB, SY.

(c) Plant a woody-based perennial (or shrub). Flower-heads white, pink or red in bud; whitish when mature.

Helichrysum sibthorpii A woody-based perennial, densely grey–white-woolly throughout. Leaves 15–60 mm long, rather *broadly* oblong-spatula-shaped; those above smaller. Flower-heads *white*, 15 mm across, few (1–3) in terminal clusters. A rare endemic of rocky cliffs on the summit of Mount Athos. GR. **H. taenari** is very similar to *H. amorginum* but with darker green, sparsely hairy leaves and smaller flower-heads. A rare endemic. PN (south).

Helichrysum amorginum A densely grey or white-downy perennial to 30 cm, slightly woody below. Leaves broadest above them middle, to 40 mm. Flower-heads 10–15 mm across, broadly bell-shaped with lax, overlapping papery bracts, *pink to bright red in bud*; whitish with yellow florets when mature. Rare. AA. **H. doerfleri** is similar but mat-forming with short, woody stems to 80 mm. Flower-heads fewer. Rocky areas; rare. CR (east).

(d) Plant a woody-based perennial (or shrub). Leaves *rather broad* (egg-shaped and more or less blunt). Flower-heads yellow.

Helichrysum orientale A tufted, white-woolly, woody-based perennial dwarf shrub with almost erect stems to 30 cm. Basal leaves rather *broad*, egg-shaped and more or less blunt; those above smaller and linear-oblong, all leaves *whitish on both sides*. Flower-heads almost spherical, 6–9 mm, borne in dense heads 12–25 mm; bracts pale lemon yellow, shiny and translucent. AA, CR, PN, TR.

(e) Plant a woody-based perennial (or shrub). Leaves *narrow* (linear or broader above the middle). Flower-heads deep yellow.

Helichrysum stoechas A very variable, spreading, much-branched dwarf shrub to 50(70) cm. Leaves white-felted, narrowly linear, untoothed and slightly aromatic (faintly curry-scented) when crushed, 5–20 mm long, and 0.5–2.1 mm wide with down-turned margins. Flower-heads 4–7 mm, yellow, borne in clusters about 30 mm across; the series of bracts (involucre) bright shiny yellow and *spherical* in flower; bracts yellowish, shiny and papery. Common on fixed sand dunes and garrigue. Throughout. **Subsp. *barrelieri*** is the form in the region, characterised by its leaves <20 mm, rather broad and scarcely aromatic, and flower-heads borne in *compact* inflorescences. **H. heldreichii** is similar but with numerous, crowded, *long* leaves 40–80 mm with down-turned margins. A very rare island endemic of limestone cliffs. CR (southwest).

Helichrysum italicum A spreading, much-branched dwarf shrub to 40(50) cm, similar to *H. stoechas* but with *smaller, narrowly linear* leaves, (6)10–40 mm long, 0.3–1.5 mm wide with down-turned margins, often spreading to recurved; often *strongly* aromatic. Flower-heads 4.5–5 mm, yellow, borne in *clusters of 15–25* about 80 mm across; the series of bracts (involucre) *dull mustard-yellow and oblong to narrowly bell-shaped*; the inner bracts at least 5 x longer than the outer. Sand dunes; rather local in the area covered. AA, CR, CY.

---

1. *Helichrysum italicum*
2. *Helichrysum stoechas* subsp. *barrelieri*
3. *Helichrysum stoechas* in fruit
4. *Phagnalon rupestre* subsp. *graecum*
5. *Phagnalon rupestre* subsp. *rupestre*
6. *Phagnalon saxatile*

Phagnalon

Grey dwarf shrubs with alternate leaves and flower-heads borne solitary at the tips of the branches; flower-heads solitary, with densely overlapping bracts. Pappus of bristles.

Phagnalon rupestre A small shrub 15–40 cm with erect to ascending, white-felted stems. Leaves *narrowly oval and more or less toothed*, small, 5–35 x 1.5–4.5 mm, green or whitish above and white-felted below, with down-turned margins. Flower-heads *solitary,* long-stalked (20–90 mm); yellowish; flower bracts brownish, membranous, somewhat hairy and closely overlapping. Fairly common on cliffs and coastal garrigue. Throughout. The following subspecies are recoginsed but their distinction and distributions are not always clear: **Subsp.** *rupestre* has triangular-oval, rather *broad, blunt bracts*. AA (east), CR, CY, GR, IA, IP, PN, TR. **Subsp.** *graecum* has *narrower*, overlapping, triangular-lance-shaped, *long-pointed bracts*. AA, CR, CY, GR, IA, PN.

Phagnalon saxatile A shrub similar to the previous species but often taller, to 50 cm with *linear* leaves 14–50 x 1.2–4 mm, sometimes broadest above the middle, *green* (not normally white-felted above) and white-felted below. Flower-heads *very long-stalked* (25 mm–13 cm); bracts with somewhat wavy margins, the tips often slightly recurved. Local in rocky and disturbed habitats, mainly in the west. AA, CR (southwest – rare), IA, PN.

Pulicaria FLEABANE

Annual or perennial herbs with simple alternate leaves. Flower-heads yellow and daisy-like, borne terminally. Achenes with scales around the pappus.

Pulicaria dysenterica COMMON FLEABANE A rhizomatous, hairy perennial herb with downy or woolly, erect, branched stems to 1 m. Basal leaves oblong, withering during flowering; stem-leaves alternate, downy and clasping at the base. Flower-heads yellow, 15–30 mm across, borne in loose clusters; ray florets exceeding the disk florets; bracts sticky and hairy with long, fine tips. Damp, grassy habitats; widespread but local. Throughout. *P. odora* is similar but lacks creeping stolons and has basal leaves *not withered* during flowering. Throughout, except many islands and the southeast (not CY). *P. arabica* subsp. *hispanica* (syn. *P. paludosa*) is similar to the previous species but with narrow, *linear*, rigid leaves, smaller flower-heads, and achenes with erect to spreading hairs. Rare and local. AA (Rhodes), CY, IP, TR (south).

Dittrichia

Annual or perennial herbs or small shrubs with sticky stems and simple, alternate leaves; very similar to *Inula*. Flower-heads borne in branched infloresences. Achenes abruptly contracted below the pappus.

Dittrichia viscosa (syn. *Inula viscosa*) A densely glandular, sticky perennial 40 cm–1.3 m with stems woody at the base. Lower leaves bright green, linear, scarcely and sparsely toothed; upper leaves stalkless and semi-clasping the stem. Flower-heads 10–15(20)mm, 5–6 mm across, bright yellow, *with ray florets to 8–12 mm long*, much exceeding the flower bracts; bracts adpressed. Common to abundant in disturbed and damp habitats. Throughout. *D. graveolens* (syn. *Inula graveolens*) is a more slender, erect annual 20–50 cm, with *flower-heads smaller, 3–5 mm across*, with linear-lance-shaped leaves and flower-heads with outward-curving bracts *and with shorter rays* (3.5–4 mm). Waste places. Throughout.

Anvillea

Rigid, branching desert shrubs. Flower-heads yellow, with only disk florets, and when mature, a woody involucre. Pappus absent.

Anvillea garcinii A grey-woolly shrub, branched from the base with rigid, intricate stems, to 30 cm. Leaves 5–12 mm, egg- to spatula-shaped, irregularly toothed or fringe-toothed, tapered and clasping at the base. Flower-heads yellow, 15–25(30) mm across, without ray florets, borne on short, thick stalks; involucre with deflexed, leafy bracts, becoming spine-like with age. Achenes 1–1.5 mm, 4-angled, without a pappus. Deserts. IP (south, especially the Arava Valley).

Inula FLEABANE

Perennial herbs, sometimes woody below. Flower-heads yellow, often showy, usually with disk and ray florets, borne few, in branched, flat-topped clusters. Pappus 1 row of hairs.

(a) Flower-heads with disk florets only, or ray florets very small.

Inula conyza PLOUGHMAN'S SPIKENARD An erect, leafy, sparingly branched perennial to 1.3 m with hairy, oval, stalked lower leaves and narrowly oval-shaped upper leaves, all finely toothed and hairy beneath. Flower-heads small, 7–12(15) mm across, *numerous on each stem*, yellow; *ray florets absent*. Pappus reddish-white. Damp habitats, mainly northern. GR, PN.

1. *Pulicaria dysenterica*
2. *Dittrichia viscosa*
3. *Anvillea garcinii*
4. *Inula conyza*

(b) Flower-heads with disk and ray florets.

Inula verbascifolia A bushy, branched *white-downy* perennial to 50 cm with oval-elliptic, finely toothed, slightly folded, pointed, grey leaves. Flower-heads bright golden yellow with disk and ray florets. Rocky habitats and cliffs. AA, GR, IA, PN, TR.

Inula salicina WILLOW-LEAVED INULA A slender, unbranched perennial to 70 cm with stiff, rough-margined leaves, hairless above, the uppermost elliptic with heart-shaped bases ½-clasping the stem and spreading horizontally; all leaves net-veined above. Flower-heads 25–45 mm, solitary or few, yellow, with long, slender rays. Moist habitats. GR (north).

Limbarda

Succulent, salt-tolerant perennials, previously included in the genus *Inula*. Pappus 1 row of hairs.

Limbarda crithmoides (syn. *Inula crithmoides*) GOLDEN SAMPHIRE An erect, *succulent, maritime perennial* to 1 m, hairless, woody-based, with *fleshy, crowded, linear leaves*, untoothed or 3-toothed at the apex. Flower-heads 15–25 mm, rather few, yellow with a golden disk, borne in flat-topped clusters. Saltmarshes, cliffs and other saline, coastal habitats. Throughout.

Aster

Annuals or perenials with oval to linear stem-leaves and flower-heads typically with yellow disk flowers and white, blue, pink or purple ray florets; bracts in several rows. Pappus of 1–2 rows of hairs. North American species traditionally classed in the genus now transferred to other genera such as *Tripolium* and *Galatella*.

Aster squamatus (syn. *Symphyotrichum squamatum*) An erect, slender, almost hairless annual 10 cm–1 m (tiny in bare habitats). Leaves 40 mm–18 cm, linear-lance-shaped, sometimes obscurely toothed. Flower-heads 7–9 mm long, with greenish bracts with broad, translucent margins, often purplish towards the tips; florets *inconspicuous*, whitish, cream, mauve or pink. Native to South America, naturalised in saline areas and sand flats. Probably throughout.

1. *Limbarda crithmoides*
2. *Aster squamatus*
3. *Erigeron acris*

Tripolium

Previously grouped with *Aster*. *Succulent*, hairless annuals to biennials in saline habitats.

Tripolium pannonicum subsp. *tripolium* (syn. *Aster tripolium*) SEA ASTER A hairless annual to biennial to 1 m with reddish, erect or ascending stems, branched from the base. Leaves *fleshy*, linear to linear-lance-shaped, rounded in cross section and clasping the stem; unstalked above. Flower-heads with bright blue–mauve rays and yellow disks, to 20 mm across, borne in large, flat-topped inflorescences. Local in maritime environments, especially in the west. Scattered throughout, except CY and the east.

Galatella

Previously grouped with *Aster*. Typically herbaceous perennials with prominently 1(3)-veined, glandular leaves, flower-heads with violet rays, bracts in 3–4 series and achenes with 1–2 ribs on each face.

Galatella linosyris (syn. *Aster linosyris*) GOLDILOCKS ASTER A hairless perennial with straggling or erect, very leafy stems. Leaves all 1-veined, linear, unstalked and rough-edged, gland-dotted, not fleshy, and pointing upwards. Flower-heads small, 12–18 mm across and *bright golden yellow, without rays*, borne in spreading clusters; bracts unequal and loosely appressed. Rare but possibly overlooked. Lagoon margins. GR (north).

Erigeron FLEABANE

Annuals or perennials with linear to oblong, entire to toothed, unstalked to shortly stalked leaves. Flower-heads with numerous ray florets in several rows, narrow and strap-like; bracts numerous and overlapping. Pappus 1 row of hairs (or with additional shorter hairs).

(a) Ray florets *lilac*, erect and a *little longer than the disk florets*.

Erigeron acris (syn. *E. acer*) BLUE FLEABANE A variable, densely grey-hairy biennial to 60 cm with rigidly erect stems. Basal leaves narrow and entire, the uppermost unstalked. Flower-heads 10–15 mm across with 2 rows of erect, very *pale lilac ray florets scarcely longer than the yellowish disk florets*, borne on upward-spreading stalks, in spreading or flat-topped clusters. GR.

(b) Ray florets white, spreading and *much longer than the disk florets*.

Erigeron annuus SWEET SCABIOUS An erect, branched, sparsely hairy biennial to 70 cm (1 m) with oval to lance-shape, deeply lobed and coarsely toothed, virtually hairless leaves. Flower-heads daisy like, with a yellow disk and *very slender pale lilac–white, outward-spreading rays*, borne in lax clusters. Native to North America, locally naturalised. GR.

(c) Flower-heads with very *inconspicuous or absent* ray florets. Previously described as *Conyza*, and still described under that genus in many floras.

Erigeron canadensis (syn. *Conyza canadensis*) CANADIAN FLEABANE A tall, sparsely hairy annual to 15 m. Leaves alternate, narrowly oblong, stalked and yellow–green, often withered below and deciduous before flowering. Flower-heads very small, 3–5 mm, whitish, borne abundantly in a *cylindrical infloresence*; outer florets exceeding the involucre; disk florets with 4-lobed corolla. Pappus cream. A weed native to North America, very commonly naturalised on waste ground in towns. Throughout (less common in the south). *E. bonariensis* (syn. *Conyza bonariensis*) is a similar annual to 60 cm, and more densely brown to reddish-grey hairy, with larger flower-heads 6–10 mm borne in an *inflorescence with long lateral branches,* overtopping the main axis; *terminal axis with ≤30 flower-heads*; outer florets *minute*; disk flowers with 5-lobed corolla. Pappus dirty-white to reddish. Native to tropical America, commonly naturalised. Throughout. **E. sumatrensis** (syn. *Conyza sumatrensis*) *is* similar to *E. bonariensis*, with flower-heads 5–7 mm borne in a *pyramidal* inflorescence; *terminal axis with >50 flower-heads*; disk flowers with a 5-lobed corolla. Pappus cream to grey. Native to the Americas, naturalised throughout.

1. *Erigeron canadensis*
2. *Erigeron bonariensis*
3. *Bellis sylvestris*
4. *Bellium minutum*
5. *Artemisia arborescens*

Bellis DAISY

Annuals or perennials with basal leaf rosettes; leaves simple. Flower-heads solitary, on long stalks; ray florets white to pink; disk flowers yellow. Pappus absent.

(a) Stems branched and leafy below.

Bellis annua ANNUAL DAISY A low, normally *soft-bristly-hairy annual* to 10 cm with short, erect, *leafy stems* (at least below). Leaves spatula-shaped, toothed or not, stalked below, *not borne in distinct rosettes*. Flower-heads white with a yellow disk, to 15 mm across, the rays sometimes tinged purple. Common in damp, grassy as well as dry open habitats. Throughout.

(b) Stems unbranched and leafless.

Bellis perennis DAISY A perennial to 12(20) cm with a *dense basal rosette*, leaves *abruptly* tapered at the base into the stalks, toothed. Flower-heads 12–25 mm across, borne on *unbranched, leafless stems*. Common on lawns, roadsides and other grassy habitats. Throughout (rare in the north). **B. sylvestris** SOUTHERN DAISY is similar to *B. perennis* but with *3-veined* leaves *gradually* tapered at the base into the stalk. Flower-heads *large*, to 40 mm across, borne on long, stout stalks to 30 cm; ray florets white but sometimes tinged purple on both sides. Achenes compressed and bristly. Woods, thickets and shady roadsides. Throughout.

Bellium

Very similar to *Bellis* but *minute*; pappus comprising a ring of hairs (rather than a pappus with 1 scale between 2 long bristles).

Bellium minutum A minute annual, similar to *Bellis annua* but much smaller. Leaves all basal, spatula-shaped. Flower-heads borne on very slender stalks just 20–50 mm; ray florets tinged purple. Achenes 1 mm; pappus formed of an outer ring of short, translucent scales and an inner ring of 4 long bristles. Dunes and rocky places, often seasonally damp. AA (central + south), CR (+ KP), CY.

Artemisia WORMWOOD

Annual or perennial herbs, or low shrubs, often strongly aromatic. Leaves alternate, entire to divided. Flower-heads small, often nodding in a lax inflorescence. Achenes lacking (or virtually lacking) a pappus.

Artemisia campestris FIELD SOUTHERNWOOD An *almost scentless*, low shrub to 75 cm with a stout, woody stock and numerous non-flowering shoots; stems ascending, slightly, but persistently *silkily hairy*. Leaves also silky when young, later hairless, 2–3-pinnately divided below, simple and stalkless above; leaf lobes *narrow, 0.3–1(1.5) mm wide*. Flower-heads 2–4 mm across, pale green, ovoid, short-stalked, erect or spreading; involucre hairless or virtually so, bracts with a wide, papery margin; corolla yellowish or reddish. Grassy habitats; absent from hot, dry areas. GR, IA.

Artemisia arborescens A branched, aromatic, white-downy shrub to 1.2 m. Leaves 1–2-pinnately divided into linear-oblong, almost blunt segments. Flower-heads borne numerously in panicles, almost spherical, somewhat nodding, rather large, 6–7 mm across; bracts with wide, translucent margins; florets yellow. AA, CR, GR, IA, PN. **A. absinthium** is similar but less shrubby, <1.5 m, with flower-heads 2.5–3.5 mm, nodding in narrow to broad panicles. Widespread across the northern mainland. GR, TR.

Artemisia vulgaris An almost hairless, *herbaceous perennial* 60 cm–1.2 m with reddish stems. Leaves 1-pinnately divided, dark green above, white downy below with segments 3–15 mm wide. Flower-heads 4 mm long, borne in wide-spreading panicles; bracts hairy; florets reddish brown. AA (rare), GR.

Achillea

Perennial herbs with alternate, simple and wavy-edged to shallowly or deeply dissected leaves. Flower-heads congested into umbel-like flat-topped clusters; disk and ray florets white, pink or yellow. Pappus absent.

(a) Florets yellow.

Achillea santolina An aromatic, woody-based perennial 25–90 cm with deeply furrowed stems. Leaves linear, to 24 cm, pinnately lobed with *distant,* simple, 3-lobed or 3-toothed linear-lance-shaped segments. Flower-heads (10)20–70(100), borne in corymbs 20–50(70) mm across; involucre ovoid to hemispherical, 30–40 mm; bracts lance-shaped to oval, rather pointed; florets pale to golden yellow, 0.5–1 mm. Mainly in deserts or dry, rocky habitats. IP (mainly south-central), SY. **A. falcata** is a similar dwarf shrub to 45 cm with linear, grey leaves 15–30 mm, with down-turned margins and overlapping segments; lobes circular. Flower-heads 3–40 borne in corymbs 15–40 mm across; involucre broadly ovoid to hemispherical, 3.5–6 mm; bracts oval-triangular to oblong, without papery margins. Rocky slopes and forests. IP (north – rare), LB, SY, TR.

Achillea holosericea A perennial herb to 50(60) cm with oblong, silver-hairy leaves to 15 cm, pinnately divided into toothed, blunt lobes. Flower-heads *golden yellow,* borne in dense, compound corymbs of 20–50. Fairly common in mountains of mainland Greece. GR, PN.

Achillea taygetea A *woody-based* perennial to 35 cm, white–grey-woolly throughout. Leaves 1–2-pinnately lobed. Flower-heads lemon yellow, borne in dense heads 8–20 mm across. PN.

(b) Florets white to cream.

Achillea millefolium YARROW An aromatic, creeping perennial with clumped, erect, downy and furrowed, unbranched stems to 80 cm. Leaves much-divided 2–3-pinnately divided with lobes diverging in 3 dimensions (feathery), to 15 cm long, the upper leaves unstalked. Flower-heads 4–7 mm across, numerous (>25–50), aggregated into flattened, umbel-like inflorescences; ray florets white or pink; disk florets yellow. Cool, grassy habitats. GR (north). **A. ligustica** is similar but with shorter, wider, more finely divided leaves with *winged stalks* and green, forward-directed segments, and smaller flower-heads 3(5) mm across. CR (west – rare), GR, IA, PN. **A. nobilis** is similar but with narrow, unwinged leaf stalks with spreading segments. GR (mainly north + centre).

Achillea cretica A dense, dwarf shrub to 50 cm with white-downy stems. Leaves linear-oblong, pinnately divided into overlapping, blunt segments. Flower-heads white, borne in dense corymbs of (5)15–30; ray florets *broad,* 3–4.5 mm, white. AA (southeast), CR.

Achillea ageratifolia A tufted perennial to 20 cm with simple, erect stems. Leaves linear or broadest above the middle, grey-downy, entire to toothed. Flower-heads *solitary* to few (1–3), white, with long white ray florets 11–20 mm; disk florets whitish. GR (north).

Achillea umbellata A short, grey-woolly, tufted perennial with flowering stems to 12 cm. Leaves oval in outline, 1–2-pinnately divided into 3–6 pairs of oblong-elliptic, blunt lobes. Flower-heads large and few (3–9); ray florets 4–6 mm, white; disk florets cream. PN.

(c) Flower-heads *only with disk florets*; florets elongated below into 2 enlarged spurs.

Achillea maritima (syn. *Otanthus maritimus*) COTTONWEED A short, *densely white-woolly*, spreading perennial to 30 cm with robust, ascending stems. Leaves oblong-lance-shaped, untoothed or blunt-toothed, fleshy and unstalked. Flower-heads few, 6–9 mm across, *yellow, rayless* and button-like, borne in lax, flat-topped clusters; flower bracts white-woolly. Fixed dunes and coastal sands; local. Throughout.

Cladanthus

A genus closely related to *Anthemis* and *Chamaemelum* (many species previously included in the latter genus).

Cladanthus mixtus (syn. *Chamaemelum mixtum*) A hairy, chamomile-like annual to 60 cm, normally with much-branched stems. Leaves oval in outline, 1–2-pinnately divided below, deeply toothed or 1-pinnately divided and *unstalked above*; leaf lobes linear-lance-shaped and toothed or not. Flower-heads daisy-like with an involucre to 45 mm and spreading white rays and a yellow disk; rays 3-toothed at the tip; flower bracts with a wide, pale brown membranous margin. Achenes 1.2–1.6 mm, ovoid and weakly ridged. Cultivated, fallow and sandy waste ground and coastal sands. AA (local), GR (west), IA, TR.

Anthemis CHAMOMILE

Slightly hairy, often aromatic herbs or dwarf shrubs with *leaves cut into linear segments*. Flower-heads usually with white or yellow rays. Achenes oval to obconical (not compressed), circular to square in cross section, usually with about 10 smooth or rough *distinct* ribs. Pappus absent. Numerous species in the area; only a subset are included here.

(a) Annuals to biennials without woody bases. Ultimate leaf segments variously (but rarely finely) toothed. Achenes *not* compressed. Flower-heads usually with *disk florets only*.

Anthemis rigida A spreading, *mat-forming* annual with prostrate to ascending, rigid stems to 20 cm. Leaves grey-hairy, 2-pinnately divided with broad segments. Flower-heads with yellow disk florets *only*; bracts pointed, with narrow, translucent margins; stalks thickening and recurved in fruit. Bare and sandy habitats. AA, CR, CY, GR (southeast), PN (east), TR (southwest).

1. *Achillea santolina*
2. *Achillea maritima*
3. *Anthemis rigida*

Anthemis ammanthus A *slender annual* to 20 cm, often purplish. Leaves rather fleshy with adpressed hairs, the lowermost blunt-lobed, the stem-leaves lobed to almost entire. Flower-heads yellow (turning purplish), without ray florets, 4–8 mm across, borne on long, slender stalks. Achenes 10-ribbed, almost *conical*. AA (south), CR (east + KP). The following are similar: ***A. filicaulis*** is shorter, to 14 cm with typically unbranched stems and leaves with thin, pointed lobes. Coastal habitats. CR (northeast + islets). ***A. glaberrima*** has *broadly cylindrical, bluntly* ribbed achenes. CR (northwest + islets). ***A. tomentella*** has long, minutely warty achenes. Inland habitats. CR (east).

(b) Annuals to bienneials without woody bases. Ultimate leaf segments variously (but rarely finely) toothed. Achenes *not* compressed. Flower-heads usually with ray and disk florets.

Anthemis chia A virtually hairless annual 10–35 cm with several ascending, simple or sparsely branched stems. Leaves 1–2-pinnately divided. Flower-heads solitary, borne on long, slender stalks; receptacle hemispherical; ray florets white; disk florets yellow; bracts with *conspicuous brown, translucent margins*. Common in a range of lowland habitats. Throughout.

Anthemis arvensis CORN CHAMOMILE A variable, hairy, aromatic annual to biennial with spreading or ascending branched, downy stems to 50 cm. Leaves to 50 mm long, oval in outline, 1–2-pinnately lobed with narrow, pointed segments, *woolly below*, especially when young. Flower-heads with white ray florets and a yellow disk to 40 mm across; bracts with brown, papery margins. Receptacle cylindrical-cone-shaped in fruit; achenes with 10 ridges. Disturbed, cultivated and fallow land; frequent. Throughout, except some islands. ***A. ruthenica*** is similar but with broader, blunter leaf segments and narrow achenes. GR (north).

Anthemis tomentosa WOOLLY CHAMOMILE A low, ascending *woolly-hairy* annual with spreading to prostrate stems and 2-pinnately divided leaves. Flower-heads 20–30 mm across with a yellow disk and white, rather short ray florets; bracts grey-haired. CR (east).

Anthemis cotula STINKING MAYWEED An erect annual, *strong and unpleasant-smelling*, with *almost hairless leaves*, slightly fleshy and irregularly 2–3-pinnately lobed into linear segments. Flower-heads with ray florets deflexed when mature. Achenes with rough ridges. Various habitats. Throughout.

1. *Anthemis chia*
2. *Anthemis tricolor*
3. *Aaronsohnia factorovskyi*

(c) Perennials with slightly woody bases. Ultimate leaf segments *not* finely toothed. Achenes *not* compressed.

Anthemis tricolor A prostrate to spreading, hairy perennial with stems to 46 cm, angular and purplish, woody below. Leaves tufted, 10–40 mm long, lobed, greyish. Flower-heads solitary and terminal with involucre 6–13 mm across with closely overlapping bracts in 3 rows, with raised, greenish mid-veins; ray florets *whitish or pinkish* with darker purplish base, 5–6 mm long; disk strongly convex and *pink or purple*. Achenes pale brown, obscurely 4-angled, rough, and with a collar 0.3 mm long. An island endemic of garrigue, chalk and limestone outcrops. CY.

Anthemis cretica A variable, often slightly *woody-based perennial* to 40 cm with stems leafless above. Leaves greyish, stalked, 1–3-pinnately divided into into linear segments (not finely toothed). Flower-heads with bracts with *dark brown margins*; ray florets white or absent; receptacle hemispherical to cone-shaped. Achenes bluntly square in cross section (not compressed), *ribbed* and greyish. AA, GR, IA (Lefkada), LB, PN, SY, TR. *A. abrotanifolia* is similar but with slender, spreading stems, and pinkish ray florets. CR (mountains).

(d) Perennials with slightly woody bases. Leaf segments very finely toothed. Achenes compressed.

Anthemis tinctoria A variable, often slightly *woody-based perennial* to 60 cm. Leaves 2–3-pinnately divided with very *finely toothed segments*. Flower-heads with oblong-lance-shaped bracts with dark brown, slightly fine-toothed tips; ray florets white, cream or yellow. Achenes *diamond-shaped* in cross section (compressed). GR, PN, TR.

(e) Annuals to bienneials without woody bases. Leaf segments very finely toothed. Achenes compressed.

Anthemis altissima An erect, virtually hairless annual to 60 cm. Leaves oval to oblong in outline, 2–3-pinnately divided with pointed segments. Flower-heads borne on stalks much thickened in fruit; receptable hemispherical; ray florets 8–12 mm, white. Achenes diamond-shaped in cross section (compressed) and with 8–10 ribs on each face; achenes soon detached when ripe. AA (local), CR, GR, PN. *A. melanolepis* is similar but with ripe achenes forming a *persistent, compact head*. AA, CR, CY, IP, SY. *A. palaestina* is scarcely distinguishable, and has pale (not bronze to purplish) scales on the receptacle when ripe. AA (Rhodes, Samos), CY, LB, SY, TR.

Cota

Differing subtly from *Anthemis* in having obconical, flattened achenes with promiment lateral ribs or with 2–10 ribs on each face.

Cota tinctoria (syn. *Anthemis tinctoria*) YELLOW CHAMOMILE An erect or ascending, branched, woolly-hairy perennial to 30(50) cm with leaves deeply twice cut into narrow, linear or narrowly oblong, toothed segments, woolly beneath. Flower-heads with 10–20 yellow ray florets 6–12 mm. Receptacle hemispherical in fruit; achenes 1.8–2.2 mm, angled. Dry, rocky and bare habitats. Throughout.

Aaronsohnia

A North African and Middle Eastern genus of herbaceous desert annuals. Leaves alternate, pinnately divided. Flower-heads solitary, with or without ray florets, and a cone-shaped involucre. Pappus whitish, papery and sheet-like (or missing).

Aaronsohnia factorovskyi An ascending desert annual 50 mm–20(30) cm. Leaves green, pinnately divided into rather fleshy, rigid, blunt broadly linear lobes. Flower-heads hemispherical, *solitary*, borne on *long*, erect, slender stalks, bright yellow, *without ray florets*; florets with 5 triangular lobes; involucre green, cone-shaped with bracts in several rows, with narrow, whitish, papery margins. Achene 1.7–2 mm, pale brown, slightly compressed; pappus a *papery*, white extension, decreasing in size on achenes from the centre to the margin. Wadis in deserts after winter rain. IP (very common in the far south and east).

Anacyclus

Anthemis-like annual or perennial herbs with alternate, 1–2-pinnately divided leaves. Outer achenes 2-*winged*, inner achenes unwinged; pappus absent.

Anacyclus clavatus A short, widely branched, hairy annual 20–50 cm, similar to *Anthemis tomentosa* but with narrower leaf segments and *winged* achenes. Leaves alternate, 2–3-pinnately lobed with linear segments. Flower-heads daisy-like with *white recurved rays and a yellow disk*; flowers borne solitary on stalks *distinctly swollen* below the fruiting head; flower bracts with a narrow whitish or purplish margin. Sandy ground and coastal rocks in the west. IA, PN.

Anacyclus radiatus A soft-hairy annual 40–60 cm with erect, rigid stems branched above. Leaves oval-oblong in outline with linear lobes. Flower-heads yellow, 25–30 mm across; bracts woolly, with white, transparent appendages. Wings of outer achenes broader than the achene body; inner achenes wingless. Sandy habitats, often near towns (introduced). Rare. IP (coastal plains).

Glebionis (*Chrysanthemum*)

Annuals with simple leaves. Flower-heads with yellow, cream or white ray florets; receptacle without scales. Achene without a pappus.

Glebionis segetum (syn. *Chrysanthemum segetum*) CORN MARIGOLD A short to tall, green, hairless, slightly fleshy annual to 60 cm with erect to ascending, branched or unbranched stems. Leaves blue–grey, alternate, narrowly oval in outline, slightly to *deeply toothed*, at least below. Flower-heads 30–70 mm across, bright yellow and daisy-like, with a flat disk. Achenes 2.5–3 mm, deeply ridged and unwinged. Common on disturbed fallow and cultivated land, roadsides, and coastal waste places. Throughout.

Glebionis coronaria (syn. *Chrysanthemum coronarium*) CROWN DAISY A slightly hairy annual to 80 cm with leaves *2-pinnately lobed*. Flower-heads 40–80 mm across, yellow and daisy-like, with rays cream–white in the upper ½ (or cream or yellow). Very common to abundant on disturbed land, roadsides, and coastal waste places. Achenes 3–3.5 mm, deeply ridged and winged. Throughout (mainly all yellow-flowered forms in IP).

Coleostephus

Similar to *Glebionis* but with leaves *regularly fine-toothed* (not deeply lobed or divided).

Coleostephus myconis A sparingly branched, virtually hairless annual. Leaves oval, *regularly fine-toothed* (*not* lobed), the lower leaves tapered gradually into a stalk; the upper leaves unstalked and semi-clasping the stem. Flower-heads yellow and daisy-like, to 22 mm across with yellow or paler rays and a yellow disk. Cultivated, fallow and damp ground; local and mainly western. AA (rare), CR, GR (west), IA, SY, TR.

Leucanthemum OX-EYE DAISY

Rhizomatous perennials. Flower-heads daisy-like, solitary or in groups of 2–3; ray florets white. Achenes with secretory canals.

Leucanthemum vulgare OX-EYE DAISY A variably hairy, clump-forming perennial to 75 cm with leafy stolons and erect, ridged stems. Leaves alternate, oblong, deeply toothed and stalked below, unstalked and clasping the stem above. Flower-heads large, 25–60(75) mm across and daisy-like, with long, white rays and a yellow disk. Cool, grassy habitats and hillsides. GR (north), TR.

Cotula

Small annual or perennial herbs with entire to deeply pinnately divided, alternate leaves. Flower-heads borne in leaf axils or terminally, without rays. Achene without a pappus.

Cotula coronopifolia A small, succulent, hairless, often patch-forming annual to 30 cm. Leaves rather fleshy; linear and entire or scarcely toothed, clasping the stem at the base. Flower-heads all yellow, 8–12 mm across, without rays; long-stalked and often nodding. Achenes strongly compressed and winged (outer disk florets) or unwinged (inner disk florets). Native to South Africa, locally naturalised in damp and seasonally flooded habitats. AA (Lesbos), GR, IA, PN.

---

1. *Glebionis coronaria*
2. *Glebionis coronaria* (eastern yellow-flowered form)
3. *Glebionis segetum*
4. *Cotula coronopifolia*

Senecio

A large and diverse genus of annual or perennial herbs and shrubs with alternate, pinnately veined leaves. Flower-heads often numerous, borne in flat-topped clusters, usually yellow. Achene with a white or greyish, hairy pappus.

(a) Some or all leaves deeply lobed; *ray florets absent*.

Senecio vulgaris GROUNDSEL A short, more or less hairy, rather succulent annual to 30 cm (usually less) with weak, sparingly and irregularly branched stems. Leaves coarsely lobed and toothed, oval in outline and short-stalked below, semi-clasping the stem at the base. Flower-heads *small, without (or with few) rays*, to 5 mm across and yellow; involucre cylindrical with black-tipped bracts. Achenes <2.5 mm with much longer pappus hairs. A very common weed in a range of disturbed natural and urban habitats. Throughout. **S. aegyptius** is similar but with numerous flower-heads borne in large terminal coymbs or cymes. Pappus hairs equalling or shorter than the achenes. CY.

(b) Some or all leaves deeply lobed; *ray florets inconspicuous*.

Senecio viscosus STICKY GROUNDSEL A small, *very sticky*, rather unpleasant-smelling annual to 60 cm (usually less) with weak, freely or sparingly branched stems. Leaves dark grey–green, densely glandular-hairy and sticky, deeply and regularly lobed, and short-stalked below, and unstalked (but not clasping) above. Flower-heads to 12 mm, with 13 pale yellow, recurved ray florets, borne in a large, irregular corymb. Achenes hairless or hairy in the grooves. Open stony and sandy places; local and absent from hot, dry areas. GR (north).

(c) Some or all leaves deeply lobed; ray florets short to spreading.

Senecio lividus A short, hairless or slightly hairy annual to 60 cm usually branched from the base, sparsely glandular-hairy. Leaves fleshy, pinnately divided with lobed segments; stalked below, unstalked above and *clasping* the stem. Flower-heads 4.5–6(7) mm with black-tipped bracts and *short* and inconspicuous, *recurved* ray florets. Achenes 3.2–3.7(4.5) mm. Dunes and fallow land; rare and local in the area. AA, GR.

Senecio leucanthemifolius An ascending, branched annual to 25(30) cm. Leaves fleshy, hairless or slightly white-hairy (sometimes purple-tinged). Flower-heads 7–8 mm with *black-tipped flower bracts*; ray florets conspicuous and spreading. Achenes 2–2.2 mm, hairless. Coastal habitats; absent from the west and north. AA, GR (southeast), CR, CY, IP. **Subsp. vernalis** (also often treated as a true species) is much taller, often with a single stem, branched above and with leaves not fleshy. CY, IP.

Senecio glaucus An erect to spreading *hairless* (to sparsely hairy), greyish-blue, sometimes fleshy annual to 40(60)cm; similar to *S. leucanthemifolius*. Leaves oblong or lance-shaped in outline, 10–50 mm (13 cm) long, the upper usually stalkless and clasping; all deeply lobed with toothed, linear segments. Flower-heads yellow, borne in lax cymes or corymbs; ray florets conspicuous; bracts *without* conspicuous black tips. Achenes 1.8–2.2 mm long. CY, IP. **Subsp. cyprius** is the endemic subspecies of Cyprus; often short and tufted (15–45 cm) with narrowly oblong, rather bright to dull green leaves. Mainly sandy or stony ground near the sea. CY. **Subsp. coronopifolius** is the eastern, mainly desert-growing subspecies, which has fleshy leaves with lobed to remote-toothed linear segments with down-turned margins. IP (south).

1. *Senecio vulgaris*
2. *Senecio leucanthemifolius*
3. *Senecio leucanthemifolius* subsp. *vernalis*
4. *Senecio glaucus* subsp. *cyprius*
5. *Senecio glaucus* subsp. *coronopifolius*
6. *Jacobaea maritima*

ASTERACEAE | 547

Jacobaea

Usually (though not exclusively) perennials with pinnately lobed leaves. Pappus of simple hairs. Previously classified under *Senecio*.

Jacobaea vulgaris (syn. *Senecio jacobaea*) RAGWORT A large biennial or weak perennial to 1.5 m. Stems reddish, robust and erect, more or less hairless or slightly woolly (floccose) and unbranched below; *furrowed*. Basal leaves large to 20 cm long, *deeply pinnately lobed with a large blunt end-lobe* and stalked, often withered during flowering; upper leaves semi-clasping the stem; all slightly hairy below. Flower-heads to 20 mm across, bright yellow and numerous, borne in *flat-topped clusters*. Involucre more or less hairless. Achenes of disk florets hairy. Pastures and grassy habitats; absent from hot, dry areas. GR (north).

Jacobaea maritima A shrub 30–80 cm with shortly white-hairy stems. Leaves more or less pinnately divided into blunt lobes, green above, white-felted below. Flower-heads borne in compound corymbs; ray florets yellow, 3–6 mm. **Subsp.** *bicolor* is native in rocky coastal habitats. AA, GR, IA, PN. **Subsp.** *maritima* is widely planted and locally naturalised. CR (probably elsewhere).

ASTERACEAE

Doronicum LEOPARD'S-BANE

Herbaceous, rhizomatous perennials. Leaves simple and alternate, the lowermost withered when in bloom. *Flower-heads large and yellow*, with a hairy receptacle. Pappus of short bristles, absent in outer florets.

Doronicum columnae A laxly tufted perennial to 30(50) cm with simple stems. Leaves persistent, almost hairless, long-stalked, oval, heart-shaped at the base. Flower-heads solitary, 30–50 mm across, yellow. Achenes silkily hairy, the innermost with a pappus of white hairs. Mountains. GR. ***D. orientale*** is similar but somewhat taller (to 75 cm) with leaves withering in summer, hairy on both surfaces, and with just 2 (not 3–5) stem-leaves. AA (local), GR, IA (Cephalonia, Corfu), LB, PN, SY, TR.

Calendula MARIGOLD

Annual or perennial, often aromatic herbs with alternate, undivided leaves. Flower-heads daisy-like with yellow or orange ray florets; bracts in 1–2 rows. *Achenes strongly curved*, without a pappus.

Calendula arvensis FIELD MARIGOLD A short, slender, thinly hairy, ascending or spreading annual to 30 cm. Leaves oblong and finely toothed. Flower-heads 10–25(27) mm across; yellow–orange throughout or with a brownish disk. Fruiting head with an outer row of beaked and strongly incurved achenes. Common on fallow ground, dunes and open garrigue. Throughout. **C. officinalis** is similar but with larger flowers, 40–70 mm across, yellow–orange with an orange or brownish disk. Widely cultivated as an ornamental, possibly naturalised throughout.

Calendula tripterocarpa A short, slender, glandular-hairy annual to 15 cm. Leaves linear-lance-shaped, entire or toothed. Flower-heads solitary, yellow, *small, 7–10(12) mm across*; involucre with 2 rows of almost equal glandular-hairy bracts. Fruiting head with markedly differing achenes, the outermost beakless and with *3 prominent wings;* the intermediate boat-shaped, the innermost transversely with stripes of small warts. Arid habitats, soon-flowering after winter rain. IP (mainly southern Negev).

Pallenis

Annuals to subshrubs. Flower-heads yellow with bracts in 2–3 rows, leafy, the outermost spineless or spine-tipped. Pappus of scales. A genus now established to include species traditionally placed in *Asteriscus*.

Pallenis spinosa A slender, softly hairy annual to 60 cm with rigid stems, woody at the base with branches overtopping the main stem. Leaves elliptic, stalked below, unstalked and semi-clasping the stem above. Flower-heads daisy-like and bright yellow with a large disk to 20 mm across. Flower *bracts spine-tipped and 2 x the length of the ray florets;* inner flower bracts papery and not spine-tipped. Achene 2–2.5 mm. Common on coastal garrigue, dry roadsides and rocky places. Throughout.

Pallenis hierochuntica (syn. *Asteriscus pygmaeus*) RESURRECTION PLANT A small, *virtually stemless* (main stem 0–7 mm) annual, similar to *Asteriscus aquaticus* in general appearace (see below), but to just 15 cm with small lateral branches. Flower-heads generally exceeded by their outermost involucral bracts. Achene 1.2–1.9 mm. Bare, arid ground. IP, SY.

Asteriscus

A genus very similar to *Pallenis* to which some species have recently been transferred.

Asteriscus aquaticus A similar species to the *Pallenis* group but an *aromatic annual* to 50 cm with erect to spreading stems, and flower bracts *greatly exceeding* the ray florets. Similar to *Pallenis spinosa* but lacking spine-tipped flower bracts. Achene 1.5–2 mm. Various damp habitats as well as garrigue. Throughout.

1. *Calendula arvensis*
2. *Calendula tripterocarpa*
3. *Pallenis spinosa*
4. *Pallenis hierochuntica*
5. *Asteriscus aquaticus*

Xanthium COCKLEBURS

Cosmopolitan annual weeds, native to the Americas. Leaves alternate. Flower-heads discoid. Fruiting head *conspicuously spiny*. Pappus absent.

Xanthium strumarium ROUGH COCKLEBUR A stiffly branched, often aromatic, spineless annual to 1.2 m without spines. Leaves alternate and heart-shaped at the base, shallowly lobed and long-stalked. Flower-heads greenish, with male and female flowers borne separately in *lateral clusters*. Fruiting heads covered in hooked spines; *those at the apex not hooked*. Naturalised in damp and disturbed habitats. Throughout. *X. spinosum* SPINY COCKLEBUR is similar but with *prominent beige spines* projecting from the leaf bases. Disturbed habitats. Throughout.

Bidens BUR-MARIGOLD

Annuals with opposite leaves. Flower-heads rounded, often solitary, often without ray florets; bracts in 2 rows, the innermost papery. Pappus of 2–5 strong, barbed bristles.

Bidens tripartita TRIFID BUR-MARIGOLD A hairless to slightly hairy annual to 75 cm. Leaves with 1–2 pairs of deep lateral lobes and a short, winged stalk. Flower-heads yellow, erect, to 25 mm across, borne in branched clusters, *usually without ray florets but with leaf-like spreading bracts beneath*. Achenes with 2–4 bristles. Locally common on riversides and other damp habitats. AA (rare), GR, IA (Corfu), PN.

SUBFAMILY CICHORIOIDEAE

Biennials and perennials, often *with a white latex,* and stem-leaves spirally alternate, or absent. Flower-heads usually yellow and ligulate (florets comprising 5-toothed ligules only).

Scolymus SPANISH OYSTER PLANT

Stout, spiny, thistle-like perennials. Flower-heads with outer bracts leaf-like and grading into the true upper leaves. Achenes flattened, not beaked, with pappus absent or composed of few rigid hairs.

Scolymus hispanicus SPANISH OYSTER PLANT A robust, spiny, biennial or perennial to 80 cm with interrupted *spiny-winged* stems. Lower leaves yellowish-green, oblong, pinnately lobed with sparse spines; upper leaves smaller and spinier; leaves with paler veins and border. Flower-heads golden-yellow, 25–40 mm across, *borne in long, narrow spike-like panicles*; florets rayed (not tubular); flower-bracts slightly hairy or hairless, with membranous margins, and narrowed into a sharp point. Frequent on fallow ground, roadsides and sandy waste ground. Throughout. *S. maculatus* is very similar but with *broader wings 2–5 mm at the narrowest), with prominent white margins*. Throughout. *S. grandiflorus* is distinguished by its *markedly hairy* oval to linear flower-bracts. A central Mediterranean species, rare in the east. AA (Rhodes).

1. *Xanthium strumarium*
2. *Xanthium spinosum*
3. *Scolymus hispanicus*
4. *Scolymus grandiflorus*
5. *Cichorium intybus*
6. *Catananche lutea*

Cichorium CHICORY

Annual or perennial herbs with a white latex when cut. Leaves toothed or lobed. Flower-heads numerous; florets all rayed and toothed at the tips. Achenes angular (not flattened); pappus a series of short scales.

Cichorium intybus CHICORY A hairless to stiffly hairy, erect, branched perennial to 1 m. Basal leaves pinnately lobed and short-stalked below, lance-shaped and clasping the stem above. Flower-heads bright sky-*blue*, 25–40 mm across, borne in narrow, leafy, branched spikes. Fairly common on fallow ground and roadsides. Throughout. *C. endivia* is similar but a blue–grey annual with less deeply lobed leaves and *flower-stalks distinctly swollen* beneath the terminal flower-heads. Widely cultivated and naturalised. Throughout. *C. spinosum* SPINY CHICORY is distinguished by its densely branched, mounded habit and spiny branches. AA, CR, CY, GR, IA (Cephalonia), PN, TR (unconfirmed).

Catananche

Rhizomatous perennials. Flower-heads solitary, each at the end of a long branch, with *shiny, papery bracts* loosely overlapping. Achenes 5–10-ribbed, not beaked; pappus of scales.

Catananche lutea An annual 10–30 cm with almost erect, simple or sparsely branched stems with adpressed hairs. Leaves linear or broadest above the middle, parallel-veined with remote teeth, or toothless and with spreading hairs. Flower-heads pale yellow, borne solitary on long, slender stalks; involucre cylindrical; bracts translucent, the innermost narrower than the outer. Achene a pappus of scales. Virtually throughout.

Gundelia

Spiny perennials with the aspect of an *Eryngium*. Flower-heads compound, with fertile and sterile florets. Achenes large; pappus absent.

Gundelia tournefortii A virtually hairless, conspicuously spiny perennial 30–60 cm (1 m), branching from the base. Leaves 60 mm–25(30) cm, alternate, very thick, rigid and leathery with prominent veins, divided into spiny-toothed lobes; the uppermost reduced and forming spiny wings at the base (decurrent). Flower-heads complex, large, 20–50 mm across (excluding bracts); florets yellow and tubular; heads becoming woody when mature; bracts spiny-margined, scarcely exceeding the flower-heads, or extended into spines to 70 mm. Achenes large, smooth, ovoid, falling together with a spiny body. Fields, plains and dry pastures. CY (southwest, northeast), IP (north), LB, SY, TR.

Tolpis

Annual or perennial herbs. Flower-heads with *very long, narrow, curved-spreading bracts*. Pappus typically with long, rough hairs.

Tolpis umbellata A slender annual 25–75 cm, sparsely branched above, similar to the western Mediterranean *T. barbata*. Basal leaves broadest above the middle with distant teeth. Flower-heads pale yellow, often with purplish-brown inner florets, borne on stalks, which thicken in fruit. Outer achenes hairy with a short pappus of bristles; inner achenes short-bristly with 2–5 long bristles. Virtually throughout, except IA and the far west.

Hedypnois

Small annuals with rosette leaves. Flower-heads yellow, with *fleshy bracts in a single row, persisting and encircling the fruiting heads*. Achenes ribbed, not flattened, not beaked; pappus of scales.

Hedypnois rhagadioloides (syn. H. cretica) A variable, more or less hairy annual to 45 cm with mostly basal leaves, and *many, branched stems*. Leaves narrowly elliptic, entire to deeply lobed, with winged stalks below; stalkless above. Flower-heads 5–15 mm across, yellow and dandelion-like, borne on stalks thickened immediately below; involucre with narrow linear-lance-shaped bracts *strongly incurved in fruit*. Fairly common in a range of dry and disturbed habitats across the region. Throughout.

Hyoseris

Flower-heads borne solitary at the end of leafless stems, arising from a basal rosette of leaves. Pappus with *unequal yellowish hairs*.

Hyoseris scabra A small *annual* with a rosette of basal leaves. Leaves pinnately divided with *backward-pointing* triangular lobes. Flower-heads borne on *ascending to prostrate* (not erect), *thickened* stalks 6–80 mm, *swollen* at, or above the middle and hollow above. Flower-heads yellow and dandelion-like, 8–10(30) mm across; florets all rayed. Achenes 6–7 mm. Throughout, except the north. *H. lucida* is similar but a *perennial* with almost erect flowering stalks to 30 cm, not conspicuously swollen in fruit. AA, CR, IA, PN (southeastern tip), TR.

---

1. *Gundelia tournefortii*
2. *Hedypnois rhagadioloides*
3. *Hyoseris scabra* (habit)
4. *Hyoseris scabra* (flower head)
5. *Hyoseris lucida* (flower head)
6. *Hyoseris lucida* (habit)

Rhagadiolus

Hairy annuals. *Achenes few, lobe-like, borne in a star-like formation*; pappus absent.

Rhagadiolus stellatus STAR HAWKBIT A rather weedy, coarsely hairy annual with branched, spreading stems to 40 cm. Leaves 25 mm–14 cm, oblong, sparsely toothed to lobed and indistinctly stalked. Flower-heads yellow, small to 10 mm across, long-stalked in lax panicles. *Fruiting head a star-shaped series of (5)7–8 long*, slender, *incurved*, lobe-like achenes 10–16 mm long. Fallow land. Common throughout. *R. edulis* is variably treated as a variety of the former (their level of genetic distinction is unclear), differing in its lyre-shaped leaves with a large, rounded terminal lobe and *5 straight or scarcely curved* achenes. Throughout the range.

Hypochaeris CAT'S-EAR

Annual or perennial herbs with rosettes of leaves and solitary or few, branched stems. Flower-heads yellow, with rayed florets only; bracts in several overlapping rows; *receptacle with scales between the florets*. Achenes finely ribbed, beaked or not; pappus 1–2 rows of brownish hairs.

Hypochaeris radicata COMMON CAT'S-EAR A perennial to 60 cm with almost *hairless, branched stems* and a basal rosette of bristly hairy, wavy-toothed, oblong, dark-tipped leaves to 25 cm. Flower-stalks *thickened* below the flower-heads; flower-heads yellow, 20–40 mm across, with a bell-shaped involucre; *bracts hairless*, dark-tipped. Central achenes 8–17 mm, beaked (marginal achenes usually also beaked). Common in grassy habitats above sea level. Probably throughout, except for hot, dry areas. *H. glabra* SMOOTH CAT'S-EAR is similar but with virtually hairless, glossy leaves and *small, partially closed flower-heads* 10–15 mm across with ray florets scarcely exceeding the bracts. Central achenes 6–9(14) mm (*marginal achenes not beaked*). Throughout.

Hypochaeris achyrophorus A rough-hairy perennial 12–14 cm with a basal rosette of leaves and, often branched, leafless stems thickened below the flower-heads. Leaves rather broad, spatula-shaped and lobed or unlobed. Flower-heads bright yellow, solitary, 10–15 mm across, and with *densely bristly, linear bracts*. Achenes 5–7 mm. Grassy habitats and maquis; common. Throughout.

Leontodon HAWKBIT

Rhizomatous perennials similar to *Hypochaeris* with unbranched stems without bracts and with forked leaves. Flower-heads solitary, with a *receptacle without scales between the florets*. Achenes finely ribbed, not or indistinctly beaked; pappus 2 rows of pale brown hairs (or scales).

Leontodon tuberosus A tuberous perennial to 30 cm with a rosette of toothed to lobed leaves with 2–3-branched bristles. Flower-heads solitary; bracts blunt, 12–15 mm in a bell-shaped capitulum; florets yellow above, brownish or purplish beneath. Achenes of 2 types: the outer with a short pappus, the inner with a long pappus of fluffy hairs. Common and widespread in various habitats. Throughout.

Leontodon hispidus ROUGH HAWKBIT A rhizomatous (*not* tuberous), rosette-forming perennial to 60 cm *covered in rough, white hairs*. Leaves oblong-lance-shaped and narrow at the base, deeply wavy-toothed and *very hairy*. Flower-heads solitary, with yellow florets longer than the bracts; bracts 11–13(15) mm. Achenes with a pappus of hairs of *2 types*: the outer composed of short hairs, the inner longer, fluffy. Grassy habitats, often above sea level. GR, PN. *L. crispus* is similar but with a pappus of 2 rows of hairs, *both* fluffy. AA (local), GR, IA, PN. *L. graecus* is very similar but with leaves with *both* stalk*less starry hairs* as well as *longer 3–7-forked hairs* (not with *only* 2–6-forked hairs). AA (local), GR, IA, PN.

Scorzoneroides

Perennials similar to *Leontodon*, without rhizomes, with simple leaves and (often) branched stems with scale-like bracts. Flower-heads often >1, with a receptacle without scales between the florets (but hairy); involucral bracts merging into stem scales. Achenes finely ribbed, not beaked; pappus 1 row of pale brown hairs.

Scorzoneroides autumnalis (syn. *Leontodon autumnalis*) AUTUMN HAWKBIT A small perennial to 60 cm, *covered in white, rough hairs*. Leaves lance-shaped, wavy-toothed and narrowed at the base. Flower stems hairy throughout, with forked hairs; flower-heads solitary, with yellow florets much longer than the bracts; bracts 6–15 mm. Grassy places above sea level. Northern. GR.

1. *Rhagadiolus stellatus*
2. *Rhagadiolus edulis*
3. *Hypochaeris radicata*
4. *Hypochaeris achyrophorus*
5. *Leontodon tuberosus*

Picris OXTONGUE

Rough-hairy biennials or perennials with lax, branched stems. Flower-heads yellow, with rayed florets, the outer often with a reddish stripe; *outer bracts lance-shaped, similar to the inner*. Achenes weakly ribbed, not, or shortly beaked; pappus 2 rows of off-white hairs.

Picris hieracioides HAWKWEED OXTONGUE A slender, *bristly* biennial with branched stems to 1 m. Leaves lance-shaped to narrowly oblong, shallowly toothed or lobed, the upper leaves clasping the stem. Flower-heads long-stalked, yellow, to 35 mm across, the *involucral bracts lance-shaped, spreading to recurved, simple hairs, dull green* (some blackish). Achenes all similar, 3–5 mm with a creamy pappus. Various habitats; strongly northern and western. GR, IA, PN.

Picris pauciflora A annual or biennial to 45 cm with bristly untoothed to sparsely toothed leaves. Flower-heads borne on erect, branched, spreading panicles; rather few and lax; involucre 8–12 mm with *bracts all linear-lance-shaped, rather concave and inwardly adpressed*, often reddish, and bristly; florets yellow. Achenes 4–5 mm, curved with a white pappus. Pastures and scrub. Probably throughout. *P. rhagadioloides* is similar but with more (not few) flower-heads, semi-clasping leaves, and flower-stalks not swollen in fruit. Throughout, except IP.

Scorzonera VIPER'S-GRASS

Perennial herbs with 1–few stems. Leaves often linear and entire. Flower-heads yellow, pink, purple or white; bracts borne in overlapping rows; florets all rayed and toothed at the tip. Achenes ribbed, with or without a tubular base, not beaked; pappus brownish-white, feathery. A taxonomically rather difficult group.

Scorzonera cretica A perennial with several erect to asending stems to 30 cm. Leaves linear and grass-like, the lowermost broadest at the base. Flower-heads bright lemon yellow; involucre 10–25 mm, bell-shaped; bracts triangular-lance-shaped and pointed. Achenes woolly with a pale *reddish-brown* pappus of long hairs (1.5–2 x as long as the achene). AA (rare and local), CR (common), PN (Kithira, Andikithira). *S. araneosa* is very similar (possibly indistinct), but with *whitish* pappus hairs. AA (central).

Scorzonera judaica (syn. *S. psychrophila*) A small *desert* perennial 30 mm–12 cm, with *ovoid-spherical tubers*. Leaves (5)8–15 mm wide with conspicuously undulate margins, dull green with whitish hairs. Flower-heads yellow with involucre ≤16 mm, borne on *spreading to prostrate* stems 30 mm–10(12) cm; ovaries *woolly-hairy*. Pappus *prominent long-woolly*; achenes 8 mm. Deserts and bare, arid plains. IP (east-central).

Scorzonera hispanica A variable, leafy, *branched* perennial 30 cm–1.2 m. Leaves narrow, variably linear to spatula-shaped, tapering at the base. Flower-heads yellow, typically 1–5, with florets 2 x as long as the involucre; involucre 20–25 mm with broad bracts 6–8 mm wide. Pastures, also an escape from cultivation. GR (northeast). *S. scyria* is similar but with *much broader*, elliptic-lance-shaped, almost *hairless* leaves 15–30 mm wide and a pappus with hairs of 2 distinct lengths. A rare island endemic chasmophyte of limestone cliffs. AA (Skiros).

Podospermum

A genus closely related to (and still often included in) *Scorzonera*. Annuals, biennials or perennials, often with pinnately lobed leaves. Achenes with a tubular base; pappus of hairs.

Podospermum laciniatum (syn. *Scorzonera laciniata*) CUT-LEAVED VIPER'S-GRASS A variable annual or biennial with 1–2-pinnately lobed leaves with linear segments in rosettes; upper leaves clasping the stem. Flower-heads yellow, 15–25 mm across, solitary, borne on thick, unbranched stalks. Flower bracts lance-shaped to oval, often with dark, recurved tips. Fruiting heads large 'dandelion clocks'. Cultivated land and waste places; common. Probably throughout.

Podospermum canum (syn. *Scorzonera jacquiniana*) A *hairy perennial* similar to *P. laciniatum*, often with *branched stems*. Flower-heads yellow, the rays often grey–green, purple or red on the reverse; flower-bracts much shorter than the rays. Saline waste places. Probably throughout.

Tragopogon GOAT'S BEARD

Annual or perennial herbs with a white latex when cut. Stems usually solitary. Leaves linear, often rush-like. Flower-heads with only ray florets. Fruiting head a large 'dandelion clock'; achenes ribbed, beaked; pappus with feathery and simple hairs.

(a) Flower-heads yellow.

Tragopogon pratensis GOAT'S BEARD An annual or perennial with erect stems to 75 cm, downy when young, hairless and greyish when mature. Lower leaves narrowly linear-lance-shaped, to 30 cm with a distinct keel; stem-leaves similar, ending in a fine point. Flower-heads solitary, long-stalked, to 50 mm across with yellow florets. Fruiting head a *large 'dandelion-clock'*. Outer achenes ‹30 mm. GR. *T. dubius* WESTERN SALSIFY is similar but with *stalks markedly swollen beneath flower-heads* and outer bracts much exceeding florets. Dry fields. AA, GR, IA, PN, TR.

---

1. *Picris hieracioides*
2. *Scorzonera cretica*
3. *Scorzonera judaica*
4. *Scorzonera judaica* (fruiting plant)

(b) Flower-heads pink (*T. porrifolius* group: conflicting accounts exist in regional floras).

Tragopogon porrifolius SALSIFY A variable annual or perennial, similar to the previous species, with broadly linear-lance-shaped leaves with broad, sheathing bases, and with *robust flower-heads of purplish florets borne on stalks markedly swollen beneath the heads*. Achene with an equally long beak. Throughout (in its wider treatment). Several subspecies have been described (some are regarded as true species by some authors), including: the widespread **subsp. *australis***, which has ray florets not more than ½ as long as the bracts; **subsp. *porrifolius*** with ray florets about as long as the bracts; and **subsp. *longirostris***, which has *narrowly* linear leaves, bracts exceeding the ray florets, and achenes with a beak 2 x as long as the body of the achene. Maquis and garrigue in the east (regional forms are described under various names). AA, CR, CY, IP.

Geropogon

Annuals similar to *Tragopogon* (still classified in this genus by many authors). Achenes ribbed, beaked; *pappus of rigid (not plumose) hairs*.

Geropogon hybridus (syn. *Tragopogon hybridus*) A slender, hairless annual to 50 cm with branched stems. Leaves long-linear, hairless and rush-like. Flower-heads borne on inflated stalks; lilac–pink, with few, purplish central florets much-exceeded by the surrounding rays, which are in turn *greatly exceeded by the narrow, long-pointed flower bracts*. Stony fallow land, widespread. Throughout.

Aetheorhiza

Herbaceous perennials *with long underground stolons* with white tubercles, from which arise leaves that are often half-buried in sand. Achenes with 4 grooves, nor beaked; pappus of several rows of white, simple hairs.

Aetheorhiza bulbosa A small, green or purplish perennial to 30(55) cm with stolons and rhizomes and few, slender stems. Leaves hairless or hairy, all basal and *held prostrate on the surface of the sand*; elliptic and entire or lobed. Flower-heads 8–15 x 3–12 mm, borne terminally; yellow and dandelion-like; involucral bracts 14–15(16)mm, with blackish hairs. Achenes 3–4.5 mm. Common on coastal sands, rarer inland. Throughout.

Sonchus SOW-THISTLE

Annual or perennial herbs with stout, hollow stems exuding a white, sticky latex when cut. Leaves pinnately lobed or wavy with spiny margins; the stem-leaves clasping at the base. Flower-heads yellow, only with ray florets. Achenes flattened, ribbed, not beaked; pappus with 2 or more rows of white, simple hairs.

Sonchus asper PRICKLY SOW-THISTLE A variably tall, erect, greyish or reddish, hairless (glandular above) annual to 1 m with simple or branched stems. Lower leaves spoon-shaped, sometimes pinnately lobed, with sharp, triangular toothed lobes, the lowest 2 *rounded* at the base and *glossy* green on the upper surface; clasping the stem above. Flower-heads golden yellow, to 25 mm across. Achenes 2–3 mm. Very common in a wide range of habitats from towns to coastal dunes. Throughout. **S. *tenerrimus*** is similar but with deeply pinnately divided leaves with all leaf lobes strongly constricted or linear at the base, and the terminal lobe equalling the laterals (not distinctly larger). Flower-stalks often white-hairy. Disturbed habitats; locally common. Throughout.

Sonchus oleraceus SMOOTH SOW-THISTLE An erect annual to 1.5 m similar to *S. asper*; greyish or reddish, hairless (glandular above). Lower leaves spoon-shaped and variously, often deeply *pinnately lobed* with triangular toothed lobes, the *end-lobe distinctly larger* than the next pair down, and the lowest 2 *pointed* (not rounded) at the base, *matt* (not glossy) green on the upper surface; leaves clasping the stem above. Flower-heads pale yellow, to 25 mm across. Achenes 2.5–3.8 mm. Very common in a range of habitats. Throughout.

Lactuca LETTUCE

Annual, biennial and perennial herbs with a white latex when cut. Flower-heads with cylindrical involucre only with ray florets. Achenes flattened and ribbed; pappus with 2 rows of white simple hairs.

(a) Flower-heads blue.

Lactuca perennis BLUE LETTUCE An erect, hairless perennial to 70 cm (1.2 m) with spineless, hairless, bluish leaves deeply pinnately cut into narrow, nearly entire lobes. Flower-heads blue or violet, borne in loose, spreading clusters, each with 12–20 flat, spreading rays (similar to flower-heads of *Cichorium* but borne on long stalks in loose, spreading clusters, not on short stalks in leaf axils). Achenes 5.5–8 mm, black and wrinkled. GR (north-central).

1. *Tragopogon porrifolius* subsp. *australis*
2. *Tragopogon porrifolius* subsp. *longirostris*
3. *Geropogon hybridus*

(b) Flower-heads yellow; leaves prickly.

Lactuca serriola PRICKLY LETTUCE A tall, greyish, stiffly erect annual or biennial to 1.5(2) m, unbranched below and branched above with whitish stems. Leaves *held stiffly erect*, oblong-lance-shaped, entire to pinnately divided with distant lobes below and sharply toothed, more or less hairless but spiny along the midrib beneath, and along the margins; all leaves waxy and greyish. Flower-heads pale yellow and small, to 13 mm across, borne in a long, lax inflorescence. Achenes olive-grey, 3–4 mm (excluding beak). Waste ground. Probably throughout. **L. virosa** is similar but taller, to 2 m, with *leaves spreading, not erect*, with rounded, clasping lobes at the base. Achenes 4.2–4.8 mm (excluding beak). Waste ground in the west. GR, IA, PN.

(c) Flower-heads yellow; leaves not prickly.

Lactuca saligna LEAST LETTUCE A slender annual to 75 cm (1 m) with *whitish stems* and greyish leaves, *not prickly*, with arrow-shaped bases clasping the stem, those below deeply pinnately lobed, those above linear-oblong and clasping. *Flower-heads borne in narrow, spike-like inflorescences with short branches*; florets pale yellow and reddish below. Achenes olive-grey, 2.8–3.5 mm with a white beak. Bare, grassy habitats. Throughout.

Lactuca viminea PLIANT LETTUCE A hairless biennial without prickles, often with numerous, erect stems 15–80 cm (1 m). Leaves dark grey–green, the lowermost pinnately lobed with linear-lance-shaped toothed lobes, the stem-leaves above lance-shaped and more or less unlobed except at the base. Flower-heads pale yellow, borne in much-branched panicles; florets (4)5–8 per head. Achenes 6.5–12 mm. Throughout.

Chondrilla

Flower-heads cylindrical with 8–10 long, inner bracts and a row of very short, leafy outer bracts; disk florets absent. Achene ribbed, beaked or beakless, with scales at the base of the beak; pappus of simple hairs.

Chondrilla juncea A greyish biennial, hairy, especially below, to 1 m. Stems normally solitary, stiff and *broom-like with few leaves*. Leaves oblong, lobed, withering below and linear and entire above. Flower-heads unstalked, in small clusters; florets all rayed. Sandy or scrubby open habitats; fairly common. Throughout.

Launaea

Shrubs and perennials of dry and arid habitats, often intricately branched, with spineless leaves that exude a latex when cut. Flower-heads with ray florets only. Pappus of feathery bristles.

Launaea nudicaulis A perennial herb 40–50 cm with basal leaf rosettes and *many, divergently branched*, spreading to ascending flowering stems. Basal leaves 12–28 cm, with wavy-toothed to lobed margins; stem-leaves few, smaller and semi-clasping the stems. Flower-heads yellow; involucre 10–16 mm long. Dry, arid habitats, flowering after winter rain. IP (south). *L. fragilis* is a similar, but *smaller* perennial 50 mm–30 cm, often with branched to *unbranched* stems and linear to lance-shaped leaves 30 mm–10 cm long, divided into narrow lobes. Flower-heads yellow; involucre 10 mm long. Bare habitats and desert sands. CY, IP (mainly northern and western Negev).

Urospermum

Flower-heads with 7–8 bracts *all in 1 row* and fused at the base; only ray florets present. Achene beaked; pappus of 2 rows of plumose hairs.

Urospermum picroides An *annual* with bristly stems 20–60 cm. Leaves bristly, rather large to 20 cm long, and sow thistle-like; toothed to lobed, oblong below and linear-lance-shaped above. Flower-heads yellow and dandelion-like, borne on stalks thickened below; involucre 25 x 28 mm, cylindrical. Pappus white and fluffy (plumose). Waste ground. Throughout.

Reichardia

Annual or perennial herbs. Involucre pitcher-shaped with bracts in several rows, each with a white margin. Achenes 4–5-angled (at least the outermost); pappus of numerous rows of soft, simple hairs.

(a) Plant a perennial, woody at the base.

Reichardia picroides A woody-based perennial with toothed to pinnately lobed leaves. Flower-heads yellow, borne on more or less leafless stalks scarcely thickened at the top; flower *bracts with narrow membranous margins* (<0.5 mm wide), the *florets all yellow* (not purplish at the base) though sometimes with a dark stripe on the outer surface. Outer achenes wrinkled, but *the inner smooth*, and appearing sterile. Cultivated and waste land. Common throughout, except the north.

---

1. *Sonchus tenerrimus*
2. *Chondrilla juncea*
3. *Launaea nudicaulis*
4. *Reichardia picroides*
5. *Reichardia tingitana*

(b) Plant typically an annual or biennial, not woody at the base.

Reichardia tingitana A variable, hairless annual or biennial with oblong, toothed, pinnately lobed leaves, often with white pimples; basal leaves broadly winged at the base, the stem-leaves few and linear, half-clasping the stem at the base. Flower-heads yellow, to 25 mm across, the *ray florets purplish at the base*, the outermost with a red stripe on the reverse; flower bracts to 15 mm long, with a membranous margin and hairless. Pappus *long, to 14 mm*. Hot, dry areas. CY, IP. *R. intermedia* is similar to *R. picroides* in habit and overall appearance, but an annual, and with *flower-bracts with membranous margins* to 1.25 mm wide, more similar to *R. tingitana*. AA, CR, CY, IP, LB, TR.

Andryala

Flower-heads many in a cluster; bracts in a single row with some additional bracts; only ray florets present. Pappus of soft hairs.

Andryala integrifolia A variable, short to tall, white-hairy annual or perennial with sparingly branched, erect, leafy stems to 50 cm (1 m). Leaves oval-lance-shaped, unlobed to lobed, semi-clasping the stem above, and densely covered in yellowish glandular hairs. *Flower-heads pale lemon yellow* (often with a darker centre), 20 mm across, borne abundantly in rather flat-topped clusters; flower bracts linear-lance-shaped, and hairy; involucre 6–10 x 5–9 mm. Achenes 1 mm. Locally common in dry, sandy and grassy habitats. AA (central), CR (central, south – rare), GR (north – rare), PN (east – rare), TR (southwest).

Taraxacum DANDELIONS

Perennial herbs with tap-roots, a basal rosette of leaves, and flower-heads borne on leafless, hollow stems that exude a milky latex. Achenes finely ribbed; pappus of white simple hairs. Only one cosmopolitan species section is described here, but *Taraxacum* in its wider sense comprises a complex aggregate of numerous microspecies, the extent of which is unclear in the region.

Taraxacum sect. Ruderalia (syn. *T. officinale*) COMMON DANDELION A variable (group of) perennial(s) with leaves in a basal rosette and leafless, hollow stems. Leaves deeply pinnately lobed, with long, winged stalks at the base. Flower-heads bright yellow, the rays striped with purple beneath, on stalks to 40 cm; outer flower bracts 9–16 mm, backwardly curved. Fruiting heads borne as a familar 'dandelion clock'; achenes brown, 2.5–4 mm. Common in grassy and waste places across the region. Throughout.

Crepis HAWK'S-BEARD

Annual or perennial herbs with spirally arranged leaves with lobes pointing backwards and erect, branched stems. Flower-heads yellow with florets all rayed; flower bracts in 2 rows, the outer row often shorter and spreading. Achenes ribbed and flattened; pappus of white, brittle hairs. Numerous, similar species in the region.

(a) Flower-heads yellow. Achene without a pominent beak; pappus white and silky.

Crepis biennis ROUGH HAWK'S-BEARD A rough-hairy biennial with stems to 1.2 m, often purple below. Basal leaves irregularly pinnately lobed with a large terminal lobe; stem-leaves similar but *semi-clasping the stem and without markedly arrow-shaped bases*. Flower-heads to 30 mm, borne in loose clusters (crowded at first). Pappus equalling or slightly exceeding the involucre, and white and soft; achenes 4–7.5 mm. GR, IA.

Crepis sancta An annual to 40 cm with sparsely bristly leaves in a basal rosette, lyre-shaped or saw-toothed to simply toothed. Flower-heads yellow, 1–few; involucre 10 mm; *receptacle with long, white hairs* subtending the florets. Achenes 6 mm, of 3 kinds, the outermost spindle-shaped and rather spiny, *winged*, the innermost *slender* and *smooth*; pappus shorter than the achene, with fine, soft white hairs. Bare habitats. Throughout, except many smaller islands.

Crepis fraasii A short-lived *perennial* to 45 cm, sparingly branched. Leaves borne in rosettes, lyre-shaped and lobed with a large broadly oval end segment. Flower-heads yellow; involucre *blackish*, 7–10 mm, glandular-hairy. Achenes more or less of 1 kind, small, 3–3.5 mm, without a clear beak; pappus with thin, white hairs. Probably throughout.

Crepis zacintha An almost hairless annual to 25 cm with few, rather divergent branches. Basal leaves in a rosette, lyre-shaped to saw-toothed; stem-leaves few, the uppermost bract-like. Flower-heads yellow, borne *stalkless in the branch axils* (as well as laterally and terminally); involucre 4–5 mm with lower bracts thickened, incurved and persistent in fruit. Achenes of 2 kinds, the outer 2–2.5 mm, compressed and triangular, the inner 2.5 mm, conical, smooth. Probably throughout.

(b) **Flower-heads yellow. At least some achenes beaked, often ribbed.**

Crepis vesicaria BEAKED HAWK'S-BEARD A robust, *hairy perennial* with leaves with broad lobes, downy all over. Flower-heads to 15 mm across, with *orange–yellow* florets, the outer striped reddish externally; outer involucral bracts spreading; heads erect in bud. Achenes with *long, slender beaks*, 5–9 mm; fruiting pappus much exceeding the involucre. Meadows and waste places. Throughout.

Crepis foetida A very variable, erect to spreading, *strong-smelling* annual or biennial to 50 cm with few branches. Leaves bristly, with saw-like divisions or 1–2-pinnately lobed. Flower-heads drooping in bud, 1–many, yellow, the ray florets purplish on the outer surface; involucre 7–16 mm long. Achenes of 2 kinds, the outermost robust, beaked or not, the inner *longer and slender*, (10)12–17 mm with a slender beak; pappus protruding from the fruiting head. Common. Throughout.

Crepis neglecta A slender annual to 30 cm with basal leaves in a rosette, lyre-shaped, saw-toothed or simply toothed; stem-leaves few, clasping, the uppermost bract-like. Flower-heads yellow, nodding in bud; involucre 4–7 mm. Achenes spindle-shaped, 10-ribbed with a variably stout beak; pappus as long as the achene. Throughout; rare in the north.

1. *Andryala integrifolia*
2. *Crepis sancta*
3. *Crepis incana*
4. *Crepis rubra*

Crepis setosa An erect annual to 70 cm, sparingly branched. Basal leaves saw-toothed; stem-leaves clasping, lance-shaped and deeply lobed. Flower-heads yellow; involucre 8–10 mm, *conspicuously long-bristly*. Achenes 5 mm including the slender beak, which equals the body of the achene. GR, IA, PN.

(c) Flower-heads pink (rarely white).

Crepis incana A small perennial to 15 cm with a rather woody stock and rigid stems. Leaves lobed, grey–green and hairy. Flower-heads pale to dark *pink*; bracts often with blackish bristles along the mid-veins. Achenes short, 4–6 mm with a white pappus. Rare, in mountains. GR (Sterea Ellas + Euboea), PN.

Crepis rubra PINK HAWK'S-BEARD An annual to 35 cm readily distinguished by its solitary or few *pink or whitish-pink flower-heads*. Achenes 10-ribbed, the outer short-beaked, the inner with long, slender beaks. More or less throughout; absent from CY, the southeast, and many smaller islands.

Hieracium HAWKWEED

Perennials with or without a basal rosette of leaves during flowering; stems often with leaves. Flower-heads yellow; involucre with bracts in several rows. Achenes not flattened or beaked. Often in mountain habitats. Numerous microspecies described (especially north of the region covered); only one, reasonably widespread species is included here.

Hieracium bracteolatum A thick-rooted perennial, often slightly woody-based, to 70 cm, hairy below, hairless above. Basal leaves broadest above the middle, shallowly lobed; upper leaves smaller and lance-shaped. Flower-heads lemon-yellow, borne sparsely on long, straight branches. Achenes pale brown, 10-ribbed; pappus 4–5 mm with hairs with short, hooked bristles. GR, IA, PN.

Arctotheca

Short, leafy annuals. Flower-heads yellow with long-spreading rays. Pappus of scales. Native to South Africa.

Arctotheca calendula An annual with spreading, white-woolly, leafy stems to 40 cm long. Leaves 70 mm–20 cm, deeply lobed, roughly hairy above and white-softly hairy beneath. Flower-heads to 50 mm across, borne on long stalks; outer involucral bracts with a membranous margin, and often with a terminal appendage; inner bracts membranous; rays long, deep yellow and pale yellow at the tip, purplish beneath; disk florets blackish green. Sandy waste ground; naturalised. CR (northwest –rare), PN (south).

SUBFAMILY CARDUOIDEAE

Plants *without* white latex, often *spiny* (thistle-like), and stem-leaves spirally alternate (or absent). Flower-heads discoid (with tubular flowers only); the outer flowers often longer with larger lobes (appearing ray-like), often *red to blue or white* (rarely yellow).

Lamyropsis

Spiny perennials with alternate, often leather leaves. Flower-heads purple; involucre with overlapping bracts. Achenes long, compressed, leathery with a distinct, raised margin surrounding a cylindrical central projection.

Lamyropsis cynaroides A thistle-like perennial to 50 cm with sparsely branched woolly-hairy stems. Leaves pinnately divided into deep lobes, almost hairless above except for the veins, densely white-downy beneath with narrowly triangular lobes tapering into spines. Flower-heads purplish-pink, 1–few; involucre 27–35 mm, almost spherical with narrowly lance-shaped bracts, some *tapering into stout deflexed spines*. Achenes 5–6 mm, smooth with an apical collar surrounding a stalked projection; pappus with fluffy hairs, (15)17–21 mm. Field margins and wasteland. CR (very common), PN, TR (southwest).

Cardopatium

Perennial, very spiny herbs, rather similar to *Eryngium* in general appearance. Leaves alternate, lobed. Flower-heads numerous, few-flowered, almost stalkless in clusters; bracts in several rows, spiny; corolla deeply 5-lobed. Achenes hairy; pappus with 5–8 scales.

Cardopatium corymbosum A conspicuous, *very spiny*, erect, perennial to 25 cm, much-branched above, about as long as high. Leaves 70 mm–35 cm, hairless, with spiny lobes, borne in rosettes, often flattish on bare ground. Flower-heads *bright blue*, *small*, just 5–10 mm across, with few (7–10) florets; corolla 10 mm; bracts 12–18 mm, spiny; heads *numerous in dome-shaped inflorescences*. Achene 3 mm with an equally long pappus. Rocky slopes, dry grassland and garrigue, and stony bare ground, often near the sea. Throughout, except the far west and southeast.

---

1. *Lamyropsis cynaroides*
2. *Cardopatium corymbosum* (immature specimen)

Echinops

Thistle-like perennials with deeply pinnately lobed leaves. Flower-heads spiny, *spherical* and >25 mm across. Pappus of scale-like bristles.

(a) Inner involucral bracts *fused* at least 1/3 from the base.

Echinops graecus A densely stemmed, very spiny perennial to 50 cm. Leaves dense, leathery, (1)2-pinnately lobed with winged stalks, grey-downy beneath; segments linear-lance-shaped with sparp spines. Flower-heads *deep blue*, solitary, terminal, spherical, 30–60 mm across; involucre with inner bracts *fused* along the basal 1/3. AA, GR.

(b) Inner involucral bracts *free* to the base. Leaves rather weakly spiny.

Echinops microcephalus A bushy perennial to 50 cm with many, ridged stems. Leaves rather leathery, lance-shaped, toothed to lobed, grey beneath and *weakly spiny* (compared with other *Echinops* spp.). Flower-heads *small*, 20–35 mm across, bright blue. AA, GR, PN, TR.

(c) Inner involucral bracts *free* to the base. Leaves conspicuously and robustly spiny.

Echinops sphaerocephalus PALE GLOBE THISTLE A stiffly erect thistle-like perennial 40–90 cm (2.5 m) with deeply pinnately lobed, spine-toothed leaves, which are *green and slightly glandular above* and white-cottony beneath; bases winged and clasping. Flower-heads spherical, terminal and *pale grey to whitish-blue*, 35–60 mm across; stamens with blue filaments. Achenes 6–10 mm. GR, TR. *E. ritro* GLOBE THISTLE A stiff, spiny, thistle-like perennial, similar to *E. sphaerocephalus* but with leaves shiny-green above (white-cottony beneath) and flower-heads 35–45 mm across, *bright blue*. Rocky, uncultivated ground. GR, PN.

Echinops spinosissimus A perennial similar to *E. sphaerocephalus* and *E. ritro* but often with rather widely spaced leaves; leaves densely woolly and glandular and 2-pinnately lobed, *dark green* above with *strong marginal spines*. Flower-heads 30–50 mm across, *whitish* to pale grey-blue *with sparse prominent spines protruding*; inner flower bracts fused to form a membranous tube. Stony pastures and fallow land. AA, CR, CY, TR.

Echinops viscosus A stiffly erect, spiny perennial to 80 cm with white or purplish, ridged stems. Leaves alternate, oblong to lance-shaped in outline, 2–3-pinnately lobed. Flower-heads rather large, to 10 cm across, white, blue or purple; anthers bluish-grey. Achenes with long straight, addressed hairs. Damp, shady habitats; rare. IP (north, especially Hula Plain).

Carlina CARLINE THISTLE

Spiny, thistle-like annuals or biennials with pinnately lobed leaves. *Flower-heads large, without ray florets, but bracts conspicuous and spreading.* Pappus of feathery hairs.

(a) Plant a greyish or whitish, *mound-forming shrub*.

Carlina tragacanthifolia A rigid, spiny, densely branched, greyish or whitish, mound-forming *shrub* to 60 cm across. Leaves 1-pinnately divided with narrowly winged stalks and linear-lance-shaped segments tapering into sharp spines. Flower-heads deep yellow, terminal and solitary; involucre 12–20 mm across with leafy outer bracts much exceeding the heads, tapering into spines; inner bracts brownish-yellow. AA (southeast; common on KP), TR (southwest: Marmaris peninsula).

1. *Echinops ritro*
2. *Echinops spinosissimus*
3. *Echinops viscosus*
4. *Carlina tragacanthifolia* PHOTO: ELEFTHERIOS DARIOTIS
5. *Carlina gummifera*
6. *Carlina corymbosa*
7. *Carlina corymbosa* subsp. *graeca*
8. *Carlina racemosa*
9. *Carlina lanata*

(b) Plant virtually *stalkless*. Flower-heads *large*, (35)60 mm–14 cm across, normally *solitary in a central leaf rosette*; yellowish, whitish or pink-purple.

Carlina gummifera (syn. *Atractylis gummifera*) A late-flowering, thistle-like, low, robust, *stemless* perennial with leaves 15–40 cm in a lax rosette. Leaves virtually hairless, oblong in outline but deeply divided into narrow, spiny segments; stalks clasping at the base. Flower-heads large, 35–70 mm across and pink–purple without ray florets, with an involucre with outer bracts bearing 3 apical spines; inner bracts with a single brown spine. Achenes 10–12 mm. Local in dry fields, roadsides and fallow land. Throughout, except the far east of the area (not CY).

Carlina acanthifolia A distinctive, stemless perennial to 10 cm with all leaves clustered into a basal rosette to 30 cm across. Leaves pinnately divided just ›½-way to the midrib into spine-toothed segments, hairy-felted beneath. *Flower-heads large, usually ›70 mm across, yellowish and solitary*; ray florets absent. Stony pastures. GR (north). *C. acaulis* is similar but with *flower-heads whitish*, owing to bright silvery-white spreading hairs on the ray-like bracts, with dull purplish florets in the central disk. GR (north).

(c) Plant a rigid, *sparsely branched annual*. Flower-heads 15–50 mm across, normally in *clusters*. Species rather similar.

Carlina corymbosa FLAT-TOPPED CARLINE THISTLE A variable, pale, whitish or bluish short to tall, erect, spiny perennial; main stems usually solitary or few. Leaves linear-lance-shaped, toothed and pinnately lobed, wavy and spiny-margined; clasping the stem. *Flower-heads stalked*, borne terminally in flat-topped clusters, to 20(50) mm across with yellow to straw-coloured florets and *dark golden-yellow* to beige ray-like inner bracts. Common and widespread on dry stony slopes, sea-cliffs and garrigue. Throughout, except CY and the southeast. Several subspecies have been described that are not easily distinguishable. The main forms in the region are: **Subsp. *corymbosa*** outer involucral bracts scarcely exceed the inner. GR (mainly northern and central). **Subsp. *graeca*** outer involucral bracts markedly exceed the inner (by 15–60 mm); leaves and bracts spiny-toothed to lobed, the margin between the main spines again *finely spine-toothed*. Throughout the range of the species in the area covered. **Subsp. *curetum*** similar to subsp. *graeca* but with the margin between the main spines *entire*. Mountains. CR (+ KP). *C. sitiensis* is scarcely distinct from the latter subspecies (possibly of hybrid origin), but with outer bracts with very long terminal spines, much exceeding the heads, pale red to silvery above, purplish beneath. CR (east + Kasos). *C. barnebyana* is similar to *C. corymbosa*, but a shorter, more branched perennial, and flower-heads with rather short, *reddish to purple–brown* (not yellow) inner bracts. CR (east + KP). *C. pygmaea* is similar and closely related, but with inner bracts rather longer and spreading, *bright pink–purple*, whitish at the base. CY.

Carlina racemosa A stiff, sparingly branched annual, similar to *C. corymbosa* but with stems branched from the base and larger, unstalked flower-heads to 15 mm, *distinctly overtopped* by 1–2 surrounding flowering branches that arise from immediately below. Dry and arid pastures. IP. *C. vulgaris* CARLINE THISTLE is also similar to *C. corymbosa* but with pale sivery-yellow ray-like bracts. GR (north).

Carlina lanata An erect, spiny annual with branched or unbranched flowering stems to 39(50) cm. Leaves white-woolly and lance-shaped in outline, cut into spiny lobes, the uppermost clasping the stem. Flower-heads with *reddish-purple ray-like bracts*, contrasting with the yellow disk florets; flower-heads exceeded by the subtending leaves. Garrigue and rocky slopes. Throughout.

1. *Atractylis cancellata*
2. *Carduus pycnocephalus*
3. *Carduus nutans*

Atractylis

Similar to *Carlina*: inflorescence with leafy bracts deeply cut into spiny teeth, the innermost bracts papery-tipped, not brightly coloured. Pappus silvery-haired.

Atractylis cancellata A slender, thistle-like perennial with leaves in a lax rosette and with stems from 30 mm to 30 cm. Leaves hairy, oblong in outline, toothed and shortly spine. Flower-heads large, pinkish-purple without ray florets, with an involucre to 20 mm, surrounded by upper leaves forming a *cage-like structure*; middle bracts with all spines similar. Common in dry fields, roadsides and fallow land. Throughout.

Carduus THISTLE

Annuals to biennials with spiny-winged stems (at least in part), and alternate, spine-toothed leaves. Flower-heads rounded or cylindrical, often shaving brush-shaped; ray florets absent. *Pappus with many rows of simple hairs.*

(a) Flower-heads ≤14 mm across (excluding flowers) and oblong-cylindrical.

Carduus pycnocephalus A variable (with regional subspecies described), erect, narrowly branched biennial to 1 m with stems winged up to the flower-heads, similar to *C. tenuiflorus* but with leaves more densely white-cottony and stems not leafy but with interrupted spiny wings below the flower-heads; *flower-heads solitary or in clusters of 1–3*; 14–20 mm long and 7–12(14) mm across. Disturbed habitats. Widespread and common. Throughout.

(b) Flower-heads large, >14 mm across (excluding flowers) and spherical.

Carduus nutans NODDING THISTLE A variable (with numerous subspecies described), erect, robust, biennial thistle to 1 m with cottony, spiny-winged stems branched above and *not winged, and sparsely spiny to spine-free below the flower-heads*. Leaves pinnately lobed into spine-tipped segments, cottony below. *Flower-heads large* and nodding, usually solitary, 16–30 mm long and 20–60 mm across; florets bright red–purple, surrounded by robust, broad, deflexed spine-tipped bracts. Thickets and waste places, usually on higher ground. AA (north), GR.

Carduus argentatus An annual 40–70 cm with erect, green, *spiny-winged stems*; wings narrow, to *2 mm* (including spines). Leaves pinnately lobed below, simple but spiny-toothed above. Flower-heads solitary, pinkish-purple; involucre 14–20 mm, 2 x as long as broad. Scattered throughout except IA; absent from many islands. ***C. candicans*** is very similar but with more broadly winged stems (5 mm) and *broadly bell-shaped* involucre. Mainly northern. GR, TR.

Cirsium THISTLE

Biennials and perennials similar to *Carduus* with or without spiny wings. Flower-heads purple (or yellow). Pappus of many rows of *branched, feathery hairs* (not rough, unbranched hairs).

(a) Flower-heads typically purple or reddish.

Cirsium vulgare SPEAR THISTLE An erect biennial to 1.5 m with cottony stems with *interrupted spiny wings*. Basal leaves to 30 cm long, deeply pinnately lobed with segments forked and spiny, and a *single long, pointed end-lobe*; upper leaves smaller, all leaves prickly. Flower-heads ovoid, 25–40 mm long and 20–50 mm across, with cottony bracts; the outer bracts with long spine-tips; florets reddish-purple. Disturbed and cultivated habitats. Throughout, except hot, dry areas.

Cirsium arvense CREEPING THISTLE A creeping perennial herb with leafy, mostly *unwinged stems* to 1.2 m and leaves not in a distinct basal rosette; leaves oblong and deeply divided with triangular, wavy, spiny lobes; upper leaves similar but clasping, sometimes forming a short wing along the stem. Flower-heads narrowly cylindrical, 10–22 mm long and 8–20 mm across, *pale pink–purple*, borne in loose clusters; *involucral bracts purple*. Grassland, usually not coastal. GR (north).

Cirsium hypopsilum A biennial 40 cm–1 m with much-branched stems. Leaves hairy beneath, with linear to triangular to linear lance-shaped lobes with rigid apical spines (5)10–20 mm. Flower-heads borne numerously in panicles with subtending leaves; involucre 20–27 mm; corolla 18–25 mm, *whitish-pink*. Mountains. GR, PN.

Cirsium creticum A slender, erect perennial to 1 (1.5)m with conspicuously *spiny-winged* stems. Leaves almost hairless, lance-shaped, lobed, grey-hairy beneath with sharp marginal spines. Flower-heads numerous, erect, pink–purple; involucre 15 mm, broadly cylindrical. Throughout.

1. *Cirsium vulgare*
2. *Notobasis syriaca*
3. *Galactites tomentosus*

Cirsium italicum An erect biennial 30–80 cm, branched above, the stem-leaves with *bases extending* down the stems over half the internodes (decurrent); leaves remotely linear-lobed and with sharp spines and bristles on the upper surface. Flower-heads exceeded by their subtending leaves; involucre 15 mm, broadly cylindrical to ovoid; bracts linear-lance-shaped, tapering into spines. Damp pastures and grassy habitats. GR, IA, TR.

(b) Flower-heads typically whitish.

Cirsium candelabrum A tall, hairless biennial 2.5–3 m with much-branched stems. Leaves leathery with undulate margins, lance-shaped to oblong with triangular, spiny segments. Flower-heads white or yellowish in clusters of 4–12; corolla 13–17 mm; bracts adpressed. Achenes 2.5–5 mm; pappus 13–16 mm. Grassy and stony ground. Mainly on mainland Greece. AA (north and west), GR, PN.

Notobasis

Similar to *Carduus* but with *flower-heads encircled by tough, spiny upper leaves.*

Notobasis syriaca SYRIAN THISTLE A thistle-like annual 20 cm–2 m with rigid stems *not* spiny-winged. Leaves alternate, dark green with paler veins; narrowly elliptic, pinnately lobed with spine-tipped, narrow triangular lobes, reduced and *clustered* around the stalkless flower-heads above. Flower-heads solitary, purple; bracts with spine-tips; involucre 18–23 mm long. Achenes 67 mm long. Field margins, fallow land and roadsides; common, sometimes abundant. Throughout.

Galactites GALACTITES

Thistle-like perennials with deeply dissected leaves. Flower-heads without ray florets, but with disk florets much spreading at the margins. Pappus long and feathery.

Galactites tomentosus GALACTITES A stiffly erect perennial 30–50 cm with conspicuously *white-veined and variegated* dark green leaves; alternate, oblong, pinnately lobed with spiny lobes, white-downy beneath. Flower-heads pale purple, borne solitary or in clusters; the outer *ray florets long, even and spreading*; flower bracts tapered abruptly into spine tips; white-downy; 12–18 mm long. Achenes 3.5–5 mm long. Common in a range of dry and disturbed habitats. AA, CR, GR, IA, PN.

Onopordum SCOTCH THISTLE

Stout perennials with *spiny-winged stems*, often with cobweb-like hairs. Leaves spiny-margined and toothed to lobed. Flower-heads large and purple or white; all florets tubular and deeply 5-lobed. Pappus of many rows of simple hairs.

(a) Plant *green* (not white or grey-hairy) and often sparsely *glandular* hairy.

Onopordum cyprium An erect biennial to perennial to 1.2 m with unbranched or sparsely branched stems with narrow, spiny-lobed wings. Leaves *green, almost hairless above*, glandular-hairy below, 20–40 cm long, deeply lobed. Flower-heads purple (rarely whitish), with an almost spherical involucre 25–35 mm across with closely overlapping bracts 20–35 mm long, abruptly narrowed into rigid, purplish spine tips, often sharply recurved. Achenes 4–5 mm, oblong, 4-angled, hairless; pappus brownish, 11 mm long. An island endemic of roadsides, waste and fallow land. CY.

Onopordum tauricum An erect, robust, *bright to dark green*, often slightly glandular biennial to 2 m, sparingly branched above with broad-winged stems. Leaves stalkless, dark green with spiny-toothed lobes. Flower-heads pinkish-purple, 30–50 mm across; bracts lance-shaped, ending in strong, spreading spines, the lowermost deflexed. AA, CR, GR (local), PN (local), SY.

(b) Plant with (variably dense) *white or grey* **woolly hairs throughout.**

Onopordum illyricum A stout perennial to 1.5 m with spiny-winged stems and grey or white-felted leaves, similar to *O. bracteatum*, with winged stems and remotely lobed leaves, and rather large, purplish flower-heads 40–50(60) mm across; all bracts *shorter* than the florets. AA (not Cylcades), CR, GR, IA, PN. *O. rhodense* is similar but slender, greenish with sparser hairs, and bracts narrowly triangular-lance-shaped, tapering into *long spines* exceeding the florets. Pappus purplish. AA (Rhodes). *O. laconicum* is similar to *O. Illyricum* but very tall, to 2.5 m, slender, sparsely hairy and with leaves conspicuously net-veined beneath. PN (+ Kithera).

Onopordum acanthium SCOTCH THISTLE A large, spiny biennial to 2.5 m with broadly winged stems. Leaves oblong in outline, and pinnately lobed with broadly triangular, spiny-toothed segments, grey–white with cottony hairs. Flower-heads solitary, 20–60 mm across with tubular pink–purple florets 20 mm long; ray florets absent; bracts sepal-like, ending in yellow spines, woolly hairy at the base. Dry hillsides in the north. GR.

Onopordum bracteatum A variable, erect, robust biennial to 2 m, more or less densely grey to white-downy throughout, sparsely branched above; stems with very spiny wings. Leaves pinnately divided into triangular, spiny-toothed lobes. Flower-heads pinkish-purple, *large*, 50–70 mm across, hairless; bracts broadly lance-shaped with smooth margins, ending in spines, the outermost deflexed, the innermost equalling the florets. Grassland and open woods. AA (east), CR (+ KP), CY, GR, PN, TR (northwest).

Cynara ARTICHOKE

Stout perennials with leaves in a basal rosette or alternate; deeply divided into spiny segments. Flower-heads borne solitary or sparingly; purplish, blue or white. Pappus of many rows of feathery hairs.

(a) Leaves uniformly grey–green or green.

Cynara cardunculus CARDOON A large, bushy, *greyish or whitish* perennial to 1.8 m with numerous basal leaves; stems white-hairy and unwinged. Leaves thick, lance-shaped in outline, deeply 1–2-pinnately lobed into *lance-shaped* (not narrow and linear), *flat* segments; toothed, shortly and sparsely hairy and green above, white-hairy beneath; lower leaves short-stalked and upper leaves stalkless; leaves with *long spines to 35 mm clustered* at the base of each segment. Flower-heads large, 35–95 mm with bracts in 5–8 rows, narrowed into spreading spine-tips, basally brown; florets violet–purple. Fallow habitats. GR (west), IA, PN. *C. scolymus* GLOBE ARTICHOKE is similar to *C. cardunculus* but with broad, oval, *blunt* (sometimes notched but *not spine-tipped*) involucral bracts. Of unknown origin but cultivated throughout.

Cynara syriaca A thistle-like perennial to 70 cm with leaves rather sparse along the stems. Leaves pinnately divided into 9–11 segments, *densely glandular* beneath and with a *basal cluster of long spines*, the longest ›6 mm. Flower-heads blue–purple; bracts in 6–8 series, stout, triangular and broad-based, strongly purple-flushed; middle bracts *deflexed, abruptly* terminating in spine-points. IP (north), LB, SY (south). *C. auranitica* is similar but taller, to 2 m, with rough leaves and bracts with a broad *dark brown band* at the base. LB, SY, TR (southeast).

Cynara cyrenaica A thistle-like perennial to 80 cm with mostly basal leaves. Leaves uniformly green, 1–2-pinnately divided into 19–33 segments with spiny-toothed and *down-turned* margins. Flower-heads bluish-purple; involucre 32–36 mm; bracts in 5–7 rows, stout, spreading; middle bracts *long-tapering* into spine tips. A rare, mainly Libyan species, extending into Crete. CR (east).

1. *Onopordum cyprium*
2. *Onopordum illyricum*
3. *Onopordum acanthium*
4. *Cynara cardunculus*
5. *Cynara cornigera*
6. *Silybum marianum*
7. *Cynara syriaca*

PHOTO: BANAN AL SHEIKH

(b) Leaves variegated.

Cynara cornigera A low, thistle-like perennial to 30 cm with mostly basal leaves. Leaves *variegated, rather leathery*, 1–2-pinnately divided into 19–28 segments with spiny-toothed margins. Flower-heads large, *whitish or cream*, usually solitary with an involucre 50 mm long; bracts in 5–8 rows, very stout, long, *spine-like and spreading*; middle bracts abruptly contracted at the tips. Rocky slopes and maquis. Strongly southern. AA (south), CR, CY, PN (south – rare).

Silybum MILK THISTLE

Robust annuals or perennials with spineless stems. Flower-heads with *filaments fused at the base to form a tube*. Pappus with many rows of simple hairs.

Silybum marianum MILK THISTLE A robust, weakly spiny biennial to 1 m. Basal leaves oblong, pinnately lobed and prominently *white-veined and variegated*, virtually hairless and stalked beneath; stem-leaves smaller, with fewer lobes and clasping the stem. Flower-heads purple, 25–40 x 50 mm (14 cm), borne solitary and terminally; bracts terminating in long, stout spines to 7 mm. Common and widespread on fallow, cultivated, waste ground and field margins. Throughout.

Centaurea CORNFLOWER, KNAPWEED

A large genus of summer-flowering annual or perennial herbs with alternate simple, entire to pinnate leaves. Flower-heads purple, pink, white or yellow; bracts with an apical appendage (bract characteristics important for identification). Pappus absent (or simple to toothed hairs or scales). A very large and complex genus (possibly comprising several genera) of which only a small subset are described.

(a) Flower-heads yellow (or cream); bracts (usually) ending with *conspicuous terminal spine(s)* and usually with smaller lateral spines.

Centaurea solstitialis ST BARNABY'S THISTLE A variable, white-hairy, much-branched annual with *winged stems* to 60 cm, *grey woolly-hairy or felted*. Lower leaves pinnately lobed or toothed, often *withered* during flowering; upper leaves linear-lance-shaped and entire with spine-tips, the base decurrent down the stems to form a wing. Flower-heads usually solitary and *yellow*; involucre spherical to ovoid, to 12 mm across; bracts tipped with *stout, apical yellow spines 10–15 mm, with shorter lateral spines palmately arranged at the base*. Fallow ground and waste places, common. Throughout. *C. idaea* is similar but a woody-based perennial with leaves persistent during flowering. CR. *C. melitensis* is similar to *C. solstitialis* but with bracts terminating in shorter spines 5–8 mm long and pappus equalling (not exceeding) the achene. Rare and local. CR (rare), AA (south). *C. hyalolepis* is a bushy, branched biennial with flower-heads like those of *C. solstitialis* but with stems *not* winged and with spines of the involucre more stout and rigid. Flower-heads yellow (sometimes whitish or pink in the east). Achenes 2.5–3 mm, with an equal pappus. Fallow land in the south and east. CY, GR (south), IP, PN, SY.

Centaurea acicularis A virtually stemless, woody-based perennial. Leaves lyre-shaped to 2-pinnately divided, downy beneath. Flower-heads bright yellow, *borne in clusters in the leaf rosettes*; involucre ovoid, 20–25 mm; bracts with long terminal spines and 2–4 lateral pairs of smaller spines; inner bracts with papery appendages. Achenes silky-hairy with a long, white pappus. AA (east), TR.

1. *Centaurea solstitialis*
2. *Centaurea hyalolepis*
3. *Centaurea acicularis* PHOTO: ARNE STRID
4. *Centaurea aegyptiaca*
5. *Centaurea calcitrapa*
6. *Centaurea pumilio*

Centaurea aegyptiaca A short, stout, *white woolly-hairy*, short-lived perennial to 25 cm, branched from the base. Leaves linear-oblong, dissected into oblong or linear, toothed to lobed segments; uppermost stalkless and clasping at the base. Flower-heads subtended by small leaves; involucre ovoid, 15 mm; bracts with stout brownish spines 4–5 x as long as the bracts (longest to 20 mm), with smaller spines diverging from the base; florets creamy-yellow and purple, 2 x as long as the involucre. Achene 3 mm, sparsely hairy and with a pappus of bristles. Deserts. IP (south), SY.

Centaurea salonitana A variable, erect, slightly woody-based perennial to 70 cm with angled stems. Leaves green, lobed. Flower-heads usually solitary, *lemon yellow*; involucre 25–30 cm; bracts broadly oval, greenish, variably spiny – usually with lateral slender teeth and a short to long terminal spine. Achenes 4–5 mm with a pappus 2 x as long. Common in parts of mainland Greece; absent or local elsewhere AA (Rhodes), GR, TR.

Centaurea benedicta (syn. *Cnicus benedictus*) A leafy, *thistle-like* annual to 20 cm with rather *soft*, hairy, toothed to lobed leaves with weak lateral spines. Flower-heads small, yellow and solitary, subtended by large leaves; involucre ovoid, 15–20 mm; bracts rigid, reddish-purple, spreading and spine-like with paired lateral smaller spines. Achenes 7–8 mm, pale brown and shiny with 20 ribs and pappus of 2 rows of bristles (distinct from other *Centaurea* species, and still often referred to by its synonym). Throughout, except IA.

(b) Flower-heads yellow; bracts with or without lateral spines and *without* a conspicuous, long terminal spine.

Centaurea ptosimopappa A *shrub to 2 m* with erect to ascending stems with few branches. Leaves rigid, leathery and with rather conspicuous lateral veins, hairless on both surfaces except the margins (which are woolly), lance-shaped and stalked below; stalkless above, the uppermost *enveloping* the flower-heads. Flower-heads yellow, without spreading florets; involucre 15–30 mm, narrowly ovoid; bracts *adpressed, hairless*, with minute terminal spines <1 mm. Achenes 4–5 mm, egg-shaped, with a pappus 5–9 mm. Rare. SY, TR. *C. ptosimopappoides* is very similar but a smaller shrub, just 20–65 cm, and with an involucre 14–25 mm. Achenes 5.9 mm, rectangular, with a pappus 6–11 mm. Rare. TR.

Centaurea argentea A *bushy perennial* to 40 cm with a much-branched, woody stock. Leaves *densely white-downy* when young, grey when mature, lyre-shaped to pinnately lobed. Flower-heads 1–few in clusters, pale lemon yellow, cream or whitish, sometimes tipped pink; involucre 10–15 mm; bracts with pale to dark brown apical borders with 8 teeth on each side. Achenes 3 mm with a slightly longer, whitish pappus. Three subspecies described. CR, PN (island of Kithira).

(c) Flower-heads pink; bracts typically with *few to many*, prominent spines.

Centaurea calcitrapa RED STAR-THISTLE A much-branched, virtually hairless perennial 50 mm–50(80) cm with grooved stems. Leaves grey when young, glandular, remotely shallowly to deeply pinnately lobed with bristle-pointed lobes, often withered below during flowering; upper leaves smaller and narrower. Flower-heads *reddish-purple*, surrounded by conspicuous *spreading, long and star-like spines 15–30 mm, >3 x as long as the longest laterals*; involucre ovoid-cylindrical, to 15 mm across. Achenes 3 mm; *pappus absent*. Waste places, roadsides, bare, sandy ground and other disturbed habitats; common. Throughout. *C. iberica* is very similar but with flower-heads with broader involucre (>15 mm) and achenes *with* pappus 2–3 mm. GR, PN.

Centaurea pumilio A tuberous perennial with a sparsely branched, woody stock. Leaves borne in rosettes, irregularly pinnately lobed-lyre-shaped, grey-hairy. Flower-heads pink, short-stalked, borne in small clusters; involucre 25 mm across, broadly ovoid, middle bracts with *broad, pale, papery margins* in the upper half, with a weak spine at the tip 4–9 mm. Achenes 5 mm with an unequal whitish pappus. Very local on maritime sands. CR (west), IA (Cephalonia), IP, PN (southwest), SY. *C. aegialophila* is similar but smaller with almost hairless leaves and very short spine tips on the bracts; florets less spreading. Rare. CR (central, east + KP), CY, TR.

Centaurea atropurpurea An erect perennial 30–60 cm, sparingly branched. Leaves lyre-shaped, white-downy beneath. Flower-heads *dark reddish-brown*; involucre broadly ovoid, 25 mm across; middle bracts triangular, dark brown with dense, slender, whitish bristle teeth and a short terminal spine 2–3 mm. Achene 4 mm, short-hairy with an equal, dirty white pappus. AA (central).

Centaurea spruneri An erect, robust perennial to 80 cm with few branches. Leaves 1–2-pinnately lobed, green, rough. Flower-heads pink, solitary; involucre egg-shaped, to 30 mm with *green* bracts with marginal teeth, terminating abruptly into long, spreading to *deflexed, brownish terminal spines*. Local. AA (west), CR (centre), GR (west), IA, PN. *C. euboica* is similar but with straw-coloured bracts, which graduate into terminal spines. Very rare, only on Euboea. *C. laconica* is similar but with 2-pinnately lobed leaves with smaller segments between the main lobes and prominently ridged stems. AA (central), PN. Closely related local species on the Greek mainland, the distinction of which is unclear, are: *C. achaia*, *C. aetolica* and *C. corinthiaca*.

(d) Flower-heads pink or purple; bracts with apical spines *small or absent*, or enlarged into a conspicuous, papery terminal section.

Centaurea paxorum A clump-forming biennial with ascending stems 20–30(50) cm long, irregularly branched from the middle. Leaves rather dense, rigid, rather *succulent*, bright green and rough, ending in a short spine-tip. Flower-heads pink–purple, solitary with a spherical involucre 15–18 mm across; bracts whitish and papery almost throughout, minutely toothed, with a central triangular, brownish rib; ending in a slender spine 1–1.5 mm. Achene 3–3.5 mm, rough or almost smooth, with equal or longer pappus. A rare endemic of the Ionian Islands, usually on inaccessible sea cliffs. IA (Corfu – rare, Paxos, Antipaxos).

Centaurea lactucifolia A robust perennial with stout stems 30–70 cm, branched from the middle. Leaves borne in large basal rosettes, lobed with 3–5 pairs of oblong segments, green and almost hairless. Flower-heads pale yellow (or reddish), *very large*, involucre 40–70 mm across; *middle bracts large, papery and whitish*, almost *circular* with fine teeth, and *without* a distinct terminal spine. Achenes 7 mm, grey, with a somewhat longer, whitish pappus. AA (Rhodes + adjacent islands).

---

1. *Centaurea aegialophila*
2. *Centaurea atropurpurea* PHOTO: ARNE STRID
3. *Centaurea paxorum*
4. *Centaurea lactucifolia* PHOTO: ARNE STRID
5. *Centaurea akamantis*

Centaurea akamantis A chasmophytic, woody-based shrub with spreading stems to 70 cm. Leaves greyish to whitish-green, to 60 mm long, pinnately lobed into linear segments. Flower-heads bright pink; involucre spherical, 10 mm long, 9 mm across with downy to cobweb-hairy bracts; outer florets spreading. Achenes 3.5–4(5) mm. A critically endangered island endemic known from just 3 sites. Rocky gorges. CY (west).

Centaurea jacea BROWN-RAYED KNAPWEED A downy perennial 20–60(80) cm with slender, rough stems, sometimes thickened below the flower-heads, unbranched, or with some branches above the middle. Leaves oval-lance-shaped and unlobed or sometimes pinnately lobed. Flower-heads solitary, 15–20 mm across, with red–purple florets, the outermost larger; *bracts rather shiny, pale brown, rounded and irregularly jagged* (not regularly toothed), *without* spines. GR.

(e) Flower-heads *bluish-purple.*

Centaurea triumfettii A rhizomatous perennial to 40 cm. Leaves often withered during flowering beneath, the middle leaves extending down the stem at the base (decurrent). Flower-heads with *conspicuous bluish, spreading florets*; involucre narrowly ovoid, 15–20 mm; bracts green with a narrow dark to blackish-brown border with silvery teeth 1.5–3 mm. Achenes 4–5 mm with a short pappus. Scattered across the mainland; absent from most islands.

Cyanus CORNFLOWER

Very similar to *Centaurea* (and frequently still included the genus); *annuals*, typically with flower-heads with spreading ray florets and bracts with slender, silvery-white teeth 2–3 x as long as the width of the bract.

Cyanus segetum (syn. *Centaurea cyanus*) CORNFLOWER A slender, sparsely branched annual to 60 cm. Leaves lyre-shaped with 1–2 pairs of lateral lobes; upper leaves linear and without lobes. Flower-heads *bright blue*, with spreading ray florets, solitary, long-stalked; involucre 10–15 mm, broadly cylindrical to bell-shaped; bracts brown or blackish with pale slender teeth 1 mm long. Achenes 3–4 mm with shorter pappus. Open woods and rocky habitats; also naturalised in fields. Probably scattered throughout. *C. depressus* (syn. *Centaurea depressa*) is similar but with oblong, undivided lower leaves and bracts with longer teeth. Rare or casual in most of the range. AA (rare), CR (rare), LB, SY, TR. *C. pinardii* (syn. *Centaurea pinardii*) is similar but smaller (to 25 cm) and with achenes *without* a pappus. Scattered and absent from most islands. GR, PN, TR.

Mantisalca

Very similar to *Centaurea*. Flower-heads with leathery, yellowish-green, spine-tipped bracts. Pappus of long, bristle-like scales.

Mantisalca salmantica A knapweed-like, erect, more or less hairy perennial to 1.5 m. Leaves crowded at the base in a rosette; oval in outline and pinnately lobed; upper leaves few, linear and toothed, to 35 cm. Flower-heads mauve, borne solitary or in lax branched clusters; involucre prominent and swollen in appearance; 13–19(21) x 6–11(15) mm, with oval, black-tipped bracts, each with a short *apical spine to 1 mm, often soon-falling*. Achene 3.2–4.7 mm with a *double pappus of bristles 2–4.5 mm*. Dry, grassy habitats; absent from much of the north and west. CY, GR, IP, PN, TR.

Crupina

Slender, erect annuals, branched above. Leaves alternate, spineless. Flower-heads purple with cylindrical to ovoid involucre; florets 5-lobed. Pappus absent in outer achenes, in 2 rows in the inner achenes.

Crupina crupinastrum A very slender, knapweed-like annual with few branches, 20–70 cm. Leaves pinnately divided into linear, toothed lobes. Flower-heads reddish-purple with rather few (9–15) regularly 5-lobed florets; involucre 15–20 mm, cylindrical, greenish or purplish. Achenes 4 mm, egg-shaped, rather compressed and keeled in the lower half with a large basal scar; pappus a series of dark purple bristles 5–6 mm, the outer short, the inner exceeding the achene. Throughout (rarer in the north and west). *C. vulgaris* is similar but with *almost entire* (not deeply lobed) leaf segments and leafy up to the branches (not just below). GR, IA (rare).

Jurinea

Perennials, sometimes woody at the base. Leaves entire or lobed. Flower-heads solitary or in inflorescences; bracts linear to lance-shaped, straight or recurved; florets tubular, pink, red or purple. Achene with a papery collar around the base of the pappus; pappus hairs in several unequal rows.

(a) Short, often stemless perennials on mountain rocks.

Jurinea taygetea A short, stemless or almost stemless perennial to 40 mm with few (2–7), grey, pinnately lobed leaves 20–35 mm long. Flower-heads pink, with spherical involucre 20–35 mm; bracts lance-shaped, straight and adpressed, green, leafy, purplish at the tips and keeled at the base. Achenes 3–7 mm, ribbed, hairy, brown. Dry rocky mountains, growing directly from rock. PN. *J. cadmea* is similar in habit but with lilac–purple to *whitish* flower-heads with involucre 14–20 mm, with bracts with greenish-purple recurved tips. Mountain rocks. AA (Chios, Samos), TR.

(b) Erect, slender perennials.

Jurinea mollis A slightly woody-based, tall perennial to 70 cm. Leaves few, variable but usually pinnately lobed, white-downy beneath; upper leaves bract-like. Flower-heads solitary with a *large*, almost spherical involucre (wider than long) 30–60 mm wide; bracts purple-tipped and recurved. Achenes cone-shaped, 4-angled with a slightly longer pappus. Rocky slopes and woods. AA (east), GR, PN, TR. *J. consanguinea* is similar but smaller and more slender, with flower-heads with involucre 20–35 mm across, with *erect-spreading* (not recurved) bracts. Distribution as for previous.

Klasea

Rhizomatous perennials with spineless, entire or deeply lobed leaves (often on the same plant). Flower-heads solitary, with oval-triangular bracts ending in a spine. Pappus of simple, persistent hairs.

Klasea cretica A slender, erect perennial to 45 cm with 1–few stems, ridged or winged. Leaves almost entire to wavy-lobed, hairless, with bases extending along the stem (decurrent). Flower-heads solitary purple–mauve; involucre ovoid, 20–25 mm long; bracts oval, abruptly contracted at the tips. Achenes hairless; pappus with several rows of hairs to 13 mm. CR (east). *K. moreana* is similar but with leaves without deccurent bases, *unwinged stems* and bracts gradually tapering at the tips. PN (southeast).

Carthamus

Very spiny, often glandular *annuals*. Leaves pinnately lobed with spiny margins. Flower-heads solitary, surrounded by spiny, leaf-like bracts, yellow to orange. Pappus absent or *a series of narrow, pointed, persistent scales*.

Carthamus lanatus A very spiny, thistle-like annual 15–60 cm (1 m) with straw-coloured stems unbranched below, branched above, covered in white-woolly hairs when young. Leaves lance-shaped, pinnately lobed with a spiny margin, withered below during flowering; clasping the stem above. Flower-heads 30 mm across, *yellow*; bracts with a spine-toothed appendage. Pappus a series of scales as long as the achene; achene 4.5–6 mm. Common in bare, dry and sandy places. Throughout. **C. dentatus** is similar but with *pink* flower-heads. AA, CR, CY, GR, PN, SY, TR.

Carthamus leucocaulos A prickly, erect, rigid annual to 40 cm with whitish or purplish stems, sparsely branched from above. Stem-leaves stalkless, rigid and shiny with 2–4 pairs of narrowly triangular lobes with *robust spines*. Flower-heads very *small, whitish-pink* (to purplish) with narrowly ovoid involucre, just 10 mm across; bracts oval-lance-shaped, the outermost 2 x as long as the inner (or longer). Achenes whitish; pappus with pale brownish scales. AA, CR (+ KP), PN (northeast + Kithira).

Carduncellus

Closely related to *Carthamus* but *woody, many-stemmed perennials*. Flower-heads yellow or purple. *Pappus a series of persistent or shedding bristles.*

Carduncellus caeruleus (syn. *Carthamus caeruleus*) A greyish, hairy perennial 20–80 cm with normally unbranched, erect, unwinged stems. Leaves rather shiny, grey–green, lyre-shaped and toothed or untoothed to pinnately lobed with bristle-tips; upper leaves semi-clasping the stem. Flower-heads blue–purple, borne solitary, surrounded by leafy bracts; involucre 16–30 x 20–30 mm; florets 14–22 mm, tubular and *deeply 5-lobed*. Fallow land and roadsides, usually coastal. AA (local, southern), CR, CY (rare), GR (west), PN (Kithira, Andikithira).

SUBFAMILY GYMNARRHENOIDEAE

Distinctive winter annuals (in the region covered) with *rosettes of monocot-like leaves* and 2 types of flower-head. Very different from other Asteraceae in the region.

Gymnarrhena

Desert herbs with rosettes of linear to lance-shaped leaves and central flower-heads, either functionally male (few florets) or female (single florets); stamens 3–4. Achenes compressed; pappus reduced or absent.

Gymnarrhena micrantha A small, *grass-like* annual herb. Leaves simple, linear to lance-shaped, borne in *rosettes*. Flower-heads congested, inconspicuous, borne centrally in the leaf rosettes, or 10–15 mm underground; florets with 3–4-lobed corolla, whitish; stamens 3–4. Achenes rather large, flattened, blackish; pappus absent or short with scale-like bristles. Rocky deserts, slopes, plains and gullies. IP (south, east, central; especially the Negev), SY (Syrian Desert).

1. *Mantisalca salmantica*
2. *Crupina crupinastrum*
3. *Jurinea taygetea*   PHOTO: ARNE STRID
4. *Carthamus lanatus*
5. *Carduncellus caeruleus*
6. *Gymnarrhena micrantha*

ADOXACEAE

A small family of perennials with opposite, toothed leaves. Inflorescences cymose; stamens 4–5; style 0 or 1. Fruit a 1–several seeded drupe (rarely an achene). This family now includes species that were previously described under the Sambucaceae or Caprifoliaceae.

Viburnum

Deciduous or evergreen shrubs with simple leaves. Flowers numerous, in compound clusters; stamens 5; style 0. Fruit a succulent 1-seeded drupe.

Viburnum tinus An evergreen shrub to 7 m with more or less hairless stems, and dark green, oval, short-stalked leaves, sparsely hairy beneath. Flowers white, pinkish in bud, borne in dense, flattened heads to 90 mm across. Fruit a berry to 7 mm, *blue–black* when ripe. Maquis and mountain woods; uncommon. GR, IA, PN. *V. lantana* is deciduous with starry hairs on the leaves. Mountains. GR.

Sambucus ELDER

Deciduous shrubs or perennials with pinnately divided leaves. Flowers borne in compound clusters; stamens 5. Fruit succulent, with 3–5 seeds.

Sambucus ebulus EUROPEAN DWARF ELDER A robust, vigorous, herbaceous perennial to 1.5(2) m with erect, grooved stems; unpleasant-smelling. Leaves pinnately divided with 7–13 narrow leaflets and conspicuous oval stipules at the base. Flowers borne in flat-topped inflorescences with 3 main rays; pinkish-white, with *purple* anthers. Fruit a black berry. Streamsides and field boundaries; common in the north. Throughout, except smaller islands. *S. nigra* is similar but a small tree, 3–8 m, and flowers with *cream* anthers. Distribution as above (not CY or IP).

CAPRIFOLIACEAE | HONEYSUCKLE FAMILY

Woody shrubs and climbers with opposite, simple leaves. Flowers solitary, paired or in showy panicles on shrubs; calyx small; corolla regular or 2-lipped, fused below to form a tube; stamens mostly 5; ovary inferior. Fruit a berry or nutlet.

Abelia

Shrubs to small trees mostly native to eastern Asia. Leaves opposite (sometimes whorled). Flowers solitary or in clusters; calyx 2-, 4- or 5-parted; corolla funnel- to bell-shaped; stamens 4, of 2 sizes. Fruit a leathery achene.

Abelia × grandiflora A semi-evergreen shrub to 1.2(3) m with reddish shoots. Leaves 25–45 mm, oval with rounded bases, toothed, glossy. Flowers borne in clusters; calyx with 2–5 lobes, 7–10 mm; corolla funnel-shaped, to 20 mm, white flushed pink, slightly fragrant. Widely planted in gardens.

1. *Abelia × grandiflora*
2. *Lonicera implexa*; (inset) fruits

Lonicera HONEYSUCKLE

Deciduous or evergreen shrubs or climbers with simple, sometimes lobed leaves. Flowers stalkless, with a zygomorphic, 5-lobed corolla; stamens 5; style long. Fruit a several-seeded berry.

Lonicera implexa HONEYSUCKLE An evergreen, much-branched, shrubby and woody climber to 3 m with bluish, hairless shoots. Leaves persistent, entire, 15–40 mm long, oval and pointed, stalkless and cone-shaped on the upper parts of the stems, hairless, dark green and shiny above, and bluish beneath. Inflorescence *stemless* with 5–6(8) flowers; corolla 32–40 mm, yellowish-white tinged red, and fragrant, particularly at night; *style woolly, stamens slightly exserted*. Fruit a red berry 6–7.5 mm. Maquis, garrigue and among shrubs in thickets; absent from many islands. GR, IA, PN, TR. *L. etrusca* is similar but deciduous with leathery leaves, usually hairy beneath, and with inflorescences with clusters of flowers in groups of 3 *on stalks 30–40 mm long*. Fruit 5.2–6.2 mm, ovoid and red. Limestone woods and thickets. Throughout.

Scabiosa SCABIOUS

Annual or perennial herbs with simple or pinnately divided leaves, the lowermost often in a rosette. Flower-heads flat or domed and long-stalked; outer flowers longer than the inner; calyx with 5 long bristles; corolla 5-lobed.

Scabiosa atropurpurea MOURNFUL WIDOW A hairy, bushy annual or biennial to 70 cm with erect, branched stems. Leaves oblong, untoothed and long-stalked below, pinnately lobed above. Flower-heads lilac to reddish-purple (often mixed in populations), 15–35 mm across, becoming oblong in fruit, the outer florets 2 x the size of the centrals; involucral bracts *not* longer than the florets. Fruits with long calyx teeth. Common in a range of dry habitats, particularly on sandy, maritime waste ground. Throughout. *S. webbiana* has *cream* flower-heads and more or less spherical fruiting heads. GR (north, east + adjacent islands), TR.

CAPRIFOLIACEAE

Lomelosia

Very similar to *Scabiosa* (and previously described in the same genus) but with a *pitted epicalyx*.

(a) Erect to spreading annuals.

Lomelosia prolifera An erect, robust annual 13–30(50) cm, mostly unbranched except in the inflorescence. Basal leaves often withered during flowering, egg- to spatula-shaped, 30 mm–10 cm long, entire or remotely and irregularly toothed. Flowers borne in rigid, divergent to spreading inflorescences; flower-heads hemispherical but flattened, 20–40 mm across, creamy white to yellowish. Fruiting head conspicuous with membranous, cup-shaped corona with 30–35 veins, 10–12 mm across. Roadsides and fallow ground or garrigue. CY, IP, SY, TR (southeast).

Lomelosia porphyroneura **(syn. *Scabiosa aucheri; S. porphyroneura*)** A herbaceous, hairy *small desert* annual with erect to ascending stems to 20 cm, branched from the base. Lower leaves 50–60 mm, oblong-lance-shaped; middle leaves 45–50 mm, oblong to egg-shaped. Flower-heads *pale yellow, not radiating*, surrounded by spreading leaves; heads 15–25 mm across (20–30 mm when mature); fruiting corona with 25–28 prominent *dark purple* or reddish veins. Deserts and other dry habitats. IP (mainly eastern and southern), LB, SY.

Lomelosia divaricata **(syn. *Scabiosa sicula*)** A short, erect or sprawling, hairy annual 10–50(70) cm. Leaves spoon-shaped to oval and toothed or not, the upper pinnately divided with narrow segments. Flower-heads mauve or reddish, 20 mm in flower, 7–19 mm in fruit, becoming spherical in fruit; *involucral bracts much exceeding the florets*. Dry, stony habitats. Virtually throughout (not IP). ***L. cyprica*** is similar (also with mauve or purplish flower-heads), but with leaves usually entire. Dry hilsides. CY.

(b) Woody-based shrubs (several similar and closely related species).

Lomelosia minoana A bushy, tufted *dwarf shrub forming mounds* to 60 cm. Leaves narrowly egg-shaped, tapering into their stalks, with *dense, silky grey* hairs. Flowers borne in flat to rounded heads, 20–35 cm across; corolla whitish to pale mauve. CR (west, centre). ***L. albocincta*** has *long straight* hairs on the leaves, especially the margins. CR (west, centre). ***L. hymettia*** has 3–5-lobed leaves with densely silvery silky hairs and pinkish-mauve flowers. GR (east + adjacent islands), PN. ***L. variifolia*** is woody, even to the terminal branches, with leaves crowded at the tips, densely silky-hairy when young, almost hairless when mature. AA (KP + Rhodes and adjacent islands).

Pycnocomon

Perennials with entire to divided leaves. Flower-heads hemispherical with 7–12 bracts fused at the base; calyx with 8–14 bristles; corolla 4-lobed.

Pycnocomon rutifolium **PYCNOCOMON** Similar to *Scabiosa* species during flowering; a robust, hairy or hairless perennial to 1 m with erect, branched stems. Leaves slightly fleshy, 1–2-pinnately lobed, reduced above, sometimes unlobed below. Flower-heads *white or yellow*, 12–28 mm across, the outer florets slightly larger than the central; bracts below the flower-heads 7–8 fused together in the lower part, forming a cup. Coastal sands; rare and local. PN.

Dipsacus

Robust, prickly biennials and biennials with entire or pinnately divided leaves. Flower-heads very dense; corolla 4-lobed; stamens 4; style 1.

Dipsacus fullonum **WILD TEASEL** A tall biennial to 2 m with robust, prickly, angled stems. Basal leaves forming a large rosette; oblong-elliptic, toothed or not *and prickly*; stem-leaves in pairs and fused together on the stem forming a water-collecting cup. Flower heads 40–80 mm, pink-purple, oblong-cylindrical with long, spiny bracts at the base. Local in damp places. AA (rare), GR, IA, PN, TR.

---

1. *Scabiosa atropurpurea*
2. *Lomelosia prolifera*
3. *Lomelosia porphyroneura*
4. *Lomelosia prolifera* fruiting head

CAPRIFOLIACEAE

Knautia

Perennials with hairy stems. Flower-heads hemispherical with 1–2 rows of free bracts at the base (none below each flower); calyx with 8 long bristles; outer florets usually longer than the inner; corolla 4-lobed; stamens 4; style 1.

(a) Plant an annual.

Knautia integrifolia A hairy annual with stems to 60 cm (1 m) arising from a basal leaf rosette; leaves deeply cut into oval segments, stem-leaves divided into linear segments, or narrowly lance shaped. Flower-heads 15–32 mm across, blue–violet, almost flat with spreading outer florets; calyx usually with 12–24 inconspicuous teeth. Fruits with numerous white hairs. Common in a range of habitats. Throughout. *K. orientalis* has entire leaves (or with basal leaves with narrow segments) and heads of just 5–10 flowers, the outermost much larger than the central; corolla pink to reddish-purple. GR, PN, TR.

(b) Plant typically a perennial.

Knautia arvensis FIELD SCABIOUS A stout biennial or perennial with rough-hairy stems to 75 cm (1.2 m). Basal leaves roughly spoon-shaped, rough-hairy and undivided, stem-leaves pinnately divided into segments with an oval-lance-shaped end-segment. Flower-heads 25–40(50) mm across, blue–violet and hemispherical, borne on long hairy stalks; calyx with 8–14 teeth. Open woods and tracks, usually inland and on higher ground. GR (north).

---

1. *Knautia integrifolia*
2. *Knautia arvensis*
3. *Centranthus calcitrapa*
4. *Centranthus ruber*

Cephalaria

Annual or perennial herbs or shrubs with leafy stems and entire to deeply lobed leaves. Flowers cosexual, borne in dense rounded or hemispherical heads; corolla 4-lobed; stamens 4; style 1. Fruit a capsule.

(a) Plant a small shrub with *leathery leaves*.

Cephalaria squamiflora A *shrub* to 90 cm with oval-lance-shaped to elliptic leaves to 18 cm, *untoothed to shallowly lobed, leathery* and tapered at the base. Flower-heads 20–32 mm across, yellow–white with prominently exserted stamens; corolla 8–12 mm; bracts oval and hairy. Limestone crevices. AA, CR (western mountains).

(b) Plant a woody-based or herbaceous perennial, leaves *not leathery*.

Cephalaria leucantha A tall perennial herb to 1.3 m, woody at the base, with erect, branched stems. Leaves to 20 cm, *pinnately lobed* into narrow segments, hairy or hairless. Flower-heads 14–30 mm across, white–yellow with prominently exserted stamens; corolla 7–9(13)mm; bracts oval and hairy. Dry, stony habitats in the west. GR, IA, PN. *C. syriaca* is similar but with blue or lilac flower-heads just 3–9 mm across; corolla 6–9 mm. CY, IP.

Cephalaria transsylvanica A tall, rather hairy perennial to 1.5 m with erect, branched stems. Leaves to 20 cm, deeply pinnately lobed into elliptic to linear segments, toothed or not. Flower-heads 5–25 mm across, blue–lilac (or white); corolla 7–11 mm; bracts lance-shaped and pointed, with ciliate margins. Fields and fallow land. GR, IA, PN (north).

former VALERIANACEAE (now considered part of CAPRIFOLIACEAE)

Centranthus

Rhizomatous annual or perennial herbs with entire to deeply pinnately lobed, stalked or unstalked leaves. Flowers small, borne in dense panicles, pink or white; corolla tubular; stamen 1; stigma 1.

Centranthus calcitrapa A small, hairless annual with simple or branched stems to 30 cm. Leaves rounded to oval, green flushed with purple, toothed or not, stalked below and stalkless above. Flowers *small*, with corolla tube 1–4.6 mm, white or pink, long- and short-spurred, borne in small terminal clusters. Fruit a feathery, persistent calyx. Common on sand dunes, low garrigue and bare waste places. Throughout (rare in the north, absent in the far southeast).

Centranthus ruber RED VALERIAN A familiar, tufted, somewhat fleshy and waxy perennial to 80 cm. Leaves bluish, oval, pointed or blunt, clasping the stem above. Flowers with tube 5–12 mm, usually dark pink–red (sometimes white), spurred, borne in large, slightly fragrant, showy panicles. Fruit 1-seeded with a feathery, persistent calyx. Cultivated as an ornamental, and naturalised throughout.

Valerianella CORNSALAD

Small annuals with symmetrical branches. Flowers borne in the branch axils and in terminal clusters; stamens and stigmas 3. Numerous species, all rather similar; ripe fruits and calyx necessary for identification.

(a) Calyx minute and inconspicuous.

Valerianella locusta COMMON CORNSALAD A small, variable, hairless annual to 15(40) cm with symmetrical, spreading branches, spoon-shaped, blunt, sometimes toothed leaves to 70 mm and dense terminal heads of pale lilac flowers; corolla 5-lobed; *calyx very small and 1-toothed*. Fruit hairless, 1.8–2.5 mm long and 1–1.5 mm across (scarcely longer than thick), shallowly grooved. Locally frequent in grassy habitats and along tracks. Throughout.

(b) Calyx conspicuous and often persistent.

Valerianella eriocarpa HAIRY-FRUITED CORNSALAD A small annual to 15(40) cm similar in appearance to *V. locusta* but with denser, compact flower clusters *and distinctive calyx nearly as broad as the fruit, strongly net-veined and deeply (2)5–6-toothed*; corolla 1–1.6 mm. Fruit 0.9–2 mm, with rigid hairs. Throughout except smaller islands (not the southeast, CY or IP). ***V. dentata*** NARROW-FRUITED CORNSALAD is similar to *V. eriocarpa* but with laxer flower clusters and calyx *just ½ as broad as the fruit and scarcely veined*. Fruit 1.5–2 mm. AA, GR, IA, PN, TR.

Valerianella discoidea A small annual similar to *V. locusta* with opposite-spreading stems and oval to narrowly spatula-shaped leaves, blunt, toothed or untoothed. Flowers with corolla 1.5–2 mm, whitish-blue or mauve; *calyx well-developed, with many (6–12) spine-tipped lobes, hairy outside and in*. Fruit 1.5–2.1 mm. Fallow land, pastures and maquis clearings; common. Throughout. ***V. vesicaria*** is distinguished by its *spherical inflated, bladder-like fruits*. ***V. echinata*** SPINY CORNSALAD is distinguished by its distinctive *fleshy, curved, spiny calyx lobes, which are even visible in bud*, and thickened fruiting stalks. Corolla 2–2.5 mm. Fruit 7–9 mm. Throughout.

Valeriana VALERIAN

Perennials with small, funnel-shaped flowers with 5 lobes; stamens and stigmas 3. The most common species are described here; other species occur at higher altitudes.

(a) Basal leaves pinnately divided or lobed.

***Valeriana italica* (syn. *V. dioscoridis*)** A more or less hairless, scarcely branched or unbranched perennial to 80 cm with spindle-shaped tubers. Basal leaves pinnately divided or lobed with 1–4 pairs of lateral lobes and a shallowly toothed end lobe; stem-leaves 1–3 pairs. Flowers white to pinkish-mauve, borne in lax inflorescences; corolla 6–8 mm, narrowly funnel-shaped with 5 almost equal lobes. Fruit 4–5 mm, hairy on one side. Throughout (not CR or KP).

Valeriana officinalis COMMON VALERIAN A tall, erect perennial to 1.5(2.3) m with single or clustered stems. Leaves pinnately divided, to 14 cm, borne in opposite pairs; the lowermost stalked, the upper almost stalkless; leaflets toothed. Flowers borne in terminal heads, with several lower clusters, corolla 4–7.8 mm, pink and funnel-shaped, 5-lobed with 3 protruding stamens. Fruit 2.6–4 mm. Aquatic habitats; absent from hot, dry areas. GR.

(b) Basal leaves mostly undivided.

Valeriana asarifolia A hairless perennial to 50 cm with a short, tuberous rhizome. Basal leaves long-stalked, the blade *entire* (or sometimes with 1–2 small basal lobes), 40 mm–12 cm across, almost circular; stem-leaves 1–2 pairs, much smaller and dissected into linear-oblong lobes. Flowers white (or flushed pink), borne in dense heads; corolla narrowly funnel-shaped, 8 mm. Fruit 4–5 mm. AA (Chalki), CR (+ KP and adjacent islets), PN (Andikithira).

Valeriana alliariifolia A large, hairless, spreading perennial to 1.5 m with branched, cylindrical rhizomes and hollow stems. Leaves *all simple*, irregularly toothed, to 20 cm long; upper leaves lance-shaped to triangular and more or less stalkless. Flowers white (to pale pink). Fruits hairless. Shady, damp habitats in mountains. GR (unconfirmed), TR.

Valeriana tuberosa **TUBEROUS VALERIAN** A rather fleshy, green, purple-flushed perennial to 70 cm with undivided, oblong basal leaves (sometimes shallowly lobed) and pinnately divided stem-leaves with deeply dissected, linear lobes. Flowers borne in dense, *rounded* clusters; corolla 5–7.2 mm, pale pink-white. Fruits 2.7–3.89 mm, with silvery hairs. Frequent on maquis and scrub. GR, PN, TR.

PITTOSPORACEAE

Trees, shrubs and lianes with simple, alternate leaves. Flowers with 5 sepals, petals and stamens; style 1. Fruit a 2–4 parted capsule.

Pittosporum

Trees and shrubs with simple (rarely lobed) leaves. Flowers with 5 sepals and petals; style 1. Fruit a woody capsule with numerous seeds.

Pittosporum tobira A dense, dark green, evergreen shrub 2–6 m. Leaves 35 mm–10 cm, alternate, oval, deep green and leathery with inrolled margins. Flowers cream, with 5 petals 10–13 mm, fragrant, borne in dense, flat-topped clusters. Fruit 15–20 mm, splitting 3 ways. Widely cultivated in towns and on roadsides. Throughout. *P. undulatum* has broad, mid-green leaves 50 mm–15 cm, narrowed at the base, with *undulate* (*not inrolled*) margins. Fruit 10–14 mm, splitting 2 ways. Occasionally planted in the region.

1. *Valerianella discoidea*
2. *Valeriana italica*
3. *Valeriana asarifolia*
4. *Pittosporum tobira* fruits
5. *Pittosporum tobira* flower

ARALIACEAE | IVY FAMILY

Woody climbers, trees and shrubs with alternate, simple, often lobed leaves. Flowers small, often borne in umbels and 5-parted; style(s) 1(5). Fruit a drupe or berry with 2–5 seeds.

Hedera IVY

Woody vines, which climb by means of rootlets. Flowers borne in terminal umbels. Fruit berry-like.

Hedera helix IVY A vigorous evergreen climber or creeper with flexuous stems with adhesive roots. Leaves glossy, hairless and dark green, palmately lobed (>½ to the base), 13 cm across; leaves on flowering stems oval and unlobed. Flowers pale yellowish-geen and borne in terminal, dense spherical umbels. Berry spherical and black when ripe, 7–8.3 mm across. Woods above sea level. Widespread but absent from arid areas. Virtually throughout.

APIACEAE | CARROT FAMILY

A large and important family. Mostly herbs with alternate leaves, often pinnately divided with sheath-like bases. Flowers normally borne in very characteristic compound *umbels*; often green, yellow or white with 5 free petals and 0 or 5 sepals; stamens 5; styles 2, often arising from a swelling. Fruit a dry, 2-parted schizocarp (composed of 2 mericarps), often with a central carpophore. Fruit characteristics are an important diagnostic.

Eryngium SEA HOLLY

Hairless, stiff, spiny herbs with simple or pinnate leaves. Flowers dense, borne in rounded, thistle-like heads, surrounded by spiny bracts. Fruit rounded and often scaly or bristly. A confused genus with conflicting accounts.

(a) Involucral bracts oval and toothed or deeply lobed. On maritime sands.

Eryngium maritimum SEA HOLLY A short, rigid perennial to 50 cm with persistent, leathery, *blue-grey, white-veined* leaves, 3–5 lobed with an undulate, spiny margin. Flowers pale blue, borne in dense, rounded heads 10–30 mm surrounded by *broadly lance-shaped, lobed, coarsely spiny bracts* with 1–3 pairs of spines; upper stems flushed bright blue. Fruit >6 mm, with hooked bristles. Very common on coastal sands and dunes. Throughout.

(b) Involucral bracts linear or lance-shaped and spiny-toothed or entire; *basal leaves pinnately lobed*.

Eryngium campestre FIELD ERYNGO A short, rigid, *much-branched* perennial 20–60(75) cm with green or yellowish, leathery, 3-lobed basal leaves, which are spine-toothed with unwinged stems; stem-leaves unstalked and clasping the stem. Flowers pale green-white or yellowish, borne in dense, rounded heads 8–15 mm; bracts 5–7, *linear-lance-shaped and entire or spiny-toothed*, spreading. Very common in dry, sandy or grassy places both coastal and inland. Throughout, except the far southeast.

Eryngium amethystinum BLUE ERYNGO A stiff, much-branched perennial 40–60 cm with long, *pinnately lobed basal leaves* with spine-toothed margins and broadly winged stalks. Flower-heads *strongly flushed blue*; bracts 7–8, blue, *linear and entire* or sparsely spine-toothed, 2–5 x as long as the flower-heads. Garrigue and pastures in the west. GR, IA (rare – Cephalonia, Lefkada). *E. creticum* is similar but with leaves broadly oval in outline, leathery, divided into 3 segments, each pinnately lobed. Throughout. *E. falcatum* has fewer, larger flower-heads with bracts with 2 slender, deflexed spines at the base. LB, SY, TR.

Eryngium amorginum A robust perennial to 1 m, leafy below, sparingly branched above. Leaves green, rather rigid, *not spiny*, and pinnately lobed into few oval, toothed segments. Flower-heads 10–15, borne in lax inflorescences; bracts 5, narrowly lance-shaped. Limestone sea cliffs; rare. AA (southeast), CR (east).

Eryngium ternatum A perennial 30–70 cm with slender stems, rather bluish. Leaves with winged stalks, divided into 3 linear-lance-shaped segments with sparse, soft spines. Flower-heads few, broadly ovoid, 12–18 mm wide, exceeded by 5–9 linear-lance-shaped bracts. A rare chasmophyte of cliffs in gorges. CR (mainly west).

---

1. *Hedera helix*
2. *Eryngium maritimum*
3. *Eryngium campestre*
4. *Eryngium amethystinum*

Hydrocotyle PENNYWORT

Aquatic, creeping perennials *with entire, circular leaves*. Flowers very small and inconspicuous. Fruit strongly laterally compressed. Unlike other genera in the family in the region covered.

Hydrocotyle vulgaris MARSH PENNYWORT A low, spreading perennial with creeping stems to 30 cm rooting at the nodes, and with long- and hairy-stalked, roughly *circular leaves* 8–35(50) mm wide with blunt, curved teeth and 7–9 radiating veins. Inflorescence small and inconspicuous, stalkless, to 3 mm across, with pinkish-green flowers obscured by the leaves. Fruit 1.8–2.3 mm, rounded. Rare, in damp, marshy habitats (more common in Northern Europe). CR (mainly west), GR (west), IA, IP, PN.

Lagoecia

Annual, hairless herbs with pinnately divided leaves. Umbels with 8 bracts and 4 bracteoles; petals minute. Fruits with very fine ribs.

Lagoecia cuminoides LAGOECIA An annual to 55 cm with pinnately divided leaves 20–70 mm with toothed segments; upper leaves yarrow-like, deeply divided into 3–13 pairs of narrow lobes. *Umbels feathery, white, rounded and dense*, 7–15 mm; petals 1 mm. Fruit 1.5 x 1 mm with brittle hairs. Widespread and common in dry, fallow habitats. Throughout.

Echinophora

Stiff, spiny perennials. Flowers borne in umbels with 5–8(12) rays. Fruits angled.

Echinophora spinosa A very spiny, much-branched, robust perennial to 70 cm with fleshy, 2–3-pinnately divided, rigid leaves 40 mm–22 cm with thick, spine-tipped lobes. Flowers white, borne in flattish umbels to 30 mm across, short-stalked, with 5–8(12) rays; bracts linear and spiny. Fruit 8 x 4 mm, egg-shaped and angled. Coastal sands. AA (very rare), GR (west), IA, PN. *E. tenuifolia* is similar, though greyish, often very densely branched and with *numerous yellow umbels*. Fallow and waste ground. AA, CR (north), CY, GR (east), LB, PN, SY, TR.

Smyrnium

Biennial herbs in which the 2–3 leaf lobes arise from a single point (ternate). Umbels yellow, the flowers lacking sepals. Fruit egg-shaped with slender ridges and oil glands, hairless.

Smyrnium olusatrum ALEXANDERS A tall, pungent, hairless biennial to 1.5(2.2) m. Stem robust, hollow when mature, bearing 1–2-pinnately *divided leaves* with triangular, toothed lobes, borne on short, somewhat inflated stalks; shiny green below, yellow above. Umbels terminal, borne in the leaf axils with 7–15(18) unequal rays; flowers yellow, to 3 mm across without sepals. Fruit ovoid, 5–7 mm and *black* when ripe, conspicuously *ridged*. Common in damp, coastal or disturbed habitats. Throughout. *S. creticum* (syn. *S. apiifolium*) is similar in habit but taller, to 1.2 m with hollow, much-branched stems and fruits 3–3.5 mm, *scarcely ridged*. AA, CR, GR, PN, TR.

Smyrnium perfoliatum PERFOLIATE ALEXANDERS A virtually hairless, tall biennial to 60 cm (1.5) m. Stem robust, angled with longitudinal narrow wings, solid and hairy at the nodes. Basal leaves with 1–2 oval, toothed lobes; *upper leaves yellowish, simple, oval, markedly clasping* the stem with heart-shaped bases. Umbels yellow, rounded with 4–8(12) rays and no bracts. Fruit 2–4 mm, broader than long, brown–black. Wooded and open grassy slopes. Throughout. *S. rotundifolium* is similar (also considered to be a subspecies of the former), but with the upper leaves entire (not toothed) and often yellow, and the stems ribbed but not winged; leaves greyish. Throughout.

Smyrnium connatum A robust, erect perennial to 1.5 m, similar to *S. perfoliatum* and *S. rotundifolium* but with the upper stem-leaves *opposite* (not alternate) and *fused* along their basal margins (not clasping with deeply heart-shaped bases). Umbels with 10–20 rays and no bracts. Fruit 3 mm long, blackish when ripe. Rocky mountainsides; uncommon and local, in the far southeast only. CY, IP, SY, TR.

Bunium

Tuberous, hairless, erect, hairless perennial herbs with divided leaves. Flowers borne in umbels with 5–20(25) equal to unequal rays; petals white or pink. Fruits hairless, with 5 primary, rounded ribs.

Bunium ferulaceum An erect to ascending perennial 20–60 cm. Basal leaves with linear, pointed lobes. Flowers borne in inflorescences with 6–15 rays; bracts 0–2; sepals absent or minute. Fruit borne on stalks *almost as thick*, 4–6 mm long, oblong, with very thick, prominent ridges. Scattered across the region, more common in the south. Throughout.

---

1. *Hydrocotyle vulgaris*
2. *Lagoecia cuminoides*
3. *Smyrnium rotundifolium*
4. *Smyrnium connatum*
5. *Astomaea seselifolia*

Astomaea

Slender perennials with *tuberous* roots, similar in aspect to *Bunium*. Flowers white. Fruits *twinned*, laterally flattened.

Astomaea seselifolia A hairless, slender perennial, branched above, 10–50 cm. Leaves oval in outline, 2–3-pinnately divided into oblong to linear lobes; lobes of upper leaves few, linear and elongated. Flowers white, borne in umbels with (5)8–12 rays 30–60 mm long. Fruit 2 mm, black with whitish ribs. Fields, often in desert highlands; locally common. IP, LB, SY.

Scandix

Slender annual herbs with thread-like, feathery leaves. Flowers with white petals and with minute sepals. *Fruits conspicuous long, needle-like and with cylindrical beaks, hairless.*

Scandix pecten-veneris SHEPHERD'S NEEDLE A small, spreading, hairless annual to 40(50) cm with leaves 2–3(4)-pinnately divided, the segments widened towards the tips. Umbels simple or with 2–3 rays; *bracteoles divided into linear lobes*; flowers small and white; petals 2–4.5 mm. Fruits highly distinctive, borne in claw-like clusters: *elongated, 50–95 mm long*, at least ½ the length comprising the seedless beak, which is *flattened and distinct*. Common on disturbed and fallow ground. Throughout.

Scandix australis An annual similar to *A. pectin-veneris* but with undivided (or bilobed) bracteoles, (2)3–5 rays not thickened in fruit, and the fruit shorter, 11–35 mm, and *scarcely compressed*, and the beak *indistinct* from the rest of the fruit. Pine forests and shady places. Throughout. *S. stellata* is similar but with long, thread-like leaf segments just 0.5 mm wide (not 1 mm wide) and scarcely spreading petals. Mainly in the east of the area; rare and local. AA, GR, IP, TR.

Crithmum

Hairless, fleshy perennials with leaves 1–2-pinnately divided. Petals yellow–green. Fruit oblong, hairless, not compressed, and rather corky (spongy when fresh).

Crithmum maritimum ROCK SAMPHIRE A somewhat bushy, greyish, hairless perennial to 45 cm woody at the base. Leaves 1–2-pinnately divided with almost cylindrical, untoothed, *fleshy segments*; the base membranous and clasping the stem; *smelling of furniture polish when crushed*. Flowers yellow–green in umbels to 60 mm across with 10–32(36) rays, sepals absent; bracts and bracteoles numerous (5–10), triangular. Fruit 3.5–5 mm, oval and corky, later purple. Common on cliff-tops, rocky shores and other maritime environments. Throughout.

Oenanthe WATER-DROPWORT

Typically aquatic, hairless annuals or perennials, often with tuberous roots. Flowers with white petals. Fruits rounded along the back, with shallow, grooved ribs.

Oenanthe pimpinelloides CORKY-FRUITED WATER-DROPWORT An erect perennial with solid (pithy), ridged stems to 1 m with 1–3-pinnately divided basal leaves with linear to oval leaflets and 1-pinnately divided stem-leaves with narrow, unlobed leaflets. Flowers white, borne in terminal umbels to 60 mm across with 6–15 rays, which *thicken in fruit* (to >0.5 mm thick); bracts and bracteoles bristle-like. Fruits 3–3.5 mm, *cylindrical, corky and swollen at the base*; styles erect and equalling the fruit. Damp grassland. Throughout (not CY). *O. silaifolia* is similar but slender, with few branches, umbels with few (3–8) rays, which are slightly thickened in fruit. GR, IA, IP, PN, TR.

Oenanthe globulosa MEDITERRANEAN WATER-DROPWORT A *short*, much-branched perennial 15–60 cm with egg-shaped tubers; stems rather bluish-grey, hollow and distinctly *grooved*. Basal leaves 2-pinnately divided with oval to linear lobes, the upper leaves less divided and narrower. Umbels white with 2–8(15) lax, disparate rays; flowers male or cosexual, sepals persistent; flower-stalks *thickened* in fruit. Fruit *spherical*, 3–5(7.5) mm, with thickened lateral ridges. Marshes and riversides; very rare in the area. AA, TR.

Oenanthe fistulosa A slender, rhizomatous, erect and hairless perennial to 80 cm. Leaves long-stalked below, 1–2-pinnately divided into linear segments. Umbels with 2–3 rays, thickened in fruit. Fruits stalkless and with *bristly, persistent styles*. Marshes. GR, IA, PN, TR.

Foeniculum FENNEL

Perennials with 3–4-pinnately divided leaves with slender lobes, *strongly aromatic*. Flowers with sepals absent, petals yellow. Fruit distinctly ridged and scarcely compressed.

Foeniculum vulgare FENNEL A late summer-flowering, robust, tall, hairless, *strongly aromatic* perennial to 2.5 m, *smelling of aniseed*. Stems *bluish*, hollow when mature. Leaves feathery, 3–4-pinnately divided with *thread-like* segments 5–40 mm, light green and with sheathing bases. Umbels yellow with 5–44 rays without any bracts; petals 1.3–1.6 mm. Fruit oblong, 3–6(9) mm long, ridged. Very common and widespread in a range of dry habitats. Throughout.

Kundmannia

Erect perennial herbs with pinnately divided leaves with large, oval segments. Flowers with yellow petals. Fruits oblong, hairless and ribbed.

Kundmannia sicula KUNDMANNIA A hairless perennial with erect stems 30 cm–1 m. Lower leaves 2-pinnately divided with an extra pair of lobes at the base of each main lobe; lobes 15–40 mm, oval, toothed; upper leaves 1-pinnately divided and coarsely toothed. Umbels yellow with 3–30 rays, bracts numerous and linear and keeled backwards. Fruit cylindrical, 6–10 mm long with slender ridges. Dry slopes and fallow ground; local and uncommon. AA, CR, IA, PN.

1. *Scandix pecten-veneris*
2. *Foeniculum vulgare*
3. *Oenanthe pimpinelloides*

Cachrys

Bushy perennials with leaves 2–4-pinnately divided into linear lobes. Flowers with yellow petals. Fruit with undulate wings and ridges.

Cachrys cristata An erect, hairless perennial with angled stems and opposite or whorled branches. Leaves oval in outline, 2–4-pinnately divided into long divergent, rigid, thread-like lobes to 15 mm, flat, 3 x branched at the tips. Flowers borne in umbels of (4)10–12 rays; bracts and bracteoles short and linear. Fruit 7–10 mm with prominent crimpled, toothed ridges (almost winged). AA, CR, GR, IA, PN, TR.

Prangos

Bushy, aromatic herbs with pinnately divided leaves with thread-like lobes. Flowers with yellow petals. Fruits large and corky.

Prangos ferulacea A robust, leafy perennial to 1.5 m with ribbed stems. Basal and lower stem-leaves 4–6 pinnately divided, 60–80 cm long with linear lobes 20–50 mm long. Umbels with 7–15(20) rays with linear-lance-shaped bracts and linear bracteoles. Fruit (10)12–25 mm, slightly laterally compressed with prominent crested or toothed wavy wings, whitish tinged pink or maroon. Mountains, almost exclusively in the east of the area. IP, PN (rare), SY, TR.

Conium

Tall, erect, hairless biennials with compound leaves. Flowers with white petals. Fruits broadly ovoid with 5 prominent ribs.

Conium maculatum HEMLOCK A tall, erect, almost hairless biennial with *purple-spotted stems* to 2 m; strong-smelling and poisonous. Lower leaves large, to 50 cm long, triangular in outline, 2–4-pinnately divided with slender leaflets. Umbels terminal and axillary, to 50 mm across with 8–10 rays and few, small, backwardly turned bracts and similar bracteoles; flowers white. Fruit 2–3.9 mm, almost spherical with wavy ridges. Locally common in damp, grassy habitats, probably throughout.

Bupleurum

Hairless annuals or perennials with *simple leaves*. Flowers pale green or yellow, borne in small umbels surrounded by petal-like bracts. Sepals absent, petals not notched. Fruits prominently ridged. Observation of the bracts and bracteoles is important for identification.

(a) Annuals or perennials <1 m high.

Bupleurum rotundifolium THROW-WAX An erect, grey–green annual to 50(85) cm with hollow stems. Leaves elliptic to *circular*, those below to 10 cm, tapering into the stalk, those above 10–60 mm, *clasping the stem entirely at the base*. Flowers borne in umbels to 30 mm across with few (3–8) rays and no bracts but conspicuously oval, fused bracteoles (*Euphorbia*-like in appearance); petals yellow. Fruit 2.5–4 mm, with ridges, smooth in between. Arable and grassy habitats. GR, PN, TR.

Bupleurum apiculatum A slender, erect annual 30–60 cm. Lower leaves stalked, the others stalkless, all narrowly lance-shaped to linear and 3–5-veined. Flowers borne in umbels of 6–8(15); bracts as long as the rays, narrow, with translucent margins; bracteoles broader, whitish, *completely translucent* and with *short* spine-like tips. Fruit 3 mm with obscure ridges. Dry, rocky habitats and limestone scrub. GR (north and northeast).

Bupleurum glumaceum A small annual similar to *B. apiculatum* but with 3–8(12) rays with bracts up to as long as the longest ray, with a wide, white, minutely toothed margin; bracteoles elliptic, translucent outside the lateral veins, contrasting the green middle part; spine-like tip *long*. Common in the west, rarer or absent further east. AA (Euboea – rare), GR, IA, PN. ***B. gracile*** is very similar to the previous species but smaller, with fewer umbels and much longer bracts and bracteoles. AA, CR, CY, GR (southeast), PN (northeast), TR (west and south). ***B. gaudianum*** is also similar to *B. glumaceum* but with 3 rays and purplish petals. A rare island endemic. CR (island of Gavdos).

Bupleurum tenuissimum SLENDER HARE'S-EAR A slender, spreading to erect annual to 50 cm with wiry stems. Leaves 13–80 mm, linear-lance-shaped to spoon-shaped, pointed. *Umbels tiny, <5 mm across,* with few (2–5) rays borne in the leaf axils, and very short-stalked. Fruit 1.5–3 mm. Saline, grassy habitats. GR, IA, LB, SY.

Bupleurum semicompositum A small annual 50 mm–12 cm, branched from the base. Leaves linear. Umbels with 4–6 very *unequal* rays; bracts 4, lance-shaped, 3-veined, finely toothed; bracteoles 5, similar, slightly exceeding the flowers. Fruits 1–1.5 mm, ellipsoid, covered in whitish protruberances (papillae). Saline habitats. Throughout, except LB.

Bupleurum trichopodum A small annual 15–30 cm. Leaves linear, long, the uppermost *clasping*. Umbels with 2–6 very slender, almost equal rays; bracts 0–3; bracteoles 2–5. Fruit 2.5–3 mm, smooth but with fine ridges. Various habitats. Throughout, except IP.

1. *Prangos ferulacea*
2. *Prangos ferulacea* fruits
3. *Bupleurum rotundifolium*
4. *Bupleurum glumaceum*
5. *Bupleurum fruticosum*

(b) Shrubs >1 m high.

Bupleurum fruticosum SHRUBBY HARE'S-EAR A large, aromatic, evergreen, perennial *shrub* to 2.5(3) m. Leaves oblong with prominent mid-vein and lateral veins, and *blue–grey beneath*. Stems reddish and much-branched. Umbels from previous years persistent and woody, giving a spiny appearance. Flowers yellow, borne in open umbels with 3–20(25) rays; bracts and bracteoles short and bristly. Fruit 5–7(8) mm. Maquis and roadsides in the west. GR, LB, PN.

Apium MARSHWORT

Aquatic, hairless perennials with pinnately divided leaves. Flowers white. Fruits laterally compressed.

Apium graveolens A strong-smelling, robust biennial to 1 m with solid stems. Leaves 1–2-pinnately divided into segments 5–50 mm long, diamond-shaped to lance-shaped, lobed and toothed. Flowers borne in umbels, short-stalked to stalkless, mostly opposite the leaves; rays 4–12; bracts and bracteoles absent. Fruit 1.5–2 mm long, broadly ovoid. Damp habitats near the coast. Throughout.

Ridolfia

Tall, erect, aromatic, hairless annuals with 3–4(6)-pinnately divided leaves with thread-like segments. Flowers yellow. Fruits laterally compressed, longer than wide.

Ridolfia segetum RIDOLFIA A tall, fennel-like, bluish-green annual to 2.1 m, unpleasant-smelling when crushed (not like aniseed). Leaves 3–4(6)-pinnately divided with thread-like lobes 5–15 mm long and inflated stalks. Umbels yellow with many (8–56) slender, curved rays and no bracts or sepals; petals 1.9–2 mm, yellow. Fruit small, 1.2–2.7 mm long with slender ridges. Cultivated and waste land; local. AA (rare), GR, IA, IP (common), PN (northwest).

Ammi

Hairless annual or biennial herbs with erect, striated stems and leaves divided into linear or lance-shaped segments. Flowers with white petals, borne in dense umbels. Fruits hairless and slightly laterally compressed.

Ammi visnaga A robust, carrot-like annual or biennial to 1 m. Lower leaves 1-pinnately divided, upper leaves 1–2-pinnately divided, all somewhat feathery with narrow to *linear, thread-like lobes*. *Flowers white*, borne in *dense* umbels; rays numerous (45–125), slender and spreading in flower but much *thickened* and *erect (bunching)* in fruit. Bracts divided, equalling or exceeding the rays. Fruit 2–2.8 mm. Uncommon, on rocky garrigue. AA (rare), GR, IA, IP (common). *A. majus* is a similar, erect annual to 1 m with broader, linear-lance-shaped leaf segments and with 20–55 rays, *slender, not bunching* in fruit. Fruit 1.5–2 mm. Common, sometimes abundant on disturbed ground. Throughout (rare in much of the Greek mainland and the north).

Angelica

Robust biennials and perennials with hollow stems and divided leaves, triangular in outline, with large lobes. Flowers borne in large umbels, with white or purlish petals. Fruits hairless with winged lateral ridges.

Angelica sylvestris A robust, erect biennial to 2.5 m, more or less hairless with thick, purple stems. Leaves green, not shiny, those below 2–3-pinnately divided into oblong-elliptic, finely toothed segments, those above variably divided into oval, toothed segments. *Umbels large, with up to 75, rather distant rays of white flowers*, often pink- or purple-flushed. Fruits 4–5 mm. Rare, in aquatic habitats; mainly northern. GR, PN, LB, SY, TR.

Ammoides

Slender annuals or biennials with divided leaves, cylindrical in outline. Flowers with white petals, borne in lax umbels. Fruits ridged and hairless.

Ammoides pusilla A weak, very *slender, hairless annual* to 50(75) cm. Leaves yarrow-like, 2-pinnately divided below with 7–12 pairs of *very short lobes*, the upper leaves with longer, *linear segments*; all leaves narrow in outline and yarrow-like. Umbels with 5–15(20) slender, *lax, unequal rays*; bracts absent or few; petals white. Fruit small, just 0.7–1.3 mm, with slender ridges; hairless. Dry habitats and field borders. GR (west).

Pastinaca PARSNIP

Slightly hairy, strong-smelling biennials. Flowers with yellow petals with cut-off, incurved tips; sepals absent. Fruit ovoid and strongly compressed.

Pastinaca sativa WILD PARSNIP An erect, downy, branched biennial to 1(2) m with hollow, ridged and angled stems (sometimes solid and unridged). Basal leaves light green, *1-pinnately divided with broad, oval, lobed and toothed segments*. Umbels to 10 cm across with 5–20 rays; bracts and bracteoles few or absent; *flowers yellow and fennel-like*. Fruit 4–7 mm, ovoid and flattened with narrowly winged edges. Damp fields and grassy habitats on the mainland. GR, PN, TR.

1. *Ammi majus*
2. *Malabaila secacul*
3. *Malabaila aurea*

Malabaila

Erect biennials to perennials. Leaves triangular in outline, the basal leaves 1–2-pinnately divided; upper leaves reduced to sheaths. Flowers borne in 4–20 rayed umbels with or without bracts; sepals absent; petals yellow. Fruit compressed, hairy to hairless, almost circular with a thickened margin.

Malabaila secacul   A white to grey-hairy perennial to 75 cm. Leaves triangular in outline, the lowermost divided into oblong, blunt-toothed segments 10–12 mm. Flowers yellow, borne in small clusters of 20, in umbels with 15–20 rays. Fruits 6–8 mm, round to egg-shaped, flattened, notched and with thick, smooth white borders (1 mm). Mainly fields and desert scrub; sometimes open oak forests. IP (east and centre), SY, TR. **M. lasiocarpa** has umbels with 10–16(30) rays. Fruits 6–12(14) mm, with white borders. Oak forests and stony slopes. TR.

Malabaila aurea   A minutely hairy annual (to biennial) with erect, scarcely branched stems 20–50 cm. Lower leaves 1-pinnately divided into broad, toothed or lobed segments; upper leaves with narrower, toothed segments. Umbels with rather few (3–9) long, almost equal rays; bracts absent; bracteoles few; petals golden yellow. Fruit compressed, 8–10 mm, circular with pale, smooth, slightly thickened margins. Maquis and woods. AA (local), GR (common), IA, PN, TR. **M. involucrata** is similar but with greyish leaves with scalloped margins, umbels with *more* (5–16) rays and 6–8 persistent bracts and bracteoles. Garrigue and woods. AA (west), GR (south), IA, PN.

Ferula

Tall, robust perennials with leaves 3–4-pinnately divided into linear lobes. Flowers with yellow petals. Fruits strongly compressed dorsally. DNA-based analysis shows that morphological traits are of limited value; accurate identification may be impossible in the field for closely related species (hence traditional floras not reliable).

(a) Leaves with long, thread-like lobes 10–40(80) mm long.

Ferula communis GIANT FENNEL A robust perennial herb to 1.5(3.3) m, bushy at the base; stems long-persisting after flowering. Leaves 3–4(6)-pinnately divided with bright green, *thread-like segments 10–40(80) mm long and just 0.5–1.3 mm wide*, and prominent sheathing bases; upper leaves reduced to sheathing bases only. *Inflorescences pyramidal and spreading*; umbels large, bright yellow–green, with *(12)20–40(50) rays*; bracts absent, bracteoles few and soon-falling. Fruit large 10–16 mm long, oval, flattened with slender dorsal wings 1–1.5 mm wide and resin canals. Common on roadsides. Throughout, except the far north and northeast.

Ferula glauca (syn. *F. communis* subsp. *glauca*; *F. chiliantha*) A robust perennial, similar to *F. communis* but with alternate (not whorled) upper branches and *broader leaf lobes* (1–3 mm wide), which are *paler and bluish beneath*. Inflorescences rather broadly pyramidal; sheaths narrower. Fruits elliptic. Gorges and cliffs, sometimes on roadsides; often confused with the previous species. AA, CR, GR (southeast), PN (Kithira).

(b) Leaves with short terminal lobes 2–4(14) mm.

Ferula tingitana Similar to *F. communis* but with *shorter, narrowly wedge-shaped (not thread-like) leaf lobes 2–4(14) mm long*, and 1.3–6 mm wide, with or without down-turned margins, blunt-tipped and stiff. Umbels with 8–16(22) rays. Fruit 10–16 mm with very narrow lateral wings. It is not clear whether specimens recorded under this name in the eastern Mediterranean correspond with those in the west (Spain and Portugal). AA (rare), IP, SY, TR (south).

(c) Leaves with short terminal lobes 2–10(30) mm. Eastern mainland only.

Ferula brevipedicellata An erect, *mountain-dwelling* perennial with leaves 5–6-pinnately divided with linear lobes (3)7–12(18) mm long and *very short flower-stalks*, just 0.5–6 mm in fruit. Rare, in rocky habitats in the Taurus range. TR. *Ferula rigidula* has leaves with *short* segments 1–5(8) mm long. TR.

Ferula biverticillata A *short*, robust, hairless perennial 20–40 cm with almost *leafless* stems, branched from below the lower third, with sheathing bracts at the base. Leaves to 35 cm, *rough*, completely *withered* during flowering, 3–4-pinnately divided; lobes short, oblong to linear, rather feathery. Flowers yellow, borne in umbels with 4–14 rays. Fruit 10–12 mm, linear-elliptic. Dry fields, plains and deserts; rare. IP, LB, SY.

Ferula sinaica A blue–grey *desert* perennial (40)80 cm–1 m, branched above. Leaves 15–20 cm, 3–4-pinnately divided, the ultimate lobes linear, with down-turned margins, *short, 3–10 mm*, and *2–3 × branched*. Flowers yellow, borne in terminal cosexual umbels with 7–12 rays, and lateral, long-stalked male umbels. Fruit 8–12 mm, yellowish-green, elliptic to almost circular, ribbed, narrowly winged. Deserts; rare. IP.

1. *Ferula communis*; (inset, top) fruit; (inset, bottom) leaves
2. *Ferula glauca*; (inset) leaves
3. *Ferula tingitana*

Ferula species: **A.** *Ferula orientalis*, **B.** *F. sinaica*, **C.** *F. biverticillata*, **D.** *F. daninii*, **E.** *F. negevensis*.

Ferula negevensis A very *tall, slender, desert* perennial to 2(3) m. Leaves rough, 10–25 cm, 3–4-pinnately divided; ultimate lobes simple or 2–4 branched, linear to oblong, blunt, 2–4 mm; leaf sheaths 20–40(60) mm. Flowers borne in umbels with 4–6(9) rays, on lax, repeatedly branched inflorescences. Fruit 6–7 mm, broadly elliptic, transversely *wrinkled*. Deserts; rare. IP (central Negev).

Ferula daninii A rather short, slender, hairless, green, *desert* perennial 30–80 cm. Leaves 20–40 cm, 3-pinnately divided; terminal lobes short and *oblong*, 10–30(50) mm; leaf sheaths inflated, rather large, 50–60(70) mm. Flowers yellow, borne in umbels with (3)6–10 rays. Fruit 7–9 mm, elliptical, whitish. Deserts; rare. IP (central Negev).

Ferula orientalis A slender, tall, hairless perennial 1–1.5(2) m. Leaves *large* and bushy, to 50 cm, flopping, 4-pinnately divided with rather long, linear ultimate lobes to 15 mm; sheaths *strongly inflated*, yellowish, 90 mm–12(15) cm. Flowers yellow, borne in cosexual umbels of 30 rays and 2 lateral, long-stalked male umbels. Fruit 10–12 mm with thick, swollen margins. Limestone cliffs; very rare. IP (Samaria and adjacent areas).

Ferula hermonis An erect, slender, fennel-like *mountain-dwelling* perennial. Leaf leaf lobes linear, 1–4(9) mm long; sheaths oval-oblong (absent above). Flowers yellow, borne in umbels with 5–16 rays to 30 mm long. Fruits elliptic-oblong with slender ridges and lateral wings just 0.5–0.9 mm. Rare, on rocky hillsides of Mount Hermon (mainly Syrian and Lebanese slopes). IP (very rare), LB, SY.

Opopanax

Robust perennials with 2-pinnately divided leaves and yellow flowers. Fruit oval, hairless and ribbed, sometimes shortly winged.

Opopanax chironium A tall perennial to 1.5(2) m superficially similar to *Ferula* in profile, but with shorter, broader inflorescences. Leaves long-stalked (18 cm) pinnately divided with *large, oval lobes* 12 x 10 cm, bristly hairy, withering during or soon after flowering. Flowers yellow, borne in rounded umbels with 7–10(20) rays. Fruit 8(10)mm, with a narrow, thick, white border. Common on the mainland. GR, IA, PN, TR. ***O. hispidus*** is similar but larger, to 3 m, with umbels with fewer (6–13) rays and fruits with a broad but thin white border. Throughout.

Ferulago

Bushy perennial herbs with hairless stems, branched above and 2–4-pinnately divided leaves. Flowers yellow, borne in umbels on alternate branches. Fruits with narrow and expanded (wing-like) ridges.

(a) Stem nodes markedly swollen.

Ferulago nodosa An erect, hairless perennial to 60(80) cm with stems *very conspicuously swollen at the nodes*. Leaves triangular in outline with short, linear lobes. Flowers yellow, borne in umbels with 10–20 rays; bracts and bracteoles oval-lance-shaped. Fruits 8–10(12) mm with narrow, undulating wings. Rather local. AA (west), CR, GR, IA, PN (common).

(b) Stem nodes *not* markedly swollen (habitat important).

Ferulago campestris A rather bright green, slender perennial to 2 m with angled stems. Leaves to 60 cm long, triangular-oval in outline with short, linear to thread-like segments. Bracts and bracteoles lance-shaped and conspicuous. Flower-heads yellow, borne in ovoid to pyramidal inflorescences. Fruit 10–12 mm, *narrowly* elliptic-oblong with slender, *blunt* ridges, the dorsal unwinged. Grassy inland habitats and hillsides in the north. GR (north-central and northwest). **F. sylvatica** is a similar, erect perennial 30 cm–1 m, branched above. Leaves *narrower* in outline (elliptical to almost linear), divided into many thread-like segments. Flowers deep yellow, borne in umbels, the central often overtopped by several smaller umbels; bracts oblong to oval and deflexed. Fruit (6)7–10 mm, elliptic with beige wings 1.5 mm wide; ridges *not* blunt. Grassy habitats. GR (northeast + Thasos), TR (northwest).

Ferulago thyrsiflora A bushy, robust, hairless perennial to 1.2 m. Lower leaves large, triangular in outline, divided several times into linear segments 30–50 mm long. Flowers not produced in some years; yellow, borne in umbels aggregated into complex panicles; bracts and bracteoles linear-lance-shaped. Fruit 8 mm, elliptic with narrow lateral wings and blunt ridges. A chasmophyte of mountain cliffs, and in gorges. CR.

1. *Opopanax chironium* fruits
2. *Ferulago nodosa*
3. *Ferulago campestris*
4. *Elaeoselinum asclepium* fruits
5. *Thapsia garganica*
6. *Torilis japonica* fruits
7. *Torilis japonica*

Ferulago syriaca A *tall*, blue–grey, hairless perennial, 1–1.5(2) m; stems erect, cylindrical, finely ridged. Lower leaves 20–40 cm, oval-oblong in outline, 2–3(4) pinnately divided into linear to oblong, rough, pointed lobes 6–10 mm long; margins down-turned. Flowers yellow, borne in umbels with (4)6–12 rays; rays 30–80 mm; bracts and bracteoles 3–7 mm, oval-triangular. Fruit 8–9(10) mm, broadly ellipsoidal, ribbed, with lateral wings 1 mm. Uncommon, in fallow fields and rocky slopes. IP (mainly Mount Carmel), LB, SY (west).

Elaeoselinum

Erect perennials with 3–4-pinnately divided leaves. Flowers with yellow petals. Fruits hairless, with prominent wings.

Elaeoselinum asclepium A perennial to 1 m with cylindrical stems. Leaves feathery, divided into short, linear, thread-like terminal segments 1.5–4.5 mm long, arranged spirally. Flowers borne in umbels of 8–25 rays; petals dull yellow. Fruit 6–12 mm, with broad, papery, shiny lateral wings. Local in rocky garrigue in hot, dry areas; strongly western and southern. AA (Rhodes), IA, PN.

Thapsia DEADLY CARROT

Robust poisonous perennials with distinct basal rosettes of pinnately divided leaves. Flowers yellow, borne in rounded to flat umbels. Fruits broadly winged.

Thapsia garganica A perennial herb to 1.4 m with a rosette of green, bushy leaves, triangular in outline, and 2-pinnately divided into linear to narrowly triangular, long, pointed lobes 50–80 mm long, somewhat flopping; hairless on the underside. Flowers yellow, borne in rounded umbels of 10–20(24) rays, carried on tall, erect, solitary stalks. *Fruits winged and large: 17–22 mm across*; wings 4–5(6) mm wide. AA, CR, GR (southeast), IA, PN (Kithira), TR (southwest).

Torilis HEDGE PARSLEY

Bristly annuals with adpressed hairs, solid stems and pinnately divided leaves. Flowers white or purplish. Fruits with *prominent protuberances*.

(a) All umbels borne opposite the leaves.

Torilis nodosa KNOTTED HEDGE PARSLEY A *short, rough-hairy annual* to 50 cm. Leaves 1–2-pinnately divided with deeply toothed segments. Flowers with pinkish-white petals, clustered on stalkless or short-stalked *leaf-opposed* umbels with 2–3 very short rays (<5 mm). Fruit egg-shaped, 2.5–3.5 mm long with *warts and straight bristles*. Common in open grassy areas and waste ground. Probably throughout.

(b) Umbels mostly terminal. (Many similar species.)

Torilis arvensis An erect, slender annual to 1 m with adpressed hairs. Leaves 1–2-pinnately divided, the ultimate segments coarsely toothed or lobed. Flowers white, borne in terminal umbels with (2)4–12 rays, usually without bracts; bracteoles numerous. Fruit 3–6 mm, ovoid, slightly compressed, covered in spines, and with persistent styles 0.4–1.5 mm. Common in grassy habitats. Throughout (in its broader treatment). *T. japonica* is similar but more robust, with 4–6 bracts and smaller fruits (≤ 3 mm) densely covered in soft spines and with *recurved persistent styles*. GR, IA, PN. *T. tenella* is similar to *T. arvensis* but very slender, to 40 cm, with leaf segments <1 mm wide (not ≥2 mm) and 5–9 rays; petals minute. Fruit *without* conspicuous persistent style (stigmas virtually *stalkless*). *T. africana* has reduced upper leaves with just 3 lobes and umbels with very few (2–4) rays. Fruits with *very short* styles, 0.1–0.4 mm. Widespread and common, possibly throughout.

Chaetosciadium

Slender annual herbs, with the aspect of *Pimpinella* and similar to *Torilis* (different in fruit bristle characteristics). Leaves 2–3-pinnately lobed. Flowers white with few rays, borne on long-stalked umbels. Fruits densely covered in *long bristles*.

Chaetosciadium trichospermum A slender annual, divergently branched from the base, 10–60 cm. Leaves oblong-oval in outline with oval to oblong segments, pinnately divided into oblong, pointed lobes. Flowers white, borne in umbels on long stalks with 3–6 rays; bracteoles many, 3–8 mm, exceeding the flower-stalks (pedicels). Fruits covered in conspicuous, dense, long, weak, red (or white or purple) bristles, 3–4(5) x as long as the seed-bearing part of the fruit. Rocky, often shady habitats. IP (common), LB, SY.

Tordylium

Normally annuals with simple, lobed or pinnately divided leaves. Flowers whitish or purplish; sepals often conspicuous and unequal. Fruit circular or oval-elliptic, strongly compressed with much-thickened margins, often lobed.

Tordylium apulum An erect, soft-hairy annual 20–45 cm, sparsely branched from the base. Leaves pinnately divided into 2–5 pairs of broadly oval, scalloped leaflets; upper leaves reduced. Flowers white, borne in umbels with few (2–8) rays; bracts and bracteoles narrow and deflexed; outer petals 2.5–9 mm, divided into 2 lobes; the inner petals smaller and unlobed. Fruit 6–10 mm with a thickened, whitish, scalloped margin. Throughout, except the far southeast.

Tordylium aegyptiacum A soft to woolly hairy annual, branched above the middle with slightly angular stems, 30–60 cm. Leaves 1–3-pinnately divided below, 1–2-pinnately divided above, with forked, oval lobes, those below with scalloped margins. Flowers white, rather *large*, borne in small clusters of 10–12 in umbels with 6–14(15) unequal, slightly rough rays up to 10 cm long, with a sterile, blackish structure in the centre; marginal flowers larger and *strongly radiating*; bracts 1–2; bracteoles 1–3, deflexed. Fruits compressed, the outer (5)8–10 mm with a thin wing and slightly thickened, wrinkled margins; the central hemispherical, 4 mm across. Deserts, roadsides and fields. CY (widespread), IP (frequent in the north), SY.

Tordylium syriacum A rough annual, sparsely branched from the base, 15–30 cm with spreading, angular stems. Lower leaves simple, round-elliptic, coarsely scalloped to toothed; upper leaves 1–2-pinnately divided with stalkless, oval to wedge-shaped segments, the terminal the largest. Flowers *small* and white, *slightly* radiating, borne in clusters of 8–12 in umbels with 5–10 very unequal rays; bracts 3–5, deflexed; bracteoles 5, very unequal, the longest 2–3 x as long as the fruiting stalks. Fruits broadly elliptic to circular, (8)10–12 mm, with *strongly wrinkled* margins. Fields, hillsides and roadsides. CY (south and east), IP (north –rare), SY.

Seseli

Biennial to perennial herbs. Leaves several times pinnately divided. Flowers white (rarely pink or yellow); bracts 0–16; sepals small or absent. Fruits oblong to ovoid (not strongly compressed) with prominent ridges. A difficult genus with many rare, local species.

Seseli gummiferum A *stout* perennial to 1 m with a robust, woody stock and few branches. Leaves rigid, hairless, greyish, 2-pinnately divided into narrowly elliptic segments 8–30 mm long. Umbels white with numerous (20–60) rays to 50 mm long. Fruit 3–4 mm, oblong, with low, prominent ridges. A conspicuous chasmophyte of rocky cliffs and gorges. AA (south, southeast), CR, TR (north). The following are widely recognised subspecies: **Subsp. *crithmifolium*** has hairless basal leaves. **Subsp. *gummiferum*** has minutely hairy leaves with lobes 1–5 mm wide and 20–35 rays. **Subsp. *aegaeum*** has minutely hairy leaves with lobes 4–7 mm wide and 30–60 rays. All three subspecies are considered by some authors to belong to one widespread taxon: *S. crithmifolium*.

Seseli tortuosum A rigidly erect, *slender* and *divergently branched* biennial to perennial to 1 m with lower leaves pinnately divided several times into narrow segments 10–25(40) mm long. Flowers white, borne in compact, hemispherical umbels with 4–10 angular rays; bracts absent; bracteoles 8–12. Fruits 2 mm, ovoid, minutely hairy, with low, prominent ridges. GR.

1. *Chaetosciadium trichospermum*
2. *Tordylium apulum*
3. *Tordylium apulum* fruits
4. *Tordylium aegyptiacum*
5. *Tordylium syriacum*
6. *Seseli gummiferum* subsp. *crithmifolium*   PHOTO: NICK TURLAND

Falcaria

Annuals or perennials with leaves with 1–2 leaflets. Flowers whitish. Fruits 3 x as long as wide, oblong, compressed, with broad, low ridges.

Falcaria vulgaris A grey–green annual to short-lived perennial to 90 cm, much-branched above. Leaves rather rigid, divided into narrowly lance-shaped segments 5–15 mm long, *sharply* and *irregularly toothed*. Flowers white, borne in umbels of 8–15 slender, almost equal rays; bracts and bracteoles several. Fruit 4 mm, laterally compressed with low, broad ridges. Roadsides and grassy habitats. AA (Samothraki only – very rare), CY, GR (north – rare), IP (north), PN (rare), TR.

Orlaya

Annuals with leaves 2–3-pinnately divided. Flowers with petals white or pink, deeply notched, often very unequal. Fruits oval or oblong with 5 *slender ribs* and 4 secondary ribs, *covered in spines*.

Orlaya grandiflora ORLAYA A branched or simple annual 10–50 cm, often hairy at the base. Leaves 2–3-pinnately divided into oval, toothed segments. Flowers white or pink, the outermost petals up to 8 x as long as the inner petals; umbels long-stalked, with 5–12 rays. Fruits 6–8 mm, egg-shaped with hooked bristles. Grassy habitats, mainly in the north. CR, GR, IA, IP, PN, TR. *O. daucoides* (syn. *O. kochii*) is similar but with umbels with 2–4(5) rays. Common. Throughout.

Pseudorlaya

Small, prostrate, densely bristly annuals in maritime habitats. Flowers white or pinkish. Fruits with prominent bristles. Possibly better-placed within the genus *Daucus*.

Pseudorlaya pumila PSEUDORLAYA A very *short, densely hairy*, rather fleshy annual 50 mm–20 cm, branched from the base. Leaves 2–3-pinnately divided with oval segments 2–5 mm. Umbels white to pale purple with 2–5(7) unequal rays of 8–12 flowers; petals more or less equal, some larger at the perimeter; bracts 2–5, linear. Fruit elliptic, 7.5–12 mm long, ridged and with hooked spines. Common on coastal dunes. Throughout.

Daucus CARROT

Summer-flowering annuals or biennials with leaves 2–3-pinnately divided. Umbels white, with pinnately lobed bracts; petals unequal, those of the outer flowers often larger. Fruit elliptic and spiny.

Daucus carota WILD CARROT A very variable, hairy or hairless annual or biennial to 2.2 m with solid, often ridged stems. Leaves feathery with linear-lance-shaped segments. Umbels with white flowers, many-rayed (9–130), often with a single purple flower in the centre, or purplish throughout; bracts pinnately lobed. Fruit oblong, 1.5–4 mm, shortly spiny; spines not distinctly fused at the base; fruiting rays becoming conspicuously incurved when dry. Common in grassy places. Throughout. **Subsp.** *carota* grows to 1.1 m, has basal leaves 1–4-divided almost to the base (pinnatisect) with linear to linear-lance-shaped lobes, and umbels 50 mm–11 cm, *strongly contracted* in fruit. Fruit 1.8–3.2 mm. Scattered. **Subsp.** *major* is similar, with leaves with oval-lance-shaped leaf lobes and umbels 50 mm–10 cm across. Scattered throughout. **Subsp.** *maximus* is *robust*, to 2 m, with basal leaves 1–3-pinnatisect and *large umbels*, 12–23(30) cm across. Fruit 1.5–2.5 mm. Throughout. **Subsp.** *drepanensis* has a spreading to ascending habit, with hairy or hairless stems to 30 cm, *fleshy*, shiny leaves with oval-lance-shaped segments; umbels *convex or scarcely contracted* in fruit. Coastal cliffs. AA, CR, IA.

Daucus involucratus An annual to 20 cm with ascending stems, branched from the base, sparsely bristly to almost hairless. Leaves 1–2-pinnately divided with oblong-lance-shaped lobes. Flower-heads with just 3–4 short lobes; bracts longer than the rays, not deflexed; petals *very small, <1 mm*. Fruit 2–5 mm. AA (common), CR (very common), CY, GR (southwest), IA (rare), PN (rare), TR.

Daucus broteri An erect to ascending annual to 50 cm, branched from the base, bristly. Leaves 2-pinnately divided with linear-oblong lobes. Flower-heads with 8–14 short rays; petals 1.2–5 mm, white or pink; bracts equalling, or shorter than the rays. Fruit 4–6 mm with strongly swollen ridges, *merging* at the base. Fields and coasts; rare. AA, CY, GR. ***D. guttatus*** is similar but with bracts *longer* than the 8–25 rays. Fruits with ridges *scarcely merging* at the base. Throughout, except the north. ***D. glaber*** is similar to the previous two species but ground-hugging and with short 3-parted bracts (not divided into linear lobes). AA (east), CY, SY, TR.

1. *Falcaria vulgaris*
2. *Pseudorlaya pumila*
3. *Daucus carota* subsp. *major*
4. *Daucus carota* subsp. *drepanensis*

Arum dioscoridis, Cyprus

Glossary

Achene, dry, 1-seeded, rather hard, indehiscent (non-splitting) fruit; there are often many achenes in a fruiting head, as in *Ranunculus*.

Actinomorphic (of a flower), with a radially symmetrical (regular) shape that has multiple axes of symmetry, e.g. a lily flower; syn. **regular**.

Alien, not native, introduced to a region.

Alternate, arising at 1 axis per node; e.g. leaves arising at different heights along the stem (not opposite or whorled).

Angiosperm, flowering plant.

Annual, completing the life cycle in one year; typically without woody parts (herbaceous).

Anther, fertile, pollen-producing part of the stamen, typically on a terminal stalk (filament).

Apex, uppermost part of a structure.

Aril, succulent covering around the seed.

Ascending, arising upwards at an angle (curving).

Awl-shaped, long, pointed spike.

Awn, long, stiff bristle, e.g. in florets of grasses (Poaceae).

Axil, point at which the leaf or leaf stalk joins the stem; adj. **axillary**.

Beak, elongated projection, usually on a fruit; adj. **beaked**.

Berry, succulent fruit, typically with >1 seed; seeds without stony coats.

Biennial, completing its life cycle in 2 years (often with leaves in a rosette in the first year and flowering in the second year).

Bifid, divided into 2 parts, typically deeply at the apex.

Blade, the main, often flattened, part of the leaf (or petal).

Bract, small, leaf-like structure, often subtending (beneath) a flower or inflorescence.

Bracteole, small secondary bract.

Bulb, underground storage organ composed of condensed stem and fleshy leaves.

Bulbil, small reproductive bulb borne in the leaf axil (sometimes among flowers, e.g. *Allium*).

Calyx, all the sepals of the flower (the outer whorls of the perianth if different from the inner).

Capitulum, a dense flower-head (inflorescence) composed of small stalkless flowers (often ray and disk florets) crowded together on a compound receptacle; typical of the daisy family (Asteraceae) and Dipsacaceae.

Carpel, female organ of the flower, comprising stigma, style and ovary.

Carpophore, a thin, sterile stalk above the pistil (typical in fruits of some Apiaceae and Caryophyllaceae).

Casual, introduced to an area and sporadic in its appearance, not persisting.

Cauliflory, flowers produced directly from the primary branch or trunk.

Ciliate, fringed along the margin with hairs.

Chasmophyte, typically grows out of crevices and fissures in rock, often on cliffs or in gorges.

Chromosome, thread-like structure in the cell's nucleus, carrying genetic information in the form of genes.

Cladode, flattened organs arising from the stem, typically resembling leaves.

Compound, made up of more than one similar part or segment (not simple), typically a leaf.

Cone, compact body of scales or bracts containing the reproductive structures in gymnosperms.

Cordate, heart-shaped (often the base of a leaf).

Corm, underground storage organ formed from a swollen stem base.

Corolla, all the petals of a flower (which may form a tube), the inner whorls of the perianth.

Corona, trumpet- or cup-shaped extension of the corolla in *Narcissus*, or fused filaments in some Apocynaceae.

Corymb, raceme in which the lower flowers have longer stalks, producing a flat-topped inflorescence; adj. **corymbose**.

Cosexual, flowers with both male and female reproductive organs, stamens and carpels, respectively.

Cupule, hardened, cup-like structure composed of bracts in Fagaceae.

Cyathium, specialised inflorescence of *Euphorbia* (Euphorbiaceae) with a cup-like structure containing a single carpellate (female) flower and several staminate (male) flowers.

Cyme, an inflorescence in which each flower terminates a branch; adj. **cymose**.

Deciduous, not persistent, for example leaves in autumn or petals after flowering.

Dehiscent, splitting.

Desiccation, drying out.

Dioecious, with male and female flowers on separate plants (individual plants of 1 sex).

Diploid, with two sets of chromosomes.

Disk floret, small actinomorphic flower, forming part or all of a capitulum (in Asteraceae).

Divided, not entire (typically a leaf); divided into teeth, lobes or leaflets.

Drupe, succulent or spongy fruit, typically with 1 seed with a stony coat.

Elliptic, flat shape (typically a leaf), widest at the middle.

Endemic, restricted to a particular country, region or island.

Entire, whole; without distinct lobes, teeth or divisions.

Epicalyx, additional whorl of sepal-like bracts beneath the true calyx (e.g. in flowers of Malvaceae).

Epichile, distal part of the lip (labellum) in some orchid species (e.g. *Epipactis*) separated from the basal part (hypochile) by a joint.

Exserted, protruding (not included, such as anthers from a corolla tube).

Falcate, sickle-shaped.

Falls, outer perianth segments (tepals) of *Iris* flowers (Iridaceae).

Family, monophyletic group of related genera, the taxonomic group between the lower rank of genus and the higher rank of order.

Floccose, covered in soft, woolly hairs.

Floret, small individual flower making up part of a dense inflorescence, e.g. a component of the capitulum (in Asteraceae) or in grasses (Poaceae).

Flower-head, a group of flowers (inflorescence) such as a capitulum (Asteraceae).

Follicle, dry, many-seeded fruit, dehiscent along 1 side; usually formed from 1 carpel.

Fruit, ripened ovary or ovaries of a flower containing 1 or more seed(s).

Garrigue, stunted sclerophyllous shrub-dominated vegetation, often on coastal limestone (see also **maquis** and **phrygana**).

Genus (pl. **genera**), monophyletic group of related species, the taxonomic group between the lower rank of species and the higher rank of family; the generic name is the first part of the scientific binomial.

Geophyte, plant that survives the dry summer as a dormant underground bulb, corm or tuber.

Glabrous, not hairy.

Gland, organ of secretion, often in sticky plants; adj. **glandular** (often referring to hairs).

Glaucous, covered in a bluish, whitish or greyish waxy bloom (rather than green).

Globose, spherical.

Glume (of grasses), bract below a spikelet.

Gymnosperm, non-flowering, seed-bearing vascular plants such as conifers.

Halophyte, salt-tolerant plant.

Head, group of flowers crowded together at the end of a stalk.

Hemiparasite, parasitic plant that gains some of its nutrition from another plant (host) but which also has chlorophyll and a root system; some hemiparasites can survive without a host (facultative hemiparasitism).

Herb, plant without woody parts; a soft and leafy annual, biennial or perennial in which aerial parts naturally die to ground level at the end of the growing season; adj. **herbaceous**.

Holoparasite, parasitic plant which gains all of its nutrition from another plant (the host) and which lacks chlorophyll and a true root system.

Hybrid, offspring of a cross between two different species, races or varieties.

Hybridisation, the formation of hybrid offspring.

Hypochile, the basal part of the lip (labellum) of some orchid species (e.g. *Epipactis*); see also epichile.

Indehiscent, non-splitting.

Inferior ovary, ovary situated beneath the point of insertion of other floral organs; syn. **epigynous ovary**.

Inflorescence, group of flowers with their floral stem (axis) and any associated bracts.

Internode, a part of the stem between two nodes.

Involucral bract, bracts surrounding a head of flowers (e.g. in Asteraceae).

Involucre, collection of involucral bracts (e.g. in Asteraceae).

Irregular, flower in which 1 or more members of the whorl, or several floral whorls, differ in form from the other members.

Keel, boat-shaped structure formed by two lower petals in the pea family (Leguminosae), or a longitudinal ridge (typically on a leaf or petal).

Labellum, lower-most petal of an orchid flower, often highly specialised, e.g. in bee orchids (*Ophrys*) and *Serapias*.

Lanceolate, narrowly ovate, spear- or lance-shaped.

Legume, term used to describe either the fruit or the plant itself in the pea family (Leguminosae).

Lemma (of grasses), the lower of the pair of bracts (lemma and palea) that subtends the floret.

Ligule (of grasses), a small membranous projection or ring of hairs at the junction of the leaf sheath and stem.

Limb, expanded portion of the calyx or corolla (distinct from the tube or throat).

Linear, long, narrow and parallel (typically the leaf of a grass).

Lip, region of the calyx or corolla sharply differentiated from the rest (see also **labellum**).

Lobe, substantial division of a leaf, calyx or corolla.

Maquis, sclerophyllous shrub-dominated vegetation typical of the Mediterranean, often on deep, acidic substrates (see also **garrigue** and **phrygana**).

Membranous, paper-like or membrane-like in consistency.

Mericarp, 1-seeded portion formed by the splitting of a 2–many-seeded fruit.

Midrib, the central or main vein.

Monoecious, with separate male and female reproductive structures on the same individual plant; contrast dioecious.

Morphology, the appearance, form or structure.

Mucronate, with a short bristle-tip.

Mycoheterotrophy, process by which a non-photosynthetic plant obtains nutrition from a fungal symbiont (or sometimes from another plant via a shared fungal symbiont) living in its root system.

Native, naturally occurring in the area.

Naturalised, not native but well established.

Node, the position on the stem from which the leaves, branches or flowers arise.

Nut, dry, indehiscent, 1-seeded fruit with a woody, hard wall; often large in size.

Nutlet, a small, woody-walled nut (see also **achene**).

Obconical, an inverted cone, attached to the stalk by the pointed end.

Oblong, elongated but wide in shape, the middle part parallel-sided (usually describing a leaf).

Obovate, ovate (oval), and narrower at the base.

Obtuse, with a point $>90°$.

Opposite, of 2 organs arising from a common node (e.g. leaves from a stem).

Ovary, part of the carpel or pistil containing the ovules, and later, the seeds.

Ovate, oval in outline, or egg-shaped.

Ovoid, solid shape, oval/ovate in side view.

Ovule, structure containing the egg, becomes the seed after fertilisation.

Palea (of grasses), the upper of the pair of bracts (lemma and palea) that subtends the floret.

Palmate, lobes or segments radiating from a common axis.

Panicle, branched, compound inflorescence.

Pappus, structure consisting of hairs, bristles or scales on the fruit (achene) of Asteraceae.

Parasite, plant obtaining nutrients from another plant (may be hemi- or holo-parasitic).

Pedicel, the stalk of an individual flower in an inflorescence or the stalk of a grass spikelet.

Peltate, leaf with stalk attached to the centre of the blade.

Perennial, living for more than two years, generally flowering every year. Often woody at the base.

Perfoliate (of a leaf or bract), with the base united (around the stem).

Perianth, all the non-sexual segments (i.e. calyx and corolla together) of a flower (see also **tepals**).

Petal, one of the segments of the inner whorl(s) of the perianth; **petaloid** is petal-like.

Petiole, the stalk of the leaf; **petiolate** (with a petiole).

Phrygana, very stunted, and partially bare vegetation in dry, rocky areas; continuous with garrigue (and used synonymously in this guide).

Phyllode, a flattened, expanded petiole that resembles and functions as a leaf.

Phytogeographic, a geographic region defined by its flora (e.g. the *Mediterranean Floristic Region*).

Pinnate (of a compound leaf), composed of leaflets arranged on opposite sides of a common axis (rachis); adj. **pinnately** (e.g. pinnately divided into leaflets).

Pinnatifid (of a leaf), pinnately divided with the lobes cut nearly (but not quite) to the mid-vein.

Pinnatisect (of a leaf), pinnately divided to the mid-vein, but lobes not contracted at the base to form discrete leaflets.

Plumose, with many fine filaments or branches, giving a feathery or 'fluffy' appearance (e.g. the pappus of some Asteraceae).

Pollinium (pl. **pollinia**), a mass of adhering pollen grains (of e.g. an orchid) that is shed and transported as a unit by a pollinator.

Polyploid, with more than two sets of chromosomes.

Prickle, a spiny outgrowth, broadened at the base.

Procumbent, trailing on the ground.

Pseudocopulation, the process by which a male insect (usually a bee or wasp) attempts to mate with the flower of a bee orchid (*Ophrys*) and in so doing brings about cross fertilisation.

Raceme, a simple unbranched inflorescence with stalked flowers borne on a single axis, the youngest flowers at the top; adj. **racemose**.

Rachis, stalk of a compound leaf or the central axis bearing the flowers.

Ray, a radiating branch of an umbel or cyathium (e.g. in Apiaceae or Euphorbiaceae).

Ray floret, a small zygomorphic flower often resembling a single petal in the inflorescence (flower-head) of Asteraceae. Contrast with disk floret.

Receptacle, the portion of the axis of a flower stalk on which the flower is borne; the thickened or expanded part of the stem from which the flowers or inflorescence arise (e.g. in Asteraceae).

Reflexed, bent downwards or backwards.

Regular (of a flower), with a radially symmetrical (regular) shape that has multiple axes of symmetry (see also **actinopmorphic**), e.g. a *Lilium* flower.

Revolute, rolled back (e.g. inrolled margins of a petal).

Rhizome, horizontal underground stem; adj. **rhizomatous**.

Rosette (of leaves), radiating from a central point on the ground.

Samara, dry, indehiscent (non-splitting) 1-seeded fruit with a membranous, wing-like extension for dispersal.

Scape, flowering stem without leaves (e.g. all leaves basal).

Schizocarp, a dry fruit splitting into 2 1-seeded portions (mericarps), e.g. fruits of Apiaceae (see also **mericarp**).

Sclerophyll, hard leathery leaf containing a high proportion of thickened cells (sclereids); adj. **sclerophyllous**.

Sepal, typically leaf-like segments of the outer whorl(s) of the perianth (sometimes like a petal and then called a tepal).

Sessile, not stalked.

Simple, structure (e.g. a leaf) that is not divided into segments or lobes (not compound).

Spadix, spike-like organ bearing tiny male and female flowers at its base and surrounded by a spathe; characteristic of Araceae.

Spathe, large, leafy bract, sometimes brightly coloured; characteristic of Araceae.

Spike, an unbranched inflorescence (raceme) of stalkless flowers.

Spikelet, the basic unit of the inflorescence of grasses (Poaceae) and sedges (Cyperaceae); each flower cluster (e.g. Plumbaginaceae).

Spur, hollow, tubular projection originating from the sepals or petals, often containing nectar.

Stamen, male reproductive organ of the flower consisting of a filament and anther (microsporangium).

Staminode, aborted or sterile stamens (not pollen-producing), sometimes resembling petals.

Standard, upper petals of flowers of species in the pea family (Leguminosae) and iris family (Iridaceae).

Stellate, star-shaped, e.g. a hair.

Steppic, referring to plant communities (used here in a Mediterranean context) typical of extreme, arid, continental environments (steppes).

Stigma, the part of the carpel or pistil that receives pollen and upon which the pollen germinates.

Stipule, small, leaf-like organ at the base of some leaf petioles (stalks), often in pairs; either simple or lobed.

Stolon, a spreading, above-ground shoot, or runner; adj. **stoloniferous**.

Style, stalk on an ovary bearing the stigma(s); sometimes absent.

Subshrub, small shrub or perennial with woody stems.

Subspecies, taxonomic subdivision of a species; usually a geographically, morphologically and/or genetically distinct race.

Succulent, fleshy or juicy, e.g. stems of cacti and succulents.

Superior ovary, ovary positioned above the attachment points of other floral organs; syn. **hypogynous ovary**.

Syconium, fleshy, hollow receptacle that develops into a multiple fruit (a fig), typically associated with symbiotic, pollinating wasps.

Taxon (pl. **taxa**), taxonomic unit of any rank, for example species, genus, subspecies or variety.

Teeth, divisions of a leaf, calyx or corolla; adj. **toothed**.

Tepals, the segments of the perianth, often used to describe segments that are not clearly petals or sepals (particularly in Liliaceae and related families).

Ternate, compound leaf with 3 leaflets.

Terminal, at the top/apex.

Throat, the opening where the tube joins the limb of the corolla or calyx.

Tree, a woody plant typically >5m with a single trunk.

Trifoliate, compound leaf with 3 leaflets.

Tube, narrow, cylindrical part of the calyx or corolla, distinct from the limb, lobes or throat; adj. **tubular**.

Tufted, clustered together, e.g. a plant with numerous stems.

Umbel, flat-topped or convex 'umbrella-shaped' inflorescence consisting of a cluster of flowers with spreading stalks (pedicels) that arise from the apex of the peduncle; typical of Apiaceae.

Undulate, wavy at the margin.

Unisexual, flowers with either male or female reproductive organs only. Contrast cosexual.

Variety, form of a species that is geographically, morphologically and or genetically distinct (but not distinct enough to warrant subspecies status).

Vascular, pertaining to the veins (conducting tissue) of an organ; all plants in this book are vascular.

Viscid, sticky.

Whorl, group of lateral organs borne >2 at each node (e.g. petals of a flower).

Woody, hard and wood-like, typically persistent.

Woolly, clothed with soft, shaggy hairs.

Xerophyte, drought-tolerant plant.

Zygomorphic (of a flower), with a bilaterally symmetrical, irregular shape, e.g. an *Antirrhinum* flower.

Selected references

Danin, A., Fragman-Sapir, O. (2016 onwards). *Flora of Israel Online*. Published at http://flora.org.il/en/plants/

Danin, A., Orshan, G. (1999). *Vegetation of Israel 1. Desert and Coastal Vegetation*. Leiden, The Netherlands: Backhuys Pubishers.

Davis, P. H. (1965–1985). *Flora of Turkey and the East Aegean Islands (Volumes 1–9)*. Edinburgh: Edinburgh University Press.

Dimopoulos, P., Raus, Th., Bergmeier, E., Constantinidis, Th., Iatrou, G., Kokkini, S., Strid, A., Tzanoudakis, D. (2013). *Vascular Plants of Greece: An Annotated Checklist*. Berlin: Botanic Garden and Botanical Museum Berlin-Dahlem; Athens: Hellenic Botanical Society.

Dimopoulos, P., Raus, T., Strid, A. (2018 onwards). Flora of Greece web. Published at http://portal.cybertaxonomy.org/flora-greece/credits

Euro+Med. (2006 onwards). Euro+Med PlantBase – the information resource for Euro-Mediterranean plant diversity. Published at http://ww2.bgbm.org/EuroPlusMed/

Fielding, F., Turland, N., Matthew, B. (eds.). (2008). *Flowers of Crete*. Royal Botanic Gardens, Kew: Kew Publishing.

Flora Ionica Working Group. (2016 onwards). Flora Ionica – An Inventory of Ferns and Flowering Plants of the Ionian Islands (Greece). Published at https://floraionica.univie.ac.at

Hand, R., Hadjikyriakou, G. N., Christodoulou, C. S. (eds.). (2011 onwards). Flora of Cyprus – A Dynamic Checklist. Published at http://www.flora-of-cyprus.eu

Meikle, R. D. (1977, 1985). *Flora of Cyprus, Volumes 1 and 2*. Royal Botanic Gardens, Kew: The Bentham-Moxon Trust.

Papiomytoglou, V. The Greek Flora. Published at http://www.greekflora.gr

Strid, A. (2016). *Atlas of the Aegean Flora, Part 1 (text & plates), Part 2 (maps)*. Berlin: Botanic Garden and Botanical Museum Berlin.

Strid, A., Tan, K. (eds.). (1997). *Flora Hellenica, Volume 1: Gymnospermae to Caryophyllaceae*. Konigstein: Koeltz Scientific Books.

Strid, A., Tan, K. (eds.). (2002). *Flora Hellenica, Volume 2: Nymphaeaceae to Platanaceae*. Ruggell: A.R.G. Gantner Verlag.

Thorogood, C. (2016). *Field Guide to the Wild Flowers of the Western Mediterranean*. Royal Botanic Gardens, Kew: Kew Publishing.

Tutin, T. G., Heywood, V. H., Burges, N. A., Moore, D. M., Valentine, D. H., Walters, S. M., Webb, D. A. (eds.). (1964–1993). *Flora Europaea (Volumes 1–5)*. Cambridge: Cambridge University Press.

Zohary, M., Feinbrun-Dothan, N. (1966–1986). *Flora Palaestina (Parts 1–4, text & plates)*. Jerusalem: Israel Academy of Sciences and Humanities.

Index of English names

Entries with a photograph are indicated by a **bold** page number.

A
acacia 214
 false **220**
adenocarpus 224
albizia 214
alder 290
alexanders 593
 perfoliate 593
alkanet 439
 eastern **438**, 439
 large blue **438**, 439
Aleppo pine 41, 42, 44, 52, **53**
allseed 377
 four-leaved 377
almond 35, 36, **278**, 279
 Cretan wild **278**, 279
amaranth 15
amaranth family 387
anemone 22, 192
 crown **193**
 wood 192
angel's trumpet 456
annual bellflower 526
annual daisy 539
annual lavatera 328
annual mercury **306**
annual pearlwort 378
annual rock rose 337
annual scorpion vetch **264**, 265
annual seablite 396
annual sea heath **364**, 365
annual yellow vetchling **236**, 237
apple 280
apple mint **502**, 503
apple of Peru 450
apple of Sodom 455
Arabian fumana **338**, 339
arrowgrass family 74
arrowhead **72**, 73
artichoke 572
 globe 572
arum 48
 dragon 26, **72**
 Italian 64, **65**
 wild 66, 67
arum family 64
arum lily 71
ash 460
 flowering **458**, 460
 narrow-leaved ash **458**, 460
ashwagandha **454**, 455
asparagus 140
 Bath 141
asparagus pea **253**, 254
asphodel 127
 hollow-leaved **128**
 white **128**
 yellow 26, **128**, 129
aster,
 Goldilocks 537
 sea 537
autumn crocus 77
autumn hawkbit 555
autumn lady's tresses 100, **101**
autumn squill 146, **147**
avocado 62, **63**

B
Balkan maple 322
balloon-vine 322, **323**
balm 489
balsam family 403
banana **156**
banana family 156
barbary nut 115, **116**
barley, wall 174
bartsia,
 red 510, **511**
 southern red 510, **511**
 yellow bartsia 510, **511**
bastard cabbage 360, **361**
Bath asparagus 141
bay laurel 62
beaked hawk's-beard 563
bean broomrape **520**, 522
bean trefoil **220**, 221
bearded iris 117
bearded oat **165**
bear's breech 476, 477
 spiny **476**, 477
bedstraw 414
 lady's 416
 round-leaved 414
beech 22
bee orchid 28, 102, **106**
beet 394
 sea 394
belladonna lily 48
bellflower 15, 523
 annual 526
 narrow-leaved 524
 spreading 524
bellflower family 523
Bermuda buttercup **292**, 293
Bermuda grass **176**, 177
berry catchfly 385
bindweed 445, 446, **447**
 great **444**, 445
 hedge 445
 mallow-leaved 446, **447**
 sea 445
bird of paradise **155**
bird's foot 255
bird's-foot restharrow 240
bird's-foot trefoil 252
bird's nest orchid **92**, 93
birthwort 62, **63**
birthwort family 57
bitter orange 36
black bog-rush 161
black bryony 76, **77**
black medick 245, **247**
black mustard 358
black nightshade **454**, 455
black pine 30
black vetch 237
bladder campion 383
bladder dock 372
bladder hibiscus 331, **332**
bladder senna 225
bladder vetch 254, **255**
bladderwort,
 greater **480**
 lesser 481

bladderwort family 480
blinks family 400
blue eryngo **590**, 591
blue fleabane **536**, 537
blue hound's tongue 440, **441**
blue-leaved wattle **215**
blue lettuce 559
blue potato bush 450, **451**
blue water-speedwell 470
blue woodruff 414
bogbean family 528
bog-rush, black 161
borage 15, **436**, 438
borage family 427
bottlebrush **315**, 316
bougainvillea 400
box **202**
box family 203
bramble 275
branched broomrape **516**
branched bur-reed 157
branched horehound 485
branched plantain 463
bristle club-rush 161
bristle-fruited silkweed 423
bristle grass 179
　green 179, **180**
broadbean 236
　purple **236**, 237
broad-leaved cudweed 529
broad-leaved everlasting pea 239
broad-leaved glaucous spurge **303**
broad-leaved helleborine 91
brome 162
　compact 164
brookweed **404**, 405
broom 221
　butcher's 41, **152**, 153
　hairy thorny **220**, 221
　Spanish **220**, 221
　white **223**, 224
broomrape 45, 47, 516, 517
　bean **520**, 522
　branched **516**
　clove-scented **519**, 521
　common 517, **518**
　ivy 518, **519**
　knapweed 521
　oxtongue 518, **519**
　slender **520**, 521
　sunflower 522
　thistle **519**, 521

　thyme 521
　yellow 521
broomrape family 510
brown-rayed knapweed 578
bryony 293
　black 76
buckler mustard 356
buck's horn plantain **462**, 463
buckthorn 281, 282
　Mediterranean **282**
　rock 281
bugle **482**, 483
　eastern 483
　southern **482**, 483
bugloss 436
　pale **436**, 437
　small-flowered 436
　viper's **436**, 437
bug orchid **97**
bullrush family 156
bumblebee orchid 104, **105**
bunch-flowered daffodil 132, **133**
bur-marigold 550
　trifid 550
bur-reed, branched 157
burnet 273
　salad **272**, 273
　thorny 273, **274**
burnet rose 273
burning bush **324**, 325
burnt orchid 96
bur-reed 157
bush vetch 237
butcher's broom 41, **152**, 153
buttercup 194
　Bermuda 293
　Jersey **195**, 196
　rough-fruited 194
　turban 48
buttercup family 192
butterwort 480

C

cabbage 14, 358
　bastard 360, **361**
　southern warty 349
　violet **351**, 352
　wild **358**, 359
cabbage family 343
cactus family 401
Calabrian soapwort 380
calamint,
　large-flowered 496

　lesser 495, **497**
Californian pepper tree 318, **319**
campion 380
　bladder **383**
Canadian fleabane **536**, 537
Canary palm **154**, 155
candelabra tree 300, **301**
candytuft 355
　evergreen **354**, 355
Cape honeysuckle 478, **479**
caper 342
　Syrian bean 212
caper family 343
caralluma **421**
cardoon 572, **573**
carline thistle 566, 568
carob 35, 36, 41
carob tree 217, **219**
carrot family 591
carrot 14, 591, 608
　deadly 605
　wild 608, **609**
cassia 218, **219**
　popcorn 218, **219**
castor oil plant **306**
cat mint 489
catchfly,
　berry 385
　Cretan 380
　Italian 384
　night-flowering 383
　small-flowered 382
cat's-ear 554
　common 554, **555**
　smooth 554
cedar 33, 50
　Cyprus 28
　Lebanon 30
cedar of Lebanon 50, **51**
celery-leaved crowfoot 196
centaury 418
　common **418**
　lesser 418
　slender 418
　spiked **418**, 419
　yellow 419
century plant **150**, 151
chamomile 541
　corn 542
　woolly 542
　yellow 543
charlock 359
chaste-tree **482**
cherry 276

cornelian **401**, 402
  prostrate 277
cherry plum **277**
chestnut 42
chicory **551**
  spiny 551
Chilean wine palm 155
Chinese privet 460, **461**
Chinese wisteria 268, **269**
chives 136
cinquefoil 275
  creeping 275
citron **324**, 325
  fingered 36
citrus 35, 36
clary 506
  whorled 508
  wild 506, **507**
classical fenugreek 244
clematis,
  fragrant 200, **201**
clover 256
  crimson 257
  hare's-foot 256, **257**
  narrow-leaved crimson 256, **257**
  purple 256, 257
  red 260, **261**
  reversed 260, **261**
  rough 258
  spiny 260
  star 260, **261**
  strawberry 260
  suffocated 257
  sulphur 260
  white 258, **259**
clove-scented broomrape **519**, 521
club rush, bristle 161
cocklebur 550
  rough 550, **551**
  spiny 550, **551**
cockscomb sainfoin 268, **269**
cock's-foot 169
columbine 201
common bird's-foot trefoil 252
common broomrape 517, **518**
common cat's-ear 554, **555**
common centaury 418
common cornsalad 588
common cudweed **528**, 529
common dandelion 562
common dodder **443**, 444
common eelgrass 75

common field-speedwell 470
common figwort 473
common fleabane 534, **535**
common fumitory **189**, 190
common glasswort 394
common grape hyacinth 149, **150**
common hibiscus 331
common juniper 54, **55**
common mallow **329**, 330
common milkwort 270, **271**
common myrtle **315**, 316
common orache 392
common passion flower 298, **299**
common poppy 182, **183**
common privet 460
common rock rose 339
common rue 324
common self-heal **489**, 490
common smilax 80, **81**
common spike-rush 161
common star of Bethlehem **142**, 143
common stork's-bill **312**
common toadflax 467
common valerian 588
common vetch 237
compact brome 164
compact dock 373
convolvulus,
  dwarf 445, **447**
  pink 446, **447**
convolvulus family 442
coral necklace **375**, 376
corky-fruited water-dropwort 594, **595**
corn chamomile 542
corncockle **379**, 380
corn crowfoot 196
cornelian cherry **401**, 402
cornflower 574, 578
corn marigold **544**, 545
cornsalad 587
  common 588
  hairy-fruited 588
  narrow-fruited 588
  spiny 588
cornus family 402
Corsican pine 52, **53**
cotton 327
cotton bush 397, **398**
cottonweed **541**
cowslip **404**, 405

creeping cinquefoil 275
creeping thistle 570
crane's-bill,
  cut-leaved **310**, 311
  dove's-foot **310**
  long-stalked 311
  shiny **310**, 311
  small-leaved 310
cress,
  hoary 356
  shepherd's 361
Cretan catchfly 380
Cretan wild almond **278**, 279
crimson clover 257
crocus,
  autumn 77
  sand 124, **125**
crosswort 413
crowfoot,
  celery-leaved 196
  corn 196
crow garlic 136
crown anemone **193**
crown daisy **544**, 545
crown vetch **264**, 265
crucianella 412
cruel vine **423**
cucumber, squirting 291, **292**
cucumber family 291
cudweed 529, 530, 531
  broad-leaved 529
  common **528**, 529
  heath 531
  Jersey 531
  marsh 531
  narrow-leaved 530
  small 530
cultivated flax 308
curry 531
curved sea hard-grass 172
cut-leaved crane's-bill **310**, 311
cut-leaved self-heal 490
cut-leaved viper's-grass 557
cyclamen 33, 35, 48, 406
  ivy-leaved 406, **407**
  Persian **408**, 409
cypress 28, 54, **55**
  Monterey 54
cypress spurge **303**
Cyprus cedar 28
cytinus 326, **327**

D
daffodil 48, 132

INDEX OF ENGLISH NAMES

bunch-flowered 132, **133**
pheasant's-eye 132
sea 44, 132
winter **130**, 131
daisy family 529
daisy 14, **539**
   annual 539
   crown **544**, 545
   ox-eye 545
   southern **538**, 539
damask rose 273
dandelion 562
   common 562
dark-red helleborine 90
deadly carrot 605
deadly nightshade 450, **451**
dead-nettle 491
   henbit **490**, 491
   large red **490**, 491
   red **490**, 491
   spotted 491
dense-flowered orchid **97**
Deptford pink **384**, 385
desert gourd 291, **292**
desert hyacinth **147**
diamond flower 348
dock,
   bladder **372**
   compact 373
   horned **372**
dock family 371
dodder 442
   common **443**, 444
   field 445
dog-rose **272**
dogwood 402
dorycnium **262**, 263
dove's-foot crane's-bill **310**
downy oak 288, **289**
downy woundwort **493**
dragon arum 26, **72**
dragon fruit **401**
drooping star-of-Bethlehem 141, **142**
Duke of Argyll's teaplant 451
dwarf convolvulus 445, **447**
dwarf fan palm **152**, 153
dwarf mallow 330
dwarf rush 157
dwarf spurge 302
Dyer's greenweed 222, **223**

E
early forget-me-not **441**

early purple orchid 96
early star-of-Bethlehem **82**, 83
eastern alkanet **438**, 439
eastern bugle 483
eastern hornbeam 291
eastern rocket 349
Eastern strawberry tree 42
edible cyperus 160
edible lotus 252, **253**
eelgrass, common 75
eelgrass family 75
Egyptian St John's-wort 295
elder 582
   European dwarf 582
elder-flowered orchid 102
elm, Mediterranean **282**, 283
elm famiy 283
emex 373, **374**
enchanter's nightshade 22
eryngo,
   blue **590**, 591
   field **590**, 591
euphorbia family 15, 300
European dwarf elder 582
European nettle tree **282**, 283
evergreen candytuft **354**, 355

F
false acacia **220**
false sainfoin 235
fat hen 393
fennel **595**
   giant **600**, 601
fenugreek 243
   classical 244
   sickle-fruited 243
   star-fruited 245, **247**
fern-grass 170, **171**
fiddle-leaf morning glory **448**, 449
field dodder 445
field eryngo **590**, 591
field marigold **548**, 549
field sage 506
field scabious **586**
field southernwood 539
field-speedwell,
   common 470
   grey **471**
field woundwort **494**
fig 35, 36, **285**
   Hottentot 397
fig family 284
figwort,

common 473
French 472
nettle-leaved 473, **474**
figwort family 472
fine-leaved sandwort 374
fingered citron 36
fingered speedwell 470
fir 33, 50
   Greek **50**, 51
   Macedonian 22, 50
   Taurus 30
fire thorn 276
flag, yellow 113, **115**
flat-topped carline thistle **567**, 568
flax,
   cultivated 308
   pale 308
   upright yellow 307, **308**
   yellow 308
flax family 307
fleabane 534, 535, 537
   blue **536**, 537
   Canadian **536**, 537
   common 534, **535**
flowering ash **458**, 460
flowering rush family 74
fly orchid 102, **105**
fodder vetch 234, **235**
forget-me-not 441
   early 441
   tufted 441
forking larkspur 200
four-leaved allseed 377
foxglove 469
   rusty **469**
foxglove tree **508**, 509
foxtail 168
fragrant clematis 200, **201**
fragrant orchid 101
French figwort 472
French lavender **504**
French oat-grass **165**, 166
French rocket **349**
French speedwell 470
friar's cowl 69, **70**
fringed rue **323**, 324
fringed water-lily 528
fritillary 86
fumana,
   Arabian **338**, 339
   procumbent 340
   thyme-leaved **338**, 339
fumitory, common **189**, 190

INDEX OF ENGLISH NAMES | 623

fungus, Maltese **210**, 211
furrowed melilot 242
Fyfield pea 239

G
galactites **570**, 571
garlic,
   crow 136
   Naples **135**, 136
   rosy **138**, 139
gentian family 418
geranium family 309
germander 483
   wall **484**, 485
   water 483
   yellow **484**
germander speedwell 470, **471**
giant fennel **600**, 601
giant orchid 100, **101**
giant reed 175, **176**
glasswort 394
   common 394
globe artichoke 572
globe thistle 566, **567**
globularia, matted 472
goat's beard 557
goat's rue 255
golden dog's tail 170, **171**
golden oak 288, **289**
golden samphire **536**
Goldilocks aster 537
good King-Henry 394
goosefoot 392
   maple-leaved 393
   sticky 394
   stinking 393
goose-grass **415**
grape, sea 371, **372**
grape family 211
grapefruit 325
grape hyacinth 149
   common 149, **150**
grass 14
   Bermuda 177
   bristle 179
   green bristle 179, **180**
   Johnson 178, **180**
   large quaking 167
   love 177
   marram 44, 167
   Mediterranean hair **165**, 166
   pampas 181
   ravenna 178, **180**
   rye 169

stalked bur 177
   torpedo 181
grass family 162
grass poly 313
grass vetchling 238
great bindweed **444**, 445
greater bladderwort **480**
greater butterfly orchid **101**
greater musk-mallow **329**, 330
greater periwinkle 420, **421**
greater plantain 464, **465**
greater sea-spurrey 378
great reedmace 156
great willowherb 316
Greek fir 50, **51**
green bristle grass 179, **180**
greenweed 222
   Dyer's 222, **223**
green-winged orchid **98**, 99
grey field-speedwell **471**
gromwell, yellow **433**, 434
ground pine **482**, 483
groundsel 546, **547**
   sticky 546
gum, river red **315**, 317
gum rock rose 336
gypsywort **504**

H
hare's-ear,
   shrubby **597**, 598
   slender 597
hairy-fruited cornsalad 588
hairy medick 249, **250**
hairy rocket 360
hairy rupture-wort **377**
hairy sea heath 365
hairy tare 234, **235**
hairy thorny broom **220**, 221
hairy trefoil 261
hairy vetchling 239
hairy violet 297
hard-grass 172
   curved sea 172
hare's-foot clover 256, **257**
hare's tail 166
Hawaiian lily **458**, 459
hawkbit 554
   autumn 555
   rough 554
   star 554, **555**
hawk's-beard 562
   beaked 563
   pink **563**, 564

   rough 562
hawkweed 564
hawkweed oxtongue **556**
hawthorn **274**, 275
heart-flowered orchid 111, **112**
heath,
   tree 37, **410**, 411
heath cudweed 531
heather 37, 411
heather family 409
hedge bindweed 445
hedge mustard 348
hedge parsley 605
heliotrope **433**
helix 590
helleborine 90, 91
   broad-leaved 91
   dark-red 90
   marsh 90
   red 92
   small-leaved 90
   sword-leaved **91**, 92
   white **91**, 92
hemlock 596
hemp **282**, 283
hemp family 283
hemp-nettle 486
   narrow-leaved 486
henbane 452, **453**
   white **452**, 453
henbit dead-nettle **490**, 491
herb paris 76, **77**
herb Robert **310**, 311
hibiscus 331
   bladder 331, **332**
   common 331
hoary cress 356
hoary mustard 359
hoary plantain 464
hoary stock **354**, 355
hollow-leaved asphodel 128
holly **522**, 523
   sea 590, **591**
holly family 523
hollyhock 331, **332**
Holm oak 41, 288, **289**
honeysuckle **583**
   Cape 478, **479**
honeysuckle family 582
honeywort **434**, **435**
   lesser 434, **435**
hop trefoil 256, **257**
horehound 485
   branched 485

white 485
hornbeam 291
  eastern 291
horned dock 372
horn-wort, spineless 181
horn-wort family 181
horsechestnut 320, **321**
horse mint 503
horseshoe vetch 266
Hottentot fig 397
hound's tongue 440
  blue 440, **441**
hyacinth,
  desert **147**
  tassel **148**
hyssop 498

I
ice plant **398**, 399
Indian bead tree 326, **327**
Indian bean tree 478
Indian laurel 284
inula, willow-leaved 536
iris 113
  bearded 117
  widow 115, **116**
iris family 113
Italian arum 64, **65**
Italian catchfly 384
Italian man orchid 94, **95**
Italian sainfoin 268, **269**
ivy 590
ivy broomrape 518, **519**
ivy family 590
ivy-leaved cyclamen 406, **407**
ivy-leaved speedwell **471**
ivy-leaved toadflax 468, **469**

J
jacaranda 478, **479**
Japanese loquat **280**
jasmine 459
  night 458
  white 459
  wild 459
Jersey buttercup **195**, 196
Jersey cudweed 531
Jersey toadflax **466**, 467
Jerusalem sage **487**
Johnson grass 178, **180**
joint pine family 56
Judas tree 218, **219**
juniper 33, 41
  common 54, **55**

Phoenicean 54, **55**
prickly 54, **55**
juniper family 54

K
Kermes oak 41, **286**, 287
knapweed 45, 574
  brown-rayed 578
knapweed broomrape 521
knotted hedge parsley 605
knotgrass, sea 371, **372**
kohlrauschia 386, **387**
kundmannia 595

L
lady orchid, **95**
lady's bedstraw 416
lady's tresses 100, **101**
  autumn 100, **101**
lagoecia **592**
lantana 481
large blue alkanet **438**, 439
large disk medick 246, **247**
large-flowered calamint 496
large Mediterannean spurge
  300, **301**
large milkwort 270
large quaking grass 167
large red dead-nettle **490**, 491
large self-heal 491
large thyme 499
large Venus's looking glass 527
large yellow restharrow 240
large yellow vetch **236**
larkspur 200
  forking 200
  violet **198**, 199
laurel,
  bay 62
  Indian 284
  spurge 334, **335**
laurel family 62
lavatera, annual 328
lavender 504
  French **504**
  sea 44, 367
  winged 367, **368**
least lettuce 560
Lebanon cedar 30
leek,
  round-headed 134, **135**
  wild 134, **135**
lemon 35, **324**, 325
leopard's-bane 548

lesser bladderwort 481
lesser calamint 495, **497**
lesser celandine **197**
lesser centaury 418
lesser honeywort 434, **435**
lesser hop trefoil 256
lesser milk-vetch 227
lesser sea spurrey 378, **379**
lesser snapdragon **465**
lesser spearwort **195**, 197
lesser swinecress 356, **357**
lesser water-plantain 73
lettuce 559
  blue 559
  least 560
  pliant 560
  prickly 560
lily 14
  arum 71
  belladonna 48
  Hawaiian **458**, 459
  Madonna 80, **81**
  St Bernard's 139, **152**, 53
lily family 80
lime **324**, 325
limodore 92
  violet **92**
limoniastrum **369**, 370
liquorice **233**
  spiny-fruited **233**
little robin 311
lizard orchid 100
London rocket 349
long-headed poppy 182, **183**
long-lipped serapias 111
long-stalked crane's-bill 311
loose-flowered orchid 99
loosestrife, purple **312**, 313
loosestrife family 313
loquat 280
  Japanese **280**
love grass 177
love-in-a-mist **198**
Lucerne 245, **247**
lupin 224
  white 224
  yellow 224
lyrata 473

M
Macedonian fir 22, 50
madder, wild 411, **412**
madder family 411
Madonna lily 80, **81**

magnolia **63**
magnolia family **62**
maize **178**
mallow **15**, **328**
  common **329**, **330**
  dwarf **330**
  marsh **331**, **332**
  musk- **329**, **330**
  rough marsh **328**
  small tree **329**, **330**
  tree **329**, **330**
mallow family **327**
mallow-leaved bindweed **446**, **447**
mallow-leaved stork's-bill **311**
Maltese fungus **210**, **211**
mandrake **456**, **457**
man orchid **93**, **94**
maple **320**
  Balkan **322**
  Montpelier **322**
  Syrian **321**, **322**
maple-leaved goosefoot **393**
marigold **549**
  corn **544**, **545**
  field **548**, **549**
marjoram,
  pot **501**, **502**
  sweet **502**
marram **167**
marram grass **44**, **167**
marsh cudweed **531**
marsh helleborine **90**
marsh mallow **331**, **332**
marsh pennywort **592**
marsh willowherb **316**
marshwort **598**
marvel of Peru **400**
mastic **2**, **318**, **319**
matted globularia **472**
meadow-rue **194**
meadow saffron **78**, **79**
medick **245**
  black **245**, **247**
  hairy **249**, **250**
  large disk **246**, **247**
  sea **248**, **250**
  small **249**, **250**
  spotted **248**
  tree **245**, **247**
Mediterranean buckthorn **282**
Mediterranean elm **282**, **283**
Mediterranean hair grass **165**, **166**

Mediterranean kidney vetch **262**, **263**
Mediterranean meadow saffron **77**
Mediterranean medlar **274**, **275**
Mediterranean sandwort **374**
Mediterranean stork's-bill **313**
Mediterranean water-dropwort **595**
Mediterranean white violet **297**
Mediterranean willow **299**
Mediterranean woundwort **493**
medlar, Mediterranean **274**, **275**
melilot **242**
  furrowed **242**
  small **242**, **243**
  tall **242**
  white **242**
membranous nettle **285**
mercury, annual **306**
mignonette,
  white **340**, **341**
  wild **341**
mignonette family **340**
military orchid **94**, **95**
milk thistle **573**, **574**
milk-vetch,
  lesser **227**
  Montpellier **226**, **227**
milkwort,
  common **270**, **271**
  large **270**
  Nice **270**
milkwort family **270**
milky orchid **96**, **97**
mint **14**, **503**
  apple **502**, **503**
  cat **489**
  horse **503**
  spear **503**
  water **502**, **503**
mint family **481**
mirror orchid **102**, **105**
mistletoe **20**, **363**, **364**
monkey orchid **93**, **94**
monkey-puzzle family **56**
Monterey cypress **54**
Montpelier maple **322**
Montpellier milk-vetch **226**, **227**
morning glory **449**
  fiddle-leaf **448**, **449**
mossy stonecrop **206**
mournful widow **583**, **584**

mouse-ear **376**
  sticky **375**, **376**
mullein,
  nettle-leaved **475**
  purple **476**
  white **475**
musk-mallow **329**, **330**
  greater **329**, **330**
musk stork's-bill **312**, **313**
mustard **359**
  black **358**
  buckler **356**
  hedge **348**
  hoary **359**
  white **358**, **359**
myrtle, common **315**, **316**
myrtle family **316**

N
Naples garlic **135**, **136**
narcissus, paperwhite **132**, **133**
narrow-fruited cornsalad **588**
narrow-leaved ash **458**, **460**
narrow-leaved bellflower **524**
narrow-leaved crimson clover **256**, **257**
narrow-leaved cudweed **530**
narrow-leaved everlasting pea **238**, **239**
narrow-leaved glaucous spurge **304**, **305**
narrow-leaved hemp-nettle **486**
narrow-leaved red vetchling **238**
narrow-leaved rock rose **336**, **337**
nasturtium family **340**
navelwort **210**
nettle,
  membranous **285**
  Roman **285**
  small **285**
  stinging **284**
nettle family **284**
nettle-leaved figwort **473**, **474**
nettle-leaved mullein **475**
Nice milkwort **270**
night-flowering catchfly **383**
night jasmine **458**
nightshade **455**
  black **454**, **455**
  deadly **450**, **451**
  enchanter's **22**
  woody **454**, **455**

nit-grass 168
nodding thistle **569**
Norfolk Island pine 56

O
oak 28, 33, 42
  downy 288, **289**
  golden 288, **289**
  Holm 41, 288, **289**
  Kermes 41, **286**, 287
oak family 287
oat 165
  bearded 165
oat-grass, French **165**, 166
oleander 420, **421**
oleaster **280**, 282
olive 35, 36, 41, 460, **461**
olive family 459
ophrys, yellow 104, **105**
opium poppy 182
orache 390
  common 392
  shrubby 390, **391**
  spear-leaved **391**, 392
orange 35
  bitter 36
  Seville **324**, 325
  sweet 325
orange bird's foot 255
orchid 15, 20, 35
  bee 28, 102, 106
  bird's nest 93
  bug **97**
  bumblebee 104, **105**
  burnt orchid 96
  dense-flowered **97**
  early purple 96
  elder-flowered 102
  fly 102, **105**
  fragrant 101
  giant 100, **101**
  greater butterfly **101**
  green-winged **98**, 99
  heart-flowered 111, **112**
  Italian man 94, **95**
  lady, **95**
  lizard 100
  loose-flowered 99
  man 93, **94**
  military 94, **95**
  milky 96, **97**
  mirror 102, **105**
  monkey 93, **94**
  pink butterfly **98**, 99

  provence 96
  pyramidal **98**, 99
  Roman 102
  sawfly 104, **105**
  sombre bee 104, **105**
  tongue 111, **112**
  woodcock **106**
orchid family 90
orlaya 608
oxalis, pink **293**, 294
oxalis family 293
ox-eye daisy 545
oxtongue 556
  hawkweed **556**
oxtongue broomrape 518, **519**

P
pale bugloss **436**, 437
pale flax 308
pale globe thistle 566
pale persicaria 371
palm,
  Canary **154**, 155
  Chilean wine 155
  dwarf fan **152**, 153
palm family 153
pampas grass 181
paperwhite narcissus 132, **133**
papyrus **159**, 160
parsley,
  hedge 605
  knotted hedge 605
parsnip 599
  wild 599
pasqueflower 194
passion flower, common 298, **299**
passion flower family 298
pea 14, 237, 239
  asparagus **253**, 254
  broad-leaved everlasting 239
  Fyfield 239
  narrow-leaved everlasting **238**, 239
  wild 239
pea family 214
peach **278**, 279
pear **278**, 279
pearlwort 378
  annual 378
  procumbent 378
pellitory-of-the-wall **286**
penny-royal 503
pennywort 592

  marsh **592**
peony family 203
pepper tree, Californian 318, **319**
pepperwort 356
perfoliate alexanders 593
periwinkle 420
  greater 420, **421**
periwinkle family 419
Persian cyclamen **408**, 409
persicaria, pale 371
petty spurge **301**, 302
pheasant's-eye daffodil 132
Phoenicean juniper 54, **55**
pigweed 388, **389**
pimpernel 404
  scarlet **404**
pine 28, 33, 52
  Aleppo 41, 42, 44, 52, **53**
  black 30
  ground **482**, 483
  Norfolk Island 56
pine family 50
pink 14, 385
  Deptford **384**, 385
  proliferous 386
  wood pink 386
pink family 374
pink butterfly orchid **98**, 99
pink convolvulus **446**, **447**
pink hawk's-beard **563**, 564
pink oxalis **293**, 294
pink water-speedwell 470
pistacio nut 318
plane tree family 202
plantain 15
  branched 463
  buck's horn **462**, 463
  greater 464, **465**
  hoary 464
  ribwort 464
  silvery 464
plantain family 463
pliant lettuce 560
ploughman's spikenard **535**
plum, cherry **277**
pokeweed **398**, 399
pomegranate 314, **315**
pond water-crowfoot **195**, 197
popcorn cassia 218, **219**
poppy 15
  common 182, **183**
  long-headed 182, **183**
  opium 182

prickly 184
  red-horned 186, **187**
  yellow horned- **187**
poppy family 182
Port St John's creeper **479**
potato bush, blue 450, **451**
potato family 450
potato vine 455
pot marjoram 501, **502**
prasium **484**, 485
prickly juniper 54, **55**
prickly lettuce 560
prickly pear 35, **401**, 402
prickly poppy 184
prickly saltwort **395**, 396
prickly sow-thistle 558
pride of Madeira 437
primrose **404**, 405
primula family 403
privet 460
  Chinese 460, **461**
  common 460
procumbent fumana 340
procumbent pearlwort 378
procumbent yellow sorrel **292**, 293
proliferous pink 386
prostrate cherry 277
provence orchid 96
pseudorlaya 608, **609**
purple broadbean **236**, 237
purple clover 256, 257
purple loosestrife **312**, 313
purple mullein 476
purple sainfoin 268, **269**
purple spurge **304**, 305
purslane 400
  sea 390, **391**
  water 314
putoria 412, **413**
pycnocomon 585
pyramidal orchid **98**, 99

Q
quince **278**, 279

R
radish 360
  wild 360, **361**
ragwort 547
ramping-fumitory, white 189
rampion 524
ravenna grass 178, **180**
red bartsia 510, **511**

red clover 260, **261**
red dead-nettle **490**, 491
red helleborine 92
red-horned poppy 186, **187**
redshank 371
red star-thistle **575**, 577
red stonecrop **208**, 209
red-topped sage 506, **507**
red valerian **586**, 587
red vetchling 238
  narrow-leaved **238**
reed, giant 175, **176**
reedmace, great 156
restharrow 239
  bird's-foot 240
  large yellow 240
  spiny 240, **241**
resurrection plant **548**, 549
reversed clover 260, **261**
rhododendron, yellow **410**, 411
ribwort plantain 464
rice 174
ridolfia 598
river red gum **315**, 317
rock buckthorn 281
rocket 348
  eastern 349
  London 349
  sea 44, **358**, 359
  tall 349
  wall 357
rock rose 15
  annual **337**
  common 339
  gum 336
  narrow-leaved 336, **337**
  sage-leaved 336, **337**
  white **338**, 339
  willow-leaved 338
rock rose family 336
rock samphire 594
roemeria 184, **185**
Roman nettle **285**
Roman orchid 102
rose 15
  burnet 273
  damask 273
rose family 271
rosemary **504**, 505
rosy garlic **138**, 139
rough clover 258
rough cocklebur 550, **551**
rough dog's-tail 170
rough-fruited buttercup 194

rough hawkbit 554
rough hawk's-beard 562
rough marsh mallow 328
round-headed leek 134, **135**
round-leaved bedstraw 414
round-leaved birthwort 57, **58**
rubber plant 284
rue 324
  common 324
  fringed **323**, 324
  goat's 255
  meadow- 194
rupture-wort 377
  hairy **377**
rush,
  black bog- 161
  dwarf 157
  flowering 74
  sea 157
  sharp 158, **159**
  soft 157
rush-like scorpion vetch **264**
rusty foxglove **469**
rye grass 169

S
saffron,
  meadow **78**, 79
  Mediterranean meadow **77**
sage 505, **507**
  field 506
  Jerusalem **487**
  red-topped 506, **507**
  silver 506, **507**
  three-lobed 505, **507**
sage-leaved rock rose 336, **337**
sainfoin 268
  cockscomb 268, **269**
  false 235
  Italian 268, **269**
  purple 268, **269**
salad burnet **272**, 273
sallowwort **424**, 425
salsify 558, **559**
  western 557
saltwort 396
  prickly **395**, 396
samphire,
  golden **536**
  rock 594
sandbur 179
sand crocus 124, **125**
sand spurrey 379
sandwort 374

fine-leaved 374
thyme-leaved 374
Mediterranean 374
satureja 495, **497**
sawfly orchid 104, **105**
saxifrage family 205
scabious 583
  field **586**
  sweet 537
scarlet pimpernel **404**
scilla 144
scorpion 264
scorpion vetch, annual **264**, 265
Scotch thistle 571, 572, **573**
Scots pine 52
sea aster 537
sea beet 394
sea bindweed **445**
seablite 395
  annual 396
  shrubby **395**
sea daffodil 44, 132
sea grape 371, **372**
seagrass **75**
sea heath, annual **364**, 365
sea heath family 365
sea holly 590, **591**
seakale 360
sea knotgrass 371, **372**
sea lavender 44, 367
  winged 367, **368**
sea medick 248, **250**
sea purslane 390, **391**
sea rocket 44, **358**, 359
sea rush 157
sea spurge **304**, 305
sea-spurrey 378
  greater 378
  lesser 378, **379**
sea squill 37, 144, **145**
sedge 158
self-heal 490
  common **489**, 490
  cut-leaved 490
  large 491
senna, bladder **225**
serapias, long-lipped 111
Seville orange **324**, 325
sharp rush 158, **159**
sheep's-bit 527
sheep's sorrel 373
shepherd's cress 361
shepherd's needle **594**, **595**
shepherd's purse 348, **349**

shiny crane's-bill **310**, 311
short-fruited willowherb 316
shrubby hare's-ear **597**, 598
shrubby orache 390, **391**
shrubby seablite **395**
shrubby sphericalia 472
shrub tobacco **458**
Sicilian snapdragon **465**
sickle-fruited fenugreek 243
sickle spurge 302
silk-vine **424**, 427
silkweed 423
  bristle-fruited 423
silver sage **506**, **507**
silver wattle 214
silvery plantain 464
skullcap 501
slender bird's-foot trefoil 254
slender broomrape **520**, 521
slender centaury 418
slender hare's-ear 597
small caltrops 212, **213**
small cudweed 530
small-flowered bugloss 436
small-flowered catchfly 382
small-flowered hairy willowherb 316
small-flowered willowherb 316
small-leaved crane's-bill 310
small-leaved helleborine 90
small medick 249, **250**
small melilot 242, **243**
small nettle **285**
small tree mallow **329**, 330
smilax, common 80, **81**
smoke tree 320, **321**
smooth cat's-ear 554
smooth sow-thistle 559
snake's-head 115, **116**
snapdragon 464, **465**
  lesser **465**
  Sicilian **465**
snowdrop 129
soapwort **379**, 380
  Calabrian 380
soft rush 157
sombre bee orchid 104, **105**
sorrel,
  procumbent yellow 292, **293**
  sheep's 373
southern bird's foot trefoil 252, **253**
southern bugle **482**, 483
southern daisy **538**, 539

southern red bartsia 510, **511**
southern warty cabbage 349
sow-thistle 558
  prickly 558
  smooth 559
Spanish broom **220**, 221
Spanish oyster plant 550, **551**
spear-leaved orache **391**, 392
spear mint 503
spear thistle **570**
spearwort, lesser **195**, 197
speedwell 470
  fingered 470
  French 470
  germander 470, **471**
  ivy-leaved **471**
  thyme-leaved 470, **471**
  wall 470
spicate Venus's looking glass 528
spiked centaury **418**, 419
spike-rush, common 161
spindle family 293
spineless horn-wort 181
spiny bear's breech **476**, 477
spiny chicory 551
spiny clover 260
spiny cocklebur 550, **551**
spiny cornsalad 588
spiny-fruited liquorice **233**
spiny restharrow 240, **241**
spotted dead-nettle 491
spotted medick 248
spreading bellflower 524
spring vetch 237
spurge 300
  broad-leaved glaucous **303**
  cypress **303**
  dwarf 302
  large Mediterannean 300, **301**
  narrow-leaved glaucous **304**, 305
  petty **301**, 302
  purple **304**, 305
  sea **304**, 305
  sickle 302
  sun **301**, 302
  tree 300, **301**
spurge laurel 334, **335**
square-stalked St John's-wort 294
square-stalked willowherb 316
squill,

autumn 146, **147**
  sea 37, 144, **145**
squirting cucumber 291, **292**
stalked bur grass 177
star clover 260, **261**
star-fruited fenugreek 245, **247**
star hawkbit 554, **555**
star-of-Bethlehem 81, 141
  common **142**, 143
  drooping 141, **142**
  early **82**, 83
St Barnaby's thistle 574, **575**
St Bernard's lily 139, **152**, 53
stemless stork's-bill 312
sternbergia 48
sticky goosefoot 394
sticky groundsel 546
sticky mouse-ear **375**, 376
stinging nettle 284
stinking goosefoot 393
stinking mayweed 542
stinking tutsan 294, **296**
St John's-wort ,
  square-stalked 294
  Egyptian 295
  perfoliate 294
stock 353
  hoary **354**, 355
stonecrop 206
  mossy 206
  red **208**, 209
  thick-leaved **208**
  white **207**, 208
stone pine 41
storax **408**, 409
storax famiy 409
stork's-bill 311
  common **312**
  mallow-leaved 311
  stemless 312
  three-lobed 311
stranglewort 425
strawberry clover 260
strawberry tree 37, **410**
  Eastern 42
suffocated clover 257
sulphur clover 260
sumach 320, **321**
summer snowflake **130**, 131
sunflower broomrape 522
sun spurge **301**, 302
swamp wattle 215
sweet alison **354**, 355
sweet chestnut 22, **286**, 287

sweet marjoram 502
sweet orange 325
sweet scabious 537
sweet violet **296**, 297
swinecress, lesser 356, **357**
sword-leaved helleborine **91**, 92
Syrian bean caper 212
Syrian maple **321**, 322
Syrian thistle **570**, 571

T
tall melilot 242
tall rocket 349
tamarisk family 365
tassel hyacinth **148**
Taurus fir 30
teasel, wild 585
tea tree 451
thick-leaved stonecrop **208**
thistle 569, 570
  carline 566, 568
  creeping 570
  flat-topped carline **567**, 568
  globe 566, **567**
  milk **573**, 574
  nodding **569**
  pale globe 566
  red star- **575**, 577
  Scotch 571, 572, **573**
  spear **570**
  St Barnaby's 574, **575**
  Syrian **570**, 571
thistle broomrape **519**, 521
thorny burnet 273, **274**
three-leaved toadflax **466**, 467
three-lobed sage 505, **507**
three-lobed stork's-bill 311
thrift family 366
throw-wax 596, **597**
thrumwort 73
thyme 498
  large 499
thyme broomrape 521
thyme-leaved fumana **338**, 339
thyme-leaved sandwort 374
thyme-leaved speedwell 470, **471**
toadflax,
  common 467
  ivy-leaved 468, **469**
  Jersey **466**, 467
  three-leaved **466**, 467
  white 467
tobacco 458

shrub **458**
tongue orchid 111, **112**
toothwort **512**
torpedo grass 181
traveller's joy 200, **201**
tree heath 37, **410**, 411
tree mallow **329**, 330
tree medick 245, **247**
tree of heaven **324**, 325
tree spurge 300, **301**
trefoil,
  bean **220**, 221
  bird's-foot 252
  common bird's-foot 252
  hairy 261
  hop 256, **257**
  lesser hop 256
  slender bird's-foot 254
  southern bird's foot 252, **253**
  woolly 258, **259**
trifid bur-marigold 550
trumpet vine **479**
tuberous valerian 589
tufted forget-me-not 441
tufted vetch 234, **235**
tulip 84
  wild 84, **85**
tunic flower 386
turban buttercup 48
Turkish pine 52, **53**
turnip 359
turpentine tree 318, **319**
twayblade **92**, 93

U
umbrella pine 52, **53**
upright yellow flax 307, **308**

V
valerian 588
  common 588
  red **586**, 587
  tuberous 589
Venus's looking glass 528
  large 527
  spicate 528
vervain **480**, 481
vetch 234
  annual scorpion 264, 265
  black 237
  bladder 254, **255**
  bush 237
  common 237
  crown **264**, 265

fodder 234, **235**
horseshoe 266
large yellow **236**
Mediterranean kidney **262**, 263
rush-like scorpion **264**
spring 237
tufted 234, **235**
yellow **235**, 236
vetchling
 grass 238
 hairy 239
 red 238
 winged 238
 yellow 238
vine,
 balloon- 322, **323**
 cruel **423**
 potato 455
 silk- **424**, 427
 trumpet **479**
violet,
 hairy 297
 Mediterranean white 297
 sweet **296**, 297
violet cabbage **351**, 352
violet family 295
violet larkspur **198**, 199
violet limodore 92
viper's bugloss **436**, 437
viper's-grass 556
 cut-leaved 557
virgin's bower **201**

W
wall barley 174
wallflower 350, **351**
wall germanium **484**, 485
wallpepper 206, **207**
wall rocket 357
wall speedwell 470
walnut 290, **292**
water-crowfoot, pond **195**, 197
water-dropwort 594
 corky-fruited 594, **595**
 Mediterranean 595
water germander 483
water-lily, fringed 528
water mint **502**, 503
water-plantain 74, **75**
 lesser 73
water purslane 314
water-speedwell,
 blue 470

pink 470
waterwort family 295
wattle,
 blue-leaved **215**
 silver 214
 swamp **215**
weld 342
western salsify 557
white asphodel **128**
white broom **223**, 224
white bryony **292**, 293
white campion 380
white clover 258, **259**
white floss silk tree 333
white helleborine **91**, 92
white henbane **452**, 453
white horehound **485**
white jasmine 459
white lupin 224
white melilot 242
white mignonette 340, **341**
white mullein 475
white mustard **358**, 359
white ramping-fumitory 189
white rock rose **338**, 339
white saxaoul 390
white stonecrop **207**, 208
white toadflax 467
whorled clary 508
widow iris 115, **116**
wild arum **66**, 67
wild asparagus 140
wild basil **496**, **497**
wild cabbage **358**, 359
wild carrot 608, **609**
wild clary 506, **507**
wild jasmine 459
wild leek 134, **135**
wild madder 411, **412**
wild mignonette **341**
wild parsnip 599
wild pea 239
wild radish 360, **361**
wild teasel 585
wild tulip 84, **85**
willow, Mediterranean 299
willow family 298
willowherb,
 great 316
 marsh 316
 short-fruited 316
 small-flowered 316
 small-flowered hairy 316
 square-stalked 316

willowherb family 314
willow-leaved inula 536
willow-leaved rock rose 338
winged sea lavender 367, **368**
winged vetchling 238
winter daffodil **130**, 131
winter savory 495
wisteria, Chinese 268, **269**
woad 348
wood anemone 192
woodcock orchid **106**
wood pink 386
woodruff, blue 414
woody nightshade **454**, 455
woolly chamomile 542
woolly trefoil 258, **259**
wormwood 539
woundwort,
 downy **493**
 field **494**
 Mediterranean **493**

Y
yarrow 540
yellow asphodel 26, **128**, 129
yellow bartsia 510, **511**
yellow bird's nest **408**, 409
yellow broomrape 521
yellow centaury 419
yellow chamomile **543**
yellow flag 113, **115**
yellow flax 308
yellow germander **484**
yellow gromwell **433**, 434
yellow horned-poppy **187**
yellow lupin 224
yellow ophrys **104**, **105**
yellow rhododendron **410**, 411
yellow vetch **235**, 236
yellow vetchling 238
 annual **236**, 237
yellow-wort 419
yucca **152**

Index of scientific names

Species entries in roman text are synonyms; entries with a photograph are indicated by a **bold** page number.

A
Aaronsohnia 543
  factorovskyi 46, **542**, 543
Abelia 582
  × *grandiflora* 582, **583**
Abies 50
  alba 50
  cephalonica 23, 50, **51**
  cilicica 30, 33
  × *borisii-regis* 50
Abutilon 328
  fruticosum **327**, 328
Acacia 214
  cyanophylla 215
  cyclops **215**
  dealbata 214
  farnesiana 214, **215**
  pycnantha 215
  raddiana 216
  retinodes **215**
  saligna **215**
  tortilis 216
ACANTHACEAE 477
Acantholimon 366
  aegaeum 366, **367**
  androsaceum 366
Acanthus 477
  dioscoridis 477
  greuterianus 477
  hirsutus **476**, 477
  hungaricus 477
  mollis **476**, 477
  spinosus **476**, 477
  syriacus 477
Acer 320
  campestre 320, **321**
  heldreichii 322
  hyrcanum 322
  monspessulanum 322
  negundo **321**, 322
  obtusifolium **321**, 322
  pseudoplatanus 322
  sempervirens **321**, 322
Aceras anthropophorum 93
Achillea 540
  ageratifolia 540

  cretica 540
  falcata 540
  holosericea 540
  ligustica 540
  maritima **541**
  millefolium 540
  nobilis 540
  santolina 540, **541**
  taygetea 540
  umbellata 540
Acinos
  arvensis 496
  grandiflorum 496
Acis ionica 131
Adenocarpus 224
  complicatus 224
Adonis 184, 186
  aestivalis **185**
  aleppica 186
  annua 184, **185**
    subsp. *cupaniana* 184
  cretica **185**, 186
  dentata **185**, 186
  flammea 186
  microcarpa **185**, 186
  palaestina **185**, 186
ADOXACEAE 582
Aegilops 173
  geniculata 173
  neglecta 39, **171**, 173
Aesculus 320
  hippocastanum 320, **321**
Aetheorhiza 558
  bulbosa 558
Aethionema 347
  saxatile 347
    subsp. *creticum* **346**, 347
    subsp. *graecum* **346**, 347
Agave 151
  americana **150**, 151
  attenuata **150**, 151
  sisalana **150**, 151
Agrimonia 271
  eupatoria 271
Agrostemma 380
  githago **379**, 380

Ailanthus 325
  altissima **324**, 325
AIZOACEAE 397
Aizoon 397
  canariense 397, **398**
  hispanicum 397, **398**
Ajuga 482
  chamaepitys 483
    subsp. *chamaepitys* **482**, 483
    subsp. *chia* 483
    subsp. *palaestina* **482**, 483
  genevensis 483
  iva **482**, 483
  orientalis 483
  reptans **482**, 483
Albizia julibrissin 214
Alcea 331
  rosea **331**, 332
Alisma 74
  gramineum 74
  lanceolatum 74
  plantago-aquatica 74, **75**
ALISMATACEAE 73
Alkanna 427
  chrysanthiana 427
  graeca 429
  hellenica 429
  oreodoxa 427, **428**
  orientalis 429
  sartoriana 427
  sfikasiana **428**
  sieberi 427, **428**
  tinctoria **428**
Allium 134
  akirense 136
  amethystinum 134
  ampeloprasum 134, **135**
  aschersonianum **138**
  atroviolaceum 134
  chamaemoly 136
  commutatum 134, **135**
  cyrillii 137
  erdelii 136, **137**
  guttatum 134

israeliticum **137**, 138
junceum 134, **135**
longisepalum 136
neapolitanum **135**, 136
negevense 136, **137**
nigrum 137
orientale 137
palaestinum **135**, 136
pallens **138**, 139
paniculatum 139
papillare 136
qasyunense 136
roseum **138**, 139
rothii **32**, **137**, 138
schoenoprasum 136
schubertii **135**, 136
scorodoprasum 134
siculum **138**, 139
sphaerocephalon 134, **135**
stamineum 139
subhirsutum **135**, 136
tel-avivense 138
tel-avivensis **137**
trifoliatum 136, **137**
vineale 136
Alnus 290
glutinosa 290
Aloe 127
maculata **126**, 127
vera **126**, 127
Alopecurus 168
arundinaceus **168**
Althaea 331
cannabina 331
hirsuta 328
officinalis 331, **332**
Alyssum 344
akamasicum 344, **345**
cypricum 345
damascenum **345**, 347
simplex **345**
strigosum **345**
troodi 345
AMARANTHACEAE 15, 387
Amaranthus 388
albus 390
blitoides 390
blitum 388
cruentus **389**
deflexus 388, **389**
emarginatus 388
graecizans 390
hybridus 389
hypochondriacus 389

muricatus 388, **389**
palmeri 389
powellii 389
retroflexus 388, **389**
viridis 388, **389**
AMARYLLIDACEAE 129
Amaryllis 131
belladonna **48**, **130**, 131
Ammi 598
majus 598
visnaga 598
Ammoides 599
majus **599**
pusilla 599
Ammophila 167
arenaria **44**, 167
Ampelodesmos 162
mauritanicus 162, **163**
Amsonia 420
orientalis 420
Amygdalus 33
Anabasis 387
articulata 31, **46**, **387**
Anacamptis 97
boryi 99
collina **98**, 99
coriophora **97**
israelitica 99
laxiflora 99
morio 99
   subsp. *syriaca* **98**, 99
papilionacea 99
   subsp. *aegaea* **98**, 99
   subsp. *palaestina* **98**, 99
pyramidalis **98**, 99
sancta **98**
ANACARDIACEAE 318
Anacyclus 544
clavatus 544
radiatus 544
Anagallis 404
arvensis **404**
foemina 404
Anagyris 221
foetida **220**, 221
Anchusa 439
aegyptiaca **438**, 439
azurea **438**, 439
cespitosa **438**, 439
cretica **438**, 439
officinalis 439
strigosa **438**, 439
undulata 439
   subsp. *hybrida* **438**, 439

variegata **438**, 439
Anchusella
cretica 439
variegata 439
Androcymbium 79
palaestinum **78**, 79
rechingeri 45, **78**, 79
Andropogon 179
distachyos 179, **180**
Andryala 562
integrifolia 562, **563**
Anemone 192, 194
apennina 192
   subsp. *blanda* 192, **193**
coronaria **8**, **47**, **193**
hortensis 193
   subsp. *heldreichii* **193**
nemorosa 192
pavonina **22**, **193**
rhodopaea 194
Angelica 598
sylvestris 598
Anthemis 541
abrotanifolia 543
altissima 543
ammanthus 542
arvensis 542
chia **542**
cotula 542
cretica 543
filicaulis 542
glaberrima 542
melanolepis 543
palaestina 543
rigida **541**
ruthenica 542
tinctoria 543
tomentella 542
tomentosa 542
tricolor **542**, 543
Anthericum 139, 153
liliago 139, **152**, 153
ramosum 139
Anthyllis 263
hermanniae **39**, **44**, **262**, 263
tetraphylla 254
vulneraria 263
   subsp. *rubriflora* **262**, 263
Antirrhinum 4, 464
majus 464
   subsp. *majus* **465**
   subsp. *tortuosum* **465**
siculum **465**
Anvillea 535

garcinii **535**
APIACEAE 14, 48, 591
Apium 598
  graveolens 598
APOCYNACEAE 419
APOCYNOIDEAE 419
Apocynum 419
  venetum 419
Aptenia cordifolia 399
AQUIFOLIACEAE 523
Aquilegia 201
  vulgaris 201
Arabis 347
  collina **346**, 347
  cypria **346**, 347
  purpurea **346**, 347
  turrita **346**, 347
  verna **346**, 347
ARACEAE 11, 64
ARALIACEAE 590
Araucaria 56
  heterophylla 56
ARAUCARIACEAE 56
Araujia 423
  sericifera **423**
Arbutus 410
  andrachne 42, **410**
  unedo 37, **42**, **410**
Arceuthobium 363
  oxycedri 20, 363
Arctotheca 565
  calendula 565
ARECACEAE 153
Arenaria 374
  leptoclados 374
  serpyllifolia 374
Arisarum 69
  vulgare 69, **70**
Aristolochia 57
  auricularia 58
  bodamae 59, 60
  bottae **32**, 59, 60, **61**
  cilicica 60
  clematitis 62, **63**
  cretica 26, 57, **58**
  elongata 57
  geniculata 58
  guichardii 30, 58, 59
  hirta 57
  incisa 59, 60
  isaurica 58
  krausei 59, 60
  maurorum 59, 60
  microstoma 62

paecilantha 59, 60, **61**
pallida 57
parviflora 59
parvifolia 57
poluninii 59, 60
pontica 59, 60
rechingeriana 58
rotunda 57, **58**
scabridula 60, **61**
sempervirens 57, **58**
ARISTOLOCHIACEAE 57
Armeria 366
  canescens 366
  icarica 367
  johnsenii 367
Arnebia 429
  decumbens **428**, 429
Artemisia 539
  absynthium 539
  arborescens **538**, 539
  campestris 539
  vulgaris 540
Arthrocnemum 395
Arthrocnemum
  fruticosum 44, 395
  perenne 395
Arum 64
  alpinum 67
  byzantinum 64, 68
  concinnatum **47**, 64, **65**, 68
  creticum **27**, 64, **65**
  cylindraceum 67
  cyrenaicum **66**, 67, 68
  dioscoridis **7**, 29, **66**, 69, 610
  elongatum 67, 68
  gratum 67, 68
  hygrophilum **43**, 64, **65**
  idaeum 64, **65**
  italicum **12**, **13**, 64, **65**
  maculatum **66**, 67
  nigrum **66**, 67, 68
  orientale 67, 68
  palaestinum **66**, 67, 68
  purpureospathum **66**, 67
  rupicola 28, 30, **66**, 68, 69
ARUNDINOIDEAE 175
Arundo 175
  donax 175, **176**
  plinii 175, **176**
ASCLEPIADOIDEAE 421
ASPARAGACEAE 139
Asparagus 140
  acutifolius **140**

aphyllus **39**, 140
horridus **140**
officinalis 140
stipularis 140
Asperula 414
  arcadiensis 414, **415**
  aristata 414, **415**
  arvensis 414
  brevifolia 414
  laevigata 414
  lutea 414
  rigida 414
ASPHODELACEAE 127
Asphodeline 129
  luburnica **128**
  lutea **25**, **26**, **27**, **128**
Asphodelus 127
  albus **128**
  fistulosus **128**
  liburnica 129
  lutea 129
  ramosus **7**, **47**, 127, **128**
  tenuifolius **128**
Aster 536
  linosyris 537
  squamatus **536**
  tripolium 537
ASTERACEAE 11, 14, 46, 48, 529
Asteriscus 549
  aquaticus **548**, 549
  pygmaeus 549
ASTEROIDEAE 529
Astomaea 594
  seselifolia **592**, 594
Astracantha 232
  cretica 232
  dolinicola 232
  parnassi 232
  thracica 232
Astragalus 226
  aleppicus **230**, 231
  angustifolius 232
    subsp. *aegeicus* 232
    subsp. *angustifolius* 232
    subsp. *balcanicus* **230**, 232
    subsp. *echinoides* 232
    subsp. *erinaceus* 232
    subsp. *odonianus* 232
    subsp. *postianus* 232
  asterias 228
  boeticus **227**
  callichrous **229**, 231
  cretaceus 226, **227**

cyprius 226, **227**
echinatus **229**, 230
epiglottis 228, **229**
hamosus 226, **227**
intercedens 226, **227**
lusitanicus 232
macrocarpus **230**, 231
  subsp. *lefkarensis* 231
monspessulanus **21**, 226, **227**
odoratus 227
palaestinus **229**, 231
pelecinus **229**, 230
sanctus **229**, 230
schimperi 228, **229**
sesameus 228
sinaicus 228
spruneri 226
stella 228
suberosus 226
tribuloides 228, **229**
Asyneuma 527
  giganteum 527
  limonifolium **526**, 527
  pichleri 527
Atractylis 569
  cancellata **569**
  gummifera 568
Atriplex 390
  glauca 390, **391**
  halimus 390, **391**
  hastata 392
  holocarpa **391**, 392
  patula 392
  portulacoides **44**, 390, **391**
  prostrata **391**, 392
  rosea 392
  semibaccata 390, **391**
  suberecta **391**, 392
Atropa 450
  belladonna 450, **451**
Aubretia 344
  deltoidea 344, **345**
  erubescens 344
  glabrescens 344
  gracilis 344
  thessala **20**, **21**, 344, **345**
Avena 165
  barbata **165**
  sterilis 165

B
Baldellia 73
  ranunculoides **72**, 73
Ballota 492

acetabulosa 492
nigra 492
pseudodictamnus 492
  subsp. *lycia* 492
BALSAMINACEAE 403
Barlia robertiana 100
Bartsia 510
  trixago 510, **511**
    var. *flaviflora* **511**
Bassia 392
  arabica 392, **393**
Bellardia trixago 510
Bellevalia 146
  ciliata **23**, 148
  densiflora 146
  desertorum **147**
  dubia 146
  eigii **147**
  flexuosa 146, **147**
  nivalis 146, **147**
  romana 148
  trifoliata 146, **147**
  warburgii 146, **147**
Bellis 539
  annua 539
  minutum **538**
  perennis 539
  sylvestris **538**, 539
Bellium 539
  minutum 539
BERBERIDACEAE 190
Berberis 190
  cretica 190, **191**
  libanotica 190
Beta 394
  maritima **393**, 394
  vulgaris 394
    subsp. *maritima* 394
Betonica officinalis 494
BETULACEAE 290
Biarum **48**, 69
  angustatum 69
  auraniticum **70**, 71
  carduchorum 69
  davisii **70**, 71
  ditschianum **70**, 71
  eximium 69
  marmarisense **70**, 71
  olivieri 69
  rhopalospadix 69
  syriacum 69
  tenuifolium 69, **70**
Bidens 550
  tripartita 550

Biebersteinia 317
  orphanidis **23**, **315**, 317
BIEBERSTEINIACEAE 317
BIGNONIACEAE 478
Biscutella 356
  didyma 356, **357**
    subsp. *dunensis* 356, **357**
Biserrula pelecinus 230
Bituminaria 234
  bituminosa **11**, 234, **235**
Blackstonia 419
  acuminata 419
  perfoliata 419
Blitum 394
  bonus-henricus 394
Boerhavia 399
  helenae **398**, 399
Bongardia 191
  chrysogonum **191**
BORAGINACEAE 427
Borago 438
  officinalis **436**, 438
Bosea 387
  cypria **387**
Bougainvillea 400
  glabra 400
  spectabilis 400, **401**
Brachychiton 333
  populneus 333
Brachypodium 162
  retusum 162, **163**
Brassica 358
  cretica **358**
    subsp. *aegaea* 358
    subsp. *cretica* 358
    subsp. *laconica* 358
  hilarionis **358**, 359
  nigra 358
  oleracea **358**, 359
  rapa 359
BRASSICACEAE **14**, **48**, 343
Briza 167
  maxima **13**, 167
  minor 167
Bromus 162
  diandrus 164
  fasciculatus **163**, 164
  hordeaceus **163**, 164
  madritensis 164
  squarrosus 164
Brugmansia **2**, 456
  arborea 456, **457**
  sanguinea 456
  versicolor 456

Bryonia 293
   cretica subsp. dioica 293
   dioica **292**, 293
Buglossoides 429
   arvensis **428**, 429
   incrassata 429
   purpurocaerulea 429
   tenuiflora **428**, 429
Bunias 349
   erucago 349
Bunium 593
   ferulaceum 593
Bupleurum 596
   apiculatum 596
   fruticosum **597**, 598
   gaudianum 597
   glumaceum **597**
   gracile 597
   rotundifolium 596, **597**
   semicompositum 597
   tenuissimum 597
   trichopodum 597
BUTOMACEAE 74
Butomus 74
   umbellatus 74
BUXACEAE 203
Buxus 203
   sempervirens **202**, 203

C

Cachrys 596
   cristata 596
CACTACEAE 401
Caesalpinia gilliesii 217
CAESALPINIOIDEAE 214
Cakile 359
   maritima 44, **358**, 359
Calamintha
   grandiflora 496
   nepeta 495
Calendula 549
   arvensis **548**, 549
   officinalis 549
   tripterocarpa **548**, 549
Calicotome 221
   villosa 2, **39**, **220**, 221
Callistemon citrinus 316
Calotropis 423
   procera **423**
Calystegia 445
   sepium 445
   silvatica **444**, 445
   soldanella **444**, 445
Campanula 523

andrewsii 524
boreosporadum 526
calaminthifolia **525**
celsii 524
cretica 526
creutzburgii 526
delicatula 526
drabifolia 526
erinus 526
euboica 524
glomerata **525**
heterophylla 526
hierapetrae 526
incurva 22, 526
kamariana 524
olivieri 526
patula 524
persicifolia 524
phrygia 524
ramosissima 524, **525**
rapunculus 524
rhodensis 526
samothracica 526
saxatilis 524
   subsp. *cytherea* 524
   subsp. *saxatilis* **43**, 524, **525**
scutellata 526
sidoniensis 524, **525**
simulans 526
topaliana 523, **525**
tubulosa 524
CAMPANULACEAE 15, 523
Campsis 479
   grandiflora 479
   radicans **479**
CANNABACEAE 283
Cannabis 283
   sativa **282**, 283
CAPPARACEAE 343
Capparis 343
   spinosa **342**, 343
   zoharyi 343
CAPRIFOLIACEAE 582
Capsella 348
   bursa-pastoris 348, **349**
Caralluma 421
   europaea **421**, 426
   sinaica 31, **421**, 426
   tuberculata 421, 426
Cardaria draba 356
Cardiospermum 322
   halicacabum 322, **323**
Cardopatium 565

corymbosum **564**, 565
Carduncellus 581
   caeruleus **580**, 581
CARDUOIDEAE 565
Carduus 569
   argentatus 570
   candicans 570
   nutans **569**
   pycnocephalus **569**
Carex 158
   divisa 158
Carlina 48, 566
   barnebyana 568
   corymbosa **567**
      subsp. *corymbosa* 568
      subsp. *curetum* 568
      subsp. *graeca* **567**, 568
   gummifera **567**
   lanata **567**, 568
   pygmaea 568
   racemosa **567**, 568
   sitiensis 568
   tragacanthifolia 24, 566, **567**
   vulgaris 568
Caroxylon aegaeum 396
Carpinus 291
   orientalis 291
Carpobrotus 397
   edulis 397
Carrichtera 362
   annua **362**
Carthamus 580
   caeruleus 581
   dentatus 580
   lanatus **580**
   leucocaulos 581
CARYOPHYLLACEAE 14, 374
Cascabela 420
   thevetia 420, **421**
Cassia 218
   artemisioides 218, **219**
   corymbosa 218
   didymobotrya 218
Castanea 287
   sativa **42**, **286**, 287
Casuarina 290
   equisetifolia 290, **292**
CASUARINACEAE 290
Catalpa 478
   bignonioides 478
Catananche 552
   lutea 552
Catapodium 170
   rigidum 170, **171**

Cedrus 50
  libani 30, 33, 50, **51**
    var. *brevifolia* 28, 50, **51**
Ceiba 333
  insignis 333
  speciosa **332**, 333
CELASTRACEAE 293
Celtis 283
  australis **282**, 283
  tournefortii 283
Cenchrus 179
  ciliaris 179, **180**
Centaurea 574
  achaia 577
  acicularis 574, **575**
  aegialophila **576**, 577
  aegyptiaca **575**
  aetolica 577
  akamantis 43, **576**, 578
  argentea 576
  atropurpurea **576**, 577
  benedicta 575
  calcitrapa **575**, 577
  corinthiaca 577
  cyanus 578
  depressa 578
  euboica 577
  hyalolepis 574, **575**
  iberica 577
  idaea 574
  jacea 578
  laconica 577
  lactucifolia **576**, 577
  melitensis 574
  paxorum 18, **576**, 577
  pinardii 578
  ptosimopappa 576
  ptosimopappoides 576
  pumilio 45, **575**, 577
  salonitana 575
  solstitialis 574, **575**
  spruneri 577
  triumfettii 578
Centaurium 418
  erythraea 418
    subsp. *erythraea* **418**
    subsp. *rhodense* **418**
  maritimum 419
  pulchellum 418
  serpentinicola 418
  spicatum 419
  tenuiflorum 418
Centranthus 587
  calcitrapa **586**, 587

ruber **586**, 587
Cephalanthera 91
  cucullata 91
  damasonium **91**, 92
  epipactoides 91
    subsp. *kurdica* 30, **91**
  longifolia **91**, 92
  rubra 92
Cephalaria 587
  leucantha 587
  squamiflora 587
  syriaca 587
  transsylvanica 587
Cerastium 376
  glomeratum **375**, 376
Ceratocephala 202
  falcata 202
Ceratonia 217
  siliqua **11**, **13**, **35**, **36**, **38**, 217, **219**
CERATOPHYLLACEAE 181
Ceratophyllum 181
  demersum 181
  submersum 181
CERCIDOIDEAE 218
Cercis 218
  siliquastrum 218, **219**
Cerinthe 434
  major 434
  minor 434
  palaestina 434
Cestrum 458
  nocturnum 458
  parqui 458
Chaetosciadium 606
  trichospermum 606, **607**
Chamaemelum mixtum 541
Chamaerops 153
  humilis **152**, 153
Chamaesyce
  canescens 305
  maculata 305
  peplis 305
  prostrata 305
  serpens 305
Chelidonium hybridum 184
Chenopodium 392
  album 393
  bonus-henricus 394
  botrys 394
  hybridum 393
  murale **393**
  vulvaria 393
CHLORIDOIDEAE 177

Chondrilla 561
  juncea **560**, 561
Chorispora 350
  purpurascens 350, **351**
Chrozophora 306
  obliqua 306
  tinctoria **306**
Chrysanthemum
  coronarium 545
  segetum 545
CICHORIOIDEAE 550
Cichorium 551
  endivia 551
  intybus **551**
  lutea **551**
  spinosum 551
Circaea lutetiana 22
Cirsium 570
  arvense 570
  candelabrum 571
  creticum 570
  hypopsilum 570
  italicum 571
  vulgare **570**
CISTACEAE 15, 336
Cistanche 514
  fissa **514**, **515**
  salsa 514
  tinctoria 514
  tubulosa 31, **46**, 513, 514, **515**
  violacea viii, 513, 514, **515**
Cistus 336
  creticus **47**, 336
    subsp. *creticus* 336, **337**
    subsp. *eriocephalus* 336, 337
  ladanifer 336
  monspeliensis 336, **337**
  parviflorus 336, **337**
  salviifolius 336, **337**
  villosus 336
Citrullus 291
  colocynthis 291, **292**
Citrus 325
  medica **324**, 325
    var. *sarcodactylis* **36**
  × aurantiifolia **324**, 325
  × aurantium **36**, **324**, 325
  × limon **324**, 325
  × paradisi 325
  × sinensis 325
Cladanthus 541
  mixtus 541
Clematis 200

cirrhosa 201
elisabethae-carolae 201
flammula 200, **201**
vitalba 200, **201**
viticella 201
Clinopodium 495
  acinos 496
  nepeta 495
    subsp. *glandulosum* 495, **497**
  vulgare **39**, 496, **497**
Clypeola 344
  jonthlaspi 344, **345**
Cnicus benedictus 575
Coccoloba 371
  uvifera 371, **372**
COLCHICACEAE 77
Colchicum 48, 77
  autumnale **78**, 79
  baytopiorum 77
  bivonae **48**, **78**, 79
  boissieri 79
  chalcedonicum 79
  cretense 77
  cupanii **77**
  euboeum 79
  graecum **78**, 79
  macrophyllum 79
  parlatoris 79
  parnassicum **79**
  pusillum 77
  stevenii 77
  triphyllum 78
  troodi 79
Coleostephus 545
  myconis 545
Colutea 225
  arborescens **225**
  cilicica 225
  insularis 225
Commicarpus helenae 399
Conium 596
  maculatum 596
Consolida 200
  ajacis 200
  arenaria 200
  glandulosa 200
  hispanica 200
  orientalis 200
  phygria 200
  regalis 200
  samia 200
  sulphurea 200
  tomentosa 200

CONVOLVULACEAE 442
Convolvulus 445
  althaeoides 446, **447**
  argyrothamnos 449
  arvensis 446, **447**
  betonicifolius 446
  cantabrica 446, **447**
  coelesyriacus 446
  dorycnium 449
  elegantissimus 446
  libanoticus 446
  lineatus 446
  oleifolius **448**, 449
  pentapetaloides 445, **447**
  scammonia 449
  siculus 445
  stachydifolius 446, **448**
  tricolor 445, **447**
Conyza
  bonariensis 537
  canadensis 537
  sumatrensis 538
Cordyline 141
  australis 141
Coris 405
  monspeliensis **404**, 405
CORNACEAE 402
Cornus 402
  mas **401**, 402
  sanguinea 402
Coronilla 264
  emerus 266
  glauca 264
  juncea **264**
  parviflora **264**
  repanda 265
  scorpioides **264**, 265
  securidaca **264**
  valentina **39**
  valentina 264
  varia **264**
  varia 265
Coronopus
  didymus 356
  squamatus 356
Cortaderia 181
  selloana 181
Corydalis 188
  cava 188
  integra 188
  solida **19**, **187**, 188
  thasia 188
  uniflora 188
Corynephorus 169

articulatus 169
divaricatus 169
Cota 543
  tinctoria 543
Cotinus 320
  coggygria 320, **321**
Cotula 545
  coronopifolia 545
Crambe 360
  hispanica 360
Crassula 206
  alata 206
  tillaea 206
  vaillantii 206
CRASSULACEAE 206
Crataegus 33, 275
  aronia 276
  azarolus **274**, 275
    var. *aronia* **274**, 276
    var. *azarolus* 276
  heldreichii 276
  monogyna **274**, 275
  orientalis 276
  rhipidophylla **274**, 276
Crepis 562
  biennis 562
  foetida 563
  fraasii 563
  incana **563**, 564
  neglecta 563
  rubra **563**, 564
  sancta **563**
  setosa 564
  vesicaria 563
  zacintha 563
Crithmum 594
  maritimum **44**, 594
Crocus 8, 48, 120
  asumaniae **122**, 123
  atticus 120, **121**
  baytopiorum **121**, 122
  biflorus 120
  boryi 123
  cartwrightianus 123
  chrysanthus 123
  flavus **8**, 123
  gargaricus **121**, 122
  goulimyi **122**, 123
  hadriaticus **122**, 123
  kotschyanus **121**, 122
  laevigatus **122**, 123
  nivalis 120
  niveus 123
  oreocreticus 123

sativus **122**, 123
sieberi 120, **121**
   subsp. atticus 120
   subsp. sublimis 120
sublimis 120, **121**
veluchensis 120, **121**
vitellinus **122**, 123
Crucianella 412
  angustifolia 412
  bithynica 412
  graeca 412
  latifolia 412, **413**
  maritima 412
Cruciata 413
  articulata **413**
  laevipes 413
Crupina 579
  crupinastrum 579, **580**
  vulgaris 579
Cucubalus baccifer 385
Cucumis 291
  prophetarum **46**, 291, **292**
CUCURBITACEAE 291
CUPRESSACEAE 54
Cupressus 54
  arizonica 54
  macrocarpa 54
  sempervirens 28, **42**, 54, **55**
Cuscuta 442
  approximata 443, 444
  atrans 444
  brevistyla **443**, 444
  campestris 445
  epithymum **443**, 444
  europaea 444
  monogyna 442
  palaestina 442, **443**
  planiflora **443**, 444
  rausii 442
  scandens 445
  suaveolens 444
Cyanus 578
  depressus 578
  pinardii 578
  segetum 578
Cyclamen 406
  cilicium 406, **407**
  confusum 406
  coum 33, **408**, 409
  creticum **407**, 408
  cyprium 406, **407**
  graecum 406, **407**
  hederifolium **12**, **13**, **35**, 406, **407**

libanoticum **407**, 408
mirabile 406, **407**
persicum **7**, **8**, **48**, **49**, **408**, 409
repandum 406
  subsp. peloponnesiacum **407**, 408
  subsp. repandum 406
  subsp. rhodense **407**, 408
rhodense 408
Cydonia 279
  oblonga **278**, 279
Cymbalaria 468
  longipes 468
  microcalyx 468
  muralis 468, **469**
Cymodocea 76
  nodosa 76
CYMODOCEACEAE 76
Cynanchum 425
  acutum 425, 426
Cynara **48**, 572
  auranitica 572
  cardunculus 572, **573**
  cornigera **48**, **573**, 574
  cyrenaica 572
  scolymus 572
  syriaca 572, **573**
Cynodon 177
  dactylon **176**, 177
Cynoglossum 440
  cheirifolium 440
  columnae 440, **441**
  creticum 440, **441**
  sphacioticum 440
  troodi 440
CYNOMORIACEAE 211
Cynomorium
  coccineum 211
Cynosurus 170
  echinatus 170
CYPERACEAE 158
Cyperus 159
  capitatus **159**
  eragrostis 160
  esculentus 160
  involucratus **159**, 160
  laevigatus **159**, 160
  longus 160
  papyrus **159**, 160
Cyprinia 427
  gracilis 427
CYTINACEAE 326
Cytinus 326

hypocistis 326, **327**
ruber 326, **327**
Cytisopsis 254
  pseudocytisus 254, **255**
Cytisus 221
  hirsutus 221
    subsp. polytrichus **220**, 221
  scoparius 221
  villosus 221

D

Dactylis 169
  glomerata 169
Dactylorhiza 102
  romana 102
  saccifera 102
  sambucina 102
Damasonium 73
  bourgaei 73
  polyspermum 73
DANTHONIOIDEAE 181
Daphne 334
  euboica 334
  gnidioides 334
  gnidium 334, **335**
  laureola 334, **335**
  oleoides 334, **335**
  sericea 334, **335**
Datura 456
  ferox 456
  innoxia 456, **457**
  stramonium 456, **457**
Daucus 608
  broteri 609
  carota 608
    subsp. carota 608
    subsp. drepanensis 608, **609**
    subsp. major 608, **609**
    subsp. maximus 608
  glaber 609
  guttatus 609
  involucratus 609
Delonix 217
  regia **216**, 217
Delphinium 199
  balcanicum 199
  bovei 199
  fissum 199
  hellenicum 199
  ithaburense 199
  peregrinum **198**, 199
  staphisagria **198**, 199

Desmazeria rigida 170
Dianthus 385
  armeria **384**, 385
  crinitus **384**, 385
  fruticosus 385
    subsp. amorginus 385
    subsp. carpathus **384**, 385
    subsp. creticus 385
    subsp. fruticosus 385
    subsp. karavius 385
    subsp. occidentalis 385
    subsp. rhodius 385
    subsp. sitiacus 385
  gracilis 386
  juniperinus 385
    subsp. aciphyllus 385
    subsp. bauhinorum **384**, 385
    subsp. heldreichii 385
    subsp. idaeus 385
    subsp. juniperinus 385
    subsp. kavusicus 385
    subsp. pulviniformis 385
  pinifolius 386
  sylvestris 386
  viscidus 386
Dictamnus 325
  albus **324**, 325
Digitalis 469
  cariensis 469
    subsp. ikarica 469
  ferruginea **469**
  laevigata 470
  lanata 469
  leucophaea 469
  trojana 469
  viridiflora 470
Dioscorea 76
  communis 76, **77**
DIOSCOREACEAE 76
Diplotaxis 357
  acris **357**
  erucoides 357
  muralis 357
Dipsacus 585
  fullonum 585
Dittrichia 534
  graveolens 534
  viscosa 534, **535**
Dombeya 333
  × cayeuxii 333
Doronicum 548
  columnae 548

  orientale 548
Dorycnium 263
  hirsutum **262**, 263
  rectum 263
Draba verna 344
Dracaena 139
  draco 139, **140**
Dracunculus 72
  vulgaris **8**, **11**, **26**, **27**, 72
Drimia 144
  aphylla 144, **145**
  maritima 37, **39**, **48**, 144, **145**
  numidica 144
Dutra innoxia 456
Dysphania 394
  ambrosioides **393**, 394
  botrys 394

E

Ebenus 232
  cretica 232, **233**
  sibthorpii 233
Ecballium 291
  elaterium 291, **292**
Echinophora 593
  spinosa 593
  tenuifolia 593
Echinops 566
  acanthifolia 568
  acaulis 568
  corymbosa 568
  graecus 566
  gummifera 568
  microcephalus 566
  ritro 566, **567**
  sphaerocephalus 566
  spinosissimus 566, **567**
  viscosus 566, **567**
Echium 436
  angustifolium **436**, 437
  arenarium 436
  candicans 437
  italicum **436**, 437
  parviflorum 436
  plantagineum **436**, 437
  vulgare **436**, 437
Einadia 388
  nutans 388, **389**
ELAEAGNACEAE 281
Elaeagnus 281
  angustifolia **280**, 281
Elaeoselinum 605
  asclepium **604**, 605
ELATINACEAE 295

Elatine 295
  alsinastrum 295
  macropoda 295
Eleocharis 161
  multicaulis 161
Elymus farctus 174
Elytrigia 174
  juncea 174
Emerus major 266
Emex 373
  spinosa 373, **375**
Eminium 71
  spiculatum **70**, 71
Ephedra 56
  aphylla 56
  distachya 56
  foeminea 56
  major 56
  nebrodensis 56
EPHEDRACEAE 56
Epilobium 314
  hirsutum 316
  montanum 314, **315**
  obscurum 316
  palustre 316
  parviflorum 316
  roseum 316
  tetragonum 316
Epipactis 90
  atrorubens 90
  condensata 90
  cretica 91
  helleborine 91
  microphylla 90
  palustris 90
  troodi 91
  turcica 91
  veratrifolia 90
Eragrostis 177
  barrelieri **176**, 177
Eremostachys 489
  laciniata **489**
Eremurus 127
  spectabilis 30, **126**, 127
Erica 37, 411
  arborea 41, **410**, 411
  manipuliflora 41, **410**, 411
  multiflora **410**, 411
ERICACEAE 409
Erigeron 537
  acer 537
  acris **536**, 537
  annuus 537
  bonariensis 537, **538**

canadensis 537, **538**
sumatrensis 538
Eriobotrya 280
  japonica **280**
Erodium 311
  acaule 312
  botrys 313
  chium 311
  cicutarium 312
  crassifolium **312**, 313
  gruinum 311
  laciniatum 311, **312**
  malacoides 311
  moschatum 312, 313
  salicaria **312**
  touchyanum **312**
Erophaca 232
  baetica
    subsp. *orientalis* 232, **233**
Erophila 344
  verna 344, **345**
Eruca 360
  sativa 360
  vesicaria 360, **361**
Erucaria 361
  pinnata **361**
Erucastrum 360
  gallicum 360
Eryngium 591
  amethystinum 590, 591
  amorginum 591
  campestre **590**, 591
  creticum 591
  falcatum 591
  maritimum **590**, 591
  ternatum 26, 591
Erysimum 350
  candicum 350
    subsp. *candicum* 350, **351**
    subsp. *carpathum* 352
  corinthium 352
  creticum 350
  horizontale 350
  krendlii 350
  mutabile 350
  naxense 352
  repandum 350
  rhodium 352
  senoneri 352
  smyrnaeum 350
  × *cheiri* 350, **351**
Erythrostemon 217
  gilliesii **216**, 217
Eucalyptus 317

camaldulensis **42**, **315**, 317
EUDICOTS 182
Euonymus 293
  europaeus 293
Euphorbia 300
  acanthothamnos 37, 300, **301**
  amygdaloides **301**, 302
  chamaesyce **304**, 305
  characias 300, **301**
  cyparissias **303**
  deflexa **301**, 302
  dendroides 300, **301**
  exigua 302
  falcata 302
  helioscopia **301**, 302
  hierosolymitana **301**
  ingens 300, **301**
  lemesiana 301
  maculata **304**, 305
  myrsinites **303**
  nicaeensis 302, **303**
  paralias **304**, 305
  peplis **304**, 305
  peplus **301**, 302
  pinea 303
  prostrata 305
  rigida **304**, 305
  segetalis 303
  serpens **304**, 305
  sultan-hassei 302
  terracina 305
  veneris 303, 305
EUPHORBIACEAE 15, 300
Evax
  aegaea 530
  contracta 530
  pygmaea 530

F
FABACEAE 14, 214
Factorovskya aschersoniana 251
FAGACEAE 287
Fagonia 213
  cretica **213**
  mollis **213**
Fagus 22
Falcaria 608
  vulgaris 608, **609**
Fallopia 373
  convolvulus 373, **375**
  dumetorum 373
Ferula 601
  biverticillata 601, 602

brevipedicellata 601
chiliantha 601
communis **7**, **600**, 601
  subsp. *glauca* 601
daninii 602, 603
glauca **27**, **600**, 601
hermonis 603
negevensis 602, 603
orientalis 602, 603
rigidula 601
sinaica 601, 602
tingitana **600**, 601
Ferulago 603
  campestris 603, **604**
  nodosa 603, **604**
  sylvatica 604
  syriaca 605
  thyrsiflora 604
Fibigia 355
  clypeata **354**, 355
Ficaria 197
  verna 197
    subsp. *calthifolia* 197
    subsp. *chrysocephala* **197**
    subsp. *ficariiformis* 197
    subsp. *verna* **197**
Ficus 284
  carica **12**, **35**, **36**, 284, **285**
  elastica 284
  microcarpa 284
Filaginella uliginosa 531
Filago 529
  aegaea 530
  contracta 530
  cretensis 530
  desertorum **528**, 529
  eriocephala 529
  gallica 530
  germanica **528**, 529
  minima 530
  pygmaea 530
  pyramidata 529
  spicata **528**
  vulgaris 529
  wagenitziana 530
Foeniculum 595
  vulgare **12**, **595**
Frankenia 365
  hirsuta 365
  pulverulenta **364**, 365
FRANKENIACEAE 365
Fraxinus 460
  angustifolia **458**, 460
  excelsior 460

ornus **458**, 460
Fritillaria 86
  acmopetala **88**
  bithynica **88**, 89
  conica 89
  davisii 86
  drenovskii 89
  ehrhartii 89
  epirotica 86
  euboeica 89
  forbesii 89
  graeca 86
  gussichiae 86
  latakiensis **88**, 89
  messanensis **87**, 88
  obliqua 89
  orientalis 88
  persica 32, **88**, 89
  pontica 86
  rhodia 89
  rhodocanakis 88
  sibthorpiana 89
  spetsiotica **87**, 88
  stribrnyi 89
Fumana 339
  arabica **338**, 339
  laevipes 339
  procumbens 340
  scoparia 340
  thymifolia **338**, 339
Fumaria 189
  bracteosa **189**, 190
  capreolata 189
  densiflora **189**, 190
  kraliki **189**, 190
  macrocarpa **189**, 190
  officinalis **189**, 190
  parviflora **189**, 190
  petteri 189
  vaillantii 190

G
Gagea 81
  amblyopetala 83
  bohemica **82**, 83
  fibrosa 83
  graeca **82**, 83
  heldreichii 83
  juliae **82**, 83
  lutea 81
  minima 81
  omalensis 83
  peduncularis 83
  pratensis 81

pusilla 81
reticulata **82**, 83
rigida 83
troodi 28
villosa 83
Galactites 571
  tomentosus **570**, 571
Galanthus 129
  elwesii 129, **130**
  ikariae 129
  peshmenii 129
  reginae-olgae 129, **130**
  samothracicus 129
  woronowii 129, **130**
Galatella 537
  linosyris 537
Galega 255
  officinalis 255
Galeopsis 486
  ladanum 486
Galium 414
  aparine **415**
  capitatum 416
  debile 416
  divaricatum 416
  floribundum 416
  incrassatum 416
  intricatum **415**, 416
  lonicum **415**
  mixtum 415
  murale 416
  rotundifolium 414
  spurium 415
  tricornutum 415
  verrucosum 415
  verticillatum 416
  verum 416
Gastridium 168
  phleoides 168
  ventricosum 168
Gaudinia 166
  fragilis **165**, 166
Genista 222
  acanthoclada 222, **223**
  acanthoclados 37
  anatolica 222
  carinalis 222
  fasselata 222, **223**
  sagittalis 222, **223**
  sakellariadis 20, **21**, 222, **223**
  tinctoria 222, **223**
GENTIANACEAE 418
GERANIACEAE 309
Geranium 309

columbinum 311
dissectum **310**, 311
gruinum 310
lucidum **310**, 311
molle 310
peloponesiacum 309
purpureum 311
pusillum 310
pyrenaicum 309
robertianum **310**, 311
rotundifolium 310
sanguineum 309, **310**
Geropogon 558
  hybridus 558
GESNERIACEAE 461
Gladiolus 124
  anatolicus 126
  atroviolaceus **126**
  communis
    var. *byzantinus* 124, **125**
  illyricus **126**
  italicus 12, **13**, 124, **125**
  triphyllus **126**
Glaucium 186
  aleppicum 186
  arabicum 186
  corniculatum 186, **187**
  flavum **187**
  grandiflorum 186, **187**
Glebionis 545
  coronaria **544**, 545
  coronopifolia **544**
  segetum **544**, 545
Globularia 471
  alypum 472
  arabica **471**, 472
  cordifolia 472
  stygia **471**
Glycyrrhiza 233
  echinata 233
  flavescens 233
  glabra 233
Gnaphalium 531
  luteoalbum 531
  sylvaticum 531
  uliginosum 531
Gomphocarpus 423
  fruticosus 423
  physocarpus **424**, 425
  sinaicus **424**, **425**, 426
Goniolimon 370
  dalmaticum 370
  heldreichii 370
  sartorii 370

Gossypium 327
  herbaceum **327**
  hirsutum 327
Gundelia 553
  tournefortii **552**, 553
Gymnadenia 101
  conopsea 101
Gymnarrhena 581
  micrantha **46**, **580**, 581
GYMNARRHENOIDEAE 581
Gymnocarpos 374
  decandrus 374, **375**
GYMNOSPERMS 50
Gynandriris
  monophylla 115
  sisyrinchium 115

H

Haberlea 461
  rhodopensis 19, **461**
Halimione portulacoides 390
Haloxylon 390
  salicornicum 390
Haplophyllum 323
  buxbaumii **323**
  suaveolens 323
Hedera 590
Hedypnois 553
  cretica 553
  lucida **552**
  rhagadioloides **552**, 553
  scabra **552**
Hedysarum 268
  coronarium 268, **269**
  spinosissimum 268, **269**
Helianthemum 338
  apenninum **338**, 339
  lavandulifolium 339
  ledifolium 338
  nummularium 339
  obtusifolium **338**, 339
  salicifolium 338
  sanguineum 338
  syriacum 339
  vesicarium **338**, 339
Helichrysum 531
  amorginum **533**
  doerfleri 533
  heldreichii 533
  italicum **532**, 533
  luteoalbum 531
  orientale 533
  sanguineum **531**, 533
  sibthorpii **531**, 533
  stoechas 533
    subsp. *barrelieri* **532**, 533
    subsp. *stoechas* **532**
  taenari 533
Heliotropium 433
  bacciferum 433
  curassavicum 434
  dolosum 433
  europaeum **433**
  lasiocarpum 433
  supinum 433
Helleborus 192
  odorus 192
  vesicarius 192
Hermodactylus tuberosus 115
Herniaria 377
  glabra 377
  hirsuta **377**
    subsp. *cinerea* **377**
    subsp. *hirsuta* 377
Hibiscus 331
  rosa-sinensis 331, **332**
  syriacus 331
  trionum 331, **332**
Hieracium 564
  bracteolatum 564
Himantoglossum 100
  affine 100
  caprinum 100
  comperianum 100
  hircinum 100
    subsp. *samariense* 100
  robertianum 100, **101**
Hippocrepis 265
  biflora 266, **267**
    subsp. *bisiliqua* 266
    subsp. *unisiliquosa* 266
  ciliata 265, **267**
  comosa 266
  emerus 266
    subsp. *emeroides* 1, 266, **267**
  multisiliquosa 266
  unisiliquosa 266
    subsp. *bisiliqua* **267**
    subsp. *unisiliquosa* **267**
Hirschfeldia 359
  incana 359
Hordeum 174
  murinum **11**, 174
  vulgare
    subsp. *spontaneum* 174
Hormuzakia 440
  aggregata 440, **441**

Hydrocotyle 592
  vulgaris **592**
Hylocereus 401
  undatus **401**
Hymenocarpos 263
  circinnatus **262**, 263
Hyoscyamus 452
  albus **452**, 453
  aureus **452**
  desertorum **452**, 453
  niger **452**, 453
  pusillus **452**, 453
  reticulatus **452**, 453
Hyoseris 553
  lucida 553
  scabra 553
Hyparrhenia 179
  hirta 179, **180**
Hypecoum 187
  dimidiatum **187**
  imberbe 188
  pendulum 188
  procumbens 188
HYPERICACEAE 294
Hypericum 294
  aegypticum 295
  empetrifolium 295
  hircinum 294, **296**
  olympicum 294, **296**
  perfoliatum 294
  perforatum 294
  tetrapterum 294
  triquetrifolium 294, **296**
Hypochaeris 554
  achyrophorus 554, **555**
  glabra 554
  radicata 554, **555**
Hyssopus 498
  officinalis 498

I

Iberis 355
  carnosa 356
  sempervirens **354**, 355
Ifloga 530
  spicata 530
Ilex
  aquifolium **522**, 523
Illecebrum 376
  verticillatum **375**, 376
Impatiens 403
  balfourii 403
Inula 535
  conyza **535**

INDEX OF SCIENTIFIC NAMES

crithmoides 536
graveolens 534
salicina 536
verbascifolia 536
viscosa 534
Ionopsidium 348
albiflorum 348
Ipomoea 449
acuminata 449
cairica **448**, 449
hederacea 449
imperati **448**, 449
indica 449
nil **448**, 449
purpurea 449
stolonifera 449
IRIDACEAE 113
Iris 113
albicans **116**, 117
atrofusca **32**, 114, 118, **119**
atropurpurea 31, 118, **119**
attica **116**, 118
bismarckiana 31, **32**, 114, **119**, 120
edomensis 114
florentina **116**, 117
haynei 114, 118, **119**
hellenica 117
hermona 114, **119**, 120
histrio 114, 117
lortetii 31, 114, **119**, 120
mariae 114, 118, **119**
mesopotamica 114, 118
nusairiensis 33, **34**, 115, **116**
palaestina 117
pamphylica 117
petrana 114, 118
planifolia **116**, 117
pseudacorus 113, **115**
reichenbachii **20**, **116**, 118
suaveolens 118
tuberosa 115, **116**
unguicularis 115
subsp. carica **115**
subsp. cretensis **27**, **115**
vartanii **116**, 117
× germanica 117
Isatis 348
lusitanica 348, **349**
tinctoria 348
tomentella 348
Isolepis 161
setacea 161

J
Jacaranda 478
mimosifolia 478, **479**
Jacobaea 547
maritima 547
subsp. bicolor 547
subsp. maritima 547
vulgaris 547
Jankaea 462
heldreichii 20, **462**
Jasione 527
heldreichii 527
Jasminum 459
fruticans 459
mesnyi 459
officinale 459
Jonopsidium 348
Jubaea 155
chilensis 155
JUGLANDACEAE 290
Juglans 290
regia 290, **292**
JUNCACEAE 157
JUNCAGINACEAE 74
Juncus 157
acutus 158, **159**
bufonius 158
capitatus 157
conglomeratus 157
hybridus 158
inflexus 157
maritimus 157
Juniper
macrocarpa 54
Juniperus 54
communis 54, **55**
drupacea **42**
foetida 33
foetidissima 54
oxycedrus 19, 20, 22, 54, **55**
oxyxedrus 41
phoenicea 28, **39**, 41, 54, **55**
var. turbinata 54, **55**
Jurinca 579
cadmea 579
consanguinea 579
mollis 579
taygetea 579, **580**

K
Kickxia 468
aegyptiaca 468, **469**
commutata 468, **469**
elatine 468

subsp. crinita 468, **469**
subsp. elatine 468
floribunda 468, **469**
lanigera 468
spuria 468
Klasea 579
cretica 579
moreana 579
Knautia 586
arvensis **586**
integrifolia **586**
orientalis 586
Kohlrauschia
prolifera 386
velutina 386
Kundmannia 595
sicula 595

L
Lactuca 559
perennis 559
saligna 560
serriola 560
viminea 560
virosa 560
Lagoecia 592
cuminoides **592**
Lagurus 166
ovatus 166
Lamarckia 170
aurea 170, **171**
LAMIACEAE 10, 14, 481
Lamium 491
amplexicaule **490**, 491
bifidum **490**, 491
garganicum 491
subsp. garganicum 491
subsp. pictum 491
subsp. striatum **490**, 491
macrodon 491
maculatum 491
moschatum 491
purpureum **490**, 491
Lamyropsis 565
cynaroides **564**, 565
Lantana camara 481
Laphangium luteoalbum 531
Lathraea 512
rhodopea **16**, **19**, **512**
squamaria 512
Lathyrus 237
annuus **236**, 237
aphaca 238
articulatus 239

cicera 238
clymenum **238**, 239
gorgoni 239
hirsutus 239
latifolius 239
nissolia 238
ochrus 238
setifolius **238**
sylvestris **238**, 239
tuberosus 239
Launaea 561
 fragilis 561
 nudicaulis **560**, 561
LAURACEAE 62
Laurentia gasparrinii 527
Laurus 62
 nobilis 41, 62, **63**
Lavandula 504
 coronopifolia **504**, 505
 pubescens **504**, 505
 stoechas 33, **504**
Lavatera
 arborea 330
 cretica 330
 trimestris 328
Legousia 527
 falcata 528
 hybrida 528
 scabra 528
 speculum-veneris 527
LEGUMINOSAE (FABACEAE) 10, 13, 14, 35, 214
LENTIBULARIACEAE 480
Leontice 191
 leontopetalum 35, **191**
Leontodon 554
 autumnalis 555
 crispus 554
 graecus 554
 hispidus 554
 tuberosus 554, **555**
Leopoldia 148
 comosa **13**, **148**
 cycladica 149
 dionysicum 149
 longipes **148**, 149
 spreitzenhoferi 27, **148**, 149
 tenuiflora 149
 weissii 149
Lepidium 356
 coronopus 356
 didymum 356, **357**
 draba 356
 spinosum 356, **357**

Leptadenia 422
 pyrotechnica 422, 426
Leucaena 217
 leucocephala **216**, 217
Leucanthemum 545
 vulgare 545
Leucojum 131
 aestivum **130**, 131
Ligustrum 460
 lucidum 460, **461**
 ovalifolium 460
 vulgare 460
LILIACEAE 14, 80
Lilium 80
 candidum 80, **81**
 carniolicum 80
 chalcedonicum 80, **81**
 rhodopeum 80
Limbarda 536
 crithmoides **536**
Limniris pseudacorus 113
Limodorum 92
 abortivum **92**
Limoniastrum 370
 monopetalum **368**, 370
Limonium 44, 367
 aegaeum 369
 antipaxorum **18**, **368**, 369
 arcuatum 369
 bellidifolium 369
 cephalonicum 369
 compactum 370
 coronense 369
 damboldtianum 369
 echioides 369
 elaphonisicum **368**, 369
 galilaeum 370
 graecum 370
 hirsuticalyx 370
 ithacense 369
 kardamylii 369
 lobatum 367
 messeniacum 369
 monolithicum 370
 narbonense **368**, 370
 palmare 370
 phitosianum 369
 proliferum 369
 roridum 369
 samium 370
 saracinatum 369
 sartorianum 370
 sieberi 369
 sinuatum 367, **368**

sitiacum 370
 vanandense 370
 virgatum 369
 xerocamposicum 369
 zacynthium 369
LINACEAE 307
Linaria 467
 chalepensis 467
 genistifolia 467
 haelava **466**, 467
 joppensis **466**, 467
 micrantha **466**, 467
 pelisseriana **466**, 467
 simplex 467
 tenuis 468
 triphylla **466**, 467
 vulgaris 467
Linum 307
 arboreum **43**, 307, **308**
 bienne 308
 corymbulosum 307
 flavum 308
 maritimum 307
 mucronatum 307, **308**
 nodiflorum 307, **308**
 pubescens 309
  subsp. pubescens **308**, 309
  subsp. sibthorpianum 309
 strictum 307, **308**
  subsp. spicatum 307, **308**
 tenuifolium 309
 usitatissimum 308
Listera ovata 93
Lithodora 434
 hispidula 434, **435**
  subsp. hispidula 434
  subsp. versicolor 434
  subsp. zahnii 434
 major **435**
 minor **435**
 palaestina **435**
 zahnii 434, **435**
Lithospermum
 arvense 429
 incrassatum 429
 purpurocaeruleum 429
 tenuiflorum 429
Lobularia 355
 libyca **354**, 355
 maritima **354**, 355
Logfia 530
 gallica 530

minima 530
Lolium 169
  multiflorum 169
  perenne 169
Lomelosia 584
  albocincta 585
  cyprica 585
  divaricata 585
  hymettia 585
  minoana 585
  porphyroneura **584**, 585
  prolifera **584**
  variifolia 585
Lonicera 583
  etrusca 583
  implexa **583**
Lotus 252
  angustissimus 254
  conimbricensis 252
  conjugatus 254
  corniculatus 252
  creticus 252, **253**
  cytisoides 252, **253**
  edulis 252, **253**
  glinoides **253**, 254
  halophilus 252, 254
  ornithopodioides **253**, 254
  palaestinus **253**, 254
  pedunculatus 252
  peregrinus 252, **253**
  tetragonolobus **253**, 254
  uliginosus 252
Lunaria 343
  annua **342**, 343
Lupinus 224
  albus 224
    subsp. *albus* 224
    subsp. *graecus* 224
  angustifolius 224
  gredensis 225
  gussoneanus 224
  hispanicus **225**
  luteus 224
  palaestinus **225**
  pilosus 224, **225**
Lycianthes 450
  rantonnei 450, **451**
Lycium 450
  barbarum 451
  chinense **451**
  europaeum 451
  schweinfurthii 450, **451**
Lycopus 504
  europaeus **504**

Lysimachia 403
  arvensis 404
  atropurpurea 403
  dubia 403
  foemina 404
  serpyllifolia 403
LYTHRACEAE 313
Lythrum 313
  borysthenicum 314
  hyssopifolia 313
  junceum 313
  portula 314
  salicaria 313
  thymifolia 313
  tribracteatum 313

M
Magnolia 62
  grandiflora 62, **63**
MAGNOLIACEAE 62
Maireana 397
  brevifolia 397, **398**
Malabaila 600
  aurea **13**, **599**, 600
  involucrata 600
  lasiocarpa 600
  secacul **599**, 600
Malcolmia 352
  africana **351**, 352
  bicolor 352
  chia 352
  flexuosa 352
    subsp. *flexuosa* 352
    subsp. *naxensis* **351**, 352
  graeca 352
  maritima **351**, 352
Malephora 398
  purpureocrocea **398**
Malope 328
  malacoides 328
Malus 280
  pumila 280
  sylvestris **280**
Malva **10**, 328
  alcea **329**, 330
  arborea **329**, 330
  cretica 330
    subsp. *cretica* 330
  moschata **329**, 330
  multiflora **329**, 330
  neglecta 330
  nicaeensis 330
  parviflora **329**
  pseudolavatera 330

punctata **9**, **329**, 330
setigera 328
sylvestris **329**, 330
trimestris 328
MALVACEAE **10**, **15**, 327
Mandragora 456
  autumnalis 456
  officinarum 456, **457**
Mantisalca 578
  salmantica 578, **580**
Marrubium 485
  peregrinum 485
  vulgare 485
Matthiola 353
  fruticulosa 354
    subsp. *fruticulosa* 354
    subsp. *valesiaca* **354**
  incana **354**, 355
  longipetala **354**
  parviflora 355
  sinuata **354**, 355
  tricuspidata **44**, **353**
Medicago 245
  aculeata 248
  arabica 248
  arborea 245, **247**
  blancheana 246
  ciliaris 248
  constricta 248, **250**
  coronata **250**, 251
  disciformis **250**, 251
  doliata 248
  hypogaea 251
  intertexta 248
  italica 249
  laciniata 249, **250**
  littoralis 248, **250**
  lupulina 245, **247**
  marina 248, **250**
  minima 249, **250**
  monspeliaca 245, **247**
  murex 251
  orbicularis 246, **247**
  polymorpha 249, **250**
  praecox 249
  radiata 246
  rigidula 248
  rotata 249
  rugosa 246
  sativa 245, **247**
  scutellata 246, **247**
  tornata 249
  truncatula 249, **250**
  turbinata 251

Melaleuca 316
　citrina **315**, 316
MELANTHIACEAE 76
Melia 326
　azedarach 326, **327**
MELIACEAE 326
Melica 164
　ciliata **163**, 164
　transsilvanica **163**, 164
Melilotus 242
　albus 242
　altissimus 242
　elegans 242
　graecus 243
　indicus 242, **243**
　italicus 242
　messanensis 242
　neapolitanus 242
　segetalis 242
　siculus 242
　spicatus 242
　sulcatus 242
Melissa 489
　officinalis **489**
Mentha 503
　aquatica **502**, 503
　longifolia 503
　pulegium 503
　spicata 503
　suaveolens **502**, 503
MENYANTHACEAE 528
Mercurialis 306
　annua **306**
　ovata 306
Mesembryanthemum 399
　cordifolium 399
　crystallinum **398**, 399
　nodiflorum **398**, 399
Michauxia 523
　campanuloides 33, **34**, **522**, 523
　pinnata **522**
Micromeria 496
　carpatha 496
　graeca 496
　　subsp. *graeca* 496, **497**
　juliana **497**, 498
　myrtifolia 496
　nervosa 496, **497**
　sinaica 496, **497**
　sphaciotica 496
Microthlaspi perfoliatum 343
Minuartia 374
　hybrida 374

　mediterranea 374
　picta 374, **375**
Mirabilis 400
　jalapa 400
Misopates 465
　orontium **465**
Monotropa 409
　hypopitys **408**, 409
Montia 400
　fontana 400
MORACEAE 284
Moraea
　mediterranea 115
　sisyrinchium 115, **116**
Moricandia 352
　arvensis **351**, 352
Morus 284
　alba **13**, **39**, 284, **285**
　nigra 284
Musa 156
　cultivar **156**
MUSACEAE 156
Muscari 149
　armeniacum 149, **150**
　botryoides 149, **150**
　commutatum **150**, 151
　comosum 148
　kerkis 149, **150**
　macrocarpum 151
　neglectum 23, 149, **150**
　parviflorum 151
　pulchellum 149
Myoporum 477
　laetum **476**, 477
　tenuifolium **476**, 477
Myosotis 441
　arvensis 442
　congesta 441
　laxa 441
　ramosissima **441**
　refracta 442
　sylvatica 442
MYRTACEAE 316
Myrtus 316
　communis **39**, **41**, **315**, 316
　　subsp. *communis* **315**, 316

N

Narcissus 48, 132
　obsoletus **48**, 132, **133**
　papyraceus 132, **133**
　poeticus 132
　tazetta 132, **133**
Nasturtiopsis 362

　coronopifolia **362**
Neatostema 434
　apulum **433**, 434
Neotinea 96
　lactea 96, **97**
　maculata **97**
　tridentata 96, **97**
　ustulata 96
Neottia 93
　nidus-avis 93
　ovata 93
Nepeta 489
　cataria 489
　italica 489
Nerium 420
　oleander **6**, 420, **421**
Neurada 333
　procumbens **332**, 333
NEURADACEAE 333
Nicandra 450
　physalodes 450
Nicotiana 458
　glauca **458**
Nigella 198
　arvensis **198**
　carpatha 198
　ciliaris 199
　damascena **198**
　degenii 198
　deserti 198
　doerfleri 198
　elata 198
　icarica 198
　orientalis 199
　oxypetala 199
　stricta 198
NITRARIACEAE 317
Nonea 435
　echioides 435
　philistaea **435**
Notobasis 571
　syriaca **570**, 571
Notoceras 362
　bicorne **362**
NYCTAGINACEAE 399
Nymphoides 528
　peltata 528

O

Ochradenus 342
　baccatus **342**
Odontites 510
　luteus 510
　vernus 510, **511**

Oenanthe 594
  fistulosa 595
  globulosa 595
  pimpinelloides 594
  silaifolia 594
Oenothera 314
  speciosa 314, **315**
Olea 460
  europaea **12**, **35**, **36**, 460, **461**
    var. *europaea* 460
    var. *sylvestris* 460
OLEACEAE 459
Omalotheca sylvatica 531
Omphalodes 440
  luciliae 440
ONAGRACEAE 314
Onobrychis 268
  aequidentata 268, **269**
  caput-galli 268, **269**
  crista-galli 268, **269**
  venosa 268, **269**
  viciifolia 268, **269**
Ononis 239, **241**
  diffusa 240
  mitissima 240, **241**
  natrix 240
  ornithopodioides 240
  pubescens 240, **241**
  pusilla 240, **241**
  ramosissima 240
  reclinata 239, **241**
  serrata 240
  sicula 240
  spinosa 240, **241**
  talaverae 240
  variegata 241
  verae 240, **241**
  viscosa 240
    subsp. *sicula* 240
Onopordum 571
  acanthium 572, **573**
  bracteatum 572
  cyprium 571, **573**
  illyricum 572, **573**
  laconicum 572
  rhodense 572
  tauricum 571
Onosma 430
  alborosea **431**, 432
  aucheriana 430
  echinata 430, **431**
  frutescens 430, **431**
  gigantea 432

  graeca 430
  heterophylla 430
  juliae 432
  lineariloba 431
  lycaonica 431
  mersinana 431
  orientalis **431**, 432
  stridii **431**
  subulifolia 431
  taurica 430
    var. *brevifolia* 430
    var. *taurica* 430
Ophrys 28, 102
  apifera **106**
  argolica 108
    subsp. *aegaea* 108
    subsp. *argolica* 108
    subsp. *lesbis* 108
    subsp. *lucis* 108
  bertolonii 103, 108
  bombyliflora 104, **105**
  ferrum-equinum **109**, 111
    subsp. *ferrum-equinum* 111
    subsp. *gottfriediana* 103, 111
  fuciflora 107
    subsp. *andria* 107
    subsp. *bornmuelleri* **29**, **106**, 107
    subsp. *candica* 107
    subsp. *fuciflora* 107
  fusca 104
    subsp. *blitopertha* 103, 104
    subsp. *cinereophila* 103, 104
    subsp. *fusca* 104, 105
      forma *creberrima* 104
      forma *cressa* 104
    subsp. *iricolor* 103, 104
  heterochila 107
  holoserica 107
  insectifera 102, **105**
  kotschyi 111
    subsp. *ariadnae* 103, 111
    subsp. *cretica* 26, **27**, **110**, 111
    subsp. *kotschyi* 103, **110**, 111
  lutea 104
    subsp. *galilaea* 104, **105**
      forma *sicula* 104
    subsp. *lutea* 104, **105**

    forma *phryganae* 104
    subsp. *melena* 104
  omegaifera 103, 104
    subsp. *fleischmannii* 103, 106
    subsp. *israelitica* 103, 106
    subsp. *omegaifera* 104
  reinholdii 103, 111
  scolopax 106
    subsp. *cornuta* 103, 107
    subsp. *heldreichii* **47**, **107**
    subsp. *heldreichii* 106
    subsp. *rhodia* 103, 107
    subsp. *scolopax* 107
  speculum 102, **105**
    subsp. *regis-ferdinandii* 102, 103
    subsp. *speculum* 102
  sphegodes **20**, 108
    subsp. *aesculapii* 110
    subsp. *cretensis* 110
    subsp. *epirotica* 110
    subsp. *gortynia* 110
    subsp. *helenae* 110
    subsp. *mammosa* 108, **109**
      forma *alasiatica* 108
      forma *herae* 110
      forma *morio* 110
    subsp. *sphegodes* 108, **109**
    subsp. *spruneri* **109**, 110
  tenthredinifera 104, **105**
    forma *villosa* 104
  umbilicata **106**, 107
    subsp. *bucephala* 103, 107
    subsp. *flavomarginata* 103, 107
    subsp. *umbilicata* 107
      forma *attica* 107
      forma *umbilicata* 107
Opopanax 603
  chironium 603, **604**
  hispidus 603
Opuntia 35, 402
  dillenii **401**, 402
  ficus-indica 402
  maxima **12**, **401**, 402
  stricta 402
ORCHIDACEAE 15, 90
Orchis 93
  anatolica 94
  anthropophora 93, **95**
  collina 99

coriophora 97
galilaea 94, **95**
intacta 97
italica 94, **95**
lactea 96
laxiflora 99
mascula 96
militaris 94, **95**
morio 99
pallens 96
papilionacea 99
pauciflora 96
provincialis 96
purpurea 94, **95**
quadripunctata 94, **95**
sancta **98**
simia 93, **95**
sitiaca 94
spitzelii 96
   subsp. *nitidifolia* 96
tridentata 96
troodi 29, 94, **95**
ustulata 96
Origanum 501
  amanum **502**
  calcaratum 502
  dayi **502**, 503
  dictamnus **502**
  majorana 502
  marjoram 501
  onites 501, **502**
  syriacum 501
  vetteri 501
  vulgare 501
Orlaya 608
  daucoides 608
  grandiflora 608
  kochii 608
Ornithogalum 141
  arabicum 143
  armeniacum 143
  atticum 143
  brachystylum 141
  collinum 143
  creticum 141
  exaratum 143
  exscapum **142**, 143
  gussonei 143
  libanoticum 33
  montanum 143
  narbonense **11**, 141, **142**
  neurostegium 143
    subsp. *eigii* **142**, 143
    subsp. *neurostegium* 143

nutans 141, **142**
oligophyllum 143
pedicellare 141, **142**
prasinantherum 141
pyrenaicum 141
sibthorpii **142**, 143
spetae 141
trichophyllum 141
umbellatum **142**, 143
wiedemannii 143
Ornithopus 255
  compressus **255**
  pinnatus 255
OROBANCHACEAE 510
Orobanche 47, 517
  alba 521
  amethystea 518, **519**
  artemisiae-campestris 518
  baumanniorum 520
  camptolepis 513, 521
  caryophyllacea **519**, 521
  cernua **520**, 522
  crenata **520**, 522
  crinita 518
  cumana 522
  cypria 513, 520
  elatior 521
  gracilis **520**, 521
  grisebachii 518
  hederae 518, **519**
  hermonis 513, 521
  lutea 521
  minor 517, **519**
  palaestina 518
  palasteana 513
  picridis 518, **519**
  pubescens 518, **519**
  reticulata **519**, 521
  sanguinea **45**, **47**, 518, **519**
  sideana 521
Oryza 174
  sativa 174, **176**
Oryzopsis 172
  coerulescens 172
  miliacea **171**, 172
Osyris 363
  alba 363, **364**
Otanthus maritimus 541
OXALIDACEAE 293
Oxalis 293
  articulata 292, 294
  corniculata **292**, 293
  debilis 294
  pes-caprae **292**, 293

P
Paeonia 203
  clusii 203
    subsp. *clusii* 203, **204**
    subsp. *rhodia* 203, **204**
  mascula 203, **204**
    subsp. *hellenica* 203
    subsp. *mascula* 203
  parnassica 203, **204**
  peregrina 205
  saueri 205
PAEONIACEAE 203
Paliurus 281
  spina-christi **39**, **280**, 281
Pallenis 549
  hierochuntica **548**, 549
  spinosa **11**, **39**, **548**, 549
Pancratium 132
  maritimum **44**, **48**, 132, **133**
  sickenbergeri 132, **133**
PANICOIDEAE 178
Panicum 181
  miliaceum 181
  repens 181
Papaver 182
  apulum 184
  argemone 184
  dubium 182, **183**
  humile 182, **183**
  hybridum **183**, 184
  lecoqii 182
  nigrotinctum **183**, 184
  purpureomarginatum 182, **183**
  rhoeas 182, **183**
  somniferum 182
    subsp. *setigerum* 182
  umbonatum 182
  virchowii 184
PAPAVERACEAE 15, 182
PAPILIONOIDEAE 220
Paracaryum 432
  aucheri 432
  lithospermifolium 432
    subsp. *cariense* 432, **433**
    subsp. *lithospermifolium* 432
  rugulosum 432, **433**
Parapholis 172
  incurva 172
Pardoglossum 440
  cheirifolium 440
Parentucellia 510
  latifolia 510, **511**

subsp. *flaviflora* 510, **511**
viscosa 510, **511**
Parietaria 286
  cretica 286
  judaica **286**
  lusitanica 286
  officinalis **286**
Paris 76
  quadrifolia 76, **77**
Paronychia 376
  argentea **375**, 376
  capitata **375**, 376
  cephalotes 376
  chionaea 376
  echinulata 376
  kapela
    subsp. *chionaea* 376
    subsp. *insularum* 376
  macrosepala 376
Passiflora 298
  caerulea 298, **299**
  passerina 335
PASSIFLORACEAE 298
Pastinaca 599
  sativa 599
Paulownia 509
  fortunei **508**, 509
  tomentosa **508**, 509
PAULOWNIACEAE 509
Peganum 317
  harmala 317
Pentatropis 422
  nivalis 422, 426
Pergularia 422
  tomentosa 422, **423**, 426
Periploca 425
  angustifolia 425
  aphylla **424**, 426, **427**
  graeca **424**, 427
  laevigata **424**, 425
PERIPLOCOIDEAE 425
Persea 62
  americana 62, **63**
Persicaria 371
  lapathifolia 371
  maculosa 371
Petromarula 523
  pinnata 523
Petrorhagia 386
  dubia 386, **387**
  illyrica 386
  prolifera 386
  saxifraga 386
Phacelia 432

tanacetifolia 432
Phagnalon 534
  rupestre 534
    subsp. *graecum* **532**, 534
    subsp. *rupestre* **532**, 534
  saxatile **532**, 534
Phelipanche 47, 516
  aegyptiaca **516**
  astragali 517
  cilicica 517
  cohenii 517
  daninii **516**, 517
  heldreichii 517
  lavandulacea 517
  mutelii **516**
  nana 516
  nowackiana 517
  oxyloba 517
  purpurea 517
  ramosa **516**
  rechingeri 517
  schultzii 517
  zosimii 517
Phelypaea 514
  boissieri 514, **515**
Phillyrea 460
  latifolia 20, 460, **461**
Phlomis **10**, 487
  armeniaca 488
  bourgaei 488
  chrysophylla 488
  cretica 488
  cypria 488
  floccosa 488
  fruticosa **487**
  herba-venti 488
  lanata 487, 488
  lunariifolia 487, 488
  lycia 488
  pichleri 488
  platystegia **487**, 488
  samia 488
Phoenix 155
  canariensis **154**, 155
    × *P. dactylifera* 155
  dactylifera **154**, 155
  theophrasti 26, **154**, 155
Phragmites 175
  australis 175
  communis 175
Physalis 453
  angulata 453
  ixocarpa 453
  philadelphica 453, **454**

Phytolacca 399
  acinosa **398**, 399
PHYTOLACCACEAE 399
Picea
  abies **19**
  brutia 41
Picris 556
  hieracioides **556**
  pauciflora 556
  rhagadioloides 556
PINACEAE 50
Pinguicula 480
  balcanica 480
  crystallina **480**
    subsp. *crystallina* 480
    subsp. *hirtiflora* 480
Pinus 52
  brutia 28, 30, 33, 52, **53**
  halepensis 41, **42**, 44, 52
  nigra **19**, 30, 52
    subsp. *nigra* 52
    subsp. *pallasiana* 23, 28, 52, **53**
  pinea 41, 44, 52, **53**
  sylvestris 52
Piptatherum 172
  coerulescens 172
  miliaceum 172
Pistacia 318
  atlantica 318, **319**
  lentiscus **12**, 37, 41, 318, **319**
  terebinthus 2, 37, **38**, 318, **319**
  vera 318
  × *saportae* 318
Pisum 239
  fulvum 239
  sativum 239
PITTOSPORACEAE 589
Pittosporum 589
  tobira **12**, **589**
  undulatum 589
PLANTAGINACEAE 4, 463
Plantago 463
  afra **462**, 463
  albicans 464
  amplexicaulis 464
  arenaria 463
  bellardii **462**, 463
    subsp. *bellardii* 464
    subsp. *deflexa* **462**, 464
  coronopus **462**, 463
  crassifolia 464, **465**
  cretica **462**, 464
  indica 463

lagopus 464
lanceolata 464
major 464, **465**
maritima subsp. *crassifolia* 464
media 464
ovata **462**, 464
phaeostoma **462**, 463
squarrosa 463
PLATANACEAE 202
Platanthera 101
  bifolia 101
  chlorantha **101**
  holmboei 101
Platanus
  orientalis 20, 43, **202**
  × *acerifolia* 202
  × *hispanica* 202
  × *hybrida* 202
Plocama 412
  calabrica **44**, 412, **413**
PLUMBAGINACEAE 366
Plumbago 366
  auriculata 366, **367**
  europaea 366, **367**
POACEAE 11, 14, 162
Podonosma orientalis 432
Podospermum 557
  canum 557
  laciniatum 557
Podranea 479
  ricasoliana 479
Polycarpon 377
  tetraphyllum 377
Polygala 270
  major 270
  monspeliaca 270, **271**
  myrtifolia 270, **271**
  nicaeensis 270
    subsp. *mediterranea* 270, **271**
    subsp. *tomentella* 270
  venulosa 270, **271**
  vulgaris 270, **271**
POLYGALACEAE 270
POLYGONACEAE 371
Polygonum 371
  lapathifolium 371
  maritimum 371, **372**
  persicaria 371
POOIDEAE 162
Populus 298
  nigra 298, **299**
  populneus **332**

Portulaca 400
  oleracea 400
PORTULACACEAE 400
Posidonia 75
  oceanica **75**
POSIDONIACEAE 75
Potentilla 275
  micrantha **274**, 275
  reptans 275
Poterium 273
  creticum 273
  sanguisorba **272**, 273
    subsp. *sanguisorba* 273
  verrucosum 273
Prangos 596
  ferulacea 596, **597**
Prasium 485
  majus **484**, 485
Primula 405
  veris **404**, 405
  vulgaris **404**, 405
    subsp. *rubra* **404**, 405
    subsp. *vulgaris* 405
PRIMULACEAE 403
Procopiania
  cretica 437
  insularis 437
Prospero 146
  autumnale 146, **147**
Prunella 490
  cretensis 490
  grandiflora 491
  laciniata 490
  vulgaris **489**, 490
Prunus 33, 276
  armeniaca 279
  avium 276, **277**
  cerasifera **277**
  communis **278**
  domestica 277
    subsp. *domestica* 277
    subsp. *insititia* 277
  dulcis **13**, 35, **36**, **278**, 279
  graeca 279
  laurocerasus 276, **277**
  persica **278**, 279
  prostrata 277
  spinosa **277**, **278**
    subsp. *dasyphylla* 277
  syriaca 278
  webbii **278**, 279
Pseudomuscari 151
  inconstrictum 151
Pseudorlaya 608

pumila 608, **609**
Psoralea bituminosa 234
Pteranthus 376
  dichotomus **377**
Pulicaria 534
  arabica
    subsp. *hispanica* 534
  dysenterica 534, **535**
  odora 534
  paludosa 534
Pulsatilla 194
  rhodopaea 194
Punica 314
  granatum 314, **315**
Putoria calabrica 412
Pycnocomon 585
  rutifolium 585
Pyracantha 276
  coccinea 276
Pyrostegia 478
  venusta 478, **479**
Pyrus 279
  communis 279
  elaeagrifolia 279
  spinosa 279
  syriaca 279

Q
Quercus 287
  alnifolia 288, **289**
  aucheri 288
  boissieri **286**, 287
    subsp. *veneris* 287
  calliprinos 33
  cerris 33, 288
  coccifera **19**, 20, **22**, 30, 37, 41, 287, **289**
    subsp. *calliprinos* 288, **289**
    subsp. *coccifera* 288
  humilis 288
  ilex 41, 288, **289**
  infectoria 33, 287
    subsp. *infectoria* 287
    subsp. *veneris* 28, **286**
  ithaburensis 288
    subsp. *ithaburensis* 288
    subsp. *macrolepis* 288, **289**
  libani 288
  petraea 287
  pubescens 19, 20, **42**, 288, **289**

R
Radiola 309
 linoides 309
RANUNCULACEAE 192
Ranunculus 194
 arvensis 196
 asiaticus **48, 195,** 196
 bullatus 196
 creticus **195**
 ficaria 197
 flammula **195,** 197
 gracilis **195,** 196
 lingua 197
 muricatus 194
 paludosus **195,** 196
 peltatus **195,** 197
 sardous 196
 sceleratus 196
 trichophyllus 197
 velutinus 196
Raphanus 360
 raphanistrum 360, **361**
Rapistrum 360
 perenne 360
 rugosum 360, **361**
RAUVOLFIOIDEAE 420
Reichardia 561
 intermedia 562
 picroides **560,** 561
 tingitana **560,** 562
Reseda 340
 alba 340, **341**
 alopecuros **341**
 decursiva 340, **341**
 lutea **341**
 luteola 342
 minoica **341,** 342
 phyteuma 340, **341**
RESEDACEAE 340
Retama 224
 monosperma 224
 raetam 46, **223,** 224
Rhagadiolus 554
 edulis 554, **555**
 stellatus 554, **555**
RHAMNACEAE 281
Rhamnus 281
 alaternus **282**
 cathartica 282
 lycioides **282**
  subsp. graeca 282
  subsp. oleoides **282**
 pichleri 282
 pumila 281

 saxatilis 281
Rhododendron 411
 luteum **410,** 411
Rhus 320
 coriaria 320, **321**
Ricinus 306
 communis **306**
Ricotia 353
 cretica **353**
 isatoides 353
 lunaria **353**
Ridolfia 598
 segetum 598
Rindera 430
 graeca 430
Robinia 220
 pseudoacacia **220**
 viscosa 220
Roemeria 184
 hybrida 184, **185**
Romulea 124
 bulbocodium 124, **125**
 columnae 124, **125**
 linaresii 124
 ramiflora 124
 tempskyana 124
Rosa 272
 arvensis 272
 canina **272**
 dumalis 272
 heckeliana 272
 phoenicia 272
 pimpinellifolia 273
 pulverulenta 273
 sempervirens **272**
 spinosissima 273
 × damascena 273
ROSACEAE 15, 33, 271
Rosmarinus 505
 officinalis **504,** 505
Rostraria 166
 cristata **165,** 166
Rubia 411
 peregrina **12, 39,** 411, **413**
 tenuifolia 412, **413**
 tinctorum 411
RUBIACEAE 411
Rubus 275
 canescens 275
 hirtus 275
 sanctus 274, 275
Rumex 371
 acetosa 373
 acetosella 373

 bucephalophorus 372
 cyprius 371, **372**
 scutatus 373
 thyrsiflorus 373
 tuberosus 373
 vesicarius 372
Ruscus 153
 aculeatus 41, **42, 152,** 153
 hypoglossum 153
 hypophyllum 153
Ruta 324
 buxbaumii 323
 chalepensis **323,** 324
  subsp. fumariifolia 323, 324
 graveolens 324
 montana 324
 suaveolens 323
RUTACEAE 323

S
Saccharum 178
 ravennae 178, **180**
Sagina 378
 apetala 378
 procumbens 378
Sagittaria 73
 sagittifolia **72,** 73
SALICACEAE 298
Salicornia 394
 europaea 394
 perennans 394
 procumbens **393,** 394
Salix 299
 alba 299
 amplexicaulis 299
 pedicellata 299
 purpurea 299
 triandra **299**
Salsola 396
 aegaea 396
 kali 44, **395,** 396
 soda 396
 tetrandra 46, **395,** 396
 vermiculata 31, 46, **395,** 396
Salvia 505
 aegyptiaca 505, **507**
 aethiopis 506
 amplexicaulis 506
 argentea 506, **507**
 bracteata **508,** 509
 dominica 505, **507**
 fruticosa **11, 12, 39,** 505, **507**
 hierosolymitana **508,** 509

indica 508
lanigera 508
napifolia 508
nemorosa 506, **507**
officinalis 505, **507**
pomifera 506
pratensis 506
samuelssonii **508**, 509
sclarea 506
tomentosa 505
triloba 505
verbenaca 506, **507**
verticillata 508
viridis 506, **507**
Sambucus 582
 ebulus 582
 nigra 582
Samolus 405
 valerandi **404**, 405
Sanguisorba
 cretica 273
 minor 273
  subsp. minor 273
  subsp. verrucosa 273
SANTALACEAE 363
SAPINDACEAE 320
Saponaria 380
 calabrica 380
 cypria **379**, 380
 officinalis **379**, 380
Sarcocornia 395
 fruticosa 395
 perennis 395
Sarcopoterium 273
 spinosum 33, 37, **38**, 273, **274**
Satureja 495
 icarica 495
 montana 495
 parnassica 495
 spicigera 495, **497**
 spinosa 495
 thymbra 495, **497**
Saxifraga 205
 hederacea 205
 rotundifolia **204**, 205
  subsp. chrysospleniifolia **204**, 205
  subsp. rotundifolia 205
 scardica 21, 22, **204**, 205
 tridactylites 205
SAXIFRAGACEAE 205
Scabiosa 583
 atropurpurea 583, **584**

aucheri 585
porphyroneura 585
sicula 585
webbiana 583
Scandix 594
 australis 594
 pecten-veneris 594
 stellata 594
Schenkia 419
 spicata **418**, 419
Schinus 318
 molle 318, **319**
 terebinthifolia 318, **319**
Schoenus 161
 nigricans 161
Scilla 144
 autumnalis 146
 bifolia 144, **145**
 bithynica 144
 cilicica 144, **145**
 hyacinthoides **145**
 messeniaca 144, **145**
 morrisii 144, **145**
 nana 144
 subnivalis 144
Scirpoides 158
 holoschoenus 158, **159**
Scolymus 550
 grandiflorus 550, **551**
 hispanicus 550, **551**
 maculatus 550
Scorpiurus 266
 muricatus 266, **267**
 vermiculatus 266, **267**
Scorzonera 556
 araneosa 556
 cretica **556**
 hispanica 557
 jacquiniana 557
 judaica **556**, 557
 laciniata 557
 psychrophila 557
 scyria 557
Scorzoneroides 555
 autumnalis 555
Scrophularia 472
 canina 472
 floribunda 472
 lucida 472
 nodosa 473
 peregrina 473, **474**
 scopolii 473
 spinulescens 472
 xanthoglossa 472, **474**

SCROPHULARIACEAE 472
Scutellaria 501
 columnae 501
 hirta 501
 orientalis 501
 sieberi 501
Securigera 265
 carinata 265
 cretica 265
 globosa 265
 parviflora 265
 securidaca 265
 varia 265
Sedum 206
 acre 206, **207**
 album **207**, 208
 amplexicaule 206
 caespitosum **208**
 cepaea 209
 creticum 209
 cyprium 209
 dasyphyllum **208**
 eriocarpum 209
  subsp. spathulifolium **208**, 209
 lampusae 209
 litoreum 206, **207**
 microcarpum 209
 microstachyum 209
 ochroleucum 206, **207**
 rubens **208**, 209
 sediforme 206, **207**
 tenuifolium 206
Senecio 546
 aegyptius 546
 glaucus 546
  subsp. coronopifolius 546, **547**
  subsp. cyprius 546, **547**
 jacobaea 547
 leucanthemifolius 546, **547**
  subsp. vernalis 546, **547**
 lividus 546
 viscosus 546
 vulgaris 546, **547**
Senna 218
 corymbosa 218, **219**
 didymobotrya 218, **219**
Serapias 111, **112**
 aphroditae **112**, 113
 bergonii 113
 cordigera 111, **112**
 lingua 111, **112**
 orientalis **112**, 113

subsp. *levantina* 112, **113**
parviflora 111, **112**
vomeracea 111
Seseli 606
  crithmifolium 607
  gummiferum 606
    subsp. *aegaeum* 607
    subsp. *crithmifolium* **607**
    subsp. *gummiferum* 607
  tortuosum 607
Setaria 179
  adhaerens 179, **180**
  viridis 179, **180**
Sherardia 416
  arvensis 416, **417**
Sideritis 486
  curvidens 486, **487**
  euboea 486
  lanata 486, **487**
  montana 486
  perfoliata 486
  purpurea 486
  scardica 486
  syriaca 486
Silene 380
  aegaea 382
  alba 380
  alexandrina 382, **383**
  apetala 382, **383**
  baccifera 385
  behen 383
  bellidifolia 382
  cerastoides 382
  colorata **381**, 382
  conica 381
  cretica 380
  damascena **381**, 382
  dichotoma **381**, 382
  discolor 382
  gallica 382
  grisebachii 381
  italica 384
  latifolia 380
  lydia 381
  macrodonta 381
  multicaulis 383
  noctiflora 383
  palaestina **381**, 382
  pentelica 382
  sartorii 381
  sclerocarpa 382, **383**
  sedoides 382
    subsp. *runemarkii* 382
    subsp. *sedoides* 382, **383**

sieberi 384
spinescens 384
subconica 381
succulenta **384**
vulgaris **383**
Silybum 574
  marianum **573**, 574
SIMAROUBACEAE 325
Sinapis 359
  alba **358**, 359
  arvensis 359
Sisymbrium 348
  altissimum 349
  erysimoides **349**
  irio 349
  officinale 348
  orientale 349
SMILACACEAE 80
Smilax 80
  aspera 13, 80, **81**
Smyrnium 593
  apiifolium 593
  connatum **592**, 593
  creticum 593
  olusatrum 593
  perfoliatum 593
  rotundifolium **592**, 593
SOLANACEAE 450
Solandra 459
  maxima **458**, 459
Solanum 455
  americanum 455
  chenopodioides 455
  dulcamara **454**, 455
  elaeagnifolium 455
  jasminoides 455
  laxum 455
  linnaeanum 455
  luteum **454**, 455
  nigrum **454**, 455
  sinaicum 455
  villosum 455
Solenopsis 527
  laurentia 527
  minuta 527
    subsp. *annua* **526**, 527
Solenostemma 422
  arghel 422, 426
Sonchus 558
  asper 558
  oleraceus 12, 559
  tenerrimus 558, **560**
Sorghum 178
  bicolor 178

halepense 178, **180**
Sparganium 157
  erectum 157
Spartium **10**, 221
  junceum **11**, **39**, **220**, 221
Spergula 378
  arvensis 378
  fallax 378, **379**
  marina 378
  media 378
  pentrandra 378
Spergularia 378
  bocconei **379**
  marina 44, **379**
  rubra 379
Spiranthes 100
  spiralis 100, **101**
Stachys 492
  arvensis **494**
  balansae **494**
  chrysantha 492, **493**
  citrina 492, **493**
  cretica **39**, **493**
  germanica **493**
  ionica **18**, **494**
  ocymastrum 492, **493**
  officinalis **494**
  spinulosa 492, **493**
  swainsonii **494**
  thirkei 493
  tournefortii 493
Sternbergia 48, 131
  clusiana 131
  colchiciflora 131
  lutea **48**, **130**, 131
Stipa 173
  capensis 46, **171**, 173
  parviflora 173
  pennata 173
  pulcherima 173
Strelitzia 155
  reginae **154**, 155
STRELITZIACEAE 155
STYRACACEAE 409
Styrax 409
  officinalis **408**, 409
Suaeda 395
  maritima 396
  vera 44, **395**
Symphyotrichum squamatum 536
Symphytum 437
  brachycalyx **436**, 437
  creticum 437

insulare 437
ottomanum 438
Syringa 459
  pubescens **458**, 459

T
TAMARICACEAE 365
Tamarix 365
  dalmatica 366
  nilotica **364**, 365
  parviflora 366
  smyrnensis **364**, 365
Tamus communis 76
Taraxacum sect. *Ruderalia* 562
Taraxacum officinale 562
Tecoma 478
  capensis 478, **479**
  stans 478, **479**
Teesdalia 361
  coronopifolia 361
Telmissa 209
  microcarpa **208**, 209
Tetraena 212
  alba 212
  coccinea 212
  dumosa 212
Tetragonolobus
  conjugatus 254
  palaestinus 254
  purpureus 254
Teucrium 483
  alpestre 483
  capitatum **39**, 483
    subsp. *capitatum* 483, **484**
  chamaedrys **484**, 485
  cuneifolium 483
  divaricatum 485
  flavum **484**
    subsp. *glaucum* 484
    subsp. *gymnocalyx* 484
    subsp. *hellenicum* **484**
  gracile 484
  massiliense 485
  microphyllum 485
  polium subsp. *capitatum* 483
  scordium 483
  spinosum 485
Thalictrum 194
  aquilegiifolium 194
  lucidum 194
  orientale 194, **195**
Thapsia 605
  garganica **48**, **604**, 605

Theligonum 417
  cynocrambe 417
Thesium 363
  humile 363, **364**
Thlaspi 343
  perfoliatum **342**, 343
Thymbra 500
  calostachya 500
  capitata 37, **500**
  capitatus 500
  spicata 500
Thymelaea 334
  hirsuta 334, **335**
  tartonraira 335
THYMELAEACEAE 334
Thymus 498
  atticus 499
  comptus 499
  integer 28, **29**, 499
  leucotrichus 498
  longicaulis 498
  pulegioides 499
  sibthorpii 498, **499**
  teucrioides 499
  thracicus 498
  zygioides 498
Tolpis 553
  umbellata 553
Tordylium 606
  aegyptiacum 606, **607**
  apulum 606, **607**
  syriacum 606, **607**
Torilis 605
  africana 605
  arvensis 605
  japonica **604**, 605
  nodosa 605
  tenella 605
Tragopogon 557
  dubius 557
  hybridus 558
  porrifolius 558
    subsp. *australis* 558, **559**
    subsp. *longirostris* 558, **559**
    subsp. *porrifolius* 558
  pratensis 557
Tragus 177
  racemosus 177
Tribulus 212
  dumosa 213
  terrestris 212, **213**
Trifolium 256
  affine 256

angustifolium 256, **257**
arvense 256, **257**
campestre 256, **257**
caudatum 260
cherleri 258, **259**
clypeatum 258, **259**
dubium 256
echinatum 260
eriosphaerum **261**, 262
fragiferum 260
hirtum 261
incarnatum 257
infamia-ponertii 256
lappaceum 258, **259**
leucanthum 260
lucanicum 258
micranthum 256
nigrescens 258
ochroleucon 260
pallidum 258, **259**
pamphylicum 256, **257**
petrisavii 258
phitosianum 256
physodes 260
pratense 260, **261**
prophetarum 256, **257**
purpureum 256, **257**
repens 258, **259**
resupinatum 260, **261**
scabrum 258
spumosum 260
squamosum 260
squarrosum 261
stellatum 260, **261**
suffocatum 257
tomentosum 258, **259**
uniflorum **261**, 262
Triglochin 74
  barrelieri 74
  bulbosa
    subsp. *barrelieri* 74
    subsp. *laxiflora* 74
  laxiflora 74
Trigonella 243
  arabica **243**, 244
  caerulea 244
  cariensis 244
  cephalotes 244
  corniculata 243
  foenum-graecum 244
  gladiata 244
  graeca **243**
  monspeliaca 245
  rotundifolia 244

schlumbergeri **243**, 245
smyrnaea 244
spicata 244
spinosa 244
spruneriana 244
stellata **243**
Tripodion 254
  tetraphyllum 254, **255**
Tripolium 537
  pannonicum
    subsp. *tripolium* 537
TROPAEOLACEAE 340
Tropaeolum 340
  majus 340
Tuberaria 337
  guttata **337**
Tulipa 84
  agenensis **32**, 86, **87**
  australis 84
  bakeri 84
  bithynica 84
  clusiana 84, **85**
  cretica **27**, 84, **85**
  cypria **28**, **29**, 86, **87**
  doerfleri 84
  goulimyi 84, **85**
  orphanidea 84, **85**
  praecox 86
  saxatilis **27**, 84, **85**
  sylvestris 84, **85**
  systola 86, **87**
  undulatifolia **23**, 84, **85**
  whittallii 84
Tunica velutina 386
Typha 156
  angustifolia 156
  domingensis 156
  latifolia 156
TYPHACEAE 156

U

ULMACEAE 283
Ulmus 283
  canescens **282**, 283
  minor **282**, 283
Umbilicus 210
  albido-opacus 211
  chloranthus 211
  coccineum **210**
  horizontalis 210
  intermedius 210
  luteus 211
  parviflorus 211
  rupestris 210

Urginea maritima 144
Urospermum 561
  picroides **13**, 561
Urtica 284
  dioica 284
  dubia 285
  membranacea 285
  pilulifera **285**
  urens **285**
URTICACEAE 284
Utricularia 480
  australis 481
  minor 481
  vulgaris **480**

V

Vachellia 216
  tortilis 216
    subsp. *raddiana* 46, **216**
Vagaria 133
  parviflora 133
Valantia 417
  aprica 417
  hispida **417**
  muralis **417**
Valeriana 588
  alliariifolia 588
  asarifolia **26**, 588, **589**
  dioscoridis 588
  italica **23**, 588, 589
  officinalis 588
  tuberosa 589
Valerianella 587
  dentata 588
  discoidea 588, **589**
  echinata 588
  eriocarpa 588
  locusta 588
  vesicaria 588
Verbascum 473
  arcturus **474**, 475
  blattaria 475
  chaixii 475
  dumulosum 473, **474**
  graecum 476
  longifolium 475
  lychnitis 475
  nigrum 473, **474**
  phlomoides 475
  phoeniceum 476
  pulverulentum 475
  sinuatum 473, **474**
  speciosum **474**, 475
  spinosum **474**, 475

undulatum 476
Verbena 481
  officinalis **480**, 481
VERBENACEAE 481
Veronica 470
  acinifolia 470
  anagallis-aquatica 470
  arvensis 470
  catenata 470
  chamaedrys 470, **471**
  hederifolia **471**
  persica 470
  polita **471**
  serpyllifolia 470, **471**
  triphyllos 470
Viburnum 582
  lantana 582
  tinus 582
Vicia 234
  benghalensis **235**
  cracca 234, **235**
  ervilia 234
  faba 236
  grandiflora **236**
  hirsuta 234, **235**
  hybrida **236**
  lathyroides 237
  lunata **29**, **236**, 237
  lutea **235**, 236
  melanops 237
  narbonensis **236**, 237
  onobrychioides 235
  pannonica 237
  parviflora 234
  pubescens 234
  sativa 237
  sepium 237
  tenuissima 234
  tetrasperma 234
  villosa 234
    subsp. *microphylla* 234, **235**
Vinca 420
  herbacea 420
  major 420, **421**
Vincetoxicum 425
  canescens 425
  creticum 425
  fuscatum 425
  hirundinaria **424**, 425
  speciosum 425
Viola 295
  alba 297
  arvensis **296**, 297

athois 298
canina 297
euboea 298
fragrans 298
heldreichiana 298
hirta 297
hymettia 297
kitaibeliana 297
odorata **296**, 297
parvula 297
phitosiana 297
rausii **22**, **296**, 297
reichenbachiana **296**, 297
riviniana 297
samothracica 298
scorpiuroides 295, **296**
sieheana **296**, 297
VIOLACEAE 295
Viscum 363
  album 363
    subsp. *abietis* 363
    subsp. *album* 363
    subsp. *austriacum* 363, **364**
    subsp. *creticum* 363
  cruciatum 363

VITACEAE 211
Vitex 482
  agnus-castus **482**
Vitis 211
  vinifera 211
    subsp. *sylvestris* 211

W
Washingtonia 154
  filifera 154
  robusta **154**
Wisteria 268
  floribunda 268
  sinensis 268, **269**
Withania 455
  somnifera **454**, 455

X
Xanthium 550
  spinosum 550, **551**
  strumarium 550, **551**
XANTHORRHOEACEAE 127
Xolantha guttata 337

Y
Yucca 152
  filamentosa **152**
  gloriosa **152**

Z
Zantedeschia 71
  aethiopica 71
Zea 178
  mays 178
Zilla 357
  spinosa **357**
Ziziphora 500
  capitata 500
  taurica 500
  tenuior 500
Zostera 75
  marina 75
  noltii 75
ZOSTERACEAE 75
ZYGOPHYLLACEAE 212
Zygophyllum 212
  fabago 212